World Soils Book Series

Series Editor

Alfred E. Hartemink
Department of Soil Science, FD Hole Soils Laboratory
University of Wisconsin–Madison
Madison, WI
USA

The World Soils Book Series publishes peer-reviewed books on the soils of a particular country. They include sections on soil research history, climate, geology, geomorphology, major soil types, soil maps, soil properties, soil classification, soil fertility, land use and vegetation, soil management, soils and humans, soils and industry, future soil issues. The books summarize what is known about the soils in a particular country in a concise and highly reader-friendly way. The series contains both single and multi-authored books as well as edited volumes. There is additional scope for regional studies within the series, particularly when covering large land masses (for example, The Soils of Texas, The Soils of California), however, these will be assessed on an individual basis.

More information about this series at https://link.springer.com/bookseries/8915

Thor Thorson • Chad McGrath •
Dean Moberg • Matthew Fillmore •
Steven Campbell • Duane Lammers •
James G. Bockheim

The Soils of Oregon

Thor Thorson
USDA Natural Resources Conservation Service
Tualatin, OR, USA

Chad McGrath
USDA Natural Resources Conservation Service
Tigard, OR, USA

Dean Moberg
USDA Natural Resources Conservation Service
Portland, OR, USA

Matthew Fillmore
USDA Natural Resources Conservation Service
Lebanon, OR, USA

Steven Campbell
USDA Natural Resources Conservation Service
Portland, OR, USA

Duane Lammers
USDA Forest Service
Philomath, OR, USA

James G. Bockheim
University of Wisconsin-Madison
Madison, WI, USA

ISSN 2211-1255 ISSN 2211-1263 (electronic)
World Soils Book Series
ISBN 978-3-030-90093-9 ISBN 978-3-030-90091-5 (eBook)
https://doi.org/10.1007/978-3-030-90091-5

This Springer imprint is published by the registered company Springer Nature Switzerland AG
The registered company address is: Gewerbestrasse 11, 6330 Cham, Switzerland

Foreword

We walk above soil, stepping lightly on spring meadow grasses that grow out of it on our way to a favorite lake. We curse and call it mud as we attempt to clean it off our shoes and our pets after a particularly wet hike through forests in the Coast Range. We rinse it from our favorite farmer's market treats, and plant our own flowers and tomatoes in it, anticipating summer's bounty. Soil is a part of everyday life for anyone who loves the outdoors or Oregon's celebrated "foodie" culture.

But soil provides so much more richness to Oregon than most of us recognize. Created from epic floods and volcanoes, decaying plants and hungry microorganisms, healthy soils benefit every part of our lives. You don't need to be a soil scientist (or even aspire to be one), to benefit from knowing more about the value soil adds to our lives. Understanding Oregon soils helps us appreciate Oregon's rich history, natural diversity, and the cultural and economic drivers that support Oregon families today.

Soils support Oregon's Indigenous Culture. Indigenous Peoples have cared for the lands and waters of Oregon and this country since time immemorial, and continue to do so today. Their resilient communities are integrally tied to the health and abundance of the natural resources derived from soil. As Indigenous Peoples have long recognized, healthy soils support the food that people, plants, fish, and wildlife need—not just for Oregonians now, but for future generations as well.

Soils support healthy ecosystems and local economies. Western Oregon's forest soils grow big trees. When managed with care, those trees create a system that can support homes for owls, habitat for salmon, and a sustainable timber harvest. Our Eastern Oregon soils grow sagebrush where birds nurture and protect their young, and ranchers raise cattle that thrive and return nutrients to those same soils. Healthy soils capture water, holding it for use by plants during Oregon's dry, Mediterranean-like summer months. And these same soils are home to countless beneficial microorganisms that are a critical food source in the web of life.

Soils grow great food and fiber. Soils are a major component of the "terroir" that we reference when sampling a great wine in the Willamette Valley, southern Oregon, or the Columbia Gorge—supporting a $3.35 billion wine economy in Oregon[1]. Soil is the essential ingredient in which Oregon farmers are able to grow over 220 agricultural products[2]—not just for our local farmer's markets, but also for national and international consumers.

Soils cool the earth. Healthy soils can mitigate the impacts of climate change by storing carbon that has been taken out of the atmosphere by plants through photosynthesis. Sustainable management of plants and soils opens up new opportunities to sequester carbon, creating a vital tool in our ability to address a changing climate.

"The Soils of Oregon" provides those of us who are not soil scientists with a broad understanding of the soils we walk on, drive by, and consume products from every day. This book is for anyone who doesn't just want to taste great food, wine, and beer. It's for people who want to better understand the soils that make these things taste so incredibly good.

---

[1]Oregon Wine Board 2015. "Oregon's Wine Industry Contributes $3.35 Billion to Oregon's Economy."

[2]Oregon Department of Agriculture. 2021. "Oregon agricultural statistics and directory 2021." Salem, OR.

Whether you've always wanted to know what a "Lickskillet stony loam" was and what types of plants prefer it, have a burning desire to impress your friends with the difference between wines that come from Jory or Willakenzie soils, or just want to know more about how epic prehistoric floods that started in Montana result in great Oregon blueberries, this is the book for you. You can carry it with you just like you would a book on roadside geology or wildflowers, read it cover-to-cover, or keep it handy as a reference. However you choose to use your new-found knowledge, a good understanding of Oregon's soils will help you better recognize why Oregon is such a unique and special place to visit or live.

Meta Loftsgaarden
Former Executive Director
Oregon Watershed Enhancement Board
Salem, Oregon, USA

Preface

This book discusses the nature, properties, genesis, classification, and use of the soils of Oregon and provides maps of dominant soil great groups in Oregon based on Soil Taxonomy. The study of soils in Oregon originated with a reconnaissance soil survey of Baker City in 1903, shortly after the Bureau of Soils, a precursor to the Soil Conservation Service and the more encompassing Natural Resources Conservation Service (NRCS), was established. Portions of all but five of the 36 counties in Oregon have received an order 2 or 3 soil survey (scale 1:24,000). Oregon has a variety of physiographic provinces that have led to the mapping of more than 1,700 soil series in the state. This study was made easier by an abundance of natural resource maps (vegetation, geology, etc.) and other technical information. We were assisted in this endeavor by Cory Owens, Oregon State soil scientist, and Whityn Owen, NRCS Oregon GIS specialist/coordinator.

This book is dedicated to the professional soil scientists from the Natural Resources Conservation Service, US Forest Service (US Department of Agriculture), Bureau of Land Management (US Department of Interior), Bureau of Indian Affairs, National Park Service, US Fish and Wildlife Service , soil and water conservation districts, and the Oregon Agricultural Experiment Station (Oregon State University), who contributed to the mapping of soils in Oregon. We would like to acknowledge the support of these agencies and other land management agencies that contributed to soil surveys. This book could not have been written without the support of NRCS database managers.

This book originated from data collected by the USDA-Natural Resources Conservation Service, including the Soils Data Mart and Web Soil Survey, and by approximately 100 research reports dealing with the soils, geology, and vegetation of Oregon. The interpretations were made solely by the authors. This book should be of interest to individuals in federal, state, county, and non-government organizations who are interested in and responsible for safeguarding Oregon's natural resources. The book will be of interest to students in soil science and allied disciplines.

Tualatin, USA — Thor Thorson
Tigard, USA — Chad McGrath
Portland, USA — Dean Moberg
Lebanon, USA — Matthew Fillmore
Portland, USA — Steven Campbell
Philomath, USA — Duane Lammers
Madison, USA — James G. Bockheim

Contents

1 Introduction 1
- 1.1 Etymology 1
- 1.2 Geography 1
- 1.3 Demographics 1
- 1.4 History 2
- 1.5 Economy 2
- 1.6 Summary 3
- References 3

2 History of Soil Studies in Oregon 5
- 2.1 Introduction 5
- 2.2 Definition of Soil 5
- 2.3 Soil Surveys 5
- 2.4 Soil Series 8
- 2.5 Soil Classification 10
- 2.6 Soil Taxonomy 10
- 2.7 General Soil Maps 15
- 2.8 Soil Research 16
- 2.9 The State Soil 17
- 2.10 Summary 17
- References 20

3 Soil-Forming Factors 21
- 3.1 Introduction 21
- 3.2 Climate 21
  - 3.2.1 Current Climate 21
  - 3.2.2 Past Climates 25
- 3.3 Vegetation 26
- 3.4 Relief 32
- 3.5 Physiographic Provinces 34
- 3.6 Geologic Structure 35
- 3.7 Surficial Geology 39
- 3.8 Time 40
- 3.9 Humans 40
- 3.10 Summary 40
- References 42

4 Elevation Gradients in Oregon Mountain Ranges 43
- 4.1 Introduction 43
- 4.2 Coast Range 43
- 4.3 Cascade Mountains 43
- 4.4 Wallowa Mountains 47
- 4.5 Blue Mountains 49

4.6 Steens Mountain 49
4.7 Fremont Mountains 49
4.8 Klamath Mountains 50
4.9 Summary 50
Reference 50

5 General Soil Regions of Oregon 51
5.1 Introduction 51
5.2 Malheur High Plateau 53
5.3 Cascade Mountains 59
5.4 Blue Mountain Foothills 60
5.5 Blue Mountains 61
5.6 Siskiyou Mountains 63
5.7 Cascade Mountains—Eastern Slope 66
5.8 Coast Range 67
5.9 Willamette Valley 70
5.10 Owyhee High Plateau 73
5.11 Columbia Plateau 74
5.12 Klamath Basin 77
5.13 Palouse Prairie 79
5.14 Sitka Spruce Belt 81
5.15 Columbia Basin 82
5.16 Snake River Plains 85
5.17 Humboldt Area 86
5.18 Coastal Redwood Belt 88
5.19 Summary 88
Reference 88

6 Diagnostic Horizons and Taxonomic Structure of Oregon Soils 89
6.1 Introduction 89
6.2 Diagnostic Horizons 89
6.3 Orders 91
6.4 Suborders 91
6.5 Great Groups 91
6.6 Subgroups 91
6.7 Families 91
6.8 Soil Series 96
6.9 Summary 96
References 101

7 Taxonomic Soil Regions of Oregon 103
7.1 Introduction 103
7.2 Haploxerolls (Soil Region 1) 110
7.3 Argixerolls (Soil Region 2) 111
7.4 Humudepts (Soil Region 3) 114
7.5 Haplargids (Soil Region 4) 116
7.6 Argidurids (Soil Region 5) 117
7.7 Vitricryands (Soil Region 6) 118
7.8 Haplocambids (Soil Region 7) 122
7.9 Palexerolls (Soil Region 8) 124
7.10 Haploxeralfs (Soil Region 9) 126
7.11 Haplodurids (Soil Region 10) 128
7.12 Dystroxerepts (Soil Region 11) 130
7.13 Durixerolls (Soil Region 12) 132

7.14 Haplohumults (Soil Region 13) . . . 133
7.15 Vitrixerands (Soil Region 14) . . . 136
7.16 Udivitrands (Soil Region 15) . . . 138
7.17 Haploxerepts (Soil Region 16) . . . 140
7.18 Palehumults (Soil Region 17) . . . 141
7.19 Dystrudepts (Soil Region 18) . . . 143
7.20 Hapludands (Soil Region 19) . . . 145
7.21 Paleargids (Soil Region 20) . . . 146
7.22 Eutrudepts (Soil Region 21) . . . 149
7.23 Fulvudands (Soil Region 22) . . . 151
7.24 Torripsamments (Soil Region 23) . . . 153
7.25 Endoaquolls (Soil Region 24) . . . 155
7.26 Torriorthents (Soil Region 25) . . . 157
7.27 Haplocryands (Soil Region 26) . . . 159
7.28 Argialbolls (Soil Region 27) . . . 161
7.29 Palexeralfs (Soil Region 28) . . . 163
7.30 Summary . . . 164

8 Mollisols . . . 165
8.1 Distribution . . . 165
8.2 Properties and Processes . . . 166
8.3 Use and Management . . . 168
8.4 Summary . . . 168
Reference . . . 173

9 Inceptisols . . . 175
9.1 Distribution . . . 175
9.2 Properties and Processes . . . 176
9.3 Use and Management . . . 178
9.4 Summary . . . 178

10 Aridisols . . . 181
10.1 Distribution . . . 181
10.2 Properties and Processes . . . 181
10.3 Use and Management . . . 185
10.4 Summary . . . 185
Reference . . . 185

11 Andisols . . . 187
11.1 Distribution . . . 187
11.2 Properties and Processes . . . 189
11.3 Use and Management . . . 192
11.4 Summary . . . 192
Reference . . . 195

12 Ultisols . . . 197
12.1 Distribution . . . 197
12.2 Properties and Processes . . . 197
12.3 Use and Management . . . 199
12.4 Summary . . . 199
References . . . 200

13 Alfisols . . . 201
13.1 Distribution . . . 201
13.2 Properties and Processes . . . 201

13.3 Use and Management ... 202
13.4 Summary ... 202
Reference ... 204

14 Entisols, Vertisols, Spodosols, and Histosols ... 205
14.1 Distribution ... 205
14.2 Properties and Processes ... 207
14.3 Use and Management ... 212
14.4 Summary ... 213
References ... 213

15 Soil-Forming Processes in Oregon ... 215
15.1 Introduction ... 215
15.2 Humification ... 215
15.3 Cambisolization ... 215
15.4 Argilluviation ... 215
15.5 Andisolization ... 216
15.6 Gleization ... 216
15.7 Silicification ... 217
15.8 Vertization ... 217
15.9 Calcification ... 217
15.10 Solonization ... 217
15.11 Salinization ... 217
15.12 Podzolization ... 217
15.13 Paludization ... 218
15.14 Soils with Minimal Soil-Forming Processes ... 218
15.15 Summary ... 218
References ... 218

16 Benchmark, Endemic, Rare, and Endangered Soils in Oregon ... 219
16.1 Introduction ... 219
16.2 Benchmark Soils ... 219
16.3 Endemic Soils ... 219
16.4 Rare Soils ... 219
16.5 Endangered Soils ... 220
16.6 Shallow Soils ... 220
16.7 Highly Represented Soil Great Groups ... 220
16.8 Summary ... 220
References ... 221

17 Land Use in Oregon ... 223
17.1 Introduction ... 223
17.2 Land Ownership and Management ... 223
17.2.1 Federal ... 223
17.2.2 Indigenous Peoples ... 225
17.2.3 State and Local Government ... 226
17.2.4 Private ... 226
17.3 Land-Use Designations ... 229
17.3.1 Developed Land ... 230
17.3.2 Farmsteads ... 230
17.3.3 Cropland ... 230
17.3.4 Pasture and Rangeland ... 232
17.3.5 Forestland ... 233

17.4 Soil Survey Management Groups . . . 236
17.4.1 Land Capability Classes . . . 236
17.4.2 National Inventory Groupings . . . 237
17.5 Oregon Land-Use Planning . . . 237
17.5.1 Zoning . . . 237
17.5.2 Oregon Land Use Act of 1973 . . . 237
17.5.3 Changes to the Oregon Land Use Act: Measures 7, 37, and 49 . . . 239
17.5.4 After Measure 49 . . . 240
17.6 Key Natural Resource Challenges Related to Land Use and Soil . . . 241
17.6.1 Climate Change . . . 241
17.6.2 Wetland Loss . . . 243
17.6.3 Flooding . . . 243
17.6.4 Landslides . . . 247
17.6.5 Volcanoes, Earthquakes, and Tsunamis . . . 248
17.6.6 Coastal Erosion . . . 249
17.6.7 Wildfires . . . 255
17.7 Summary . . . 258
References . . . 259

18 Yields, Soil Conservation, and Production Systems . . . 265
18.1 Introduction . . . 265
18.2 Yields of Oregon Working Lands . . . 265
18.2.1 Cropland Yields . . . 265
18.2.2 Grazing Land Yields . . . 268
18.2.3 Forestry Yields . . . 269
18.2.4 Productivity Indices and Yield Modeling . . . 269
18.3 Conservation of Soil and Related Resources . . . 270
18.3.1 Background . . . 270
18.3.2 State Regulations . . . 270
18.3.3 Federal Regulations . . . 273
18.3.4 Local, State, and Federal Funding for Conservation . . . 275
18.3.5 Practices . . . 275
18.3.6 Soil Health . . . 275
18.4 Cropland Management Systems . . . 277
18.4.1 Hay and Haylage . . . 277
18.4.2 Grain Crops . . . 278
18.4.3 Conservation Reserve Program . . . 280
18.4.4 Seed Crops . . . 281
18.4.5 Vegetables . . . 282
18.4.6 Orchards . . . 283
18.4.7 Corn for Silage or Grain . . . 287
18.4.8 Christmas Trees . . . 288
18.4.9 Nursery Crops . . . 289
18.4.10 Hemp and Marijuana . . . 291
18.4.11 Berries . . . 292
18.4.12 Grapes . . . 294
18.4.13 Small Acreage Vegetables and Specialty Crops . . . 296
18.5 Pasture and Grazed Rangeland Management Systems . . . 297
18.5.1 Pasture . . . 298
18.5.2 Grazed Rangeland . . . 300

18.6 Forestry Management Systems 301
18.6.1 Western Oregon Forests 302
18.6.2 Eastern Oregon Forests 305
18.7 First Foods of Indigenous Peoples 306
18.7.1 Camas 307
18.7.2 *Lomatium* Species 307
18.7.3 Oregon White Oak 309
18.7.4 The Huckleberries 310
18.7.5 Water and Salmon 311
18.8 Summary 314
References 315

19 Summary 323

Appendix A: Soil-Forming Factors for Soil Series in Oregon with an Area of 50 km^2 or More 325

Appendix B: Thicknesses (cm) of Diagnostic Horizons in Soil Series with an Area of 50 km^2 and Greater in Oregon 365

Appendix C: Area and Taxonomy of Soil Series in Oregon 379

Appendix D: Benchmark, Endemic, Rare, and Endangered Soil Series in Oregon 433

Appendix E: Land Use, Yield, and Key Soil Characteristics Influencing Yield in Oregon 465

Index 539

About the Authors

Thor Thorson was the USDA Oregon Natural Resources Conservation Service (NRCS) State Soil Scientist from 2012 to 2015. He is a native of Wisconsin and attended the University of Wisconsin-Madison, graduating with a Bachelor's degree in Soil Science in 1974. In 1974 he began his career with USDA as a field Soil Scientist in California. He worked on four California soil surveys prior to transferring to Oregon in 1980 as the Oregon State Soil Correlator. From 1983 to 2012 he held positions as Oregon Assistant State Soil Scientist and as a Regional Soil Data Quality Specialist covering parts of Oregon, Idaho, and Washington. He retired in 2015 from government service and resides in the Portland, Oregon area. He enjoys golfing, hunting, and fishing and continues to provide soil expertise to the Oregon NRCS as a volunteer.

Chad McGrath received his B.S. degree in Forest Resource Management from the University of Idaho in 1966. He then served as an officer in the US Navy. In 1975, he received his M.S. degree in Forest Soils from the University of Idaho. He worked as a Soil Scientist for the Idaho Soil Conservation Commission from 1975 to 1976. In 1976, he was hired as a Soil Scientist by the Soil Conservation Service: 1976–1977 Power County Area, Idaho; 1977–1978 Oneida County Area, Idaho; 1978–1983 Soil Survey Project Leader for four surveys in southeastern Idaho; completed Soil Survey Reports for two of those surveys; 1983–1987 Area Soil Scientist for Southeastern Idaho; 1987–1995 Soil Correlator for Idaho. In 1995 he moved to Portland, Oregon as State Soil Scientist and Leader for MLRA Regional Office One which covered parts of Oregon, Idaho, and most of Washington. As State Soil Scientist he provided leadership for the soil survey program in Oregon. MLRA Regional Office provided technical assistance and guidance for all aspects of the soil surveys within the area of responsibility. In 2012, he retired from the Natural Resources Conservation Service.

Dean Moberg joined the USDA-Natural Resources Conservation Service (NRCS) in 1984 and worked for NRCS in Oregon, Wisconsin, and Michigan. His final position was to serve as the Basin Resource Conservationist in northwest Oregon. Prior to NRCS, Dean taught high school and worked on vegetable farms, dairy farms, and a maple syrup operation. He obtained his B.S. in Plant Science from the University of California, Davis, a M.A. in Teaching from Cornell University, and a Ph.D. in Environmental Science and Engineering from Oregon Health & Sciences University. Dean and his wife Sara have two daughters. His hobbies include fly fishing, scuba diving, skiing, and gardening.

Matthew Fillmore attended Oregon State University (OSU) in Corvallis where he received B.S. degrees in Soil Science and Wildlife Science in 1976. He began his more than 37-year career with USDA Soil Conservation Service in Oregon as a field Soil Scientist in 1977 working in Linn County. In 1983 he was transferred to Baker County in northeastern Oregon. In 1987 he was promoted to Soil Survey Project Leader for the Curry County survey project in southwestern Oregon. In 1994 he moved back to Corvallis and was stationed in the Soil Science Department at OSU to begin the initial MLRA-based update for Oregon in Benton

County under the newly named USDA-Natural Resources Conservation Service. In 2008 he became the initial MLRA Soils Office Leader for all of western Oregon based in Salem. He mapped over 1 million acres throughout Oregon and authored three soil survey manuscripts (Curry, Benton, and Tillamook counties) during his career. He retired in 2013 from government service and lives in Lebanon, Oregon. He enjoys woodworking, baseball and softball, fishing, traveling, and photography in retirement.

Steven Campbell obtained B.S. degree in Forest Management from Washington State University in 1976. He has been a soil scientist with the Natural Resources Conservation Service from 1976 to the present, including being a Project Member and Project Leader for soil survey areas in Washington State, Resource Soil Scientist in the Spokane Washington Area Office, Soil Scientist at the NRCS State Office in Portland, Oregon, Soil Data Quality Specialist in the Pacific Northwest Soil Survey Regional Office in Portland, and is at present Soil Scientist at the West National Technology Support Center in Portland. His current responsibilities include providing training and technical support to the West Region States on a wide variety of soils-related topics.

Duane Lammers received a B.S. degree in Agricultural Science (Soils) from Montana State University in 1967 and a Ph.D. in Soils from Utah State University in 1975. He was then employed by the Soil Conservation Service in Utah 1975–1976—Alton Pipeline Corridor Soil Survey, 1976–1983 Project Leader on three soil survey areas in southeastern Utah, 1983-1985 Monitoring and Evaluation Team Leader, Uinta Basin Colorado River Salinity Project. In 1985 he moved to Corvallis, Oregon, and a U.S. Forest Service position in an acidic deposition study—the Direct-Delayed Response Project. For this project, Duane lead the soil mapping effort of watersheds selected for study in the Southern Blue Ridge, Mid-Appalachian, and NE United States. In 1990 he accepted a job as Soil Correlator for Region Six (WA and OR) of the Forest Service. His work from 1990 to 2008 included field review, correlation, and classification of soils on NF System Lands in Oregon and Washington. Most of his work on forests in Oregon was on the east side of the Cascade Range where soils have been influenced by tephra from Mt. Mazama. He considers this to be one of the "big" stories about soils in Oregon.

James G. Bockheim was a professor of Soil Science at the University of Wisconsin from 1975 until his retirement in 2015. He has conducted soil genesis and geography studies in many parts of the world, including five field seasons along the Oregon coast. Jim lives in Oregon, Wisconsin. He enjoys writing, reading, biking, photography, and traveling.

Acronyms and Abbreviations

ac	Acre
AFO	Animal Feeding Operation
Al	Aluminum
AUM	Animal unit month
B	Boron
B&B	Balled and burlapped nursery stock
BLM	Bureau of Land Management
°C	Degrees Celsius
C	Carbon
Ca	Calcium
$CaCO_3$	Calcium carbonate
CAFO	Concentrated Animal Feeding Operation (federal acronym)
CAFO	Confined Animal Feeding Operation (Oregon acronym)
CBD	Cannabidiol
CEC	Cation exchange capacity
cfs	Cubic feet per second
CH_4	Methane
cm	Centimeter
CO_2	Carbon dioxide
CO_2e	Carbon dioxide equivalent
CRP	Conservation Reserve Program
CSA	Community supported agriculture
Cu	Copper
CWA	Clean Water Act (United States)
CWPP	Community Wildfire Protection Plan
DLCD	Department of Land Conservation and Development (Oregon)
DOD	Department of Defense (United States)
DOGAMI	Department of Geology and Mineral Industries (Oregon)
DOI	Department of Interior (United States)
dS	Decisiemens
EC	Electrical conductivity
EFU	Exclusive farm use
EPA	Environmental Protection Agency (United States)
ESA	Endangered Species Act
ESD	Ecological site description
Fe	Iron
FEMA	Federal Emergency Management Agency
FIA	Forest Inventory and Analysis
FSA	Farm Service Agency
FSG	Forage suitability group
GHG	Green house gas

GWP	Global warming potential
ha	Hectare
HCl	Hydrochloric acid
HEL	Highly erodible land
IET	Integrated Erosion Tool
in	Inch
IPCC	Intergovernmental Panel on Climate Change
K	Potassium
ka	Thousands of years ago
KCl	Potassium chloride
kg	Kilogram
km	Kilometer
KOH	Potassium hydroxide
LCC	Land capability class
LCDC	Land Conservation and Development Commission
M	Magnitude (when used in connection with earthquakes)
m	Meter
M	Molar (when used in connection with chemical solutions)
MAAT	Mean annual air temperature
MAP	Mean annual precipitation
Mg	Magnesium
MLRA	Major land resource area
mm	Millimeter
MMBF	Million board feet
Mn	Manganese
Mo	Molybdenum
Mt	Million metric tonnes
N	Nitrogen
N_2O	Nitrous oxide
NaF	Sodium fluoride
NaOH	Sodium hydroxide
NASIS	National Soil Information System
NASS	National Agricultural Statistics Service
NCS	Natural climate solutions
NFIP	National Flood Insurance Program
NFSAM	National Food Security Act Manual
NH_4-N	Ammonium nitrogen
NIPF	Non-industrial private forestland
NO_3-N	Nitrate nitrogen
NOAA	National Oceanic and Atmospheric Administration
NPS	National Park Service
NRCS	Natural Resources Conservation Service (United States)
NRI	National Resource Inventory
NWOS	National Woodland Owner Survey
O&C	Oregon and California (railroad)
OCCRI	Oregon Climate Change Research Institute
OCMP	Oregon Coastal Management Program
ODA	Oregon Department of Agriculture
ODF	Oregon Department of Forestry
OPOA	Oregon Property Owners Associatoin
OSSPAC	Oregon Seismic Safety Policy Advisory Commission
OSU	Oregon State University

OWRD	Oregon Water Resources Department
P	Phosphorus
PRISM	Parameter-elevation Regressions on Independent Slopes Model
PSNT	Pre-sidedress nitrate test
RCP	Representative concentration pathway
REIT	Real estate investment trust
RUSLE2	Revised Universal Soil Loss Equation
S	Sulfur
SI	Site index
SMR	Soil moisture regime
SNC	Swiss needle cast
SOC	Soil organic carbon
SOD	Sudden oak death
SSURGO	Soil Survey Geographic Database
STATSGO2	Digital General Soil Map of the United States
STR	Soil temperature regime
SWCD	Soil and water conservation district
t	Metric tonne
T	Soil loss tolerance
THC	Tetrahydrocannabinol
TIMO	Timber investment management organization
UGB	Urban growth boundary
USDA	United States Department of Agriculture
USFS	United States Forest Service
USFWS	United States Fish and Wildlife Service
USGS	United States Geological Survey
USLE	Universal Soil Loss Equation
W	Watt
WEPS	Wind Erosion Prediction System
WSS	Web Soil Survey
WUI	Wildland-urban interface
yr	Year
Zn	Zinc

1 Introduction

1.1 Etymology

The origin of the word "Oregon" is disputed. However, three suggestions suggest that it originates from (1) the French Canadian word, *ouragan*, which means "windstorm" after the chinook winds on the lower Columbia River; (2) *orejón*, meaning "big ear," after Indigenous Peoples living in the area; and (3) the spice, orégano, which grew in the territory (http://www.netstate.com).

1.2 Geography

Oregon is bounded by the Pacific Ocean on the west, the Columbia River on the north, the Snake River and the 117°W longitude on the east, and the 42nd parallel on the south. At 254,800 km^2 (98,380 mi^2), Oregon is the ninth largest state in the United States. The state is 475 km (295 mi) north to south and 635 km (395 mi) east to west. Elevation varies from 0 m along the Pacific Coast to 3,429 m (11,249 ft) at the summit of Mt. Hood. Oregon has some of the most diverse terrain in the United States, including an eroded ancient mountain chain (Coast Range), a tectonically active mountain range (Cascade Mountains), an area with substantial serpentinite (Klamath Mountains), a glaciated mountain range in the northeast (Blue Mountains), a valley (Willamette) filled with multiple layers of sediments from catastrophic Pleistocene floods, high plateaus (Deschutes–Umatilla Plateau, Owyhee and High Lava Plains), and alternating mountain ranges and valleys resulting from faulting due to tectonic activity (Basin and Range).

Major water bodies in Oregon are shown in Fig. 1.1. Major rivers that drain from the Coast Range or Klamath Mountains to the Pacific Ocean include, from north to south, the Nehalem, Nestucca, Siletz, Yaquina, Siuslaw, Umpqua, Millicoma, Coos, Coquille, Elk, Rogue, Pistol, Chetco, and Winchuck. The Smith and Klamath Rivers (Williamson, Sprague, and Lost), in southwest Oregon, flow into California, then to the Pacific Ocean. Major rivers draining into the Columbia River include, from west to east, the Willamette (Clackamas), Sandy (Salmon), Hood, Deschutes (Crooked), John Day, and Umatilla (Butte). Rivers draining into the Snake River include, from north to south, Grande Ronde (Wallowa and Lostine), Imnaha, Powder, Burnt, Malheur, and Owyhee. Crater Lake, a drowned crater from the eruption of Mt. Mazama, is the deepest lake in the United States at 594 m (1,949 ft) and second largest (53 km^2) natural water body in Oregon. Smaller lakes in the High Cascades include Diamond, Waldo, and Odell.

Major lakes in Oregon that are remnant from larger pluvial lakes from the late Wisconsinan include Upper Klamath Lake (249 km^2), Malheur Lake (201 km^2), Harney Lake (107 km^2) and lakes in Lake County (Goose Lake, 380 km^2; Lake Abert, 150 km^2; Summer Lake, 109 km^2; and the Warner Lakes) (Fig. 1.1). Major reservoirs include Fern Ridge (410 km^2), Lake Wallula (157 km^2), Hells Canyon Reservoirs (61 km^2), and Wickiup Reservoir (45 km^2).

1.3 Demographics

With a population of 4.2 million, Oregon was the 27th most populous US state in 2019. Major cities and 2019 populations include Portland (653,000), Salem (174,000; also the capital city), Eugene (173,000), Gresham (109,000), Hillsboro (109,000), Beaverton (99,000), Bend (100,000), Medford (83,000), Springfield (63,000), Corvallis (59,000), Aloha (55,000), Tigard (56,000), and Albany (55,000) (United States Census Bureau 2021). Oregon is divided into 36 counties that range in size from Multnomah County, which is the smallest (1,204 km^2) to Harney County, which is the largest (26,490 km^2) (Fig. 1.2) (Oregon Secretary of State 2021).

T. Thorson et al., *The Soils of Oregon*, World Soils Book Series,
https://doi.org/10.1007/978-3-030-90091-5_1

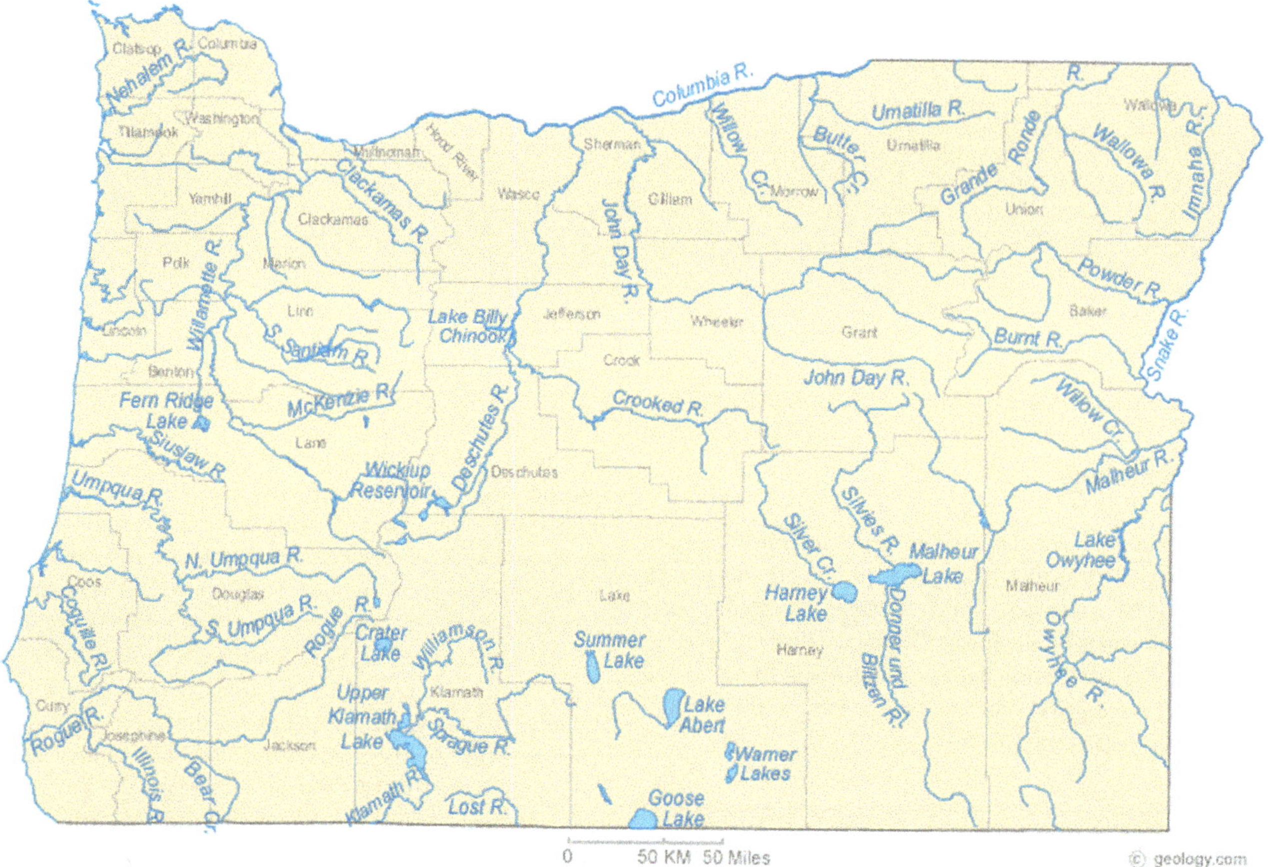

Fig. 1.1 The rivers and lakes of Oregon. *Source* http://geology.com

1.4 History

Artifacts from Indigenous Peoples[1] who lived in Oregon 15,000 years ago have been recovered throughout the state. By the sixteenth century, a number of indigenous tribes lived in Oregon, including the Chinook, Coquille, Bannock, Chasta, Kalapuya (Calapooya), Klamath, Klickitat, Molalla, Nez Perce, Takelma, Killamuk, Neah-kah-nie, Umatilla, and Umpqua.

The first Europeans to visit Oregon were Spanish explorers led by Juan Rodriguez Cabrillo in 1543. Sir Francis Drake, sailing in the Golden Hind, sheltered near Cape Arago in 1579. In 1778, James Cook explored the Oregon coast. French Canadian trappers and missionaries arrived in Oregon in the late 1700s and early 1800s. The Lewis and Clark Expedition Corps of Discovery camped the winter of 1805–1806 at Fort Clatsop, arriving and returning via the Columbia River. Representing the North West Company, David Thompson may have been the first European to navigate the entire Columbia River. John Jacob Aster financed the establishment of Fort Astoria, an outpost of the Pacific Fur Company, in 1811 at the mouth of the Columbia River. This was the first permanent European establishment in Oregon.

In the War of 1812, the British gained control of all Pacific Fur Company posts. The Treaty of 1818 gave joint custody of the region from the Rocky Mountains to the Pacific Ocean to the United States and Great Britain. The first wagon trains on the Oregon Trail arrived in the Willamette Valley in 1842. This trail provided a major conduit for emigration for more than a decade. Oregon was admitted to the US as the 33rd state in 1859.

1.5 Economy

In 2018, the gross domestic product for Oregon was $214 billion, ranking as the sixth wealthiest state in the United States. Manufacturing accounts for 47% of the GDP, including high-technology industries located in the "Silicon Forest," the industrial corridor between Beaverton and Hillsboro. Agriculture, forestry, and fishing account for

[1] Terminology used here follows, as much as possible, guidance set forth in Gregory Younging's Elements of Indigenous Style (2018).

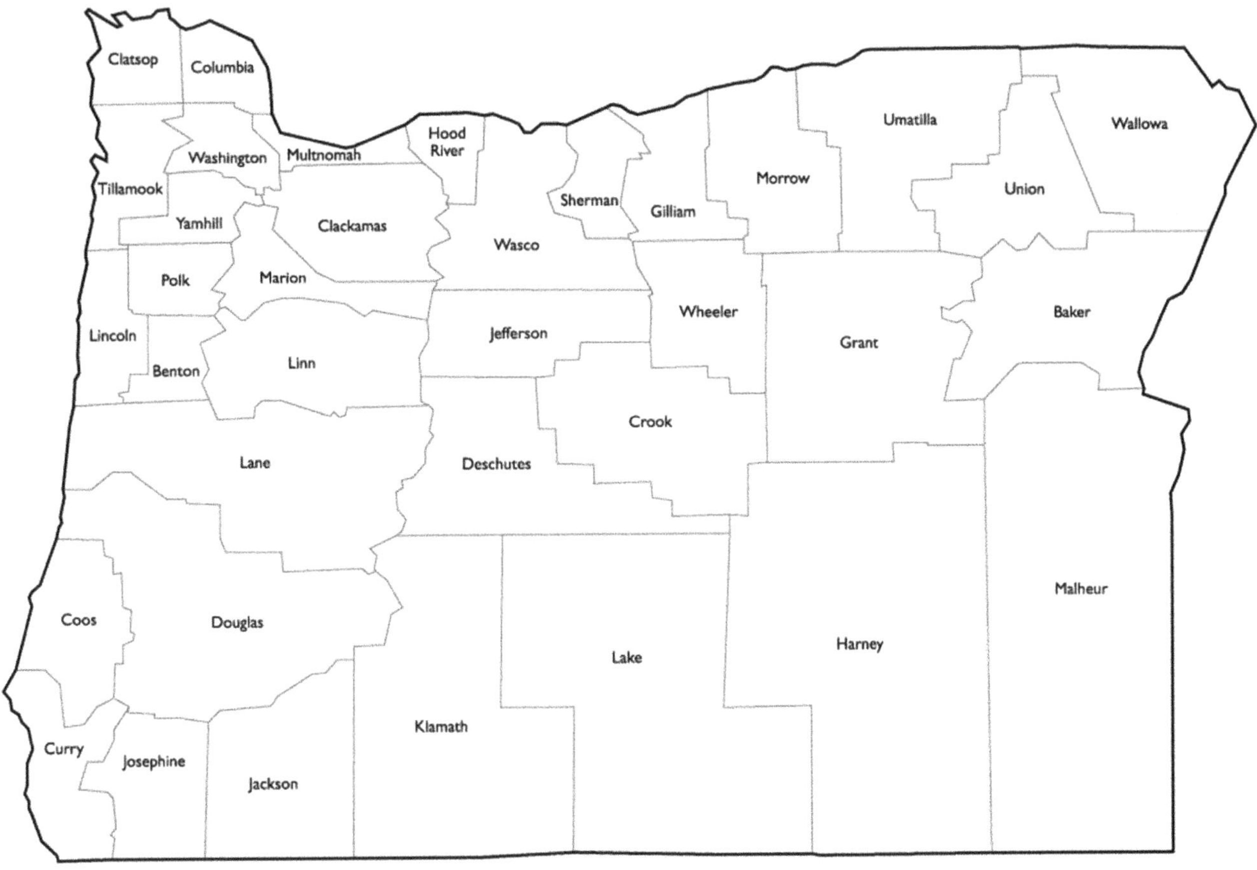

Fig. 1.2 Oregon counties. *Source* GIS Geography

4.2% of the GDP. The top five agricultural industries are greenhouse and nursery products, hay, cattle and calves, milk, and grass seed. Oregon is the top producer of softwood lumber in the contiguous United States and has one of the largest salmon fisheries in the world.

1.6 Summary

The origin of the word "Oregon" may originate from the French Canadian word, ouragan, which means "windstorm," orejón, meaning "big ear," after Indigenous Peoples living in the area; or the spice, orégano, which grew in the territory. Oregon is the ninth largest state in the United States and is one of the most diverse states in the United States, in terms of elevation, rock types, physiographic provinces, climate, and vegetation. Oregon is the 27th most populous state in the United States, with 70% of the population living in the Willamette Valley. The history of Oregon includes Indigenous Peoples, Spanish explorers, French Canadian trappers and missionaries, the Lewis and Clark Voyage of Discovery in 1804–1806, and settlers who arrived along the Oregon Trail in 1842–1843. In 2018, Oregon was ranked sixth in the USA in gross domestic products, primarily from manufacturing. Agriculture, forestry, and fishing are important industries in the state.

References

Oregon Secretary of State (2021) Oregon blue book. Retrieved May 27, 2021, from https://sos.oregon.gov/blue-book/Pages/default.aspx

U.S. Census Bureau (2021) Explore census data. Retrieved May 27, 2021, from https://data.census.gov/cedsci/

Younging G (2018) Elements of Indigenous style: a guide for writing by and about Indigenous Peoples. Brush Education Inc

History of Soil Studies in Oregon

2

2.1 Introduction

Oregon has a rich and long history of soils investigations that began at the turn of the twentieth century with 1:63,360-scale soil surveys of Baker City (Jensen and Mackie 1903) and Salem (Jensen 1903) by the Bureau of Soils under the auspices of Milton Whitney. The first countywide soil survey was of Yamhill County during World War I (Kocher et al. 1917). As of 2020, all or a large portion of 32 of the 36 counties in Oregon have been mapped, generally at a scale of 1:24,000. Soil mapping continues at present, largely by the Natural Resources Conservation Service (NRCS), with the assistance of federal agencies such as the US Forest Service (USFS), the Bureau of Land Management (BLM), the US Fish and Wildlife Service, the National Park Service, and Indigenous tribes. This work has been complemented with soils research by university, the Natural Resources Conservation Service (NRCS), BLM, and USFS personnel over the past 55 years.

2.2 Definition of Soil

There are many definitions for soil ranging from the utilitarian to a description that focuses on material. Soil has been recognized as (i) a natural body, (ii) a medium for plant growth, (iii) an ecosystem component, (iv) a vegetated water-transmitting mantle, and (v) an archive of past climate and processes. In this book, we follow the definition given in the *Keys to Soil Taxonomy* (Soil Survey Staff, 2014, p. 1) that the soil "is a natural body comprised of solids (minerals and organic matter), liquid, and gases that occurs on the land surface, occupies space, and is characterized by one or both of the following: horizons, or layers, that are distinguishable from the initial material as a result of additions, losses, transfers, and transformations of energy and matter or the ability to support rooted plants in a natural environment."

2.3 Soil Surveys

Under the auspices of Milton Whitney, the Bureau of Soils conducted the first soil surveys in Oregon at Baker City (Fig. 2.1) and Salem (Fig. 2.2). The first countywide soil survey in Oregon was of Yamhill County Area in 1917 (Fig. 2.3).

Soil mapping in Oregon increased from 1900 to 1930 and then proceeded slowly from 1930 to 1970 as the United States endured the Great Depression, World War II, and the Vietnam War (Fig. 2.4). The cumulative number of soil surveys increased sharply from 1970 to 2010 and has since tapered off. These trends also are reflected by the cumulative area of the state that has been mapped. Nearly two-thirds (64%) of the soil mapping to date occurred between 1975 and 1990 (Fig. 2.5). These trends reflect the development of a new soil classification system, the *Seventh Approximation*, when soil mapping began in earnest in Oregon. The last archived soil survey in Oregon was of Tillamook County in 2013 (Table 2.1). At the present time, mapping is reported digitally on Web Soil Survey, and about 68% of the state has received detailed soil mapping (Fig. 2.6). Only Alaska and Idaho have greater proportions of areas unmapped in the United States. About 26% of Oregon is undergoing initial surveying, particularly in Malheur, Grant, Wheeler, and Crook Counties and in portions of Linn, Lane, and Klamath Counties. When these areas are complete, only about 6% of the state will remain unmapped.

Table 2.1 provides a listing for all of Oregon's archived published soil surveys and their publication date. All of the survey text material is available online. In about 2005, the USDA phased out printing soil survey reports and made the Web Soil Survey the official source for soil survey information. In most cases, published soil survey reports are available for reference and information at local NRCS offices and public libraries. However, copies for distribution are not available. Official soil survey information for all of the Oregon soil survey areas shown in Fig. 2.6 is available online using Web Soil

T. Thorson et al., *The Soils of Oregon*, World Soils Book Series,
https://doi.org/10.1007/978-3-030-90091-5_2

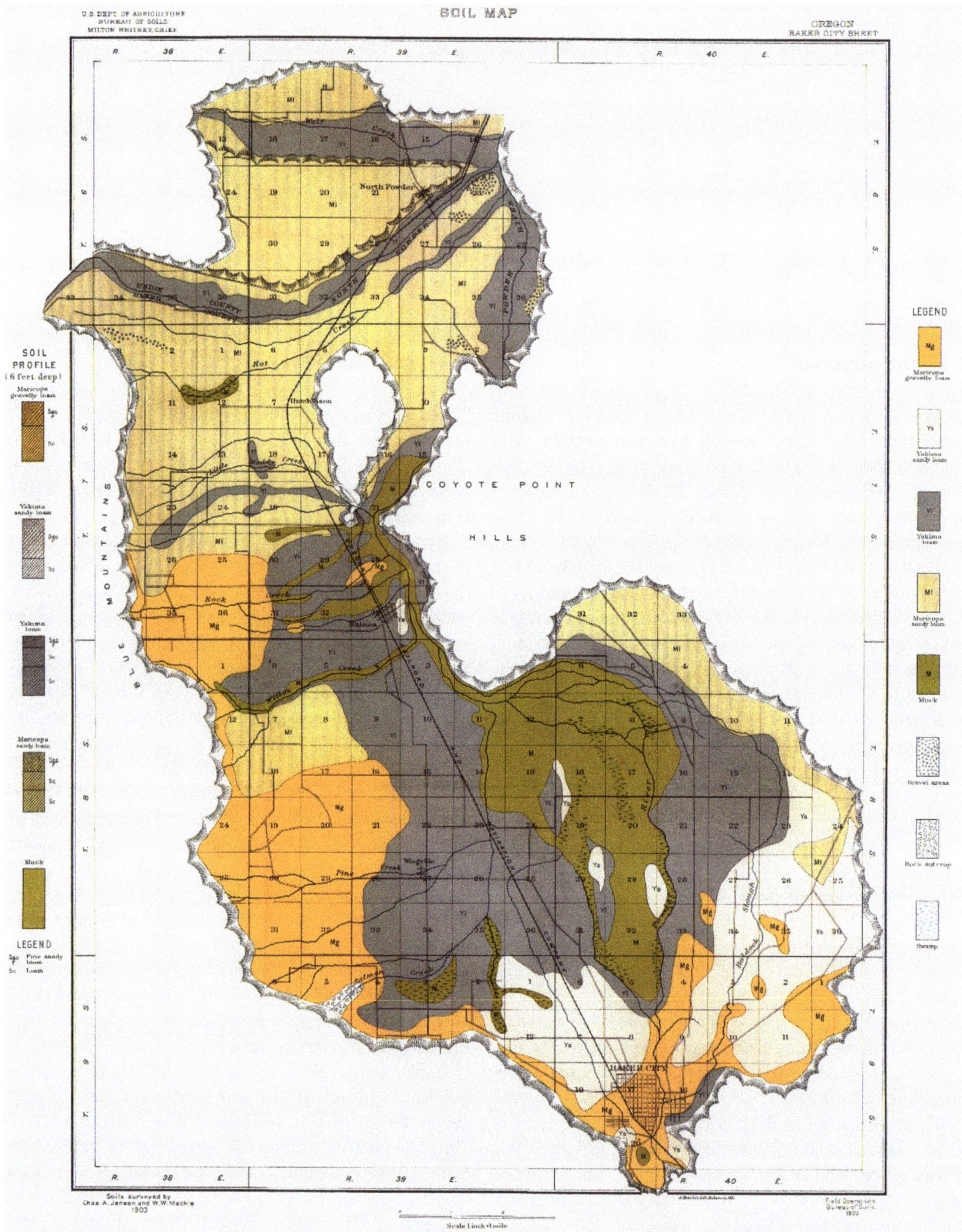

Fig. 2.1 Soil map of Baker City, Oregon, published in 1903 at a scale of 1:63,360. *Source* USDA Bureau of Soils, 1903

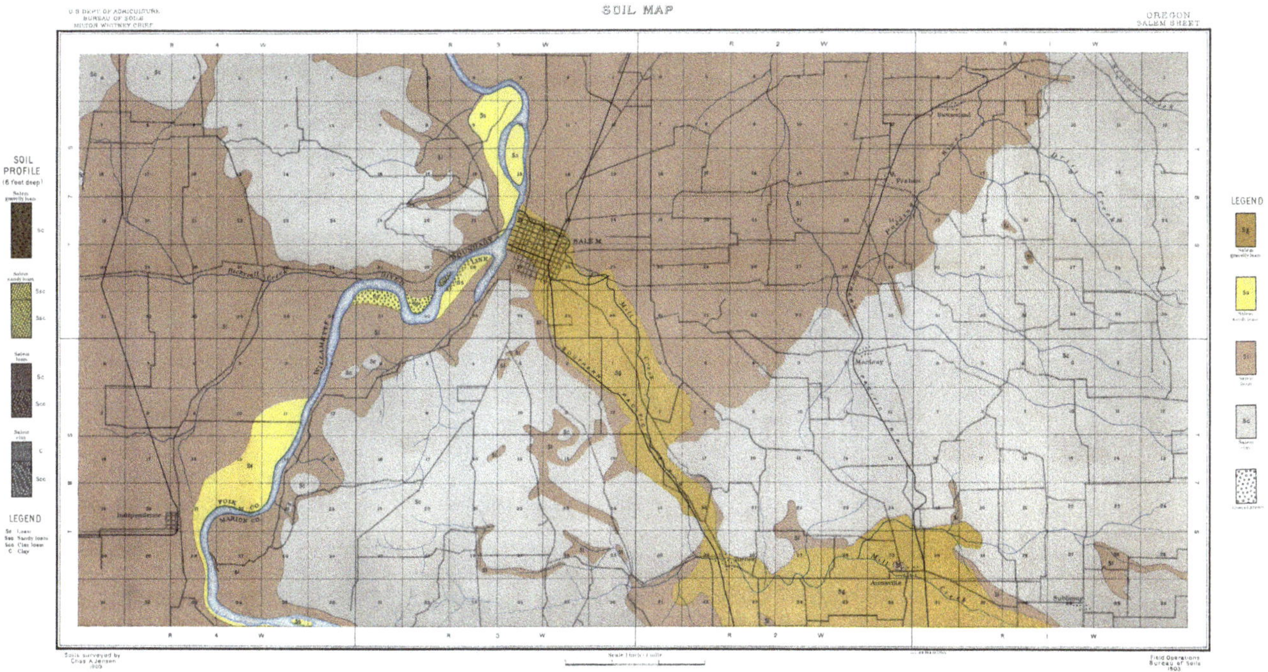

Fig. 2.2 Soil map of Salem, Oregon, published in 1903 at a scale of 1:63,360. *Source* USDA Bureau of Soils, 1903

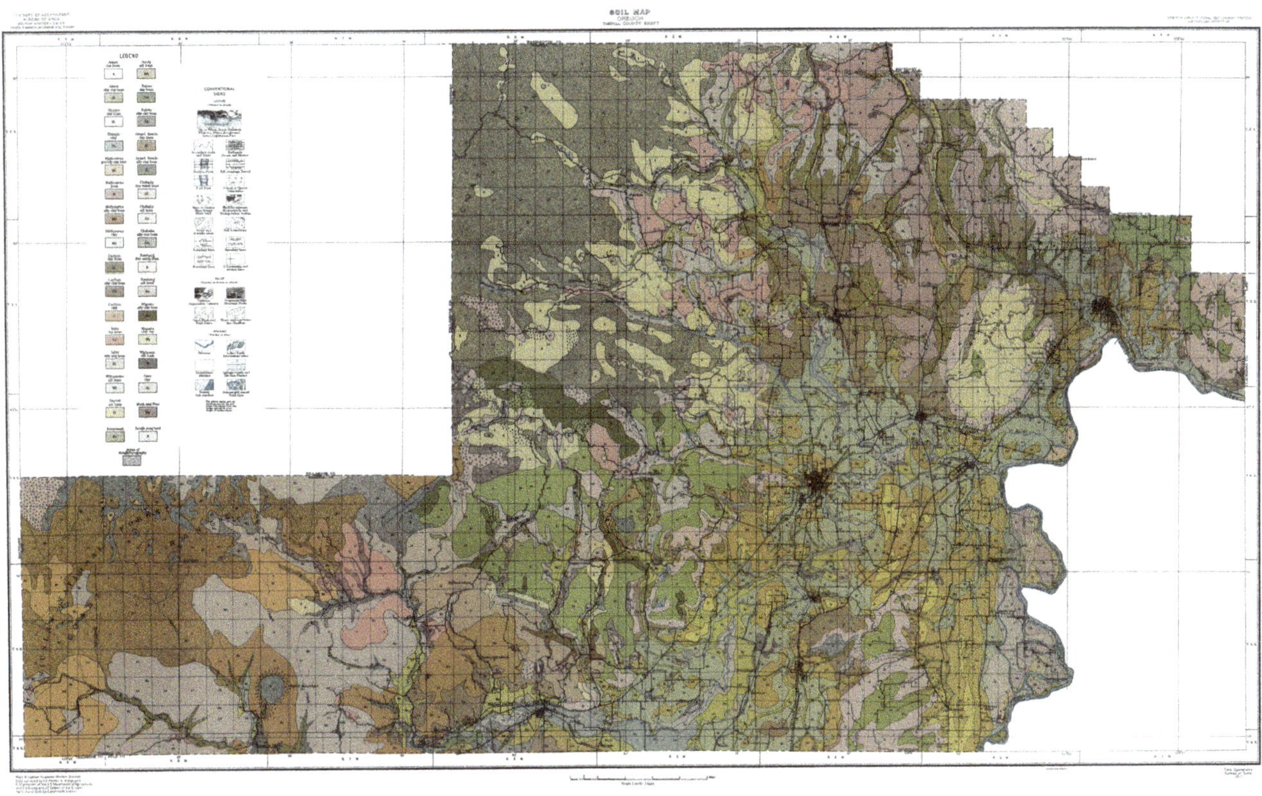

Fig. 2.3 General soil map of Yamhill County, Oregon, published in 1917 at a scale of 1:63,360. *Source* USDA Bureau of Soils, 2017

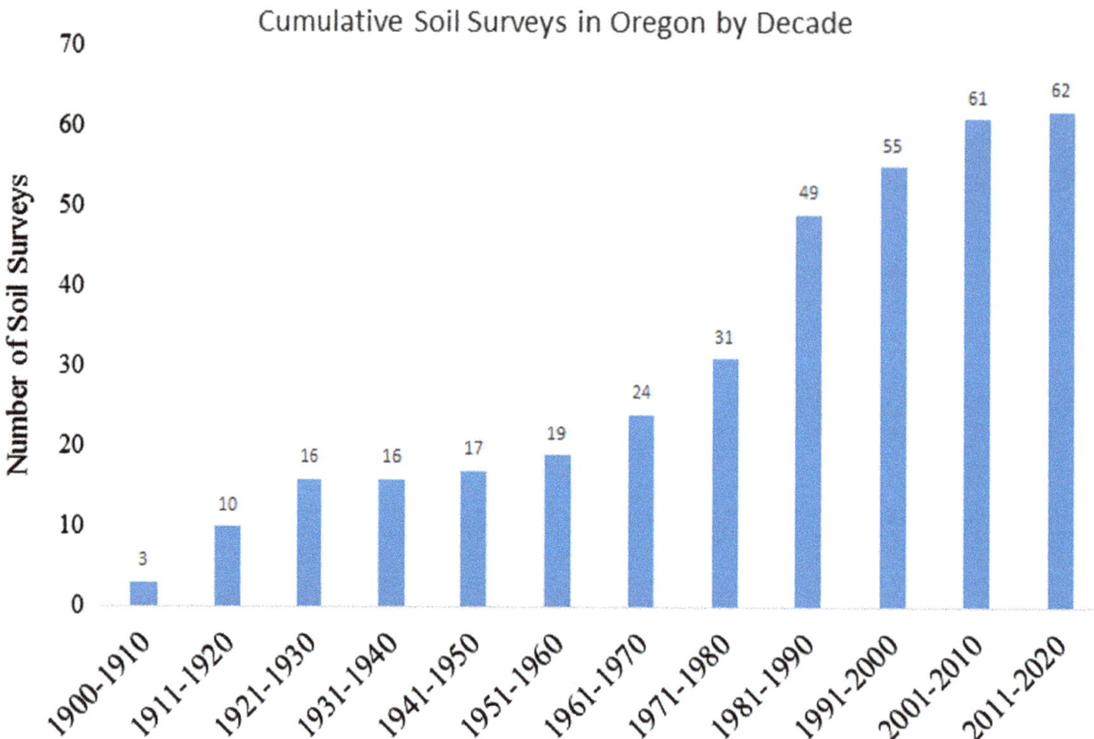

Fig. 2.4 Cumulative soil surveys in Oregon by decade

Fig. 2.5 Cumulative area mapped in Oregon by decade

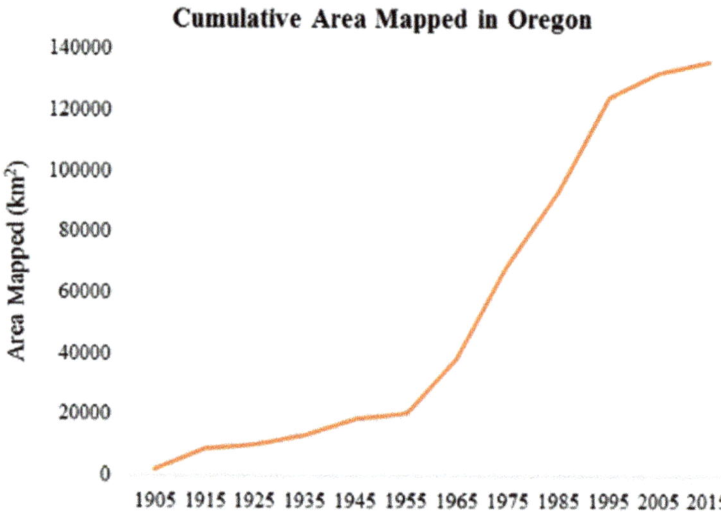

Survey at https://websoilsurvey.sc.egov.usda.gov. To start using Web Soil Survey, click on one of the links under Browse by Subject or on the large green button that says start WSS. The amount of soil information for any given survey will depend on the completeness of the survey. Areas in green in Fig. 2.6 are very complete, areas in yellow are ongoing and data is less complete, and areas in gray have little or no information.

2.4 Soil Series

The early concept of the soil series is that all occurrences of a series would have the same, or very nearly the same, chemical composition because they were derived from the same rocks. Maps completed between 1903 and 1914

Table 2.1 Archived Oregon soil survey publications

Soil survey name	Publ. date
Alsea Area	1973
Astoria Area	1949
Baker Area	1954
Baker City Area	1903
Baker County Area	1997
Benton County Area	1975
Benton County	1920
Benton County	2009
Clackamas County Area	1985
Clackamas County	1926
Clatsop County	1985
Columbia County	1925
Columbia County	1986
Coos County	1989
Crater Lake National Park	2009
Curry Area	1970
Curry County	2005
Deschutes Area	1958
Douglas County Area	2003
Eugene Area	1930
Gilliam County	1984
Grand Ronde Valley Area	1930
Grant County, Central Part	1981
Harney County Area	2006
Hood River County Area	1981
Hood River-White Salmon River Area, OR-WA	1914
Jackson County Area	1993
Josephine County	1923
Josephine County	1983
Klamath County, Southern Part	1985
Klamath Reclamation Project	1908
Lake County, Northern Part	2012
Lake County, Southern Part	1999
Lane County Area	1987
Lincoln County Area	1997
Linn County Area	1987
Linn County	1924
Malheur County, Northeastern Part	1980
Marion County Area	1972
Marion County	1927
Marshfield Area	1911
Medford Area	1913
Morrow County	1983
Multnomah County	1919

(continued)

Table 2.1 (continued)

Soil survey name	Publ. date
Multnomah County	1983
Polk County	1928
Polk County	1982
Prineville Area	1966
Salem Area	1903
Sherman County	1964
Sherman County	1999
South Umpqua Area	1973
Tillamook Area	1964
Tillamook County	2013
Trout Creek- Shaniko Area	1975
Umatilla Area	1948
Umatilla County Area	1988
Union Area	1985
Upper Deschutes River Area	1999
Wallowa Area	2007
Warm Springs Indian Reservation	1998
Wasco County, Northern Area	1982
Washington County	1919
Washington County	1982
Yamhill Area	1974
Yamhill County	1917

showed 7 or fewer soil series and 15 or fewer soil types. Those created from 1917 to 1929 showed between 14 and 20 soil series, 27–41 soil types, and 2–6 land types (Table 2.2). In *Soils of Oregon: Their Classification, Taxonomic Relationships, and Physiography*, Huddleston (1979) reported about 800 soil series, of which over 100 series are no longer recognized in Oregon, and over 900 series have been added. Many of the early maps, particularly those of Benton County in 1920 and Marion County in 1927, were truly works of art.

By 2019 more than 1,700 soil series had been identified in Oregon, 73% of which are identified only in Oregon (Fig. 2.7). The historical trends in the number of soil surveys (Fig. 2.4) and soil series (Fig. 2.8) illustrate the marked growth in both surveys and identified series between 1961 and 2010.

2.5 Soil Classification

Early soil classification schemes developed in the United States were not used in soil survey reports. The 1938 zonal soil classification system developed by Baldwin and others was used in Oregon from 1954 to 1966. After more than a decade of development, in 1960 the *Seventh Approximation*, a new classification scheme and a precursor to *Soil Taxonomy* (Soil Survey Staff 1999) was published. Beginning in the Curry County Area in 1970, *Soil Taxonomy* was used exclusively in the state.

2.6 Soil Taxonomy

All soils in Oregon are now classified using *Soil Taxonomy*, which is used throughout this book. The *Keys to Soil Taxonomy* (Soil Survey Staff 2014) is an abridged companion document that incorporates all the amendments that have been approved to the system since publication of the second edition of *Soil Taxonomy* in 1999, in a form that can be used easily in a field setting. *Soil Taxonomy* is a hierarchical classification system that classifies soils based on the properties of diagnostic surface and subsurface horizons. Both *Soil Taxonomy* and *Keys to Soil Taxonomy* are available online and readers are directed to those publications for detailed explanations of concepts.

For classification purposes, the upper limit of the soil is defined as the boundary between the soil (including organic

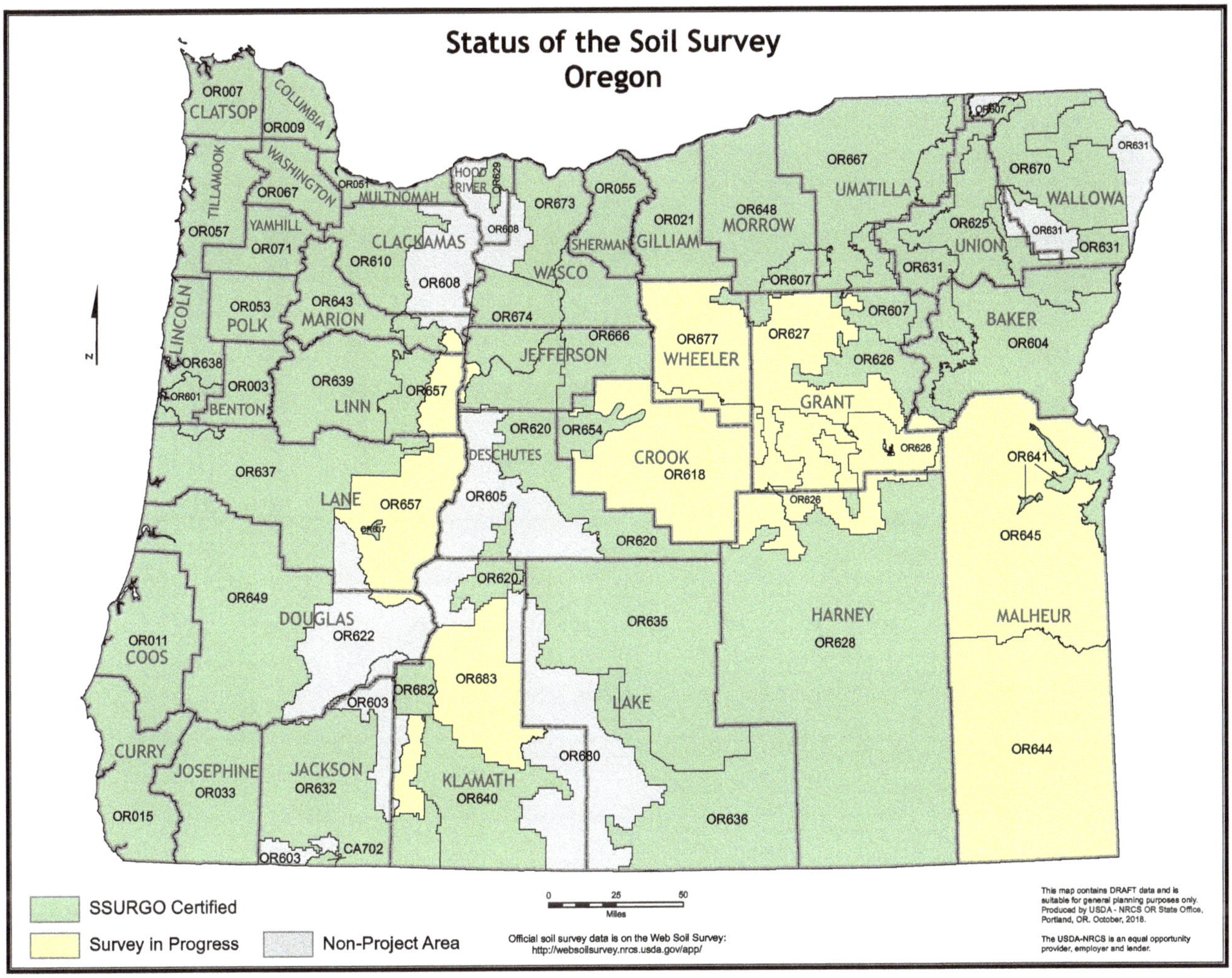

Fig. 2.6 Status of soil surveying in Oregon as of 2018. *Source* NRCS

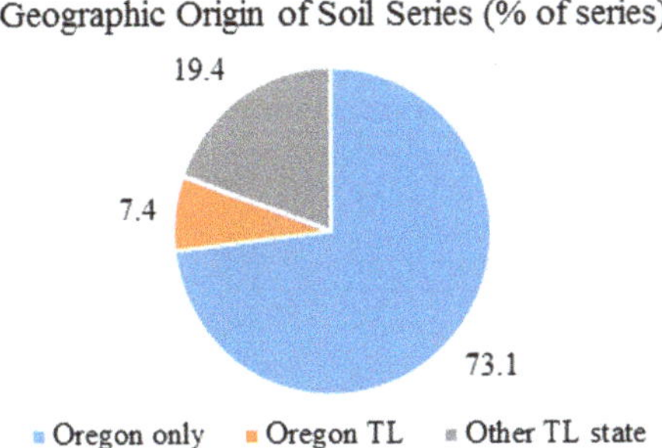

Fig. 2.7 Proportion of soil series that occur only in Oregon, elsewhere but where Oregon is the lead state (TL), and elsewhere where Oregon is not the lead state (Other TL)

horizons) and the air above it. The lower limit is arbitrarily set at 200 cm. The definition of the classes (taxa) is quantitative and uses well-described methods of analysis for the diagnostic properties. The assumed genesis of the soil is not used in the system and the soil is classified "as it is" using morphometric observations in the field coupled with laboratory analysis and other data. The nomenclature in *Soil Taxonomy* is mostly derived from Greek and Latin sources, as is done for the classification of plants and animals.

Soil Taxonomy classifies soils, from broadest to narrowest levels, into orders, suborders, great groups, subgroups, and families. *Soil Taxonomy* is similar to the taxonomic classification of living organisms, from broadest to narrowest levels, into kingdom, phylum, class, order, family, genus, and species. Casual readers may be unfamiliar with soil taxonomic terms, which often employ abbreviations based on Latin or Greek words. Those readers will benefit from a glossary of terms, such as the online Soil Formation and Classification provided by the NRCS (2022) or Chap. 7, Nomenclature, of Soil Taxonomy (1999). Introductory soil science textbooks, such as Weil and Brady's *The Nature and Properties of Soils* (2016), also explain soil taxonomy, as

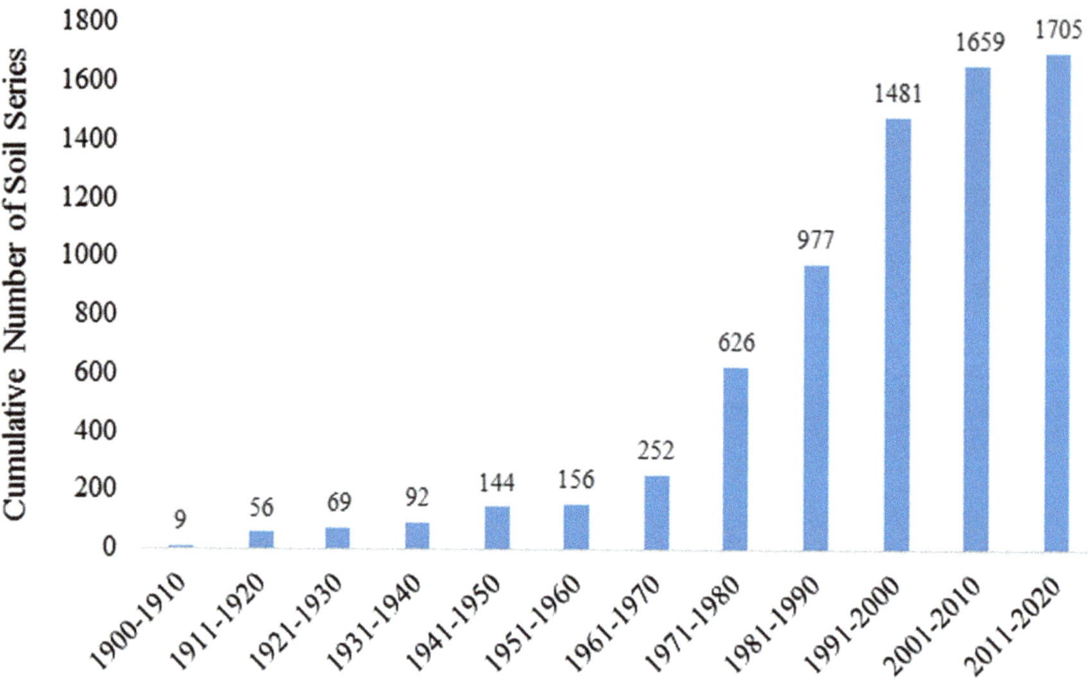

Fig. 2.8 Cumulative number of soil series recognized in Oregon by decade

Table 2.2 Number of soil series, soil type, and land types recognized in 1:62,500-scale soil surveys in Oregon from 1908 to 1929

Year	Area	No. soil Series	No. soil Types	No. land Types	Total
1908	Klamath Reclamation Project	2	9	1	10
1911	Marshfield County Area	7	15	5	20
1913	Medford County Area	23	42	4	46
1914	Hood R.-White Salmon R	7	16	2	18
1917	Yamhill Co. Area	15	29	4	33
1919	Multnomah County Area	14	27	6	33
1920	Benton County Area	16	30	2	32
1923	Josephine County Area	14	37	5	42
1926	Clackamas County Area	19	33	4	37
1927	Marion County Area	20	41	4	45
1928	Polk County Area	16	31	2	33
1929	Columbia County Area	16	34	4	38

well as other aspects of soil science. In any case, the soil taxonomic system is logical and consistent, and reveals a wealth of information about soils in a concise manner.

Soil taxonomic names are constructed in an order that progresses from narrow to broad, unlike biological classifications, which progress from broad to narrow. For example, the taxonomic classification of Douglas-fir progresses from kingdom (broad) to species (narrow) as follows: Plantae (kingdom)-Pinopsida-Pinales-Pinaceae-*Pseudotsuga-menziesii* (species). In contrast, a soil taxonomic classification for the Puderbaugh soil series to the subgroup level is "Pachic Argixeroll." In this compact and efficient system, the following terms are used, progressing from narrow to broad:

- "Pachic" describes a soil subgroup (narrow classification) with a thick epipedon,
- "Argi" describes a soil great group with a horizon that has accumulated clays from an overlying horizon,
- "xer" describes a soil suborder with an annual dry season, and
- "oll" describes a soil in the order of Mollisols (broad classification).

Using this system, taxonomic names may be applied at various levels of detail. For example, "Pachic Argixeroll" describes a subgroup, "Argixeroll" describes a great group, "Xeroll" describes a suborder, and "Mollisol" is an order. Soil associations, soil complexes (composed of two or three major soil series and miscellaneous area components), and consociations (comprising a single major component) constitute the primary soil map units, which are shown on soil maps. For example, Puderbaugh–Puderbaughridge complex, 15–30% slopes, is a soil map unit.

Eight diagnostic surface horizons (epipedons) are defined in soil taxonomy and five of them occur in Oregon: histic, melanic, mollic, umbric, and ochric (Table 1.3). The histic epipedon contains primarily organic materials and is saturated for prolonged periods during the year. The melanic epipedon has a dark color (value 2.5 or less and chroma 2 or less throughout), abundant organic carbon (6% or more), and a melanic index of 1.70 or less. The mollic and umbric epipedons occur in mineral soils and are thick, dark-colored, and enriched in organic matter. The mollic epipedon is enriched in base cations, such as calcium, magnesium, and potassium, while the umbric epipedon contains low amounts of these cations. The ochric epipedon is thin, commonly light-colored, and often low in organic matter content.

Ten of the 20 diagnostic subsurface horizons identified in soil taxonomy are present in the soils of Oregon (Table 2.3). The albic horizon is composed of materials from which clay and/or free iron oxides have been removed by eluviation to a degree that primary sand and silt particles impart a light color to the horizon. The argillic horizon is enriched in clay that has moved down the profile from percolating water. The calcic horizon features a significant accumulation of

Table 2.3 Definitions of diagnostic horizons present in Oregon soils*

Diagnostic surface horizons (epipedons)	
Histic	greater than 20 cm thick; organic matter content 16% or more; saturated for more than 30 days in normal years unless artificially drained
Melanic	at least 30 cm thick; black (moist color value and chroma 2 or less); organic C 6% or more; melanic index 1.70 or less
Mollic	at least 18 cm thick; dark-colored; organic C 0.6% or more; base saturation 50% or more
Ochric	an altered horizon that fails to meet the requirements of other epipedons; lacks rock structure or finely stratified fresh sediments; includes underlying eluvial horizons such as albic
Umbric	at least 18 cm thick; dark-colored; organic C 0.6% or more; base saturation less than 50%
Diagnostic subsurface horizons	
Albic	a light-colored eluvial horizon 1 cm or more in thickness; composed of albic materials
Argillic	an illuvial horizon that gives evidence of translocation of clay, based on the ratio of that in the clay-enriched horizon to an overlying eluvial horizon, the presence of clay films (argillans)
Calcic	a non-cemented horizon of secondary carbonate accumulation with at least 15% calcium carbonate equivalent in a horizon, that is at least 15 cm thick, and has at least 5% more carbonate than an underlying layer
Cambic	an altered horizon that shows color and/or structure development, is at least 15 cm thick, and has a texture of very fine sand, loamy very fine sand, or finer
Duripan	a horizon that is cemented in more than 50% of the volume by opaline silica; air-dry fragments do not slake in water of HCL but do slake in hot concentrated KOH; restricts rooting of plants except in vertical cracks that have a horizontal spacing of 10 cm or more
Fragipan	subsoil horizon 15 cm or more thick; high bulk density; brittle when dry and firm or very firm when moist; not effervescent in HCl; air-dry fragments slake in water
Natric	meets the requirements of an argillic horizon but also has prismatic, columnar or blocky structure, an exchangeable sodium percentage of 15 or more, or a sodium adsorption ratio of 13 or more
Ortstein	a layer of spodic materials 25 mm or more in thickness; more than 50% is cemented
Salic	a horizon of accumulation of salts more soluble than gypsum that is at least 15 cm thick; the electrical conductivity (EC) is at least 30dS/m for 90 consecutive days; the product of horizon thickness in cm and the EC is 900 or more
Spodic	an illuvial layer containing at least 85% spodic materials; 2.5 cm or more thick; spodic materials must have a pH value in 1:1 water of 5.9 or less; 0.6% organic C or more; an optical density of oxalate extract (ODOE) of 0.25 or more and that value that is at least twice that of an overlying eluvial horizon; an Al_o + ½ Fe_o percentage of 0.50 or more and that value is at least twice that of an overlying eluvial horizon

*Revised from Buol et al. (2011)

secondary calcium carbonates; this horizon must be 15 cm or more thick, have a 5% or more $CaCO_3$ equivalent, and not be cemented. The cambic horizon shows minimal development other than soil structure and color.

A duripan is a silica-cemented subsurface horizon with or without auxiliary cementing agents. The duripan is cemented in more than 50% of the volume of some horizon and shows evidence of the accumulation of opal or other forms of silica, such as laminar caps, coatings, lenses, partly filled interstices, bridges between sand-sized grains, or coatings on rock fragments. Less than 50% of the volume of air-dry fragments slakes in 1 M hydrochloric acid (HCl), but more than 50% slakes in concentrated potassium hydroxide (KOH) or sodium hydroxide (NaOH). Roots can only penetrate the duripan in vertical fractures with a horizontal spacing of 10 cm or more. Secondary calcium carbonate is often an accessory cementing agent in duripans. Duripans are very common in the soils of southeastern Oregon due in large part to the presence of soluble volcanic glass in soil parent materials.

A fragipan is a subsurface layer that is 15 cm or more in thickness, shows evidence of pedogenesis, has a high bulk density, air-dry fragments that slake in water, a firm or firmer rupture resistance class and a brittle manner of failure at field capacity, and is not effervescent in HCl. The natric horizon is a type of argillic horizon, which shows evidence of clay illuviation that has been accelerated by the dispersive properties of sodium. The spodic horizon is an illuvial layer containing at least 85% spodic material and is at least 2.5 cm thick. Spodic materials contain illuvial amorphous materials of organic matter and Al, with or without Fe. Ortstein is a layer of spodic materials more than 25-mm thick, in which more than 50% is cemented.

A salic horizon features the accumulation of salts that are more soluble than gypsum in cold water. This horizon must be 15 cm or more thick, have an electrical conductivity (EC) for 90 consecutive days or more of 30 dS/m or more in the water extracted from a saturated past; and have a product of the EC (dS/m) and thickness (cm) of 900 or more. Photographs of subsurface horizons are given in chapters describing soils in each of the orders represented in Oregon.

Soil orders are defined primarily on the basis of diagnostic soil characteristics and diagnostic surface and subsurface horizons. Ten of the 12 orders in soil taxonomy occur in Oregon: Mollisols, Inceptisols, Aridisols, Andisols, Ultisols, Alfisols, Entisols, Vertisols, Spodosols, and Histosols (Table 2.4). Mollisols are dark-colored, base-enriched grassland soils. Inceptisols are juvenile soils that contain an epipedon and either a cambic horizon, a salic horizon, or a high exchangeable sodium percentage. Aridisols are dry soils that feature the accumulation of clay or some salts. Andisols are soils having andic properties, with the upper part of the solum developed in volcanic ejecta and/or volcaniclastic materials, or in materials with abundant organic carbon, iron and aluminum, a low bulk density, and high phosphate retention. Ultisols have an argillic horizon and a base saturation less than 35% at a depth of 180 cm. Alfisols are base-enriched forest soils with an argillic horizon. Entisols are very poorly developed, recent soils that may have only an anthropic or ochric epipedon. Vertisols are derived from abundant swelling clays that lead to cracks and slickensides. Histosols are organic soils.

Table 2.4 Simplified key to soil orders in Oregon*

Histosols	Soils that do not have andic soil properties in 60% or more of the upper 60 cm and have organic soil materials in two-thirds or more of the total thickness
Spodosols	Other soils with a spodic horizon within a depth of 200 cm
Andisols	Other soils with andic soil properties in 60% or more of the upper 60 cm
Vertisols	Other soils with a layer 25 cm or more thick containing either slickensides or wedge-shaped peds, have more than 30% clay in all horizons between depths of 18 and 50 cm or a root-limiting layer if shallower, and have cracks that open and close periodically
Aridisols	Other soils with either an aridic soil moisture regime and some diagnostic surface and subsurface horizons or a salic horizon accompanied by both saturation within 100 cm of the soil surface and dryness in some part of the soil moisture control section during normal years
Ultisols	Other soils with an argillic horizon and a base saturated percentage at pH 8.2 less than 35 at a depth of 180 cm
Mollisols	Other soils with a mollic epipedon and a base saturation (by ammonium acetate at pH 7) of 50% or more in all depths above 180 cm
Alfisols	Other soils with an argillic or natric horizon
Inceptisols	Other soils with an umbric or mollic epipedon, or a cambic horizon, or a salic horizon, or a high exchangeable sodium percentage which decreases with increasing depth accompanied by ground water within 100 cm of the soil surface
Entisols	Other soils

*Revised from Buol et al. (2011)

Suborders are distinguished on the basis of soil climate for six of the ten orders occurring in Oregon: the Alfisols, Andisols, Inceptisols, Mollisols, Ultisols, and Vertisols. Soil parent materials are used to differentiate suborders of Vertisols and Entisols; and the amount of clay or types of salts are used to differentiate suborders of Aridisols. There are 41 suborders of soils in Oregon. Great groups are distinguished from a variety of soil characteristics; there are 110 great groups of soils in Oregon.

2.7 General Soil Maps

We have been able to locate three general soil maps of Oregon. The first map, published in 1973, is at a scale of 1:2 million and shows 19 physiographic regions that represent hundreds of different soils (Fig. 2.9). A revision of this map was published in 1975 showing soil-order associations for the 19 physiographic regions (Figs. 2.10 and 2.11). The 1986 General Soil Map of the State of Oregon divided the state into 25 general groups and 136 general soil map units. The groups were identified by the letters A through Y, each having a unique color (Fig. 2.12). The map units were identified by a two-character symbol; the first character indicated the group it occurred in and the second character by an Arabic number (examples A1, M12). The groups were based on a combination of soil-forming factors, notably climate and landform.

In 1997, the NRCS used STATSGO data to prepare a 1:7.5 million-scale map of "dominant soil orders" of the conterminous United States. The Oregon portion of this map shows Mollisols as being dominant (dark green color), followed by Inceptisols (brown), Andisols (red), and Aridisols (cream color) (Fig. 2.13). Spodosols (pink), Alfisols (olive

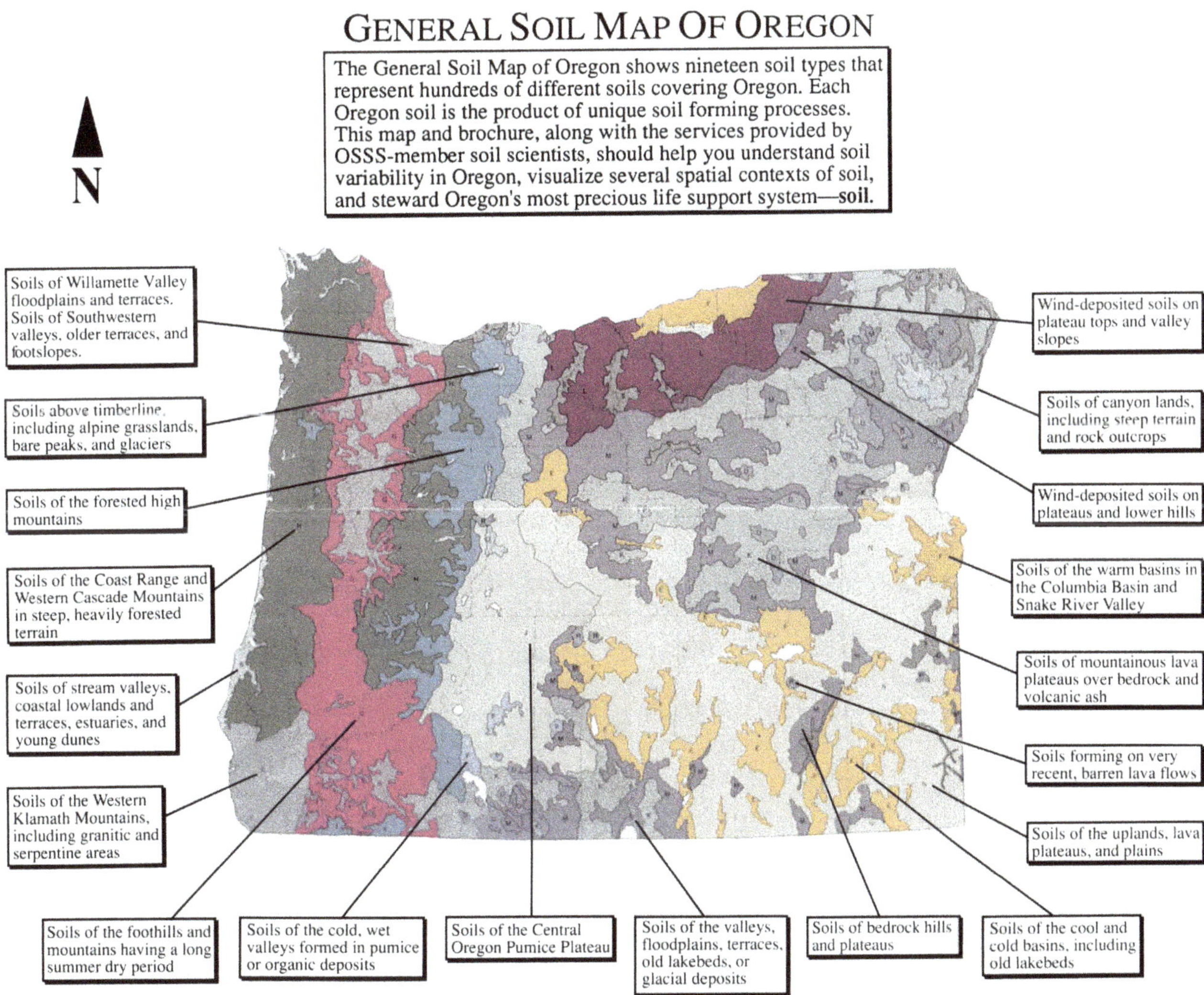

Fig. 2.9 General soil map of Oregon produced in 1973. *Source* NRCS, 1973

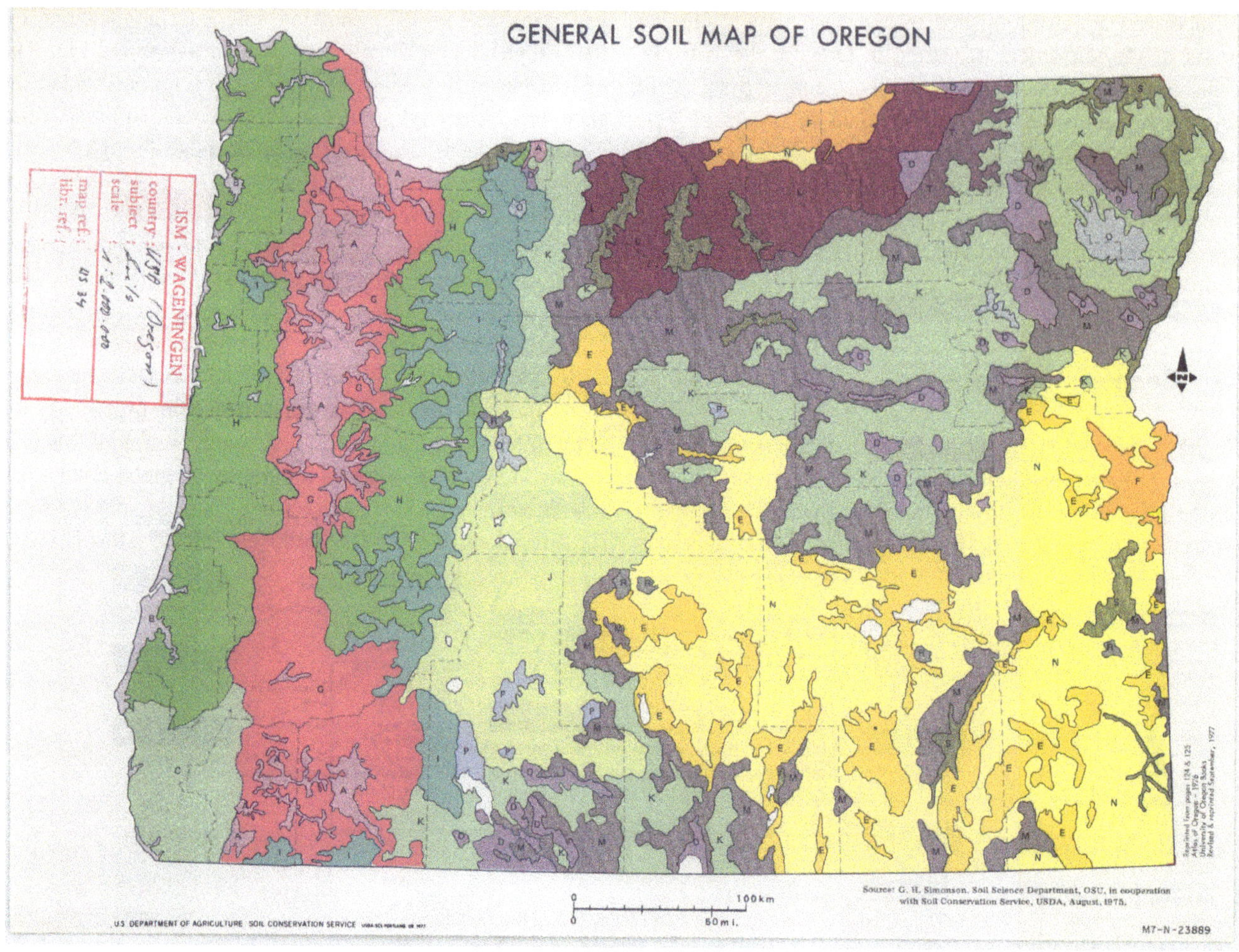

Fig. 2.10 General soil map of Oregon produced in 1975. The legend follows on the next sheet. *Source* G.H. Simonson, Soil Science Department, Oregon State University, in cooperation with Soil Conservation Service

green), Ultisols (orange), Entisols (blue), and Vertisols (yellow) occupy smaller areas. Exposed bedrock and lava flows are shown in black.

Using the STATSGO2 and SSURGO "dominant conditions" databases, Whityn Owen prepared a soil order map of Oregon at a scale of 1:3.2 million (Fig. 2.14). This map shows a substantial increase in the area of Aridisols because of more complete mapping in Lake, Harney, and Malheur Counties.

2.8 Soil Research

Oregon has benefited from considerable soil research by university and NRCS investigators over the past 65 years. Four key areas of soil research have been conducted in the state, including (i) recognition of soil-geomorphic surfaces by Roger B. Parsons, C.A. Balster, G.F. Kling, J.R. Glasmann, R.L. Herriman, and O.A. Ness; (ii) general soil

Of the ten soil orders which have worldwide distribution, Oregon includes nine. *Aridisols* are light-colored soils of dry regions; *Mollisols* have dark surfaces formed under grass; *Alfisols* are produced, under forests, by downward leaching of clay; *Ultisols* are similar but their bases have been more completely depleted; *Spodosols* reflect translocation of iron or aluminum and organic matter under forest; *Inceptisols* represent early stages of soil formation; *Vertisols* are dominated by swelling clays; *Histosols* are formed in organic deposits; and *Entisols* lack genetic horizons.

These soil orders may be further divided as follows: *Xeric* soils develop under summer-dry climates; *udic* soils under wet climates, and *aridic* soils where the climate is dry all year. The map illustrates 19 kinds of soil landscape, combining soil orders that occur in similar climates and terrain.

XERIC SOILS OF MODERATE RAINFALL WESTERN OREGON REGIONS

Mollisols, Alfisols, Vertisols and Ultisols of Valleys (A)

Soils of the Willamette Valley floodplain and terraces are dominantly deep, silty, moderately dark and somewhat acid. Poorly-drained soils are common. Soils of the southwestern valleys are generally less silty, more variable in depth and lighter-colored, although dark clay Vertisols are common. Reddish, strongly-leached Ultisols occur on older terraces and footslopes.

Alfisols, Ultisols, Inceptisols, Mollisols and Vertisols of Foothills and Mountains (G)

These uplands have a pronounced summer dry period. The soils are highly variable, but mostly dark at the surface, clayey and moderately acid in the north, becoming lighter-colored, less acid and less clayey in the south.

UDIC SOILS OF HIGH RAINFALL WESTERN OREGON REGIONS

Spodosols, Inceptisols and Entisols of Valleys and Coastal Lowlands (B)

These soils are strongly acid, mainly dark and deep. Poorly drained, clayey soils are common in the stream valleys and estuaries. Gravelly soils predominate in western Cascade valleys. Sandy Spodosols, commonly with iron-cemented subsoils, occur on lower coastal terraces; Entisols on younger dunes.

Inceptisols and Ultisols of the Coast Range and Western Cascade Mountains. (H)

These areas are steep and heavily forested. The soils are mostly dark, loamy or clayey. Shallow, stony soils are common with deeper, reddish, clayey Ultisols on smoother, more stable slopes.

Inceptisols, Alfisols and Ultisols of the Western Klamath Mountains (C)

These soils are dominantly light-colored, medium to slightly acid, loamy, and commonly stony and shallow. Granitic areas have sandy soils and serpentine areas have mostly shallow, reddish, clayey soils.

Inceptisols, Spodosols and Entisols of High Mountains (I)

Soils of the forested higher elevations are mostly dark, moderately to strongly acid, shallow and stony. Light-colored, coarse-textured Entisols and Spodosols occur on glacial deposits.

XERIC SOILS OF MODERATE RAINFALL EASTERN OREGON FORESTED REGIONS

Entisols and Spodosols of the Central Oregon Pumice Plateau (J)

Soils formed on coarse pumice are mostly light-colored, coarse-textured, and moderately to slightly acid. Strongly acid Spodosols with subsoils of iron accumulation have formed at higher elevations. Poorly-drained soils occur in depressions.

Mollisols and Inceptisols of the Mountains (K)

Soils of the mountainous lava plateaus are mostly dark, moderately to slightly acid, loamy, moderately deep or shallow and stony. Tuffaceous bedrock areas have clayey soils. Light-colored, silty Inceptisols from volcanic ash occur on broad plateaus and northerly slopes.

XERIC SOILS OF MODERATELY LOW RAINFALL EASTERN OREGON GRASSLAND REGIONS

Mollisols of the Valleys (D)

These soils are formed on floodplains and terraces, old lakebeds, or glacial deposits, and their textures vary widely. Most are dark, slightly acid and deep, but some have hardpans. Poorly-drained and salt-affected soils are common.

Mollisols, Histosols and Inceptisols of Cold, Wet Valleys (P)

These valleys have both dark and light-colored, mostly poorly drained soils formed in pumice or organic deposits.

Moderately Dark-Colored Mollisols of Loess-Mantled Plateaus (L)

Deep, silty soils with free lime in the lower subsoil predominate at lower elevations. Organic matter and depth of leaching gradually increase with elevation. Loess depth decreases to the south with moderately deep soils on plateau tops and shallow, stony soils on valley slopes.

Dark-Colored Mollisols of Loess-Mantled Plateaus (T)

These silty, slightly acid soils are mostly deep on lower hills and variable on steep terrain.

Mollisols of Bedrock Hills and Plateaus (M)

These soils are dark or moderately dark and slightly acid or neutral. The soils are mostly shallow and stony. Some areas have deeper soils formed from clayey sedimentary rock or wind deposits.

ARIDIC SOILS OF LOW RAINFALL EASTERN OREGON SHRUB-GRASSLAND REGIONS

Aridisols and Entisols of Warm Basins (F)

These soils are mainly light-colored calcareous or neutral, and deep although hardpans are common. Sandy soils predominate in the Columbia Basin. Most soils of the Snake River Valley are silty, and some have clayey subsoils over hardpans. Many of the soils on floodplains are poorly drained and salt-affected.

Aridisols of Cool and Cold Basins (E)

Many semi-arid basins formerly held large pluvial lakes and the soils are formed in deposits ranging from sand or gravel to clay. They are light-colored and neutral or calcareous. Many are poorly drained and salt-affected, and many have shallow hardpans.

Aridisols of Uplands (N)

Most of these soils are on lava plateaus or plains. They are shallow, stony and light-colored with clayey or loamy subsoils.

MISCELLANEOUS AREAS WHERE RELIEF OR BEDROCK DOMINATE SOIL FORMATION

Mollisols, Entisols and Inceptisols above Timberline (O)

These high mountain areas include alpine grasslands, bare peaks and glaciers. The soils are mainly dark, coarse-textured and very cold.

Barren Lava Flows (R)

Soils have not yet formed on these recent lava flows.

Mollisols, Aridisols and Inceptisols of Canyon Lands (S)

Canyon lands include major areas of steep, dissected terrain. The soils are shallow and stony with many rock outcroppings.

Bibliography

Soil Taxonomy. USDA Agr. Handbook 436, 1975, Soil Conservation Service, USDA.

Oregon's Long-Range Requirements for Water: Appendix I 1-15, General Soil Map Reports, 1969, State Water Resources Board, Salem.

M7-N-23889-I

Fig. 2.11 Legend for general soil map of Oregon produced in 1975. *Source* G.H. Simonson, Soil Science Department, Oregon State University, in cooperation with Soil Conservation Service

classification and genesis studies by Gerald H. Simonson, J. H. Huddleston, and E.G. Knox; (iii) soil–plant relationships by E.W. Anderson and K.W. Davies; (iv) mass wasting by F. J. Swanson, J.J. Roering, W.E. Dietrich, R.L. Sidle, D.N. Swanston, and S.C. Burns; and (v) forest soils by C.T. Youngberg, C.T. Dyrness, K.K. Cromack, R.F. Miller, M.E. Harmon, K. Laitha, and J.A. Hatten. G.J. Retallack has studied lithified paleosols in Oregon; J.G. Bockheim and H. M. Kelsey used soils to study tectonic uplift rates of marine terraces.

2.9 The State Soil

The Jory soil series, a fine, mixed, active, mesic Xeric Palehumults, was approved as the official representative soil of Oregon in 2011 (Fig. 2.15). The Jory soil occurs on over 900 km^2 of Willamette Valley foothills. The soil has formed in colluvium and residuum from sedimentary and basic igneous rocks on hillslopes under Douglas-fir and Oregon white oak vegetation. The Jory soil is used for orchards, Christmas trees, vineyards, cane berries, grass seed, timber production, wildlife habitat, and watershed health.

2.10 Summary

The first soil survey in Oregon was completed in 1903. Soil mapping increased exponentially from the early 1960s to 2011, and currently about 68% of the state has been mapped. Soil surveys in Oregon reflect historical changes in soil map units in the United States, progressing from a limited number of soil series prior to 1938, mapping of zonal great soil groups until 1968, and the use of *Soil Taxonomy* thereafter.

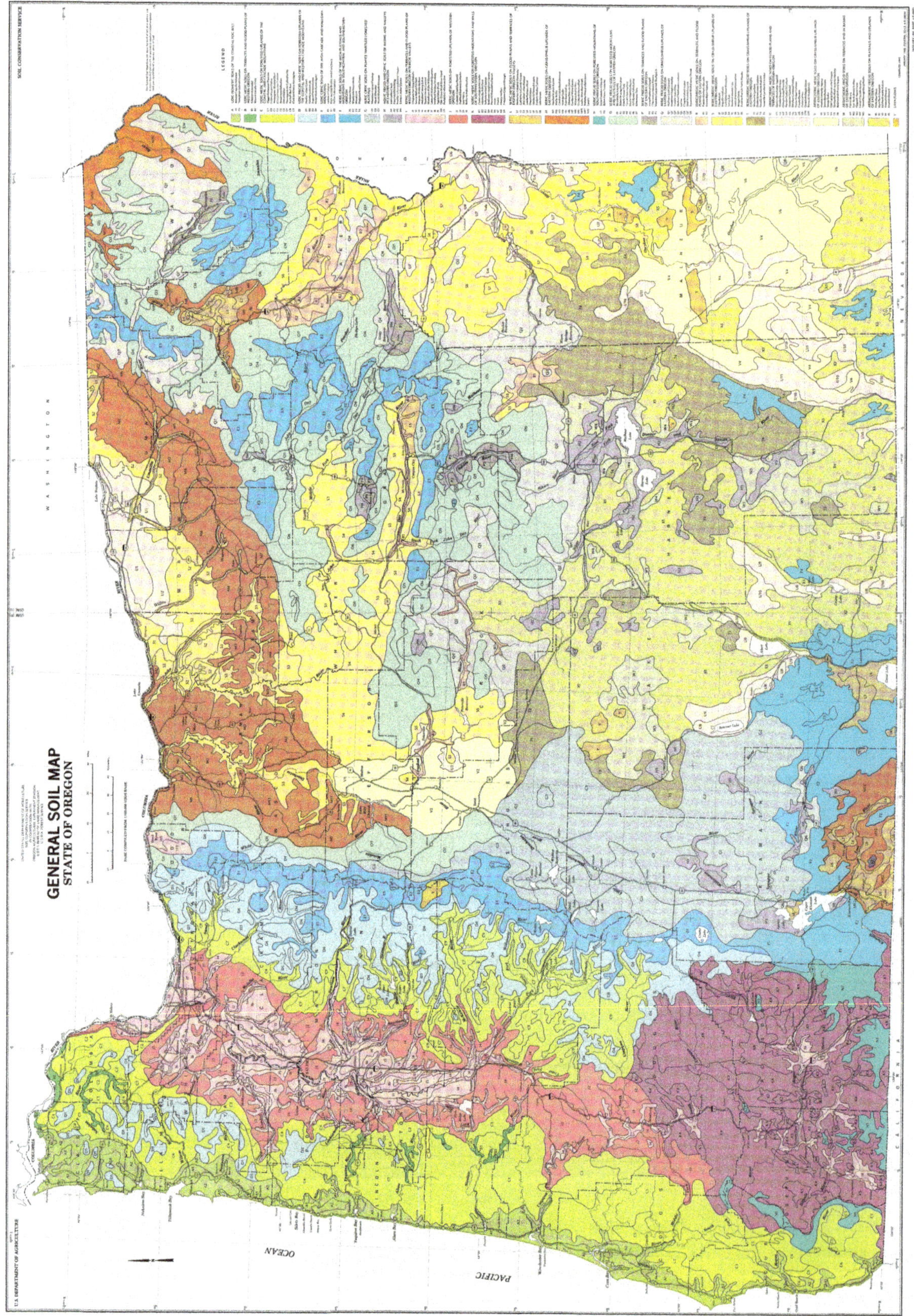

Fig. 2.12 General soil map of Oregon produced in 1986. *Source* Soil Conservation Service, et al., 1986

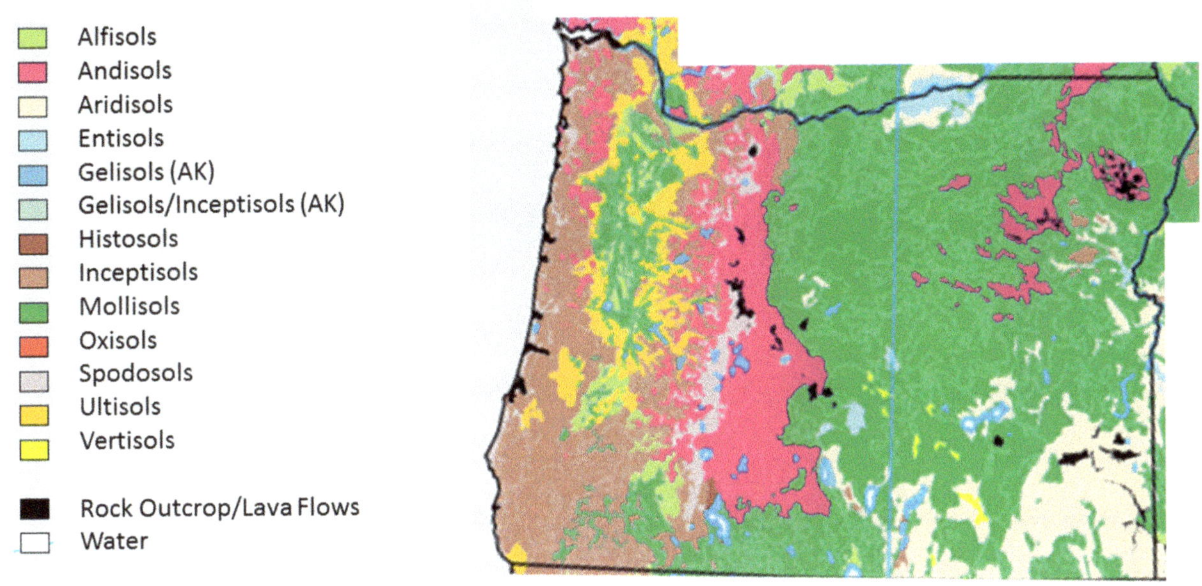

Fig. 2.13 Soil order map of Oregon. *Source* NRCS, 1997

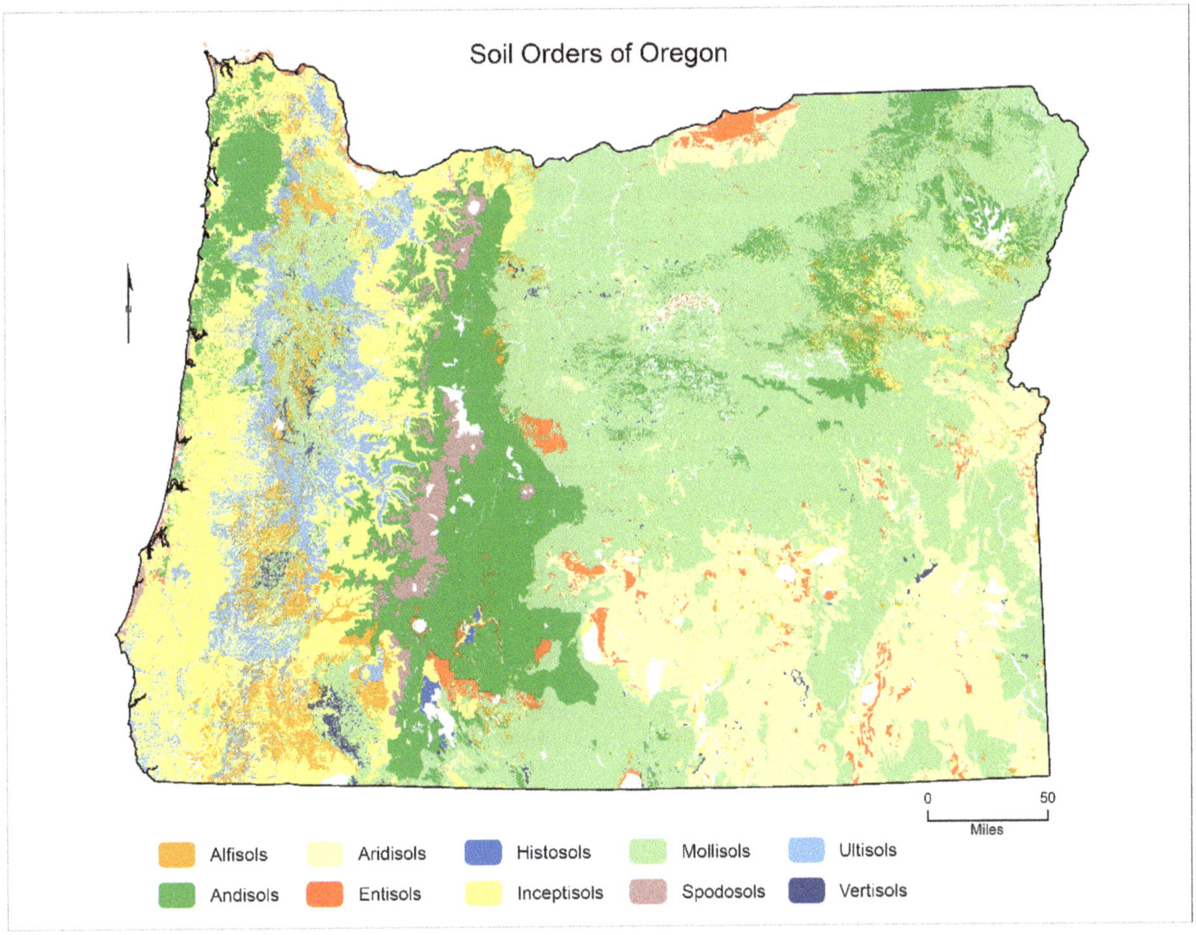

Fig. 2.14 Current (2021) soil order map of Oregon. *Source* Prepared by Whityn Owen

Fig. 2.15 The Jory soil series, a fine, mixed, active, mesic Xeric Palehumults, is the official state soil of Oregon. *Source* NRCS

The number of soil series recognized in Oregon increased markedly from 1960 to 2011. About 81% of the soil series recognized in Oregon occur only in the state.

References

Baldwin M, Kellogg CE, Thorp J (1938) Soil classification. Soils and Men. US Dep Agric Yearbook US Govt Print, Washington, DC, pp 979–1001

Buol SW, Southard RJ, Graham RC, McDaniel PA (2011) Soil genesis and classification. 6th edn. Wiley & Blackwell, West Sussex, UK

Huddleston JH (1979) Soils of Oregon: their classification, taxonomic relationships, and physiography. Oregon State Univ Exten Serv Spec Rep 535. p 121

Jensen CA (1903) Soil survey of the Salem Area, Oregon. Field Oper Bur Soils 1903:1171–1182

Jensen CA, Mackie WW (1903) Soil survey of the Baker City Area, Oregon. Field Oper Bur Soils 1903:1151–1170

Kocher AK, Carpenter EJ, Ruzek CV, Cotter JK (1917) Soil survey of Yamhill County, Oregon. Field Oper Bur Soils 1917:2259–2320

Natural Resources Conservation Service (2022) Soil Formation and Classification. http://www.nrcs.usda.gov/wps/portal/nrcs/detail/soils/edu. Retrieved March 9, 2022

Soil Survey Staff (1999) Soil taxonomy: a basic system of soil classification for making and interpreting soil surveys. 2nd edn. Agric Handb 436. US Govt Print Office, Washington, DC. p 869. https://www.nrcs.usda.gov/wps/portal/nrcs/detail/soils/survey/class/taxonomy/?cid=nrcs142p2_053580

Soil Survey Staff (2014) Keys to soil taxonomy. 12th edn. US Dept Agric Natural Resour Conserv Serv, Lincoln, NE. https://www.nrcs.usda.gov/wps/portal/nrcs/detail/soils/survey/class/taxonomy/?cid=nrcs142p2_053580

Weil R, Brady N (2016) The nature and properties of soil. 15th edn. Pearson

Soil-Forming Factors

3

3.1 Introduction

The expression of a soil results from five factors operating collectively: climate, organisms, relief, parent material, and time. The factors interact and cause a range of soil processes (e.g., illuviation) that result in a diversity of soil properties (e.g., high clay content in the subsoil). Human activities cause soil changes and are often considered a sixth factor. Following the "Russian school of soil science," in 1941, Hans Jenny published *Factors of Soil Formation* in which he described soil (s) as the result of climate (cl), organisms (o), topography (r), parent material (p) and time (t); the acronym CLORPT known to most pedologists today. The following is a review of the role of soil-forming factors in the development of Oregon's soils. Soil-forming factor data for soil series with an area of 50 km^2 or more are given in Appendix A.

3.2 Climate

3.2.1 Current Climate

Oregon lies in the northwestern United States and is bordered on the west by the Pacific Ocean, on the north by one of America's largest rivers, the Columbia, and on the east by the Snake River. Two north–south trending mountain ranges, the Coast Range adjacent to the Pacific Ocean and Cascades in the west-central part of the state, strongly influence moisture distribution from the prevailing winds off the ocean. Topography creates a remarkable climatic diversity across the state.

The lowest mean annual air temperatures are in the Cascade, Wallowa, and Blue Mountains, and the highest temperatures are along the Columbia River, in the Willamette Valley, and throughout southwestern Oregon (Fig. 3.1). The mean annual air temperature varies from 12 °C in Brookings to −0.6 °C in the Wallowa Mountains of northeastern Oregon. Along the coast, the range between high and low temperatures is moderated by maritime air and coastal fog, whereas dry regions in the south-central part of the state have the greatest range. The mean annual precipitation is greatest at upper elevations in the Coast Range and lowest in the deserts of eastern Oregon (Fig. 3.2). Total precipitation varies from 165 mm at Fields near the Alvord Desert to over 3,500 mm in the Coast Range. Because storm fronts off the ocean rise against the Coast Range and Cascades, precipitation is greater west of the mountains. A rain shadow, an area of lesser rain, extends nearly 160 km east of the Cascades. The Willamette Valley has a Mediterranean climate, and it receives about 1,000 mm annual precipitation but very little during summer months. The PRISM model has enabled prediction of mean annual precipitation in mountainous areas of the Western United States (Daly et al. 1994). Mountainous areas of Oregon receive copious amounts of snow, particularly in the Cascades and Wallowa Mountains (Fig. 3.3). Crater Lake National Park holds the state record of 22 m of snow.

Oregon soils have a diversity of temperature and moisture regimes. Soil temperature regimes (STRs) are determined by the mean temperature at a depth of 50 cm. The regimes in Oregon, from coldest to warmest, are cryic, frigid, and mesic. Because air temperature is moderated within the coastal fog belt, which occurs about 15 km in width along the Pacific Coast, STRs in that region are isomesic or isofrigid. These "iso" regimes have a difference of less than 5 °C between summer and winter soil temperature. Isofrigid STRs have a mean annual soil temperature greater than 0 °C but less than 8 °C, and isomesic STRs have mean annual soil temperature between 8 and 15 °C (Fig. 3.4). The Coast Range, Willamette Valley, Deschutes–Umatilla Plateau,

T. Thorson et al., *The Soils of Oregon*, World Soils Book Series,
https://doi.org/10.1007/978-3-030-90091-5_3

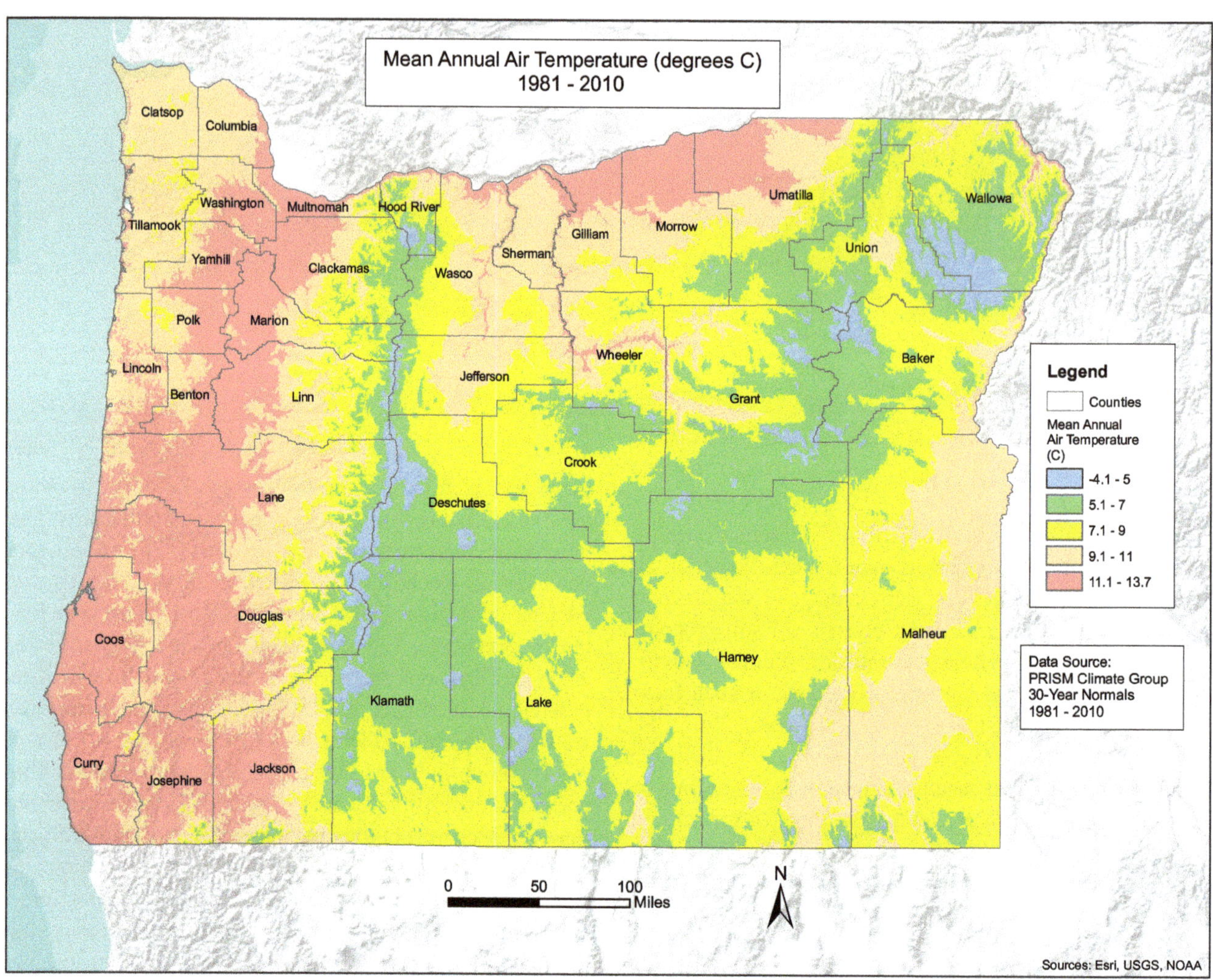

Fig. 3.1 Mean annual air temperature in Oregon. *Source* Prepared by Steve Campbell

High Lava Plains, and Owyhee Uplands have frigid or mesic STRs, in which the mean annual soil temperature ranges between 8 °C and 15 °C. In areas of the High Lava Plains that have a thick mantle of pumice from Mount Mazama, the pumice mantle coupled with cold air drainage to low areas in the landscape create an anomaly where frost may occur every month of the year. In the vicinity of Chemult to LaPine, this phenomenon restricts plant survival. Lodgepole pine (*Pinus contorta*) occupy low-lying areas and ponderosa pine (*Pinus ponderosa*) dominate the landscape above frost pockets. The Basin-and-Range and southern part of the Lava Plains have a frigid STR, in which the mean annual soil temperature is lower than 8 °C.

Soil moisture regimes (SMRs) are classified according to the presence of water in the "soil moisture control section," which corresponds to the rooting depths of many crops. Technically, the upper boundary of the soil moisture control section is the depth to which 2.5 cm of water will moisten a dry soil within 24 h, and the lower boundary is the depth to which 7.5 cm of water will moisten the soil within 48 h (Soil Survey Staff, 1999). Oregon has four SMRs: aquic, udic, xeric, and aridic, in order from most moist to most dry (Fig. 3.5). The term "torric" denotes the same dry conditions as "aridic," but the terms are used in different categories of soil taxonomy.

An aquic SMR, in which reducing conditions occur from saturation by water, occurs sporadically in the Willamette Valley and elsewhere in the state. A udic SMR exists along the coast, in the Coast Range, Cascade Mountains, and Blue Mountains; the soil moisture control section in these regions

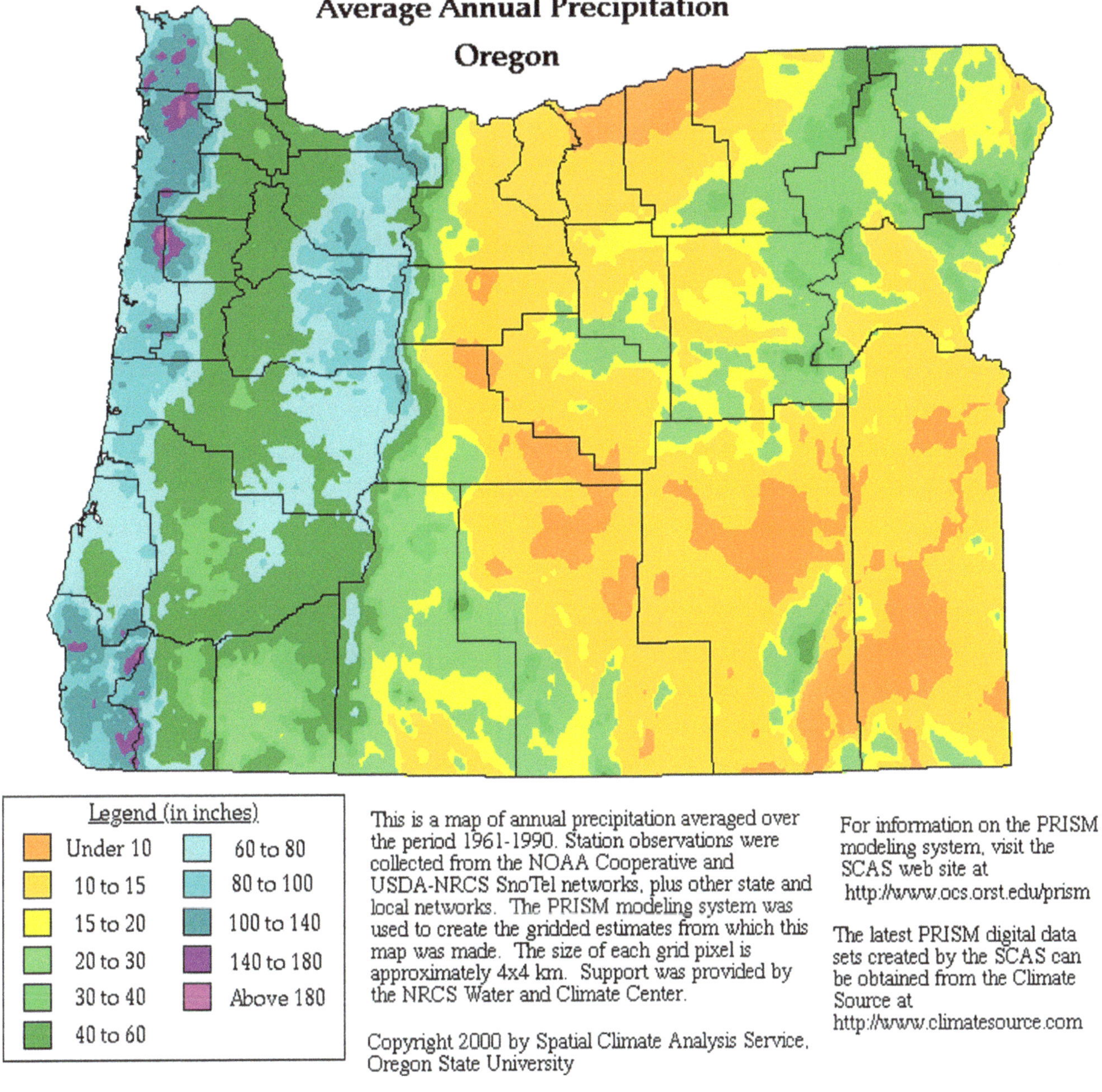

Fig. 3.2 Average annual precipitation in Oregon. *Source* Spatial Climate Analysis Service, Oregon State Univ. 2000

is not dry in any part for 90 or more cumulative days and dry less than 45 consecutive days in normal years. Most of the remainder of the state has a xeric SMR, where winters are moist and cool and summers are warm and dry and the mean annual precipitation is typically 300 cm or more. Udic and xeric moisture regimes can generally be discerned by elevation, but in mountainous regions, slope position and aspect influence soil moisture. Research in the Coast Range helped establish limits between udic and xeric SMRs (Thomas et al. 1973). Basins in the Basin and Range province and along the Columbia River have an aridic (torric) SMR, in which the soil moisture control section is dry in all parts for more than half of the cumulative days per year when the soil temperature is above 5 °C and moist in some or all parts for less

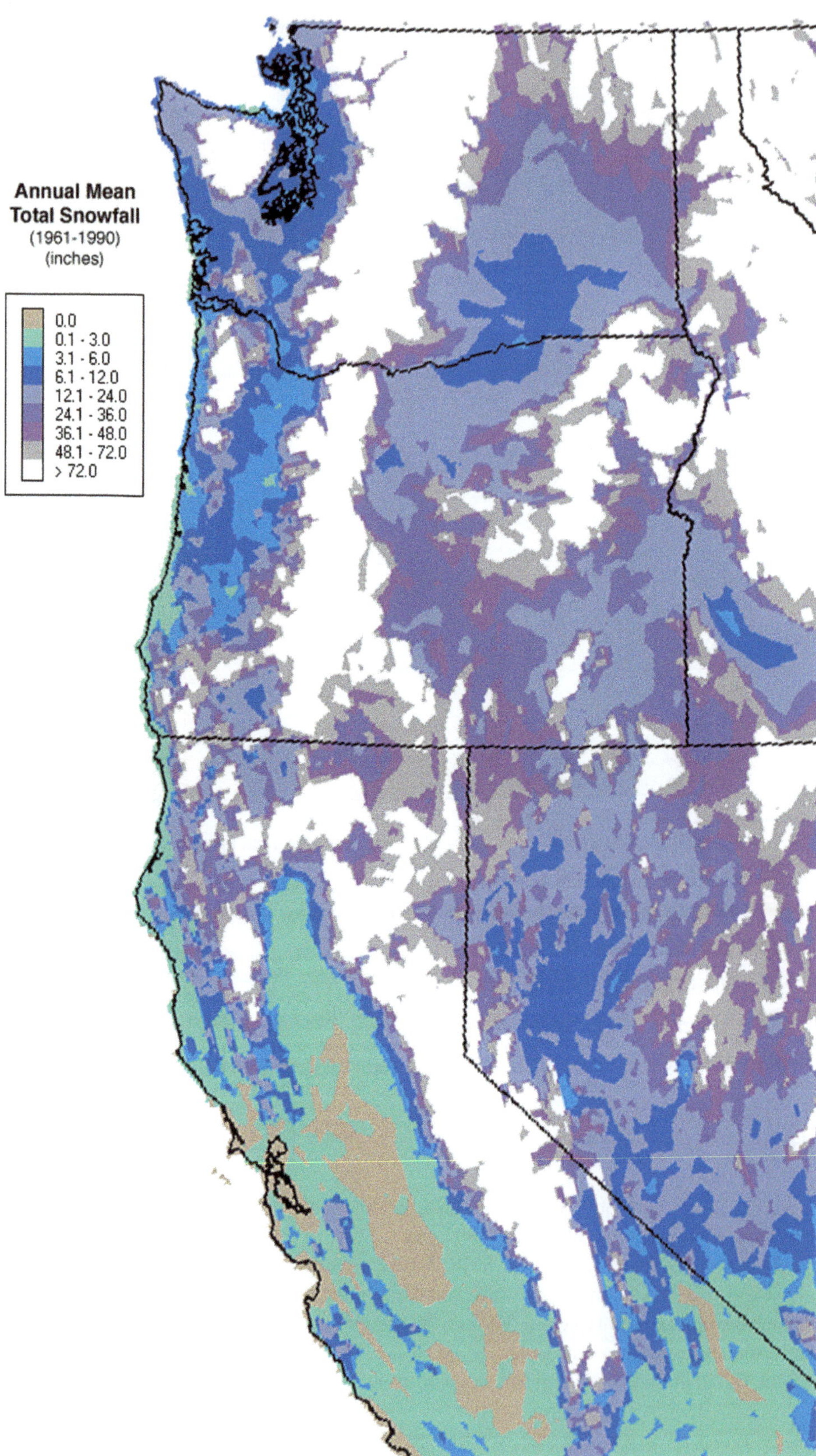

Fig. 3.3 Mean annual snowfall in the Pacific Northwest. *Source* Skimountaineer.com

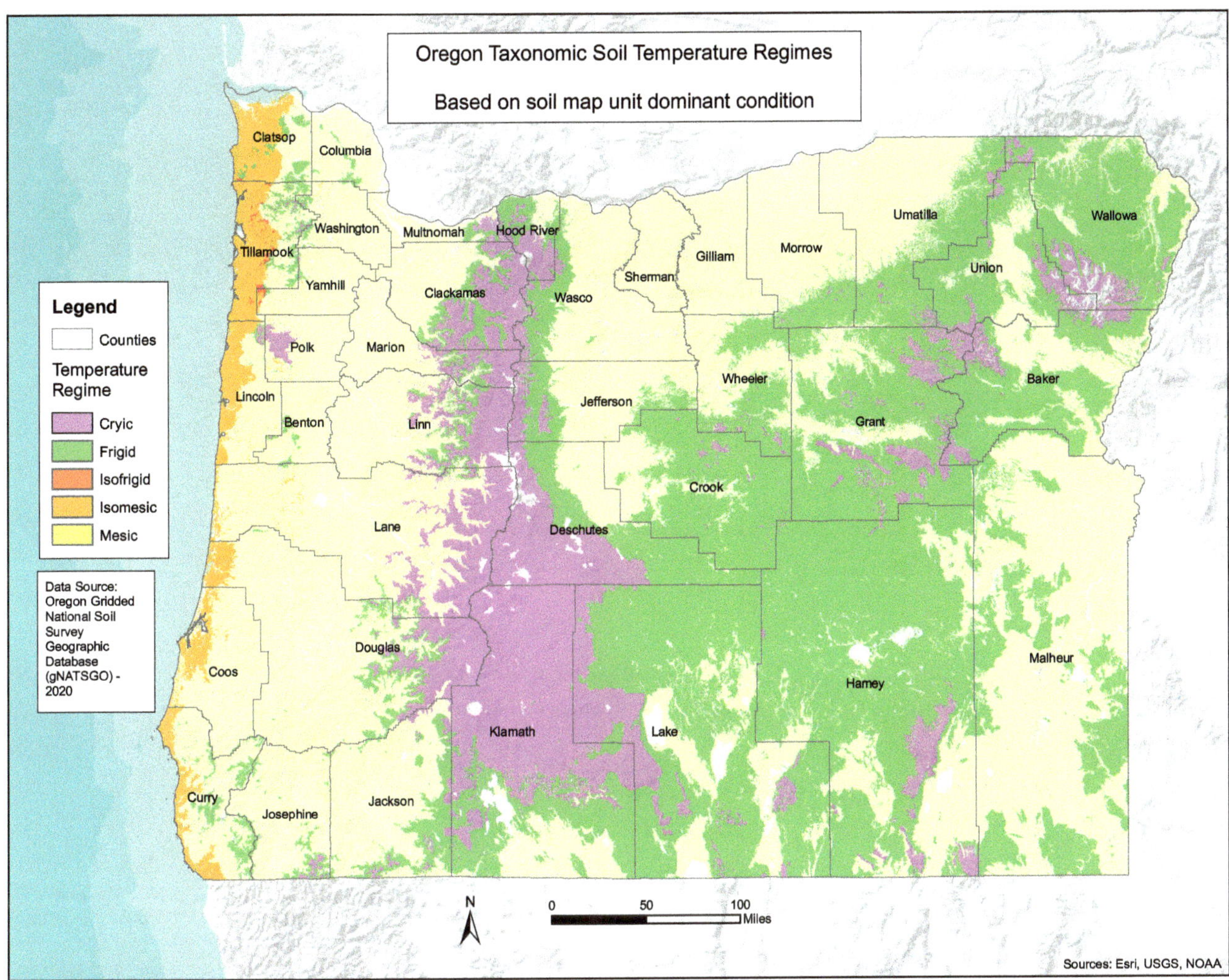

Fig. 3.4 Soil temperature regimes in Oregon. The isofrigid soil temperature regime exists in localized areas of the Coast Range, primarily in Tillamook County. *Source* Prepared by Steven Campbell

than 90 consecutive days when the soil temperature is above 8 °C. The mean annual precipitation for aridic (torric) SMRs is typically 300 mm or less.

Differences in atmospheric and soil temperature regimes are reflected in the plant hardiness zone map of Oregon (Fig. 3.6). The most favorable growing conditions (generally the longest growing season) occur to the west of the Cascade Range, particularly along the southwest coast; the least favorable conditions occur at the higher elevations east of the Cascade Divide.

3.2.2 Past Climates

The climate of Oregon has varied considerably over geologic time, particularly over the past 1.8 million years with the onset of the Pleistocene glaciers. Although the Laurentide Ice Sheet did not extend into Oregon, alpine glaciers were active in the Cascade and Wallowa Mountains and Steens Mountain during the Pleistocene. Small glaciers may have developed in the Oregon Klamath Mountains during the Pleistocene (Orr and Orr 2012). Glacial Lake Missoula was a

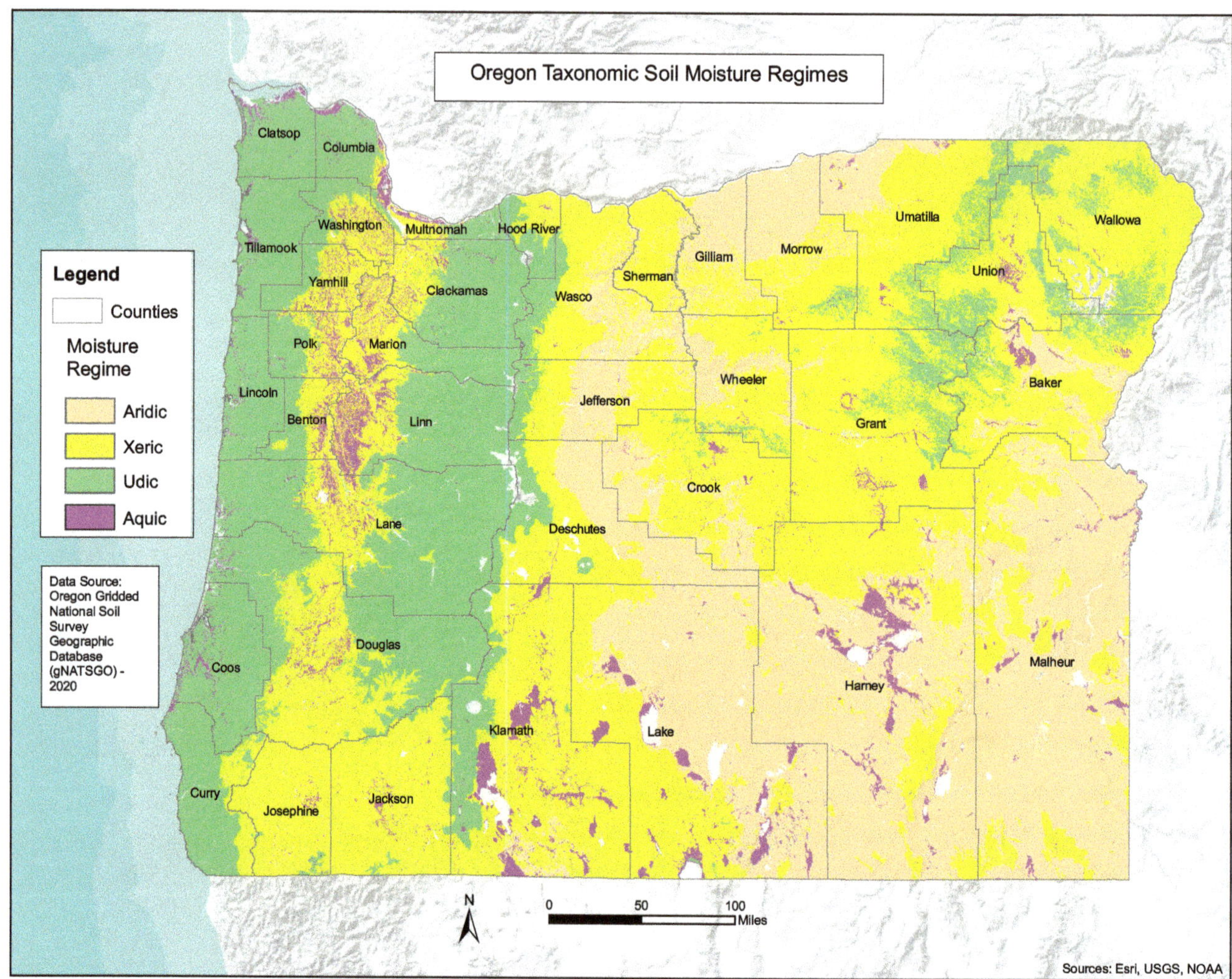

Fig. 3.5 Soil moisture regimes of Oregon. *Source* Prepared by Steve Campbell

prehistoric proglacial lake in western Montana that existed periodically at the end of the last ice age between 15,000 and 13,000 years ago. The lake measured about 7,770 km^2 and released floodwaters down the Columbia River Gorge and into the Willamette Valley approximately 40 times (e.g., Baker and Nummedal 1978). A series of pluvial lakes existed in the Basin and Range province; however, these were much smaller than those in Utah and Nevada. The "Little Ice Age," a global cool period, existed in Oregon from about 700 years ago to 150 years ago.

3.3 Vegetation

The information contained herein is drawn from the book *Natural Vegetation of Oregon and Washington* (Franklin and Dyrness 1988) and "Ecological Provinces of Oregon: a treatise on the basic ecological geography of the state" (Anderson et al. 1998). The historic vegetation map of Oregon, published in 1938 and uploaded by Tobalske and Osborne-Gowey (2002), divides the vegetation of the state

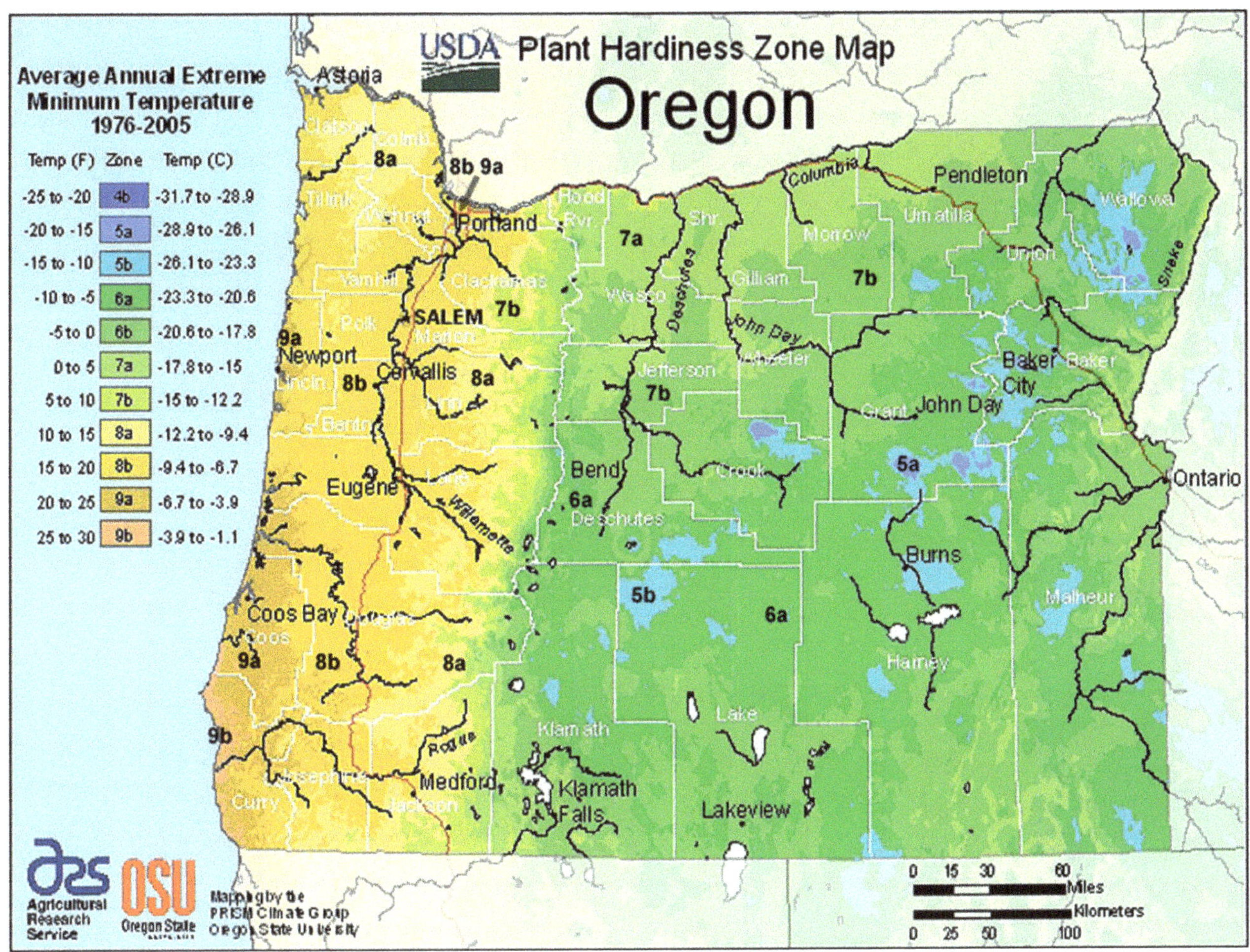

Fig. 3.6 USDA Plant hardiness zone map for Oregon. *Source* PRISM Climate Group, Oregon State Univ. undated

into seven broad categories, including from greatest to least in area: temperate evergreen needle-leaf forest (green), temperate shrubland (pink), grassland (yellow), subalpine forest (dark blue), temperate evergreen needle-leaf woodland (orange), maritime evergreen needle-leaf forest (black), and temperate warm mixed forest (purple) (Fig. 3.7). Water bodies were shown in light blue.

According to a map of Oregon's forests by the Oregon Department of Forestry (2004), approximately 48% of Oregon is forested (Fig. 3.8). Common forest types include ponderosa pine (*Pinus ponderosa*) (rust color); Douglas-fir (*Pseudotsuga menziesii*) (orange); Klamath mixed conifer (light green); lodgepole pine (*Pinus contorta*) forests (black); western juniper (*Juniperus occidentalis*) (blue); subalpine fir (*Abies lasiocarpa*) (dark green); and Sitka spruce (*Picea sitchensis*)–western hemlock (*Tsuga heterophylla*) (red).

The Klamath-mixed conifer forests contain Douglas-fir, incense cedar (*Calocedrus decurrens*), Port Orford cedar (*Chamaecyparis lawsoniana*), tanoak (*Notholithocarpus densiflorus*), ponderosa pine, and sugar pine (*Pinus lambertiana*). Mixed species forests contain Douglas-fir, western redcedar (*Thuja plicata*), western hemlock, ponderosa pine, and grand fir (*Abies grandis*). Subalpine fir forests also contain white fir (Abies concolor), noble fir (*Abies procera*), and Engelmann spruce (*Picea engelmannii*). Photographs of key forest types are provided in Figs. 3.9, 3.10, and 3.11.

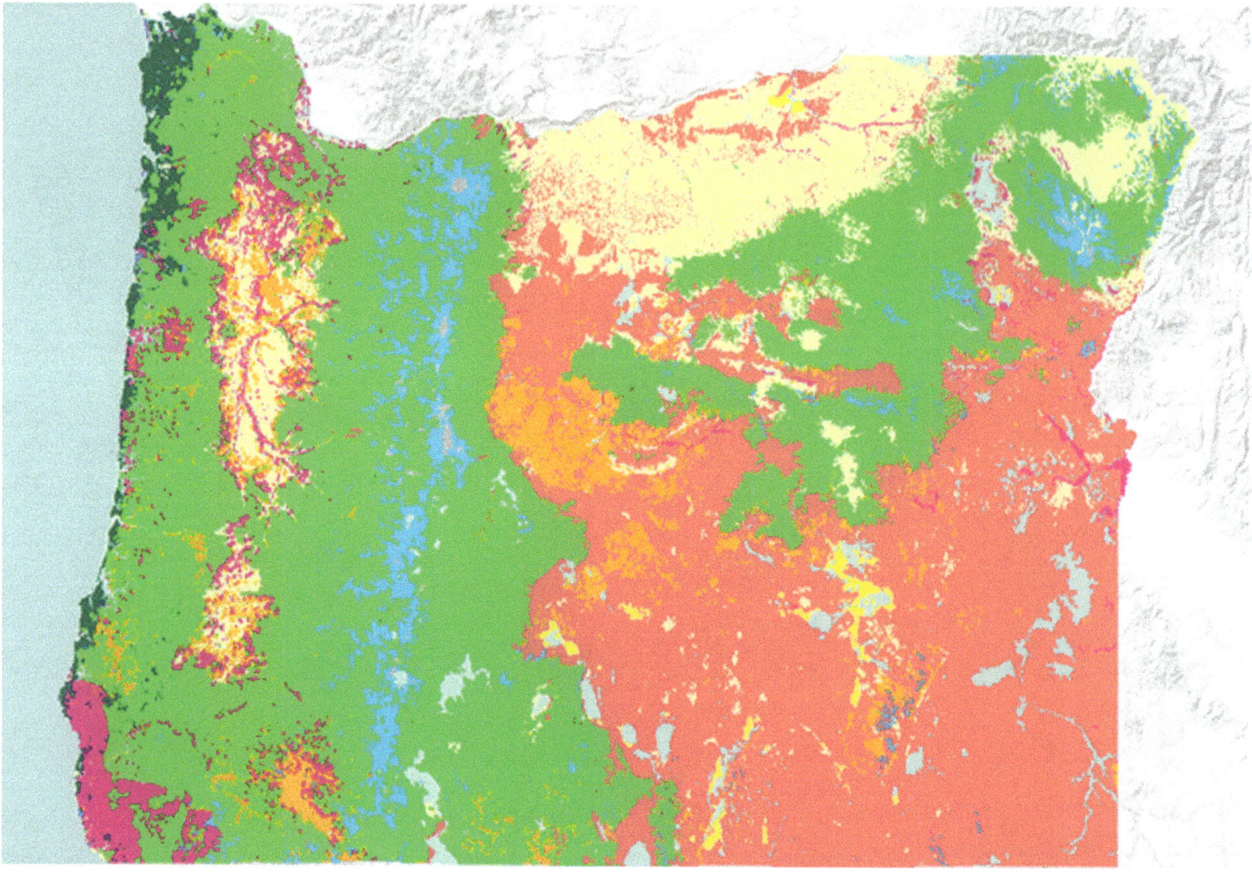

Fig. 3.7 Historic (1938) vegetation of Oregon (Oregon Natural Heritage Program). Temperate shrubland is shown in pink; temperate coniferous forest in green; subalpine forest in blue, maritime coniferous forest in dark green, temperate warm mixed forest in purple, temperate coniferous woodland in orange, and grasslands in light and dark yellow. *Source* Tobalske and Osborne-Gowey 2002

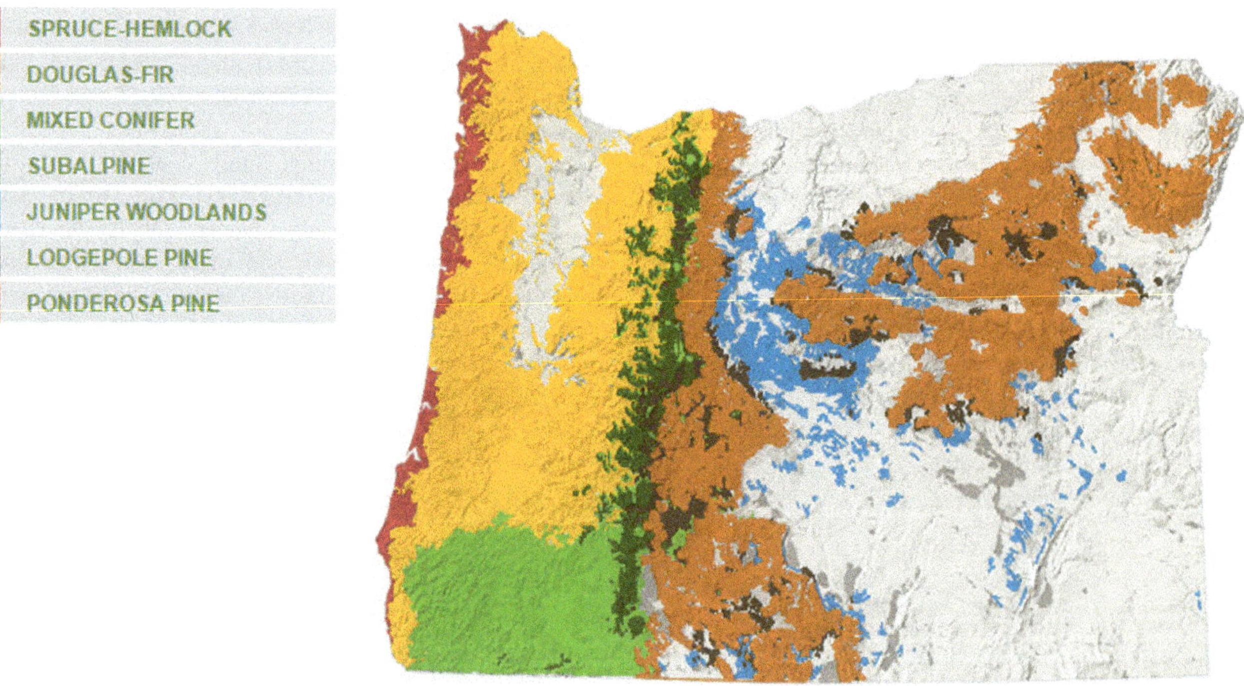

Fig. 3.8 Distribution of forest cover types in Oregon. *Source* Oregon Department of Forestry

Fig. 3.9 Major forest types of Oregon, including ponderosa pine-Douglas-fir forest in Crater Lake National Park (upper, left); Douglas-fir-western hemlock forest in Tillamook County (upper, right); mixed evergreen forest composed of noble fir, Pacific silver fir, and western hemlock in Benton County (lower, left); and lodgepole pine forest in Crater Lake National Park (lower, right). *Source* NRCS photos

The remaining 52% of the state contains shrublands, shrub-grasslands, and grasslands. Grasslands traditionally contain less than 10% canopy cover by shrubs (Anderson et al. 1998). Shrublands are dominated by Wyoming big sagebrush (*Artemisia tridentata wyomingensis*), mountain big sagebrush (*A. tridentata* subsp. *vaseyana*), low sagebrush (*A. arbuscula*), western juniper (*Juniperus occidentalis*), black greasewood (*Sarcobatus vermiculatus*), antelope bitterbrush (*Purshia tridentata*), green rabbitbrush (*Chrysothamnus viscidiflorus*), and gray or rubber rabbitbrush (*Ericameria nauseosa*). Common grasses are bluebunch wheatgrass (*Pseudoroegneria spicata*), Idaho fescue (*Festuca idahoensis*), Sandberg bluegrass (*Poa secunda*), Thurber's needlegrass (*Achnatherum thurberianum*), basin wildrye (*Leymus cinereus*), Indian ricegrass (*Achnatherum hymenoides*), inland saltgrass (*Distichlis spicata*), and cheatgrass (*Bromus tectorum*). Shrub-grasslands and grasslands from eastern Oregon are depicted in Fig. 3.12.

The Institute for Natural Resources at Oregon State University has developed an integrative map for Oregon Ecological Systems. The map was created in 2010 and integrates all available 1:24,000 vegetation maps and coverages.

The US Environmental Protection Agency has identified 64 ecoregions in Oregon (Fig. 3.13). The map emphasizes

Fig. 3.10 Additional common forest types in Oregon, including Klamath mixed conifer forest composed of Port Orford cedar and incense cedar in Trinity National Forest (USFS photo) (upper, left); Sitka spruce-western hemlock forest in Tillamook County (upper, right); western juniper forest in Harney County (lower, left, distance); and subalpine forest dominated by mountain hemlock in Crater Lake National Park (lower, right). *Source* All photos by NRCS except where indicated

Fig. 3.11 Common vegetation types in the Williamette Valley, including Oregon white oak (*Quercus garryana*) (upper) and Oregon white oak-grassland savannah in Benton County. *Source* NRCS photos

Fig. 3.12 Shrub-grasslands in eastern Oregon, including bluebunch wheatgrass (*Pseudoregneria spicata*) (US Forest Service photo) (upper); mountain big sagebrush-Idaho fescue-bluebunch wheatgrass community in Harney County (middle); and basin wildrye (*Leymus cinereus*) (lower). *Source* Photo by Natural History Museum of Utah, University of Utah

Oregon's nine physiographic provinces, and only nine of the ecoregions contain broad vegetation names, e.g., redwood, prairie, montane forest, oak savanna, etc.

3.4 Relief

Relief is a measure of surface roughness or quantitatively the measurement of elevational change in a landscape. Relatively smooth terrain may be described as having low relief, whereas areas deeply dissected and with steep slopes have high relief. Relief at a state-wide scale is depicted in a shaded relief map in Fig. 3.14. Mountainous physiographic provinces: Coast Range, Klamath Mountains, Cascades, and Blue Mountains appear to have the highest relief, while the other provinces are mostly areas of lesser relief. At a closer look, areas along the coast, crest of the Cascades and plateaus in the Blue Mountains are relatively flat. In comparison, ranges in the Basin and Range, cinder cones on the High Lava Plains and canyons that dissect the Deschutes-Umatilla Plateau express greater relief than surrounding terrain. Canyons of the Grande Ronde, the Imnaha, and the Snake rivers in the Blue Mountains have astonishing elevational change. Hells Canyon, the deepest river gorge in North America, drops more than 1.6 km below its west rim. Relief gives rise to slope morphometry (e.g. length, gradient, aspect, position, and shape), which in-turn influence erosional, gravitational, and depositional processes. South-facing mountain slopes are usually steeper than north-facing slopes, where soils are deeper to bedrock, cooler, and moister. For example, along the west slope of the Cascades, south slopes are hotter and drier in summer months and have more frequent fires. Mount Mazama blanketed the Blue Mountains with 30 cm of volcanic ash 7,700 years ago. Predictably, the thickest undisturbed ash remains on low-relief north slopes, protected from erosion. A cirque basin near Mt Howard in the Wallowas still holds 40 cm of ash at the soil surface.

Figure 3.14 shows the relief in Oregon. The lowest elevations are along the Pacific Coast and in the Willamette and Columbia River Valleys. The highest elevations are in the Cascade Mountains, with Mt. Hood being the highest at 3,429 m. High peaks also exist in the Blue and Wallowa Mountains in the northeast and Steens Mountain in the southeast.

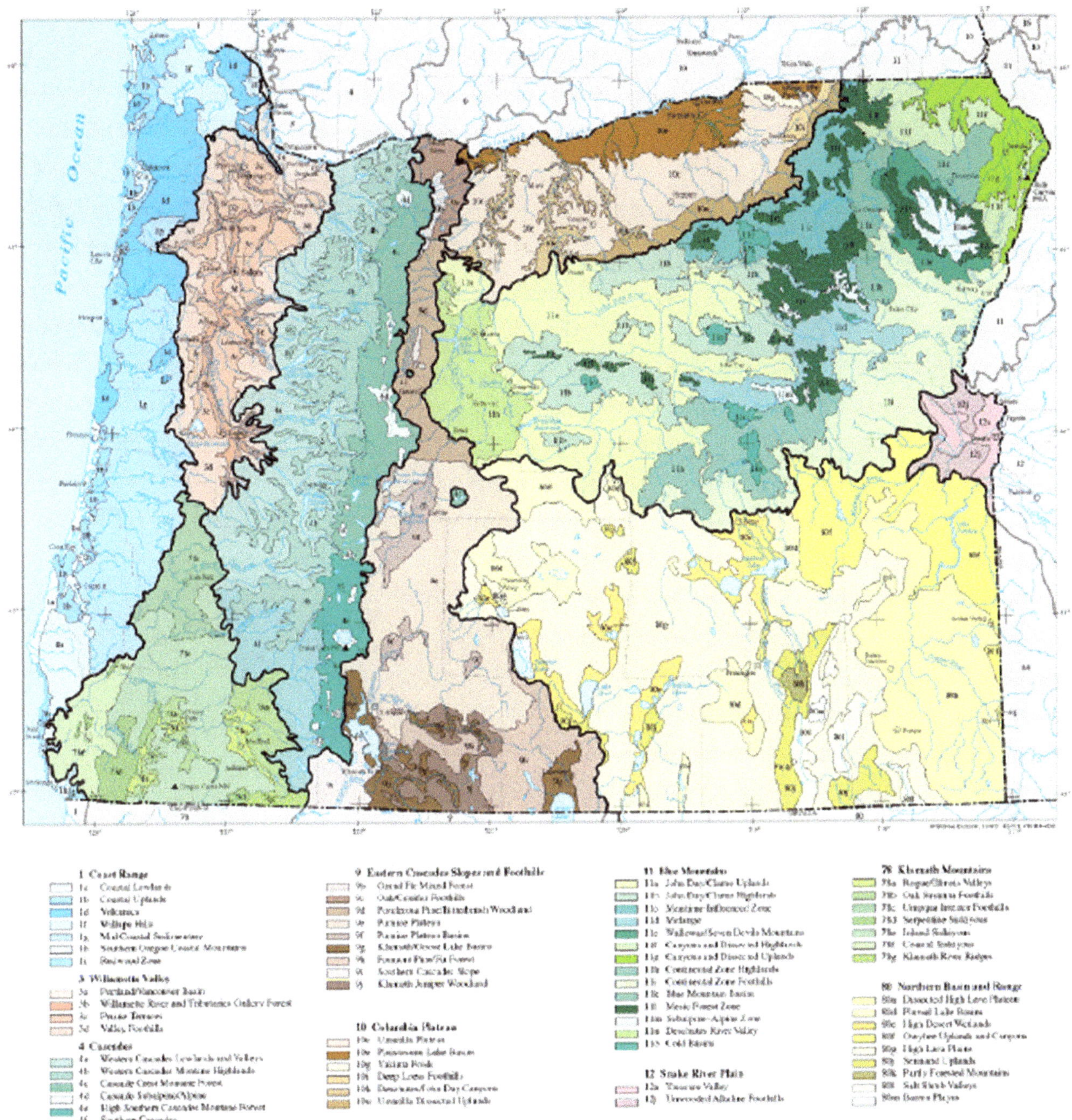

Fig. 3.13 Ecoregions of Oregon. *Source* http://epa.gov/

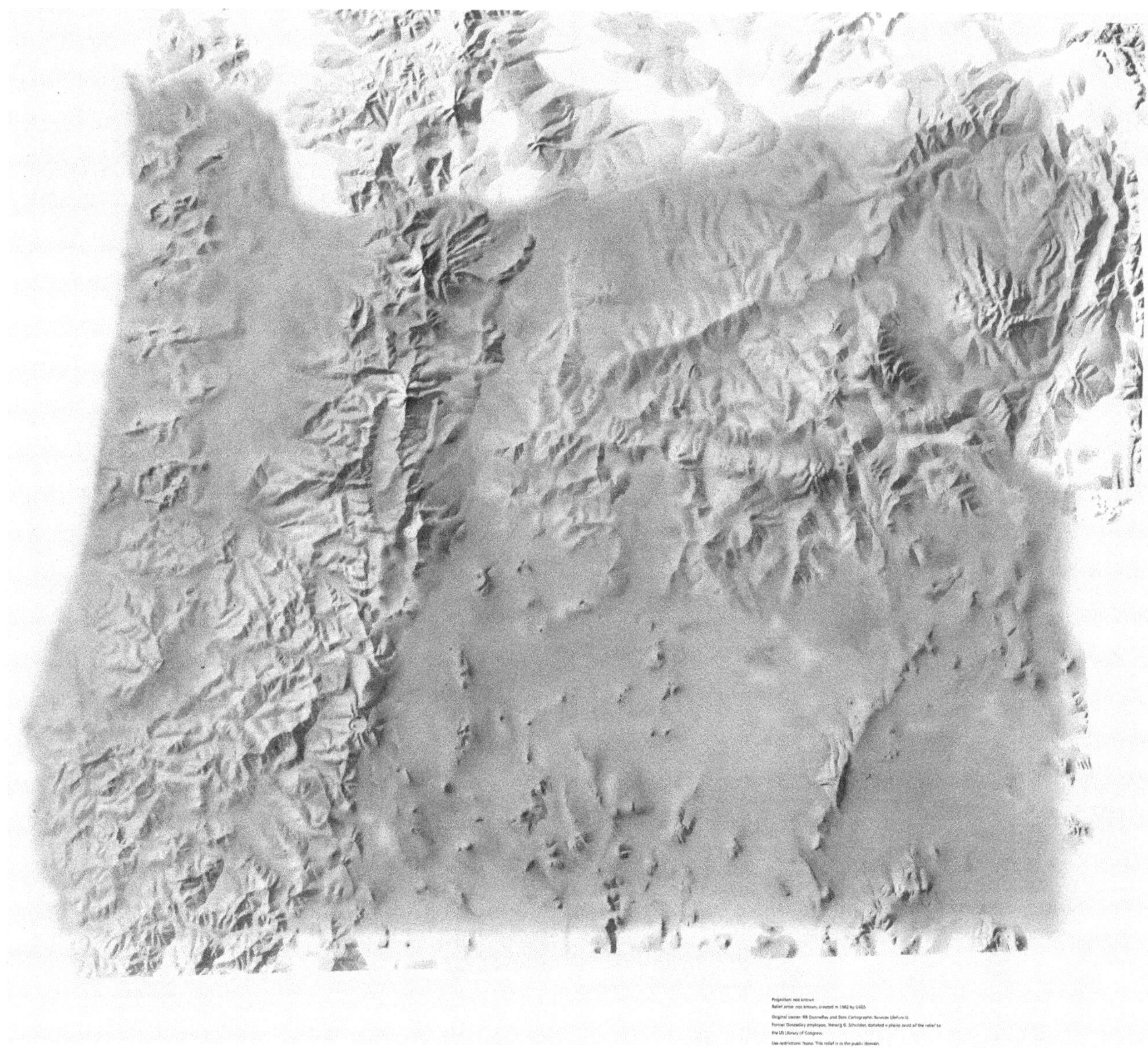

Fig. 3.14 Shaded relief map of Oregon. *Source* http://shadedreliefarchive.com

3.5 Physiographic Provinces

Oregon has been divided into nine principal physiographic provinces (Fig. 3.15). The largest of these is the Blue Mountains province, which comprise 58,600 km^2, or 23% of the state. This province is bounded by the Deschutes–Umatilla Plateau province to the northwest, the Owyhee province to the southeast, and the High Lava Plains province to the south. The highest elevation in the Blue Mountain province is Sacajawea Peak (3,000 m) in the Wallowa Mountains. The Blue Mountain province is part of the Columbia Plateau and is composed largely of basalt. The core of the Wallowas is formed in granite of the Wallowa batholith.

The Basin and Range province in Oregon has an area of 43,300 km^2 or 17% of the state and represents the northwestern most part of the Basin and Range province. This province is characterized by abrupt changes in elevation,

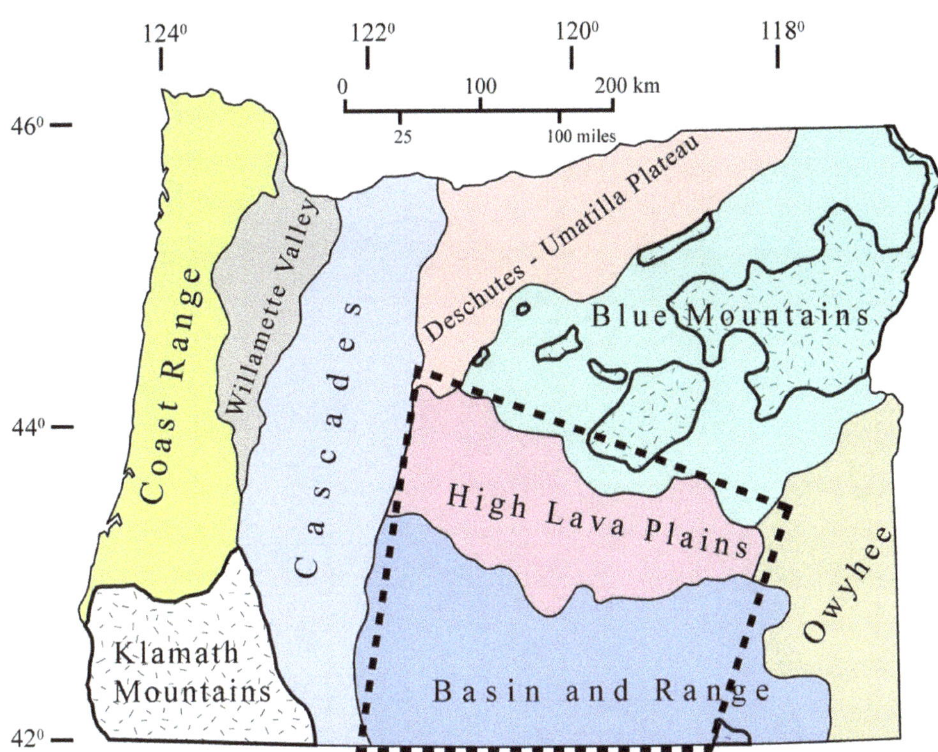

Fig. 3.15 Physiographic provinces of Oregon. *Source* www.science.oregonstate.edu

alternating between narrow faulted mountains chains and flat valleys or basins. Steens Mountain, composed of stacked basalts, is the highest part of the Basin and Range in Oregon at 2,968 m. The Cascades province extends from British Columbia to northern California. The Oregon portion has an area of 35,700 km^2, or 14% of the state. Mt. Hood is the highest peak at 3,429 m. The Cascade Range province is composed of strato-volcanoes. The Coast Range province has an area of 28,000 km^2 or 11% of the state and extends 320 km from the Columbia River in northwest Oregon to the middle fork of the Coquille River and ranges between 50 and 100 km in width. The elevation of the Coast Range averages 460 m; Marys Peak has the greatest elevation (1,250 m). This range is composed mainly of volcanic and sedimentary rocks.

The High Lava Plains province has an area of 25,200 km,^{2} which is 9.9% of the state. It is composed of basalt flows and rhyolitic tuff that represent a high plain at 1,200 m above sea-level (Fig. 3.15). The Deschutes–Umatilla Plateau province has an area of 22,400 km^2 or 8.8% of the state and is part of the Columbia Plateau that is composed mainly of basalt. The Klamath Mountains province has an area of 16,300 km^2 or 6.4% of the state and is located in southwestern Oregon. It has a varied geology that includes substantial areas of serpentinite and marble and a climate with cold winters and heavy snowfall and warm, very dry summers. Mt. Ashland (2,297 m) is the highest peak in the Oregon portion of the Klamath Mountains. The Owyhee province has an area of 14,200 km^2 or 5.6% of the state. It is composed mainly of rhyolite tuff and Quaternary basalt and andesite and averages 1,200 m in elevation. The Willamette Valley province is a 240-km long valley containing the Willamette River. It has an area of 9,900 km^2 or 3.9% of the state.

3.6 Geologic Structure

The geologic structure of Oregon is strongly related to the physiographic provinces (Fig. 3.15). The Deschutes–Umatilla Plateau, Blue Mountains, High Lava Plains, Owyhee, and Basin and Range provinces are predominantly basalts (Fig. 3.16). The Cascades province is composed of volcanic rocks, primarily rhyolite, andesite, and basalt. The Klamath

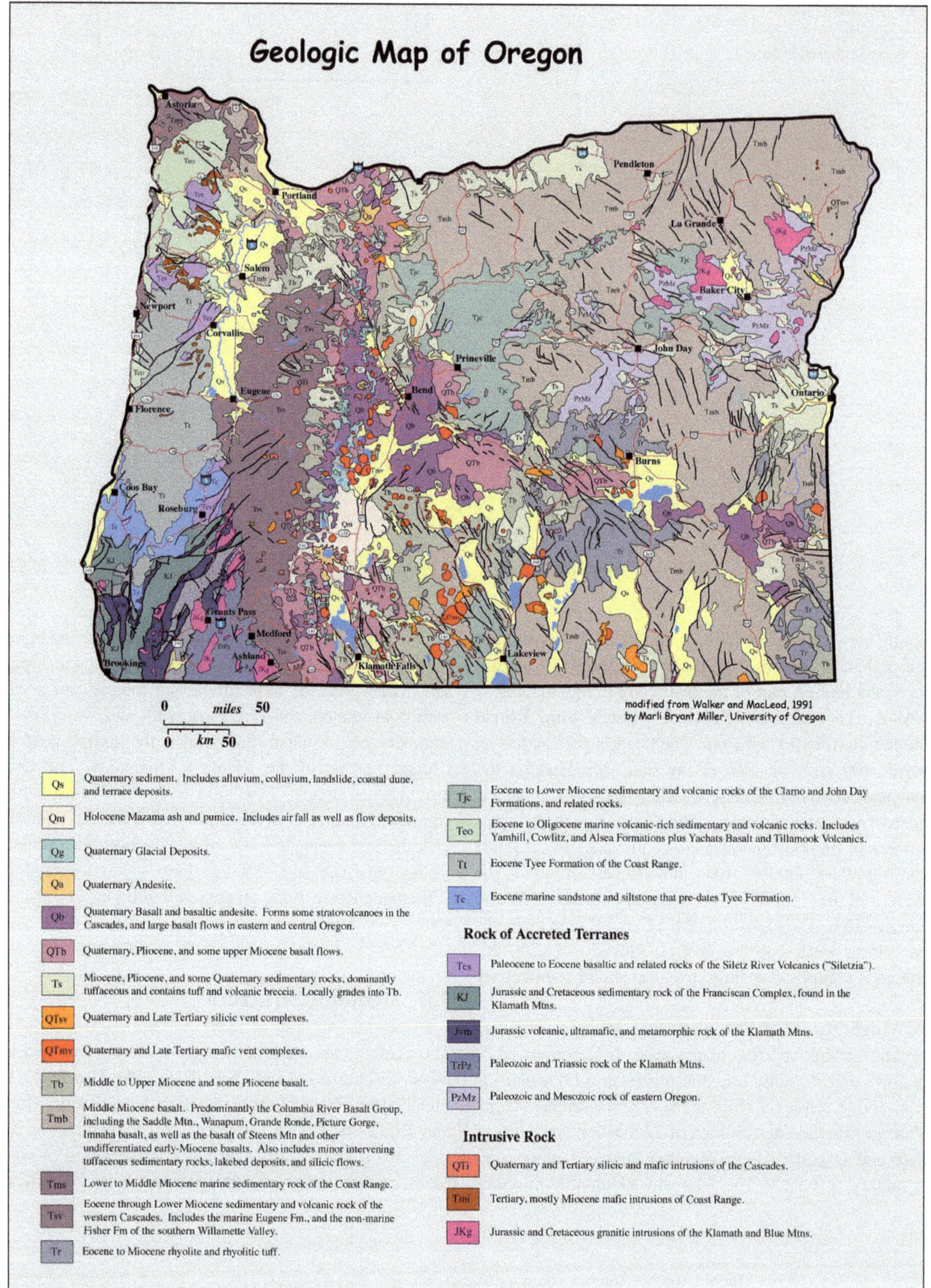

Fig. 3.16 Structural geology map of Oregon. *Source* http://marlimillerphoto.com

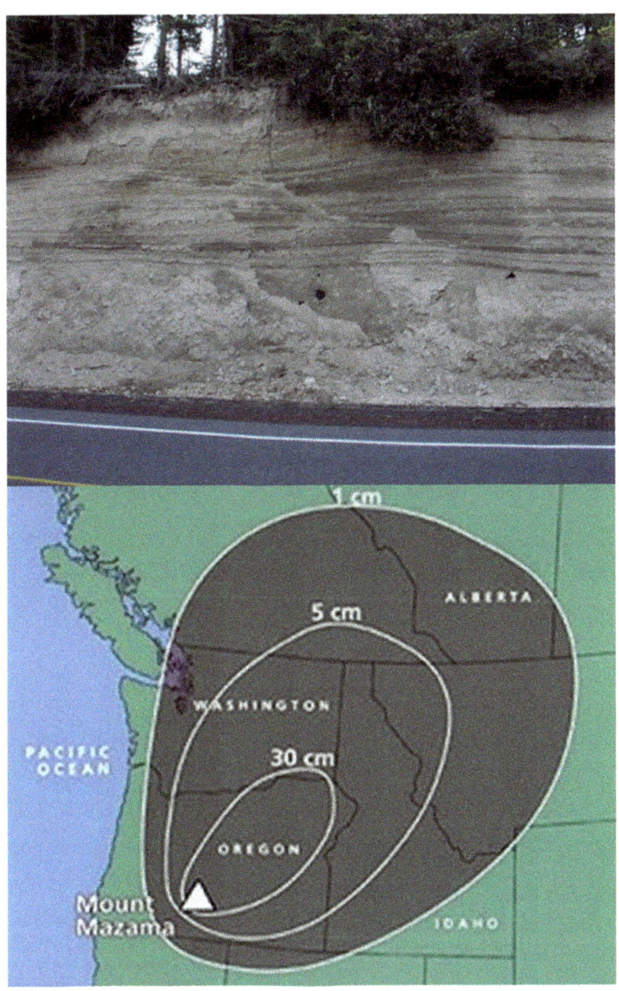

Fig. 3.17 Road cut in Crater Lake National Park showing the 7,700 year-old Mazama ash (upper panel). *Source* Photo by Stu Garrett, Earth Science Picture of the Day, University Space Research Institute, Jan. 14, 2004 and the distribution of the ash across western US and Canada (lower panel)

Fig. 3.18 Missoula Outburst Flood deposits 13–15,000 years in age in Marion County (upper panel). *Source* Photo by Roy Haggerty, Geosciences, Oregon State University) and a playa in the Alvord Desert in Harney County (lower panel) NRCS photo

Mountains province have a varied geology but have substantial areas of serpentinite and intrusions of granite. The Coast Range province is composed of volcanic and sedimentary rocks primarily basalt, sandstone, siltstone, and graywacke. The Willamette Valley province contains Quaternary sediments from the Missoula Floods.

The complexity of the geologic structure of Oregon is captured in Figs. 3.17, 3.18, 3.19, 3.20, 3.21 and 3.22. The eruption of Mt. Mazama 7,700 years ago created Crater Lake in Klamath County and left ash deposits all across the central and eastern portions of the state (Fig. 3.17). Deposits of late Quaternary age (ca. 13–15,000 years ago) include those from the Missoula Outburst Floods and in playas from drying up of lakes in arid and semiarid regions (Fig. 3.18). As stated previously, basalt is the most common rock type in Oregon. The basalts are commonly of middle Miocene age (Fig. 3.19), but range from Quaternary basalt flows to Eocene-aged basaltic rocks in accreted terranes. The upper panel of Fig. 3.19 shows the Grand Ronde Formation in

Fig. 3.19 Example of Middle Miocene basalts, including the Grand Ronde Basalt in Tillamook County (upper panel) and the Imnaha Basalt in Wallowa County (lower panel). *Source* Photos by NRCS

Tillamook County and the lower panel shows the Imnaha Formation in Wallowa County. Sedimentary rocks commonly range from middle Miocene to Eocene in age but may be older in accreted terranes. Figure 3.20 shows tilted siltstones of the Roseburg Formation in Douglas County (upper panel), tuffaceous siltstone of the Trask River Formation in Tillamook County (middle), and graywacke of the Tyee Formation in Benton County (lower). Volcanic rocks range from the lower Miocene to Eocene and Oligocene in age. Figure 3.21 shows the John Day Formation in the John Day Fossil Beds National Monument in Grant County (upper, left), the Siletz River Formation in Benton County (upper, right), the Tillamook Formation in Tillamook County (lower, left), and the Yamhill Formation in Tillamook

Fig. 3.20 Examples of sedimentary rocks in Oregon, including siltstones of the Roseburg Formation in Douglas County (upper panel), tuffaceous siltstones of the Trask River Formation in Tillamook County (middle panel), and graywacke of the Tyee Formation in Benton County. *Source* Photos by NRCS

Fig. 3.21 Examples of volcanic rocks in Oregon, including the John Day Formation of Eocene to lower Miocene age in the John Day Fossil Beds National Monument (upper, left); the Siletz River Volcanics of Eocene age in accreted terrane of Benton County (upper, right); the Tillamook Volcanics of Eocene to Oligocene age in Tillamook County (lower, left); and the Yamhill Volcanics of Eocene to Oligocene age in Tillamook County. *Source* NRCS photos

County (lower, right). Figure 3.22 shows other rock types that are less common including Jurassic and Cretaceous granite intrusions in the Blue Mountains of Wallowa County (upper) and ultramafics of Jurassic age in accreted terrane in the Klamath Mountains of Josephine County (lower).

3.7 Surficial Geology

The surficial geology of Oregon is dominated by colluvium and residuum (Fig. 3.23). The Cascades province is mainly volcanic materials, including ash and pumice. The Columbia

Fig. 3.22 Strongly weathered granite intrusions of Cretaceous age in the Idaho Batholith directly east of Wallowa County (upper panel) (NRCS photo) and ultramafic rocks in accreted terrane of Jurassic age in the Siskiyou National Forest (lower panel). *Source* Photo by John E. Roth, Siskiyou Field Institute

River Basin and the Willamette Valley contain alluvium and lacustrine materials from the Missoula Floods. Playa and pluvial lake deposits are shown in blue in the Basin and Range province. The scattered areas shown in yellow are small areas with alluvium and lacustrine deposits.

Many soil series in Oregon are derived from two or more parent materials, i.e., they contain lithologic discontinuities or a mixture of deposits. An examination of the 900 most extensive soil series, which constitute 94% of the mapped soil area, revealed that colluvium was present in at least 42% of the soils, followed by residuum (25%), volcanic ash (25%), and alluvium (18%) (Fig. 3.24).

3.8 Time

The soils of Oregon range from late Holocene to the Pliocene or earlier (Parsons and Herriman 1976; Parsons 1978; McDowell 1991; Haugland and Burns 2006; Lindeburg et al. 2013).

3.9 Humans

Humans have influenced soil development in Oregon through urbanization, cultivation, irrigation, logging, and practices that accelerated soil erosion, flooding, and mass wasting.

3.10 Summary

The soils of Oregon have been influenced by six factors operating collectively: climate, organisms, relief, parent material, time, and humans. The mean annual air temperature varies from 12 °C in Brookings to −0.6 °C at high elevations in the Wallowa Mountains of northeastern Oregon. The mean annual precipitation varies from 165 mm in the Alvord Desert to over 3,500 mm in the Coast Range and western Cascade Mountains. Oregon has a diversity of soil climates. The soil temperature regime is isomesic along the Pacific Coast, mesic in the Coast Range, Willamette Valley, Deschutes–Umatilla Plateau, and Owyhee Uplands, frigid in the Basin and Range and southern Lava Plains provinces, and cryic in the Cascades and Blue Mountains. Oregon has four soil moisture regimes, including udic along the coast, in the Coast Range, and in the Cascades and Blue Mountains; xeric throughout much of the rest of the state, and aridic in basins of the Basin and Range. The areas around the Malheur National Wildlife Refuge and the Klamath Basin have an aquic SMR.

These differences in atmospheric and soil climate have produced a variety of vegetation types in the state. Approximately 48% of Oregon is forested, including ponderosa pine, Douglas-fir, mixed conifers, lodgepole pine, juniper shrubland, subalpine forest, Sitka spruce, western hemlock and coastal redwood in southwest Oregon near the border with California. Desert shrublands and steppe feature sagebrush and bunchgrasses. Much of Oregon is mountainous, with the Cascades being the highest, followed by the

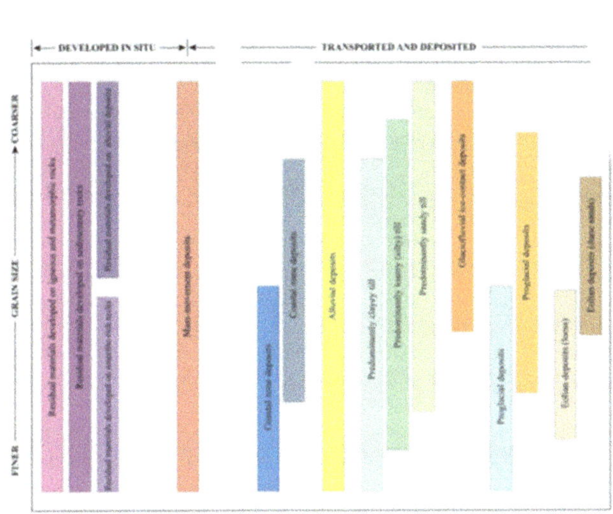

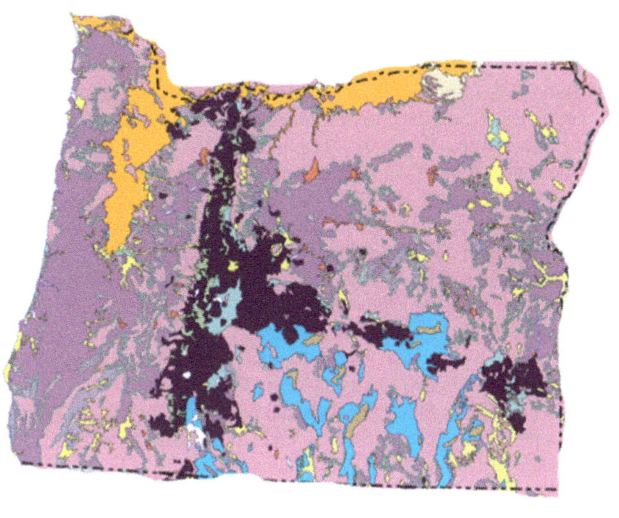

Fig. 3.23 Surficial geology of Oregon. Surficial deposits include from left to right in the legend: residual from igneous and metamorphic rocks, residual from sedimentary rocks, and residual from smectitic rocks (purple); colluvium (red); marine (blue), alluvium (yellow); clayey till, loamy or silty till, and sandy till (blue-green and green); outwash (dark orange); glaciolacustrine (cyan and light orange); and eolian-loess and eolian-dune (brown). *Source* From Soller and Reheis (2004)

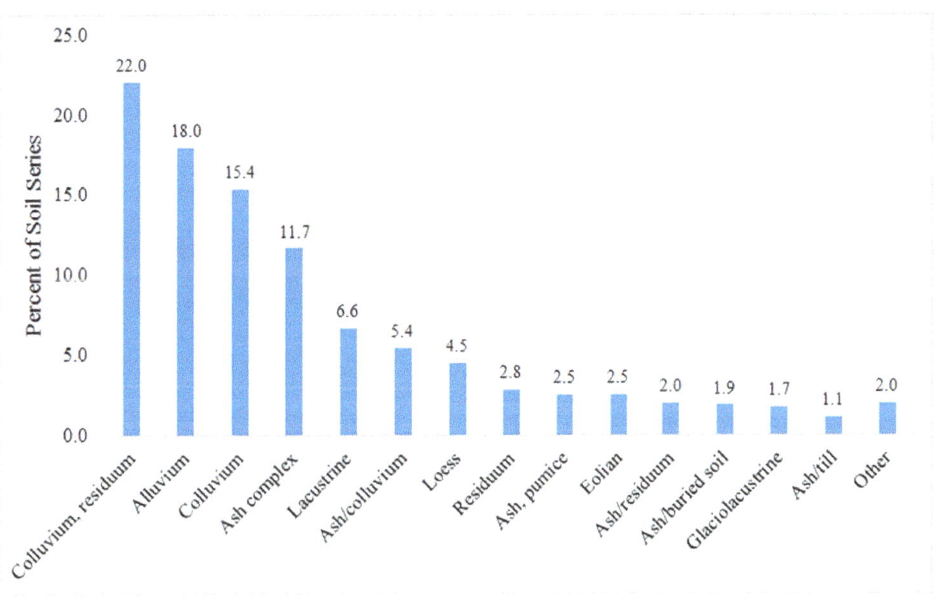

Fig. 3.24 Frequency distribution (percent) of parent materials in Oregon soil series

Blue Mountains, and the Klamath Mountains and Coast Range. Oregon has a complex geologic history. Basalts and other volcanic rocks are predominant, but graywackes, sandstones, and metasedimentary rocks also are common. Surficial deposits are primarily colluvium, residuum, volcanic ash, loess, and lacustrine sediments. The eruption of Mount Mazama 7,700 years ago deposited volcanic ash throughout central and eastern Oregon.

The soils of Oregon range from late Holocene to Pliocene in age. Humans have influenced soil development in Oregon through urbanization, cultivation, irrigation, logging and practices that accelerated soil erosion, flooding, and mass wasting.

References

Anderson EW, Borman MM., Krueger WC (1998) The ecological provinces of Oregon: a treatise on the basic ecological geography of the state. Spec Rep 990, Oregon State University, Agricultural Experiment Station, Corvallis, Oregon, pp 136

Baker VR, Nummedal D (1978) The Channeled Scabland. A guide to the geomorphology of the Columbia Basin, Washington. Planet Geol Program, Off Space Sci, NASA, p 186

Daly C, Nielson RP, Phillips DL (1994) A statistical-topographic model for mapping climatological precipitation over mountainous terrain. J Appl Meteor 33:140–158

Franklin JF, Dyrness CT (1988) Natural vegetation of Oregon and Washington. Oregon State University Press, Corvallis

Haugland JE, Burns SF (2006) Soils and geomorphology in the Oregon Cascades: a comparative study of Illinoian-aged and Wisconsin-aged moraines. Phys Geogr 27(4):363–377

Jenny H (1941) Factors of soil formation: a system of quantitative pedology. Dover Publ, NY

Lindeburg KS, Almond P, Roering JJ, Chadwick OA (2013) Pathways of soil genesis in the coast range of Oregon, USA. Plant Soil 367:57–75

McDowell PF (1991) Quaternary stratigraphy and geomorphic surfaces of the Willamette Valley, Oregon. In: Morrison RB (ed) Quaternary nonglacial geology: conterminous US, pp 156–164

Orr EL, Orr WN (2012) Oregon geology, 6th edn. Oregon State University Press, Corvallis, pp 304

Parsons RB (1978) Soil-geomorphology relations in mountains of Oregon, USA. Geoderma 21:25–39

Parsons RB, Herriman RC (1976) Geomorphic surfaces and soil development in the upper Rogue River Valley. Oregon Soil Sci Soc Am J 40:933–938

Soller DR, Reheis MC (2004) Surficial materials in the conterminous United States. US Geol Surv

Soil Survey Staff (1999) Soil taxonomy: a basic system of soil classification for making and interpreting soil surveys, 2nd edn. Agric. Handbook, vol 436. US Govt Print Office, Washington, DC, pp 869

Thomas BR, Simonson GH, Boersma L (1973) Evaluation of criteria for separating soils with xeric and udic soil moisture regimes. Soil Sci Soc Am J 37:738–741

Tobalske C, Osborne-Gowey J (2002) Oregon's historic vegetation (1938)—reclassified to functional type. Or Nat Heritage Program

4 Elevation Gradients in Oregon Mountain Ranges

4.1 Introduction

There are eight main mountain ranges in Oregon with pronounced elevation changes in climate, vegetation, and soils that are the focus of this chapter. These mountains include the Coast Range, Western Cascades, Klamath, High Cascade, Wallowa, Fremont, Steens, and Blue Mountains (Fig. 4.1). The Western Cascades and High Cascade Ranges will be considered together, as the Cascades; they form the main backbone of Oregon. The crest of the Cascades separates western Oregon from eastern Oregon. The Klamath Mountains are not identified on Fig. 4.1, however, they include much of southwestern Oregon in and around the city of Medford. The approach taken here uses soil temperature regimes to distinguish among elevation zones.

4.2 Coast Range

The Coast Range extends along the Pacific Ocean 320 km from the Columbia River in the north to the middle fork of the Coquille River in the south. The width of the range varies between 48 and 97 km. The elevation averages 460 m but reaches 1,248 m on Marys Peak in the central portion of the range. The Coast Range is composed of volcanic and sedimentary rocks, primarily basalt, sandstone, siltstone, and graywacke.

The Coast Range receives the greatest mean annual precipitation (MAP) in the state, with amounts increasing from 1,500 mm at elevations below 150 m to 2,400 mm at about 700 m, to 3,400 mm above 1,000 m (Table 4.1). Despite these apparent changes in elevation, there is no statistical relation between MAP and elevation for western Oregon, which includes the Coast Range, Cascades, and Klamath Mountains. (Fig. 4.2).

The mean annual air temperature (MAAT) decreases in the Coast Range from 9.9 °C below 700 m to 6.7 °C at 900 m, to 5.8 °C above 1,000 m (Table 4.1). This is consistent with the general decrease in MAAT with elevation for western Oregon (Fig. 4.3). The adiabatic lapse rate is 0.69 °C per 100 m, which is comparable to the commonly cited 0.66 °C per 100 m.

Elevation changes in Coast Range vegetation are subtle. Douglas-fir and western hemlock are the dominant tree species regardless of elevation in the range (Table 4.1). With an increase in elevation, Pacific silver fir and noble fir become more prevalent. Near the coast in the "fog belt" and characterized by an isomesic soil temperature, Sitka spruce dominates in the northern part of the range south to the city of Brookings. Oregon coastal redwood is dominant from about Brookings south and into California.

Dystrudepts, Humudepts, Fulvudands, and Hapludands are the major great soil groups at the lower elevations; Hapludands are common at the middle elevations; and Dystrocryepts and Fulvicryands are common at the higher elevations (Table 4.1).

4.3 Cascade Mountains

In Oregon, the north–south trending Cascade Mountains normally are divided into the ancient, highly eroded Western Cascades and the more recent High Cascades. The Cascades extend 435 km from the Columbia River to northern California and range from 50 to 125 km in width (Fig. 4.4). The Cascade Mountains are of volcanic origin and represent a series of strato-volcanoes that include Mt. Hood (3,428 m), Mt. Jefferson (3,200 m), Three Sisters (3,158 m), Mt. McLoughlin (2,895 m), and Crater Lake (Mount Mazama, 1883 m).

As with the Coast Range to the west, the Cascade Mountains force moisture-laden marine air masses from the west to ascend, depositing large amounts of precipitation on the western slopes and lesser amounts on the eastern slopes. The mean annual precipitation commonly varies from 1,755 to 1,900 mm on the western slopes and from 535 to 800 mm on the eastern slopes of the Cascades (Table 4.1). There is no statistical relation between elevation and mean annual

T. Thorson et al., *The Soils of Oregon*, World Soils Book Series,
https://doi.org/10.1007/978-3-030-90091-5_4

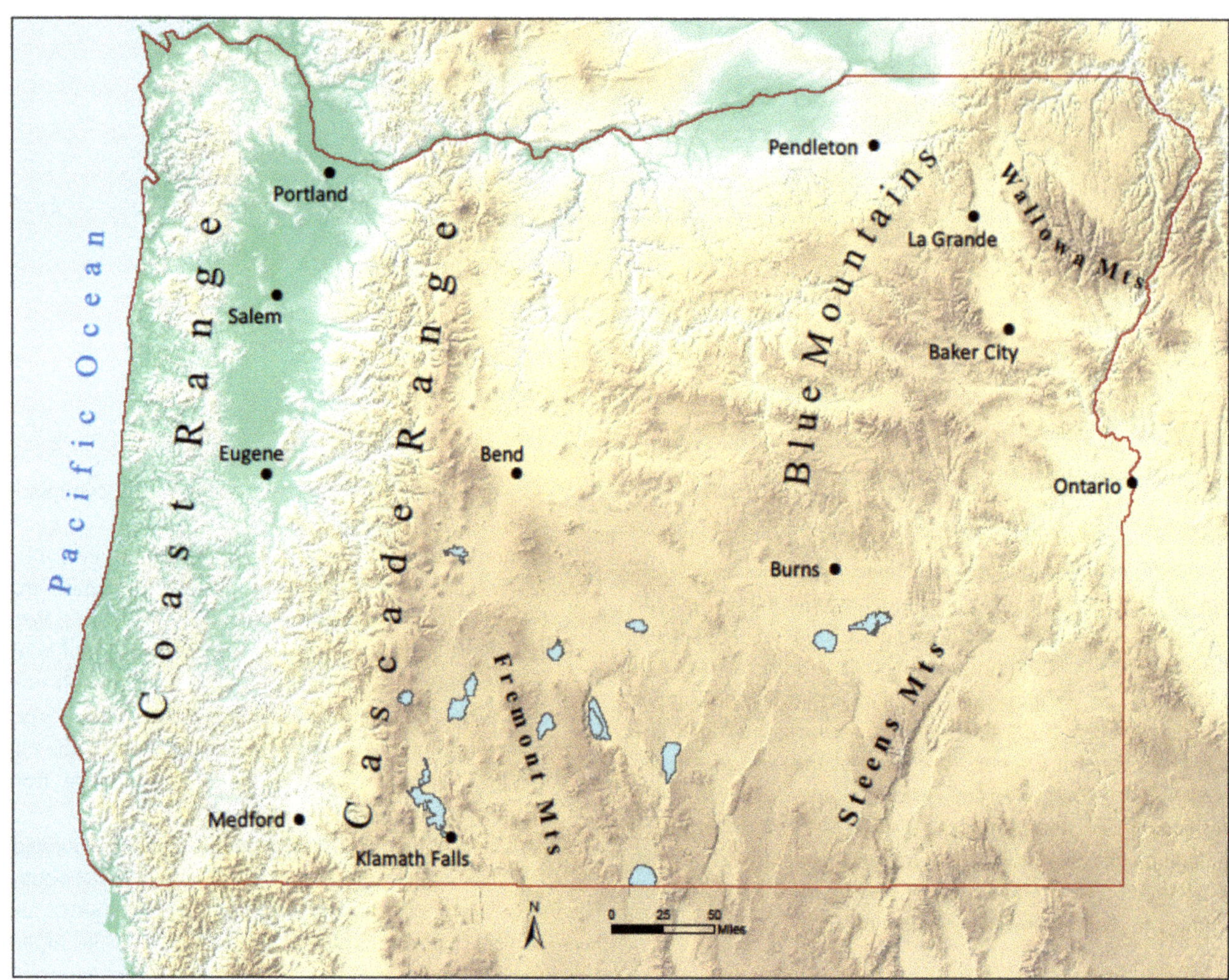

Fig. 4.1 Mountain ranges in Oregon. *Source* Prepared by Steve Campbell

Table 4.1 Climate, vegetation, and dominant soil great groups in Oregon mountain ranges

Mountain range[1]	Soil temperature class	Elev. range (m)	MAAT (° C)	MAP (mm)	Vegetation	Dominant soil great groups
Coast range	Cryic	1250–680	5.8 ± 1.3	3430 ± 0	Noble fir, Pacific silver fir, w hemlock, Douglas-fir	Dystrocryepts, Fulvicryands
	Frigid	900–465	6.7 ± 0	2445 ± 480	Douglas-fir, western hemlock, red alder	Hapludands
	Mesic	685–85	9.9 ± 0.8	2115 ± 400	Douglas-fir, western hemlock, red alder, bigleaf maple	Dystrudepts, Fulvudands, Hapludands, Humudepts
	Isomesic	400–50	10 ± 1.0	2000 ± 50	Sitka spruce, coastal redwood, Douglas-fir, red alder, western hemlock	Humudepts, Fulvudands, Haplohumults, Dystrudepts
	Isofrigid	915–480	6.7 ± 0	2750 ± 250	Sitka spruce, noble fir, western hemlock, Douglas-fir, red alder	Fulvudands
Cascade Mtns	Cryic	1915–1180	4.8 ± 1.4	1880 ± 510	Mtn hemlock, noble fir, lodgepole pine, Douglas-fir	Haplocryands, Vitricryands
(western)	Frigid	1310–760	5.3 ± 1.0	1780 ± 180	Douglas-fir, western hemlock	Udivitrands, Humudepts

(continued)

Table 4.1 (continued)

Mountain range[1]	Soil temperature class	Elev. range (m)	MAAT (° C)	MAP (mm)	Vegetation	Dominant soil great groups
	Mesic	925–165	9.6 ± 0.6	1910 ± 250	Douglas-fir, western hemlock, western redcedar	Humudepts, Palehumults
Klamath Mtns	Cryic	1990–1525	5.4 ± 0.5	1400 ± 375	White fir, Shasta red fir, Douglas-fir	Humicryepts, Haplocryands
	Frigid	1615–940	6.2 ± 0.2	1600 ± 700	White fir, Douglas-fir	Dystroxerepts
	Mesic	1175–230	10 ± 0.8	1010 ± 220	Douglas-fir, ponderosa pine, Pacific madrone, CA black oak	Dystroxerepts, Haploxeralfs, Haploxerepts, Palexerults
Cascades	Cryic	1815–1225	5.6 ± 0.6	820 ± 245	Noble fir, mtn hemlock, Douglas-fir, white fir	Vitricryands, Cryaquands
(High)	Frigid	1495–730	6.5 ± 0.6	600 ± 205	Ponderosa pine, Douglas-fir, grand fir, western juniper	Vitrixerands
	Mesic	970–285	8.9 ± 0.6	535 ± 140	Ponderosa pine, Douglas-fir, OR white oak	Haploxerolls, Argixerolls
Blue Mtns	Cryic	2150–1710	2.1 ± 2.1	1030 ± 320	Subalpine fir, Engelmann spruce, ponderosa pine, mtn big sagebrush	Humicryepts, Vitricryands, Haplocryepts
	Frigid	1720–955	6.1 ± 1.3	585 ± 215	Ponderosa pine, Douglas-fir, mtn big sagebrush	Argixerolls, Vitrixerands, Udivitrands, Haploxerolls
	Mesic	1230–685	8.7 ± 0.8	295 ± 45	WY big sagebrush, mtn big sagebrush, western juniper, wheatgrass, ID fescue	Argixerolls, Haploxerolls
Fremont Mtns	Cryic	2080–1720	5.4 ± 1.0	735 ± 235	Ponderosa pine, whitebark pine, lodgepole pine	Vitricryands, Cryaquolls
	Frigid	1825–1305	6.6 ± 1.0	570 ± 185	Ponderosa pine, white fir, Douglas-fir, ID fescue	Haploxerands, Haploxerolls, Argixerolls
	Mesic	1535–1280	8.6 ± 0.3	335 ± 45	Mtn big sagebrush, wheatgrass, ID fescue	Haploxerolls, Argixerolls, Humaquepts
Steens Mtn	Cryic	2625–1690	5.0 ± 0	480 ± 95	Mtn big sagebrush, ID fescue, wheatgrass	Haplocryolls
	Frigid	1660–1300	6.8 ± 0.4	265 ± 35	WY big sagebrush, basin big sagebrush, wheatgrass, bluegrass, ID fescue	Durixerolls, Haploxerolls, Palexerolls, Haplocambids
	Mesic	1680–1295	7.6 ± 1.6	320 ± 220	WY big sagebrush, basin big sagebrush	Haplocambids, Haplodurids
Wallowa Mtns	Cryic	2200–1530	2.8 ± 1.2	1005 ± 185	Subalpine fir, Engelmann spruce, lodgepole pine	Haplocryepts, Humicryepts, Vitricryands
	Frigid	1740–1150	4.8 ± 1.1	680 ± 105	Douglas-fir, grand fir, ponderosa pine, lodgepole pine	Argixerolls, Haploxerolls, Vitrixerands, Udivitrands
	Mesic	1220–550	8.3	610	Western juniper, WY big sagebrush, ID fescue	Haploxerolls, Argixerolls

[1]Only includes MLRAs with pronounced elevation gradients

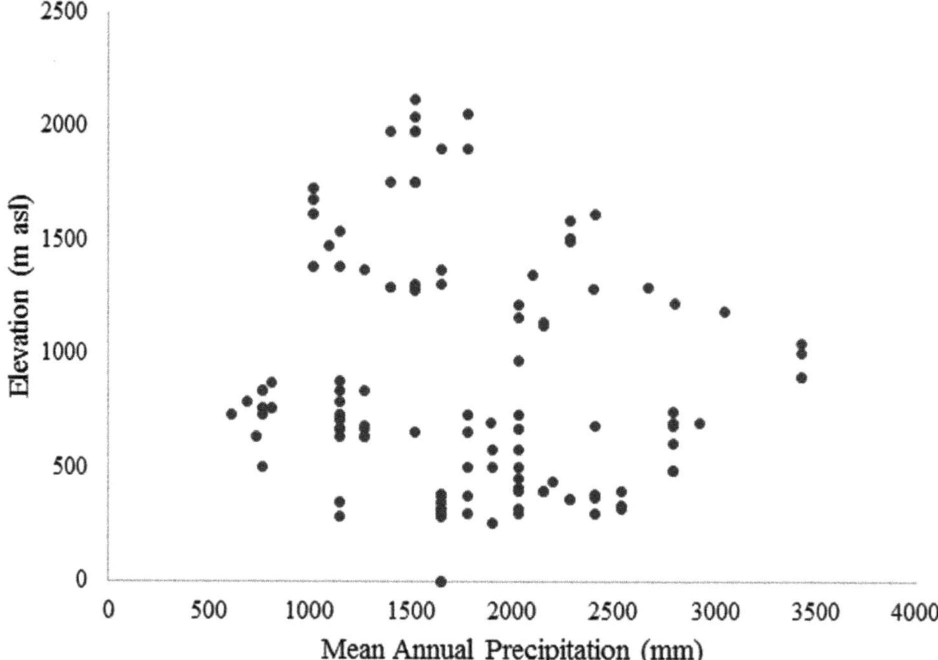

Fig. 4.2 Relation between mean annual precipitation and elevation (m above sea level) in western Oregon

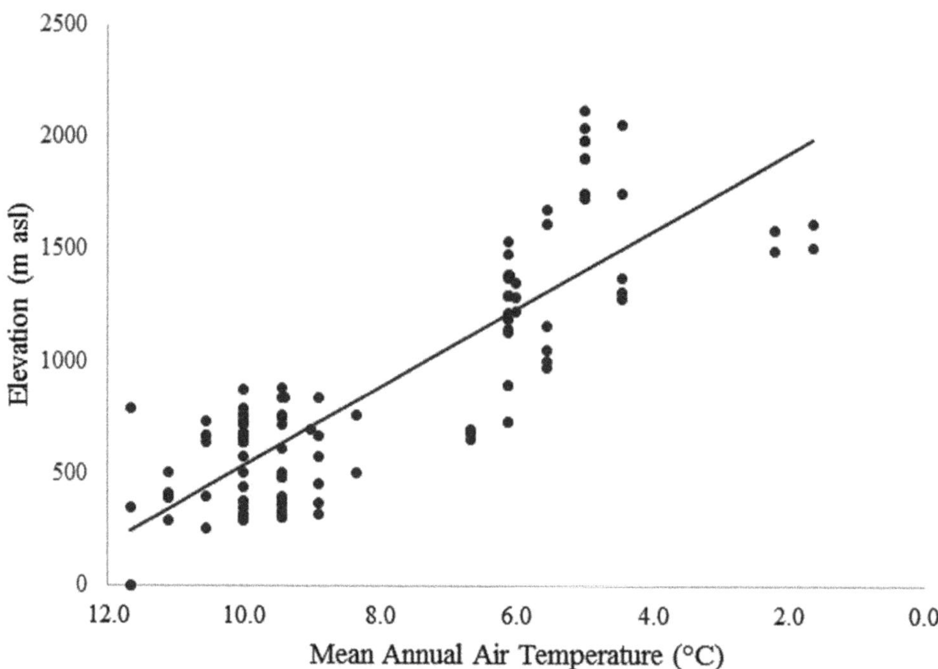

Fig. 4.3 Relation between mean annual air temperature and elevation in western Oregon

precipitation for eastern Oregon (Fig. 4.5). On western slopes, the mean annual air temperature (MAAT) varies from 9.6 °C below about 650 m, to 5.3 °C at 1,300 m, and 4.8 °C at 1,500 m (Table 4.1). On the eastern slopes, the MAAT is 8.9 °C at 1,000 m, 6.5 °C at 1,300 m, and 5.6 °C at 1,500 m. These changes are consistent with those recorded for eastern Oregon (Fig. 4.6). The adiabatic lapse rate is greater (0.78 °C per 100 m) than for western Oregon (Fig. 4.3).

On the western slopes of the Cascades, Douglas-fir and western hemlock are dominant at the lower and middle elevations, and mountain hemlock, noble fir, Douglas-fir, and lodgepole pine are common at the higher elevations (Table 4.1). On the High Cascades eastern slopes, ponderosa pine, and Douglas-fir are present at all elevations. However, Oregon white oak is common at the lower elevations near the Columbia River, grand fir at the middle elevations, and white fir, noble fir and mountain hemlock at the higher

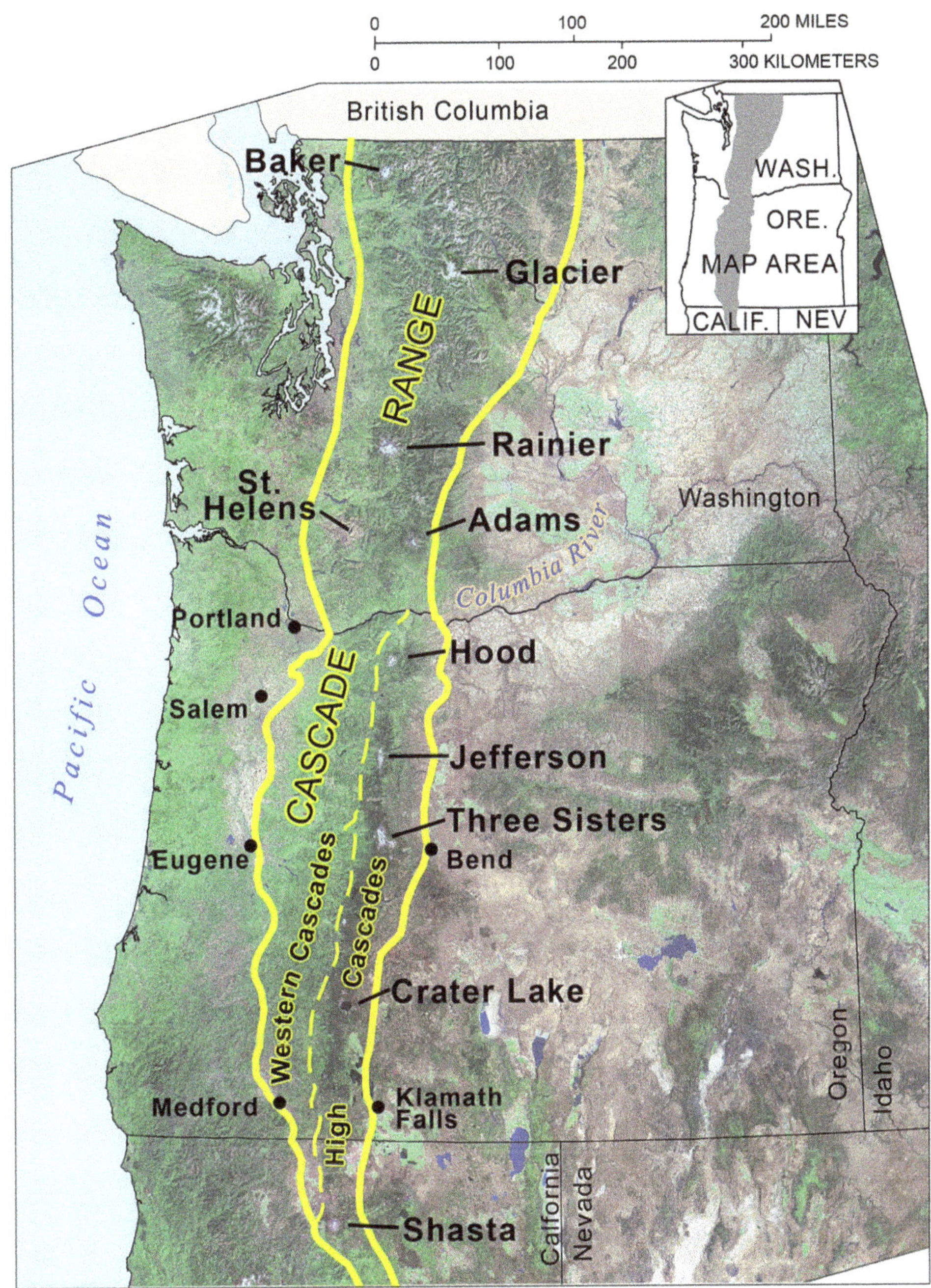

Fig. 4.4 The Cascade range in Oregon and adjoining states. *Source* The Oregon encyclopedia, a project of the Oregon Historical Society

elevations. Lodgepole pine and ponderosa pine are dominant species on the ash and pumice mantled basalt plateau in the city of LaPine area of central Oregon (Fig. 4.7).

On the western slopes, Humudepts and Palehumults are common at the lower elevations, Udivitrands and Humudepts at the middle elevations, and Haplocryands and Vitricryands at the higher elevations (Table 4.1). On the eastern slopes of the High Cascades, Haploxerolls and Argixerolls are common at the lower elevations, Vitrixerands at the middle elevations, and Vitricryands and Cryaquands at the higher elevations.

4.4 Wallowa Mountains

The Wallowa Mountains are located on the Columbia Plateau in Wallowa County in the northeastern corner of Oregon (Fig. 4.1). The range extends 64 km northwest to southeast, and the highest point is Sacajawea Peak at 3,000 m. The core of the Wallowa Mountains is composed of the granodioritic Wallowa Batholith surrounded by Columbia River basalt.

In the Wallowa Mountains, the mean annual precipitation is around 300 mm below 1,000 m, 585 mm to 1,700 m, and

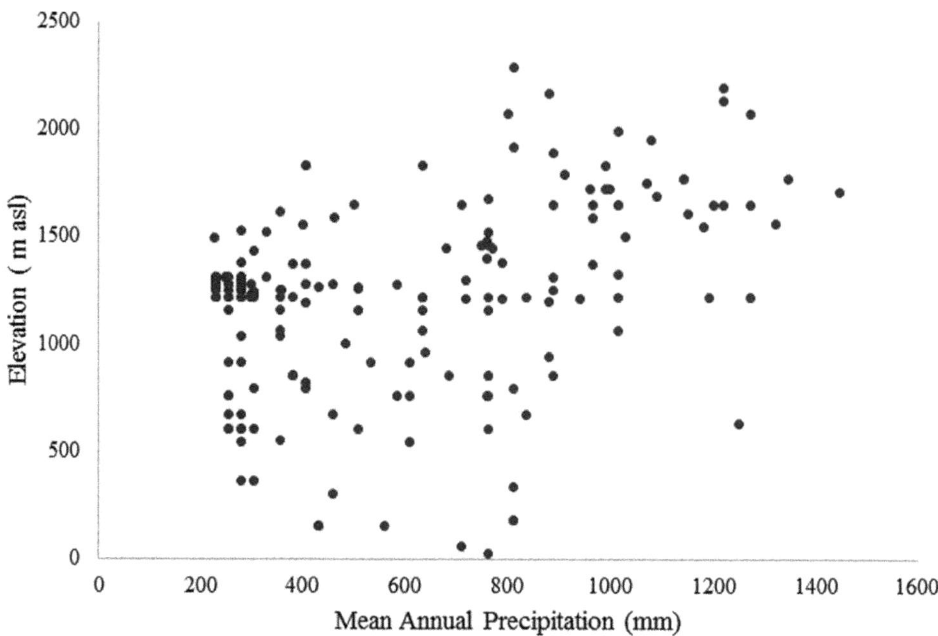

Fig. 4.5 Relation between mean annual precipitation and elevation in eastern Oregon

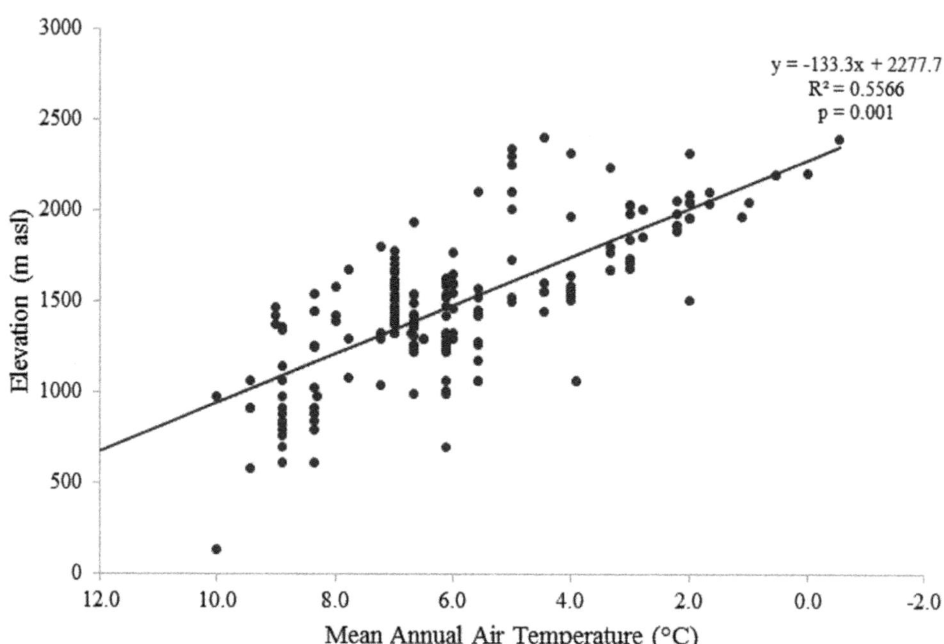

Fig. 4.6 Relation between mean annual air temperature and elevation in eastern Oregon

over 1,000 mm at 1,900 m (Table 4.1). The mean annual air temperature is 8.7 °C below 1,000 m, 6.1 °C at 1,700 m, and 2.1 °C at 1,900 m.

At the lower elevations, plant communities are dominated by western juniper, Wyoming big sagebrush, and Idaho fescue (Table 4.1). At middle elevations trees are grand fir, lodgepole pine, ponderosa pine, and Douglas-fir. At the higher elevations, subalpine fir, Engelmann spruce, and lodgepole pine forests are prevalent. Mountain big sagebrush may occur at any elevation. There is an apparent alpine zone above 2,700 m at Eagle Cap in the Wallowa Mountains (Allen and Burns 2000).

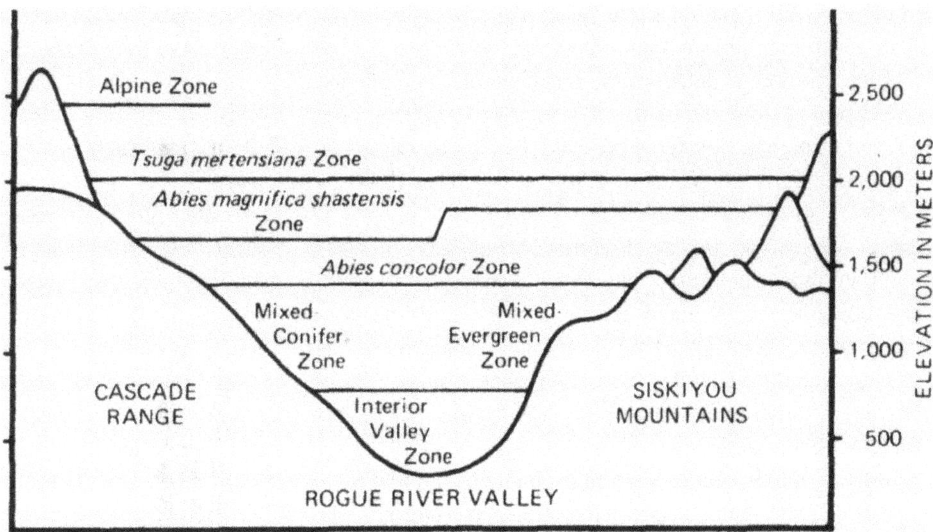

Fig. 4.7 Relation between vegetation and elevation in the Cascade Range and the Klamath (Siskiyou) Mountains. *Source* Fig. 99, Franklin and Dyrness, 1988

In the Wallowa Mountains, Argixerolls and Haploxerolls are the dominant soils at the lower elevations. At middle elevations Haploxerolls, Argixerolls, Vitrixerands, and Udivitrands are the predominant soils. At the higher elevations Vitricryands, Haplocryepts, and Humicryepts are predominant (Table 4.1). Allen and Burns (2000) reported Lithic Haplocryods above 2,700 m in the Wallowa Mountains.

4.5 Blue Mountains

The Blue Mountains are located in northeastern Oregon west of the Wallowa Mountains (Fig. 4.1). Rock Creek Butte is the highest peak at 2,776 m. The Blue Mountains are uplift mountains composed of basalt that are part of the Columbia River Plateau.

In the Blue Mountains, the mean annual precipitation ranges from 600 to 680 mm at elevations below 1,500 m, but may exceed 1,000 mm at higher elevations (Table 4.1). The mean annual air temperature is 8.3 °C at elevations below 1,000 m, 4.8 °C at 1,500 m, and 2.8 °C above 1,700 m.

At the lower elevations, plant communities are dominated by western juniper, Wyoming big sagebrush, mountain big sagebrush, Idaho fescue, and bluebunch wheatgrass. At middle elevations mountain big sagebrush, ponderosa pine and Douglas-fir are prevalent. At the higher elevations, subalpine fir, Engelmann spruce, ponderosa pine and mountain big sagebrush are dominant (Table 4.1).

At the lower elevations Haploxerolls and Argixerolls are common. At middle elevations Argixerolls, Haploxerolls, Vitrixerands, and Udivitrands are the predominant soils. The dominant soils at higher elevations are Haplocryepts, Humicryepts, and Vitricryands. (Table 4.1).

4.6 Steens Mountain

Steens Mountain is an isolated peak located on the Malheur High Plateau in Harney County, southeastern Oregon (Fig. 4.1). It is a large fault-block composed of basalt. The highest point has an elevation of 2,968 m.

On Steens Mountain, the mean annual precipitation ranges from 265 to 320 mm below 1,700 m and 480 mm above this elevation (Table 4.1). The mean annual air temperature decreases from 7.6 °C below 1,300 m to 5.0 °C at the highest elevations.

Plant communities at the lower and middle elevations include Wyoming and basin big sagebrush, bluebunch wheatgrass, Sandberg bluegrass, and Idaho fescue. Mountain big sagebrush, bluebunch wheatgrass, and Idaho fescue are common at the highest elevations (Table 4.1). These data suggest that trees, especially conifers, are uncommon on Steens Mountain. However, in some snow pockets and along drainageways quaking aspen (*Populus tremuloides*) is prevalent.

The predominant soil great groups on Steens Mountain are Haplocambids and Haplodurids at the lower elevations, Haploxerolls, Durixerolls, Palexerolls, and Haplocambids at the middle elevations, and Haplocryolls at the higher elevations (Table 4.1).

4.7 Fremont Mountains

The Fremont Mountains are a 137 km long, north–south trending mountain range in south-central Oregon (Fig. 4.1). The highest point is 2,690 m. The Fremont Mountains are composed of a thick section of volcanic, volcaniclastic, and sedimentary rocks that represent the northwest portion of the Basin and Range physiographic province.

In the Fremont Mountains, the mean annual precipitation is 335 mm below 1,300 m, 570 mm at 1,800 m, and 735 mm at higher elevations (Table 4.1). The mean annual air temperature is 8.6 °C below 1,300 m, 6.6 °C at 1,800 m, and 5.4 °C at higher elevations.

The vegetation in the Fremont Mountains includes mountain big sagebrush, bluebunch wheatgrass, Sandberg bluegrass, and Idaho fescue below 1,300 m, ponderosa pine, white fir, Douglas-fir, and Idaho fescue at 1,800 m, and lodgepole pine, whitebark pine (*Pinus albicaulis*), and ponderosa pine at higher elevations (Table 4.1).

Soils in the Fremont Mountains include Haploxerolls, Argixerolls, and Humaquepts below 1,300 m, Haploxerolls, Argixerolls, and Haploxerands at 1,800 m, and Vitricryands and Cryaquolls at higher elevations (Table 4.1).

4.8 Klamath Mountains

The Siskiyou Mountains, the portion of the Klamath Mountains in southwestern Oregon, are composed of rifted fragments of volcanic, basaltic, gabbroic, granitic, and serpentinitic rocks. At 2,300 m, Mt. Ashland is the highest point in the Siskiyou Mountains.

The mean annual precipitation ranges between 1,000 and 1,600 mm (Table 4.1). The mean annual air temperature is 10 °C at 700 m, 6.2 °C at 1,275 m, and 5.4 °C at 1,750 m.

Douglas-fir is the dominant tree species at all elevations in Oregon's Klamath Mountains (Table 4.1). Ponderosa pine and Oregon white oak accompany Douglas-fir at the lower elevations; and white fir and Shasta red fir co-exist with Douglas-fir at the middle and higher elevations.

In the Klamath Mountains, Dystroxerepts, Haploxeralfs, Haploxerepts, and Palexerults occur at the lower elevation; Dystroxerepts are common at the middle elevations; and Humicryepts and Haplocryands are predominant at the higher elevations (Table 4.1).

4.9 Summary

There are eight main mountain ranges in Oregon with pronounced elevation changes in climate, vegetation, and soils, including the Coast Range, Western Cascade, Klamath, High Cascade, Wallowa, Fremont, Steens, and Blue Mountains. There is no statistically significant correlation between mean annual precipitation and elevation in western or eastern Oregon. However, there is a strong correlation between mean annual air temperature and elevation in western and eastern Oregon, with adiabatic lapse rates of 0.69 and 0.78 ° C/100 m, respectively. In western Oregon, the lower and middle elevations have Douglas-fir, western hemlock, red alder, western redcedar, and bigleaf maple; and the higher elevations have white fir, mountain hemlock, noble fir, Shasta red fir, and Pacific silver fir. In eastern Oregon, the lower elevations have western juniper, sagebrush, and bunchgrasses; the middle elevations have Douglas-fir, ponderosa pine, and grand fir; the subalpine zone has subalpine fir, Engelmann spruce, and lodgepole pine; and the alpine zone, which is of limited extent, has forbs and grasses.

In western Oregon, Dystroxerepts, Palexerults, Humudepts, Fulvudands, and Hapludands dominate the lower and middle elevations, and Cryepts and Cryands are most common at the higher elevations. In eastern Oregon, Haploxerolls and Argixerolls dominate the lower and middle elevations, and Cryepts, Cryands, and Cryolls are most common at the higher elevations.

Reference

Allen CE, Burns SF (2000) Characterization of alpine soils, Eagle Cap, Wallowa Mountains. Oregon Phys Geog 21:212–222

5 General Soil Regions of Oregon

5.1 Introduction

The Natural Resources Conservation Service (2006) has delineated about 270 Major Land Resource Areas (MLRAs) in the United States, including Puerto Rico, the US Virgin Islands, St. Thomas, and St. John . MLRAs are geographically associated land areas that serve as a framework for organizing soil surveys and information about land as a resource for farming, ranching, forestry, recreation, and other uses. Oregon contains 17 MLRAs ranks sixth nationally in number of MLRAs behind Texas (36), Alaska (27), Oklahoma (22), California (19), and South Dakota (18) (Fig. 5.1). There are

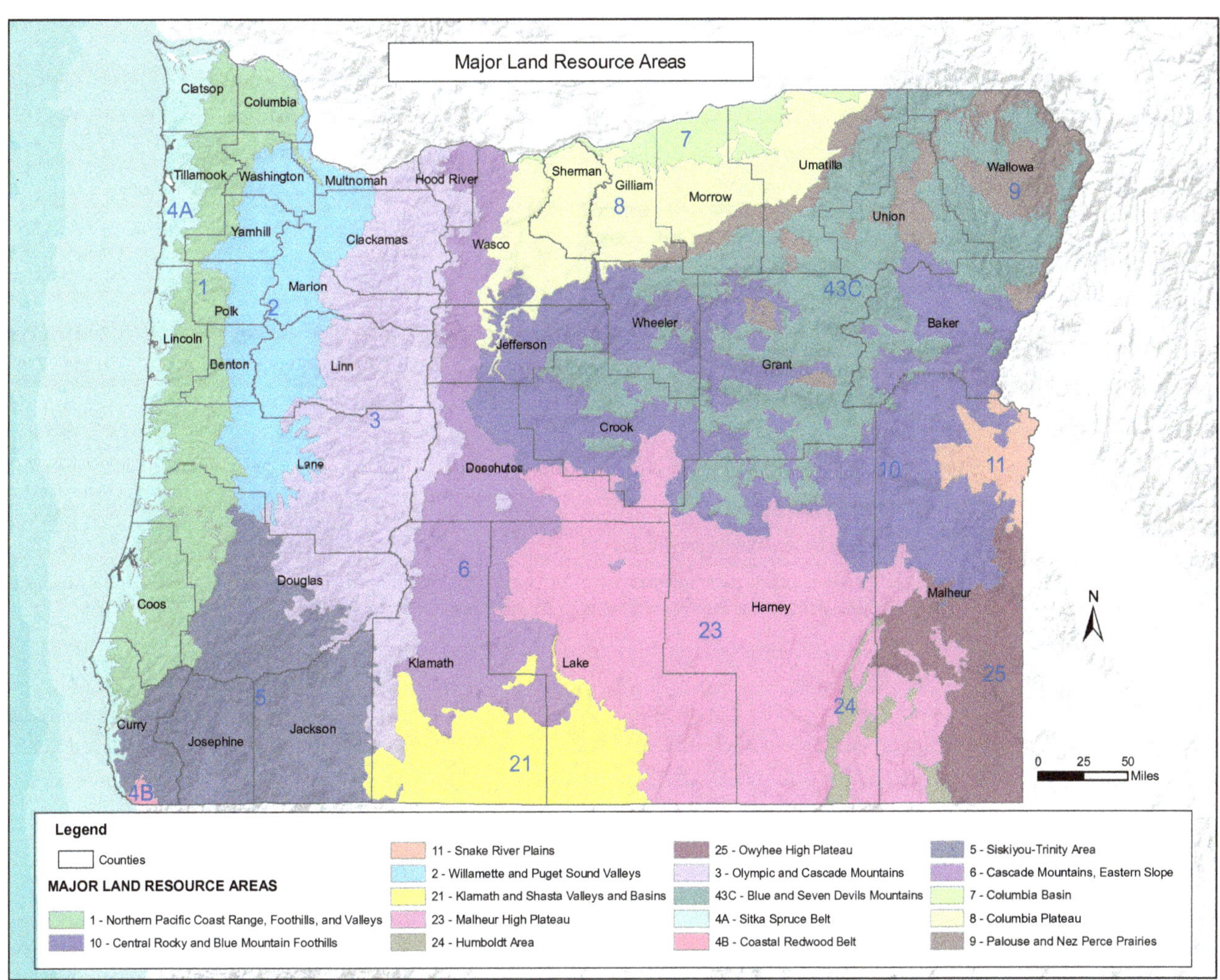

Fig. 5.1 Major Land Resource Areas in Oregon. *Source* Prepared by Steven Campbell

T. Thorson et al., *The Soils of Oregon*, World Soils Book Series,
https://doi.org/10.1007/978-3-030-90091-5_5

Table 5.1 A comparison of major land resource areas and physiographic provinces in Oregon

Physiographic province	Major land resource area
Coast Range	Northern Pacific Coast Range, Foothills, and Valleys; Sitka Spruce Belt
Willamette Valley	Willamette and Puget Sound Valleys
Klamath Mountains	Coastal Redwood Belt; Siskiyou-Trinity Area
Cascades	Olympic and Cascade Mtns.; Cascade Mountains, Eastern Slope; Klamath and Shasta Valleys & Basins
Deschutes-Umatilla Plateau	Columbia Basin; Columbia Plateau; Palouse and Nez Perce Prairies
Blue Mountains	Palouse and Nez Perce Prairies; Central Rocky and Blue Mtn. Foothills; Blue and Seven Devils Mtns
Owyhee	Owyhee High Plateau
Basin and range	Snake River Plains; Klamath and Shasta Valleys and Basins; Malheur High Plateau; Humboldt Area
High Lava Plains	Cascade Mtns., Eastern Slope; Malheur High Plateau

Table 5.2 Abbreviations for major land resourse areas in Oregon

Major land resource areas	MLRAs abbreviated OR	Number[a]
Northern Pacific Coast Range, Foothills, and Valleys	Coast Range	1
Willamette and Puget Sound Valleys	Willamette Valley	2
Olympic and Cascade Mountains	Cascade Mountains	3
Sitka Spruce Belt	Sitka Spruce Belt	4A
Coastal Redwood Belt	Coastal Redwood Belt	4B
Siskiyou-Trinity Srea	Siskiyou Mountains	5
Cascade Mountains—Eastern Slope	Cascade Mountains—Eastern Slope	6
Columbia Basin	Columbia Basin	7
Columbia Plateau	Columbia Plateau	8
Palouse and Nez Perce Prairies	Palouse Prairie	9
Central Rocky and Blue Mountain Foothills	Blue Mountain Foothills	10
Snake River Plains	Snake River Plains	11
Klamath and Shasta Valleys and Basins	Klamath Basin	21
Malheur High Plateau	Malheur High Plateau	23
Humboldt Area	Humboldt Area	24
Owyhee High Plateau	Owyhee High Plateau	25
Blue and Seven Devils Mountains	Blue Mountains	43C

[a]Although numbers are employed in USDA Handbook 296, they are not used here because of likely changes. The MLRA names are shortened so as to be applicable to Oregon

from one to four MLRAs in each physiographic province in Oregon (Table 5.1). In this chapter, we discuss each of Oregon's MLRAs, focusing on the major soil great groups. Because of anticipated changes in number assignments, an abbreviated descriptive name for each MLRA is used herein (Table 5.2).

5.2 Malheur High Plateau

The Malheur High Plateau is the largest MLRA in Oregon, comprising nearly 40,000 km,[2] which is 15% of the state area (Table 5.3). About two-thirds (67%) of this MLRA is in Oregon, with the remainder in Nevada (25%) and California (8%). The Malheur High Plateau is contained within the High Lava Plains and the Basin and Range physiographic provinces. Elevations range from 1,285 to 1,915 m in most of the area. Maximum slopes range from 5 to 55%. The Malheur High Plateau is composed of basalt fault blocks and lava plateaus with basins composed of Quaternary alluvium and colluvium (Table 5.4). Figure 5.2 shows the Malheur High Plateau near Frenchglen, Oregon.

The mean annual air temperature ranges between 5.5 and 8 °C, and the mean annual precipitation ranges from 215 to 335 mm (Table 5.3). The vegetation is low sagebrush, Wyoming big sagebrush and bunchgrasses such as Thurber's needlegrass, bluebunch wheatgrass, and Sandberg bluegrass at low to middle elevations (Fig. 5.2) and mountain big sagebrush, rabbitbrush, snowberry (*Symphoricarpos* spp.), Idaho fescue, and bunchgrasses at the higher elevations. Dry basins feature black greasewood, basin wildrye, alkali sacaton (*Sporobolus airoides*), alkali bluegrass (*Poa secunda* ssp. *juncifolia*), and inland saltgrass. Wet basins have sedges, rushes and inland saltgrass. (Table 5.4).

Common soil orders in this MLRA are Aridisols and Mollisols. The soils in the area typically have a mesic or frigid soil temperature regime, an aridic or xeric soil moisture regime, and mixed or smectitic mineralogy. Predominant soils are shallow Argidurids (Actem series) formed in residuum and colluvium on hills and tablelands (Table 5.5). Moderately deep Argidurids (Brace series) have formed in residuum and alluvium on structural benches and foothills. Shallow Haplargids (Anawalt and Coztur series) and Durixerolls (Moonbeam series) have developed in residuum on hills, mountains, and plateaus. Moderately deep Haplocambids (Felcher and Lonely series) and Haploxerolls (Westbutte series) and shallow Haplodurids (Raz series) and Durixerolls (Goodtack series) have formed in residuum and colluvium on mountains and plateaus. Moderately deep Haplocryolls (Baconcamp series) occur in colluvium on hills and mountain slopes. Shallow Argixerolls (Ninemile series) have formed in residuum on hills, plateaus, and mountain slopes. Moderately deep Palexerolls (Carryback series) have developed in colluvium and residuum on plateaus. Very deep Endoaquolls (Fury and Ozamis series) and Halaquepts (Reese series) have formed in lacustrine sediments on lake plains. Very deep Haplocambids (Catlow, Reallis, and Enko series), Natrargids (Ausmus and Poujade series), and Paleargids (Spangenburg series) occur in alluvium on lake terraces.

Figure 5.3 shows upland soils, the Harcany, Newlands, Floke, Brace, and Ratto soil series on a fault-block plateau in Lake County. Figure 5.4 shows the contact between a fault-block and intermontane lacustrine sediments in Harney

Table 5.3 Site factors of major land resource areas in Oregon

Description	% in OR	Area (km²)	% of Total	MAAT (°C)	MAP (mm)	Mean elev range (m)	Max. slope (%)
Malheur High Plateau	67	39,744	15.0	6.8 ± 1.1	275 ± 60	1915–1285	30 ± 25
Cascade Mountains	61	38,534	14.5	8.7 ± 2.7	2025 ± 43	1155–325	80 ± 8.5
Blue Mountain Foothills	71	32,223	12.2	7.7 ± 1.5	340 ± 55	1590–940	75 ± 30
Blue Mountains	84	30,471	11.5	4.5 ± 2.1	845 ± 205	1905–1140	75 ± 20
Siskiyou Mountains	38	19,842	7.5	10 ± 1.4	1070 ± 220	1230–245	75 ± 15
Cascade Mountains, Easterm Slope	52	17,732	6.7	6.7 ± 1.0	585 ± 155	1635–980	60 ± 10
Coast Range	65	17,339	6.5	9.7 ± 3.7	2200 ± 220	810–75	85 ± 8.5
Willamette Valley	43	13,605	5.1	11 ± 0.26	1165 ± 70	470–70	55 ± 35
Owyhee High Plateau	16	11,994	4.5	6.7 ± 1.3	305 ± 60	2125–1270	50 ± 20
Columbia Plateau	25	11,989	4.5	9.3 ± 0.70	305 ± 40	1060–220	60 ± 30
Klamath Basin	35	10,427	3.9	7.1 ± 1.4	470 ± 165	1770–1155	40 ± 30
Palouse Prairies	28	6391	2.4	8.5 ± 1.3	515 ± 75	1435–435	90 ± 25
Sitka Spruce Belt	45	6183	2.3	9.5 ± 0.8	2110 ± 255	430–10	60 ± 40
Columbia Basin	22	3769	1.4	10 ± 0.60	205 ± 30	600–90	60 ± 8.5
Snake River Plains	6	2561	1.0	10 ± 1.1	240 ± 45	830–520	30 ± 20
Humboldt Area	6	1971	0.7	8.0 ± 0.90	225 ± 30	1610–1245	15 ± 15
Coastal Redwood Belt	2	242	0.1	10 ± 0.39	1590 ± 630	615–30	40 ± 50

Table 5.4 Additional site factors of major land resource areas in Oregon

Description	Vegetation	Parent materials	Bedrock	Landforms
Coast Range	Douglas-fir, western hemlock, bigleaf maple, red alder, western redcedar (north, central) Douglas-fir, Pacific madrone, incense cedar, tanoak (south)	Colluvium, residuum	Sandstone, silt-stone, basalt	Mtn slopes
Willamette Valley	Oregon white oak, Douglas-fir, Oregon ash, poison oak, grasses	Glaciolacustrine, alluvium, colluvium, residuum	Basalt	Lake terraces, foothills
Cascade Mountains	Douglas-fir, western hemlock, red alder, bigleaf maple (lower montane) mountain hemlock, noble fir, lodgepole pine (upper montane)	Colluvium, residuum, volcanic ash	Igneous, basalt	Mtn slopes
Sitka Spruce Belt	Sitka spruce, western hemlock, red alder, Douglas-fir, western redcedar, shore pine	Colluvium, residuum, alluvium, marine	[Variable]	Mtn slopes, alluvial plains
Coastal Redwood Belt	Coastal redwood, Douglas-fir, western red cedar, Port Orford cedar, red alder	Colluvium, residuum, alluvium, marine	Quaternary sediments	Mtn slopes, alluvial plains
Siskiyou Mountains	Douglas-fir, incense cedar, Pacific madrone, ponderosa pine, sugar pine, white fir (moist) Douglas-fir, Oregon white oak (dry) Idaho fescue, Thurber needlegrass (prairie openings)	Colluvium, residuum	Sedimentary, igneous, volcanic	Mtn slopes, hillslopes
Cascade Mountains, Eastern Slope	Douglas-fir, white fir, sugar pine, lodgepine pine (high elevations) ponderosa pine, lodgepole pine, antelope bitterbrush, mountain big sagebrush, Idaho fescue (middle elevations) Western juniper, antelope bitterbrush, mountain big sagebrush, Idaho fescue (lower elevations)	Volcanic ash, colluvium	[Variable]	Lava plateaus, uplands
Columbia Basin	WY big sagebrush, Indian ricegrass, wheatgrass, bluegrass, needlegrass	Eolian, loess, lacustrine	Basalt	Lake terraces, hillslopes, dunes
Columbia Plateau	WY big sagebrush, bluegrass, wheatgrass, Idaho fescue	Loess, colluvium, residuum	Basalt	Uplands, hillslopes
Palouse Prairies	Wheatgrass, Idaho fescue, bluegrass	Loess, colluvium, residuum	Basalt	Mtn slopes, hillslopes
Blue Mountain Foothills	Western juniper, WY big sagebrush, antelope bitterbrush, Idaho fescue (lower elevations) mountain big sagebrush, wheatgrass, bluegrass, needlegrass (middle and higher elevations)	Colluvium, residuum, loess	Basalt	Mtn slopes, lava plateaus
Snake River Plains	WY big sagebush, needlegrass, bluegrass, wheatgrass	Lacustrine, alluvium, loess	Basalt	Lake terraces, alluvial fans
Klamath Basin	Low sagebrush, wheatgrass, Idaho fescue, basin big sagebrush, antelope bitterbrush, needlegrass (low elevations) Ponderosa pine, western juniper, mtn mahogany (middle elevations) Douglas-fir, ponderosa pine, white fir, sugar pine (higher elevations) Bullrushes, tules, lilies, cattails (wet basins)	Colluvium, residuum	Volcanic	Mtn slopes, basins

(continued)

Table 5.4 (continued)

Description	Vegetation	Parent materials	Bedrock	Landforms
Malheur High Plateau	Low sagebrush, WY big sagebrush, rabbitbrush, Thurber needlegrass, bluebunch wheatgrass, Sandberg bluegrass (lower elevations) Black greasewood, alkali sacaton, alkali bluegrass, wildrye, inland saltgrass (dry basins) Mtn big sagebrush, rabbitbrush, snowberry, Idaho fescue, bunchgrasses (higher elevations) Sedges, rushes, inland saltgrass (wet basins)	Colluvium, residuum, alluvium	Basalt	Lava plateaus, alluvial plains
Humboldt Area	WY big sagebrush, rabbitbrush, spiny hopsage Bud sagebrush, shadscale blg greasewood, basin big sagebrush, wildrye, saltgrass	Lacustrine, alluvium	Volcanic	Lakes basins & terraces, alluvial fans & plains
Owyhee High Plateau	WY big sagebrush, wheatgrass, rabbitbrush, Idaho fescue, bluegrass, wildrye, Thurber needlegrass	Colluvium, residuum	Basalt, tuff	Lava plateaus
Blue Mountains	Western juniper, mountain big sagebrush, bunchgrasses (lower elevations) Ponderosa pine, Douglas-fir (middle elevations) Engelmann spruce, whitebark pine, western larch, lodgepole pine (high elevations)	Volcanic ash, loess, till, residuum, colluvium	Basalt, granite	Mtn slopes

Fig. 5.2 The Malheur High Plateau near Frenchglen, Oregon. *Source* Photo by Bureau of Land Management

Table 5.5 Dominant soils of major land resource areas in Oregon

Description	Dominant soil series
Coast Range	Hapludands (Hemcross, Klistan, Slickrock), Dystrudepts (Bohannon, Rinearson, and Blachly), Eutrudepts (Digger and Umpcoos); Humudepts (Klickitat and Preacher); Haplohumults (Peavine); Palehumults (Honeygrove)
Willamette Valley	Argialbolls (Amity), Albaqualfs (Dayton), Argixerolls (Woodburn and Dixonville), Haplohumults (Bellpine and Nekia), Palehumults (Jory); Haploxerepts (Ritner); Fragixerepts (Cascade, Kinton, and Powell)
Cascade Mountains	Vitricryands (Castlecrest), Dystrudepts (Aschoff, Zygore, Kinney, Klickitat, and Blachly), Humudepts (McCully), Palehumults (Honeygrove), Haplohumults (Peavine)
Sitka Spruce Belt	Fulvudands (Necanicum, Tolovana, and Klootchie), Dystrudepts (Templeton and Nehalem), Fluvaquents (Coquille), Humudepts (Reedsport and Skipanon)
Coastal Redwood Belt	Haploxerults (Josephine), Udifluvents (Bigriver), Haplohumults (Winchuck and Loeb), Dystrudepts (Dulandy)
Siskiyou Mountains	Dystroxerepts (Beekman, Atring, Jayar, and Vermisa), Haploxeralfs (Speaker and Vannoy), Haploxerults (Josephine), Argixerolls (McNull), Haploxerolls (McMullin and Medco) Haplohumults (Bellpine and Windygap)
Cascade Mountains, Eastern Slope	Haploxerepts (Wamic), Vitricryands (Lapine, Steiger, and Shanahan), Vitrixerands (Smiling, Wanoga, and Maset)
Columbia Basin	Haplocalcids (Adkins and Sagehill), Haplocambids (Shano and Warden), Torripsamments (Quincy)
Columbia Plateau	Haploxerolls (Condon, Mikkalo, Ritzville, Walla Walla, Wrentham, Valby, Bakeoven, Lickskillet), Argixerolls (Morrow), Durixerolls (Willis)
Palouse Prairies	Haploxerolls (Palouse, Athena, Bocker, and Rockly), Argixerolls (Gwinly and Waha)
Blue Mountain Foothills	Haploxerolls (Bakeoven, Lickskillet, Rockly, and Westbutte), Palexerolls (Simas), Argixerolls (Tub, Madras, Merlin, Waterbury, Ateron, Ruckles, and Vitale)
Snake River Plains	Haplocalcids (Sagehill), Argidurids (Virtue), Haplodurids (Frohman and Nyssa)
Klamath Basin	Argixerolls (Lorella, Royst, and Woodcock), Haploxerolls (Fordney), Palexerolls (Booth), Humaquepts (Tulana), Haplohemists (Lather), Haploxerands (Pokegema)
Malheur High Plateau	Argidurids (Actem and Brace), Haplargids (Anawalt and Coztur), Haplocambids (Felcher, Lonely, Catlow, Enko, and Reallis), HaploPalexerolls (Carryback), Endoaquolls (Fury and Ozamis), Natrargids (Ausmus and Poujade), Halaquepts (Reese), Paleargids durids (Raz), Haplocryolls (Baconcamp), Argixerolls (Ninemile), (Spangenburg), Haploxerolls (Westbutte), Durixerolls (Goodtack and Moonbeam)
Humboldt Area	Haplocambids (Reallis, Catlow, Enko, and Actem), Argidurids (Deppy and Tumtum), Argixerolls (Reluctan), Aquicambids (Droval and Alvodest), Paleargids (Berdugo, Spangenburg), Humaquepts (Crump), Aquisalids (Icene), Haplargids (Norad), Haplocalcids (Outerkirk), Halaquepts (Reese)
Owyhee High Plateau	Argidurids (Brace), Haplargids (Anawalt and Gumble), Argixerolls (Durkee, Gaib, and Ninemile), Haplocalcids (Enko)
Blue Mountains	Palexerolls (Hankins), Haploxerolls (Hall Ranch and Umatilla), Humicryepts (Muddycreek), Dystrocryepts (Angelbasin), Vitricryands (Bucketlake, Helter, and Lackeyshole), Argialbolls (Lookingglass and Cowsly), Argixerolls (Gwinly)

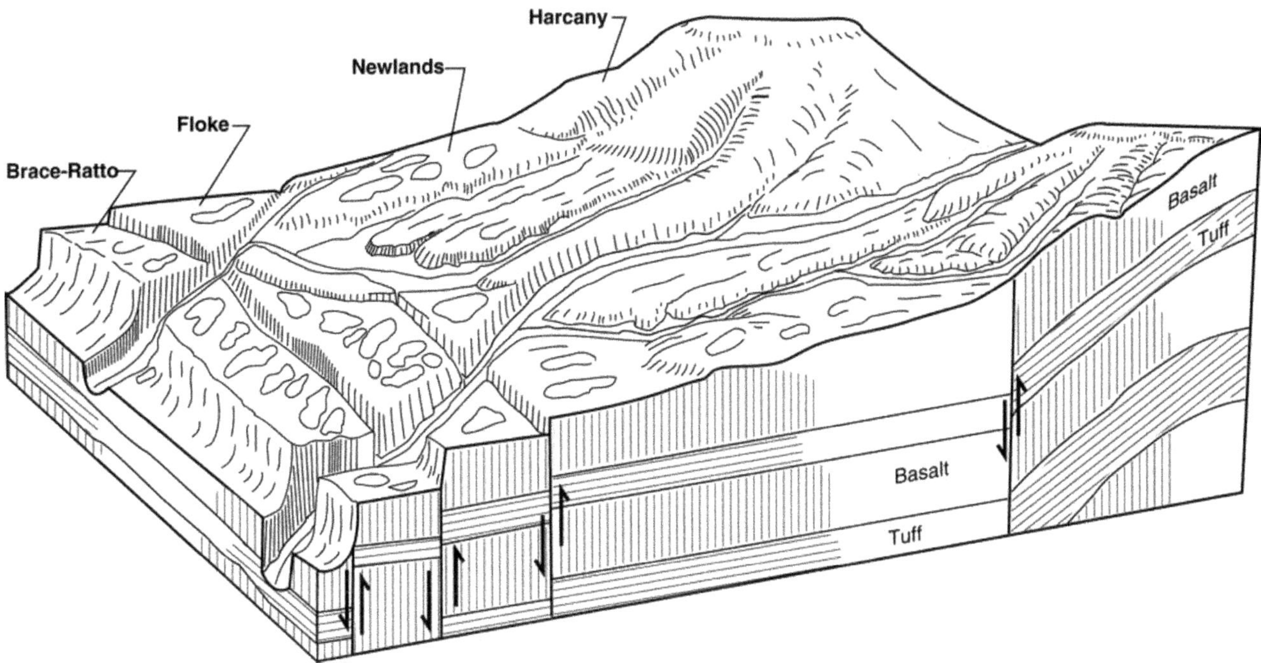

Fig. 5.3 Upland soils on a fault-block Malheur High Plateau in Lake County, including the Harcany (Haplocryolls), Newlands (Argicryolls), and Floke, Brace, and Ratto (Argidurids) soil series. *Source* NRCS

County. The extensive Ninemile soil series is on the plateau, the Felcher along the canyon sides, the Hart Camp on rock pediments, the Windybutte along high terraces and the Fury and other soil series on lower terraces. Figure 5.5 shows soil and landscape relationships on lacustrine terraces in Catlow Valley in Harney County. Upper terraces contain the Enko and Catlow (Haplocambids) and Berdugo (Paleargids) soil series from an ice age lake that filled the valley to a depth of 60 m. (Fig. 5.6). The Spangenburg (Paleargids) and Norad (Haplargids) soil series formed on lower terraces, and the Boulder Lake (Epiaquerts) soil series formed in depressions on lake plains.

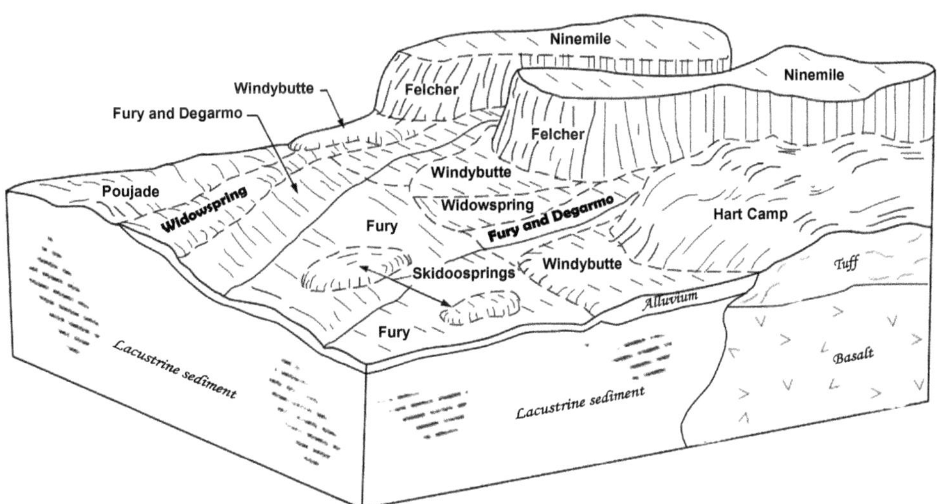

Fig. 5.4 Block diagram of a catena of soils near Malheur Lake on the Malheur High Plateau. The Ninemile (Argixerolls) and Felcher (Haplocambids) series have formed in colluvium and residuum on volcanic plateaus. The Hart Camp series (Argixerolls) has formed in tuff residuum. The Windybutte (Argixerolls), Widowspring (Haploxerolls), Skidoosprings (Halaquepts), Poujade (Natrargids), and Fury (Endoaquolls) series have formed in lacustrine deposits or alluvium on lake or stream terraces. *Source* NRCS

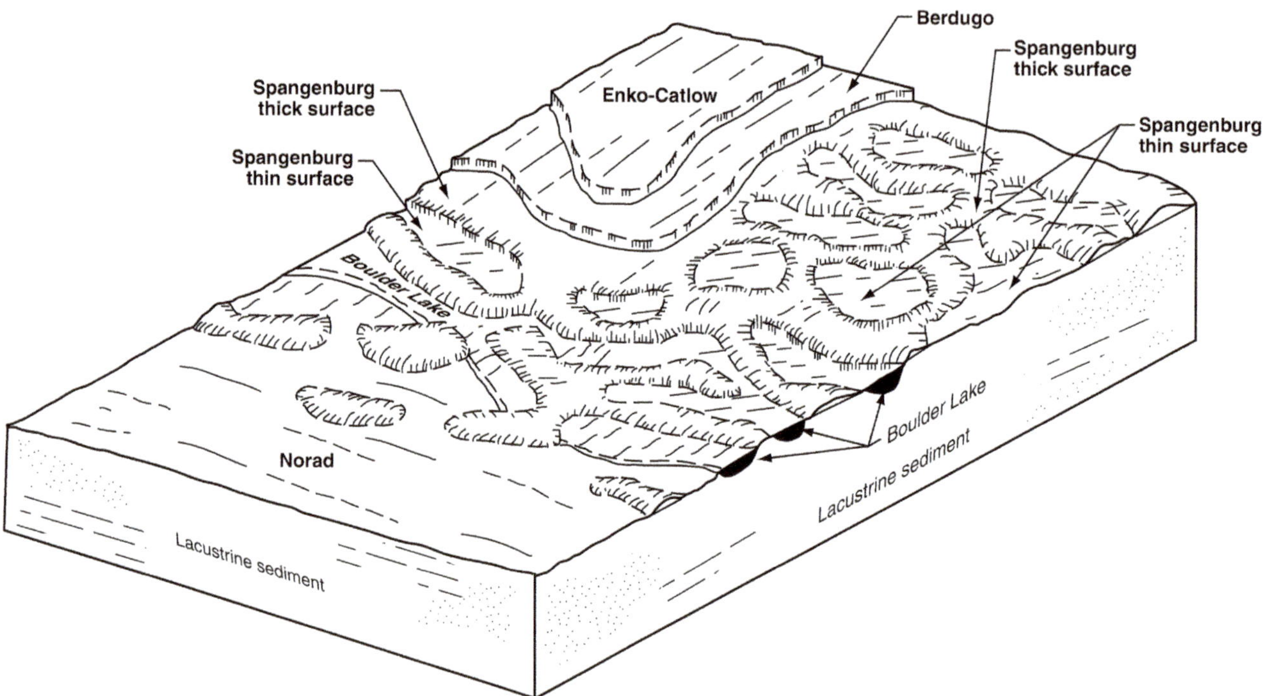

Fig. 5.5 Soil and landscape relationships on lacustrine terraces in Catlow Valley, Harney County. Upper terraces contain the Enko and Catlow (Haplocambids) and Berdugo (Paleargids) soils from an ice age lake that filled the valley to a depth of 60 m. The Spangenburg (Paleargids) and Norad (Haplargids) series have formed on lower terraces, and the Boulder Lake (Epiaquerts) soil has formed in depressions on lake plains. *Source* NRCS

Fig. 5.6 Pleistocene lake levels up to 60 m above base level on the wall of the Catlow Valley in Harney County. *Source* NRCS photo

5.3 Cascade Mountains

The Cascade Mountains MLRA is the second largest in Oregon, accounting for 38,500 km^2 14.5% of the area (Table 5.3). Nearly two-thirds (61%) of this MLRA is in Oregon, with the remainder (39%) in Washington. This MLRA corresponds with the Cascade physiographic province (Table 5.1). Elevations range from 325 m in the foothills of the Willamette Valley to 3,429 m at the summit of Mt. Hood. The Cascade Mountains typically have sharp alpine summits of accordant height and some isolated volcanic cones. These volcanic peaks rise more than 1,000 m above the surrounding mountains. Maximum slopes average 80 ± 8.5% (Table 5.3). The Oregon Cascades consist primarily of andesite and basalt flow and some tuffs. Thin deposits of alluvium are at the lower elevations along the major streams draining the Cascades and colluvium occurs throughout the range. The Cascades have been glaciated, and isolated remnants of till and outwash are at the lower elevations on the flanks of the mountains. With the exception of Crater Lake National Park, only the soils at the lower elevations have been mapped in Oregon. Figure 5.7 shows the Cascades, including the snow-capped South Sister Mountain from Hosmer Lake.

Fig. 5.7 The Cascade Mountains showing South Sister from Hosmer Lake. *Source* Chad McGrath photo

The mean annual air temperature likely ranges from about 3 °C along the Cascade crest to 11 °C at the lowest elevations. The mean annual precipitation is exceptionally high in the western Cascades, averaging 2,000 mm (Table 5.3). As will be described in Sect. 5.7, the orographic effect of the High Cascades markedly reduces precipitation on the eastern slopes. At the lower elevations of the western Cascades, Douglas-fir is the dominant tree species, along with western hemlock, red alder (*Alnus rubra*), and bigleaf maple (*Acer macrophyllum*) (Table 5.4). In the upper montane and subalpine regions, mountain hemlock, noble fir, and lodgepole pine are dominant.

Common soil orders in this MLRA are Andisols, Inceptisols, Spodosols, and Ultisols (Table 5.5). The soils in the area typically have a mesic, frigid, or cryic soil temperature regime and a udic soil moisture regime. They generally are moderately deep to very deep, well drained, ashy, medial, loamy, or clayey and occur on mountain slopes and ridges. Haplocryods (Moolack and Winopee series) and Vitricryands (Castlecrest series) have formed in volcanic ash, pumice, and cinders. Some Dystrudepts (Aschoff and Zygore series) have formed in colluvium mixed with volcanic ash. Other Dystrudepts have formed in colluvium over residuum (Kinney and Klickitat series) or in colluvium (Blachly series). Palehumults (Honeygrove series) have formed in colluvium over residuum weathered from sandstone. Haplohumults (Peavine series) have formed in colluvium over residuum weathered from siltstone and shale.

5.4 Blue Mountain Foothills

The Blue Mountain Foothills is the third largest MLRA in Oregon at 32,500 km^2 12.2% of the state area (Table 5.3). Nearly three-quarters (71%) of this MLRA is in Oregon, with the remainder (29%) in Idaho. This MLRA is contained within the Blue Mountains physiographic province (Table 5.1). Elevations generally range from 950 to 1,600 m. Maximum slopes are 75 ± 30%. This MLRA is a dissected volcanic plateau in a mountain complex. Lithologies include basalt, rhyolite, schist, granite, graywacke, limestone, sandstone, and tuff. Valleys contain Quaternary alluvial sediments. Figure 5.8 shows the central Rocky Mountains near Baker City, Oregon.

Fig. 5.8 The Blue Mountains near Baker City, Oregon. *Source* Photo by NRCS

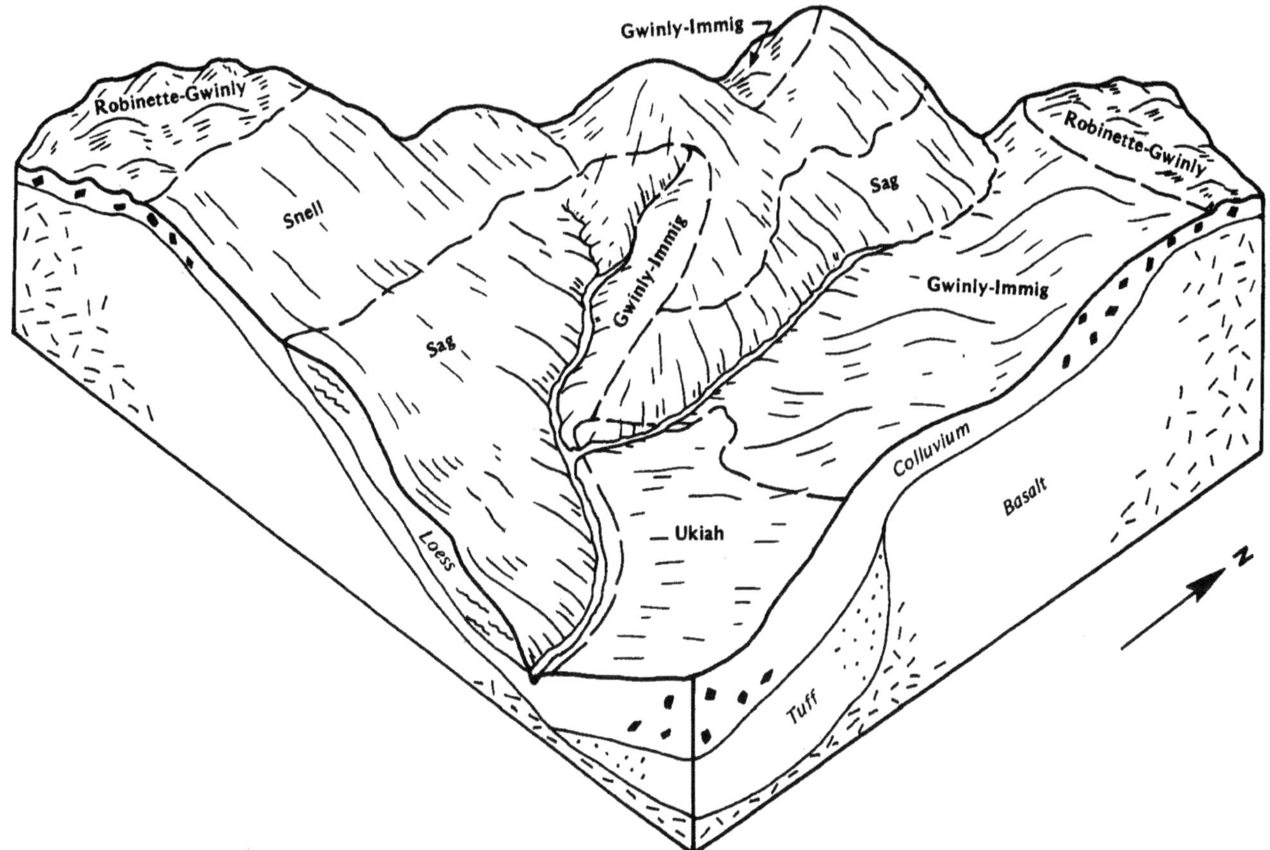

Fig. 5.9 The Gwinly–Immig–Snell soil association in Baker County. All of the soils are derived from loess over colluvium and are Argixerolls. *Source* NRCS

The mean annual air temperature ranges from 6 to 9 °C, and the mean annual precipitation ranges from 300 to 400 mm (Table 5.3). The dominant vegetation is western juniper, Wyoming big sagebrush, antelope bitterbrush, and bunchgrasses at lower elevations (Table 5.4). Mountain big sagebrush, western juniper, and bunchgrasses occupy the middle and higher elevations.

The predominant soil order in this MLRA is Mollisols; Aridisols are of minor extent (Table 5.5). The soils in the area have a mesic or frigid soil temperature regime, a xeric or aridic soil moisture regime, and mixed or smectitic mineralogy. They are very shallow to very deep, well drained, and clayey or loamy. Haploxerolls have formed in residuum and colluvium (Lickskillet, Rockly, and Westbutte series) on hills, plateaus, and mountains. Palexerolls and Argixerolls (Simas and Tub series) have formed in mixed loess and colluvium on hills. Haploxerolls have developed in volcanic ash (Deschutes series), in eolian sediments (Madras series), residuum (Merlin and Waterbury series), and residuum mixed with colluvium, or loess (Ateron, Ruckles, and Vitale series) on hills, plateaus, and mountains. Figure 5.9 shows the Gwinly–Immig–Snell soil series, all Argixerolls, on colluvium influenced by loess overlaying tuff and basalt in Baker County.

5.5 Blue Mountains

The Blue Mountains is the fourth largest MLRA in Oregon, with an area of 30,500 km^2 and 11.5% of the state area (Table 5.3). The majority (84%) of this MLRA is in Oregon, with the remainder in Idaho (11%) and Washington (5%). These mountains are contained within the Blue Mountains physiographic province (Table 5.1). Elevations commonly range from 1,140 to 1,905 m, with a maximum of 3,990 m on top of Mt. Sacajawea. Maximum slopes generally range from 55 to 95%. The Blue Mountains are composed of fault blocks and deep canyons with a complexity of bedrock types, including sedimentary, metasedimentary, and volcanic rocks. Figure 5.10 shows Chief Joseph Mountain in the Wallowa Mountains of the Blue Mountains MLRA in Wallowa County.

The mean annual air temperature ranges from 2.5 to 6.5 ° C, and the mean annual precipitation varies between 650 and 1,000 mm (Table 5.3). The vegetation includes western juniper, mountain big sagebrush and bunchgrass at the lower elevations; ponderosa pine and Douglas-fir at the middle elevations; and subalpine fir, Engelmann spruce, whitebark

Fig. 5.10 Chief Joseph Mountain in the Wallowa Mountains of the Blue Mountains. *Source* Photo by NRCS

pine, western larch, and lodgepole pine at the higher elevations (Table 5.4).

Common soil orders in this MLRA include Mollisols, Inceptisols, and Andisols (Table 5.5). The soils in the area have a mesic soil temperature regime at the lower elevations and a frigid or cryic soil temperature regime at the higher elevations. They have a xeric or udic soil moisture regime. Most of the soils have a component of volcanic ash from Mt. Mazama. The soils are shallow to very deep and well drained to very poorly drained. The soils at the lower elevations in the drier parts of the area are typically Argixerolls (Gwinly series) and Palexerolls (Hankins series). The soils at the slightly higher elevations or in the more protected areas retain the influence of volcanic ash and are Haploxerolls

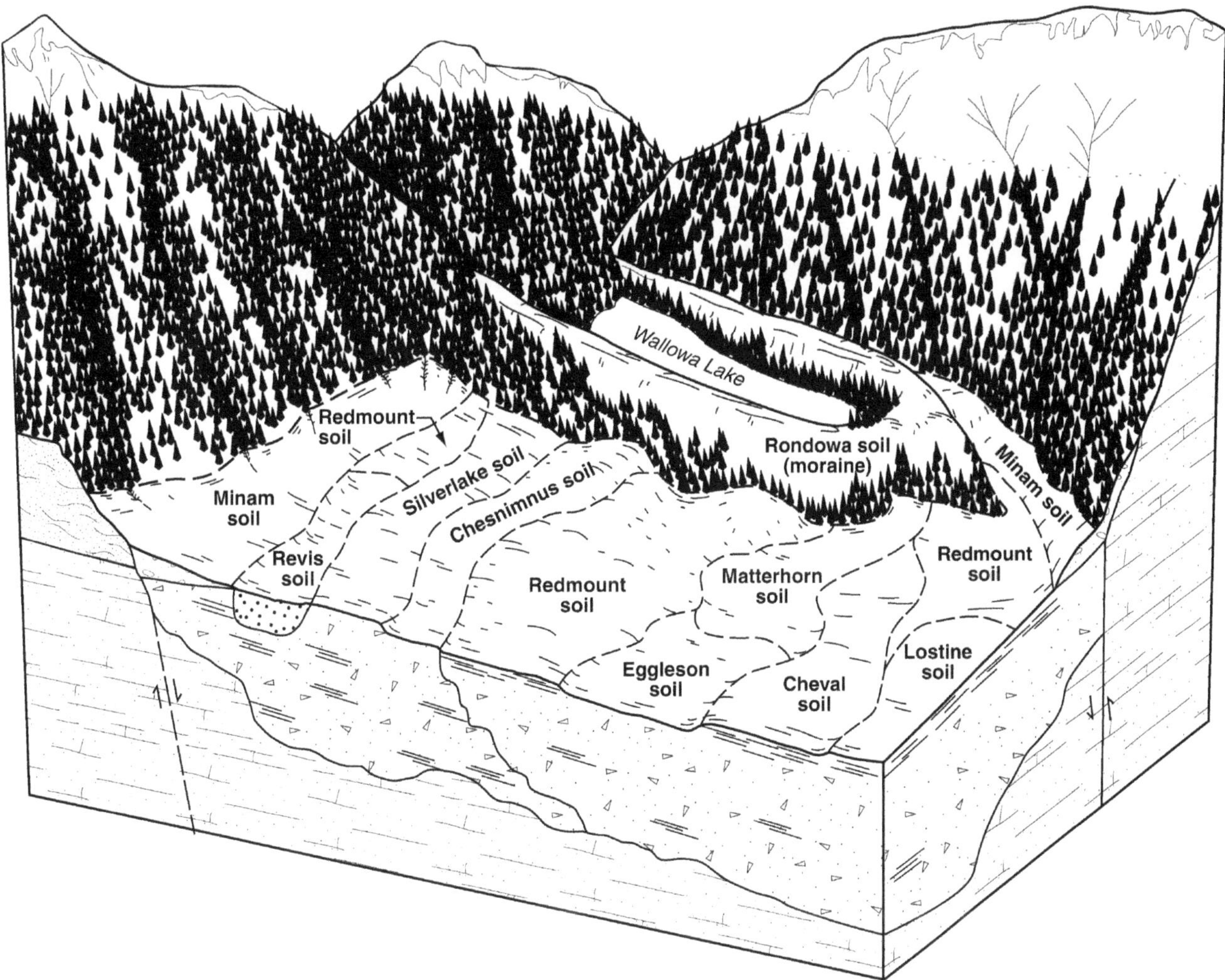

Fig. 5.11 Soils formed on a moraine and a piedmont fan covered with volcanic ash and loess in the Wallowa Mountains. The Minam, Redmount, Lostine, and Rondowa series are Haploxerolls; the Silverlake and Chesnimnus series are Argixerolls; and the Matterhorn series is a Calcixerolls. The Revis series is no longer recognized. *Source* NRCS

(Hall Ranch and Umatilla series). The volcanic plateaus in the northeastern part of the area have Argialbolls (Lookingglass and Cowsly series). Humicryepts (Muddycreek series), Dystrocryepts (Angelbasin series), Vitricryands (Bucketlake, Helter, and Lackeyshole series) occur at the higher elevations.

Figure 5.11 shows a catena of soils on a piedmont fan in a fault-block portion of Wallowa County containing primarily Haploxerolls (Minam, Redmount, and Lostine series) and Argixerolls (Chesnimnus and Silverlake series). The soils have been influenced by volcanic ash and loess. Wallowa Lake is contained by a moraine with the Rondowa series, a Pachic Haploxerolls.

5.6 Siskiyou Mountains

The Siskiyou Mountains occupy nearly 20,000 km^2 in Oregon, which is 7.5% of the state area (Table 5.3). About one-third (38%) of this MLRA is in Oregon, with the remainder (62%) in California. This MLRA corresponds with the Klamath Mountains physiographic province (Table 5.1). Elevations commonly range from 245 to 1,230 m; and maximum slopes range from 60 to 90%. The Klamath Mountains represent an uplifted peneplain composed of marine sandstones and shales intruded by granodiorite and other intrusive rocks. Ultramafic (serpentinitic) rocks occur in this area. Soil parent materials are

Fig. 5.12 Hellgate Canyon in the Siskiyou Mountains of southwestern Oregon. *Source* Photo by Bureau of Land Management

predominantly colluvium and residuum. Figure 5.12 shows Hellgate Canyon (Rogue River) in the Siskiyou Area of southwest Oregon.

The mean annual air temperature ranges from 9.5 to 11 °C, and the mean annual precipitation ranges from 850 to 1,300 mm (Table 5.3). The vegetation is primarily Douglas-fir, Pacific madrone (*Arbutus menziesii*), ponderosa pine, incense cedar, sugar pine, and white fir on moist sites and Oregon white oak on dry sites (Table 5.4). The grassland foothills adjacent to the Rogue River and the Umpqua River floodplains and terraces are dominated by Idaho fescue and Thurber's needlegrass.

Common soil orders in this MLRA are Alfisols, Inceptisols, Mollisols, and Ultisols (Table 5.5). The soils in the area have a mesic soil temperature regime, a xeric soil moisture regime, and mixed mineralogy. They generally are moderately deep or deep, well drained, and loamy and occur on mountain slopes and hills. Dystroxerepts (Beekman, Atring, and Jayar series) have formed in colluvium or in colluvium over residuum. Shallow Dystroxerepts (Vermisa series) have formed in residuum. Haploxeralfs (Speaker and Vannoy series) and Haploxerults (Josephine series) have formed in colluvium over residuum. Argixerolls (McNull series), Haploxerolls (McMullin and Medco series), and

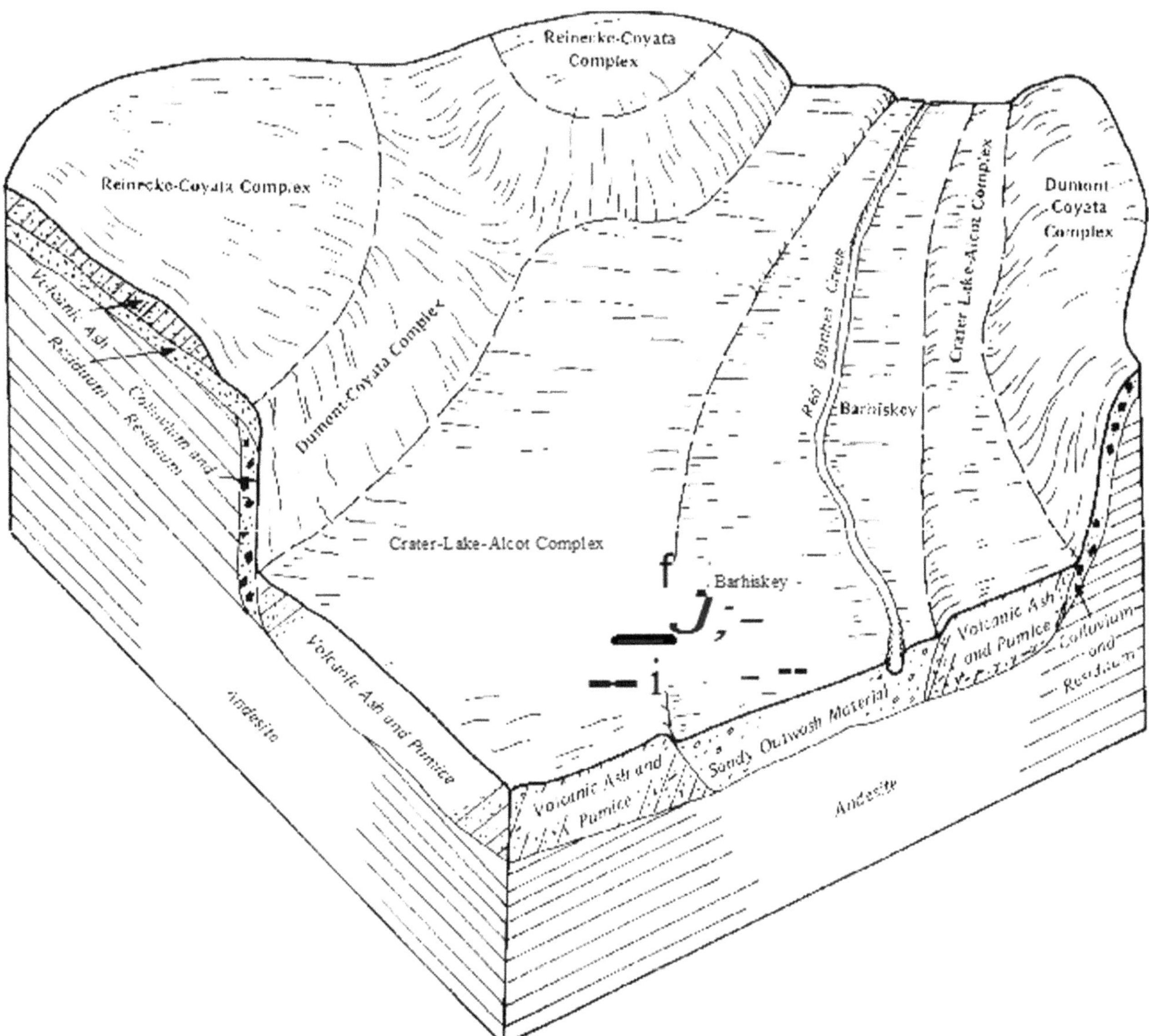

Fig. 5.13 Soils formed in colluvium, residuum and alluvium influenced by volcanic ash and pumice in Jackson County. The Reinecke series (Vitrixerands) is formed from volcanic ash over andesite residuum. The Coyata (Dystroxerepts) and Dumont (Palexerults) series are formed in colluvium and residuum from basalt. The Crater Lake and Alcot series (Vitrixerands) are formed in volcanic ash and pumice; and the Barhiskey (Dystroxerepts) is formed in alluvium strongly influenced by volcanic ash and pumice. *Source* NRCS

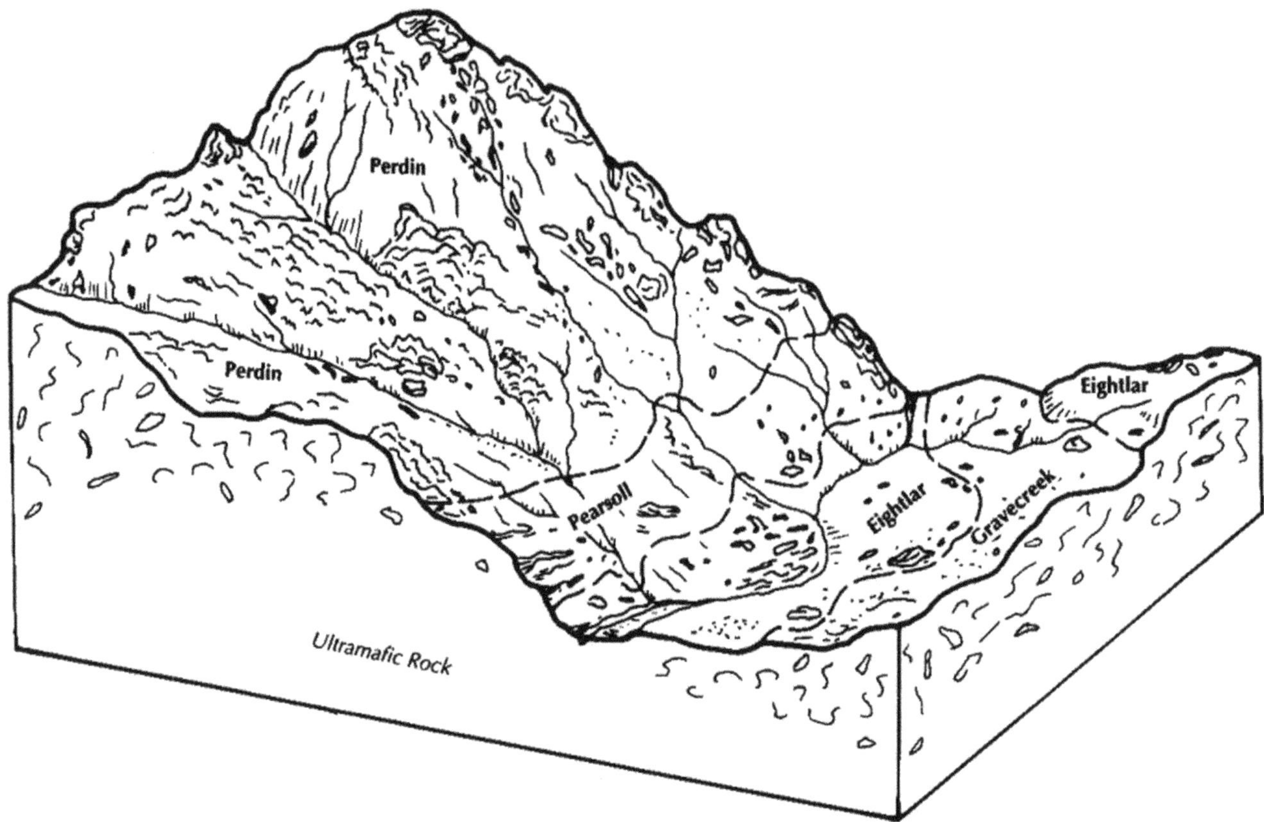

Fig. 5.14 Soils formed from ultramafic rocks in Curry County. The upland soils (Perdin series) are Haploxeralfs and the soils along the slopes (Pearsoll and Eightlar and Gravecreek series) are Dystroxerepts. *Source* NRCS

Haplohumults (Bellpine and Windygap series) have formed in colluvium and residuum on low hills.

Two soil-block diagrams are shown here to represent the MLRA. Figure 5.13 shows three soil complexes formed in volcanic ash, pumice, colluvium, and residuum over andesite in Jackson County. The Reinecke, Crater Lake, and Alcot series are Vitrixerands, and the Coyata series is a Dystroxerepts. Figure 5.14 is the Perdin–Pearsoll–Eightlar association formed in colluvium and residuum derived from ultramafic rocks. Whereas the more stable Perdin series is a Haploxeralfs, the Pearsoll and Eightlar series are Dystroxerepts. Palexerults (Dumont series) have formed in colluvium and residuum from igneous and metasedimentary rocks.

5.7 Cascade Mountains—Eastern Slope

The Cascade Mountains—Eastern Slope comprises nearly 18,000 km^2 in Oregon, which is 6.7% of the state area (Table 5.3). About one-half (52%) of this MLRA is in Oregon, with the remainder (48%) in Washington. This MLRA constitutes the eastern portion of the Cascade physiographic province (Table 5.1). Elevation commonly range from 980 to 1,635 m, but some peaks approach 3,050 m. Maximum slopes range from 50 to 70%. The northern half of this MLRA consists of Pre-Cretaceous metamorphic rocks intruded by igneous rocks; and the southern part is covered with volcanic ash and pumice from

Fig. 5.15 The Cascade Mountains from the eastern slope. *Source* Photo by Chad McGrath

Mt. Mazama. Figure 5.15 shows the eastern slope of the Cascade Mountains in Deschutes County.

The mean annual air temperature commonly ranges from 5.5 to 7.5 °C, and the mean annual precipitation ranges from 430 to 750 mm, which is substantially less than reported for the western slope of the Cascades in Sect. 5.3 (Table 5.3). The vegetation includes western juniper, antelope bitterbrush, mountain big sagebrush, and Idaho fescue at the lower elevations; ponderosa pine, lodgepole pine, mountain big sagebrush, antelope bitterbrush and Idaho fescue at the middle elevations; and Douglas-fir, white, fir, sugar pine, and lodgepole pine at the higher elevations (Table 5.4).

Common soil orders in this MLRA are Alfisols, Andisols, Inceptisols, and Mollisols (Table 5.5). The Andisols and Inceptisols in this area are similar to the soils to the west, and the Mollisols in the area are more typical of the soils to the east. Soils in this MLRA have a mesic, frigid, or cryic soil temperature regime, a xeric soil moisture regime, and mixed or glassy mineralogy. Haploxerepts have formed in a mixture of ash and loess over alluvium or colluvium on uplands (Wamic series). Vitricryands have formed in ash and pumice (Lapine series), ash (Steiger series), and ash over loamy material (Shanahan series) on plateaus. Vitrixerands (Smiling series) have formed in colluvium mixed with loess and ash. They are on mountain slopes. Moderately deep Vitrixerands (Maset and Wanoga series) formed in ash over residuum on benches and hills.

5.8 Coast Range

In Oregon, the Coast Range has an area of more than 17,000 km^2, which is 6.5% of the state area (Table 5.3). Nearly two-thirds (65%) of this MLRA is in Oregon, with the remainder (35%) in Washington. Along with the Sitka Spruce and Coastal Redwood Belts, this MLRA constitutes

Fig. 5.16 The Coast Range in Benton County. *Source* NRCS photo

the Coast Range physiographic province (Table 5.1). Elevations commonly range from 75 to 800 m; and maximum slopes range from 70 to 95%. The Coast Range is composed of sedimentary, marine basalts, and metasedimentary rocks. Figure 5.16 shows the Coast Range in Benton County.

The mean annual air temperature ranges from 6 to 12° C; and the mean annual precipitation ranges from 2,000 to 2,500 mm, the greatest for any MLRA in the state (Table 5.3). The vegetation includes Douglas-fir, western hemlock, bigleaf maple, red alder, and western redcedar in the north and central portions of the Coast Range and Douglas-fir, Pacific madrone, incense cedar, and tanoak in the southern portion (Table 5.4).

The predominant soil orders in this MLRA are Andisols, Inceptisols, and Ultisols (Table 5.5). Soils in the area have a mesic or frigid soil temperature regime and a udic soil moisture regime. They are generally shallow to very deep, well drained, medial, and loamy or clayey, occurring on foothills and mountain slopes and ridges. Hapludands (Hemcross, Klistan, and Slickrock series) have formed in colluvium over residuum. Dystrudepts (Bohannon, Preacher, Rinearson, and Blachly series) have formed in colluvium derived from sedimentary rocks; Eutrudepts (Digger and Umpcoos series) have developed in colluvium and residuum derived from sedimentary rocks; and Haplohumults (Peavine series) and Palehumults (Honeygrove series) have formed in colluvium derived from sedimentary rocks.

Three soil-block diagrams are shown to represent this MLRA. Figure 5.17 shows soils on mountains over dissected sandstone and basalt in the Coast Range in Benton County. The Apt (Haplohumults), Blachly (Dystrudepts) and Honeygrove (Palehumults) are very deep, while the Bohannon (Humudepts), and Peavine (Haplohumults) are moderately deep soil series formed in colluvium and residuum. Figure 5.18 constitutes the northern coast (Yamhill County) and shows Humudepts on basalt (Klickitat and Hembre series) and on siltstone and shale (Astoria series). Figure 5.19 is from Coos County in the southern Coast Range and shows the Umpcoos–Digger–Preacher series,

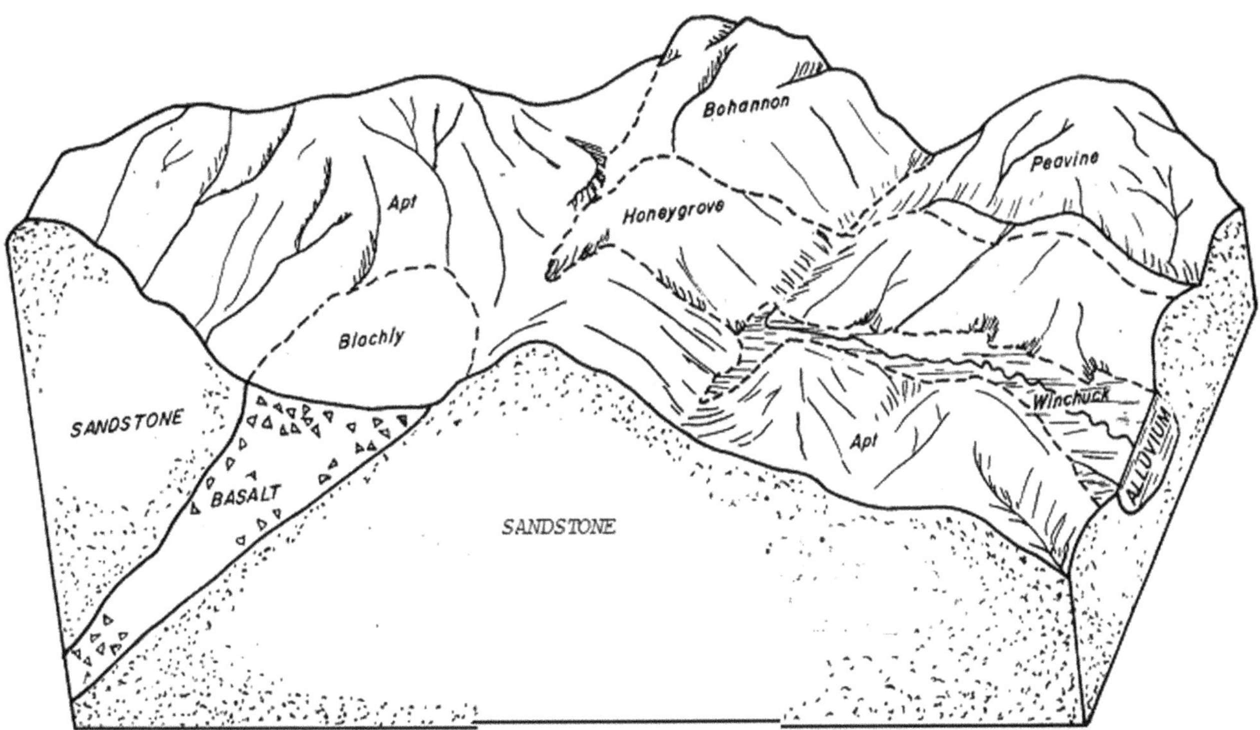

Fig. 5.17 Soils on mountains over dissected sandstone and basalt in the Coast Range in Benton County. The Apt (Haplohumults), Blachly (Dystrudepts), and Honeygrove (Palehumults) are very deep and the Bohannon (Humudepts), and Peavine (Haplohumults) are moderately deep soil series formed in colluvium and residuum. *Source* NRCS

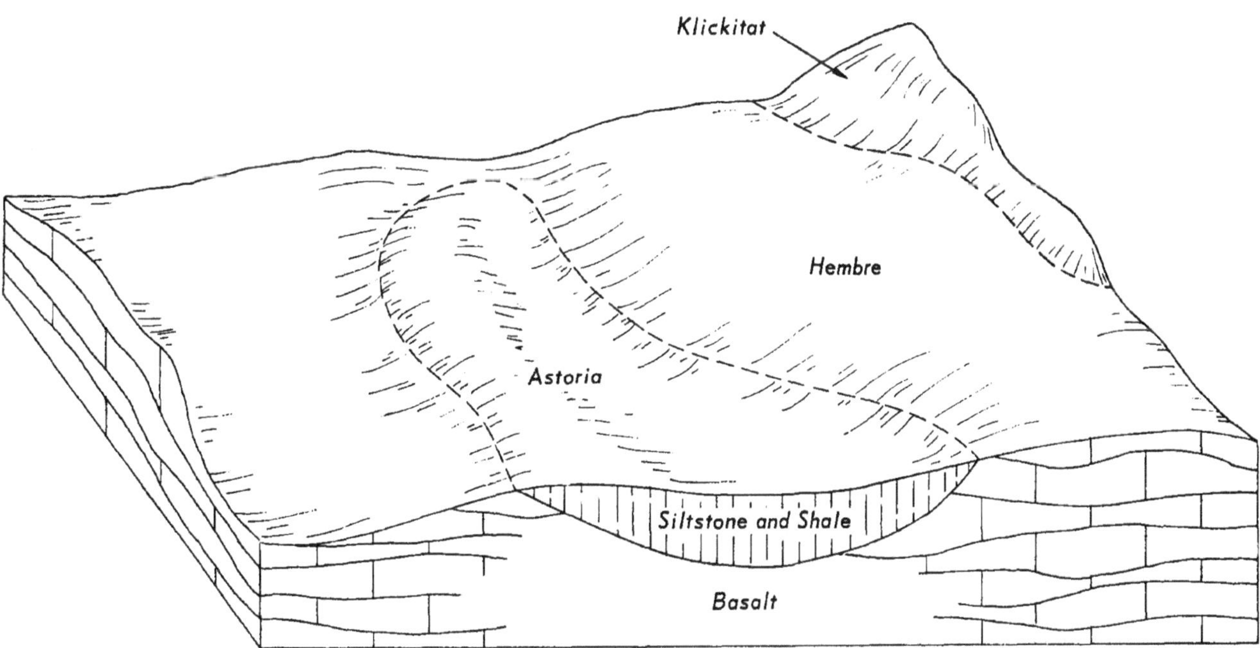

Fig. 5.18 Humudepts derived from siltstone-shale and basalt in Yamhill County, in the northern part of the Coast Range. *Source* NRCS

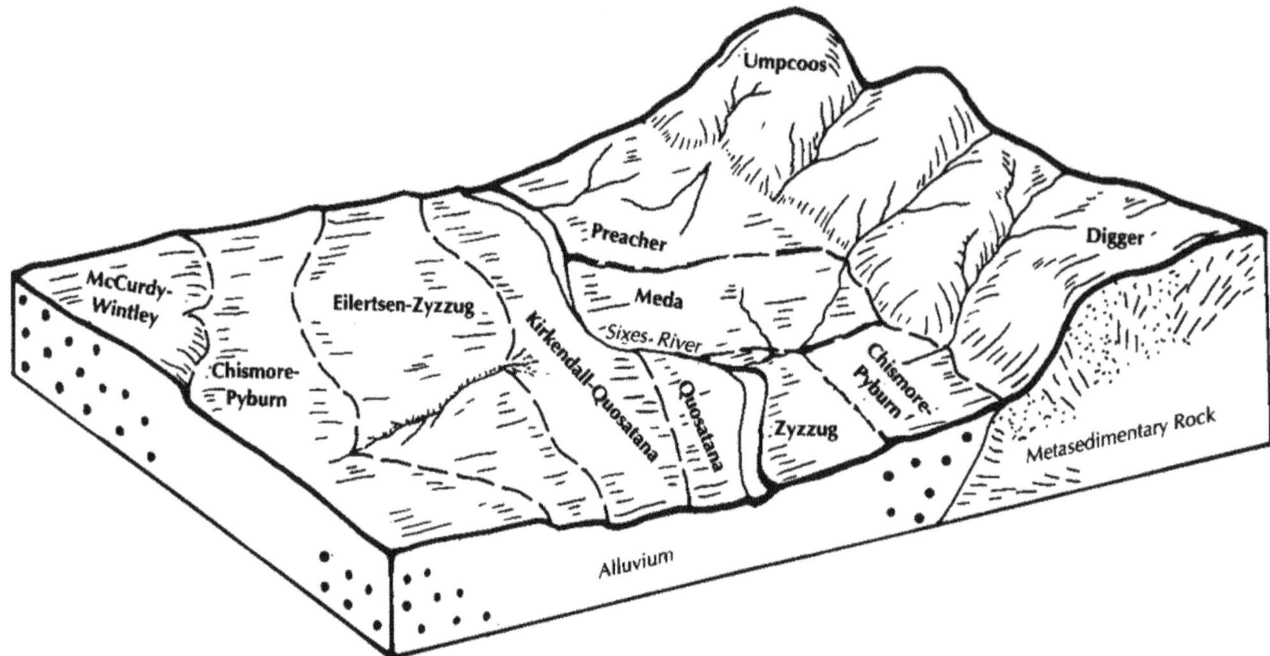

Fig. 5.19 Soils derived from metasedimentary rocks and alluvium in Curry County in the southern part of the Coast Range. The higher elevation soils, formed in colluvium and residuum from metasedimentary rocks, are Eutrudepts (Umpcoos and Digger series) and Humudepts (Preacher and Meda series). At the intermediate elevations, Palehumults (McCurdy and Chismore series), Haplohumults (Wintley series), and Umbraquults (Pyburn series) have formed in alluvium on high river terraces. At the lower elevations, Humaquepts (Quosatana and Zyzzug series), Hapludalfs (Eilertsen series), and Humudepts (Kirkendoll series) have formed in alluvium on low terraces and floodplains of rivers. *Source* NRCS

classified as Humudepts, formed in colluvium and residuum from metasedimentary rocks and soils derived from alluvium of the Sixes River.

5.9 Willamette Valley

The Willamette Valley occupies 13,600 km^2 in Oregon, which is 5.1% of the state area (Table 5.3). Less than one-half (43%) of this MLRA is in Oregon and the remaining 57% is in Washington. This MLRA corresponds directly with the Willamette Valley physiographic province (Table 5.1). Elevations range from 70 to 810 m; and maximum slopes range between 20 and 80%. The Willamette Valley is part of the Puget Trough and received meltwater from the Columbia River during the late Pleistocene Missoula Floods. For this reason, it contains glaciolacustrine deposits and alluvium that are adjacent to the Willamette Valley foothills. Figure 5.20 shows the Willamette Valley with the Coast Range in the background in Benton County.

The mean annual air temperature is 11 ± 0.26 °C; and the mean annual precipitation is between 1,100 and 1,250 mm (Table 5.3). The vegetation is Oregon white oak, Douglas-fir, Oregon ash *(Fraxinus latifolia)*, poison oak (*Toxicodendron diversilobum*), and various grasses (Table 5.4). As this MLRA contains most of the state's population, most of the native vegetation has been cleared for agriculture.

The predominant soil orders in this MLRA are Alfisols, Inceptisols, Mollisols, and Ultisols (Table 5.5). The soils in the area have a mesic soil temperature regime, a xeric soil moisture regime, and mixed mineralogy. They generally are moderately deep to very deep, poorly drained to well drained, and loamy or clayey. Nearly level, somewhat poorly drained Argialbolls (Amity series), poorly drained Albaqualfs (Dayton series), and moderately well-drained Argixerolls (Woodburn series) formed in lacustrine sediments on terraces. Moderately deep and deep, moderately well-drained and somewhat poorly drained Fragixerepts (Cascade, Kinton, and Powell series) are on foothills in the

Fig. 5.20 The Willamette Valley with the Coast Range in the background in Benton County. *Source* NRCS photo

Portland area. Well-drained Dystroxerepts (Ritner series) have formed in clayey colluvium derived from basalt. Gently sloping to steep, well-drained Haplohumults (Bellpine and Nekia series), Palehumults (Jory series), and Agrixerolls (Dixonville series) have formed in colluvium and residuum on foothills.

Three soil-block diagrams are shown here to represent this MLRA. Figure 5.21 is from the Willamette Valley foothills of Benton County and shows a catena of soils formed in colluvium and residuum over sandstone. The Jory series, Oregon's state soil, is a Palehumult, the Bellpine series is a Haplohumult, and the Hazelair series is a Haploxeroll. Figure 5.22 is a catena of soils on the Willamette glaciolacustrine silts in Yamhill County. The Willamette and Woodburn series (Argixerolls) are well drained and moderately well drained, respectively; and the Amity (Argialbolls) and Dayton (Albaqualfs) series are somewhat poorly drained and poorly drained, respectively. Figure 5.23

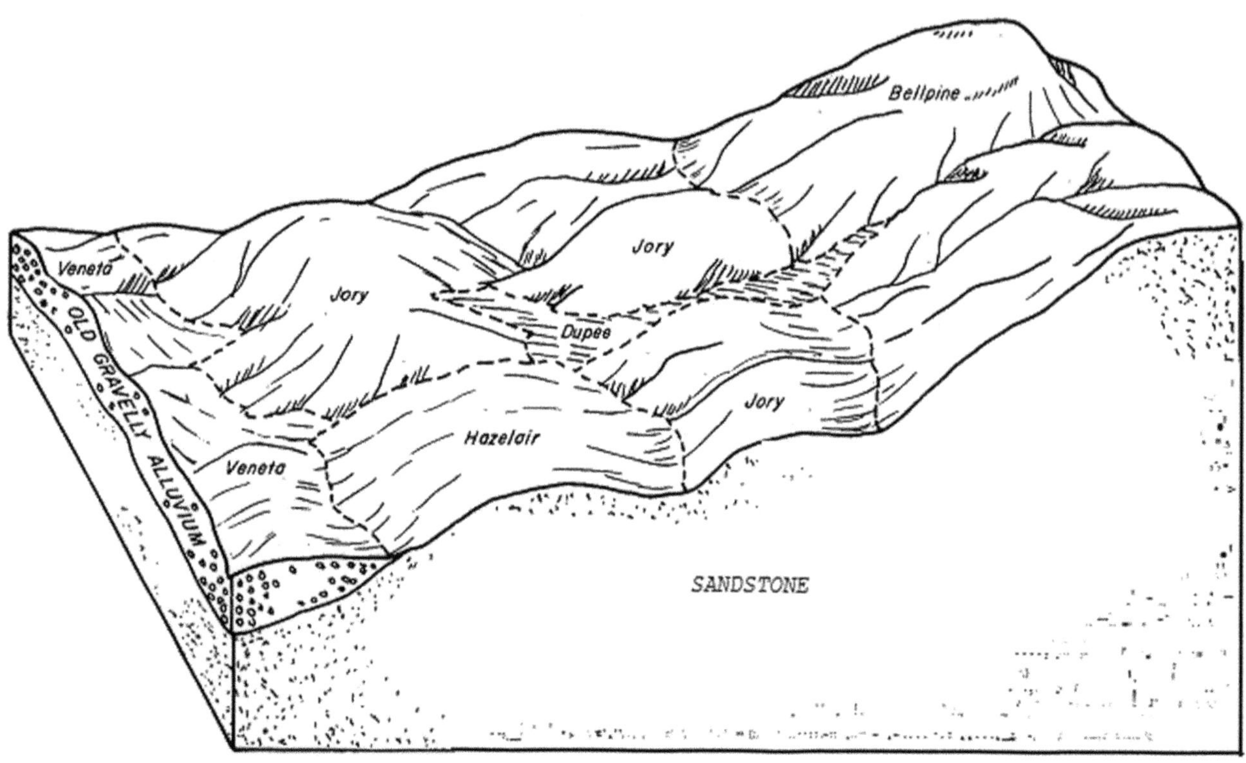

Fig. 5.21 Soils derived from sandstone in Benton County, in the foothills of the Willamette Valley. The Jory series, Oregon's state soil, is a Palehumults; the Bellpine series is a Haplohumults; and the Hazelair series is a Haploxerolls. *Source* NRCS

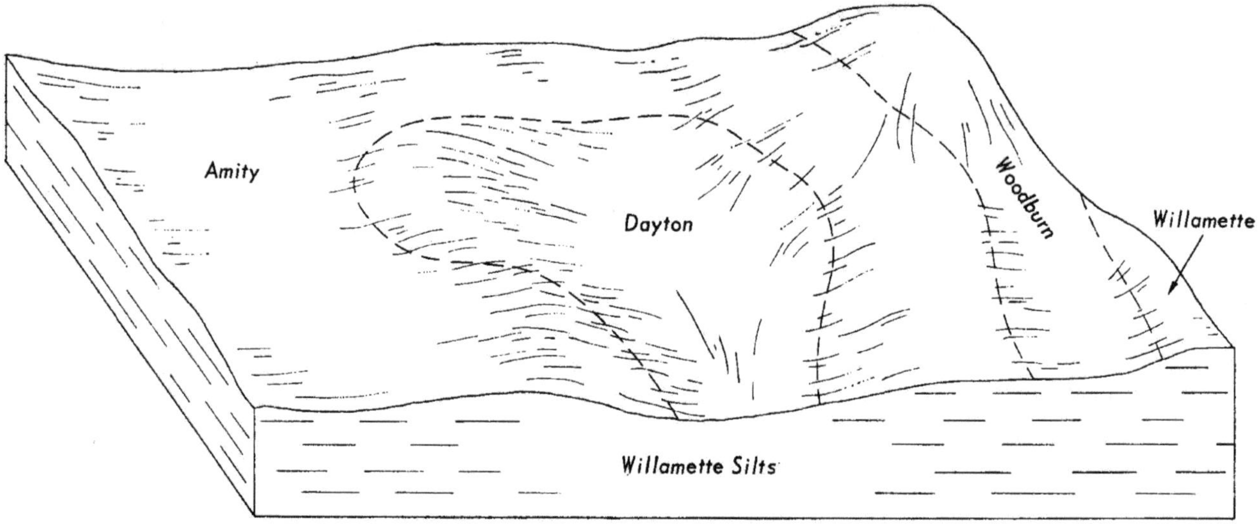

Fig. 5.22 Soil catena developed in the Willamette Silts in the Willamette Valley. The soils are very deep and include the well-drained Willamette series and moderately well-drained Woodburn series (Argixerolls), the somewhat poorly drained Amity series (Argialbolls), and the poorly drained Dayton series (Albaqualfs). *Source* NRCS

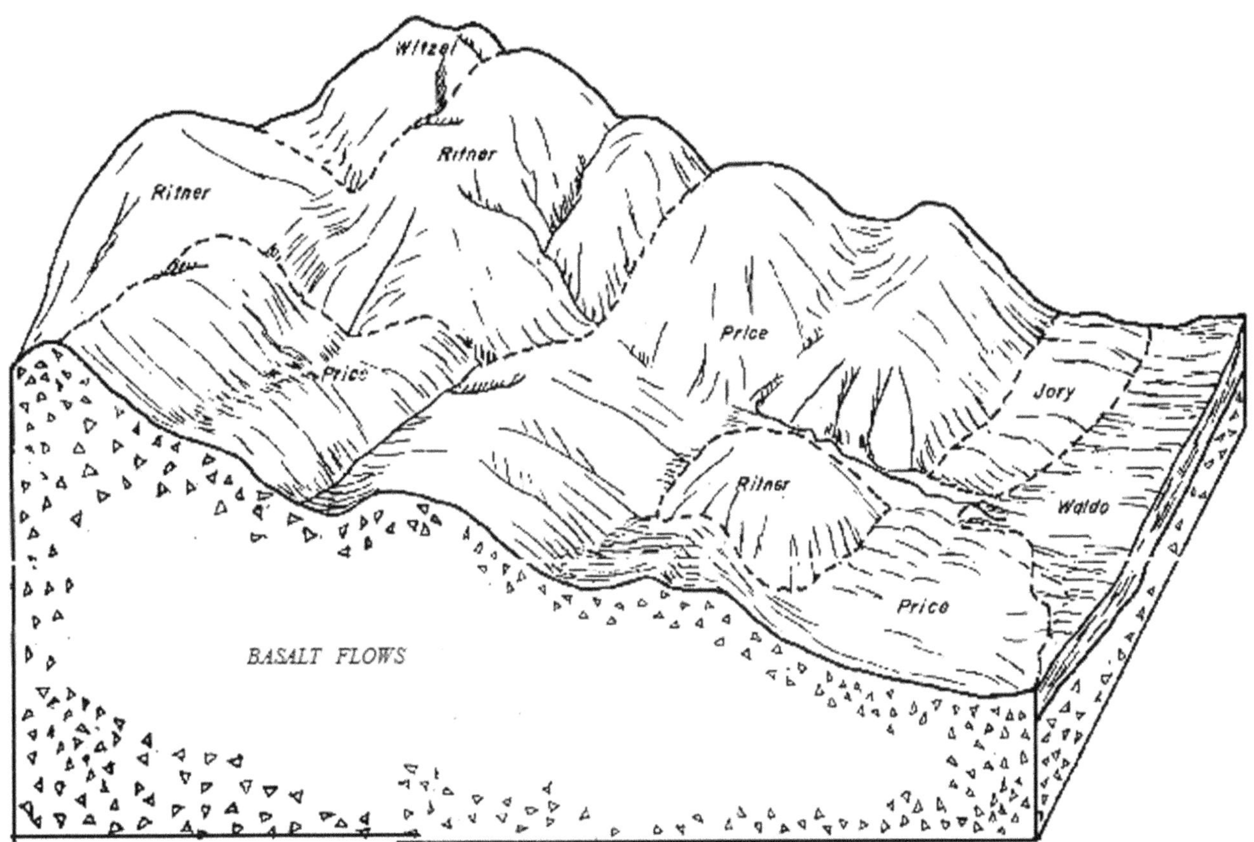

Fig. 5.23 Soils on basalt flows in the foothills in the Willamette Valley. The Waldo series (Endoaquolls) occurs in alluvium with the Witzel (Haploxerolls), Ritner and Price (Haploxerepts), and Jory (Palehumults) series have formed in colluvium from basalt. *Source* NRCS

shows the Waldo series (Endoaquolls) on alluvium with the Witzel (Haploxerolls), Ritner and Price (Haploxerepts), and Jory (Palehumults) soil series in the surrounding foothills underlain by basalt flows.

5.10 Owyhee High Plateau

The Owyhee High Plateau has an area in Oregon of 12,000 km^2, which is 4.5% of the state (Table 5.3). Only 16% of this MLRA is in Oregon; the remainder is in Nevada (52%), Idaho (29%), and Utah (3%). Along with the Snake River Plains, it is part of the Owyhee physiographic province (Table 5.1). Elevations range between 1,270 and 2,125 m, and maximum slopes range between 30 and 50%. The Owyhee High Plateau is composed of volcanic rocks, including andesite, basalt, and rhyolite. The plateau is composed of colluvium and residuum, and valleys contain alluvium. Figure 5.24 shows the Owyhee High Plateau in southeastern Oregon.

The mean annual air temperature ranges between 5.5 and 8 °C; and the mean annual precipitation is between 245 and 365 mm (Table 5.3). The vegetation is predominantly Wyoming big sagebrush, low sagebrush, rabbitbrush, bluebunch bunchgrass, Idaho fescue, Sandberg bluegrass, Thurber's needlegrass, and basin wildrye (Table 5.4).

Common soil orders in this MLRA are Aridisols and Mollisols (Table 5.5). The soils in the area typically have a mesic or frigid soil temperature regime, an aridic or xeric soil moisture regime, and mixed or smectitic mineralogy.

Fig. 5.24 The Owyhee High Plateau in southeastern Oregon. *Source* Photo courtesy of Bob Wick, Bureau of Land Management

They generally are well drained, clayey or loamy, and shallow or moderately deep. Argidurids (Brace series) have formed in colluvium and residuum derived from welded rhyolite tuff and basalt. Haplargids (Anawalt and Gumble series) have formed in residuum and colluvium containing some loess and volcanic ash on hills, mountain slopes, and plateaus. Some shallow Argixerolls (Durkee, Gaib, and Ninemile series) have formed in residuum and colluvium influenced by loess and volcanic ash on hills, plateaus, and mountain slopes. Haplocalcids (Enko series) have developed in alluvium with a component of loess and volcanic ash on fan remnants, inset fans, and fan skirts and aprons.

5.11 Columbia Plateau

The Oregon portion of the Columbia Plateau has an area of 12,000 km^2, which is 4.5% of the state area (Table 5.3). Only 25% of this MLRA is in Oregon, with the remainder (75%) in Washington. The Columbia Plateau is part of the Deschutes–Umatilla Plateau physiographic province (Table 5.1). Elevations range between 220 and 1,060 m; and maximum slopes are from 30 to 90%. The plateau is composed of Miocene basalt covered with loess, colluvium, and residuum. Figure 5.25 shows wheat country on the Columbia Plateau in Umatilla County.

The mean annual air temperature ranges between 8.5 and 10 °C; and the mean annual precipitation is between 265 and 345 mm (Table 5.3). The vegetation is predominantly Wyoming big sagebrush, Sandberg bluegrass, bluebunch wheatgrass, and Idaho fescue (Table 5.4). Much of this area is cultivated for the production of winter wheat under a wheat—fallow rotation.

The predominant soil order in this MLRA is Mollisols (Table 5.5). The soils in the area typically have a mesic soil temperature regime, a xeric soil moisture regime, and mixed mineralogy. They generally are moderately deep to very deep, well drained, and loamy. Haploxerolls have formed in loess (Condon, Mikkalo, Ritzville, and Walla Walla series),

Fig. 5.25 Wheat on the Columbia Plateau in Umatilla County. *Source* Photo by Shanna Hamilton, Oregon Wheat League

loess over colluvium (Wrentham series), ash over loess (Valby series), mixed materials (very shallow Bakeoven and shallow Kuhl series), and colluvium (shallow Lickskillet series) on uplands. Argixerolls (Morrow series) have formed in loess on plateaus and hills. Durixerolls (Willis series) formed in ash over loess on terraces and uplands. A soil-block diagram from Sherman County shows the Walla Walla (Haploxerolls) and Starbuck (Haplocambids) series on deep loess over basalt and the Hermiston series (Haploxerolls) in silty alluvium on low terraces (Fig. 5.26). A soil-block diagram shows a catena of soils formed in a thin layer of loess, residuum, and colluvium over basalt bedrock in Sherman County. The Condon series has formed in loess, the extensive Bakeoven and Lickskillet series from residuum on steep slopes, and the Hermiston series on alluvium in valleys (Fig. 5.27).

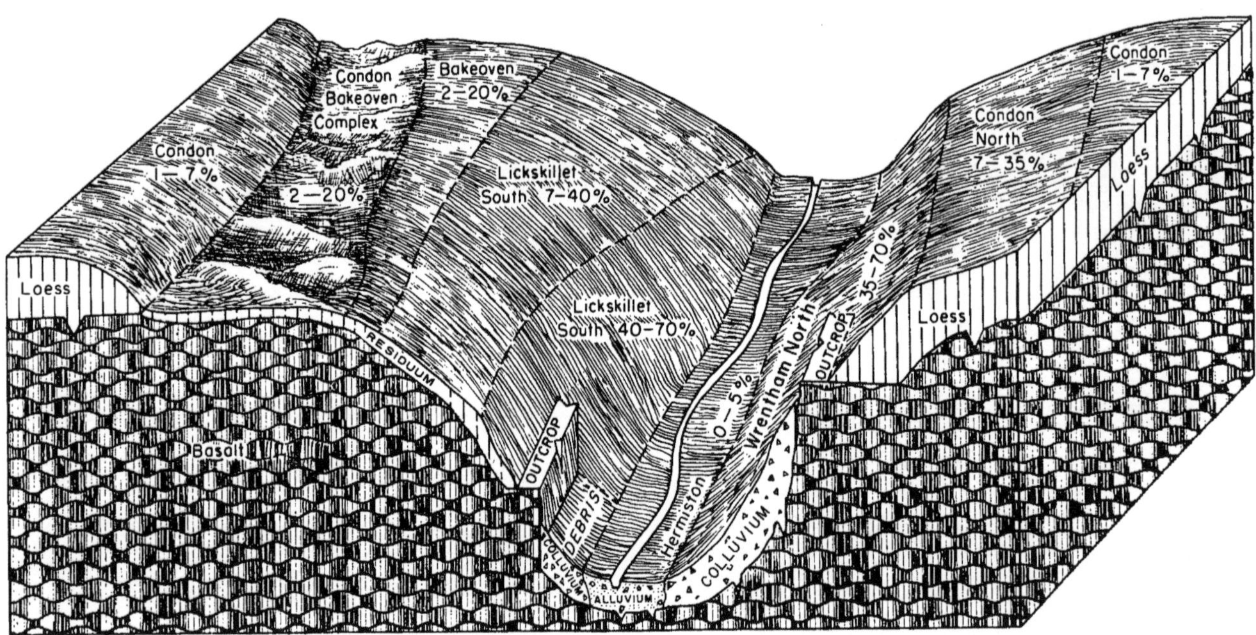

Fig. 5.26 Soils derived from loess, basalt colluvium and residuum, and alluvium on the Columbia Plateau. All of the soil series are Haploxerolls. *Source* NRCS

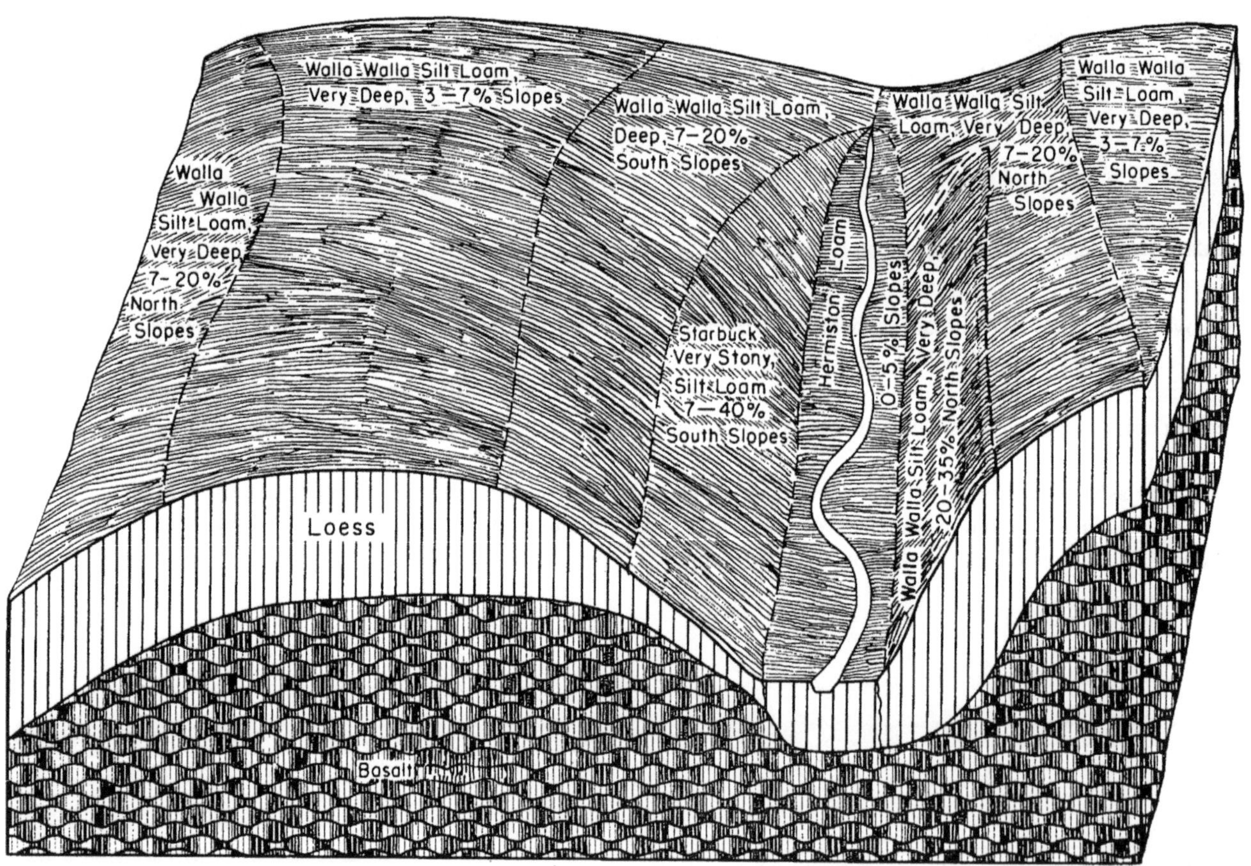

Fig. 5.27 A soil-block diagram from Sherman County shows the Walla Walla (Haploxerolls) and Starbuck (Haplocambids) series on deep loess over basalt and the Hermiston series (Haploxerolls) in silty alluvium on low terraces on the Columbia Plateau. *Source* NRCS

5.12 Klamath Basin

The Klamath Basin has an area in Oregon of 10,500 km^2, which is 3.9% of the state area (Table 5.3). About one-third (35%) of this MLRA occurs in Oregon and (65%) is in California. Despite its name, this MLRA is partly in the Cascades and mostly in the Basin and Range physiographic province (Table 5.1). Elevations commonly range from 1,155 to 1,770 m, and maximum slopes range between 10 and 70%. The geology of this MLRA is very complex. The uplands contain Cenozoic volcanic rocks composed of basalt, rhyolite, and andesite and pre-Cenozoic metamorphic and sedimentary rocks capped by colluvium and residuum. The basins and valleys contain late Pleistocene alluvium, lacustrine deposits, and playa deposits. Figure 5.28 shows the Sprague River valley in the Klamath Basin.

The mean annual air temperature ranges between 5.7 and 8.5 °C; and the mean annual precipitation is between 300 and 635 mm (Table 5.3). At the lower elevations, the vegetation is low sagebrush, basin big sagebrush, antelope bitterbrush, Idaho fescue, and bluebunch wheatgrass (Table 5.4). Western juniper, mountain mahagony (*Cercocarpus* spp.), and ponderosa pine occur at middle elevations, while Douglas-fir, ponderosa pine, white fir, and sugar pine occur at the higher elevations. Wet basins contain bulrushes, tules, lilies, and cattails.

Fig. 5.28 The Sprague River valley in the Klamath Basin in south central Oregon. The photo shows rangeland in the foreground and forestland in the background. *Source* NRCS photo

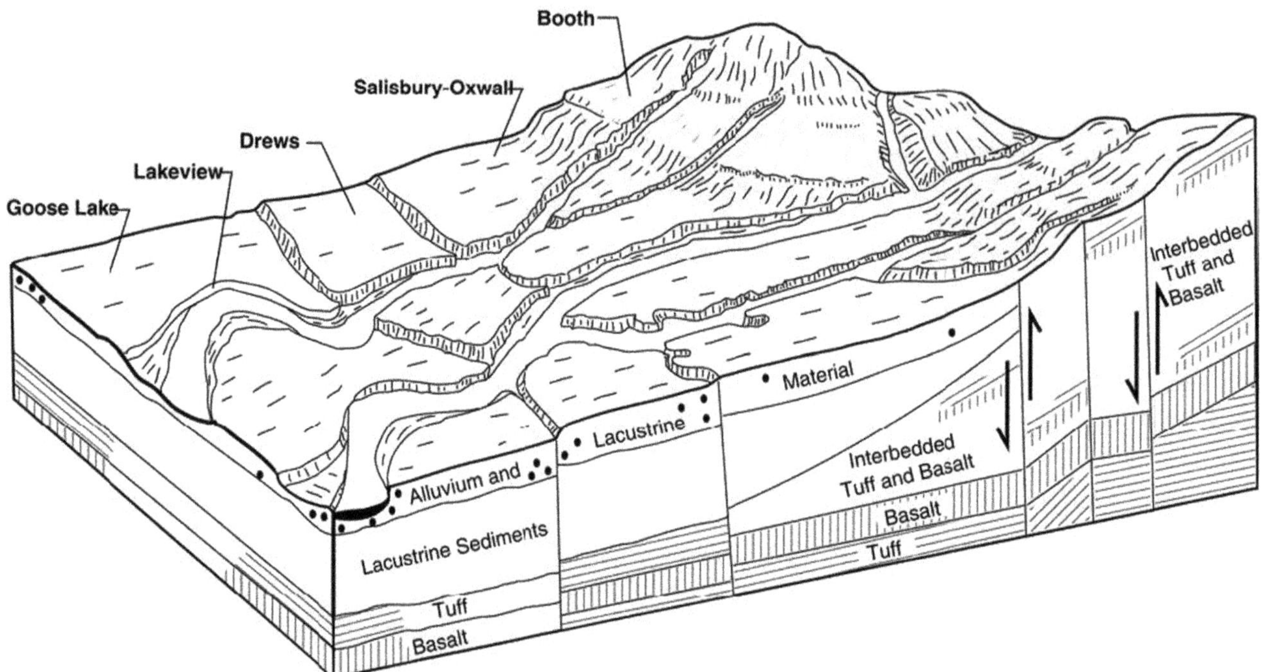

Fig. 5.29 A soil-block diagram of soils on upland volcanic rocks, lacustrine sediments, and alluvium Goose Lake Valley in Lake County. The Booth series (Palexerolls) occupies colluvium and residuum in uplands composed of interbedded tuff and basalt. The Salisbury and Oxwall series (Durixerolls) occur on high lake terraces; and the Drews (Argixerolls), Lakeview (Haploxerolls), and Goose Lake (Argialbolls) series occur on alluvium and lacustrine sediments. *Source* NRCS

The dominant soil order in this MLRA is Mollisols (Table 5.5). Small areas of Inceptisols and Histosols are in the basins. The soils in this area typically have a mesic or frigid soil temperature regime, a xeric soil moisture regime, and mixed or smectitic mineralogy. They generally are well drained, but they may be poorly drained or very poorly drained in the basins. They generally are loamy, clayey, or sandy and are shallow to very deep. Argixerolls formed in residuum (Lorella series) and in residuum mixed with loess and/or volcanic ash (Royst soil series) on plateaus, hills, and mountains. Haploxerolls (Fordney series) formed in sandy alluvium on terraces. Palexerolls (Booth series) formed in colluvium on plateaus, hills, and mountains. Haploxerands (Pokegema and Woodcock series) are on plateaus containing mudflow deposits derived from andesitic rocks and volcanic ash. Humaquepts (Tulana series) formed in lacustrine sediments on lacustrine bottoms. Haplohemists (Lather series) formed in organic material in marshes.

Figure 5.29 is a block diagram of soils on upland volcanic rocks, lacustrine sediments, and alluvium and lacustrine sediments in the Lake County portion of this MLRA. The Booth series (Palexerolls) occupies colluvium and residuum in uplands composed of interbedded tuff and basalt. The Salisbury and Oxwall series (Durixerolls) occur on high lake terraces; and the Drews series (Argixerolls), Lakeview series (Haploxerolls), and Goose Lake series (Argialbolls) occur on alluvium and lacustrine sediments.

5.13 Palouse Prairie

The Palouse Prairie occupies 6,400 km^2 in Oregon, which is 2.4% of the state area (Table 5.3). About 28% of this MLRA occurs in Oregon and the remainder of this MLRA is contained in Washington (52%) and Idaho (20%). This MLRA is contained within the Deschutes–Umatilla Plateau physiographic province (Table 5.1). Elevations range from 435 to 1,435 m; and maximum slopes range between 65 and 115%. This MLRA is underlain primarily by Miocene basalt flows overlain by loess. Figure 5.30 shows the Zumwalt Prairie in Wallowa County.

The mean annual air temperature ranges between 7.2 and 9.8 °C; and the mean annual precipitation is between 440 and 580 mm (Table 5.3). The predominant vegetation is Idaho fescue, bluebunch wheatgrass, and Sandberg bluegrass (Table 5.4). Much of this MLRA is cultivated for annual cropping of winter wheat.

The predominant soil order in this MLRA is Mollisols (Table 5.5). The soils in the area commonly have a mesic or frigid soil temperature regime, a xeric soil moisture regime, and mixed mineralogy. They are generally deep or very deep, well drained or moderately well drained, and loamy. Haploxerolls formed in loess (Palouse and Athena series) on hills and colluvium mixed with loess and ash (very shallow Bocker series and shallow Rockly series) on uplands. Argixerolls formed in loess-mantled colluvium and residuum (Gwin and Waha series).

Figure 5.31 shows the Palouse–Waha–Gwin association in Umatilla County. The loess thickness is greater than 150 cm for the Palouse series, 100 cm for the Waha series, and 45 cm or less for the Gwin series.

Fig. 5.30 Zumwalt Prairie in the Palouse Prairies of Wallowa County. *Source* NRCS photo

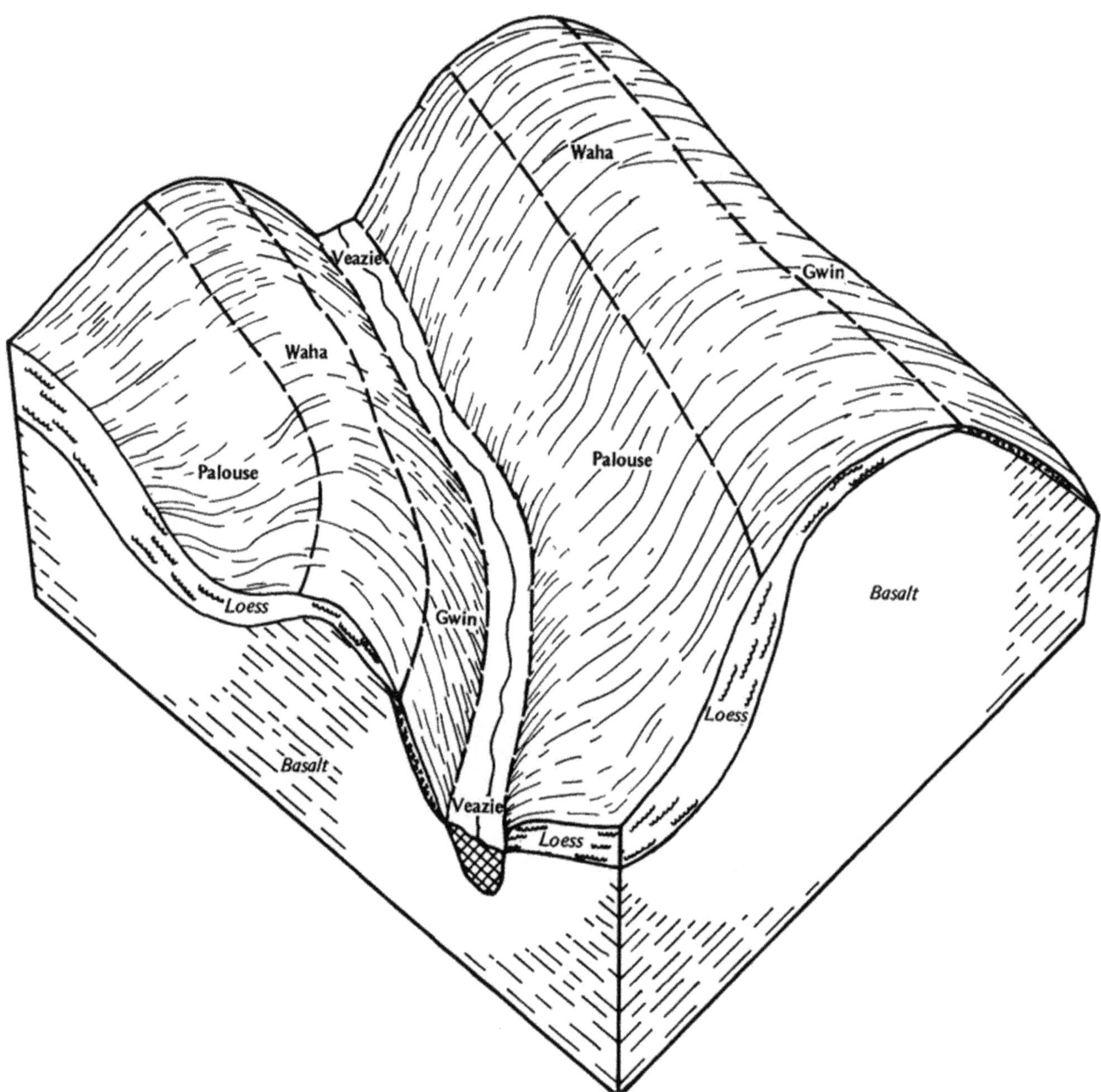

Fig. 5.31 A soil-block diagram for the Palouse and Nez Perce Prairie shows the Palouse-Waha-Gwin association in Umatilla County. The loess thickness is greater than 150 cm for the Palouse (Haploxerolls) series, 100 cm for the Waha (Argixerolls) series, and 45 cm or less for the Gwin (Argixerolls) series. *Source* NRCS

5.14 Sitka Spruce Belt

The Sitka Spruce Belt occupies a narrow belt along the Pacific Coast in Washington and Oregon. The Oregon portion (45%) has an area of 6,200 km^2 (2.3% of state area) (Table 5.3). This MLRA is part of the Coast Range physiographic province. Elevations range between 0 and 430 m; and maximum slopes usually are between 50 and 70%. The Oregon portion of this MLRA contains primarily marine and estuarine sediments, beach dune deposits, and headlands composed of volcanic rocks. Figure 5.32 shows fog moving inland in the Sitka Spruce Belt in Tillamook County.

The mean annual air temperature ranges between 8.5 and 10 °C; and the mean annual precipitation is between 1,845 and 2,365 mm (Table 5.3). The vegetation is comprised of Sitka spruce, western hemlock, red alder, Douglas-fir, western redcedar, and shore pine (*Pinus contorta* var. *contorta*) (Table 5.4).

The predominant soil orders in the MLRA are Andisols, Inceptisols, Spodosols, and Entisols (Table 5.5). The soils have an isomesic or isofrigid soil temperature regime. They have a udic soil moisture regime. They are acid throughout; most are very strongly acid or strongly acid. The hilly to extremely steep uplands are dominated by Andisols and Inceptisols. These soils are shallow to very deep and are well drained. They have ferrihydritic or isotic mineralogy. Fulvudands (Necanicum, Tolovana, and Klootchie series), Dystrudepts (Templeton series), and Humudepts (Reedsport and Skipanon series) dominate the uplands. The undulating to hilly marine terraces are dominated by Andisols and Spodosols. These soils are shallow or moderately deep to cemented materials or are deep or very deep. They are poorly drained to well drained. They have ferrihydritic or isotic mineralogy. Fulvudands (Lint series), Haplorthods (Netarts and Yaquina series), and Duraquods (Depoe series) dominate the terraces. The soils on the nearly level flood plains and in the estuaries are primarily Entisols and Inceptisols with minor areas of Histosols. These soils are very deep and typically are very poorly drained or poorly drained. They have mixed mineralogy. Fluvaquents (Coquille series), Humaquepts (Brenner and Clatsop series), Dystrudepts (Nehalem series), and Haplohemists (Brallier and Bragton soil series) dominate the flood plains and estuaries.

Fig. 5.32 Fog moving from the Pacific Coast inland in the Sitka Spruce Belt in Tillamook County. *Source* NRCS photo

5.15 Columbia Basin

The Columbia Basin comprises 3,800 km^2, which is 1.4% of Oregon (Table 5.3). About 22% of this MLRA occurs in Oregon with nearly three-quarters (78%) of this MLRA occurring in Washington. This MLRA is part of the Deschutes-Umatilla Plateau physiographic province (Table 5.1). Elevations range from 90 to 600 m; and maximum slopes are between 50 and 70%. This MLRA is underlain by Miocene basalt flows covered with loess, eolian sand, and lacustrine deposits. Figure 5.33 shows the Columbia Basin in north central Oregon, which is heavily used for irrigated agriculture and wind power generation.

The mean annual air temperature ranges between 9 and 11 °C; and the mean annual precipitation is between 175 and 235 mm, one of the driest of the MLRAs in Oregon (Table 5.3). The vegetation includes Wyoming big sagebrush, Indian ricegrass, bluebunch wheatgrass, Sandberg bluegrass, and Thurber's needlegrass (Table 5.4).

The predominant soil orders in this MLRA are Aridisols and Entisols (Table 5.5). The soils in the area have a mesic soil temperature regime, an aridic soil moisture regime, and mixed mineralogy. They generally are moderately deep to very deep, well drained to excessively drained, and loamy. Haplocalcids formed in eolian deposits on hills (Adkins soil series) and in loess over lacustrine deposits on stream terraces (Sagehill series). Haplocambids formed mixtures of loess, glaciolacustrine deposits, and colluvium on hills, plateaus, benches, and terraces (Shano and Warden soil series). Haplodurids formed in loess and glaciolacustrine sediments (Burke series) in uplands. Torriorthents formed in glaciofluvial deposits or alluvium (Burbank series) on terraces. Torripsamments formed in sandy eolian material on dunes (Quincy series).

Figure 5.34 is a schematic diagram of Benton County, Washington, showing the relationship of elevation, precipitation, and parent materials on the distribution of major soil series in the Columbia Basin. The Lickskillet, Walla Walla, and Ritzville series (Haploxerolls) occur in areas above 600 m which receive more than 275 mm of precipitation and which tend to be ridges with loess over basalt. The Prosser and Starbuck series (Haplocambids), which occur on alluvium from the Yakima and Columbia Rivers, are below 150 m and receive less than 175 mm of precipitation. The Quincy (Torripsamments) and Hezel series (Torriorthents) have formed in eolian sand on dunes and dissected terraces in areas receiving less than 180 mm of precipitation and occurring at the middle elevations. The Lickskillet (Haploxerolls), Willis (Durixerolls), and Kiona (Haplocambids) series is formed in colluvium and residuum from basaltic materials. The Warden series (Haplocambids) is formed in lacustrine or glaciolacustrine materials on terraces. The Ritzville and Walla Walla (Haploxerolls) and Shano series (Haplocambids) are formed in loess.

Fig. 5.33 The Columbia River Basin in north central Oregon. The upper panel is an ArcScene prepared by Steve Campbell showing central-pivot irrigation circles, and the lower panel shows alluvial terraces along the Columbia in the foreground and wind turbines in the background. *Source* Photo provided by Kelley Paup-Lefferts, NRCS, Pasco, WA

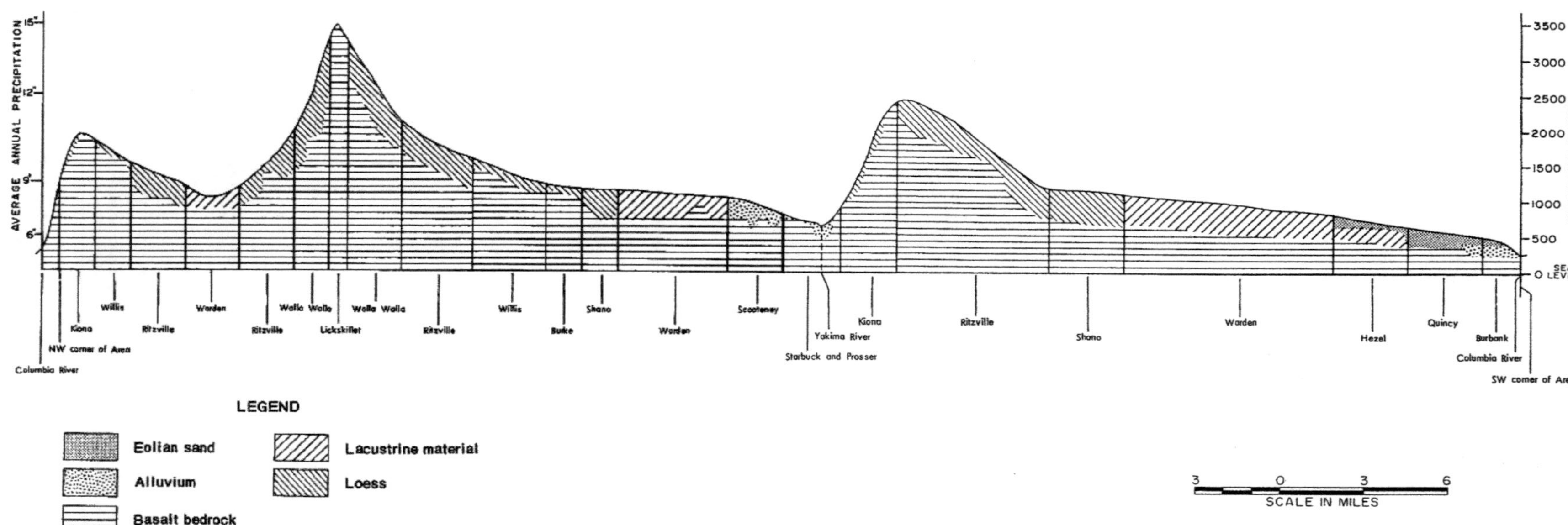

Fig. 5.34 Schematic diagram of Benton County, Washington, showing the relationship of elevation, precipitation, parent materials on the distribution of major soil series in the Columbia Basin. The Lickskillet, Walla Walla, and Ritzville series (Haploxerolls) occur in areas above 600 m which receive more than 275 mm of MAP and which tend to be ridges with loess over basalt. The Prosser and Starbuck series (Haplocambids), which occur on alluvium from the Yakima and Columbia Rivers, are below 150 m and receive less than 175 mm MAP. The Quincy (Torripsamments) and Hezel series (Torriorthents) have formed in eolian sand on dunes and dissected terraces in areas receiving less 180 mm MAP and occurring at the mid-elevations. The Lickskillet (Haploxerolls), Willis (Durixerolls), and Kiona (Haplocambids) series is formed in colluvium and residuum from basaltic materials. The Warden series (Haplocambids) is formed in lacustrine or glaciolacustrine materials on terraces. The Ritzville and Walla Walla (Haploxerolls) and Shano series (Haplocambids) are formed in loess. *Source* NRCS

5.16 Snake River Plains

Only a small portion (6%) of this MLRA occurs in Oregon (2600 km^2); the majority (94%) exists in Idaho (Table 5.3). This MLRA is part of the Owyhee physiographic province (Table 5.1). Elevations range 520 to 830 m; and maximum slopes are between 10 and 50%. The MLRA consists of lava plains formed from local basalts from fissures created as the North American Plate drifted southwest over a hot spot in the Earth's crust and the Columbia River basalts. The majority of the plain is covered with late Pleistocene loess, lacustrine deposits, and alluvium of varying depths. Figure 5.35 shows the Snake River Plains in Lake Owyhee State Park.

The mean annual air temperature ranges between 9 and 11 °C, and the mean annual precipitation is between 195 and 285 mm (Table 5.3). The vegetation includes Wyoming big sagebrush, bluebunch wheatgrass, Sandberg bluegrass, and Thurber's needlegrass (Table 5.4). Much of this area is used for irrigated agriculture.

The predominant soil order in this MLRA is Aridisols (Table 5.5). The soils in the area generally have a mesic soil temperature regime, an aridic soil moisture regime, and mixed or smectitic mineralogy. They are shallow to very deep and are generally well drained. They are silty, loamy, or clayey. Haplocalcids have formed in lacustrine deposits mantled with loess (Sagehill series). Argidurids have formed in mixed loess and ash over alluvium and lacustrine deposits (Virtue series). Haplodurids (Frohman and Nyssa series) have formed in silty lacustrine material and very gravelly alluvium on terraces.

Fig. 5.35 The Snake River Plains in Lake Owyhee State Park. *Source* Photo by Bureau of Land Management

5.17 Humboldt Area

The Humboldt Area lies mostly in Nevada (94%), with only 1970 km^2 or about 6% occurring in Oregon (Table 5.3). The Humboldt Area is part of the Basin and Range physiographic province (Table 5.1). Elevations range from 1,245 to 1,610 m; and maximum slopes are between 0 and 30%. This MLRA consists of wide valleys filled with lacustrine deposits and alluvium from adjacent mountain ranges composed of volcanic rocks. The Alvord Desert, a barren saline-sodic playa about 22,000 ha in size, is in this MLRA. The Humboldt Area in Harney County is depicted in Fig. 5.36. The snow-capped Steens Mountain forms the backdrop.

The mean annual air temperature ranges from 7 to 9 °C; and the mean annual precipitation is between 180 and 255 mm (Table 5.3). This MLRA has the lowest precipitation in Oregon. The vegetation includes Wyoming big sagebrush, rabbitbrush, black greasewood, inland saltgrass, basin wildrye, basin big sagebrush, spiny hopsage (*Grayia spinosa*), shadscale (*Atriplex confertifolia*), and bud sagebrush (*Picrothamnus desertorum*, also known as *Artemisia spinescens*) (Table 5.4).

The predominant soil orders in this MLRA are Aridisols and Inceptisols (Table 5.5). The soils in the area have a mesic soil temperature regime, an aridic soil moisture regime, and mixed mineralogy. They generally are well drained to poorly drained, loamy, and very deep. Moderately well-drained Aquisalids (Icene series), somewhat poorly drained Aquicambids (Droval and Alvodest series), and poorly drained Halaquepts (Reese series) have formed in alluvium and lacustrine deposits on flood plains and terraces.

Fig. 5.36 The Humboldt Area in Harney County. *Source* NRCS photo

Fig. 5.37 The Coastal Redwood Belt in southwestern Oregon. *Source* Photo by Thor Thorson

Haplocalcids formed in alluvium on alluvial fans and lake terraces (Outerkirk series). Shallow Argidurids formed in alluvium on fans and terraces (Deppy and Tumtum series).

5.18 Coastal Redwood Belt

The Coastal Redwood Belt is mostly in California (98%), with only 242 km^2 or about (2%) in Oregon (Table 5.3). This MLRA is contained in the Klamath Mountains physiographic province (Table 5.1). Elevations range from 0 to 630 m, and maximum slopes are between 10 and 90%. This MLRA is composed of contorted metamorphic rocks. These rocks are covered with uplifted marine terrace deposits and alluvium. Figure 5.37 shows the Coastal Redwood Belt in southwestern Oregon.

The mean annual air temperature ranges narrowly from 9.6 to 10.4 °C, and the mean annual precipitation is between 1,000 and 2,200 mm (Table 5.3). The vegetation includes coastal redwood, Douglas-fir, grand fir, western redcedar, Port Orford cedar, and red alder (Table 5.4).

Common soil orders in the MLRA are Entisols, Inceptisols, and Ultisols (Table 5.5). The soils have an isomesic soil temperature regime, a udic soil moisture regime, and mixed mineralogy. They generally are deep or very deep, well drained, and loamy or clayey, and occur on mountain slopes and hills in addition to coastal terraces. The common parent material is residuum weathered from sandstone. The soils of predominant extent include Haplohumults (Winchuck, and Loeb series) and Dystrudepts (Dulandy series). The soils on flood plains are Udifluvents (Bigriver series).

5.19 Summary

Oregon contains 17 Major Land Resource Areas (MLRAs), ranking the state sixth nationally. The MLRAs vary distinctly in mean annual air temperature, mean annual precipitation, elevation range, average maximum slope percent, vegetation, parent materials, bedrock composition, landforms, and predominant soil great groups.

Reference

Natural Resources Conservation Service. 2006. Land Resource Regions and Major Land Resource Areas of the United States, the Caribbean, and the Pacific Basin. U.S. Dept. Agric. Handbook 296.

6 Diagnostic Horizons and Taxonomic Structure of Oregon Soils

6.1 Introduction

Five of the eight epipedons (diagnostic surface horizons) in *Soil Taxonomy* (1999; 2014) and 11 of the 20 diagnostic subsurface horizons occur in soils of Oregon. The state of Oregon contains soils representative of 10 of the 12 soil orders, 40 of the 67 suborders, 114 of the 270 great groups, 389 subgroups, 1,080 families, and 1,707 soil series. Data on diagnostic horizon thicknesses are provided for soil series with an area of 50 km^2 or greater in Appendix B.

6.2 Diagnostic Horizons

Diagnostic surface horizons, or epipedons, and subsurface horizons are important in classifying the soils of Oregon. Based on occurrence in soil series, diagnostic surface horizons can be ranked: mollic (44%), ochric (41%), umbric (14%), histic (0.8%), and melanic (0.2%) (Table 6.1). The ranking of occurrence of epipedons in Oregon soils is comparable to that of soils elsewhere in the western US (Blackburn et al. 2020).

Thicknesses of epipedons for Oregon soil series are ranked: histic 106 ± 61 cm, mollic 49 ± 27 cm, melanic 47 ± 5.9 cm, umbric 44 ± 21, and ochric 19 ± 12 cm (Table 6.1). Histograms show that mollic and umbric epipedons are most commonly in the 30–40 cm thickness class (Fig. 6.1, upper). Soils with mollic or umbric epipedons greater than 50 cm are identified in pachic and cumulic subgroups. There are 117 soil series in Oregon that are in pachic subgroups and 63 in cumulic subgroups; most of these soil series are Mollisols, but some are Inceptisols or Andisols. Soils in pachic subgroups have a progressive accumulation of soil organic carbon; in contrast, soils in cumulic subgroups occur on slopes less than 25%, receive "new" organic materials at the surface from slope wash, and have an irregular distribution of soil organic carbon with depth. Oregon soils in pachic and cumulic subgroups support Wyoming big sagebrush and bunchgrasses such as bluebunch wheatgrass and Idaho fescue. Soils with a mollic epipedon are common under mixed grasses, mountain big sagebrush, ponderosa pine, western juniper, low sagebrush and grasses, and Douglas-fir with Oregon white oak. The umbric epipedon is most common in Humudepts but also occurs in Udands, Haplohumults and Humicryepts. The vegetation on soils with an umbric epipedon is primarily Douglas-fir, western hemlock, and western redcedar forest.

Diagnostic subsoil horizons can be ranked by occurrence in soil series: cambic (39%), argillic (38%), duripan (7.5%), calcic (2.5%), albic (1.8%), and natric (1.1%) (Table 6.1). Spodic, fragipan, ortstein, and salic horizons each occur in less than 1% of the soil series. Nearly one-quarter (22%) of the soil series in Oregon lack a diagnostic subsurface horizon, and 22 soil series contained more than two diagnostic subsurface horizons. Thickness of diagnostic subsurface horizons of Oregon soil series (more than three occurrences) can be ranked: spodic, argillic, and fragipan (58–62 cm), cambic, calcic, and ortstein (49–52 cm), duripan (39 cm), natric (29 cm), and albic (13 cm). Histograms show that cambic horizons are most commonly in the 20–30 cm thickness class and argillic horizons in the 20–50 cm classes (Fig. 6.1, lower). The right-skewed histograms for epipedons and diagnostic subsurface horizons are due to the lower limit requirements for each of the horizons. The right-edge peak for mollic, argillic, and cambic horizons is due to the artificial upper limit of 100 cm. The mean thickness of the duripan horizon is underestimated, because about 25% of the duripans continue beyond the limit of the profile. The same is true for 7 of 10 fragipans and 9% of the argillic horizons.

Cambic horizons are most common in Haploxerolls, Humudepts, and Haplocambids, but they occur in many other soil great groups. They form in a variety of parent materials and support diverse vegetation ranging from temperate rainforest in western Oregon to desert shrubs in

T. Thorson et al., *The Soils of Oregon*, World Soils Book Series,
https://doi.org/10.1007/978-3-030-90091-5_6

Table 6.1 Diagnostic horizon thickness for Oregon soil series

Horizon	Mean (cm)	Std. dev. (cm)	Percentage of soil series
Surface			
Mollic	49	26	44
Ochric	20	12	41
Umbric	44	21	14
Histic	106	61	0.8
Melanic	47	5.9	0.2
Subsurface			
Cambic	51	29	39
Argillic	58	38	36
Duripan	39	31	7.7
Calcic	49	28	2.2
None			22
Albic	13	7.9	1.9
Natric	29	12	1.1
Spodic	62	28	0.8
Fragipan	63	24	0.6
Ortstein	52	30	0.4
Salic	63	23	0.1
Glossic	28	–	0.1

Note many Oregon soil series contain multiple subsurface horizons

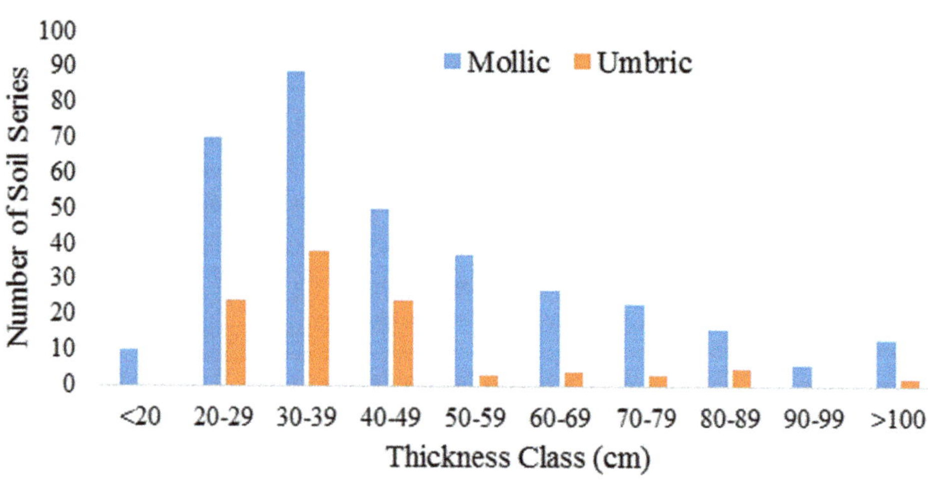

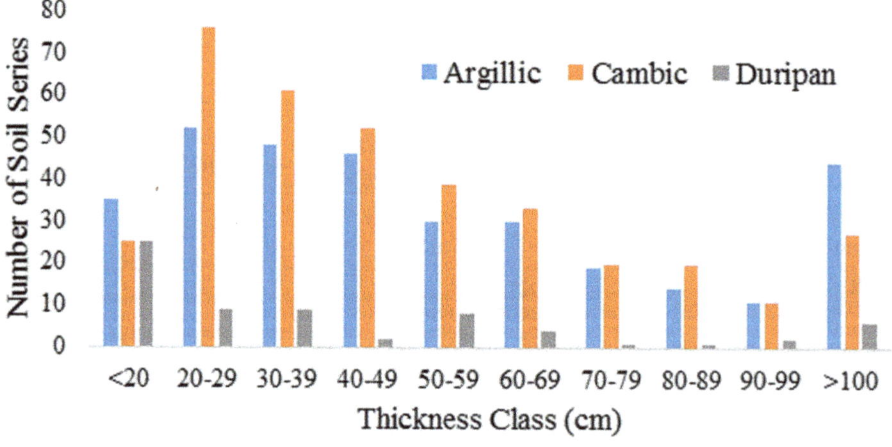

Fig. 6.1 Histograms of epipedon and diagnostic subsurface horizon thickness classes.

southeastern Oregon. Argillic horizons are most common in Argixerolls, Haploxeralfs, and Argidurids, but they occur in many other soil great groups. They form mainly from colluvium and residuum. Soils with argillic horizons support desert shrub and bunchgrass communities and forests dominated by Douglas-fir, ponderosa pine, and Pacific madrone. Duripans are found in Durixerolls, Argidurids, and Haplodurids. Soils with duripans form from alluvium on alluvial fans, colluvium and residuum, on terraces and plateaus, and lacustrine deposits in basins. These soils commonly are influenced by volcanic ash.

Calcic horizons are mainly in Argixerolls that support sagebrush-bunchgrass communities and are derived from colluvium and residuum that have been influenced by loess. Albic horizons occur most commonly in Haplorthods, Argialbolls, and Duraquods that support coniferous forest and, in the case of Argialbolls, Oregon white oak or ponderosa pine and/or Douglas-fir. Spodic horizons occur mainly in Haplorthods, Haplocryods, and Duraquods formed in marine sands along the Pacific Coast and in colluvium and till in the mountains. The dominant vegetation on Spodosols is Sitka spruce along the Pacific Coast and upper montane and subalpine conifers in the mountains. Great groups most often lacking a diagnostic subsurface horizon include the Torripsamments, Torriorthents, and Xerorthents.

6.3 Orders

Although soil series representing 10 orders occur in Oregon, six accounts for 97% of the soil area, including Mollisols (42%), Aridisols (17%), Inceptisols (16%), Andisols (13%), Ultisols (5.0%), and Alfisols (4.1%) (Table 6.2, Fig. 6.2).

6.4 Suborders

The state of Oregon contains soils representative of 40 of the 67 suborders. Nine suborders account for about 83% of the soil area of Oregon, including the Xerolls (39%), Udepts (9.8%), Durids (6.7%), Argids (6.3%), Cryands (5.2%), Xerepts (4.4%), Humults (4.0%), Cambids (3.9%), and Xeralfs (3.5%) (Fig. 6.3).

6.5 Great Groups

Oregon contains soils representative of 112 of the 270 great groups. Fourteen great groups account for 74% of the soil area in Oregon, including the Haploxerolls (18%), Argixerolls (16%), Humudepts (6.7%), Haplargids (4.3%), Argidurids (4.3%), Vitricryands (4.1%), Haplocambids (3.5%), Palexerolls (2.9%), Haploxeralfs (2.7%), Dystroxerepts (2.3%), Haplodurids (2.3%), Durixerolls (2.3%), Haplohumults (2.2%), and Vitrixerands (2.2%) (Fig. 6.4).

6.6 Subgroups

There are 389 soil subgroups in Oregon. Twelve subgroups categories account for 70% of the total in Oregon, including Typic (20%), Xeric (9.5%), Vitrandic (7.5%), Pachic (5.2%), Lithic (4.9%), Ultic (3.6%), Aridic (3.5%), Humic (3.5%), Andic (3.5%), Vitritorrandic (3.2%), Cumulic (3.1%), and Vertic (2.7%) (Fig. 6.5).

Subgroups follow one of three options: the central concept (e.g., Typic and Haplic); interagrades that have specific properties that differ from the Typic or Haplic and have one or more characteristics similar to another order, suborder, or great group; and extragrades, which identify soils with properties that do not clearly intergrade toward specifically defined categories (e.g., Lithic or Calcic). Two-thirds (66%) of Oregon soils are in intergrade subgroups, 14% are in extragrade subgroups, and 20% follow the central concept (Fig. 6.5b).

About 35% of the soil series in Oregon have a lithic or paralithic contact in the upper 100 cm.

6.7 Families

There are 1,080 soil families in Oregon. Nearly one-half (44%) of Oregon's soils are in loamy particle-size classes, followed by clayey (28%), ashy and medial (16%), silty (10%), and sandy (1%) (Fig. 6.6). The predominant mineral classes are mixed (48% of soil area), smectitic (20%), isotic (14%), and glassy (6.1%) (Fig. 6.7). In view of the fine soil textures of Oregon soils and the abundance of smectite, it is not surprising that 47% of the soil area is in the superactive

Table 6.2 Taxonomic structure of Oregon soils

Order	Suborder	Great group	No. soil series	Area (km^2)
Alfisols	Aqualfs	Albaqualfs	2	464
	Aqualfs	Endoaqualfs	3	138
	Aqualfs	Epiaqualfs	1	15
	Cryalfs	Glossocryalfs	1	6
	Cryalfs	Haplocryalfs	2	52
	Udalfs	Hapludalfs	5	234
	Xeralfs	Durixeralfs	3	83
	Xeralfs	Fragixeralfs	2	103
	Xeralfs	Haploxeralfs	57	4282
	Xeralfs	Natrixeralfs	1	20
	Xeralfs	Palexeralfs	15	1015
	Xeralfs	Rhodoxeralfs	1	30
			93	6442
Andisols	Aquands	Cryaquands	4	228
	Aquands	Endoaquands	2	21
	Aquands	Epiaquands	1	3
	Aquands	Melanaquands	1	3
	Cryands	Duricryands	3	346
	Cryands	Fulvicryands	7	155
	Cryands	Haplocryands	15	1147
	Cryands	Vitricryands	48	6518
	Torrands	Vitritorrands	3	55
	Udands	Fulvudands	19	1720
	Udands	Hapludands	18	2572
	Udands	Melanudands	2	17
	Vitrands	Udivitrands	23	3202
	Xerands	Haploxerands	4	989
	Xerands	Melanoxerands	1	32
	Xerands	Vitrixerands	52	3449
			203	20,457
Aridisols	Argids	Calciargids	2	33
	Argids	Haplargids	55	6851
	Argids	Natrargids	12	881
	Argids	Paleargids	17	2054
	Argids	Petroargids	1	48
	Calcids	Haplocalcids	7	656
	Cambids	Aquicambids	7	536
	Cambids	Haplocambids	61	5537
	Durids	Argidurids	33	6749
	Durids	Haplodurids	22	3577
	Durids	Natridurids	3	212
	Salids	Aquisalids	2	205
			222	27,339

(continued)

Table 6.2 (continued)

Order	Suborder	Great group	No. soil series	Area (km^2)
Entisols	Aquents	Fluvaquents	5	90
	Aquents	Psammaquents	2	68
	Fluvents	Torrifluvents	2	55
	Fluvents	Udifluvents	1	1
	Fluvents	Xerofluvents	1	5
	Orthents	Torriorthents	24	1222
	Orthents	Udorthents	2	16
	Orthents	Xerorthents	4	59
	Psamments	Cryopsamments	1	29
	Psamments	Torripsamments	14	1477
	Psamments	Udipsamments	4	110
	Psamments	Xeropsamments	2	26
			62	3158
Histosols	Fibrists	Cryofibrists	1	3
	Hemists	Cryohemists	1	8
	Hemists	Haplohemists	5	135
	Saprists	Cryosaprists	1	3
	Saprists	Haplosaprists	2	18
			10	167
Inceptisols	Aquepts	Cryaquepts	3	23
	Aquepts	Endoaquepts	11	289
	Aquepts	Epiaquepts	1	6
	Aquepts	Fragiaquepts	3	42
	Aquepts	Halaquepts	11	637
	Aquepts	Humaquepts	17	418
	Cryepts	Dystrocryepts	9	277
	Cryepts	Haplocryepts	7	349
	Cryepts	Humicryepts	19	402
	Udepts	Dystrudepts	45	2614
	Udepts	Eutrudepts	12	1868
	Udepts	Fragiudepts	2	293
	Udepts	Humudepts	84	10,637
	Xerepts	Durixerepts	3	76
	Xerepts	Dystroxerepts	33	3565
	Xerepts	Fragixerepts	3	289
	Xerepts	Haploxerepts	39	2852
	Xerepts	Humixerepts	7	201
			309	24,838

(continued)

Table 6.2 (continued)

Order	Suborder	Great group	No. soil series	Area (km^2)
Mollisols	Albolls	Argialbolls	9	1029
	Aquolls	Argiaquolls	6	208
	Aquolls	Calciaquolls	1	46
	Aquolls	Cryaquolls	9	302
	Aquolls	Duraquolls	6	79
	Aquolls	Endoaquolls	37	1442
	Aquolls	Epiaquolls	3	167
	Aquolls	Natraquolls	1	35
	Cryolls	Argicryolls	11	195
	Cryolls	Duricryolls	2	6
	Cryolls	Haplocryolls	12	927
	Udolls	Argiudolls	2	6
	Udolls	Hapludolls	6	38
	Xerolls	Argixerolls	236	25,846
	Xerolls	Calcixerolls	1	3
	Xerolls	Durixerolls	50	3550
	Xerolls	Haploxerolls	279	27,727
	Xerolls	Paleoxerolls	35	4514
			706	66,120
Spodosols	Aquods	Duraquods	4	42
	Aquods	Endoaquods	1	14
	Cryods	Haplocryods	2	78
	Cryods	Humicryods	1	5
	Orthods	Durorthods	1	40
	Orthods	Haplorthods	4	180
			13	359
Ultisols	Aquults	Paleaquults	1	8
	Aquults	Umbraquults	2	11
	Humults	Haplohumults	26	3499
	Humults	Palehumults	22	2780
	Udults	Hapludults	3	75
	Udults	Paleudults	4	304
	Xerults	Haploxerults	3	530
	Xerults	Palexerults	5	644
			66	7851
Vertisols	Aquerts	Endoaquerts	8	393
	Aquerts	Epiaquerts	4	166
	Uderts	Hapluderts	1	7
	Xererts	Durixererts	1	1
	Xererts	Haploxererts	9	439
			23	1006
	Total		1707	157,737

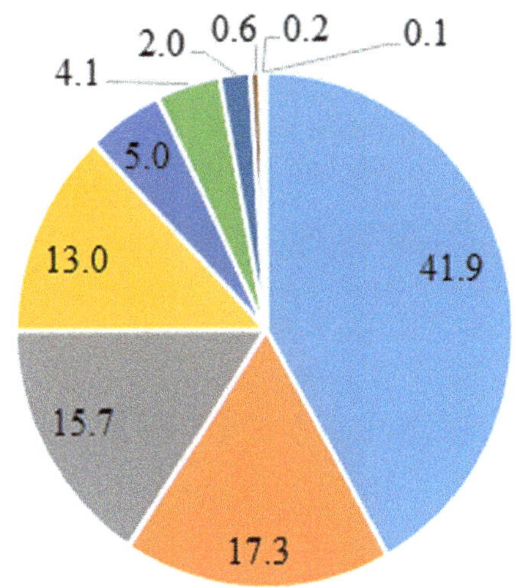

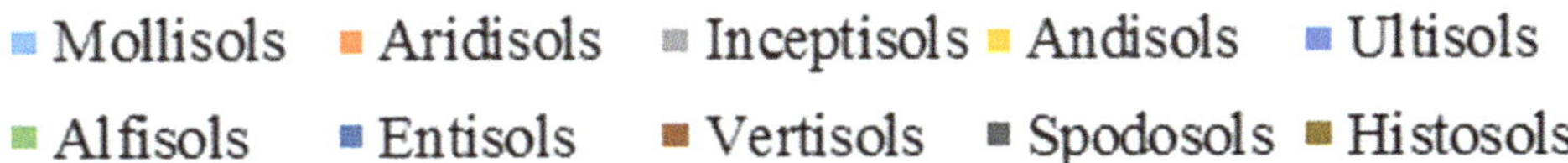

Fig. 6.2 Distribution of soil orders in Oregon (percent by area).

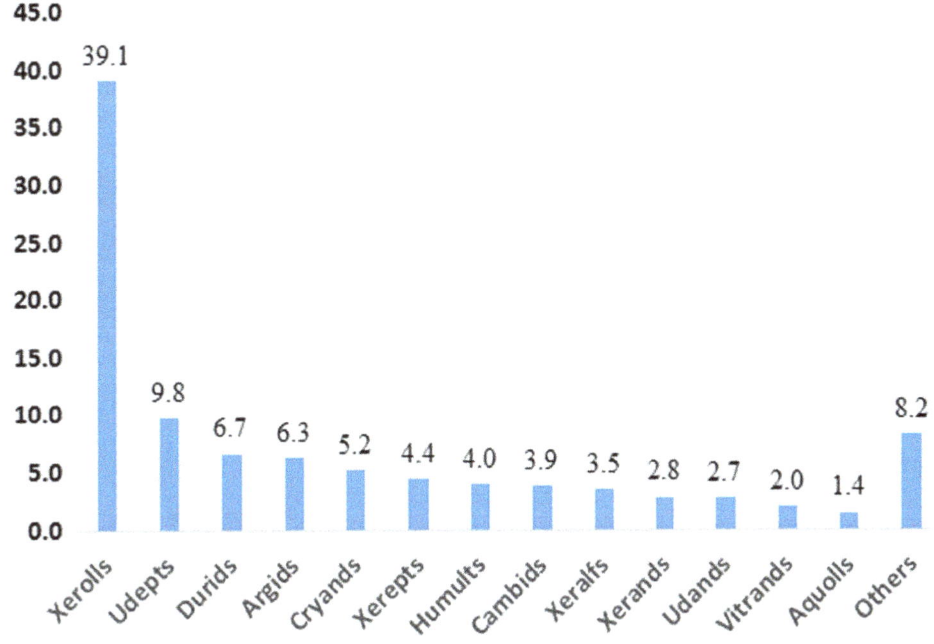

Fig. 6.3 Distribution of soil suborders in Oregon (percent by area).

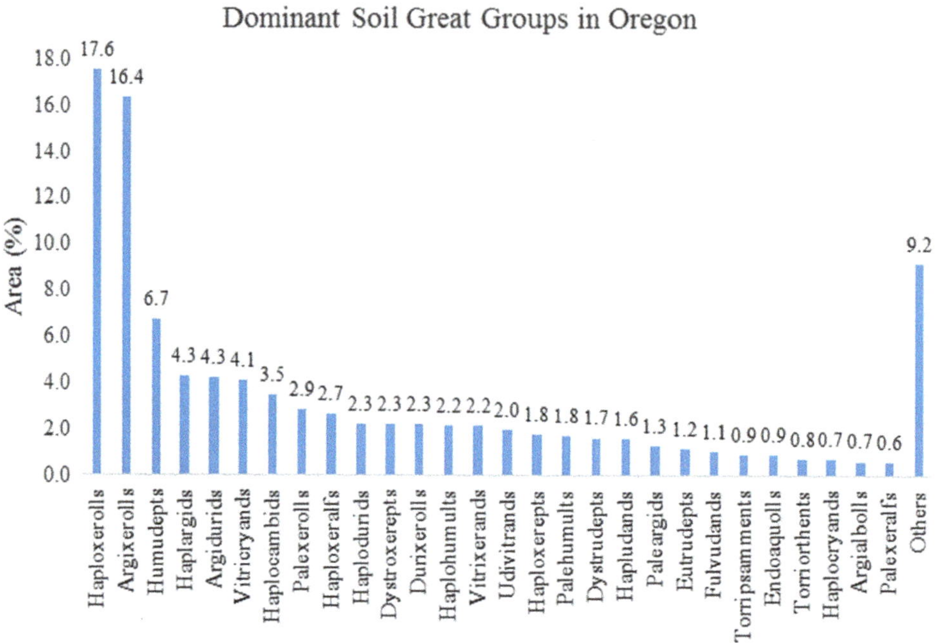

Fig. 6.4 Distribution of soil great groups in Oregon (percent by area).

and active cation-exchange activity classes (Fig. 6.8). More than one-half (55%) of the soil area in Oregon has a mesic soil temperature regime, followed by frigid (35%), cryic (7.0%), isomesic (3.0%), and isofrigid (0.05%) (Fig. 6.9). Nearly one-half (49%) of Oregon's soils are in the xeric soil moisture class; 26% are aridic, 22% are udic, and 3.6% are aquic (Fig. 6.10).

6.8 Soil Series

Oregon contains 1,707 established soil series (Table 6.2), of which 73% occur in Oregon only (Fig. 6.11). The majority (53%) of soil series in Oregon are in the 10–99 km^2 area class, followed by <10 km^2 class (25%), the 100–499 km^2 class (19%), the 500–999 km^2 class (2.3%), and the >1000 km^2 class (1.2%) (Fig. 6.12). This contrasts with the area distribution of soil series in the US as a whole, where 56% of the soil area contains "mega" (>1000 km^2) soil series.

6.9 Summary

Soils of Oregon contain primarily ochric and mollic epipedons (85% of soil area) overlying cambic and argillic horizons (78% of soil area). On an area basis, the predominant orders represented in Oregon are Mollisols, Inceptisols, Aridisols, Andisols, Ultisols, and Alfisols (97%). The remaining 3% are Vertisols, Spodosols, Histosols, and Entisols. The predominant suborders are Xerolls, Udepts, Durids, Argids, Cryands, Xerepts, and Humults and the predominant great groups are Haploxerolls, Argixerolls, Humudepts, Haplargids, Argidurids, and Vitricryands. One-fifth (20%) of the soil series are in Typic subgroups. Predominant families include the loamy particle-size class (44% of soil area), the mixed mineralogy class (48%), the superactive and active cation-exchange activity classes (47%), the mesic soil temperature regime (55%), and the xeric soil moisture regime (56%). There are more than 1,700 soil series in Oregon, of which 73% occur only in the state.

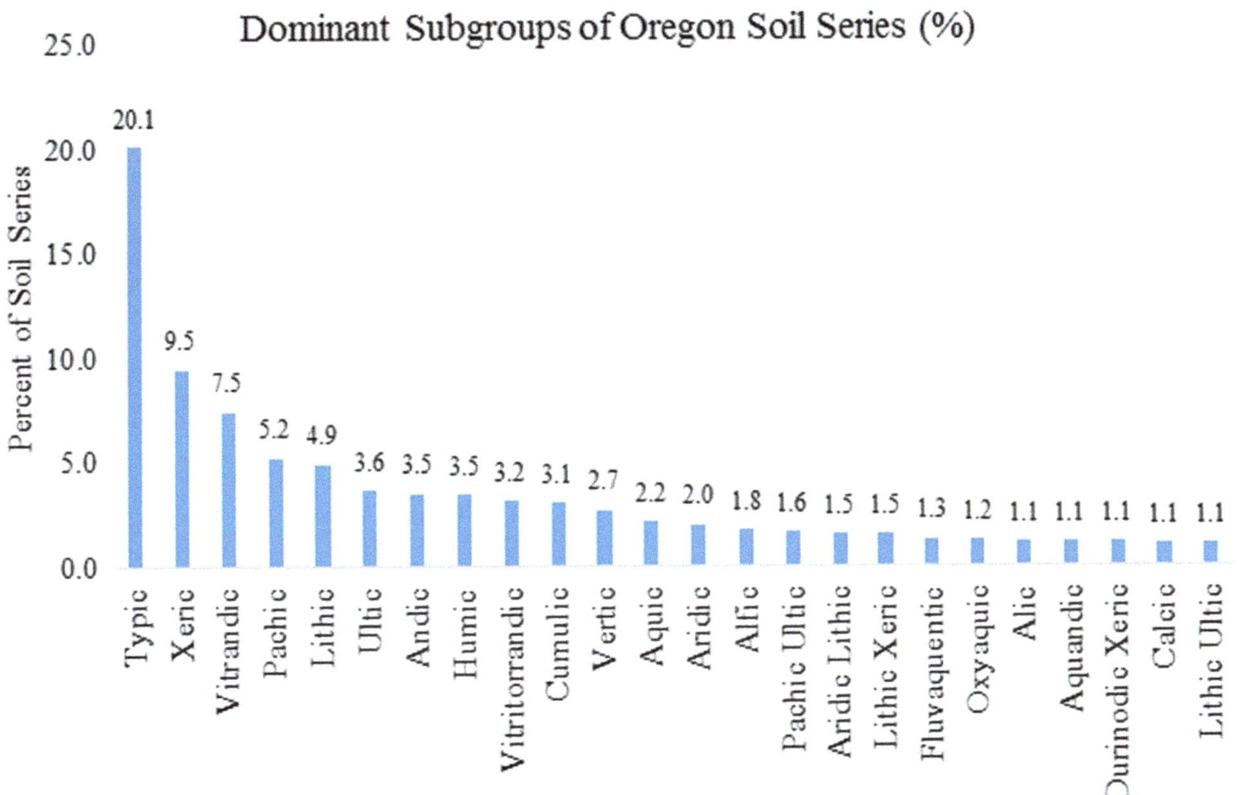

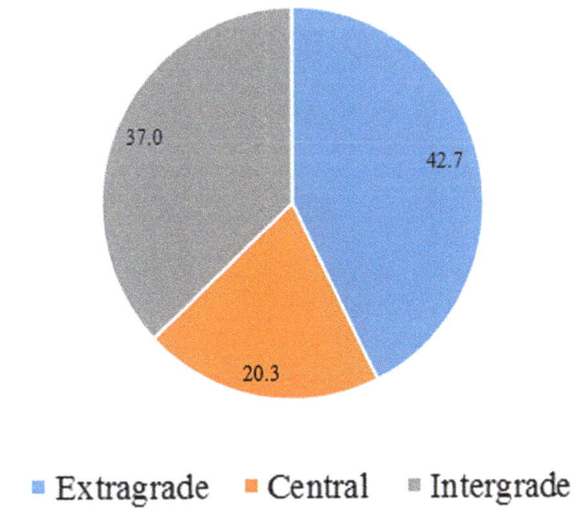

Fig. 6.5 Dominant soil subgroups in Oregon (percent by area) (upper); subgroup category (percent of soil series) (lower).

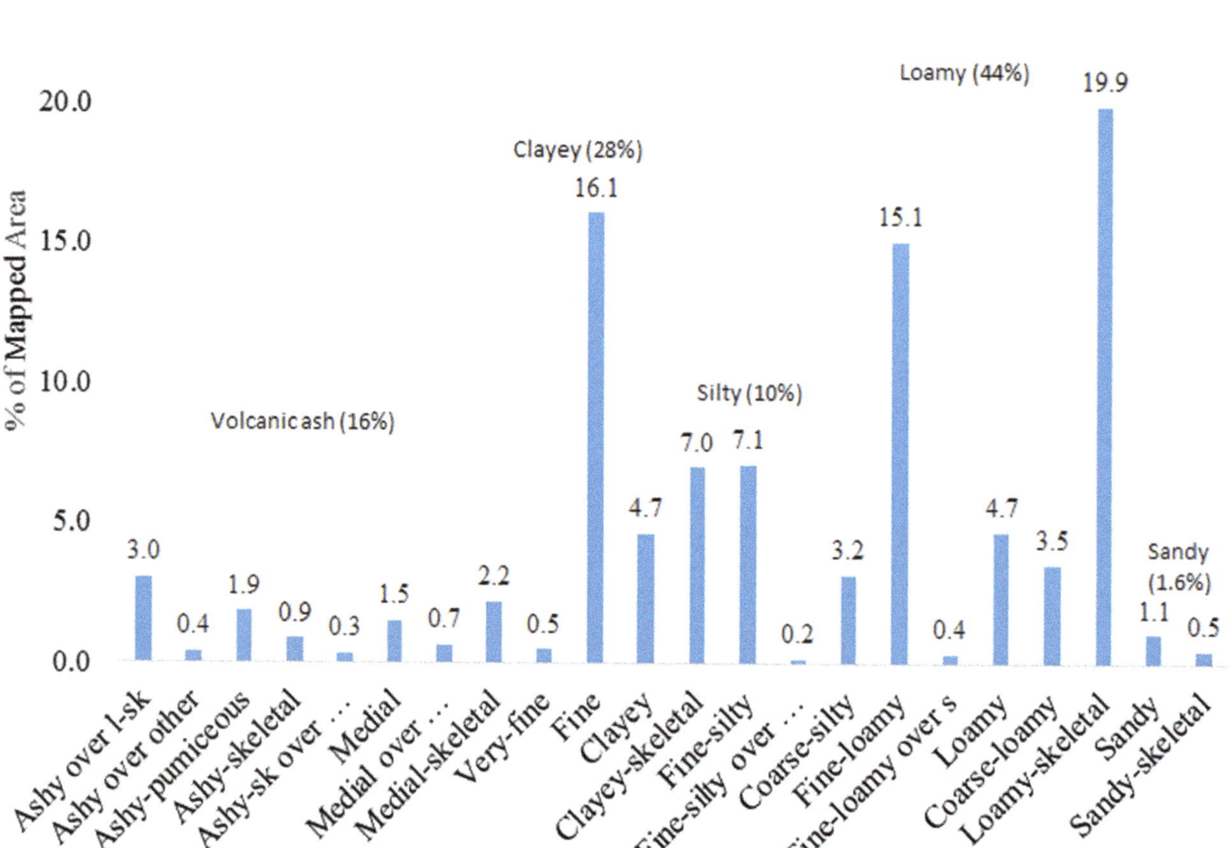

Fig. 6.6 Distribution of Oregon soil series by particle-size class.

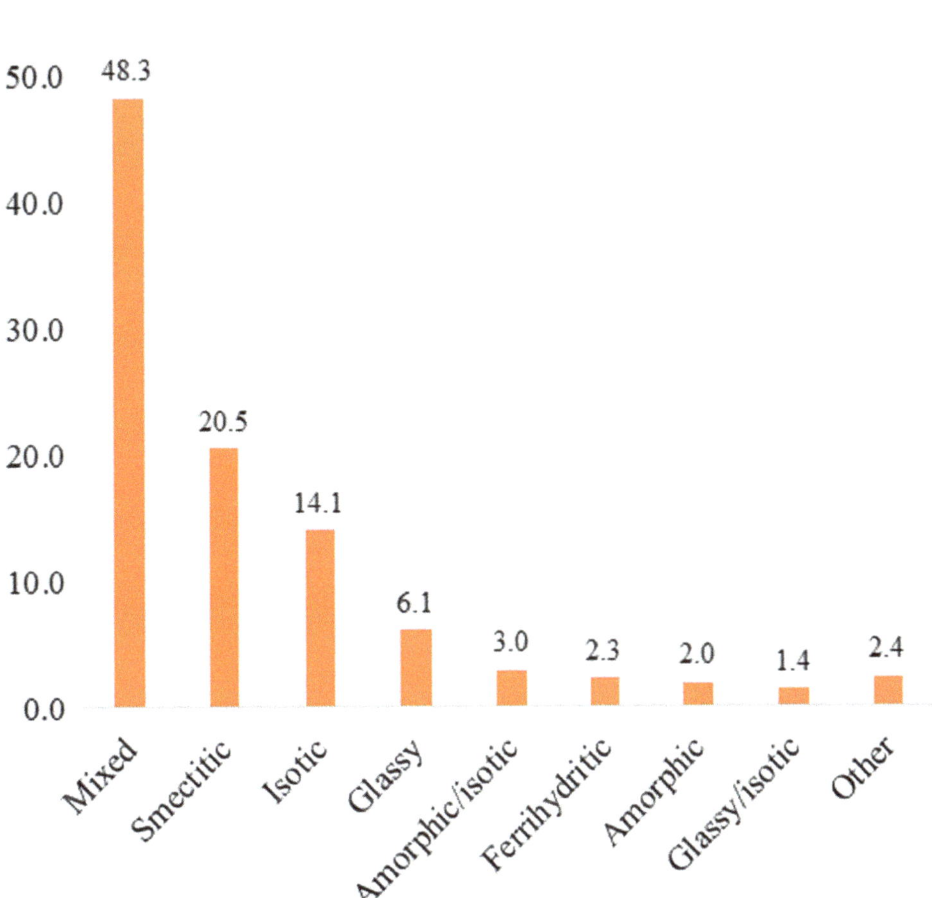

Fig. 6.7 Distribution of Oregon soil series by soil mineral class.

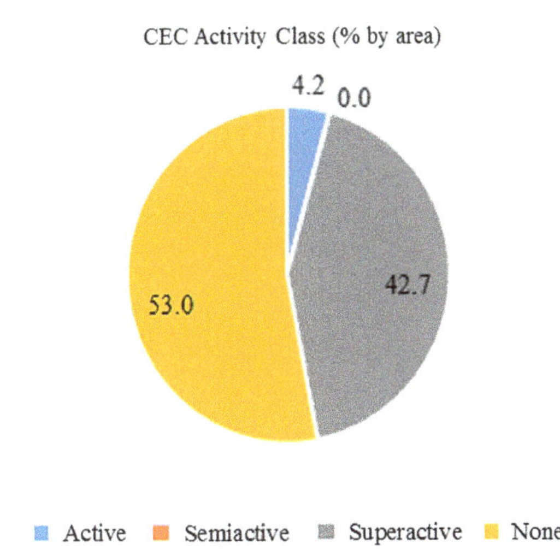

Fig. 6.8 Distribution of Oregon soil series by cation-exchange capacity activity class.

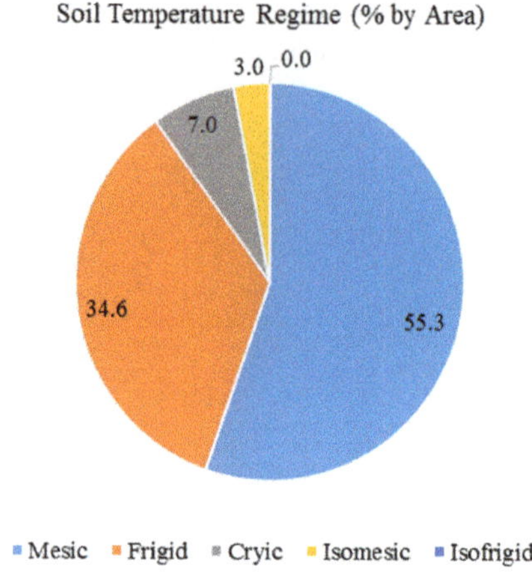

Fig. 6.9 Distribution of Oregon soil series by soil temperature class.

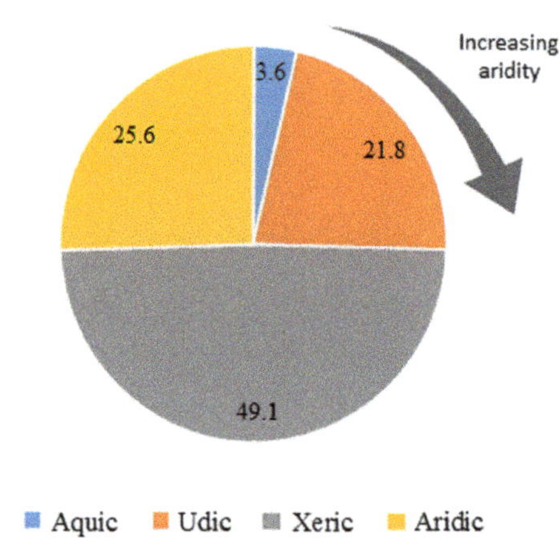

Fig. 6.10 Distribution of Oregon soil series by soil moisture regime.

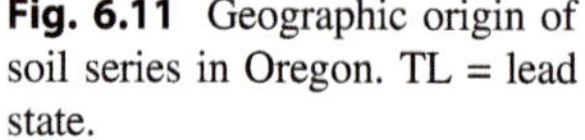

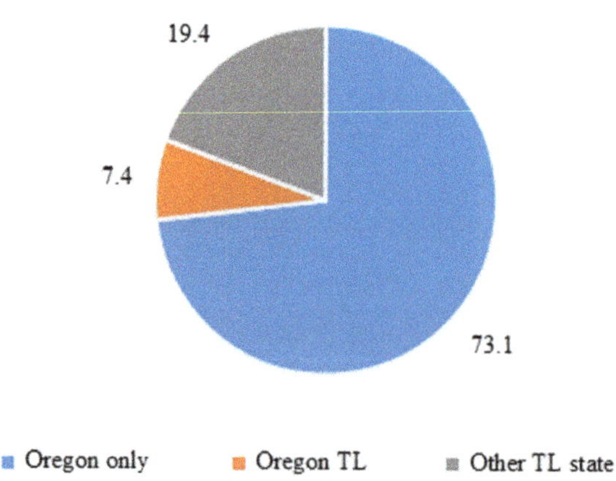

Fig. 6.11 Geographic origin of soil series in Oregon. TL = lead state.

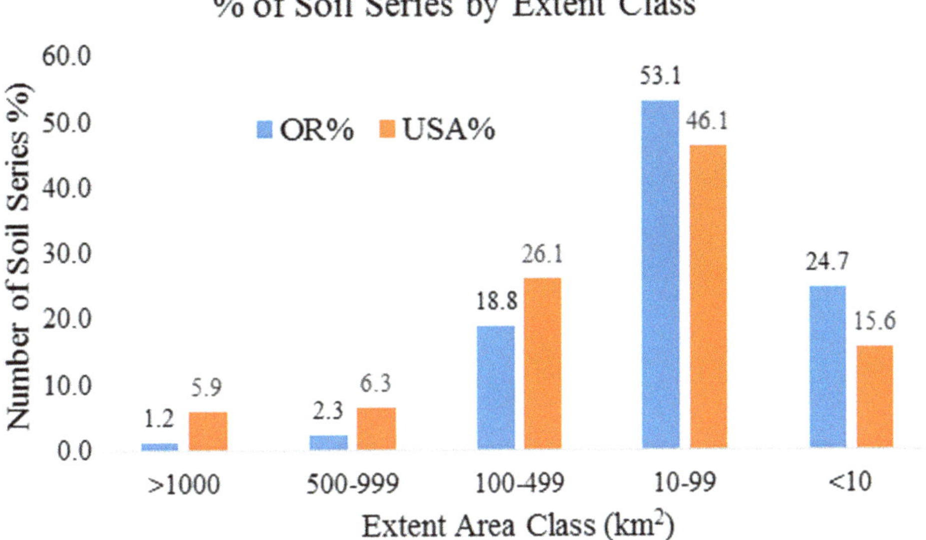

Fig. 6.12 Soil series extent classes for Oregon soil series (area basis).

References

Blackburn PW, Fisher JB, Dollarhide WE, Merkler DJ, Chiaretti JV, Bockheim JG (2020) The Soils of Nevada. Springer

Soil Survey Staff (1999) Soil taxonomy: a basic system of soil classification for making and interpreting soil surveys, 2nd edn, vol 436. AgriculturalHandbook. U.S. Govt. Print. Office, Washington, DC, 869 pp

Soil Survey Staff (2014) Keys to soil taxonomy, 12th edn. U.S. Dept. Agric. Natural Resour Conserv Serv, Lincoln, NE

7 Taxonomic Soil Regions of Oregon

7.1 Introduction

Oregon is divided here into 28 soil regions based on the relative abundance of great groups. Site factors are summarized in Table 7.1 and thicknesses of diagnostic horizons are provided in Table 7.2. The methods for preparing soil great group maps in this chapter and soil order maps in Chaps. 8–14 are described below (Table 7.3).

Soil mapping in Oregon has three levels of status: (i) SSURGO (Soil Survey Geographic Database) Certified, (ii) Initial Survey in Progress, and (iii) Non-Project Area (Fig. 2.6). The best available soil mapping in these three status levels was used to create the soil order and soil great group maps. All of the SSURGO certified data was used. The detailed mapping in the initial surveys in progress was used. STATSGO2 (State Soil Geographic Database) data was used for non-project areas and for areas within the initial surveys in progress that lacked detailed mapping (Fig. 7.1), which shows areas where STATSGO2 and where SSURGO data and initial mapping were used in creating the soil order and great group maps.

Soil survey areas that are SSURGO Certified cover about 68% of the State of Oregon. The mapping is at a scale of 1:24,000 and all soil information is contained in Web Soil Survey located at https://websoilsurvey.sc.egov.usda.gov. All soil series identified in these areas have an established status. The soil mapping units are composed of one to three major soil components. The components are typically a soil series name such as Jory, but can be a miscellaneous land type such as Rock Outcrop, or a higher soil taxa such as Argixerolls.

Soil survey areas that are Initial Survey in Progress cover about 16% of the State. The mapping is at a scale of 1:24,000 and all soil information is contained in Web Soil Survey. However, there are areas within these initial soil survey areas that are not yet mapped. These not yet mapped areas are identified as NOTCOM (not completed) on the soil maps in Web Soil Survey. All soil series identified in the initial survey areas have either an established or tentative status. Tentative series are those soil series that are newly recognized soils and have not been identified in past mapping. Tentative series will change status to established upon completion of all soil mapping in the initial mapping areas. The soil mapping units within these initial survey areas have the same kinds of components as the mapping units in SSURGO certified soil survey areas.

Soil survey areas that are Non-Project Areas cover about 16% of the State. These survey areas do not have detailed soil mapping. Soil mapping for all these areas is available in STATSGO2. The information can be assessed at https://www.nrcs.usda.gov/wps/portal/nrcs/detail/soils/survey/geo. The STATSGO2 product was a national effort in 1985 to produce a general soil map of the United States. The scale of mapping is 1:250,000. The soil series may have an established or tentative status. The soil map units can be composed of one to twenty-one soil components. These components are the same kinds of components as mapping in the SSURGO and Initial mapping areas.

Soil order and soil great group maps were developed using all soil mapping from SSURGO certified and initial soil survey areas in progress. For all remaining areas not covered by the certified and initial soil mapping, the STATSGO2 database was used. The following procedure was used in assigning a soil map unit to a soil order and a soil great group. Certified soil map units and initial soil survey map units, as previously mentioned, can have up to three major components although many map units have only one major component. Each component has an area percentage and each component has a soil classification.

EXAMPLE map unit with one soil component—Jory silty clay loam, 2–20 percent slopes.

Jory—composition 90%, soil order Ultisols, and soil great group is Palehumults.

In this example 90% of the map unit area is Utisols and all areas are included on the Ultisols order map, and all areas are included on the Palehumults great group map.

EXAMPLE map unit with three soil components—Olot-Klicker-Anatone complex, 20–50 percent slopes.

T. Thorson et al., *The Soils of Oregon*, World Soils Book Series,
https://doi.org/10.1007/978-3-030-90091-5_7

Table 7.1 Relation of soil great groups in Oregon to soil-forming factors

Soil region	Soil great group	Area (km^2)	No. soil series	MAP (mm/yr)	MAAT (°C)	Soil-moisture regime	Soil temperature regime	Vegetation type	Lower elev. (m)	Upper elev. (m)	Upper slope (%)	Parent materials	Composition	Landforms
1	Haploxerolls	27,727	279	450 ± 275	8.4 ± 1.7	Aridic, xeric	Mesic, frigid	WY big sagebrush, Idaho fescue, bluebunch, wheatgrass	650	1,400	57 ± 33	Colluvium, residuum, loess, ash	Basalt	Mountain slopes, hill-slopes, lava plateaus
2	Argixerolls	25,846	236	485 ± 285	8.1 ± 1.7	Aridic, xeric	Mesic, frigid	Wheatgrass, Idaho fescue, sagebrush, bluegrass	700	1,400	63 ± 26	Colluvium, residuum, loess	Basalt	Mountain slopes, lava plateaus
3	Humudepts	10,637	84	2,100 ± 200	9.2 ± 1.1	Udic	Isomesic, mesic, frigid	Douglas-fir, western hemlock, western redcedar	100	825	86 ± 9.0	Colluvium, residuum/lithic	Sandstone, volcanic	Hillslopes
4	Haplargids	6851	55	265 ± 35	7.3 ± 1.8	Aridic	Mesic, frigid	Sagebrush, Idaho fescue, wheatgrass, needlegrass	1,140	1,650	30 ± 29	Colluvium lacustrine	Basalt, tuff	Mountain slopes lake terraces
5	Argidurids	6749	33	240 ± 40	7.8 ± 0.8	Aridic	Mesic, frigid	Sagebrush, wheatgrass, needlegrass	1,150	1,600	24 ± 15	Alluvium alluvium (slope)		Alluvial fans lava plateaus
6	Vitricryands	6518	48	965 ± 255	4.0 ± 1.6	Xeric	Cryic	Subalpine fir, Engelmann spruce, lodgepole pine	1,450	2,160	72 ± 17	Ash/colluvium or till	Basalt	Mountain slopes, lava plateaus
7	Haplocambids	5537	61	250 ± 20	6.8 ± 0.9	Aridic	Mesic, frigid	Sagebrush, wheatgrass, needlegrass	1,250	1,850	30 ± 25	Colluvium/lithic lacustrine	Basalt, tuff	Mountain slopes lake terraces
8	Palexerolls	4514	35	350 ± 105	7.1 ± 1.0	Aridic, xeric	Mesic, frigid	Sagebrush, wheatgrass, Idaho fescue	1,065	1,625	53 ± 25	Colluvium, residuum	Tuff, basalt	Lava plateaus, hillslopes
9	Haploxeralfs	4282	57	1,125 ± 180	11 ± 0.8	Xeric	Mesic, frigid	Douglas-fir, ponderosa pine, Pacific madrone	160	900	60 ± 13	Colluvium, residuum/paralithic	Metavolcanics, metasedimenary	Foothills
10	Haplodurids	3577	22	250 ± 30	8.9 ± 1.9	Aridic	Mesic, frigid	Wyoming big sagebrush, wheatgrass, needlegrass	1,050	1,400	17 ± 7.0	Colluvium, residuum lacustrine	Basalt	Llava plateaus lake terraces
11	Dystroxerepts	3565	33	1,200 ± 100	9.7 ± 1.8	Xeric	Mesic, frigid	Douglas-fir, ponderosa pine, Pacific madrone	250	1,300	93 ± 5.2	Colluvium, residuum/paralithic or lithic	Basalt	Mountain slopes
12	Durixerolls	3550	50	295 ± 35	7.8 ± 1.4	Aridic, xeric	Frigid, mesic	Sagebrush, Idaho fescue, wheatgrass, bluegrass	1,035	1,430	25 ± 8.7	Ash, residuum/lithic	Basalt	Lava plateaus
13	Haplohumults	3499	26	1,450 ± 350	11 ± 1.0	Udic, xeric	Isomesic, mesic, frigid	Douglas-fir, western hemlock, grand fir, Pacific madrone	100	725	72 ± 13	Colluvium, residuum/paralithic or lithic	Basalt, sedimentary	Hillslopes, foothills

(continued)

Table 7.1 (continued)

Soil region	Soil great group	Area (km^2)	No. soil series	MAP (mm/yr)	MAAT (°C)	Soil-moisture regime	Soil temperature regime	Vegetation type	Lower elev. (m)	Upper elev. (m)	Upper slope (%)	Parent materials	Composition	Landforms
14	Vitrixerands	3449	52	725 ± 350	6.6 ± 0.5	Xeric	Frigid, mesic	Ponderosa pine, Douglas-fir, antelope bitterbrush	840	1,415	64 ± 11	Ash/colluvium	Basalt, tuff	Uplands, mountain slopes
15	Udivitrands	3202	23	860 ± 405	5.2 ± 1.0	Udic	Mesic, frigid	Grand fir, western larch, lodgepole pine	910	1,800	72 ± 21	Ash/colluvium		Mountain slopes
16	Haploxerepts	2852	39	875 ± 400	8.9 ± 2.3	Xeric	Mesic, frigid	Douglas-fir, ponderosa pine, Oregon white oak	500	1,125	68 ± 31	Colluvium/lithic	Basalt, andesite	Mountain slopes, hillslopes
17	Palehumults	2780	22	1,775 ± 525	10 ± 1.2	Udic, xeric	Mesic, isomesic, frigid	Douglas-fir, western hemlock, red alder, western redcedar	100	775	68 ± 26	Colluvium, residuum	Basalt	Hillslopes
18	Dystrudepts	2614	45	2,250 ± 500	10 ± 0.8	Udic	Mesic, frigid, isomesic	Douglas-fir, grand fir, bigleaf maple, red alder, western hemlock	100	900	81 ± 10	Colluvium, residuum/paralithic or lithic	Basalt, tuff	Mountain slopes
19	Hapludands	2572	18	2,425 ± 200	8.3 ± 1.5	Udic	Frigid, mesic	Douglas-fir, western hemlock, western redcedar	225	750	87 ± 7.0	Colluvium, residuum/lithic	Basalt	Mountain slopes
20	Paleargids	2054	17	260 ± 40	8.3 ± 1.5	Aridic	Frigid, mesic	Wyoming big sagebrush, needlegrass, bluegrass	1,225	1,675	20 ± 18	Lacustrine ecolluvium	Basalt, tuff	Take terraces plateaus
21	Eutrudepts	1868	12	1,880 ± 665	9.0 ± 2.8	Udic	Mesic, frigid	Douglas-fir, bigleaf maple, red alder	320	1,050	83 ± 15	Colluvium, residuum	Sandstone, siltstone	Mountain slopes
22	Fulvudands	1720	19	2,380 ± 370	9.0 ± 1.6	Udic	Isomesic	Sitka spruce, western hemlock, red alder, Douglas-fir	10	900	88 ± 6.5	Ash, colluvium, residuum	Basalt	Mountain slopes
23	Torripsamments	1477	14	240 ± 25	8.1 ± 1.3	Torric	Mesic, frigid	Sagebrush, Idaho fescue	1,200	1,400	29 ± 14	Ash, eolian, lacustrine	Mixed	Lava plateaus, dunes
24	Endoaquolls	1442	37	850 ± 500	10 ± 1.8	Aquic	Mesic, frigid	Oregon white oak, black cottonwood, Oregon ash, sedges	425	925	3 ± 1	Alluvium	Mixed	Alluvial plains
25	Torriorthents	1222	24	295 ± 160	9.4 ± 1.6	Aridic	Mesic, frigid	Sagebrush, needlegrass, bluegrass, wheatgrass	1,585	825	46 ± 26	Colluvium, residuum	Mixed	Hillslopes, terraces, alluvial fans

(continued)

Table 7.1 (continued)

Soil region	Soil great group	Area (km^2)	No. soil series	MAP (mm/yr)	MAAT (°C)	Soil-moisture regime	Soil temperature regime	Vegetation type	Lower elev. (m)	Upper elev. (m)	Upper slope (%)	Parent materials	Composition	Landforms
26	Haplocryands	1147	15	2,300 ± 675	5.1 ± 1.5	Udic	Cryic	Noble fir, mountain hemlock, Douglas-fir, Pacific silver fir	1,685	925	78 ± 16	Colluvium, residuum	Basalt, andesite	Mountain slope
27	Argialbolls	1029	9	850 ± 325	9.6 ± 1.8	Xeric, aquic	Mesic	Ponderosa pine, Douglas-fir, Oregon white oak, grasses	600	385	14 ± 13	Glaciolacustrine, alluvium, loess	Mixed	Terrace, plateau
28	Palexeralfs	1015	15	850 ± 500	9.7 ± 2.4	Xeric	Mesic, frigid	Douglas-fir, Oregon white oak	475	1,075	51 ± 45	Colluvium	Mixed, siltstone	Hillslopes, foothills

Table 7.2 Thicknesses (cm) of diagnostic horizons in major great groups of Oregon

No	Great group	Mollic	Ochric	Umbric	Cambic	Argillic	Duripan	Albic	Calcic	None (%)
1	Haploxerolls	53 ± 29			43 ± 24 (48%)					52
2	Argixerolls	47 ± 24				53 ± 33				
3	Humudepts			44 ± 18	60 ± 33 (82%)					18
4	Haplargids		14 ± 9.3			36 ± 24				
5	Argidurids		16 ± 8.1			30 ± 14	38 ± 34			
6	Vitricryands		22 ± 18 (90%)	37 ± 3.4 (8.3%)	50 ± 27 (44%)	79 ± 17 (15%)				50
7	Haplocambids		18 ± 9.5		38 ± 20					
8	Palexerolls	45 ± 16				65 ± 33				
9	Haploxeralfs		24 ± 13			72 ± 38				
10	Haplodurids		16 ± 6.4		28 ± 10 (50%)		38 ± 34		15 ± 0 (14%)	
11	Dystroxerepts		19 ± 9.4		49 ± 27 (94%)					6
12	Durixerolls	33 ± 15			35 ± 14 (14%)	33 ± 15 (78%)	41 ± 30			
13	Haplohumults		14 ± 4.2 (23%)	41 ± 17 (77%)		72 ± 32				
14	Vitrixerands	36 ± 11 (25%)	23 ± 16 (65%)	43 ± 11 (10%)	58 ± 28 (73%)	69 ± 29 (27%)				14
15	Udivitrands		19 ± 16		51 ± 15 (87%)	52 ± 17 (48%)				
16	Haploxerepts		19 ± 11		47 ± 23 (100%)					
17	Palehumults		28 ± 15 (68%)	32 ± 8.1 (32%)		108 ± 40				
18	Dystrudepts		21 ± 10		73 ± 37 (100%)					
19	Hapludands		28 ± 15 (28%)	48 ± 26 (72%)	72 ± 36 (100%)					
20	Paleargids		16 ± 6.7			46 ± 24				
21	Eutrudepts		17 ± 12		42 ± 21 (100%)					
22	Fulvudands			60 ± 36	56 ± 35 (89%)					11
23	Torripsamments		18 ± 6.8							100
24	Endoaquolls	61 ± 32			49 ± 19 (19%)					81
25	Torriorthents		17 ± 3.2							100
26	Haplocryands		24 ± 13 (60%)	60 ± 29 (40%)	61 ± 30 (87%)					
27	Argialbolls	39 ± 17				62 ± 24		15 ± 5.3		13
28	Palexeralfs		19 ± 6.5			98 ± 44				

Table 7.3 Horizons and soil properties that distinguish among soil great groups in Oregon

Soil region	Great group	Epipedon surface horizon	Diagnostic sub-	Soil mois-ture regime	Other definitive properties
1	Haploxerolls	Mollic	Cambic (53%), none (30%)	Xeric, aridic	
2	Argixerolls	Mollic	Argillic	Xeric, aridic	
3	Humudepts	Umbric	Cambic (96%)	Udic	
4	Haplargids	Ochric	Argillic	Aridic	
5	Argidurids	Ochric	Argillic, duripan	Aridic	
6	Vitricryands	Ochric or umbric	Cambic (44%), argillic (15%), none (50%)	Xeric, udic	Andic properties, cryic STR
7	Haplocambids	Ochric	Cambic	Aridic	
8	Palexerolls	Mollic	Argillic	Xeric, aridic	Pale-
9	Haploxeralfs	Ochric	Argillic (100%), cambic (13%)	Xeric	
10	Haplodurids	Ochric	Duripan (100%), cambic (57%)	Aridic	
11	Dystroxerepts	Ochric	Cambic (92%), none (8%)	Xeric	Dystr-
12	Durixerolls	Mollic	Duripan (100%), argillic (72%),cambic (16%)	Xeric	
13	Haplohumults	Umbric or ochric	Argillic	Udic, xeric	
14	Vitrixerands	Ochric or mollic	Cambic (69%) and/or argillic (38%), none (14%)	Xeric	Andic properties
15	Udivitrands	Ochric	Cambic (82%) and/or argillic (59%)	Udic	Andic properties
16	Haploxerepts	Ochric	Cambic (100%)	Xeric	
17	Palehumults	Ochric or umbric	Argillic	Udic, xeric	Pale-
18	Dystrudepts	Ochric	Cambic (100%)	Udic	Dystr-
19	Hapludands	Umbric or ochric	Cambic (100%)	Udic	andic properties
20	Paleargids	Ochric	Argillic	Aridic	Pale-
21	Eutrudepts	Ochric	Cambic (100%)	Udic	Eutr-
22	Fulvudands	Umbric	Cambic (85%), none (15%)	Udic	andic properties
23	Torripsamments	Ochric	None (100%)	Torric	
24	Endoaquolls	Mollic	Cambic (24%), none (76%)	Aquic	
25	Torriorthents	Ochric	None (100%)	Aridic	
26	Haplocryands	Ochric or umbric	Cambic (100%)	Udic	Andic properties, cryic STR
27	Argialbolls	Mollic	Argillic	Xeric	Albic
28	Palexeralfs	Ochric	Argillic (100%), cambic (17%)	Xeric	Pale-

Olot—composition is 50%, soil order is Andisols, and great group is Vitrixerands.

Klicker—composition is 30%, soil order is Mollisols, and great group is Argixerolls.

Anatone—composition is 10%, soil order is Mollisols, and great group is Haploxerolls.

In this example 50% of the area is in the Andisol soil order and 40% of the area is in the Mollisol soil order. In assigning this map unit to a soil order for map development, all of its map unit polygons are identified as Andisols on the Andisol order map, and Mollisols on the Mollisol order map. Thus, the area this map unit covers will appear on both the Andisol and Mollisol order maps.

In this same example, 50% of the area is in the Vitrixerands great group, 40% is in the Argixerolls great group and 10% is in the Haploxerolls great group. In assigning this map unit to a great group, all of its map unit polygons are identified as Vitrixerands on the Vitrixerands great group map, Argixerolls on the Argixerolls great group map. No areas of this map unit will be shown on the Haploxerolls great group map as this component only accounts for 10% of the area. A component composition of 20 percent or more

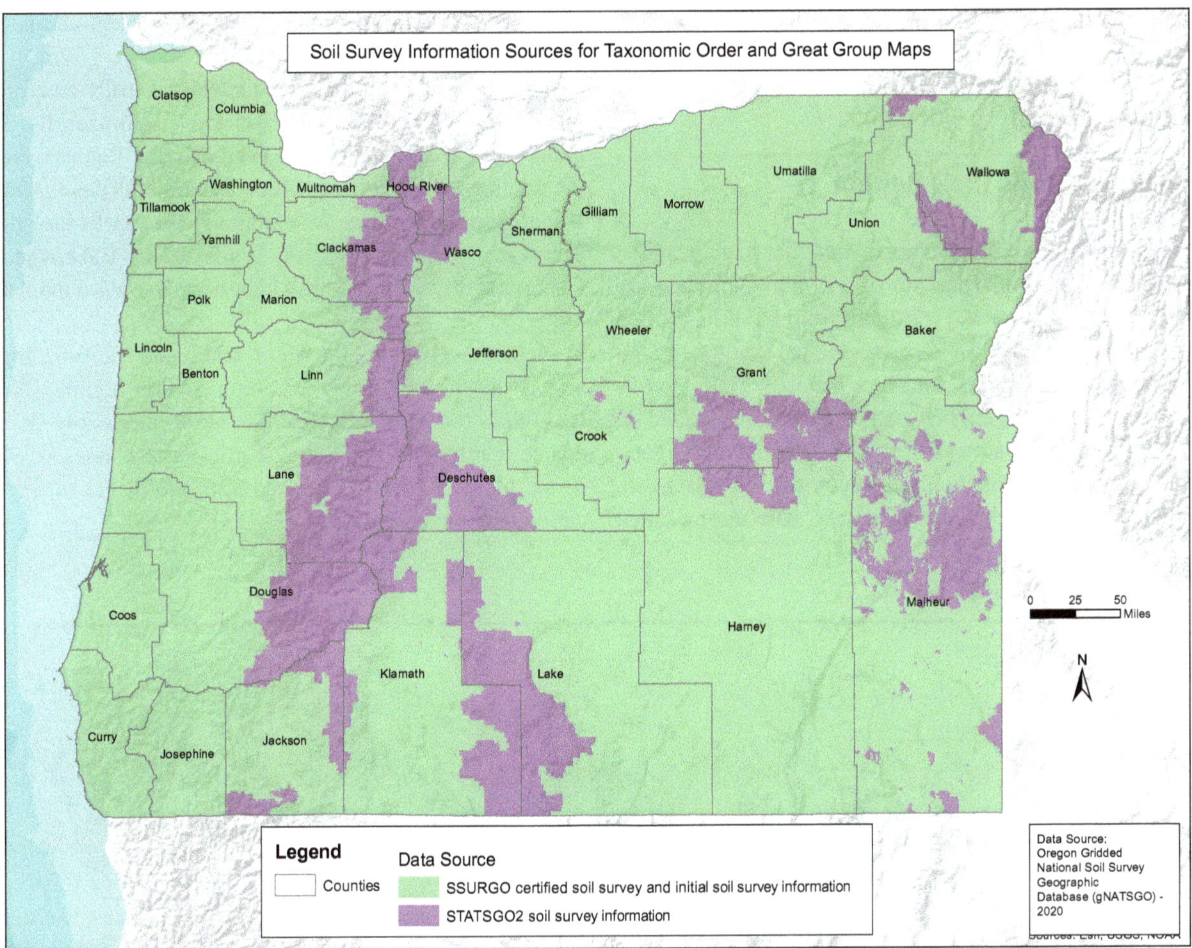

Fig. 7.1 Soil survey information sources for taxonomic order and great group maps. *Source* Prepared by Whityn Owen

was selected as a minimum threshold value in order for the map unit to be included on the order and great group maps. Thus, the area this map unit covers will only appear on two great group maps.

STATSGO2 map units can have up to twenty-one components as opposed to one to three components for the certified and initial mapping map units. Each component has a percent composition and soil classification same as the certified and initial mapping map units. However, in this case the dominant soil order and dominant soil great group of all the components was used to assign the STATSGO2 map unit to a soil order and soil great group map.

EXAMPLE map unit for soil order assignment-if the STATSGO2 map unit was composed of 18 components;

6 of those components classified as Andisols with a combined percentage of 35%,

6 of the components classified as Spodosols with a combined percentage of 20%,

3 of the components classified as Inceptisols with a combined percentage of 25%,

2 of the components classified as Alfisols with a combined percentage of 15%,

1 of the components classified as Histosols with a percentage of 5%.

In this example the STATSGO2 map unit was assigned to the Andisol order as it was the soil order having the highest combined percentage of the 18 components, also known as dominant condition. The same placement procedure of dominant condition was used for assigning the map unit to a soil great group. So, of the 18 components, the dominant great group (great group with highest percent composition) was used to assign the map unit to a great group. In using

STATSGO2 mapping the individual map unit area will only appear on one soil order map and one soil great group map.

7.2 Haploxerolls (Soil Region 1)

Haploxerolls are weakly developed soils that have a mollic epipedon over a cambic horizon or a C-horizon. They are derived from colluvium and residuum from basalt and often contain loess and volcanic ash. Major landforms are mountain slopes, hillslopes, and lava plateaus. Haploxerolls occur at elevations ranging between 650 and 1,400 m and on maximum slopes averaging 57 ± 33%. The native vegetation on Haploxerolls commonly is Wyoming big sagebrush, Idaho fescue, and bluebunch wheatgrass. The mean annual air temperature is 8.4 ± 1.7 °C, and the mean annual precipitation is 450 ± 275 mm.

There are 279 soil series in the Haploxerolls great group in Oregon which covers 27,700 km^2. Haploxerolls occur mainly in the Blue Mountain Foothills, the Palouse Prairie, the Malheur High Plateau, the Columbia Plateau, and the Willamette Valley (Fig. 7.2). Major Haploxerolls include the Lickskillet, Anatone, Bocker, Walla Walla, Bakeoven, and Condon soil series, each of which occupies more than 1,000 km^2.

Haploxerolls have a xeric or aridic soil moisture regime and a mesic or frigid soil temperature regime. Nearly two-thirds (60%) of Haploxerolls are in the loamy-skeletal, coarse-loamy, and fine-loamy particle-size classes, 76% are in the mixed mineral class, and 72% are in the superactive cation-exchange activity class.

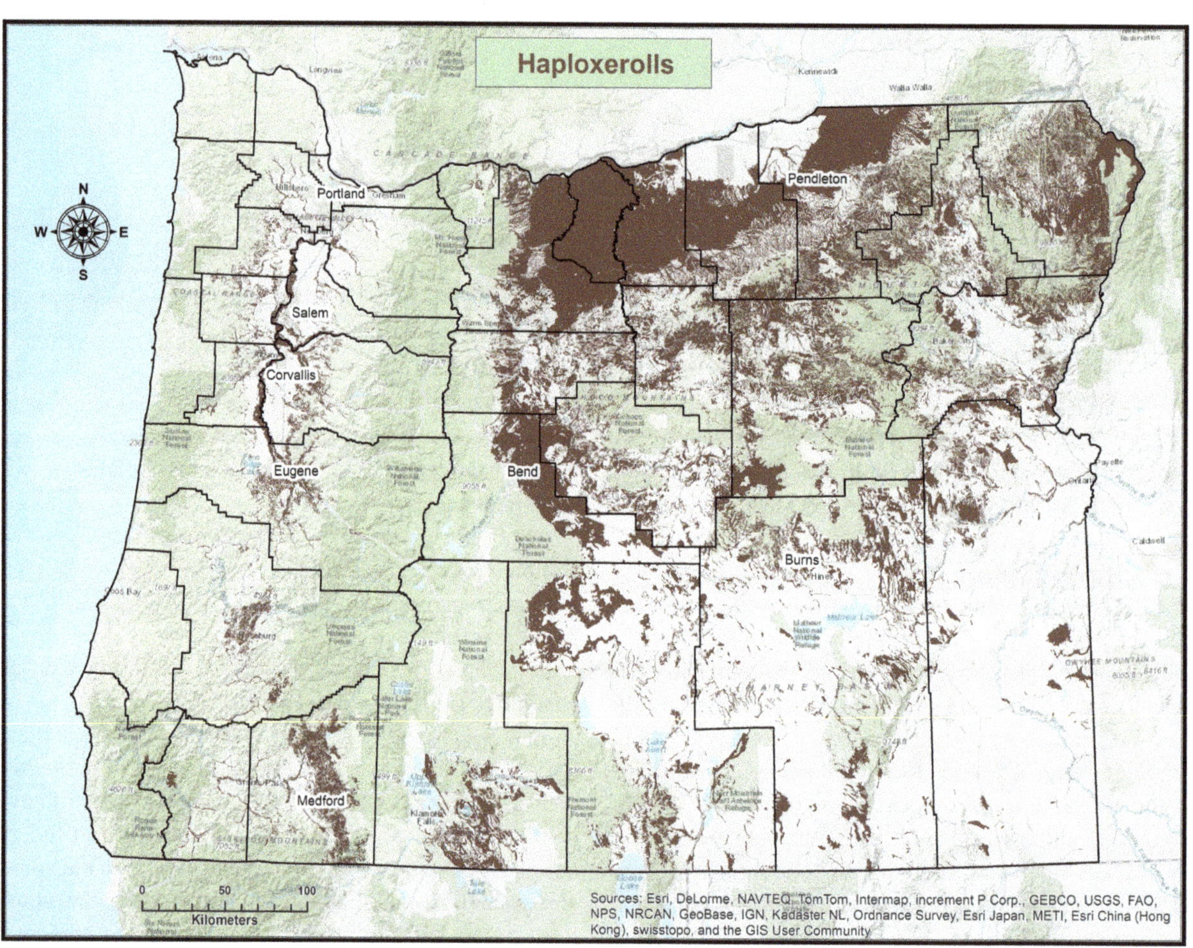

Fig. 7.2 Distribution of Haploxerolls in Oregon. *Source* Prepared by Whityn Owen

Fig. 7.3 The Josset soil series, a coarse-loamy over sandy or sandy-skeletal, mixed, superactive, frigid Cumulic Haploxerolls from Wallowa County, Oregon. The soil is derived from alluvium on a flood plain. The scale is in feet. *Source* NRCS photo

In Haploxerolls, the mollic epipedon averages 53 ± 29 cm and the cambic horizon 43 ± 24 cm in thickness. About one-half (52%) of Haploxerolls lack a diagnostic subsurface horizon. A coarse-loamy over sandy or sandy-skeletal, mixed, superactive, frigid Cumulic Haploxerolls, the Josset soil series has formed in mixed alluvium on floodplains in the Wallowa Lake area of northeastern Oregon (Fig. 7.3). The soil contains a dark brown mollic epipedon that is 27 cm thick over dark grayish brown and multicolored C-horizon materials.

7.3 Argixerolls (Soil Region 2)

Argixerolls are moderately well-developed soils that contain a mollic epipedon over an argillic horizon and sometimes a calcic horizon. They are derived from colluvium and residuum from basalt and often contain loess and/or volcanic ash. Major landforms are mountain slopes and lava plateaus. Argixerolls occur at elevations ranging between 700 and 1,400 m and on maximum slopes averaging 63 ± 26%.

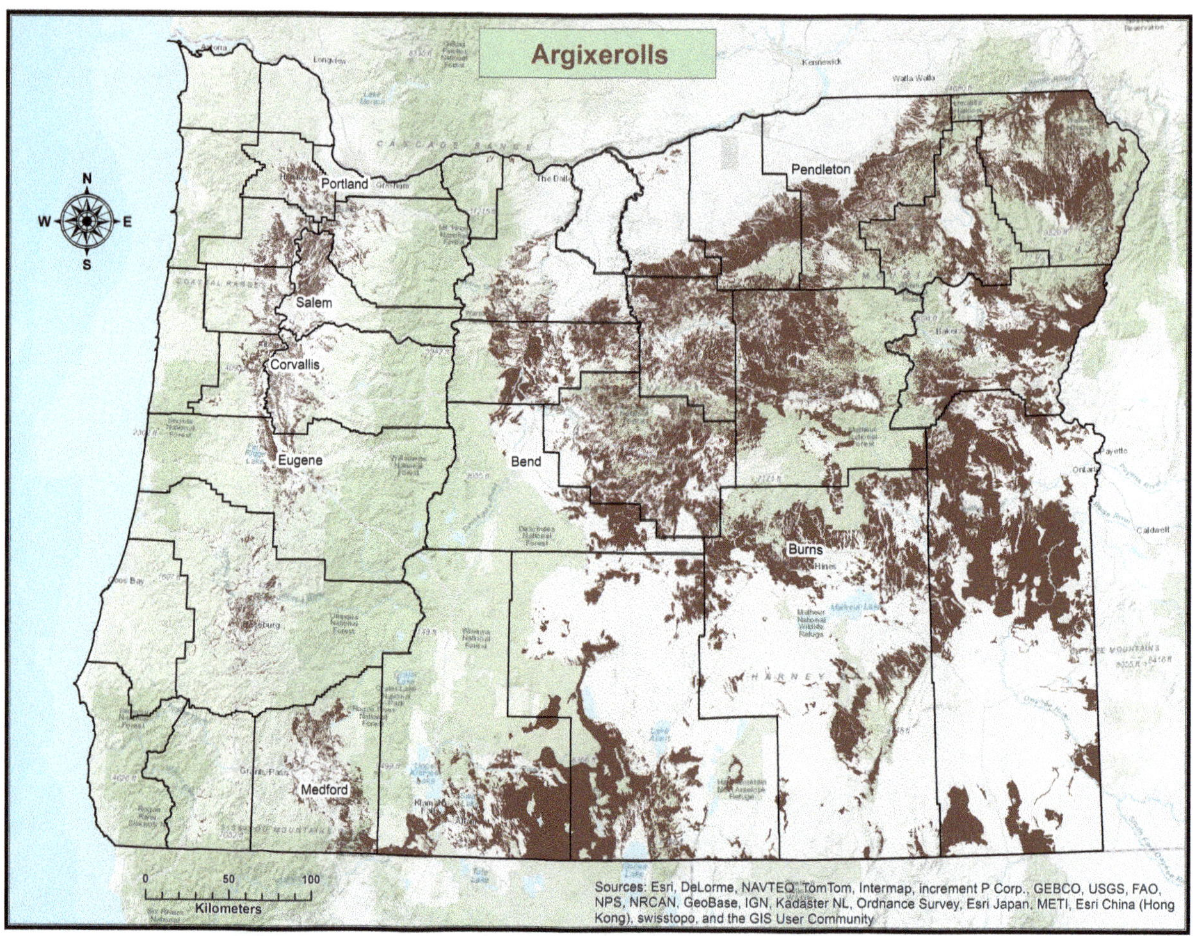

Fig. 7.4 Distribution of Argixerolls in Oregon. *Source* Prepared by Whityn Owen

The native vegetation on Argixerolls commonly is sagebrush, bluebunch wheatgrass, Idaho fescue, Sandberg bluegrass, and ponderosa pine in eastern Oregon. In western Oregon the native vegetation is Oregon white oak, Douglas-fir, and grasses. The mean annual air temperature is 8.1 ± 1.7 °C, and the mean annual precipitation is 485 ± 285 mm.

There are 236 soil series in the Argixerolls great group in Oregon which covers nearly 25,800 km^2. Argixerolls are most common in the Blue Mountain Foothills, the Palouse Prairie, the Malheur High Plateau, the Klamath Basin, the Siskiyou Mountains, the Willamette Valley, and the Owyhee High Plateau (Fig. 7.4). Major Argixerolls include the Ninemile, Ateron, Klicker, Gwinly, Merlin, and Woodburn soil series, each of which occupies more than 900 km^2.

Argixerolls typically have a xeric soil moisture regime but can have an aridic soil moisture regime and a mesic or frigid soil temperature regime. More than three-quarters (83%) of Argixerolls are in the clayey-skeletal, fine-loamy, fine, and loamy-skeletal particle-size classes; 87% are in the mixed or smectitic mineral class; and 45% are in the superactive cation-exchange activity class.

In Argixerolls, the mollic epipedon averages 47 ± 24 cm and the argillic horizon averages 43 ± 24 cm in thickness. About one-quarter of Argixerolls also have an accumulation of secondary calcium carbonates or a

Fig. 7.5 The Dixonville soil series, a fine, mixed, superactive, medic Pachic Ultic Argixerolls from Benton County, Oregon. The soil is derived from colluvium and residuum over a basalt paralithic contact. The scale is in inches. *Source* NRCS photo

calcic horizon that averages 40 ± 35 cm in thickness. A fine, mixed, superactive, mesic Pachic Ultic Argixerolls, the Dixonville soil series is formed in clayey colluvium and residuum derived from basalt on hillslopes in the southern Willamette Valley (Fig. 7.5). The soil contains a dark brown mollic epipedon that is 86 cm thick over a reddish brown argillic horizon that is 56 cm thick, overlying moderately cemented basalt.

7.4 Humudepts (Soil Region 3)

Humudepts are weakly developed soils that contain an umbric epipedon over a cambic horizon. They are derived from colluvium and residuum over bedrock. Major landforms are hillslopes. Humudepts occur at elevations ranging between 100 and 825 m and on maximum slopes averaging 86 ± 9%. The native vegetation on Humudepts commonly is Douglas-fir, western hemlock, and western redcedar. The mean annual air temperature is 9.2 ± 1.1 °C, and the mean annual precipitation is 2,100 ± 200 mm.

The Humudepts great group in Oregon has 84 soil series which covers over 10,600 km^2. Humudepts are most common in the Coast Range, the Cascade Mountains, and the Sitka Spruce and Coastal Redwood Belts (Fig. 7.6). Major Humudepts are the Bohannon, Preacher, Klickitat, Kinney, Templeton, and Rinearson soil series, each of which occupies more than 400 km^2.

Humudepts have a udic soil moisture regime and an isomesic, mesic, or frigid soil temperature regime. Nearly two-thirds (60%) of Humudepts are in the fine-loamy or loamy-skeletal particle-size classes; and 83% are in the isotic mineral class.

In Humudepts the umbric epipedon averages 44 ± 18 cm and the cambic horizon 60 ± 33 cm in thickness. About 18% of Humudepts lack a diagnostic subsurface horizon. A fine-silty, isotic, isomesic Andic Oxyaquic Humudepts, the Walluski soil series has formed in alluvium on terraces along the northern Oregon coast (Fig. 7.7). The soil contains a dark grayish brown umbric epipedon that is 36 cm thick

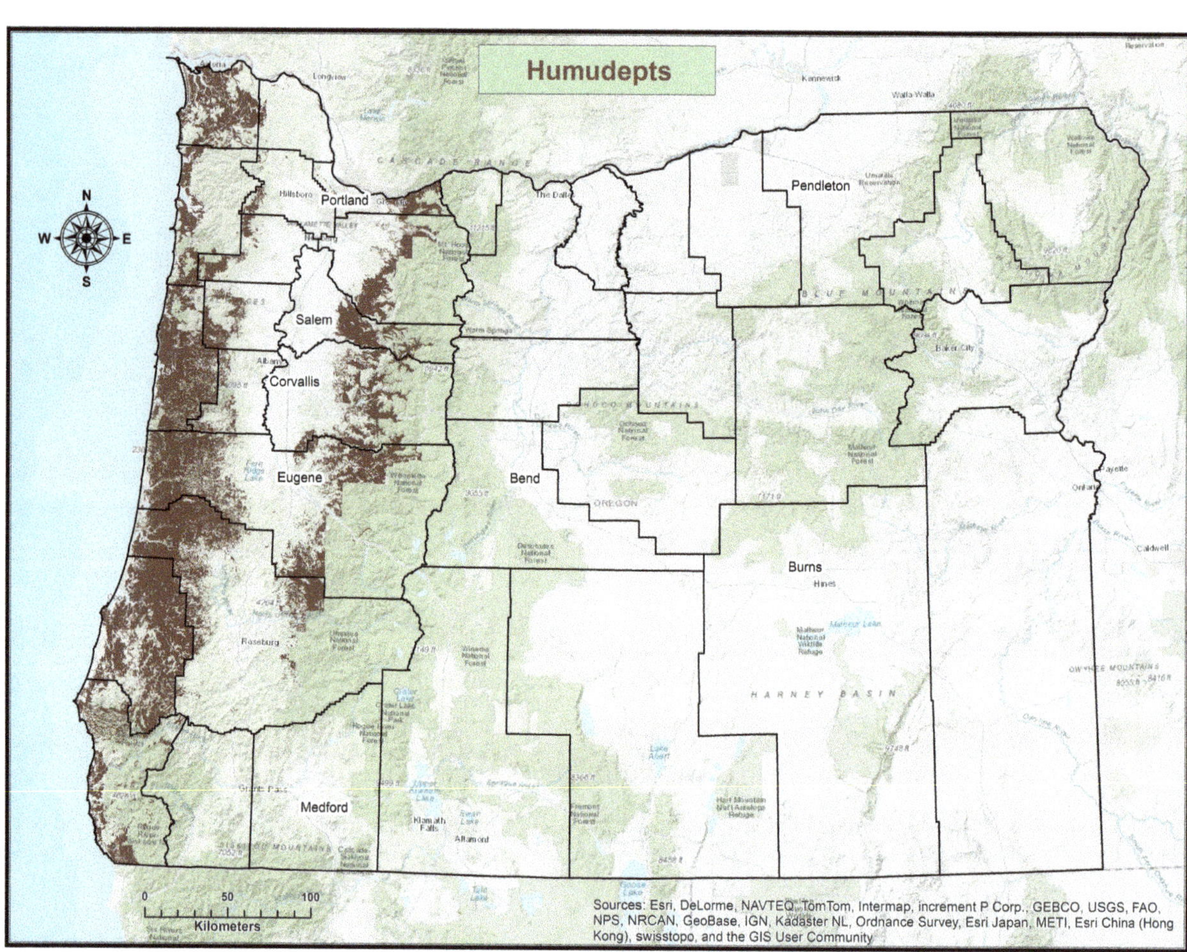

Fig. 7.6 Distribution of Humudepts in Oregon. *Source* Prepared by Whityn Owen

Fig. 7.7 The Walluski soil series is a fine-silty, isotic, isomesic Andic Oxyaquic Humudepts from Tillamook County, Oregon. The soil is derived from alluvium on terraces. The scale is in feet. *Source* NRCS photo

over a pale brown to light yellowish brown cambic horizon that is 84 cm thick.

7.5 Haplargids (Soil Region 4)

Haplargids are moderately well-developed soils containing an ochric epipedon over an argillic horizon. Haplargids are derived from colluvium and residuum on mountain slopes and lacustrine deposits on lake terraces. They support Wyoming big sagebrush, Idaho fescue, bluebunch wheatgrass, and Thurber's needlegrass. The mean annual air temperature for Haplargids is 7.3 ± 1.8 °C, and the mean annual precipitation is 265 ± 33 mm. Haplargids are predominantly at elevations between 1,140 and 1,650 m. The maximum slope averages 30 ± 29%.

The Haplargids great group contains 55 soil series in Oregon which covers 6,800 km^2. Haplargids occur mainly on the Malheur High Plateau and the Owyhee High Plateau (Fig. 7.8). The most extensive Haplargids are the Anawalt, Nevador, Robson, and Igert soil series, each of which occupies more than 500 km^2.

Haplargids have an aridic soil moisture regime and a mesic or frigid soil temperature regime. Haplargids are in a variety of particle-size classes, particularly loamy-skeletal, clayey-skeletal, and fine. They are equally distributed in the mixed and smectitic mineral classes, and about half of the Haplargids are in the superactive cation-exchange activity class.

Haplargids have an ochric epipedon over an argillic horizon that averages 36 ± 24 cm in thickness. In nearly two-thirds (62%) of Haplargids, a lithic or paralithic contact occurs in the upper 100 cm.

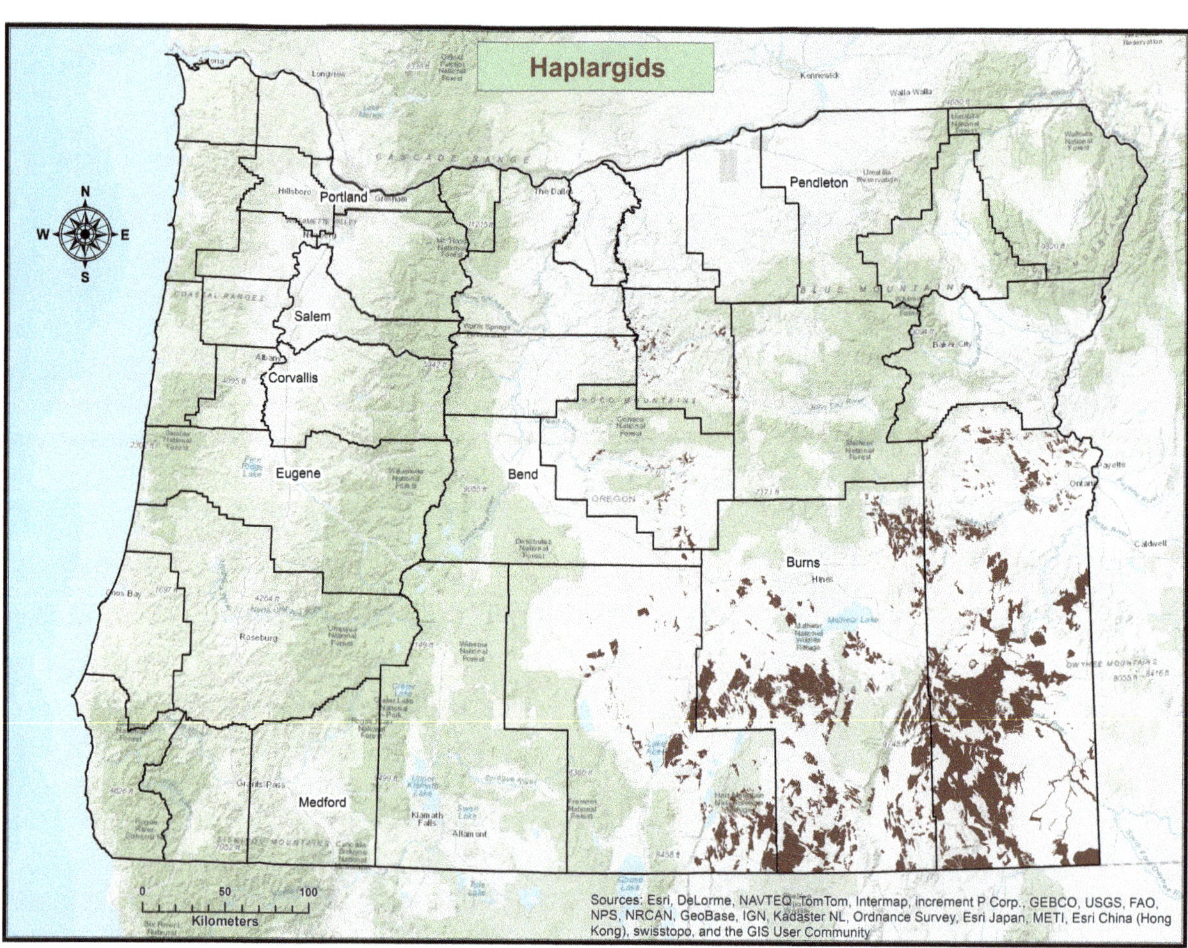

Fig. 7.8 Distribution of Haplargids in Oregon. *Source* Prepared by Whityn Owen

7.6 Argidurids (Soil Region 5)

Argidurids are well-developed soils with an ochric epipedon and an argillic horizon over a duripan. They have formed in alluvium on alluvial fans and colluvium on lava plateaus. Argidurids occur at elevations ranging between 1,150 and 1,600 m and on maximum slopes averaging 24 ± 15%. The native vegetation on Argidurids is Wyoming big sagebrush, bluebunch wheatgrass, and Thurber's needlegrass. The mean annual air temperature is 7.8 ± 0.8 °C, and the mean annual precipitation is 240 ± 40 mm.

The Argidurids great group in Oregon has 33 soil series which cover 6,700 km^2. Argidurids are most common on the Malheur High Plateau and the Owyhee High Plateau (Fig. 7.9). Major Argidurids include the Brace, Deppy, Floke, Ratto, and Actem soil series.

Argidurids have an aridic soil moisture regime and a mesic or frigid soil temperature regime. Argidurids occur in a variety of fine and loamy particle-size classes, mixed or smectitic mineral classes, and more than one-half (57%) are in the superactive cation-exchange activity class.

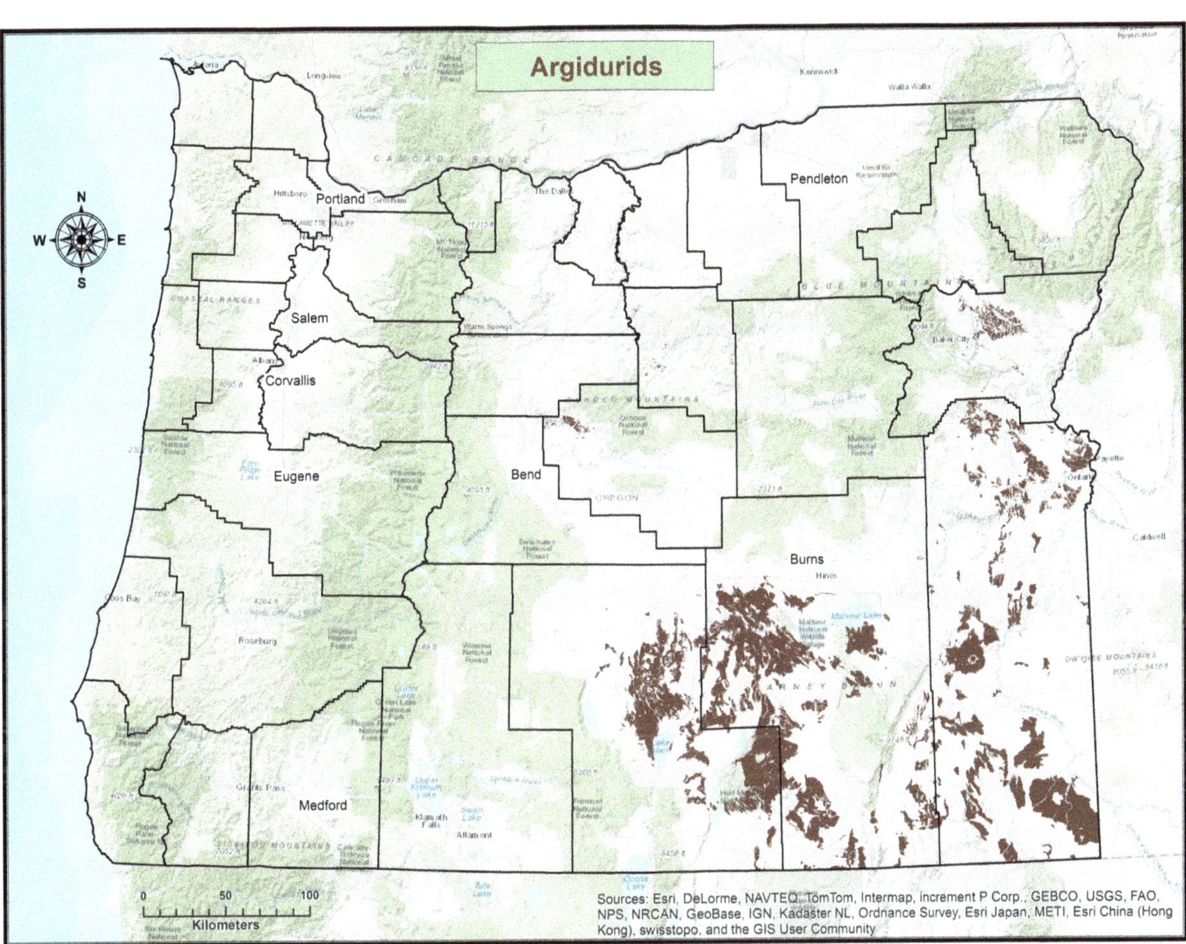

Fig. 7.9 Distribution of Argidurids in Oregon. *Source* Prepared by Whityn Owen

Fig. 7.10 The Chilcott soil series, a fine, smectitic, mesic Abruptic Xeric Argidurids, is derived from loess and volcanic ash over alluvium and occurs in Malheur County. The profile contains a light brownish gray E horizon (ochric epipedon) over a brown argillic horizon to about 75 cm that are underlain by an indurated duripan with siliceous laminae. The scale is in decimeters. *Source* Photo by Chad McGrath

Argidurids have an ochric epipedon over an argillic horizon that averages 30 ± 14 cm and a duripan that averages 38 ± 34 cm in thickness. The Chilcott soil series, a fine, smectitic, mesic Abruptic Xeric Argidurids, is derived from loess and volcanic ash over alluvium (Fig. 7.10). The profile contains a light brownish gray E horizon (ochric epipedon) over a brown argillic horizon to about 75 cm that are underlain by an indurated duripan with siliceous laminae.

7.7 Vitricryands (Soil Region 6)

Vitricryands are moderately well-developed soils that contain an ochric epipedon and a cambic subsurface horizon. They are derived from volcanic ash and pumice over colluvium or till on mountain slopes and lava plateaus. Vitricryands occur at elevations ranging between 1,450 and

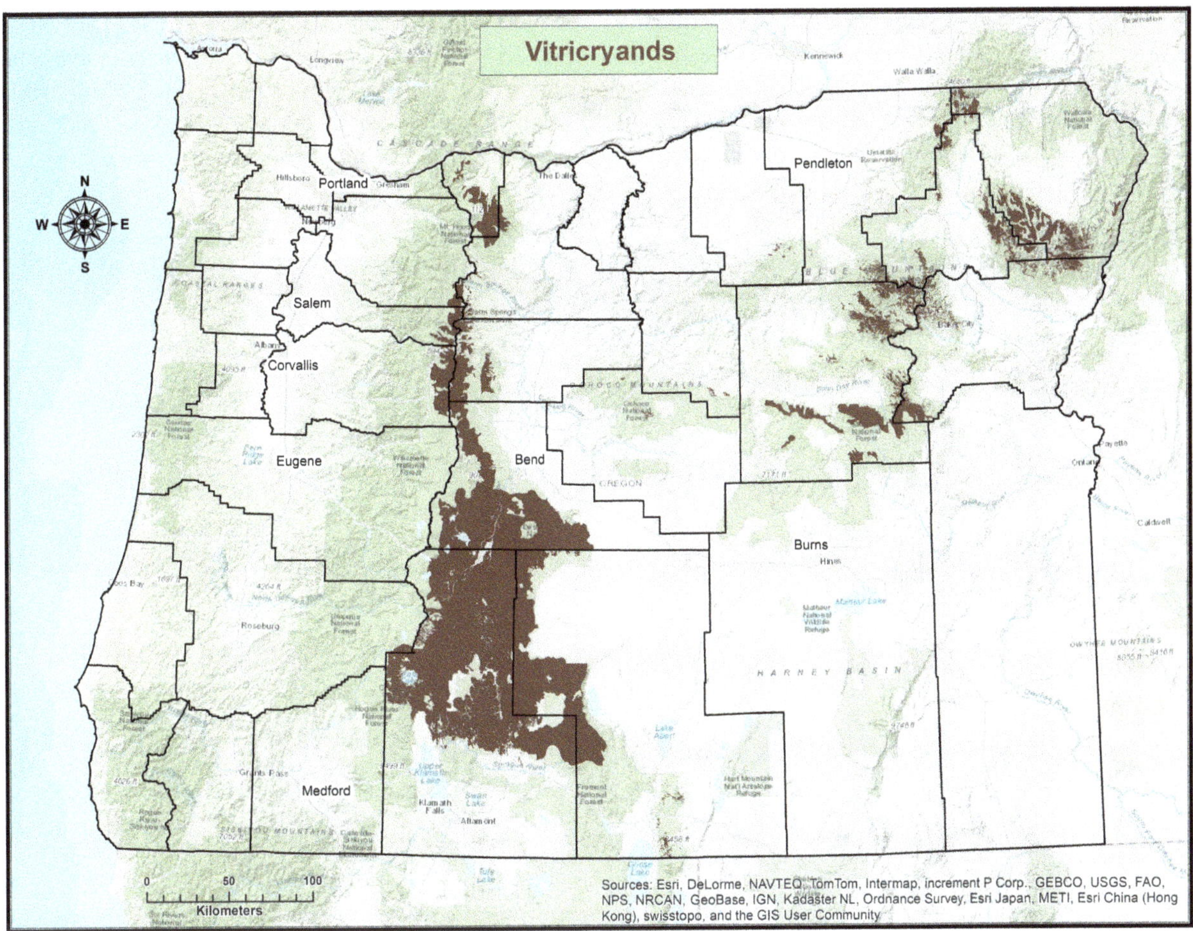

Fig. 7.11 Distribution of Vitricryands in Oregon. *Source* Prepared by Whityn Owen

2,160 m and on maximum slopes averaging 72 ± 17%. The native vegetation on Vitricryands is subalpine fir, Engelmann spruce, ponderosa pine, and lodgepole pine. The mean annual air temperature is 4.0 ± 1.6 °C, and the mean annual precipitation is 965 ± 255 mm.

The Vitricryands great group in Oregon has 48 soil series which cover 6,500 km^2. Vitricryands are most common in the Cascade Mountains—Eastern Slope, and the Blue Mountains (Fig. 7.11) The most extensive Vitricryands in Oregon is the Lapine soil series (2,500 km^2); the Shukash, Steiger, Collier, and Shanahan soil series each occupies more than 300 km^2.

Vitricryands have a xeric or udic soil moisture regime and a cryic soil temperature regime. More than three-quarters (88%) of Vitricryands are in ashy over loamy or loamy-skeletal, ashy, and ashy-skeletal particle-size classes. They are entirely in amorphic, glassy, and amorphic over isotic mineral classes.

An ochric epipedon over a cambic horizon occurs in 44% of Vitricryands, averaging 50 ± 27 cm. About 15% of the Vitricryands have an argillic horizon, i.e., are in the Alfic subgroup, that averages 79 ± 15 cm in thickness. About 50% of Vitricryands lack a diagnostic subsurface horizon. The Lapine soil series, an ashy-pumiceous, glassy Xeric

Vitricryands, has formed in deep Mazama ash and pumice deposits in Deschutes and Klamath Counties. The profile contains an ochric epipedon and Bw horizons composed of ashy loamy coarse sand over a buried soil (Fig. 7.12). The Castlecrest soil series, an ashy, amorphic Typic Vitricryands, has formed in Mazama ash near Crater Lake, Oregon (Fig. 7.13). The profile contains very dark grayish brown ochric epipedon (0–7.5 cm), a brown Bw horizon that is too coarse to be a cambic horizon, and very dark gray C horizons.

Fig. 7.12 The Lapine soil series, an ashy-pumiceous, glassy Xeric Vitricryands, is derived from grayish Mazama ash in Crater Lake National Park. A buried soil exists at a depth of 1 m. The scale is in meters. *Source* NRCS photo

Fig. 7.13 The Castlecrest soil series, an ashy, amorphic Typic Vitricryands, is derived from thick Mazama ash at Crater Lake National Park in Oregon. The scale is in feet. *Source* NRCS photo

7.8 Haplocambids (Soil Region 7)

Haplocambids are weakly developed soils that contain an ochric epipedon over a cambic horizon. They are derived from colluvium over bedrock or lacustrine materials on lake terraces. They occur at elevations between 1,250 and 1,850 m and on maximum slopes averaging 30 ± 25%. Haplocambids feature bunchgrass prairie and desert shrub vegetation. Haplocambids in Oregon receive 250 ± 20 mm of annual precipitation and have a mean annual air temperature 6.8 ± 0.9 °C.

There are 61 Haplocambids soil series in Oregon which cover 5,500 km^2. Haplocambids are most common on the Malheur High Plateau, the Owyhee High Plateau, the Snake River Plain, the Humboldt Area, and the Columbia Plateau (Fig. 7.14). The most common soil series are the Felcher, Enko, Lonely, Warden, Shano, and McConnel.

Haplocambids have an aridic soil moisture regime and a mesic or frigid soil temperature regime. Nearly two-thirds (61%) of the Haplocambids are in the coarse-loamy, fine-loamy, and loamy-skeletal particle-size classes; 88% are in the mixed mineral class; 78% are in the superactive cation-exchange activity class.

Haplocambids have an ochric epipedon over a cambic horizon that averages 38 ± 20 cm in thickness. The Shano soil series, a coarse-silty, mixed, superactive, mesic Xeric Haplocambids, is derived from loess and occurs in the Columbia Basin and on the Columbia Plateau. The photograph of the Shano soil series shows a dark grayish brown ochric epipedon to 20 cm and a brown cambic horizon to 84 cm (Fig. 7.15). The subsoil is enriched in secondary carbonates.

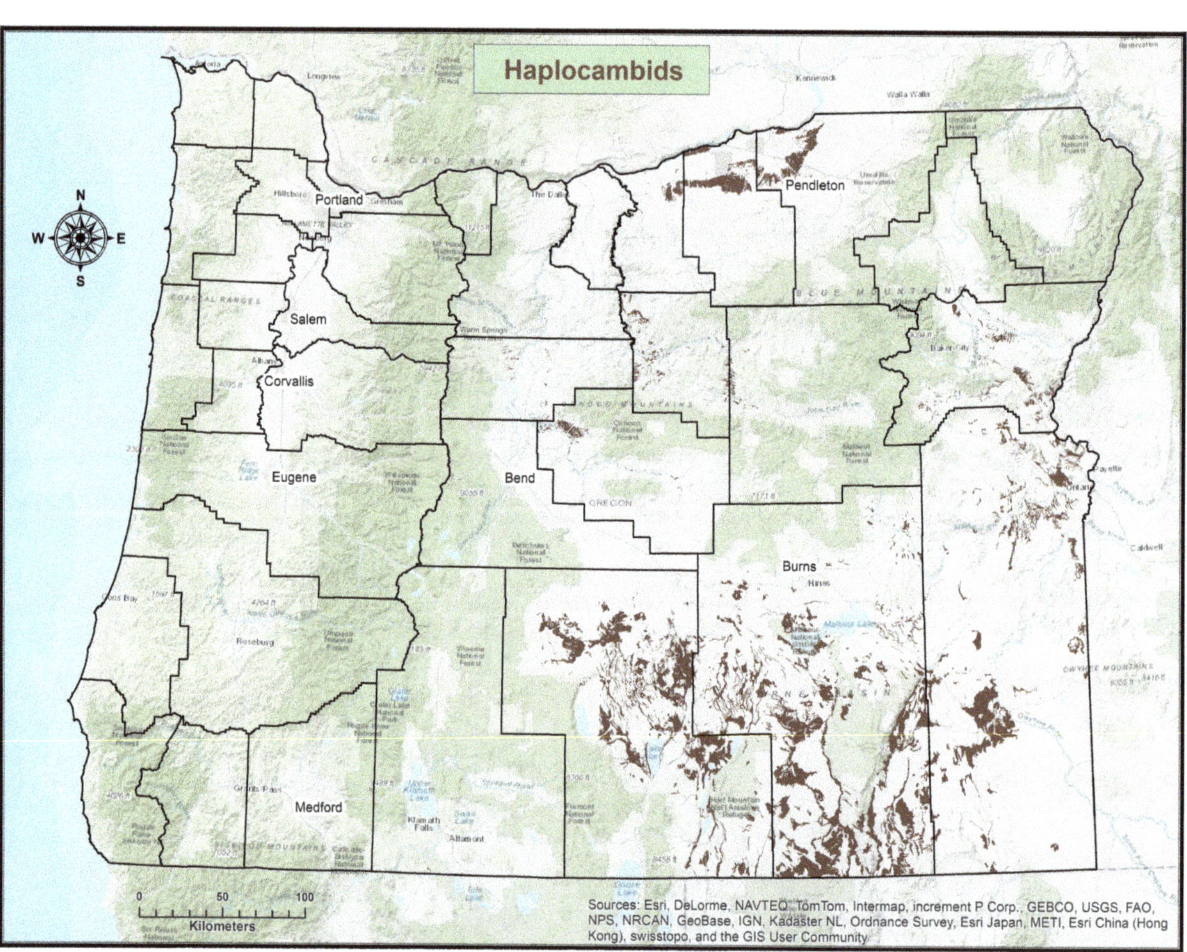

Fig. 7.14 Distribution of Haplocambids in Oregon. *Source* Prepared by Whityn Owen

Fig. 7.15 The Shano soil series, a coarse-silty, mixed, superactive, mesic Xeric Haplocambids, is derived from loess and occurs in the Columbia Basin and Columbia Plateau MLRA's. The Shano soil series has a dark grayish brown ochric epipedon to 8 in. (20 cm) and a brown cambic horizon to 33 in (84 cm). The soil is derived from loess on loess plateaus. The scale is in inches. *Source* NRCS photo

7.9 Palexerolls (Soil Region 8)

Palexerolls are well-developed soils with a mollic epipedon over an argillic horizon. They are derived from colluvium and residuum from tuff on mountain slopes and lava plateaus. Palexerolls occur at elevations ranging between 1,065 and 1,625 m and on maximum slopes averaging 53 ± 25%. The native vegetation on Palexerolls is sagebrush, bluebunch wheatgrass, and Idaho fescue. The mean annual air temperature is 7.1 ± 1.0 °C, and the mean annual precipitation is 350 ± 105 mm.

The Palexerolls great group in Oregon has 35 soil series which covers 4,500 km^2. Palexerolls are most common in the Palouse Prairie, Blue Mountain Foothills, and Malheur High Plateau (Fig. 7.16). Major Palexerolls include the Carryback, Booth, Mahoon, Zumwalt, and Simas soil series.

Palexerolls have a xeric or aridic soil moisture regime and a mesic or frigid soil temperature regime. Palexerolls are entirely in the fine and clayey-skeletal particle-size classes and the smectitic mineral class.

In Oregon Palexerolls the mollic epipedon averages 45 ± 16 cm, and the argillic horizon averages 65 ± 33 cm in thickness. The Yoncalla soil series, a fine, smectitic, mesic Aquic Palexerolls, is formed in colluvium derived from basalt on hills in the Siskiyou Mountains (Fig. 7.17). This soil features a very dark grayish brown to brown mollic epipedon to 36 cm overlying an argillic horizon to depths greater than 150 cm.

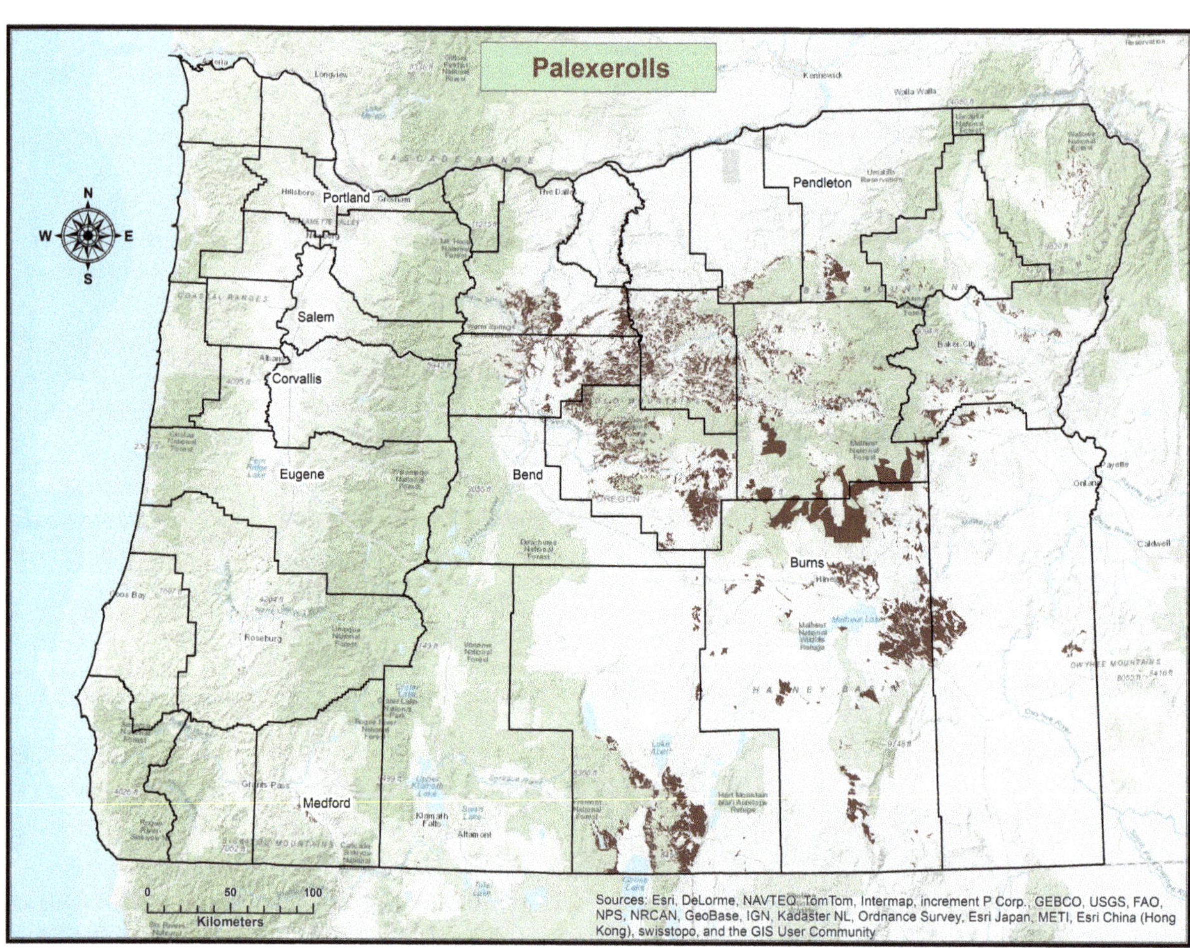

Fig. 7.16 Distribution of Palexerolls in Oregon. *Source* Prepared by Whityn Owen

Fig. 7.17 The Yoncalla soil series, a fine, smectitic, mesic Aquic Palexerolls, is formed in colluvium derived from basalt on hills in the Klamath Mountains (Fig. 7.17). This soil features a very dark grayish brown to brown mollic epipedon to 36 cm overlying an argillic horizon to depths of greater than 150 cm. *Source* Photo by Matthew Fillmore

7.10 Haploxeralfs (Soil Region 9)

Haploxeralfs are moderately well-developed soils that contain an ochric epipedon over an argillic horizon. They have formed in colluvium and residuum over a lithic or paralithic contact on hillslopes. Haploxeralfs occur at elevations ranging between 160 and 900 m and on maximum slopes averaging 60 ± 13%. The native vegetation on Haploxeralfs commonly is Douglas-fir, ponderosa pine, and Pacific madrone. The mean annual air temperature is 11 ± 0.8 °C, and the mean annual precipitation is 1,125 ± 180 mm.

The Haploxeralfs great group in Oregon has 57 soil series which cover 4,200 km^2. Haploxeralfs are most common in the Siskiyou Mountains, the Cascade Mountains—Eastern Slope, the Willamette Valley, and the Blue Mountains (Fig. 7.18). Major Haploxeralfs include the Speaker, Vannoy, Freezener, Willakenzie, Tamarackcanyon, and Dubakella soil series.

Haploxeralfs have a xeric soil moisture regime and a mesic or frigid soil temperature regime. More than two-thirds (69%) of Haploxeralfs are in the fine-loamy or fine particle-size classes; 83% are in the mixed or isotic mineral class; and 64% are in the superactive or active cation-exchange activity classes.

Haploxeralfs have an ochric epipedon over an argillic horizon that averages 72 ± 38 cm in thickness. A clayey-skeletal, magnesic, mesic Mollic Haploxeralfs, the Dubakella soil series has formed in material weathered from ultrabasic rocks with a large amount of serpentinitic minerals (Fig. 7.19). The soil contains a dark reddish brown ochric epipedon that is 27 cm thick over a dark yellowish brown argillic horizon that is 25 cm thick. Serpentinized bedrock occurs at 64 cm.

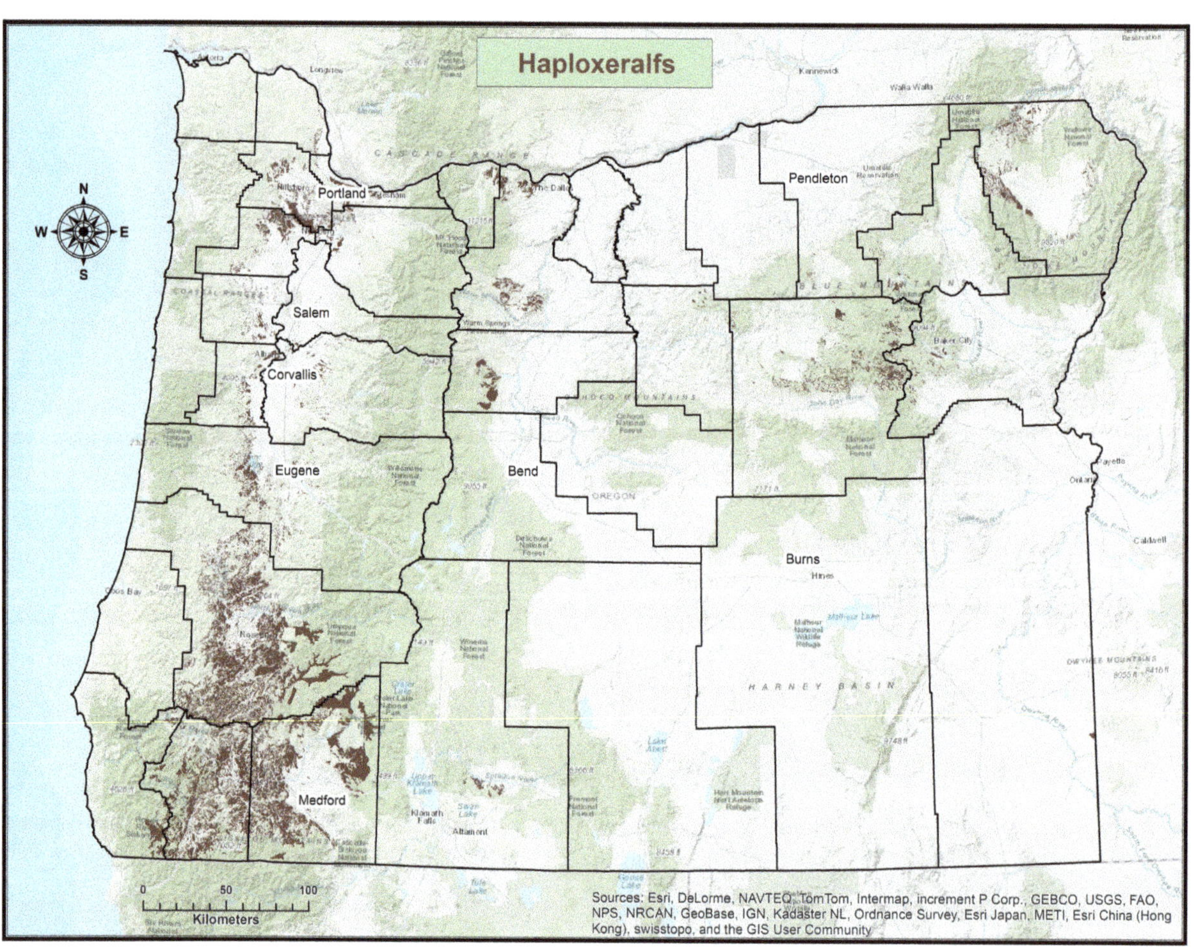

Fig. 7.18 Distribution of Haploxeralfs in Oregon. *Source* Prepared by Whityn Owen

Fig. 7.19 The Dubakella soil series, a clayey-skeletal, magnesic, mesic Mollic Haploxeralfs, has formed in material weathered from ultrabasic rocks with a large amount of serpentinitic minerals in Douglas County. The scale is in feet and rests on serpentinized bedrock. *Source* NRCS photo

7.11 Haplodurids (Soil Region 10)

Haplodurids are moderately well-developed soils with an ochric epipedon over a duripan. They are derived from colluvium and residuum on lava plateaus and lacustrine materials on lake terraces. Haplodurids occur at elevations ranging between 1,050 and 1,400 m and on maximum slopes averaging 17 ± 7.0%. The native vegetation on Haplodurids is Wyoming big sagebrush, bluebunch wheatgrass, and Thurber's needlegrass. The mean annual air temperature is 8.9 ± 1.9 °C, and the mean annual precipitation is 250 ± 30 mm.

The Haplodurids great group in Oregon has 22 soil series which cover 3,500 km^2. Haplodurids are most common on the Malheur High Plateau, the Snake River Plains, Owyhee High Plateau, and the Columbia Plateau (Fig. 7.20). The Raz soil series comprises 65% of the Haplodurids area in Oregon. Other extensive Haplodurids include the Rabbithills, Minveno, Taunton, and Frohman soil series.

Haplodurids have an aridic soil moisture regime and a mesic or frigid soil temperature regime. Haplodurids tend to be in the loamy particle-size classes, the mixed mineral class, and the superactive cation-exchange activity class. About 40% of the Haplodurids soil area is shallow.

Haplodurids have an ochric epipedon over a duripan that averages 38 ± 34 cm. About 50% of the Haplodurids have a cambic horizon averaging 28 ± 10 cm, and 14% have secondary calcium carbonate or a calcic horizon averaging 15 ± 0 cm in thickness. The Taunton soil series, a coarse-loamy, mixed, superactive, mesic Xeric Haplodurids, has formed in alluvium on structural benches and fan terraces on the Columbia Plateau. This soil has a dark gray brown ochric epipedon to 23 cm, a brown cambic horizon to 46 cm, and a pale brown duripan to 100 cm or more (Fig. 7.21).

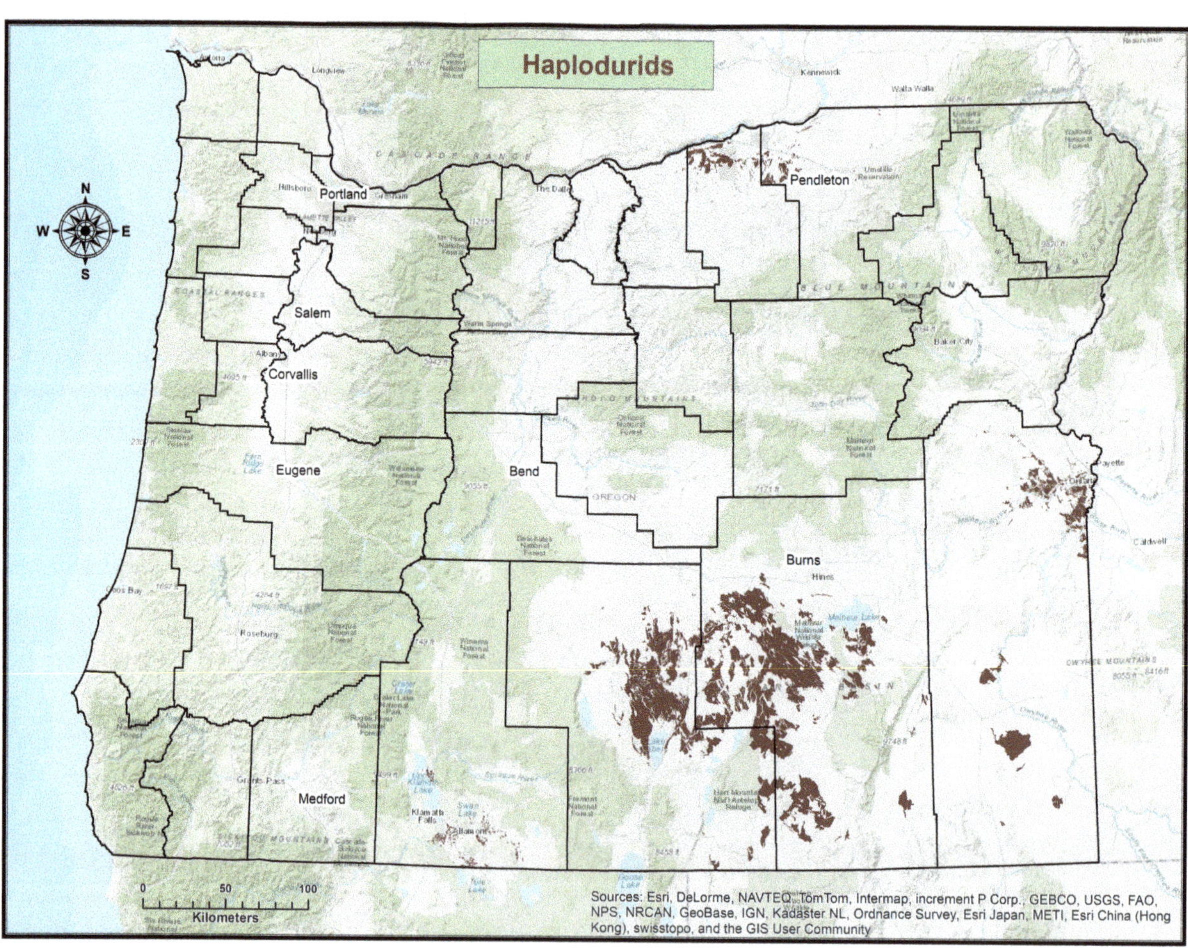

Fig. 7.20 Distribution of Haplodurids in Oregon. *Source* Prepared by Whityn Owen

Fig. 7.21 The Taunton soil series, a coarse-loamy, mixed, superactive, mesic Xeric Haplodurids, is formed from alluvium on structural benches and fan terraces on the Columbia Plateau. This soil has a dark gray brown ochric epipedon to 23 cm, a brown cambic horizon to 46 cm, and a pale brown duripan to 100 cm or more. *Source* NRCS photo

7.12 Dystroxerepts (Soil Region 11)

Dystroxerepts are weakly developed soils with an ochric epipedon over a cambic horizon. They are derived from colluvium and residuum over basalt bedrock on mountain slopes. Dystroxerepts occur at elevations ranging between 250 and 1,340 m and on maximum slopes averaging 93 ± 5.2%. The native vegetation on Dystroxerepts commonly is Douglas-fir, ponderosa pine, and Pacific madrone. The mean annual air temperature is 9.7 ± 1.8 °C, and the mean annual precipitation is 1,200 ± 100 mm.

The Dystroxerepts great group in Oregon has 33 soil series which cover 3,500 km^2. Dystroxerepts are most common in the Siskiyou Mountains (Fig. 7.22). Major Dystroxerepts include the Beekman, Vermisa, Atring, Kanid, and Pearsoll soil series.

Dystroxerepts have a xeric soil moisture regime and a mesic or frigid soil temperature regime. Nearly three-quarters (74%) of Dystroxerepts are in the loamy-skeletal or fine-loamy particle-size classes, 94% are in the mixed mineral class, and 88% are in the superactive or active cation-exchange activity class.

Dystroxerepts have an ochric epipedon over a cambic horizon that averages 49 ± 27 cm in thickness. The Brokeoff soil series, a loamy-skeletal over fragmental, isotic, frigid Typic Dystroxerepts, is formed in tephra and colluvium over residuum (Fig. 7.23). This photograph is from the Lassen Volcanic National Park in northern California, but it is similar to the Beckman soil series, which occurs in Oregon.

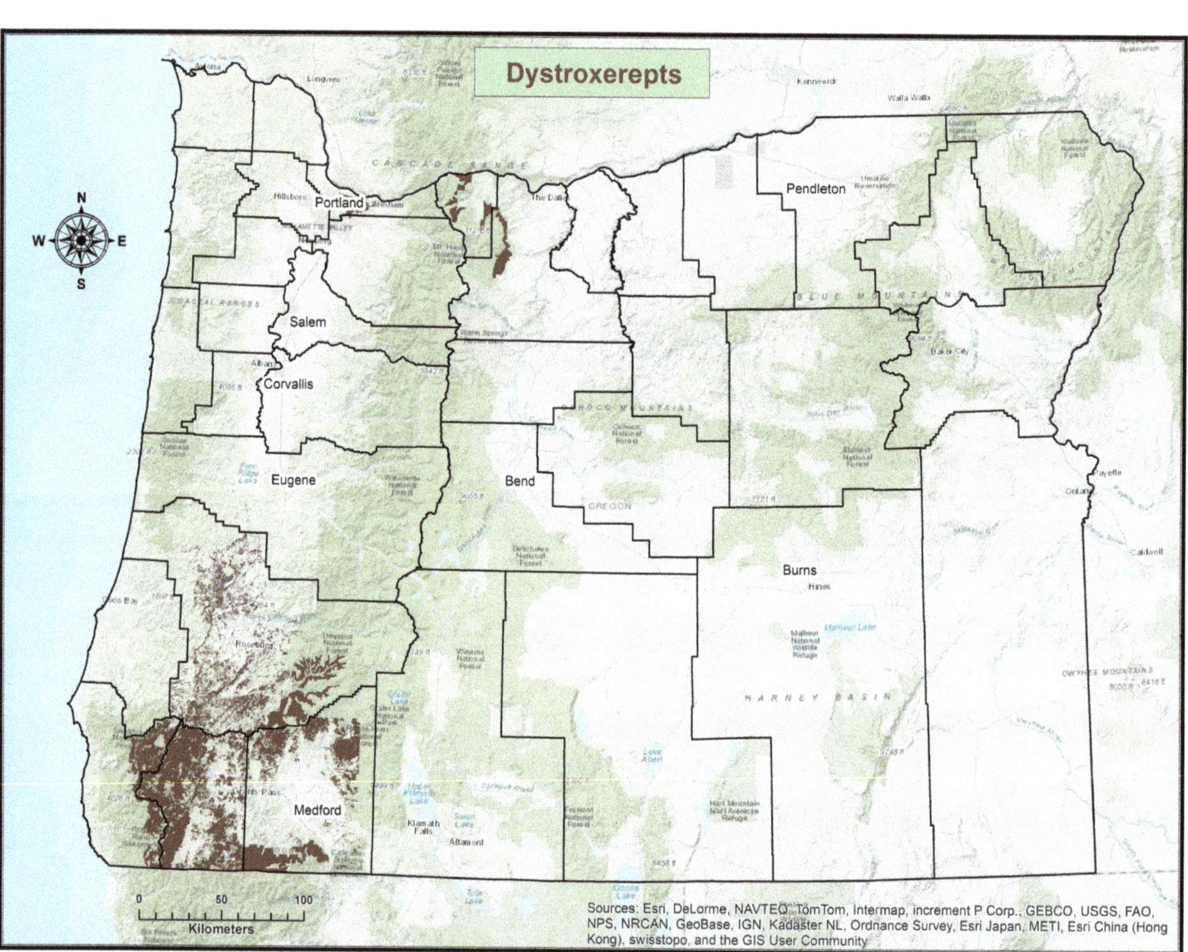

Fig. 7.22 Distribution of Dystroxerepts in Oregon. *Source* Prepared by Whityn Owen

Fig. 7.23 The Brokeoff soil series, a loamy-skeletal over fragmental, isotic, frigid Typic Dystroxerepts, is formed in tephra and colluvium over residuum. Although this photograph is from the Lassen Volcanic National Park in northern California, the Brokeoff is similar to the Beekman soil series in Oregon. The soil is derived from tephra and colluvium over indurated volcanic rocks on mountain slopes. The scale is metric down to one meter. *Source* NRCS photo

7.13 Durixerolls (Soil Region 12)

Durixerolls are moderately well-developed soils containing a mollic epipedon over a duripan. An argillic horizon is common in Durixerolls. They are derived from residuum over basalt bedrock on lava plateaus and old alluvium on terraces and fans. Durixerolls occur at elevations ranging between 1,035 and 1,430 m and on maximum slopes averaging 25 ± 8.7%. The native vegetation on Durixerolls is Wyoming big sagebrush, Idaho fescue, bluebunch wheatgrass, and Sandberg bluegrass. The mean annual air temperature is 7.8 ± 1.4 °C, and the mean annual precipitation is 295 ± 35 mm.

The Durixerolls great group in Oregon has 50 soil series which cover 3,500 km^2. Durixerolls are most common on the Malheur High Plateau and the Blue Mountain Foothills (Fig. 7.24). The most extensive Durixerolls in Oregon are the Goodtack and Moonbeam soil series. Other common series include the Ayres, Drewsgap, Gradon, Oxbow, Willis, Pilot Rock, and Stampede.

Durixerolls have a xeric or aridic soil moisture regime and a frigid or mesic soil temperature regime. Nearly two-thirds (65%) of Durixerolls are in fine, fine-loamy, ashy, and clayey particle-size classes; they most commonly are in the smectitic or mixed mineral class; and 39% are in the superactive cation-exchange activity class.

In Oregon Durixerolls, the mollic epipedon averages 33 ± 15 cm, the duripan averages 41 ± 30 cm, and the argillic horizon, which occurs in 78% of Durixerolls, averages 33 ± 15 cm. The Stampede soil series, a fine, smectitic, frigid Vertic Durixerolls, has formed in eolian material and alluvium derived from volcanic rock (Fig. 7.25). The profile contains very dark grayish brown mollic epipedon (0–24 cm), a brown argillic horizon (30–71 cm), and a yellowish brown duripan (71–94 cm).

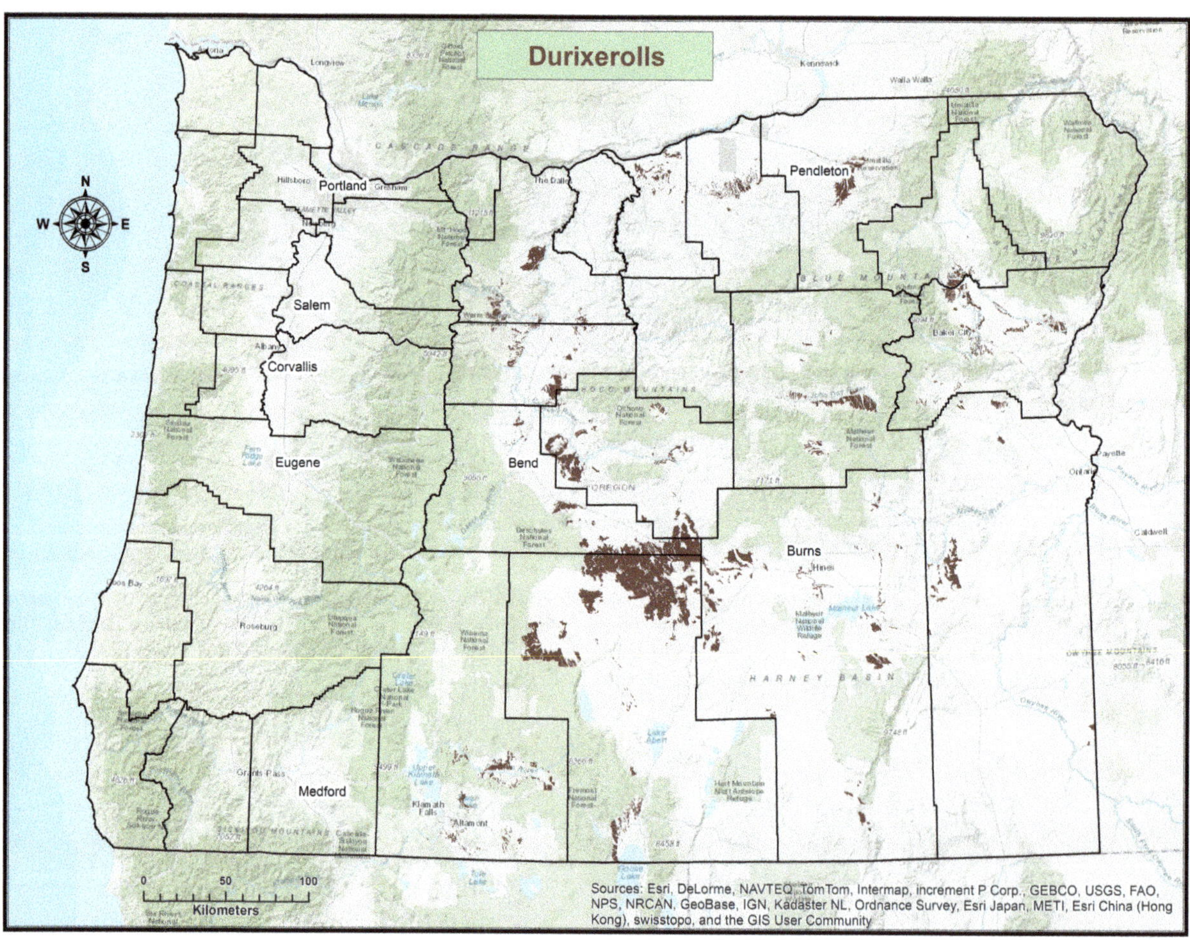

Fig. 7.24 Distribution of Durixerolls in Oregon. *Source* Prepared by Whityn Owen

Fig. 7.25 The Stampede soil series, a fine, smectitic, frigid Vertic Durixerolls, has formed in eolian material and alluvium derived from volcanic rock in Harney County. The scale is decimeters (left) and feet (right). *Source* NRCS photo

7.14 Haplohumults (Soil Region 13)

Haplohumults are well-developed soils with an ochric or umbric epipedon over an argillic horizon. They are derived from colluvium and residuum over paralithic or lithic bedrock originating from sedimentary rocks or basalt on hillslopes. Haplohumults occur at elevations ranging between 100 and 725 m and on slopes averaging 72 ± 13%. The native vegetation on Haplohumults is Douglas-fir, western hemlock, grand fir, and Pacific madrone. The mean annual air temperature is 11 ± 1.0 °C, and the mean annual precipitation is 1,450 ± 350 mm.

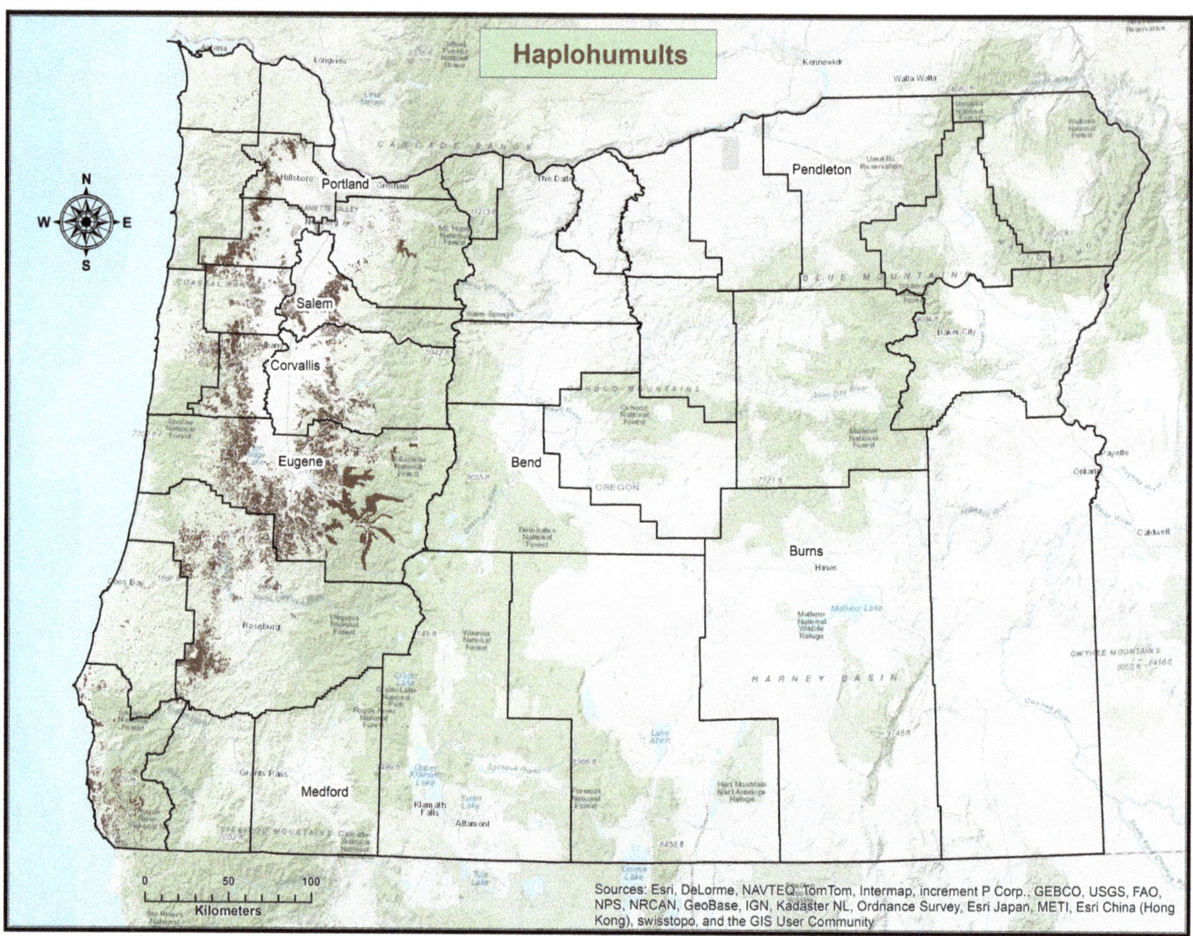

Fig. 7.26 Distribution of Haplohumults in Oregon. *Source* Prepared by Whityn Owen

The Haplohumults great group in Oregon has 26 soil series which cover 3,500 km^2. Haplohumults are most common in the Coast Range, the Willamette Valley, the Cascade Mountains, and the Coastal Redwood Belt (Fig. 7.26). Major Haplohumults include the Peavine, Bellpine, Nekia, Windygap, Loeb, and Olyic soil series.

Haplohumults have a xeric or udic soil moisture regime and a mesic, isomesic, or frigid soil temperature regime. Three-quarters (77%) of Haplohumults are in the fine particle-size class; 100% are in the mixed or isotic mineral class; and 50% are in the active or superactive cation-exchange activity class.

In Oregon Haplohumults the umbric epipedon averages 41 ± 17 cm (75% of soil series), and the argillic horizon averages 72 ± 32 cm. The Peavine soil series, a fine, mixed, active, mesic Typic Haplohumults, is derived from clayey colluvium and residuum derived from sandstone and other materials (Fig. 7.27). The dark brown umbric epipedon extends from the surface to 25 cm and the yellowish red argillic horizon is from 25 to 90 cm and is underlain by hard bedrock.

Fig. 7.27 The Peavine soil series, a fine, mixed, active, mesic Typic Haplohumults, is derived from clayey colluvium and residuum derived from sandstone and other materials. *Source* Soil Survey of Benton County, Oregon). The photographs show the upper 125 cm. NRCS photo

7.15 Vitrixerands (Soil Region 14)

Vitrixerands are moderately well-developed soils that contain an ochric epipedon over a cambic, argillic, or no diagnostic subsurface horizon. They are derived from volcanic ash over colluvium on mountain slopes. Vitrixerands occur at elevations from 850 to 1,425 m and on maximum slopes of 64 ± 11%. The native vegetation is ponderosa pine, Douglas-fir, antelope bitterbrush, and Idaho fescue. The mean annual air temperature is 6.6 ± 0.5 °C, and the mean annual precipitation is 725 ± 350 mm.

The Vitrixerands great group in Oregon has 52 soil series which cover 3,500 km^2. Vitrixerands are most common in the Cascade Mountains—Eastern Slope, and the Blue Mountains (Fig. 7.28). Major Vitrixerands include the Allingham, Circle, Threebuck, Tolo, Olot, Wanoga, Smiling, and Maset soil series.

Vitrixerands have a xeric soil moisture regime and a frigid or mesic soil temperature regime. The Vitrixerands are in ashy, ashy over loamy or loamy-skeletal, and ashy-skeletal particle-size classes and in glassy, glassy over isotic, and amorphic over isotic mineral classes.

Vitrixerands have an ochric epipedon in 65% of soil series, or a mollic epipedon that averages 36 ± 11 cm in 25% of soil series, or an umbric epipedon that averages 43 ± 11 cm in 10% of the soil series. Vitrixerands have a cambic horizon that averages 58 ± 28 cm (73% of soil soil series) and/or an argillic horizon (i.e., Alfic subgroup) that averages 69 ± 29 cm (27% of soil series). The Threebuck soil series, an ashy over clayey-skeletal, glassy over smectitic, frigid Alfic Vitrixerands, is derived from volcanic ash over clayey colluvium derived from basalt (Fig. 7.29). The dark grayish brown ochric epipedon extends from the surface to 10 cm and is underlain by a pinkish gray cambic

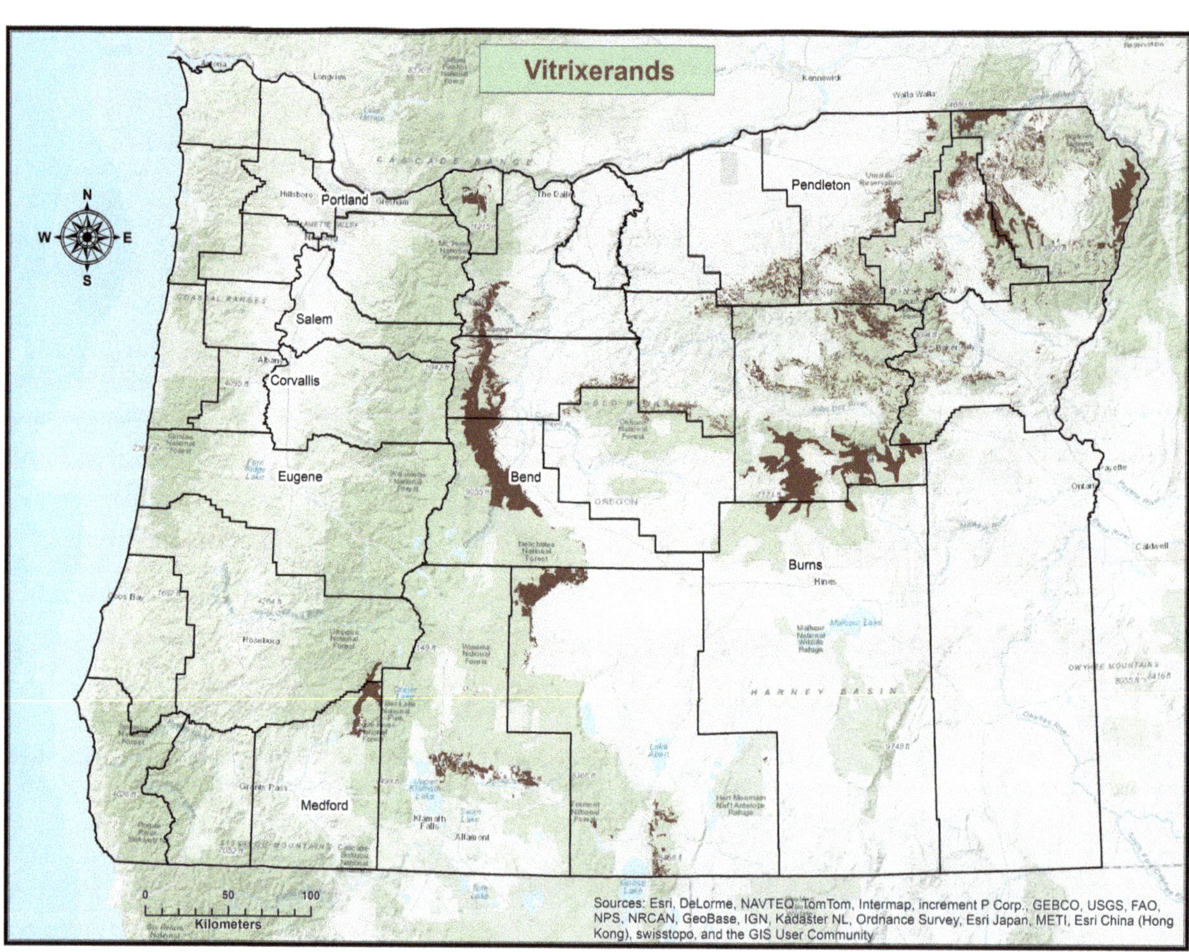

Fig. 7.28 Distribution of Vitrixerands in Oregon. *Source* Prepared by Whityn Owen

Fig. 7.29 The Threebuck soil series, an ashy over clayey-skeletal, glassy over smectitic, frigid Alfic Vitrixerands, is derived from volcanic ash over clayey colluvium derived from basalt in northeastern Oregon. The scale is in feet. *Source* NRCS photo

horizon to 38 cm and a brown argillic horizon to 122 cm, and basalt bedrock.

7.16 Udivitrands (Soil Region 15)

Udivitrands are moderately well-developed soils that contain a mollic, umbric, or ochric epipedon over a cambic and/or argillic horizon. They are derived from volcanic ash over colluvium on mountain slopes. The dominant vegetation is grand fir, western larch, and lodgepole pine forest. Udivitrands in Oregon receive 850 ± 405 mm of annual precipitation and have a mean annual air temperature of 5.2 ± 1.0 °C. Elevations range from 910 to 1,800 m, and maximum slopes are 72 ± 21%.

The Udivitrands great group in Oregon has 23 soil series, which cover 3,200 km^2. Udivitrands are most prevalent in the Cascade Mountains—Eastern Slope, and the Blue Mountains (Fig. 7.30). The Limberjim, Syrupcreek, Tamara, Gutridge, Tamara, and Howash soil series are the extensive Udivitrands in Oregon.

Udivitrands have a udic soil moisture regime and a mesic or frigid soil temperature regime. Udivitrands are primarily in ashy over loamy-skeletal and ashy-skeletal over loamy-skeletal particle-size classes and the amorphic over isotic mineralogy class.

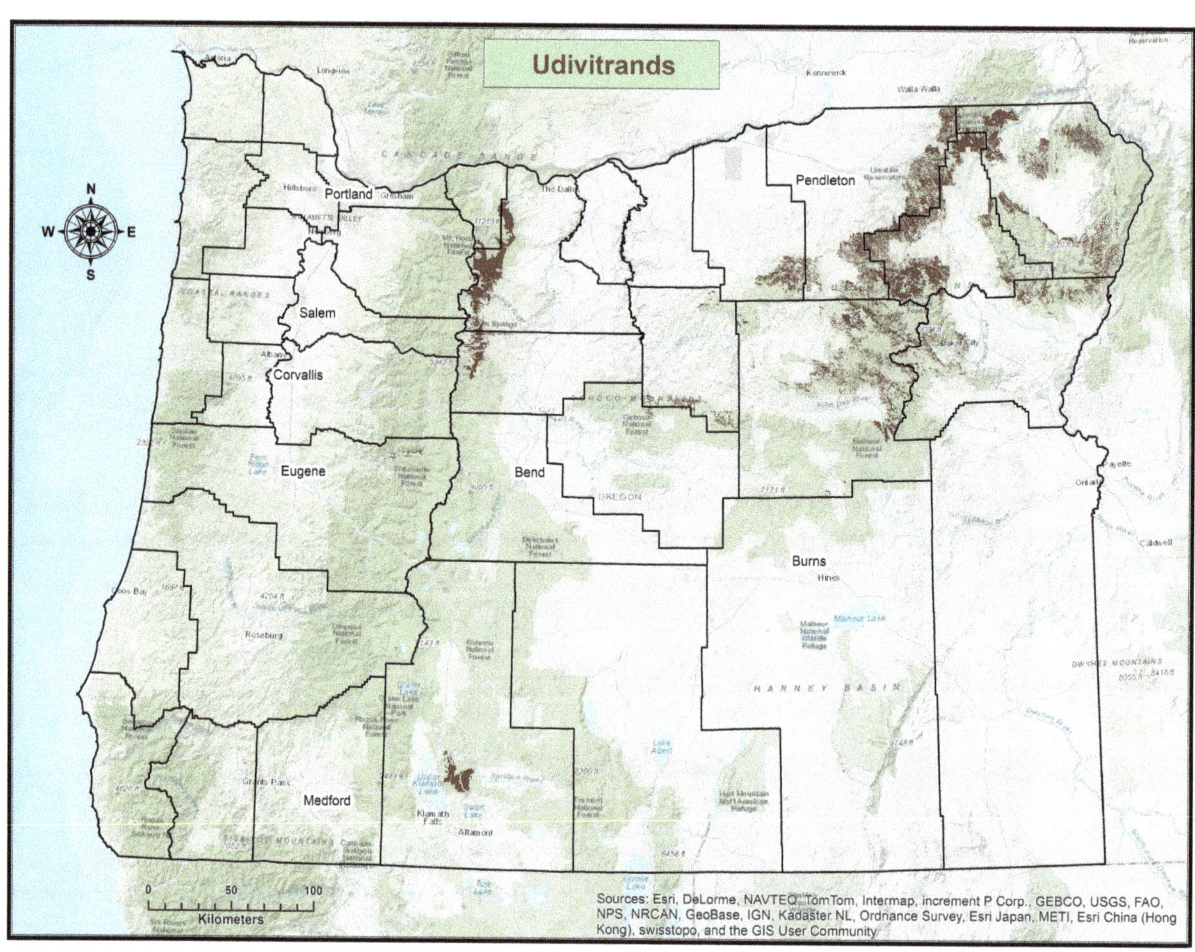

Fig. 7.30 Distribution of Udivitrands in Oregon. *Source* Prepared by Whityn Owen

Fig. 7.31 The Tamara soil series, an ashy over loamy, amorphic over isotic, frigid Alfic Udivitrands, is formed in a mantle of volcanic ash overlying material derived from a mixture of loess, colluvium, and residuum from basalt in northeastern Oregon. The scale is in feet. *Source* NRCS photo

About 82% of the Udivitrands in Oregon have an ochric epipedon. About 87% of the Udivitrands have a cambic horizon averaging 51 ± 15 cm and 48% have an argillic horizon averaging 52 ± 17 cm in thickness. The Tamara soil series, an ashy over loamy, amorphic over isotic, frigid Alfic Udivitrands, has formed in a mantle of volcanic ash overlying material derived from a mixture of loess, colluvium, and residuum from basalt (Fig. 7.31). The Tamara has a yellowish brown ochric epipedon over a light yellowish brown to pale brown cambic horizon to 56 cm, which cover a reddish brown buried soil with an argillic horizon.

7.17 Haploxerepts (Soil Region 16)

Haploxerepts are weakly developed soils with an ochric epipedon over a cambic horizon. They have formed in colluvium over bedrock on mountain slopes. Haploxerepts occur at elevations ranging between 500 and 1,125 m and on maximum slopes averaging 68 ± 31%. The native vegetation on Haploxerepts is Douglas-fir, ponderosa pine, and Oregon white oak. The mean annual air temperature is 8.9 ± 2.3 °C, and the mean annual precipitation is 875 ± 400 mm.

The Haploxerepts great group in Oregon has 39 soil series which cover 2,800 km^2. Haploxerepts are most common in the Siskiyou Mountains, the Willamette Valley, the Cascade Mountain—Eastern Slope, and the Blue Mountains (Fig. 7.32). The most extensive Haploxerepts soil series are the Ritner, Analulu, Kamela, Farva, Caris, Wamic, and Aloha.

Haploxerepts have a xeric soil moisture regime and a mesic or frigid soil temperature regime. More than three-quarters (79%) of the Haploxerepts are in the loamy-skeletal, coarse-loamy, or fine-loamy particle-size

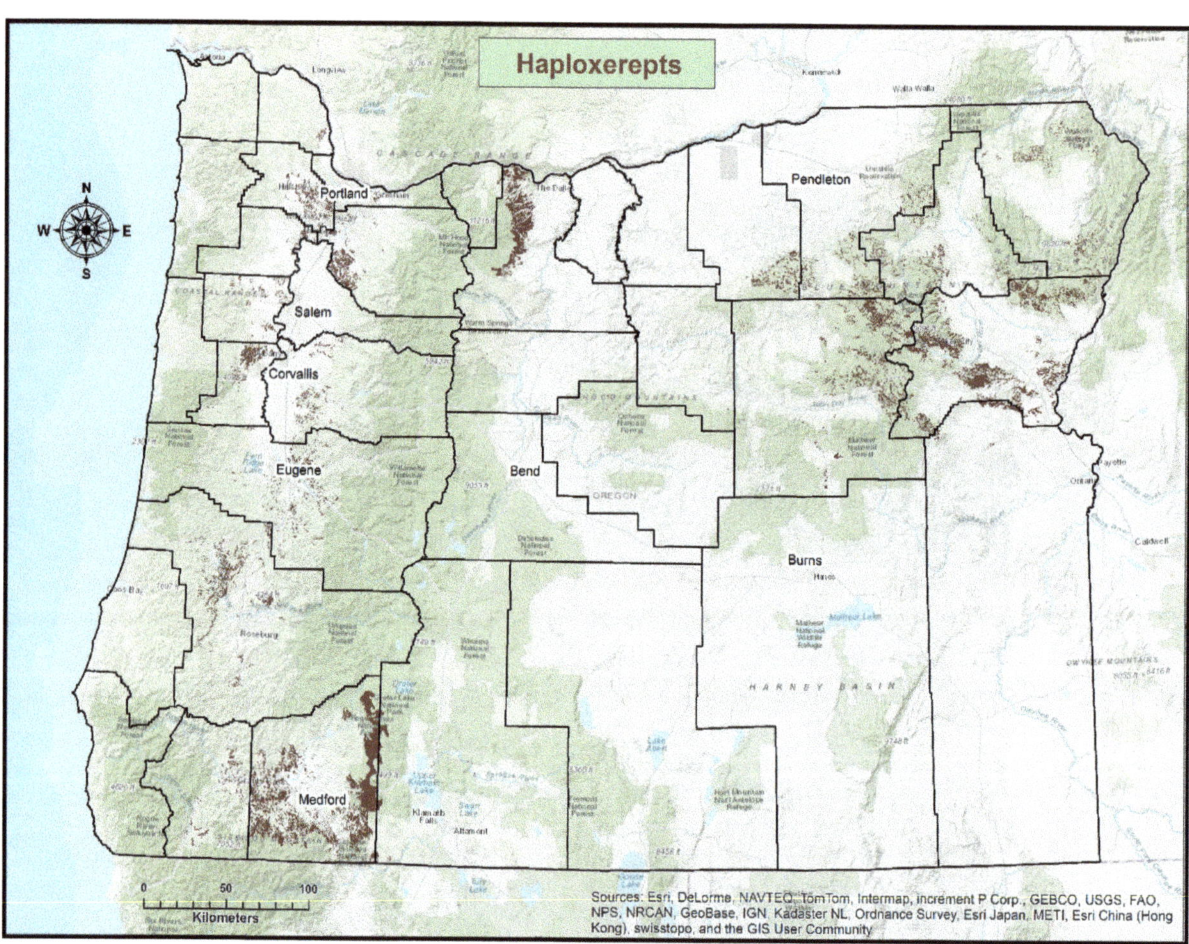

Fig. 7.32 Distribution of Haploxerepts in Oregon. *Source* Prepared by Whityn Owen

classes; they are most common in the mixed or isotic mineral classes, and 56% are in the superactive cation-exchange activity class. Haploxerepts have an ochric epipedon over a cambic horizon that averages 47 ± 23 cm.

7.18 Palehumults (Soil Region 17)

Palehumults are well-developed soils that contain an ochric epipedon over an argillic horizon that exceed 100 cm in depth. They are derived from colluvium and residuum on hillslopes and in old alluvium on marine terraces. Palehumults occur at elevations ranging between 100 and 775 m and on maximum slopes averaging 68 ± 26%. The native vegetation on Palehumults is Douglas-fir, western hemlock, red alder, Oregon white oak, Sitka spruce, coastal redwood, and western redcedar. The mean annual air temperature is 10 ± 1.2 °C, and the mean annual precipitation is 1,775 ± 525 mm.

The Palehumults great group in Oregon has 22 soil series which cover 2,800 km^2. Palehumults are most common in the Coast Range, the Cascade Mountains, the Willamette Valley, and the Siskiyou Mountains (Fig. 7.33). The most extensive Palehumults in Oregon are the Jory, Honeygrove, Orford, Bacona, Cumley, and Salkum soil series.

Palehumults have a xeric or udic soil moisture regime and a mesic, isomesic, or frigid soil temperature regime. More than three-quarters (82%) of Palehumults are in the fine

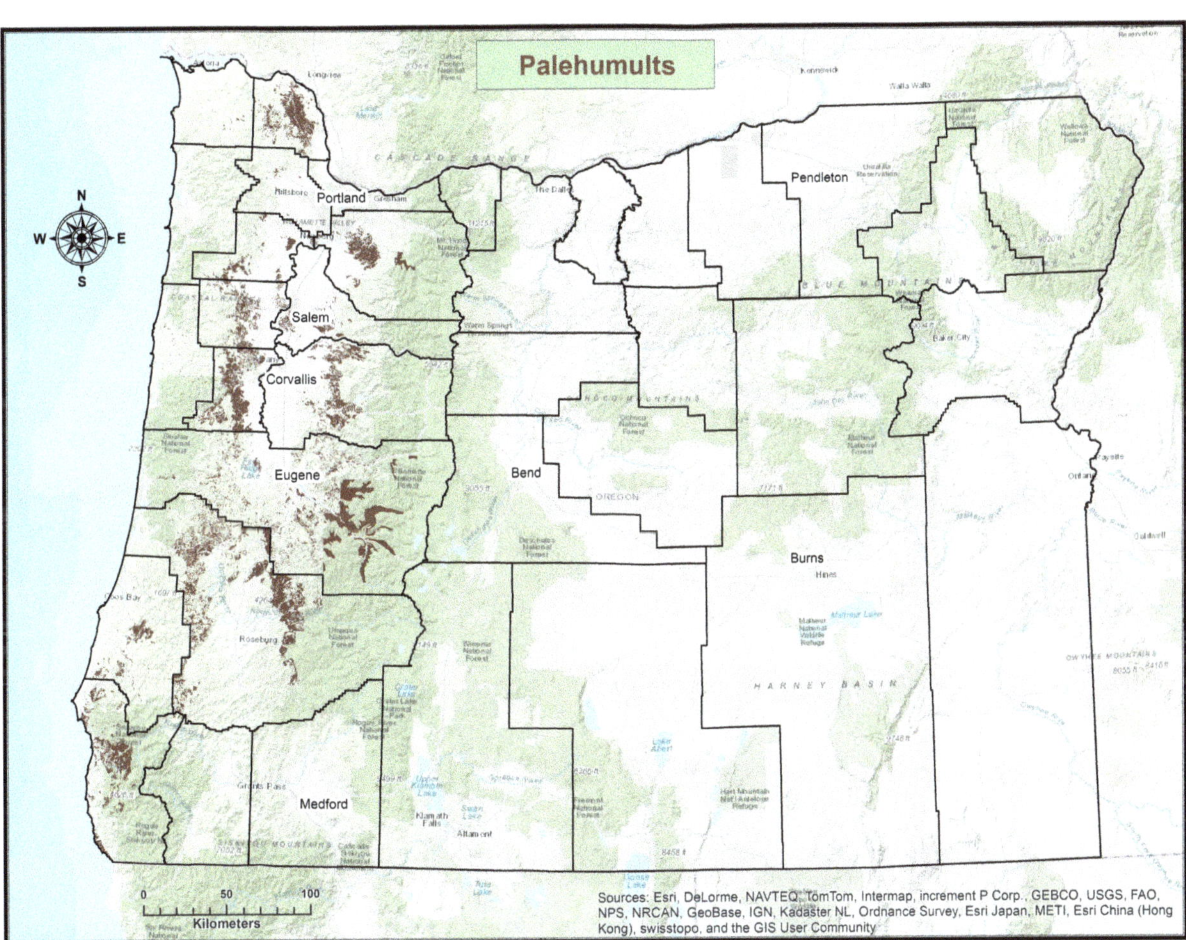

Fig. 7.33 Distribution of Palehumults in Oregon. *Source* Prepared by Whityn Owen

Fig. 7.34 The Jory soil series, a fine, mixed, active, mesic Xeric Palehumults, is formed in colluvium and residuum derived from sedimentary and basic igneous bedrock. The scale is metric down to 1.50 meters. *Source* Soil Survey of Benton County, Oregon

particle-size class; they are mostly in the mixed and isotic mineral classes; and 64% are in the active and superactive cation-exchange activity classes.

Palehumults have an ochric epipedon in 68% of the soil series, or an umbric epipdon averaging 32 ± 8.1 cm in 32% of the soil series, and an argillic horizon averaging 108 ± 40 cm in thickness. The Jory soil series, a fine, mixed, active, mesic Xeric Palehumults, has formed in colluvium and residuum derived from sedimentary and basic igneous bedrock (Fig. 7.34). The profile contains reddish brown ochric epipedon (0–40 cm) over a reddish brown argillic horizon to 254 cm.

7.19 Dystrudepts (Soil Region 18)

Dystrudepts are weakly developed soils that contain an ochric epipedon over a cambic horizon. They are derived from colluvium and residuum over a paralithic or lithic contact on hillslopes. Dystrudepts occur at elevations ranging between 100 and 900 m and on maximum slopes averaging 81 ± 10%. The native vegetation on Dystrudepts is Douglas-fir, grand fir, bigleaf maple, western hemlock, and red alder. The mean annual air temperature is 10 ± 0.8 °C, and the mean annual precipitation is 2,250 ± 500 mm.

The Dystrudepts great group in Oregon has 45 soil series which cover 2,600 km^2. Dystrudepts are most common in the Coast Range, the Cascade Mountains, and the Siskiyou Mountains (Fig. 7.35). The most extensive Dystrudepts in Oregon are the Blachly, Remote, Fritsland, Etelka, Cassiday, Chamate, Leopold, and Scaponia soil series.

Dystrudepts have a udic soil moisture regime and a mesic, isomesic, or frigid soil temperature regime. Nearly three-quarters (73%) of Dystrudepts are in the loamy-skeletal or fine-loamy particle-size class; 73% are in the isotic mineral classes; and 27% are in the active and superactive cation-exchange activity classes.

Dystrudepts have an ochric epipedon over a cambic horizon that averages 73 ± 37 cm in thickness. The Chamate soil series, a loamy-skeletal, isotic, mesic Typic Dystrudepts, has formed in colluvium derived from welded tuff (Fig. 7.36). The profile contains a light brownish gray ochric epipedon (0–40 cm) over a light gray to pale brown cambic horizon.

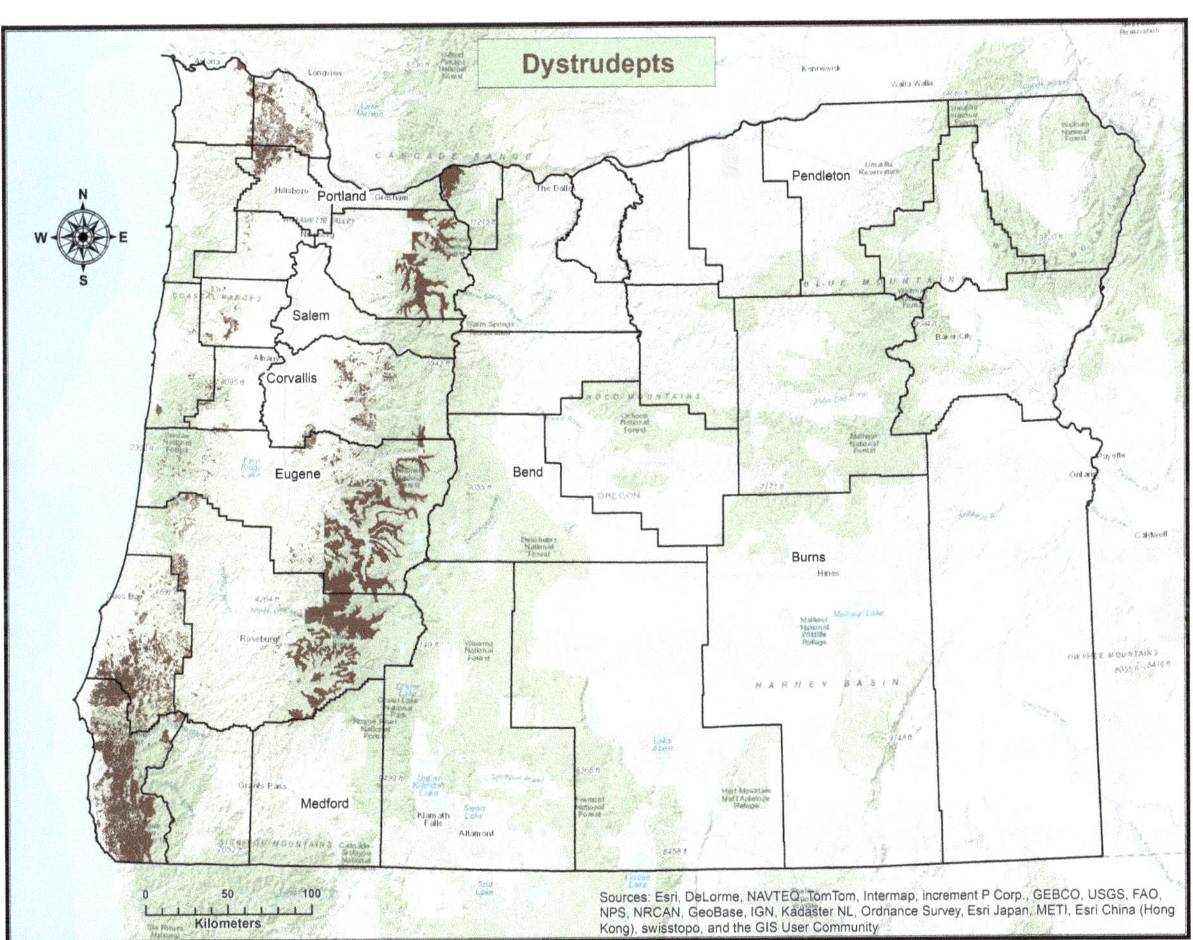

Fig. 7.35 Distribution of Dystrudepts in Oregon. *Source* Prepared by Whityn Owen

Fig. 7.36 The Chamate soil series, a loamy-skeletal, isotic, mesic Typic Dystrudepts, has formed from colluvium derived from welded tuff. The scale is in feet. *Source* Soil Survey of Douglas County, Oregon

7.20 Hapludands (Soil Region 19)

Hapludands are moderately well-developed soils with an umbric epipedon over a cambic horizon. They are formed in colluvium and residuum over bedrock on hillslopes. Hapludands occur at elevations ranging between 225 and 750 m and on maximum slopes averaging 87 ± 7%. The native vegetation on Hapludands is Douglas-fir, western hemlock, and western redcedar. The mean annual air temperature is 8.3 ± 1.8 °C, and the mean annual precipitation is 2,425 ± 200 mm.

The Hapludands great group in Oregon has 18 soil series which cover 2,500 km^2. Hapludands are most common in the Coast Range (Fig. 7.37). The most extensive Hapludands in Oregon are the Hemcross, Klistan, Slickrock, Caterl, and Murtip soil series.

Hapludands have a udic soil moisture regime and a frigid or mesic soil temperature regime. The Hapludands are all in the medial, medial-skeletal, or medial over loamy particle-size classes; and most are in the ferrihydritic mineral class.

Hapludands have an umbric epipedon averaging 48 ± 26 cm (72% of soil series) or an ochric epipedon (21% of soil series) over a cambic horizon averaging 72 ± 36 cm in thickness. The Harslow soil series, a medial-skeletal, ferrihydritic, mesic Alic Hapludands, has formed in loamy colluvium and residuum derived from basalt (Fig. 7.38). The profile contains a brown umbric epipedon (5–33 cm) over a brown cambic horizon (50–66 cm).

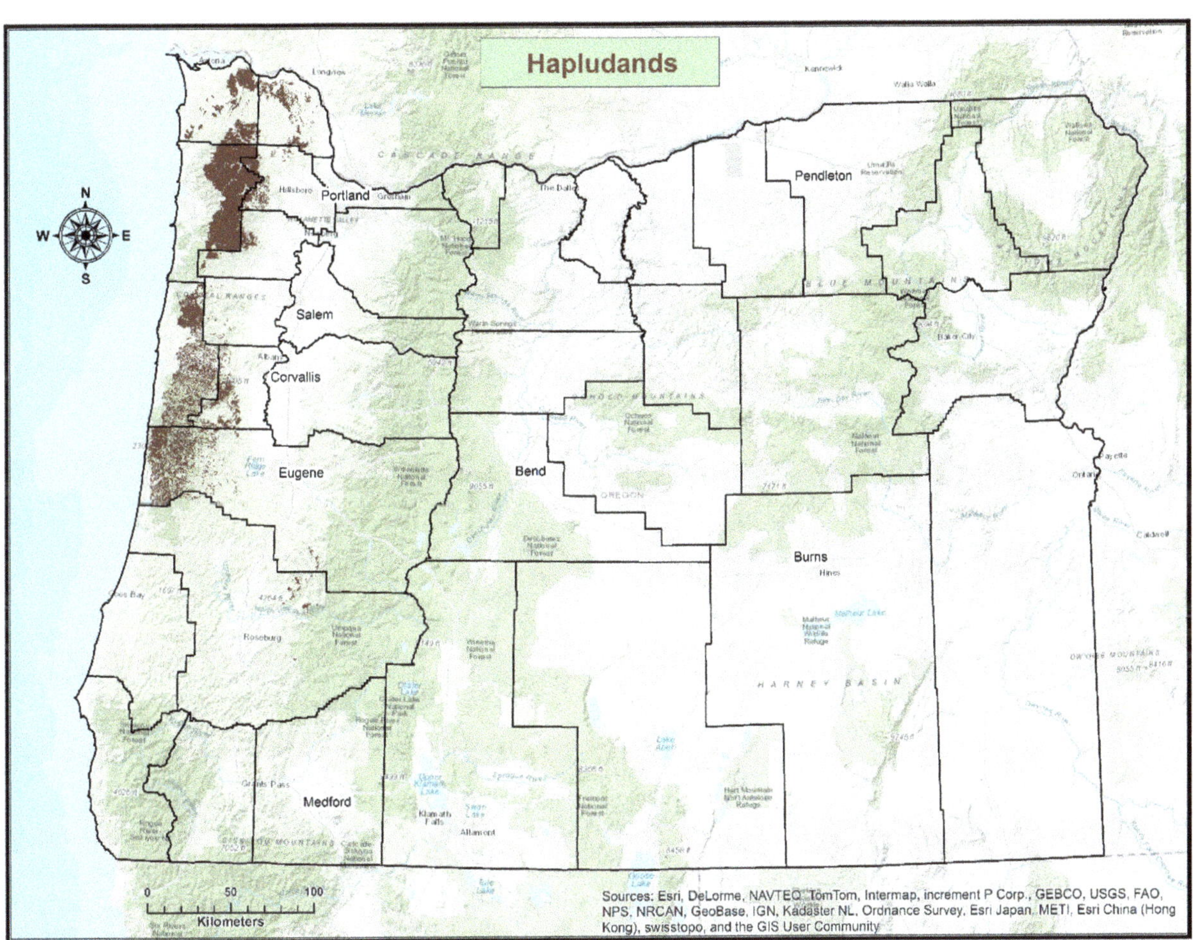

Fig. 7.37 Distribution of Hapludands in Oregon. *Source* Prepared by Whityn Owen

Fig. 7.38 The Harslow soil series, a medial-skeletal, ferrihydritic, mesic Alic Hapludands, has formed in loamy colluvium and residuum from basalt. The bottom of the profile is at 36 inches (90 cm) on basalt bedrock. *Source* Soil Survey of Benton County, Oregon

7.21 Paleargids (Soil Region 20)

Paleargids are well-developed soils containing an ochric epipedon over an argillic horizon. They are formed from lacustrine materials on lake terraces or colluvium on plateaus. The dominant vegetation is Wyoming big sagebrush, Thurber's needlegrass, and bluebunch wheatgrass. Paleargids in Oregon receive 260 ± 40 mm of annual precipitation and have a mean annual air temperature 8.3 ± 1.5 °C. Elevations range from 1,225 to 1,675 m, and maximum slopes are 20 ± 18%.

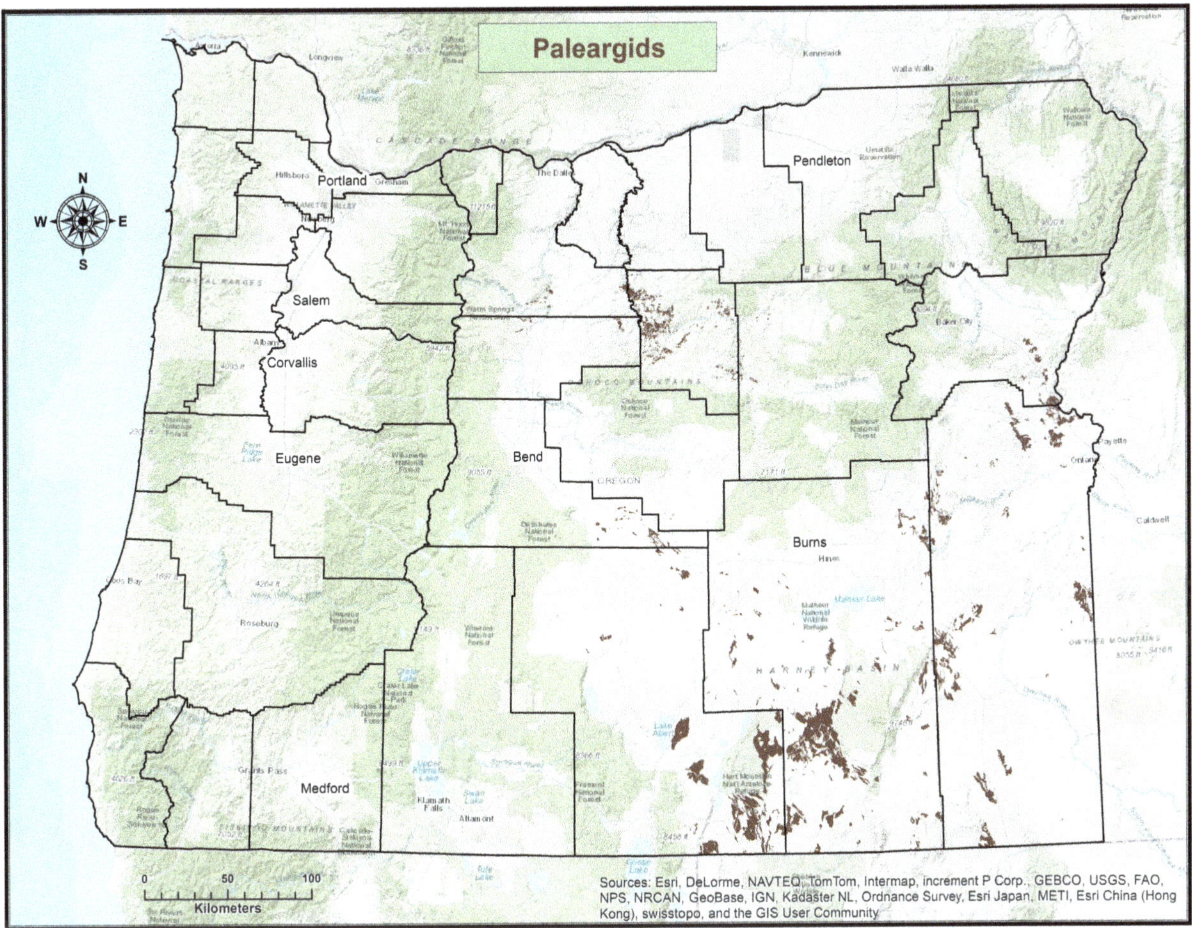

Fig. 7.39 Distribution of Paleargids in Oregon. *Source* Prepared by Whityn Owen

The Paleargids great group of Oregon has 17 soil series, which cover 2,000 km^2. Paleargids are most prevalent on the Malheur High Plateau and the Blue Mountain Foothills (Fig. 7.39). The Spangenburg and Freznik soil series are the extensive Paleargids in Oregon. Paleargids have an aridic soil moisture regime and a mesic or frigid soil temperature regime. In Oregon, 77% of the Paleargids are in the fine particle-size class; 92% are in the smectitic mineralogy class; and 64% are in the superactive or active cation-exchange activity classes.

Paleargids have an ochric epipedon over an argillic horizon that averages 46 ± 24 cm in thickness. The Gooding soil series, a fine, smectitic, mesic Vertic Paleargids, is derived from mixed alluvium and loess and occurs on alluvial fan terraces and basalt plains and buttes (Fig. 7.40). This soil contains an argillic horizon from 23 to 86 cm, a calcic horizon from 86 to 124 cm, and a duripan from 124 to 157 cm, underlain by basalt bedrock with silica capping.

Fig. 7.40 The Gooding soil series, a fine, smectitic, mesic Vertic Paleargids, is formed in mixed alluvium and loess on alluvial fan terraces on basalt plains and buttes. This soil contains an argillic horizon from 23 to 86 cm, a calcic horizon from 86 to 124 cm, and a duripan from 124 to 157 cm, underlain by basalt bedrock with silica capping. This soil occurs in Malheur County. The scale is in decimeters. *Source* NRCS photo

7.22 Eutrudepts (Soil Region 21)

Eutrudepts are weakly developed soils that contain an ochric epipedon and a cambic horizon. They are formed from colluvium and residuum derived from sandstone and siltstone or serpentinitic materials. The dominant vegetation is Douglas-fir, bigleaf maple, and red alder forest. Eutrudepts in Oregon receive 1,880 ± 665 mm of annual precipitation and have a mean annual air temperature 9.0 ± 2.8 °C. Elevations range from 320 to 1,050 m, and maximum slopes are 83 ± 15%.

There are 12 soil series in the Eutrudepts great group of Oregon that covers 1,800 km^2. Eutrudepts are most prevalent in the Coast Range (Fig. 7.41). The Digger, Umpcoos, Braun, and Whobrey soil series are the extensive Eutrudepts in Oregon. In Oregon, 75% of the Eutrudepts are in the loamy-skeletal particle-size class; they are either in the magnesic or isotic mineralogy class.

Eutrudepts have a udic soil moisture regime and a mesic or frigid soil temperature regime. Eutrudepts have an ochric epipedon over a cambic horizon that averages 42 ± 21 cm in thickness. The Digger soil series, a loamy-skeletal, isotic, mesic Dystric Eutrudepts, has formed in loamy colluvium and residuum weathered from sandstone and siltstone (Fig. 7.42). The profile contains a dark grayish brown to brown ochric epipedon over a cambic horizon and is underlain by weathered sandstone bedrock.

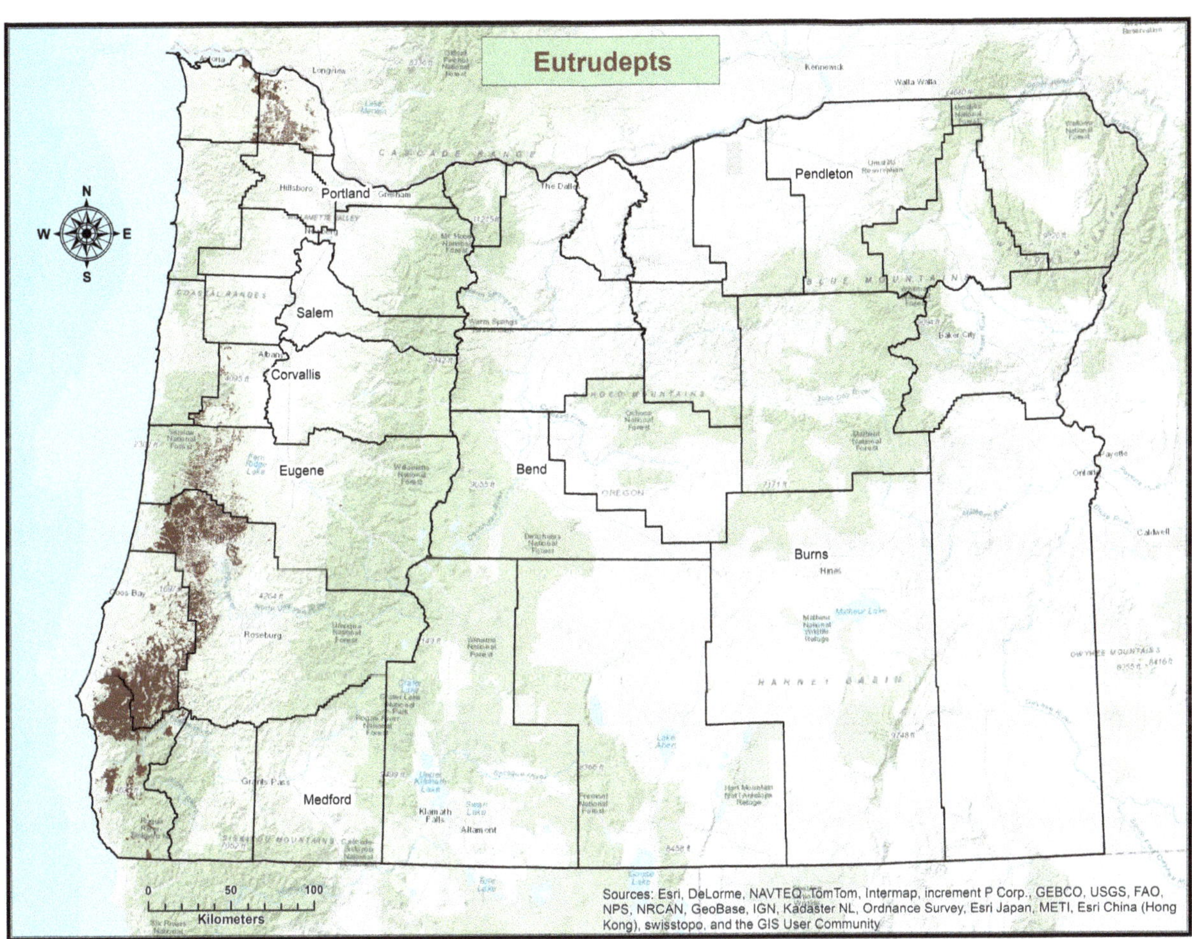

Fig. 7.41 Distribution of Eutrudepts in Oregon. *Source* Prepared by Whityn Owen

Fig. 7.42 The Digger soil series, a loamy-skeletal, isotic, mesic Dystric Eutrudepts, has formed in loamy colluvium and residuum weathered from sandstone and siltstone. The photograph shows the upper 120 cm. *Source* Soil Survey of Benton County, Oregon

7.23 Fulvudands (Soil Region 22)

Fulvudands are moderately well-developed soils that contain an umbric epipedon over a cambic horizon. They are formed in colluvium, and residuum derived from basalt or sedimentary rocks on mountain slopes. The dominant vegetation is Sitka spruce, western hemlock, red alder, and Douglas-fir. Fulvudands receive 2,380 ± 370 mm of annual precipitation and have a mean annual air temperature of 9.0 ± 1.6 °C. The cool, moist climate and coniferous plant communities are conducive for the accumulation of organic carbon and the development of aluminum–humus complexes. Elevations range from 10 to 900 m, and maximum slopes are 88 ± 6.5%.

The Fulvudands great group in Oregon has 19 soil series, which cover 1,700 km^2. Fulvudands occur mainly in the Sitka Spruce Belt (Fig. 7.43). The Klootchie, Necanicum, and Tolovana soil series each occupy more than 200 km^2.

Fulvudands are in the medial, medial-skeletal, or medial over other materials particle-size class and the ferrihydritic or ferrihydritic over isotic mineralogy class. They have an isomesic or isofrigid soil temperature regime and a udic soil moisture regime. They meet the color, thickness, and organic carbon requirements but lack the melanic index requirement for a melanic epipedon. In Fulvudands, the umbric epipedon averages 60 ± 36 cm thick over a cambic horizon (89% of soil series) with a thickness of 56 ± 35 cm.

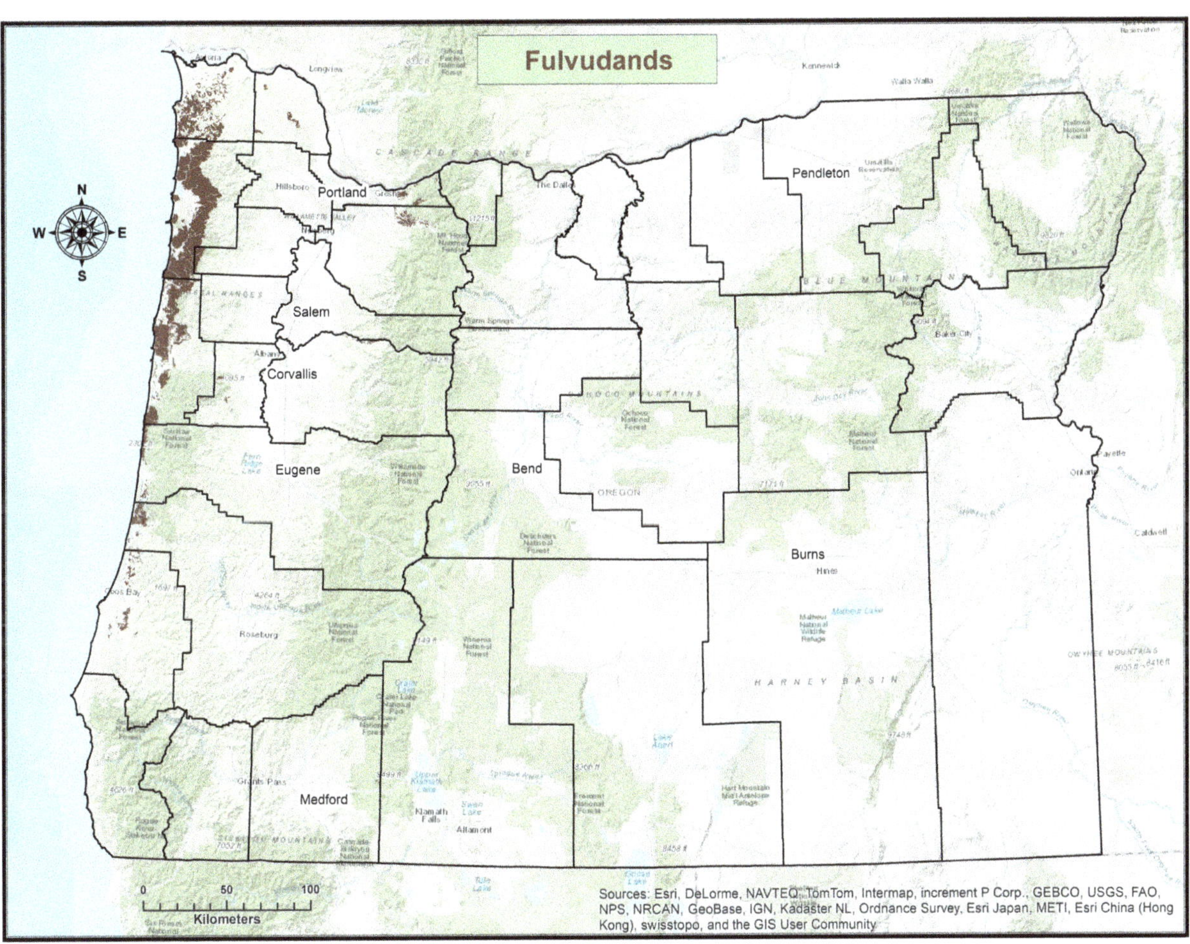

Fig. 7.43 Distribution of Fulvudands in Oregon. *Source* Prepared by Whityn Owen

Fig. 7.44 The Klootchie soil series, a medial, ferrihydritic, isomesic Typic Fulvudands, has formed in colluvium and residuum on mountain slopes along the northern Oregon coast. The scale is in feet. *Source* NRCS photo

The Klootchie soil series, a medial, ferrihydritic, isomesic Typic Fulvudands, has formed in colluvium and residuum from basalt breccia. This soil contains an umbric epipedon to 35 cm and a cambic horizon to 69 cm, over fractured, weathered basalt breccia, and are very strongly or strongly acid with high amounts of potassium chloride-extractable aluminum (Fig. 7.44).

7.24 Torripsamments (Soil Region 23)

Torripsamments are among the most weakly developed soils in Oregon, having only an ochric epipedon. They are derived from a mixture of volcanic ash, eolian sand, and lacustrine materials and occur on dunes or plateaus. The dominant vegetation includes Wyoming big sagebrush, bluebunch wheatgrass, and Sandberg bluegrass. Torripsamments in Oregon receive 240 ± 25 mm of annual precipitation and have a mean annual air temperature of 8.1 ± 1.3 °C. Elevations range from 1,200 to 1,400 m, and maximum slopes are 29 ± 14%.

The Torripsamments great group in Oregon has 14 soil series, which cover 1,400 km^2. Torripsamments are most prevalent on the Malheur High Plateau, the Columbia Basin, and the Blue Mountain Foothills (Fig. 7.45). The Quincy, Gosney, Morehouse, and Kewake soil series are the extensive Torripsamments in Oregon. Torripsamments have an aridic (torric) soil moisture regime and a mesic or frigid soil temperature regime. Torripsamments are in the sandy or ashy particle-size classes and the mixed or glassy mineralogy classes.

In Oregon Torripsamments, the ochric epipedon averages 18 cm. The Quincy soil series, a mixed, mesic Xeric Torrispsamments, is formed in sands on dunes and terraces. This soil contains an ochric epipedon over relatively sandy unaltered materials (Fig. 7.46).

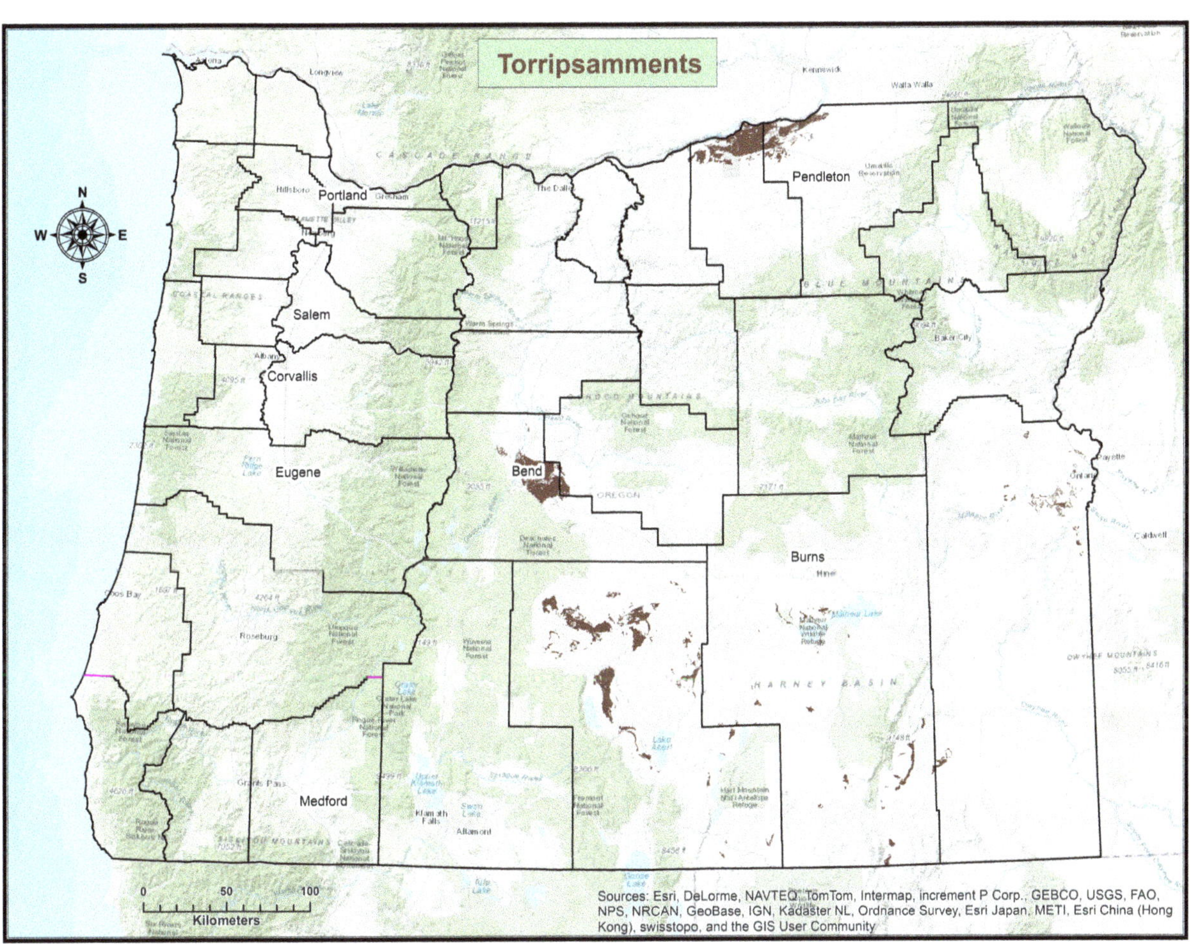

Fig. 7.45 Distribution of Torripsamments in Oregon. *Source* Prepared by Whityn Owen

Fig. 7.46 The Quincy soil series, a mixed, mesic Xeric Torripsamments, is formed in sands on dunes and terraces along the Columbia River. This soil contains an ochric epipedon over relatively unaltered sandy material. The scale is in inches. *Source* NRCS photo

7.25 Endoaquolls (Soil Region 24)

Endoaquolls are weakly developed soils that have a mollic epipedon and a cambic subsurface horizon. They are derived from alluvium, occur in floodplains or on alluvial fans, and feature willows and sedges. Endoaquolls in Oregon receive 850 ± 500 mm of annual precipitation and have a mean annual air temperature 10 ± 1.8 °C. Elevations range from 425 to 925 m, and maximum slopes are 3 ± 1%.

The Endoaquolls great group in Oregon has 37 soil series, which cover 1,400 km^2. Endoaquolls are most prevalent in the Willamette Valley, the Malheur High Plateau, and the Klamath Basin (Fig. 7.47). The Algoma, Housefield, Welch, Ozamis, Wapato, Waldo, and Sauvie soil series are the most extensive Endoaquolls in Oregon.

Endoaquolls have aquic conditions (aquic soil moisture regime) and mesic or frigid soil temperature regime. In Oregon, 76% of the Endoaquolls are in the fine-silty, fine-loamy, and fine particle-size classes; 73% are in the isotic and 27% are in the mixed mineralogy class; 27% are in the active and superactive cation-exchange activity classes.

In Endoaquolls, the mollic epipedon averages 61 ± 32 cm. Only 19% of the Endoaquolls have a cambic horizon, which averages 49 ± 19 cm; the remaining 81% lack a diagnostic subsurface horizon. The Wapato soil series, a fine-silty, mixed, superactive, mesic Fluvaquentic Endoaquolls, formed in loamy mixed alluvium on flood plains (Fig. 7.48). The soil has a very dark grayish brown mollic epipedon (0–16 inches) over a dark grayish brown, gleyed

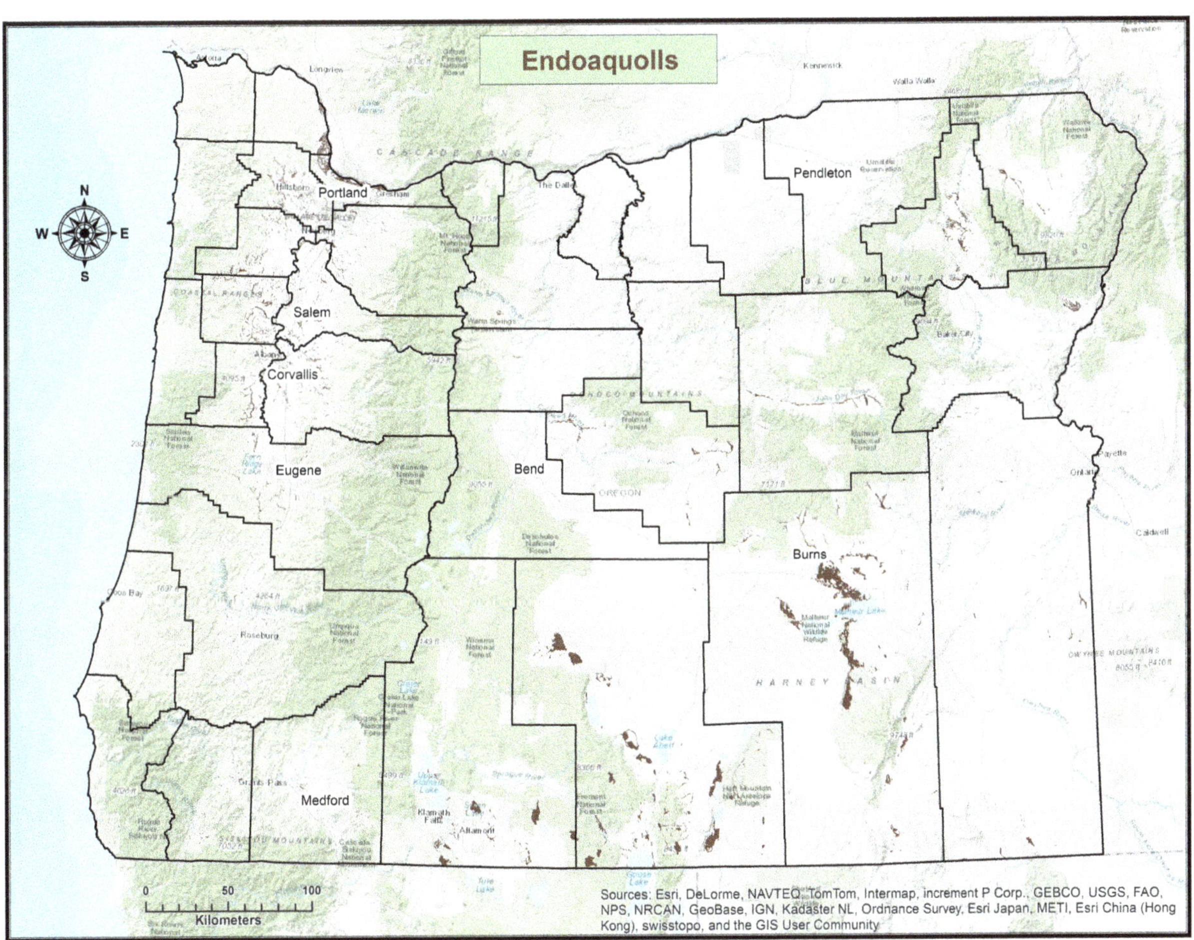

Fig. 7.47 Distribution of Endoaquolls in Oregon. *Source* Prepared by Whityn Owen

Fig. 7.48 The Wapato soil series, a fine-silty, mixed, superactive, mesic Fluvaquentic Endoaquolls, formed in loamy mixed alluvium on flood plains. The scale is in inches. *Source* Soil Survey of Benton County, Oregon

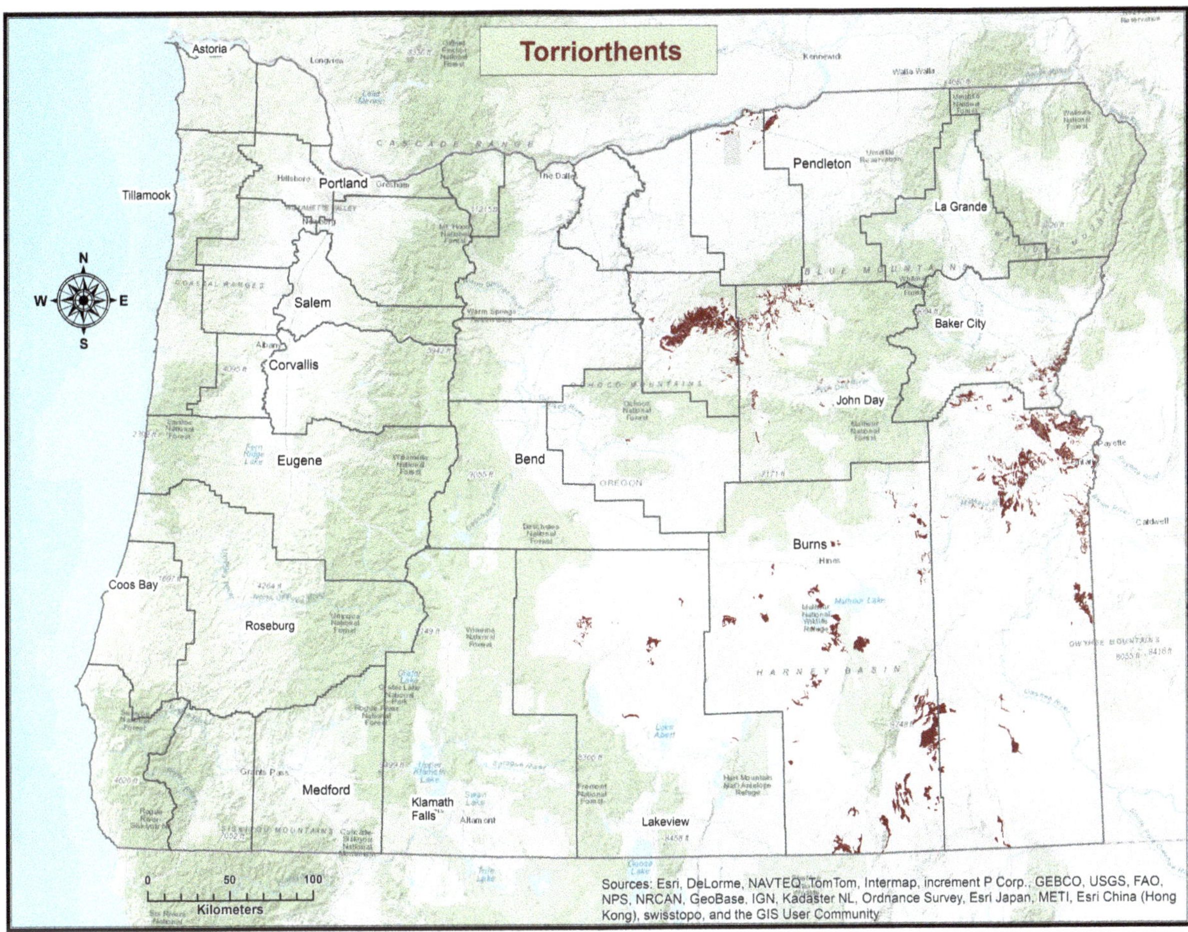

Fig. 7.49 Distribution of Torriorthents in Oregon. *Source* Prepared by Whityn Owen

cambic horizon (16–32 inches). Analytical data for the Wapato soil series are provided in Table 8.1.

7.26 Torriorthents (Soil Region 25)

The Torriorthents are among the most weakly developed soils in Oregon and only have an ochric epipedon averaging 18 cm in thickness. They are formed primarily in colluvium and alluvium on hillslopes and alluvial terraces derived from a variety of rock types. The vegetation on Torriorthents is predominantly sagebrush and grasses. The mean annual precipitation is 300 ± 167 mm, and the mean annual air temperature is 9.2 ± 1.6 °C. Torriorthents are most common between the elevations of 800 and 1500 m and on maximum slopes of 50 ± 24%.

There are 24 Torriorthents soil series, which comprise 1400 km^2 in Oregon. These soils occur mainly on the Malheur High Plateau, the Snake River Plains, the Blue Mountain Foothills, and the Humboldt Area (Fig. 7.49). The most extensive Torriorthents are the Haar, Exfo, and Skedaddle soil series.

Torriorthents most commonly have a mesic soil temperature regime, coarse-loamy, ashy, or loamy-skeletal particle-size class, and a mixed or glassy mineralogy. More than one-half of the Torriorthents are in the superactive cation-exchange activity class.

Torriorthents have an ochric epipedon averaging 18 cm in thickness and lack a diagnostic subsurface horizon. The Garbutt soil series, a coarse-silty, mixed, superactive, calcareous, mesic Typic Torriorthents, is formed in loess and silty alluvial on fan terraces, basalt plains, and alluvial fans (Fig. 7.50). This soil has a light brownish gray ochric

Fig. 7.50 The Garbutt soil series, a coarse-silty, mixed, superactive, calcareous, mesic Typic Torriorthents, is formed in loess and silty alluvium on fan terraces, basalt plains, and alluvial fans. This soil has a light brownish gray ochric horizon 12 cm thick over relatively unaltered, light gray sandy loam parent materials. *Source* NRCS photo

horizon 12 cm thick over relatively unaltered, light gray sandy loam parent materials.

7.27 Haplocryands (Soil Region 26)

The Haplocryands are weakly developed and contain an ochric epipedon over a cambic horizon. They are formed in colluvium and residuum derived from volcanic rocks, and they occur on mountain slopes. The vegetation on Haplocryands includes conifer forests with Douglas-fir, noble fir, Pacific silver fir, and mountain hemlock. The mean annual precipitation is 2,300 ± 675 mm, and the mean annual air temperature of 5.1 ± 1.5 °C. Haplocryands occur at elevations between 900 and 1,700 m and on maximum slopes of 78 ± 16%.

There are 15 soil series in the Haplocryands great group, which compose 1,100 km^2. These soils occur almost exclusively in the Cascade Mountains (Fig. 7.51). The most extensive Haplocryands include the Keel, Valsetz, Pinhead, and Oatman soil series.

Haplocryands have a cryic soil temperature regime and a udic or xeric soil moisture regime. They have a medial-skeletal or medial particle-size class and an amorphic, or less commonly, a ferrihydritic mineralogy.

Haplocryands have either an ochric epipedon in 60% of the soil series or an umbric epipedon averaging 60 ± 29 cm (40% of soil series) over a cambic horizon averaging

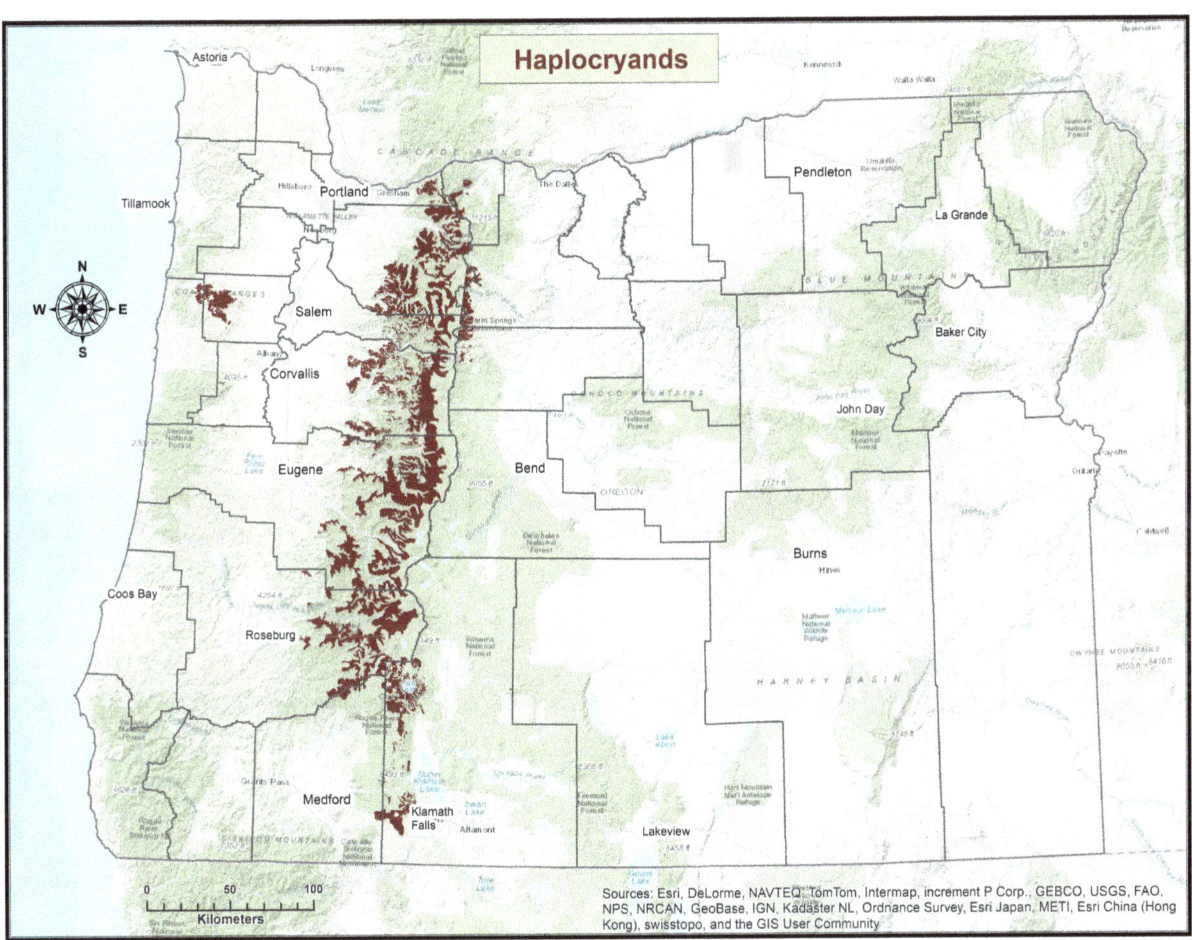

Fig. 7.51 Distribution of Haplocryands in Oregon. *Source* Prepared by Whityn Owen

Fig. 7.52 The Oatman series, a medial-skeletal, amorphic Typic Haplocryands, is formed in glacial deposits composed of andesite, scoriaceous cinders, and basalt and occurs near Klamath Falls. This soil has a dark brown ochric epipedon to 48 cm over a dark brown cambic horizon to 114 cm. The scale is metric down to 1.80 meters. *Source* NRCS photo

61 ± 30 cm (87% of soil series). The Oatman soil series, a medial-skeletal, amorphic Typic Haplocryands, is formed in till derived from andesite, scoriaceous cinders, and basalt on volcanic cones (Fig. 7.52). This soil has a dark brown ochric epipedon from 0 to 48 cm and a dark brown cambic horizon to 114 cm.

7.28 Argialbolls (Soil Region 27)

The Argialbolls are moderately well-developed soils in Oregon and contain mollic, albic, and argillic horizons. They are derived most commonly from glaciolacustrine and alluvial materials on lake terraces and loess, ash, and basalt residuum on lava plateaus. The vegetation on Argialbolls include Oregon white oak, Douglas-fir, and mixed grasses. The mean annual precipitation is 850 ± 325 mm, and the mean annual air temperature is 9.6 ± 1.8 °C. Argialbolls occur at elevations ranging between 390 and 600 m and on maximum slopes of 14 ± 13%.

There are nine Argialbolls soil series, which cover 1,000 km^2 in Oregon. These soils occur primarily in the Willamette Valley and the Palouse Prairie (Fig. 7.53). The most extensive Argialbolls include the Amity, Cowsly, and Lookingglass soil series.

Argialbolls have xeric or aquic soil moisture regimes and mesic or frigid soil temperature regimes. They occur in fine or fine-silty particle-size classes and in smectitic or less commonly mixed mineralogy classes. About one-third of the Argialboll soil series are in the superactive cation-exchange activity class.

Argialbolls have a mollic epipedon averaging 39 ± 17 cm, an albic horizon averaging 15 ± 5.3 cm, and an argillic horizon averaging 62 ± 24 cm. The Wilkins soil series, a fine, smectitic, frigid Xerertic Argialbolls, is formed in loess and minor amounts of volcanic ash over alluvium weathered from basic igneous rocks (Fig. 7.54). The mollic epipedon is black to dark grayish brown and extends from 0 to 19 inches; the albic horizon is light gray to pale brown and occurs to 25 inches, and the argillic horizon is brown and extends to 130 cm.

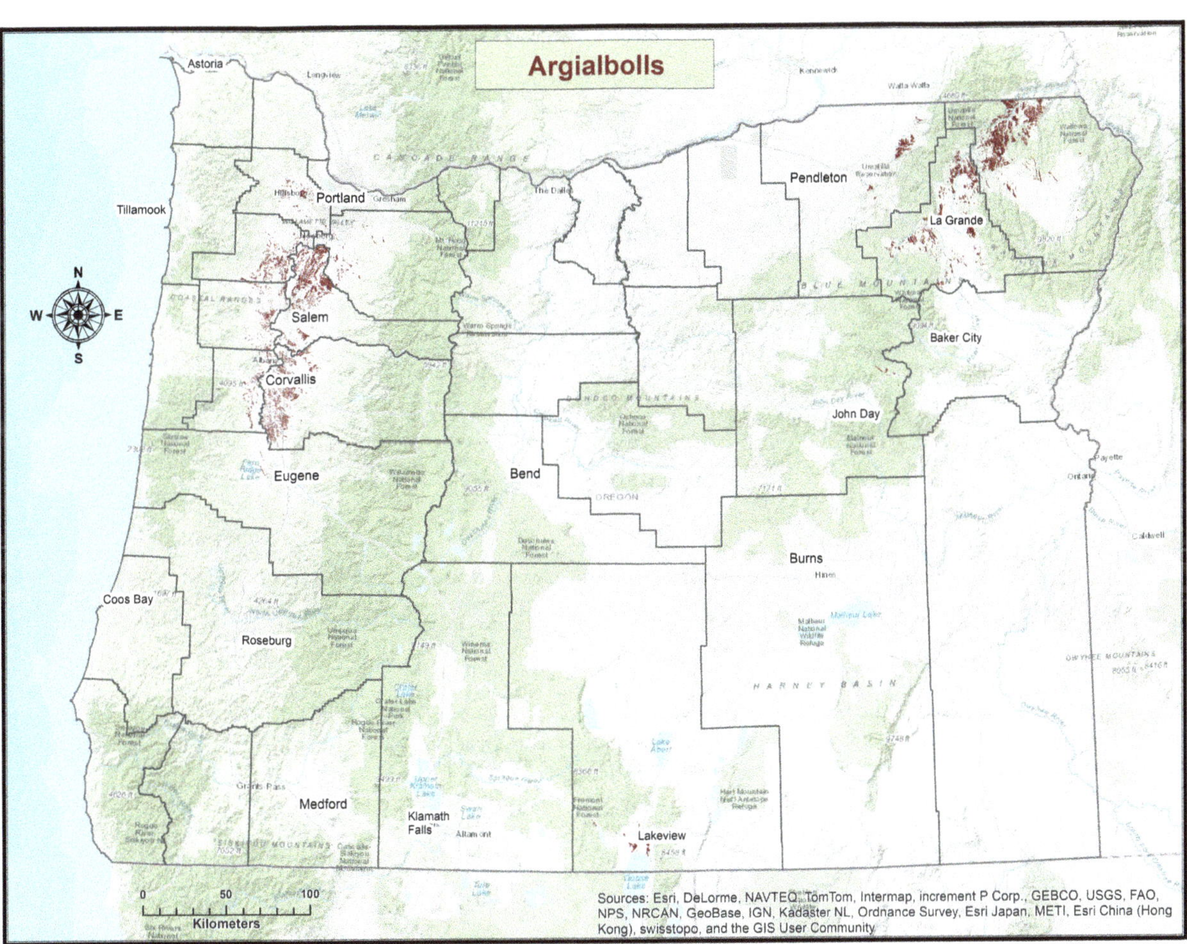

Fig. 7.53 Distribution of Argialbolls in Oregon. *Source* Prepared by Whityn Owen

Fig. 7.54 The Wilkins series, a fine, smectitic, frigid Xerertic Argialbolls, is formed in loess with minor amounts of volcanic ash over alluvium in Wallowa County in northeastern Oregon. This series has a very dark gray mollic epipedon to 48 cm, a light gray albic horizon to 64 cm, and a brown argillic horizon to 132 cm. *Source* NRCS photo

7.29 Palexeralfs (Soil Region 28)

Palexeralfs are well-developed soils with an ochric epipedon over an argillic horizon. They are formed from colluvium on hillslopes and in foothills. The dominant vegetation is Douglas-fir and Oregon white oak. Palexeralfs in Oregon receive 850 ± 500 mm of annual precipitation and have a mean annual air temperature 9.7 ± 2.4 °C. Elevations range from 475 to 1,075 m, and maximum slopes are 20 ± 18%.

The Palexeralfs great group in Oregon has 15 soil series, which cover 1,000 km^2. Palexeralfs are most prevalent in the Siskiyou Mountains, the Willamette Valley, and the Malheur High Plateau (Fig. 7.55). The Swalesilver, Bateman, Saum, and Melbourne soil series are the extensive Palexeralfs in Oregon. Palexeralfs have a xeric soil moisture regime and a mesic or frigid soil temperature regime. In Oregon, 75% of the Palexeralfs are in the fine particle-size class. They are nearly equally divided between the mixed and smectitic mineralogy classes, and 50% are in the superactive or active cation-exchange activity classes. Palexeralfs have an ochric epipedon over an argillic horizon that averages 98 ± 44 cm in thickness. The Ruch series, a fine-loamy, mixed, superactive, mesic Mollic Palexeralfs, is formed in mixed alluvium on high stream terraces, alluvial fans, and foot slopes in valleys within the Klamath Mountains. The soil features a dark brown ochric epipedon to 18 cm overlying an argillic horizon to depths of greater than 150 cm (Fig. 7.56).

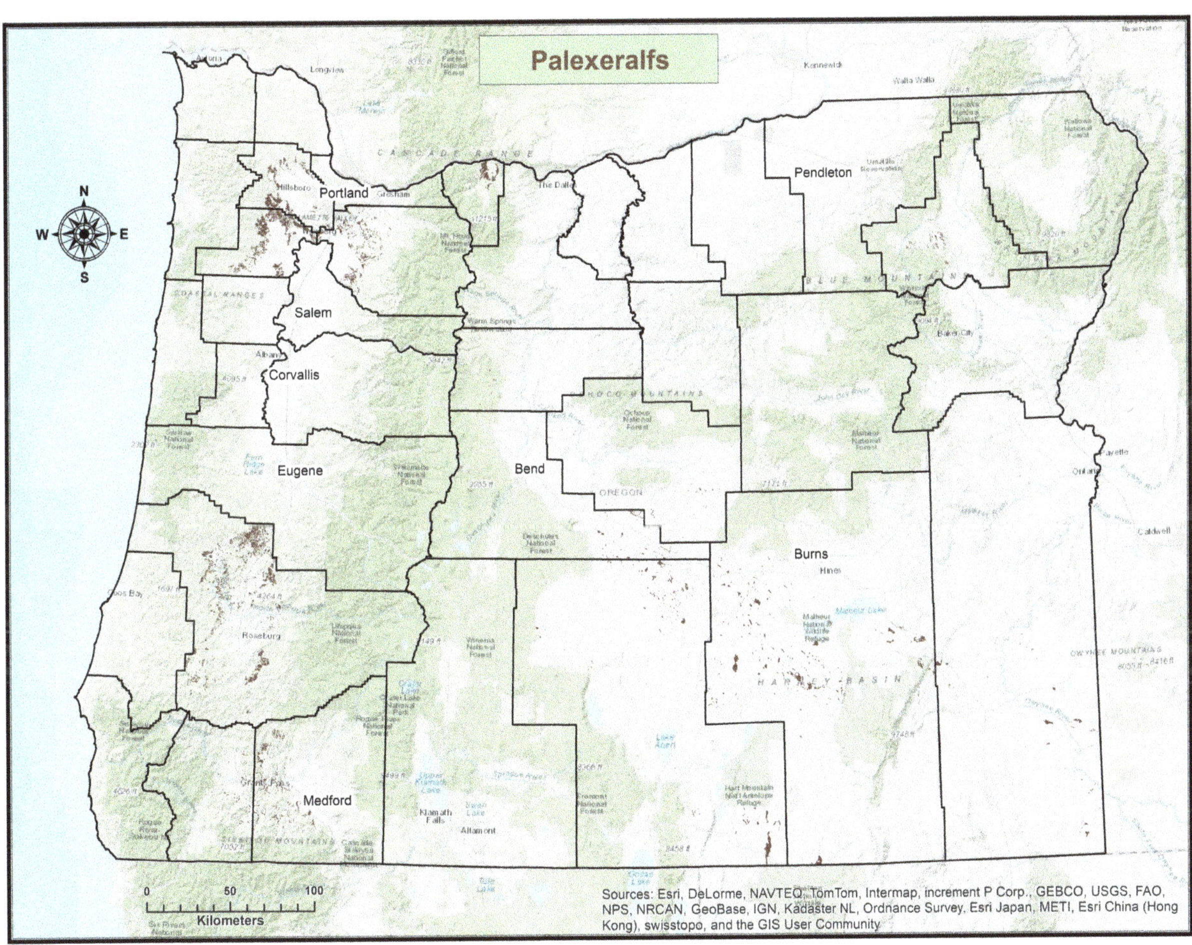

Fig. 7.55 Distribution of Palexeralfs in Oregon. *Source* Prepared by Whityn Owen

Fig. 7.56 The Ruch soil series, a fine-loamy, mixed, superactive, mesic Mollic Palexeralfs, is formed in mixed alluvium on high stream terraces, alluvial fans, and foot slopes in valleys within the Klamath Mountains. This soil features a dark brown ochric epipedon to 18 cm overlying an argillic horizon to depths of greater than 150 cm. *Source* Photo by M. Fillmore

7.30 Summary

Oregon is divided here into 28 soil regions based on the relative abundance of great groups. The great groups are ranked according to decreasing abundance: Haploxerolls, Argixerolls, Humudepts, Haplargids, Argidurids, Vitricryands, Haplocambids, Palexerolls, Haploxeralfs, Haplodurids, Dystroxerepts, Durixerolls, Haplohumults, Vitrixerands, Udivitrands, Haploxerepts, Palehumults, Dystrudepts, Hapludands, Paleargids, Eutrudepts, Fulvudands, Torripsamments, Endoaquolls, Torriorthents, Haplocryands, Argialbolls, and Palexeralfs.

8 Mollisols

8.1 Distribution

Mollisols are the most abundant soil order in Oregon, in terms of numbers of soil series (42%) and soil area (40%) (Table 6.2). Mollisols occur in all 17 Major Land Resource Areas in Oregon, but are most common on the Columbia Plateau, Klamath Basin, Blue Mountain Foothills, Palouse Prairie, Malheur High Plateau, and Willamette Valley (Fig. 8.1). Xerolls are the most abundant suborder in Oregon and comprise 90% of the Mollisol order area. More than three-quarters (78%) of the Mollisol soil area are Haploxerolls and Argixerolls (Table 6.2). In northwestern Oregon, subgroups of Argixerolls are strongly controlled by topographic position (Gelderman and Parsons 1972).

Mollisols exceeding 1,000 km^2 include the Lickskillet, Ninemile, Walla Walla, Condon, Ritzville, Bakeoven, and Carryback soil series. Mollisols covering between 500 and 1,000 km^2 each in Oregon include the Woodburn, Merlin, Klicker, Morrow, Westbutte, Simas, Tub, Anatone, Ateron, Gwinly, and Goodtack soil series. Chapter 7 contains photographs of major great groups of Mollisols, including a Haploxerolls (Josset soil series; Fig. 7.3), Argixerolls (Dixonville series; Fig. 7.5, Palexerolls (Yoncalla series; Fig. 7.17), Durixerolls (Stampede series; Fig. 7.25), Endoaquolls (Wapato series; Fig. 7.48), and Argialbolls (Wilkins soil series; Fig. 7.54).

Photographs of additional Mollisols are given here. The Chapman soil series, a fine-loamy, superactive, mesic Cumulic Ultic Haploxerolls, has developed in alluvium on terraces in the Umpqua River valley of southwestern Oregon (Fig. 8.2). The Packard soil series is not extensive but is an example of Pachic Haploxerolls from Douglas County (Fig. 8.3). This series is formed in alluvium in valleys of the Siskiyou Mountains and features a dark brown mollic epipedon to 80 cm, which contains the cambic horizon in the lower part. The Pilot Rock soil series, a coarse-silty, mixed, superactive, mesic Haploxerollic Durixerolls, has formed in loess over a very gravelly duripan on fan terraces. This soil has a 50-cm thick mollic epipedon with a cambic horizon in the lower one-half, and a duripan from 70 to 115 cm (Fig. 8.4). The Ninemile series is the second most common Argixerolls in Oregon and is formed in colluvium and residuum derived from volcanic rocks on plateaus in the Malheur High Plateau, and Owyhee High Plateau. The Ninemile series has a dark grayish brown mollic epipedon to 20 cm and an argillic horizon from 5 to 26 cm in depth (Fig. 8.5).

The Conser soil series (Vertic Argiaquolls) is an example of a poorly drained Mollisol formed in alluvium in depressions of stream terraces in the Willamette and Umpqua valleys in western Oregon. This soil has a grayish brown mollic epipedon to 70 cm, which includes the upper part of the argillic (Btg) horizon which extends to 100 cm (Fig. 8.6). The argillic horizon has strong brown masses of oxidized iron and grayish brown iron depletions.

Xerolls are predominant on plateaus and in mountain ranges throughout the state; Aquolls occur in scattered river valleys and basins, such as the Willamette Valley and the Malheur High Plateau; and Cryolls occur in the Steens and Hart Mountains in south-central Oregon.

The mean annual air temperature for Mollisols in Oregon is commonly between 6 and 10 °C. Mollisols in Oregon form under cool temperatures. The mean annual precipitation for Mollisols in Oregon ranges broadly between 200 and 850 mm/yr. Mollisols receiving about 300 mm/yr or less have an aridic soil moisture regime and those receiving more

T. Thorson et al., *The Soils of Oregon*, World Soils Book Series,
https://doi.org/10.1007/978-3-030-90091-5_8

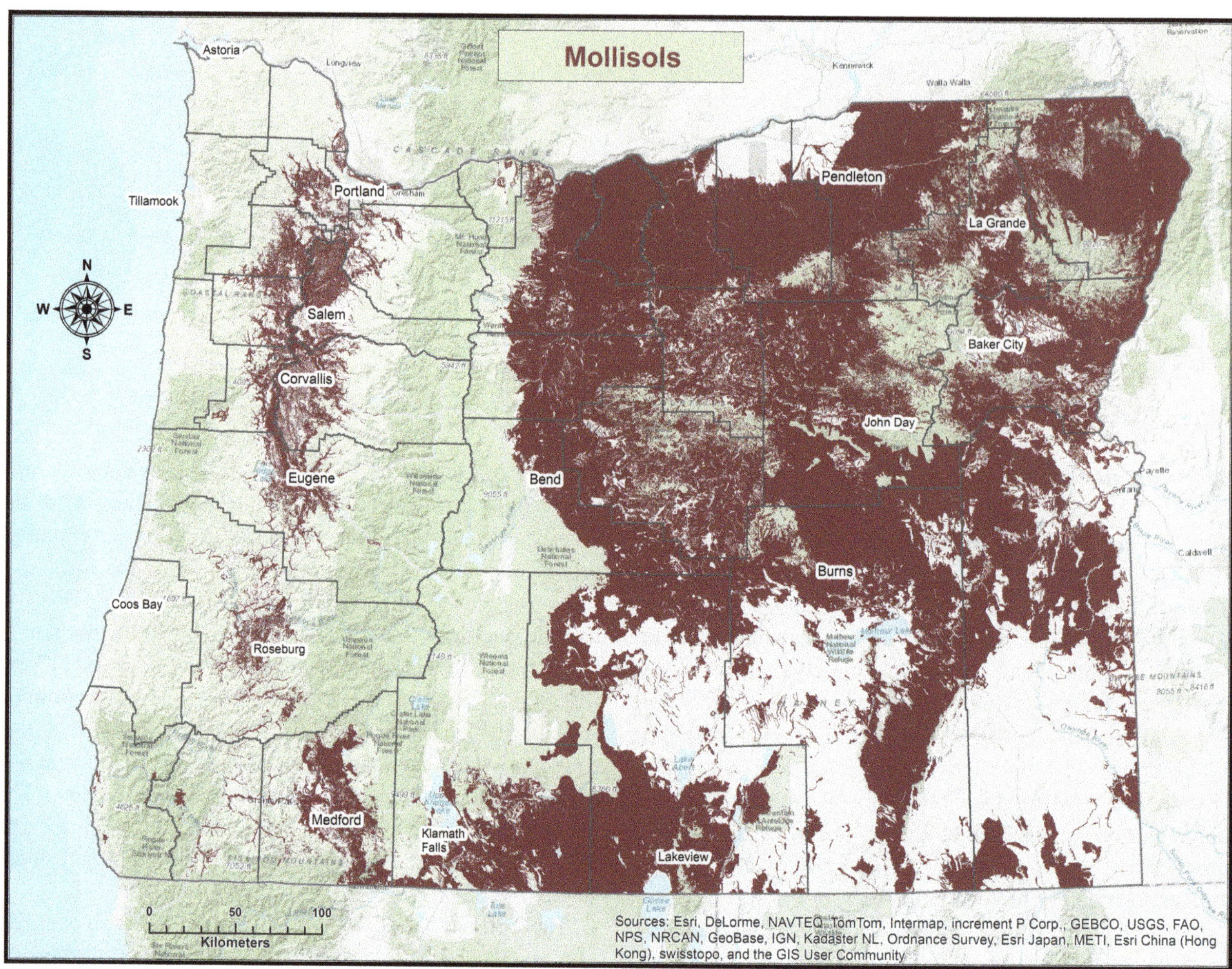

Fig. 8.1 Distribution of Mollisols in Oregon. *Source* Prepared by Whityn Owen

than 300 mm/yr have a xeric soil moisture regime. Mollisols support a variety of plant communities, but are dominated by bunchgrasses, particularly Idaho fescue, bluebunch wheatgrass, Sandberg bluegrass, Thurber's needlegrass, and wildrye. These grasses are commonly associated with sagebrush (*Artemisia* spp.) and are common in the understory of communities containing Douglas-fir, ponderosa pine, Oregon white oak, and western juniper. Mollisols in Oregon are composed of colluvium and residuum often covered by or mixed with volcanic ash and loess, alluvium, and lacustrine or glaciolacustrine materials. These materials generally are derived from basalt or tuff, but may originate from a variety of materials. The most common landforms featuring Mollisols are plateaus, followed by alluvial plains or terraces, mountain slopes, hillslopes, and lake terraces.

8.2 Properties and Processes

The key properties of Mollisols in Oregon are the presence of a mollic epipedon over an argillic (45% of Mollisols) or cambic horizon (41%). About 16% of Mollisols in Oregon lack a diagnostic subsurface horizon. The mean thickness of the mollic epipedon in Oregon soil series is 45 ± 25 cm. More than three-quarters (82%) of the mollic epipedons in Oregon soil series are greater than 25 cm (Fig. 8.7).

Fig. 8.2 The Chapman soil series, a fine-loamy, mixed, superactive, mesic Cumulic Ultic Haploxerolls, has formed in mixed alluvium on terraces in the Umpqua River valley. This soil has a thick (23 in. or 58 cm) mollic epipedon over a 19-inch thick (48 cm) cambic horizon. The scale is in feet. *Source* NRCS photo

Fig. 8.3 The Packard soil series, a loamy-skeletal, mixed, superactive, mesic Pachic Haploxerolls from Douglas County is formed in alluvium in valleys of the Siskiyou Mountains and features a dark brown mollic epipedon to 31 inches (80 cm), which contains the cambic horizon in the lower part. The scale is in feet. *Source* NRCS photo

Mollisols tend to be in loamy (49% of soil series), skeletal (39%), and clayey (30%) particle-size classes. They have a mixed (57%) or smectitic (27%) mineralogy; and they often occur in the superactive cation-exchange activity class (55% of soil series). More than one-half (57%) of the Mollisol soil series have a mesic soil temperature regime, with the remaining 47% in the frigid or cryic regimes. More than three-quarters (88%) of Mollisol soil series have a xeric soil moisture regime. Mollisols are shallow to very deep; however, 11% of the Mollisols are in Lithic subgroups and 53% have a paralithic or lithic contact within 1 m of the surface.

The six Mollisols shown in Table 8.1 have a mollic epipedon that ranges from 23 to 51 cm in thickness. The Amity, Woodburn, and Simas series have an argillic horizon that ranges from 38 to 64 cm; the Willis series has a duripan at 69 cm; and the Wapato and Lickskillet series have cambic horizons.

The dominant soil-forming processes in Mollisols are melanization and biological enrichment of bases. Ancillary processes are cambisolization, argilluviation, gleization, and calcificiation. These are discussed more fully in Chap. 15.

8.3 Use and Management

Mollisols often are highly fertile soils and are used for agriculture, livestock grazing, and forest products.

8.4 Summary

Mollisols are the most common soil order in Oregon, accounting for 40% of the soil series and 42% of the soil area. They are most common in the Palouse Prairie, Columbia Plateau, Klamath Basin, Blue Mountain Foothills, and the Malheur High Plateau. On an area basis, Xerolls account for 90% and Aquolls for 7.7% of the Mollisols. Mollisols in Oregon form under cool and dry conditions. The mean annual precipitation for Mollisols in Oregon is commonly between 200 and 850 mm/yr. Over half (57%) of the Mollisol soil series have a mesic soil temperature regime with the remaining having a cryic or frigid soil temperature regime, and 88% have a xeric soil moisture regime. The dominant vegetation on Mollisols is sagebrush (*Artemisia* spp.) and a variety of bunchgrasses. Western juniper, Douglas-fir, and ponderosa pine are the most common tree

Fig. 8.4 The Pilot Rock soil series, a coarse-silty, mixed, superactive, mesic Haploxerollic Durixerolls, has formed in loess over a very gravelly duripan on fan terraces in Umatilla County. This soil has a 50-cm thick mollic epipedon with a cambic horizon in the lower half, and a duripan from 70 to 115 cm. The photograph shows the upper 2 m of the section. *Source* NRCS photo

species on Mollisols in Oregon. Mollisols in Oregon generally occur on colluvium and residuum overlain or containing volcanic ash and loess on plateaus and mountain slopes. The key properties of Mollisols in Oregon are the presence of a mollic epipedon over an argillic horizon (45% of Mollisols) or a cambic horizon (41% of Mollisols). Mollisols are used extensively for agriculture and livestock grazing.

Fig. 8.5 The Ninemile series, a clayey, smectitic, frigid Aridic Lithic Argixerolls, is formed in colluvium and residuum derived from volcanic rocks on plateaus in the Malheur High Plateau and the Owyhee High Plateau MLRAs. This soil has a dark grayish brown mollic epipedon to 20 cm and an argillic horizon from 5 to 26 cm in depth. The scale is in decimeters on the left and feet on the right. *Source* NRCS photo

Fig. 8.6 The Conser soil series, a fine, mixed, superactive, mesic Vertic Argiaquolls, is formed in alluvium in depressions of stream terraces in the Willamette and Umpqua Valleys of western Oregon. This soil has a grayish brown mollic epipedon to 70 cm, which includes the upper part of the argillic (Btg) horizon that extends to 100 cm. The argillic horizon has strong brown masses of oxidized Fe and grayish brown Fe depletions. The scale is in feet. *Source* NRCS photo

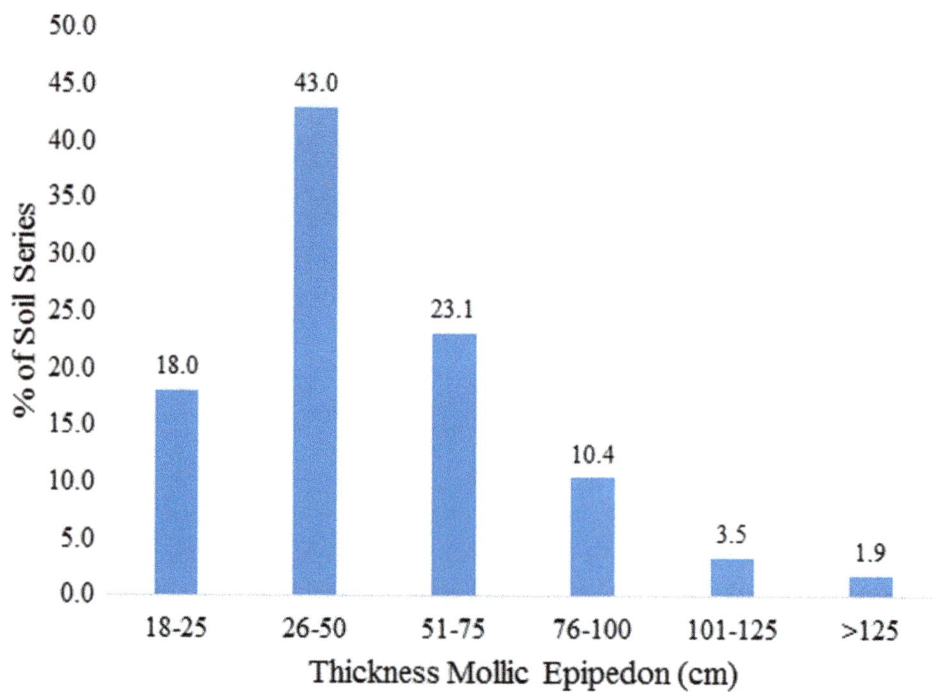

Fig. 8.7 Frequency distribution of mollic epipedon by thickness class. *Source* Prepared by current authors using NRCS databases

Table 8.1 Analyical properties of some Mollisols found in Oregon

Horizon	Depth (cm)	Clay (%)	Silt (%)	Sand (%)	SOC (%)	CEC7 (cmol$_c$/kg)	Base sat (%)	pH H_2O	$CaCO_3$ (%)	$CaSO_4$ (%)	EC (dS/m)	Ex. Na (%)	SAR	Tot. salts (%)	1.5 mPa H_2O/clay
Amity; fine-silty, mixed, superactive, mesic Argiaquic Xeric Argialbolls; Marion, OR; pedon 06N0240															
A	0–37	22.2	65	12.8	3.1	26.1	100	7.2							0.59
E	37–58	21.9	68	10.5	1.9	21.2	73	6.5							0.50
Bt1	58–96	22.2	66	11.8	0.6	16	81	6.2							0.50
Bt2	96–122	18.8	69	12.5	0.2	21.8	99	6.3							0.77
BCt	122–150	17.6	60.6	21.8	0.1	22.1	100	6.6							0.80
Woodburn; fine-silty, mixed, superactive, mesic Aquultic Argixerolls; Marion, OR; pedon 11N1291															
Ap	0–25	21.6	72.2	6.2	2.1	17.6	79	5.7				tr			0.50
A	25–51	22.9	72	5.5	1.9	17.6	77	5.9				–			0.40
2Bt	51–89	20.4	73	6.4	0.7	14.4	92	6.1				tr			0.43
3BCt1	89–112	19.7	70	10.4	0.1	21.4	100	6.4		tr		tr			0.70
3BCt2	112–152	17.0	72	10.8	0.1	21.2	100	6.6				1			0.75
Willis; ashy over loamy, mixed, superactive, mesic Vitritorrandic Durixerolls; Adams, WA; pedon 92P0075															
A1	0–10	5.8	54.1	40.1	0.78	9.7	76	7.2			0.13	1			0.81
A2	10–23	6.0	54.6	39.4	0.57	9.7	84	6.6			0.07	3			0.80
Bw1	23–38	5.7	52.6	41.7	0.54	11.0	88	6.7			0.06	3			1.07
Bw2	38–53	6.2	61.0	32.8	0.40	12.7	97	8.0			0.10	4			1.06
Bk	53–69	7.1	56.7	36.2	0.35	14.6	100	8.7			0.59	17	14	0.1	0.89
Bkqm	69–85	7.6	52.4	40.0	0.22	11.6	100	9.3	14.0		0.94		20	0.1	0.76

(continued)

Table 8.1 (continued)

Horizon	Depth (cm)	Clay (%)	Silt (%)	Sand (%)	SOC (%)	CEC7 (cmol$_c$/kg)	Base sat (%)	pH H_2O	$CaCO_3$ (%)	$CaSO_4$ (%)	EC (dS/m)	Ex. Na (%)	SAR	Tot. salts (%)	1.5 mPa H_2O/clay
Wapato; fine-silty, mixed, superactive, mesic Fluvaquentic Endoaquolls; Jackson, OR; pedon 00P0570															
A1	0–20	28.9	62.6	8.5	2.19	36.0	79	5.5				1			0.59
A2	20–43	31.8	61.7	6.5	2.19	37.7	87	6.0				2			0.63
Bg1	43–64	39.2	55.5	5.3	2.21	42.9	86	6.3				2			0.57
Bg2	64–104	41.4	50.6	8.0	1.32	42.2	86	6.1	tr			1			0.55
2C	104–152	7.7	6.5	85.8	0.16	12.5	86	6.5				3			0.65
Lickskillet; loamy-skeletal, mixed, mesic Lithic Haploxerolls															
A	0–10	10.8	35.6	53.6	0.50		89	6.9			0.30			tr	0.75
B1	10–23	16.1	35.8	48.1	0.50		91	7.2			0.30			tr	0.59
B21	23–36	25.0	30.1	44.9	0.51		79	6.8			0.30			tr	0.58
B22	36–56	25.2	25.5	49.3	0.44		78	6.5			0.30			tr	0.57
R	56														
Simas; fine, montmorillonitic, mesic Aridic Palexerolls															
Ap1	0–10	29.1	32.1	38.8	1.86	28.1	100	7.5			0.91	1	tr	tr	0.53
2Bt11	10–23	54.0	22.3	23.7	1.01	46.8	100	6.9			0.41	3	1	tr	0.51
2Bt12	23–38	60.9	19.6	19.5	0.88	54	100	7.3			0.46	4	2	tr	0.56
2Bt13	38–48	63.6	20.5	15.9	0.69	55.5	100	7.8	2		0.48	4	3	tr	0.52
2Bt21	48–59	62.0	24.1	13.9	0.68	53.5	100	8.1	12		0.51	6	3	tr	0.52
2Bt22	59–71	59.6	26.1	14.3	0.43	50.2	100	8.0	15		0.79	7	4	tr	0.51
2Bt23	71–102	63.8	25.0	11.2		54.1	100	7.9	13		1.32	8	4	0.1	0.51

Bold-face text identifies mollic epipedon (A, Ap, and B1); albic horizon (E), duripan (Bkqm); argillic horizon (Bt)

Reference

Gelderman FW, Parsons RB (1972) Argixerolls on late Pleistocene surfaces in northwestern Oregon. Soil Sci Soc Am Proc 36:335–341

9 Inceptisols

9.1 Distribution

Inceptisols are the third most abundant soil order in Oregon in terms of area, accounting for 24,800 km^2 or 16% of the land area in the state (Table 6.2). However, Inceptisols are ranked second in the number of soil series, accounting for 18% of the total. Inceptisols occur in 13 of the 17 MLRAs in Oregon. They are most common in the Coast Range, the Cascade Mountains, and the Siskiyou Mountains, but also occur in the Willamette Valley, Blue Mountains, the Sitka Spruce Belt, and the Coastal Redwood Belt (Fig. 9.1). They are absent from the Columbia Basin, Columbia Plateau, and the Owyhee High Plateau. The predominant suborder within the Inceptisols are the Udepts (58% of Inceptisol soil area), followed by Xerepts (30%), Aquepts (6.3%), and Cryepts (4.5%). The predominant great groups are Humudepts (40% of Inceptisol soil area), Dystroxerepts (16%), Haploxerepts (12%), Dystrudepts (10%), and Eutrudepts (7.3%).

Inceptisols with areas greater Fig. 9.1. Distribution of Inceptisols in Oregon Fig. 9.1. Distribution of Inceptisols in Oregon Fig. 9.1. Distribution of Inceptisols in Oregon than 1,000 km^2 in Oregon include the Bohannon, Preacher, and Klickitat soil series. Soil series with areas between 500 and 775 km^2 include the Digger, Templeton, Kinney, Umpcoos, Beekman, Jayar, Vermisa, and Blachly. Photographs of major great groups of Inceptisols are provided in Chap. 7, including a Humudepts (Walluski series; Fig. 7.7), Dystroxerepts (Brokeoff series; Fig. 7.23), Dystrudepts (Chamate series; Fig. 7.36), and Eutrudepts (Digger series; Fig. 7.42).

Photographs of three additional great groups, which are less common in Oregon than those reported above, include the Humaquepts (522 km^2), Dystrocryepts (304 km^2), and Durixerepts (81 km^2). The Hebo series, a Typic Humaquepts, has formed in alluvium of mixed materials in the Sitka Spruce Belt and the Coastal Redwood Belt (Fig. 9.2). This soil has a dark gray umbric epipedon to 25 cm and a light to dark gray cambic horizon (Bg) to 65 cm. The cambic horizon has prominent strong brown iron-manganese masses and prominent black manganese coatings on peds. The Lurnick series, an Andic Dystrocryepts, has formed in colluvium and residuum on summits in the Coast Range (Fig. 9.3). This soil has a grayish brown ochric epipedon to 25 cm over a light yellowish brown to very pale brown cambic horizon to 80 cm. Figure 9.4 is a photograph of a Durixerepts in the Lassen Volcanic National Park in northern California that is somewhat comparable to the Agate series in southern Oregon. The Agate series is a fine-loamy, mixed, superactive, mesic Typic Durixerepts that is formed in stratified alluvium. This soil has a dark colored ochric epipedon to 10 cm, with dark brown and light olive brown cambic horizons to 70 cm, which are underlain by an indurated duripan.

The mean annual air temperature for Inceptisols in Oregon ranges between 4 and 10 °C, and the mean annual precipitation ranges between 800 and 2,200 mm. The maximum slope for Inceptisols in Oregon is 77 ± 24%. Udepts support Douglas-fir forest, bigleaf maple, red alder, western hemlock, and western redcedar. Sitka spruce and western hemlock occur on Udepts along the Pacific Coast in the Sitka Spruce Belt. Xerepts support Douglas-fir, Pacific madrone, ponderosa pine, white fir, grand fir, and other species. Aquepts often support cattails, rushes, sedges, and willows, but black greasewood-inland saltgrass communities occur in saline basins. Cryepts feature subalpine fir, Engelmann spruce, and lodgepole pine communities. Inceptisols are composed primarily of colluvium and residuum, but volcanic ash and loess are often present as a mantle or mixed

T. Thorson et al., *The Soils of Oregon*, World Soils Book Series,
https://doi.org/10.1007/978-3-030-90091-5_9

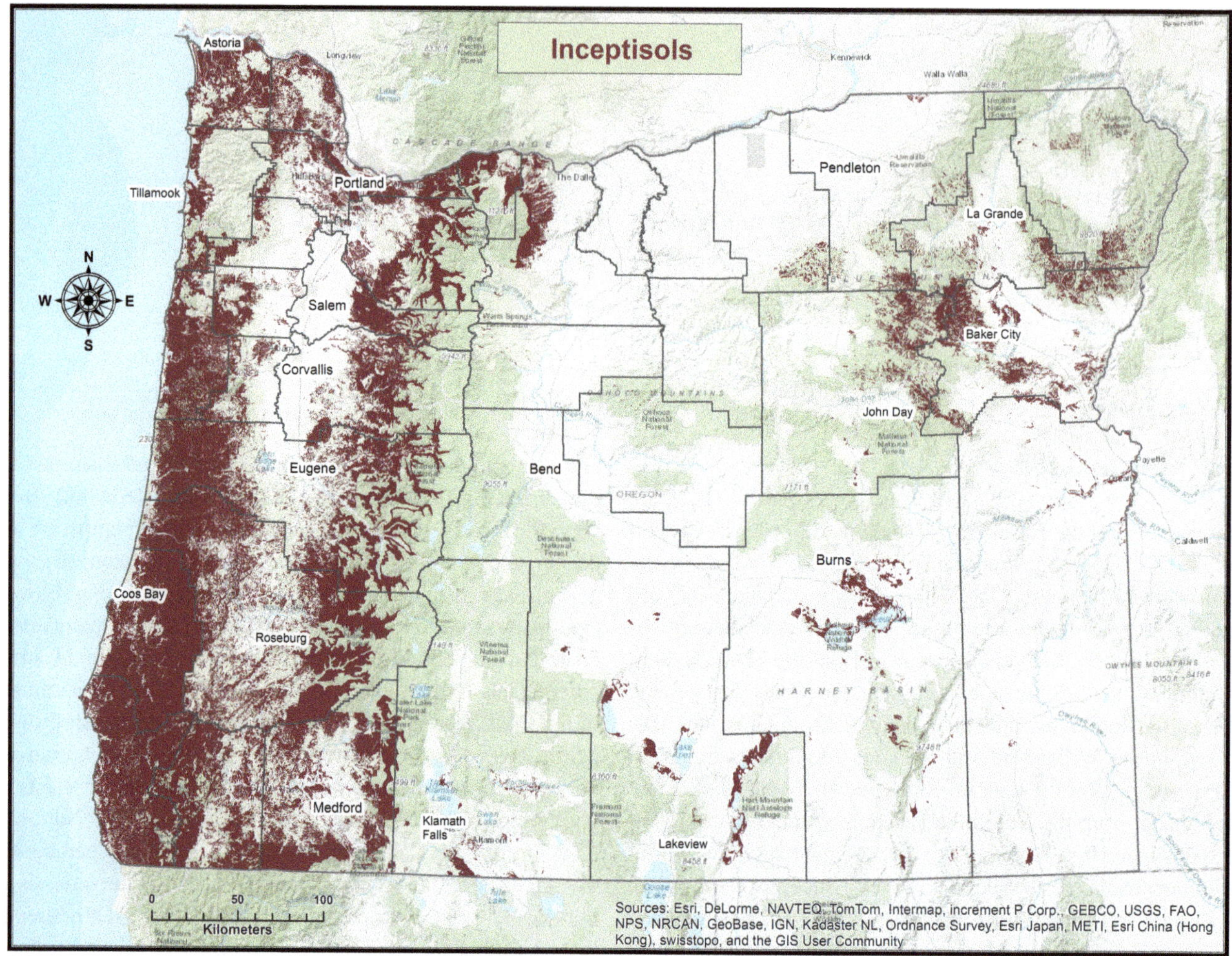

Fig. 9.1 Distribution of Inceptisols in Oregon. *Source* Prepared by Whityn Owen

within the soil profile. These materials are derived from a wide array of rock types, with sandstone–siltstone, basalt, and granite being most common. Nearly 90% of Inceptisols occur on mountains and hills.

9.2 Properties and Processes

Inceptisols usually contain an ochric (58%) or an umbric (39%) epipedon over a cambic horizon (91%), which average 20 ± 8.6 cm, 40 ± 15 cm, and 54 ± 28 cm in thickness, respectively. More than one-third (40%) of Oregon Inceptisols are in the loamy-skeletal particle-size class; 72% are in loamy (loamy, fine-loamy, coarse-loamy, and loamy-skeletal) particle-size classes and 44% are in skeletal classes (loamy-skeletal, clayey-skeletal, and sandy-skeletal). More than one-half (53%) of Inceptisols are in the isotic mineral class and 41% are in the mixed class. About 40% of Inceptisols are in the active or superactive cation-exchange activity class. More than one-half (48%) of Inceptisols have a mesic soil temperature regime, 37% are frigid or cryic, and the remaining 15% are isomesic. More than one-half (52%) of Inceptisols have a udic soil moisture regime, 33% are xeric, and 15% are aquic. Nearly two-thirds (63%) of the Inceptisols in Oregon have a lithic or paralithic contact within the upper 100 cm.

Fig. 9.2 The Hebo series, a fine, isotic, acid, isomesic Typic Humaquepts, has formed in alluvium of mixed materials in the Sitka Spruce Belt. This soil has a dark gray umbric epipedon to 25 cm and a light to dark gray cambic horizon (Bg) to 65 cm. The cambic horizon has prominent strong brown iron-manganese masses and prominent black manganese coatings on peds. Scale is in feet. *Source* NRCS photo

Analytical data for five representative Inceptisols are given in Table 9.1. Each of these soils has an ochric epipedon, except the Preacher series, which has an umbric epipedon. Each has a cambic horizon that ranges from more than 36–94 cm. The dominant processes in Inceptisols are cambisolization, humification, and gleization.

Fig. 9.3 The Lurnick series, a clayey-skeletal, isotic Andic Dystrocryepts, has formed in colluvium and residuum on summits in the Coast Range. This soil has a grayish brown ochric epipedon to 25 cm over a light yellowish brown to very pale brown cambic horizon to 80 cm. A Cr horizon begins with the abundant siltstone fragments. The photograph shows the upper 75 cm of the soil profile. *Source* NRCS photo

9.3 Use and Management

Inceptisols in Oregon are managed for forest products, recreation, and watershed protection.

9.4 Summary

Inceptisols are the second most abundant soil order in Oregon, accounting for 18% of the soil series and 16% of the land area in the state. Inceptisols occur in 13 of the MLRAs in Oregon. They are most common in the Coast Range, the Cascade Mountains, and the Siskiyou Mountains. They are absent from the Columbia Basin, Columbia Plateau, and the Owyhee High Plateau. The dominant suborders within the Inceptisols are the Udepts, Xerepts, Aquepts, and Cryepts. The predominant great groups are Humudepts, Dystroxerepts, Haploxerepts, and Dystrudepts. Inceptisols with areas greater than 1,000 km^2 in Oregon include the Bohannon, Preacher, and Klickitat soil series.

The mean annual air temperature for Inceptisols in Oregon ranges between 4 and 10 °C and the mean annual precipitation ranges between 800 and 2,200 mm. The maximum slope for Inceptisols in Oregon is 77 ± 24%. Udepts support Douglas-fir bigleaf maple, red alder, western hemlock, western redcedar, and Sitka spruce. Xerepts support Douglas-fir, Pacific madrone, ponderosa pine, white fir, and grand fir. Aquepts often have cattails, rushes, sedges, and willows, but black greasewood-inland saltgrass communities occur in saline basins. Cryepts feature subalpine fir, Engelmann spruce, and lodgepole pine communities. Inceptisols

Fig. 9.4 This photograph of a Durixerepts in the Lassen Volcanic National Park in northern California is somewhat comparable to the Agate series in southern Oregon in the Siskiyou Mountains. The Agate series is a fine-loamy, mixed, superactive, mesic Typic Durixerepts that is formed in stratified alluvium. The Agate soil series has a dark colored ochric epipedon to 10 cm, dark brown and light olive brown cambic horizons to 70 cm, which are underlain by an indurated duripan. The scale is in centimeters. *Source* NRCS photo

Table 9.1 Analyical properties of some Inceptisols found in Oregon

Horizon	Depth (cm)	Clay (%)	Silt (%)	Sand (%)	SOC (%)	CEC7 (cmol$_c$/kg)	Base sat (%)	pH H_2O	$CaCO_3$ (%)	$CaSO_4$ (%)	EC (dS/m)	Ex. Na (%)	SAR	Tot. salts (%)	1.5 mPa H_2O/clay
Multnomah; coarse-loamy over sandy or sandy-skeletal, mixed, superactive, mesic Humic Dystroxerepts; Multnomah, OR;															
pedon no. 70C0065															
Ap	0–20	10.4	56.8	32.8	2.98	21	40	5.6				tr			1.05
B1	20–41	9.4	58.9	31.7	1.77	17.8	51	6.1				1			1.09
B2	41–63	7.3	58.5	34.2	0.43	13.8	50	6.1				1			1.27
C	63–99	6.5	52.8	40.7	0.11	15.8	51	6				1			1.37
2C	99–140	4.5	9.2	86.3		21.1	64	6.1				1			1.89

(continued)

Table 9.1 (continued)

Horizon	Depth (cm)	Clay (%)	Silt (%)	Sand (%)	SOC (%)	CEC7 ($cmol_c$/kg)	Base sat (%)	pH H_2O	$CaCO_3$ (%)	$CaSO_4$ (%)	EC (dS/m)	Ex. Na (%)	SAR	Tot. salts (%)	1.5 mPa H_2O/clay
Blachly; fine, isotic, mesic Humic Dystrudepts; Lane, OR; pedon no. 12N7672															
A2	6–18	27.4	42.7	3.7	4.91	33.5	21	5.2							0.83
BA	18–29	33.9	24.6	6.1	2.83	27.8	29	5.6							0.65
2Bw1	29–60	43.1	17.7	15.2	1.70	25.6	23	5.3							0.55
2Bw2	60–79	55.3	14.1	29.2	0.74	24.4	16	5.2							0.46
2Bw3	79–132	53.4	14	25.7	0.34	23.0	14	5.4							0.49
2BC	132–200	48.5	16.6	17.9	0.42	20.1	28	5.5							0.53
Digger; loamy-skeletal, isotic, mesic Dystric Eutrudepts; Benton, OR; pedon no. 01N0323															
A	3–10	19.3	32.7	48.0	3.07	23.8	66	5.9				2			0.76
BA	10–41	22.2	35.7	42.1	1.98	21.8	59	5.6				3			0.65
Bw1	41–76	25.2	40.9	33.9	1.02	27.6	61	5.5				1			0.60
Bw2	76–97	22.7	40.7	36.6	0.62	33.7	69	5.4				2			0.64
Lostbasin; loamy-skeletal, mixed, superactive, frigid Typic Haploxerepts; Baker, OR; pedon no. 85P0835															
A	0–13	18.0	34.6	47.4	1.17	10.4	100	7.0							0.39
Bw1	13–33	37.3	32.9	29.8	0.85	19.5	100	6.5				1			0.32
Bw2	33–58	30.6	20.8	48.6	0.70	22.3	100	7.0				1			0.43
Bw3	58–71	21.7	24.1	54.2	0.43	19.9	100	7.0				1			0.45
Preacher; fine-loamy, isotic, mesic Andic Humudepts; Lincoln, OR; pedon no. 84P0898															
A	0–13	26.3	35.3	38.4	10.64	49.9	23	4.5				1			0.86
AB	13–33	21.9	41.6	36.5	4.43	32.6	7	5.0				1			0.74
Bw1	33–81	17.3	28.2	54.5	0.25	23.9	3	4.9							0.94
Bw2	81–127	13.0	25.5	61.5	0.16	22.2	2	4.9							1.12
BC	127–147	13.8	27.6	58.6	0.22	25.5	1	4.9							0.86
2Bwb	147–170	22.2	31.5	46.3	0.30	25.0	3	4.9				tr			0.78

Bold-face text identifies cambic horizon (B1, B2, Bw)

are composed primarily of colluvium and residuum derived from a wide array of rock types, with sandstone–siltstone, basalt, and granite being most common. They often have a mantle of volcanic ash and loess which also may be mixed within the soil profile. Nearly 90% of Inceptisols occur on mountains and hills.

Inceptisols usually contain an ochric or an umbric epipedon over a cambic horizon that average 20 ± 8.6 cm, 40 ± 15 cm, and 54 ± 28 cm in thickness, respectively. Inceptisols are most common in loamy and skeletal particle-size classes, the isotic or mixed mineral class; the mesic, frigid, and cryic soil temperature regimes, and the udic and xeric soil moisture regimes. Nearly two-thirds of the Inceptisols in Oregon have a lithic or paralithic contact within the upper 100 cm. The dominant processes in Inceptisols are cambisolization, humification, and gleization.

10 Aridisols

10.1 Distribution

Aridisols are the second most abundant soil order in Oregon in terms of land area (27,300 km^2; 17% of total) and are ranked third in number of soil series (13%) (Table 6.2). Aridisols occur in 7 of the 17 Major Land Resource Areas in Oregon but are most common in the Malheur High Plateau, Owyhee High Plateau, Snake River Plains, and Columbia Basin (Fig. 10.1). On an area basis, 39% of the Aridisol soil series are Durids, followed by Argids (29%), Cambids (28%), Calcids (3.0%), and Salids (1.0%) (Table 6.2). Oregon ranks third nationally in area of Durids. Four great groups account for 79% of the Aridisols soil area, including the Haplocambids (25%), Argidurids (20%), Haplodurids (17%), and Haplargids (17%).

The Raz, Brace, and Anawalt soil series each cover more than 1,000 km^2 in Oregon, followed by the Felcher and Actem soil series, which occupy between 500 and 800 km^2. Photographs of major great groups of Aridisols are given in Chap. 7, including an Argidurids (Chilcott series; Fig. 7.10), Haplocambids (Shano series; Fig. 7.15), Haplodurids (Taunton series; Fig. 7.21), and Paleargids (Gooding series; Fig. 7.40).

The Ausmus soil series, an Aquic Natrargids, has formed in alluvium and lacustrine deposits from volcanic rocks and volcanic ash in the Harney Basin in Harney County. This soil has a dark grayish brown ochric epipedon to 5 cm, a dark grayish brown natric horizon to 23 cm, and a calcic horizon from 5 to 72 cm (i.e., the calcic horizon includes natric horizon) (Fig. 10.2).

Durids occur on plateaus; Argids, Cambids, and Calcids are found mainly on Pleistocene lake terraces; and Salids are restricted to basin floors. Aridisols in Oregon have a mean annual air temperature ranging between 6.5 and 9.5 °C and commonly receive from 200 to 300 mm of water-equivalent precipitation per year. They support sagebrush communities dominated by Wyoming big sagebrush, basin big sagebrush, bud sagebrush, low sagebrush, or mountain big sagebrush, along with bunchgrasses, such as basin wildrye, Sandberg bluegrass, Indian ricegrass, bluebunch wheatgrass, and Thurber's needlegrass. Saline basins contain black greasewood, shadscale, spiny hopsage, and inland saltgrass. Aridisols form on slopes ranging between 0 and 50%. Aridisols in Oregon are derived from lacustrine sediments, colluvium, residuum, and alluvium, often with a mixed mantle of eolian sand or loess. Typical landforms include lake terraces, plateaus, and alluvial fans. Aridisols may form in less than 6,700 years (Alexander and Nettleton, 1977).

10.2 Properties and Processes

By definition, Aridisols have an aridic soil moisture regime, in which the soil control section is dry in all parts for more than half of the cumulative days per year, when the soil temperature at a depth of 50 cm below the soil surface is

T. Thorson et al., *The Soils of Oregon*, World Soils Book Series,
https://doi.org/10.1007/978-3-030-90091-5_10

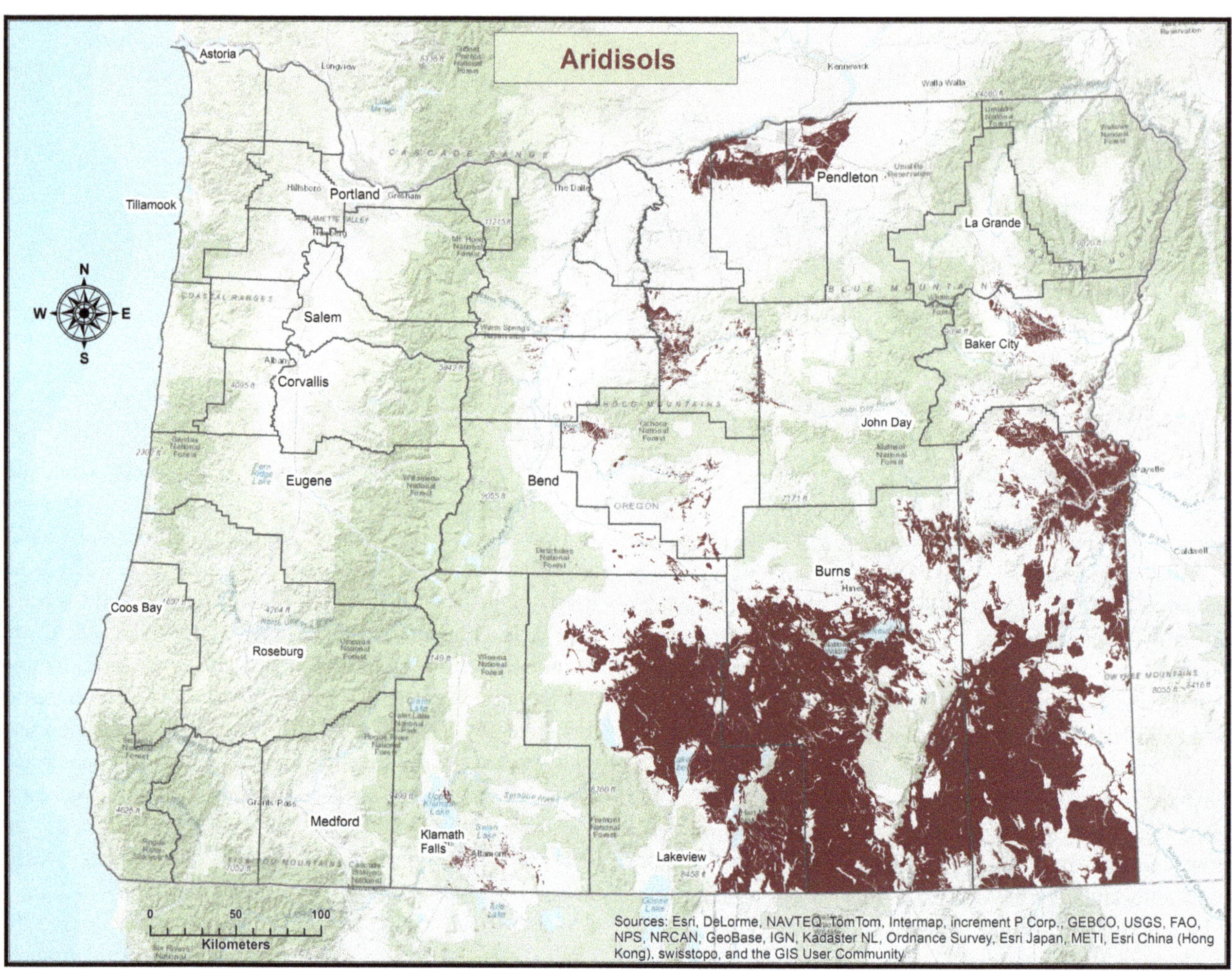

Fig. 10.1 Distribution of Aridisols in Oregon. *Source* Prepared by Whityn Owen

above 5 °C, and moist in some or all parts for less than 90 consecutive days, when the soil temperature at a depth of 50 cm below the soil surface is above 8 °C.

The key properties of Aridisols in Oregon are the presence of an ochric epipedon over a duripan, cambic, argillic, calcic or salic horizon. The ochric epipedon averages 17 ± 6.4 cm in thickness, the cambic horizon 40 ± 25 cm, the argillic horizon 33 ± 19 cm, the duripan 26 ± 17 cm, the calcic horizon 62 ± 43 cm, and the salic horizon 62 ± 23 cm.

Aridisols in Oregon tend to be in loamy (54%) or clayey (25%) particle-size classes, have a mixed (65%) or smectitic (25%) mineralogy, and are in the superactive cation-exchange activity class (61%). Aridisols commonly

Fig. 10.2 The Ausmus soil series, a fine-silty, mixed, superactive, frigid Aquic Natrargids, is formed in alluvium and lacustrine deposits derived from volcanic rocks and volcanic ash in the Malheur Lake region. This soil has a dark grayish brown ochric epipedon to 5 cm, a dark grayish brown natric horizon to 23 cm, and a calcic horizon from 5 to 72 cm (i.e., the calcic horizon includes the natric horizon). The scale is in centimeters. *Source* NRCS photo

have a mesic (65% of soil series) or frigid (37%) soil temperature regime. Although only 16% of the Aridisols are in the shallow family class, 29% have a paralithic or lithic contact within the upper meter.

Because of their aridity, most Aridisols in Oregon have a high base saturation, an alkaline pH, and low soil organic carbon concentrations (Table 10.1). Diagnostic subsurface horizons are designated in bold face in Table 10.1. The selected soils have an argillic (Brace, Nevador, and Sorf series), duripan (Brace and Henley series), cambic (Abert series), calcic (McBain series), or a salic (Icene series). Gypsids and Cryids have not been identified in Oregon.

Table 10.1 Analytical properties of some Aridisols found in Oregon

Horizon[1]	Depth (cm)	Clay (%)	Silt (%)	Sand (%)	SOC (%)	CEC7 (cmol$_c$/kg)	Base sat (%)	pH H_2O	$CaCO_3$ (%)	$CaSO_4$ (%)	EC (dS/m)	Ex. Na (%)	SAR	Tot. salts (%)	1.5 mPa H_2O/clay
Brace; fine-loamy, mixed, superactive, frigid, shallow Xeric Argidurids; Harney, OR; pedon no. 03N0327															
A	0–8	17.7	33.7	48.6	1.22	17.9	93	6.9	2						0.60
Bt1	8–18	27.8	27.3	44.9	0.68	23.3	93	7.1				2			0.47
Bt2	18–41	36.5	21.0	42.5	0.59	29.6	93	7.2	5			3			0.46
Btkqm	41–66														
2R	66														
Nevador; fine-loamy, mixed, superactive, mesic Durinodic Xeric Haplargids; Malheur, OR; pedon no. 10N0420															
A1	0–7	8.1	30.1	61.8	0.8	10.4	88	5.9			0.49	tr	1	0.1	0.64
A2	7–14	12.9	34.7	52.4	0.4	11.9	92	6.5			0.13	2			0.40
AB	14–24	12.0	32.3	55.7	0.2	11.2	91	7.2			0.06	7			0.43
Bt	24–40	28.5	27.0	44.5	0.4	30.7	98	7.8			0.18	15			0.48
Bkq1	40–57	18.0	29.0	53.0	0.2	36.5	100	8.3	1		0.66	16	11	0.1	0.95
Bkq2	57–100	4.4	10.0	85.6	0.1	19.2	100	8.4	3		0.76	20	13	0.1	1.75
Bkq3	100–128	5.0	15.2	79.8	tr	13.7	100	8.5	1		0.55	20	15	tr	1.08
Bk	128–160	6.6	14.9	78.5	tr	13.3	100	8.6	1		0.58	22	17	tr	0.86
Abert; ashy, glassy, frigid Sodic Xeric Haplocambids; Lake, OR; pedon no. 79P0431															
A	0–7	14.7	30.3	55.0	1.36	24.8	100	7.6	6			1			0.82
Bw1	7–20	13.8	28.5	57.7	0.78	25.4	100	7.9	tr			2			0.96
Bw2	20–42	10.4	25.6	64.0	0.64	23.6	100	8.2	3			3			1.24
Bk1	42–67	9.2	27.7	63.1	0.61	29.5	100	8.7	4			18			1.96
Bk2	67–106	8.4	46.5	45.1	0.33	35.5	100	8.5	2			54			2.98
C	106–164	5.7	40.9	40.8	0.20	26.3	100	7.8		6		79			1.16
Henley; coarse-loamy, mixed, superactive, mesic Aquic Haplodurids; Klamath, OR; pedon no. 67C0034															
Ap	0–18	17.2	28.0	54.8	1.49	36.0	100	7.7	2.0		0.8	3	1		0.93
B11	18–33	16.3	23.3	60.4	0.82	34.8	84	8.1	tr			8			0.82
B12	33–53	15.7	34.4	49.9	0.8	30.0	100	8.4	2.0			13			0.85
B21	53–71	14.1	30.2	55.7	0.67	36.6	100	8.3	3			11			1.13
B22	71–91	12.3	27.3	60.4	0.54	40.3	100	8.2	5.0			15			1.48
2Bqm1	91–112	6.8	17.8	75.4	0.34	41.4	100	8.3	7			17			3.00
2Bqm2	112–127	5.2	16.1	78.7	0.13	43.2	100	8.6	4.0			32			2.73
Sorf; fine, smectitic, mesic Vertic Paleargids; Wheeler, OR; pedon no. 99P0247															
A	0–12	13.5	26.5	60.0	0.5	19.5	100	7.6	tr		0.21	6			0.61
2Bt	12–23	56.0	19.7	24.3	0.4	63.5	100	8.0	1		0.33	9	4	tr	0.56
2Btss	23–30	50.9	28.5	20.6	0.2	67.7	100	8.4	8		0.41	11	4	tr	0.60
2Bk1	30–53	48.3	33.6	18.1	0.1	71.3	100	8.1	5		0.98	12	10	0.1	0.66

(continued)

Table 10.1 (continued)

Horizon[1]	Depth (cm)	Clay (%)	Silt (%)	Sand (%)	SOC (%)	CEC7 ($cmol_c$/kg)	Base sat (%)	pH H_2O	$CaCO_3$ (%)	$CaSO_4$ (%)	EC (dS/m)	Ex. Na (%)	SAR	Tot. salts (%)	1.5 mPa H_2O/clay
McBain; fine-loamy, mixed, superactive, frigid Sodic Xeric Haplocalcids; Harney, OR; pedon no 96P0560															
Akzn	0–13	17.6	64.1	18.3	2.5	26.8	100	9.0	13		30.10	92	125	1.4	0.56
Abkzn	13–28	17.8	59.1	23.1	1.4	30.4	100	8.9	23		8.63	53	50	0.4	0.93
Bk1	28–55	18.6	57.9	23.5	0.9	31.1	100	8.4	25		8.22	27	14	0.3	0.96
Bk3	67–93	7.9	48.1	44.0	0.1	29.3	100	8.0	tr		3.05	10	5	0.1	1.51
Icene; fine-loamy, mixed, superactive, mesic Typic Aquisalids; Lake, OR; pedon no. 86P0989															
Akn	0–13	4.0	11.1	84.9	0.8			9.1	tr		1.95		19	0.1	1.48
2Akn	13–25	29.1	26.9	44.0	0.6			9.2	tr		9.79		77	0.4	0.62
3Bknz	25–58	14.8	36.6	48.6	0.6			8.6	10		20.50		109	2.4	1.66
3Bknz	58–71	12.6	68.6	18.8	0.3			8.4	5		37.50		97	2.7	2.59
4Cknz	71–165	38.4	56.0	5.6	0.2			8.0	2		24.70		59	2.1	0.92

[1]Horizons in bold-face are diagnostic: Bt = argillic; Bw = cambic; Bqm = duripan; Bk = calcic; Bz - salic

The dominant soil-forming processes in Aridisols are silicification, argilluviation, cambisolization, calcificiation, and salinization, which are discussed fully in Chap. 15.

10.3 Use and Management

As with all of the orders in *Soil Taxonomy*, there is considerable variation in the nature and properties of Aridisols. However, the key property that links all of the Aridisols is the aridic soil moisture regime. For this reason, some Aridisols may be cultivated, especially the Cambids and some Argids, but irrigation is a prerequisite for most agriculture. Some of the Aridisols are shallow, including 50% of the Durids, which have a duripan. Most Aridisols are used for livestock grazing and wildlife.

10.4 Summary

Aridisols are the second most common soil order in Oregon, accounting for 17% of the land area. They occur in 7 of the 17 MLRAs, but are most common on the Malheur High Plateau, the Owyhee High Plateau, the Snake River Plains, and in the Columbia Basin. On an area basis, Aridisols in Oregon can be ranked: Durids > Argids, Cambids > Calcids, Salids. Aridisols in Oregon commonly receive from 200 to 300 mm of precipitation per year, occur at low- to middle elevations under desert shrubs, usually are derived from lacustrine materials or alluvium, and may form in less than 6,700 years. The key properties of Aridisols in Oregon are the presence of an ochric epipedon over a duripan or a cambic, argillic, calcic, or salic horizon.

Reference

Alexander EB, Nettleton WD (1977) Post-Mazama Natrargids in Dixie Valley, Nevada. Soil Sci Soc Am J 41:1210–1212

11 Andisols

11.1 Distribution

Andisols are the fourth most abundant soil order in Oregon in terms of land area (13%) and the number of soil series (12%) (Table 6.2). Andisols occur in 10 of the 17 MLRAs but are most common in the Cascade Mountains—Eastern Slope, the Coast Range, the Cascade Mountains, and the Blue Mountains (Fig. 11.1). However, because of incomplete detailed mapping in the Cascade Mountains, the Blue Mountains, and the Klamath Basin, it is likely that Andisols are far more abundant in Oregon. Andisols have formed primarily from the eruption of Mt. Mazama in the Crater Lake area 7,700 years ago (Fig. 3.17), but andic properties also occur in cool, humid climates of the Coast Range and Sitka Spruce Belt in materials with abundant organic carbon and high iron and aluminum without the influence of volcanic glass. The predominant suborders include the Xerands (29%), Udands (28%), Cryands (28%), and Vitrands (13%) (Table 6.2). Five great groups account for 91% of the Andisols on an area basis, including the Vitrixerands (23%), Vitricryands (19%), Hapludands (17%), Udivitrands (13%), Fulvudands (11%), and Haplocryands (7.7%).

The most extensive Andisols include the Tolo, Woodcock, Lapine, Klootchie, Syrupcreek, and Limberjim soil series, each of which occupies an area between 450 and 665 km^2. Photographs of major great groups of Andisols are given in Chap. 7, including Vitricryands (Lapine series, Fig. 7.12 and the Castlecrest, soil series Fig. 7.13); Vitrixerands (Threebuck series, Fig. 7.29); Udivitrands (Tamara series, Fig. 7.31); Hapludands (Harslow series, Fig. 7.38); Fulvudands (Klootchie series, Fig. 7.44), and Haplocryands (Oatman soil series, Fig. 7.52).

Additional photographs are provided here of important but less common Andisols, including Fulvudands, Duricryands, Fulvicryands, Melanudands, and Haploxerands. Fulvudands are common in the Sitka Spruce Belt. The Lebam series, a medial over clayey, ferrihydritic over isotic, isomesic Typic Fulvudands, is formed in residuum from tuffaceous sedimentary rocks on slopes of the Sitka Spruce Belt. This soil has a dark brown umbric epipedon to 33 cm and a deep, yellowish brown cambic horizon to 157 cm (Fig. 11.2).

Duricryands occur only in the Crater Lake region of the Cascade Mountains and have a total area of only 44 km^2. The Grousehill series, a medial-skeletal, amorphic Oxyaquic Duricryands, is formed in volcanic ash over till on ridges and benches. This soil has a dark brown ochric epipedon to 25 cm, a dark grayish brown to dark brown cambic horizon to 100 cm, and a dark gray, moderately cemented duripan below (Fig. 11.3). Fulvicryands occupy 164 km^2 in the Coast Range, the Cascade Mountains, and the Siskiyou Mountains of western Oregon. The Newanna series, a medial-skeletal, ferrihydritic typic fulvicryands, is formed in colluvium and residuum over fractured basalt. This soil has a dark reddish brown umbric epipedon to 30 cm and a strong brown cambic horizon over fractured basalt at 65 cm (Fig. 11.4).

Melanudands occupy only 32 km^2 in Oregon, entirely in the Sitka Spruce Belt in Tillamook County. The Quillamook series, a medial, ferrihydritic, isomesic Pachic Melanudands, is formed in alluvium on stream terraces. This soil has a dark grayish brown, smeary melanic epipedon (Pachic subgroup) to 53 cm, followed by a light yellowish brown cambic horizon to 147 cm (Fig. 11.5). Figure 11.6 is an Aquic

T. Thorson et al., *The Soils of Oregon*, World Soils Book Series,
https://doi.org/10.1007/978-3-030-90091-5_11

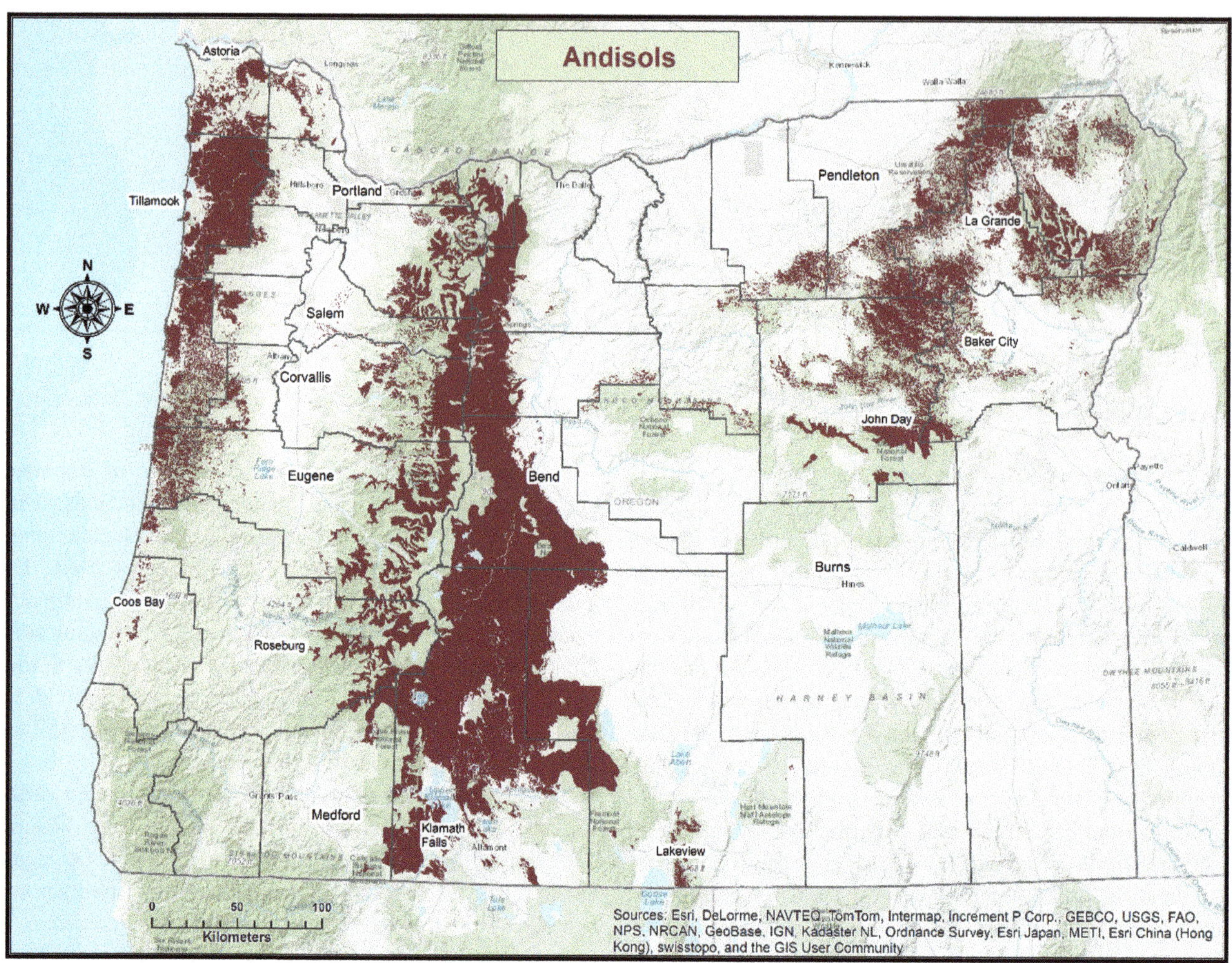

Fig. 11.1 Distribution of Andisols in Oregon. *Source* Prepared by Whityn Owen

Haploxerands from Lassen Volcanic National Park that contains a mollic epipedon over a cambic horizon. This soil is somewhat comparable to the Hot Lake series in northeast Oregon, Union County, which also contains a mollic epipedon over a cambic horizon.

Andisols in Oregon have a mean annual air temperature ranging between 3.5 and 8 °C and receive between 650 and 2,200 mm/yr of precipitation. The vegetation on Andisols is predominantly Douglas-fir forests with bigleaf maple, grand fir, western hemlock, red alder, Sitka spruce, and western redcedar; subalpine forests of Engelmann spruce, subalpine fir, and lodgepole pine; and ponderosa pine forests with lodgepole pine, grand fir, and white fir. Maximum slopes commonly range from 40 to 90%. Parent materials include volcanic ash, pumice, and cinders or colluvium and residuum with a high amount of organic matter and extractable iron and aluminum in a cool, humid climate. Typical landforms are mountain slopes, lava plateaus, and volcanic cones. Many Andisols are of early Holocene age, having originated from the Mount Mazama ash and pumice deposits 7,700 years ago.

Fig. 11.2 The Lebam soil series, a medial over clayey, ferrihydritic over isotic, isomesic Typic Fulvudands, is formed in residuum from tuffaceous sedimentary rocks on slopes of the northern Coast Range. This soil has a dark brown umbric epipedon to 33 cm and a deep, yellowish brown cambic horizon to 157 cm. The scale is in inches. *Source* NRCS photo

11.2 Properties and Processes

Andisols are distinguished from other soils by having andic soil properties from the weathering of volcanic glass or the accumulation of organic matter complexed with aluminum and iron, a low bulk density, a high phosphate retention, and a high aluminum plus one-half iron content (by ammonium oxalate). Baham and Simonson (1985) delineated soils with andic properties along the Oregon coast that helped lead to the development of the Andisol soil order. Andisols in Oregon contain either an ochric (52% of soil series), umbric

Fig. 11.3 The Grousehill soil series, a medial-skeletal, amorphic Oxyaquic Duricryands, is formed in volcanic ash over till on ridges and benches in the Crater Lake area. This soil has a dark brown ochric epipedon to 25 cm, a dark grayish brown to dark brown cambic horizon to 100 cm, and a dark gray, moderately cemented duripan below. The scale is in feet. *Source* NRCS photo

(33%), mollic (15%), or melanic (<1%) epipedon over a cambic horizon (72%) and/or argillic horizon (18% of soil series). Nearly one-quarter (18%) of Andisols lack a diagnostic subsurface horizon. The ochric, umbric, and mollic epipedons average 19 ± 15 cm, 49 ± 27 cm, and 36 ± 13 cm, respectively. The cambic and argillic horizons average 57 ± 28 cm and 57 ± 31 cm. Less than one-quarter (15%) of Andisols in Oregon are endemic, i.e., are the only soil series in a family.

Andisols of Oregon are in ashy or medial particle size classes that are often skeletal. Nearly one-half (47%) of Andisols have strongly contrasting particle size classes, indicating that the ash has buried soils with a loamy-skeletal particle size. Andisols of Oregon are most commonly in amorphic, glassy, and ferrihydritic soil mineralogy classes. In cases of burials, the underlying soils most often have an isotic mineralogy. More than three-quarters (78%) of Andisols in Oregon have a cryic or frigid soil temperature regime. The predominant soil moisture classes represented in Andisols are xeric (48%) and udic (45%). Andisols in Oregon tend to be deep and seldom limit plant rooting.

Table 11.1 contains analytical data for six Andisol great groups. Andisols have a high pH when measured in sodium fluoride (NaF), abundant volcanic glass, low bulk densities, relatively high aluminum and iron contents, and a high phosphorus retention capacity. Andisols with an amorphic mineralogy tend to have a high cation-exchange capacity. Andisols lack the increase in $Al_o + ½\ Fe_o$ from the eluvial to the B horizon that is typical of Spodosols. Andisols and Spodosols have similar chemical characteristics. When andic soil properties criteria were being developed, spodic horizon criteria also were typically met. Upon approval of the Andisol criteria, the Spodosol criteria were changed to include the morphological criteria lacking in Andisols and Spodosols were placed before Andisols in the key to the orders in *Soil Taxonomy*. The dominant processes in Andisols are andisolization, melanization, and cambisolization, with gleization and silicification as subsidiary processes (Chap. 15).

Fig. 11.4 The Newanna soil series, a medial-skeletal, ferrihydritic Typic Fulvicryands, is formed in colluvium and residuum over fractured basalt in the northern Coast Range. This soil has a dark reddish brown umbric epipedon to 30 cm and a strong brown cambic horizon over fractured basalt at 65 cm. The photograph shows the upper 55 cm of the soil profile. *Source* NRCS photo

Fig. 11.5 The Quillamook soil series, medial, ferrihydritic, isomesic Pachic Melanudands, is formed in alluvium on stream terraces in the Sitka Spruce Belt. This soil has a dark grayish brown, smeary melanic epipedon (Pachic subgroup) to 53 cm, followed by a light yellowish brown cambic horizon to 147 cm. The scale is in feet. *Source* NRCS photo

11.3 Use and Management

Andisols are used primarily for timber production, wildlife, recreation, and watershed protection. Some areas in the Hood River Valley are used for orchards. When used for agriculture, phosphorus is added to the soil because of the high phosphate retention typical in Andisols.

11.4 Summary

Andisols are the fourth most abundant soil order in Oregon in terms of land area (13%) and the number of soil series (12%). Andisols occur in 10 of the 17 MLRAs but are most common in the Cascade Mountains—Eastern Slope, the Coast Range, and the Blue Mountains. However, because of incomplete detailed mapping in the Cascade Mountains, Blue Mountains, and the Klamath Basin, it is likely that Andisols are far more abundant in Oregon. The predominant suborders include the Xerands, Udands, Cryands, and Vitrands. Five great groups account for 91% of the Andisols on an area basis, including the Vitrixerands, Vitricryands, Hapludands, Udivitrands, Fulvudands, and Haplocryands. The most extensive Andisols include the Tolo, Woodcock, Lapine, Klootchie, Syrupcreek, and Limberjim soil series.

Andisols have a mean annual air temperature ranging between 3.5 and 8 °C and receive between 650 and 2,200 mm/yr of precipitation. The vegetation on Andisols is predominantly Douglas-fir forests with bigleaf maple, grand fir, western hemlock, red alder, Sitka spruce, and western redcedar; subalpine forests of Engelmann spruce, subalpine fir, and lodgepole pine; and ponderosa pine forests with lodgepole pine, grand fir, and white fir. Maximum slopes commonly range from 40 to 90%. Parent materials include volcanic ash, pumice, and cinders or colluvium and residuum with high amounts of organic matter and extractable iron and aluminum in a cool, humid climate. Typical landforms are mountain slopes, lava plateaus, and volcanic cones. Many Andisols are of early Holocene age, having originated from the Mount Mazama ashfall 7,700 years ago.

Fig. 11.6 An Aquic Haploxerands in the Lassen Volcanic National Park in northern California that is comparable to the Hot Lake series in southern Oregon. Both soils contain a mollic epipedon over a cambic horizon. The scale is in centimeters. *Source* NRCS photo

Table 11.1 Analyical properties of some Andisols found in Oregon

Horizon	Depth (cm)	Clay (%)	Silt (%)	Sand (%)	SOC (%)	CEC7 ($cmol_c$/kg)	Base sat (%)	pH H_2O	Al_o + 1/2Fe_o (%)	pH NaF	Melanic index (%)	Volc. glass (%)	Bulk density (g/cm^3)	NZ P retent (%)	1.5 mPa H_2O/clay
Tolovana; medial over loamy, ferrihydritic over isotic, isomesic Typic Fulvudands; Lincoln, OR; pedon no. 84P0899															
A	0–18	22.6	43.7	33.7	11.21	58.6	24	4.4	0.34	9.8	96		0.33		0.52
AB	18–46	16.4	46.0	37.6	4.81	37.2	9	4.9	2.70	11.0	100		0.68		0.38
Bw	46–81	14.5	51.7	33.8	2.25	28.2	6	5.0	2.46	10.8	100		0.85		0.71
BC	81–117	17	38.3	44.7	0.59	30.1	5	4.8	1.48	10.2	96				1.08
C	117–152	18.1	41.7	40.2	0.6	29.5	5	4.8	1.57	10.3	97		0.96		0.99
Keel; medial, amorphic Typic Haplocryands; Lane, OR; pedon no. 75C0077															
A1	0–15	32.4	39.0	28.6	3.84	80.1	21	5.1		10.3			0.41		1.25
A2	15–31	8.3	63.8	27.9	1.72	78.2	19	5.0		10.6			0.61		4.66
Bw1	31–48	33.7	40	26.3	1.25	77.6	11	5.1		10.7			0.76		1.16
Bw2	48–71	31.1	36.4	32.5	0.56	75.0	9	4.9		10.5			0.76		1.28
BC	71–89	20.7	30.8	48.5	0.34	76.8	9	4.9		10.2					1.99
Cr	89–102	9.8	34.1	56.1	0.42	73.6	12	4.8		10.1					3.86
Hemcross; medial, ferrihydritic, mesic Alic Hapludands; Washington, OR; pedon no. 91P0932															
Oi	0–6					74.6		5.6			3				
A	6–20	12.1	60.5	27.4	48.66	34.7	18	5.5	3.18		94	2			1.60
Bw1	20–56	11.3	57.7	31.0	4.07	23.7	10	5.6	2.65		97	2			1.44
Bw2	56–103	13.7	54.4	31.9	1.13	21.4	19	5.5	1.75		95	1			1.09
Bs3	103–150	10.5	50.7	38.8	0.26	20.7	11	5.4	1.95		94				1.36
Bw4	150–186	13.2	50.9	35.9	0.16	26.1	16	5.4	1.95		99	1			1.27
R	186	2.1	19.5	78.4	0.18	22.6	4	5.5	2.89		91				5.57
Syrupcreek; ashy over loamy-skeletal, amorphic over isotic, frigid Alfic Udivitrands; Umatilla, OR; pedon no. 97P0552															
Oi	0–4				23.10	19.6		4.9		7.2					
Bw1	4–18	8.6	63.9	27.5	0.90	15.4	50.0	5.8	0.70	9.8	46.0		0.82	46	1.15
Bw2	18–41	10.5	62.2	27.3	0.41	15.3	63.0	6.1	0.34	8.8	29.0	16	1.03	29	0.90
2Eb	41–54	10.9	61.1	28.0	0.30	16.2	75.0	6.1	0.35	8.3	26.0	8	1.19	26	0.92
2Btb	54–84	14.9	66.6	18.5	0.18	19.2	80	6.0	0.39	8.3	26.0		1.22	26	0.78
Lapine; ashy-pumiceous, glassy Xeric Vitricryands; Klamath, OR; pedon no. 72C0044															
A	0–8	3.4	26.3	70.3	9.50	33.6	65	5.8			> 70		0.91		5.82
AC	8–23	3.3	23.1	73.6	0.89	13.1	58	6.4			> 70		0.99		2.36
C1	23–53	5.3	23.8	70.9	0.47	10.0	82	6.6			> 70		0.95		1.15
C2	53–114	2.8	22.1	75.1	0.10	4.5	73	7.0			> 70				2.29
C3	114–142	0.1	3.3	96.6	0.02	1.5	100	6.6			> 70				0.86
C4	142–152	0.6	5.3	94.1	0.03	1.8	100	7.0			> 70				
2Bb	152–167	8.5	44.5	47.0	0.20	26.8	67	6.6					1.74		1.74

(continued)

Table 11.1 (continued)

Horizon	Depth (cm)	Clay (%)	Silt (%)	Sand (%)	SOC (%)	CEC7 (cmol$_c$/kg)	Base sat (%)	pH H_2O	Al_o + 1/2Fe_o (%)	pH NaF	Melanic index (%)	Volc. glass (%)	Bulk density (g/cm^3)	NZ P retent (%)	1.5 mPa H_2O/clay
Tolo; ashy over loamy, amorphic over isotic, frigid Alfic Vitrixerands; Union, OR; pedon no. 40A5473															
A	0–3	12.3	72.1	15.6	5.32	24.1	87	6.1			> 60				1.06
Bw1	3–18	12.8	71.0	16.2	1.14	14.8	72	5.7			> 60				0.73
Bw2	18–43	12.8	70.7	16.5	0.70	14.1	72	5.9			> 60				0.66
Bw3	43–71	12.8	73.1	14.1	0.37	12.1	81	6.0			> 60				0.59
2Eb/Bb	71–84	17.0	72.1	10.9	0.26	13.6	93	6.1							0.53
2Eb	84–102	19.6	70.6	9.8	0.21	15.3	92	5.9							0.46
2Btb1	102–137	28.3	63.2	8.5	0.18	21.4	99	6.0							0.50
2Btb2	137–160	31.1	60.1	8.8	0.12	25.1	100	6.6							0.54

Reference

Baham J, Simonson GH (1985) Classification of soils with andic properties from the Oregon coast. Soil Sci Soc Am J 49:777–780

12 Ultisols

12.1 Distribution

Ultisols are the fifth most abundant soil order in Oregon in terms of land area (5.0%) and the number of soil series (3.9%) (Table 6.2). Ultisols occur in 6 of the 17 MLRAs, all of which are west of the Cascade Mountain crest (Fig. 12.1). Suborders are ranked by area within Ultisols: Humults (80%), Xerults (15%), Udults (4.8%), and Aquults (0.2%). Two great groups account for 80% of the Ultisols on an area basis, including the Haplohumults (44%) and the Palehumults (36%) (Table 6.2).

The most extensive Ultisols include the Jory, Peavine, Bellpine, Honeygrove, Nekia, and Josephine soil series, each of which occupies an area between 500 and 900 km^2. Photographs of major great groups of Ultisols are given in Chap. 7, including a Haplohumults (Peavine series; Fig. 7.26), and a Palehumults (Jory series; Fig. 7.33). Figure 12.2 is of the Burnthill series, a fine-loamy, siliceous, superactive, isomesic Typic Palehumults, found on the uplifted Poverty Ridge marine terrace in Curry County (Bockheim et al. 1996). This terrace is the highest and oldest that has been reported along the Oregon coast and may be 500,000 years in age or older. The soil contains a very dark grayish brown umbric horizon to 28 cm and a brown to reddish brown argillic horizon to 110 cm.

Ultisols in Oregon have a mean annual air temperature ranging between 9.5 and 11 °C and receive between 1,100 and 2,000 mm/yr of precipitation. The vegetation on Ultisols is predominantly Douglas-fir forests with bigleaf maple, western hemlock, red alder, and western redcedar. On drier sites, Douglas-fir is accompanied by Oregon white oak, ponderosa pine, sugar pine, and Pacific madrone. Maximum slopes commonly range from 55 to 90%. Parent materials are predominantly colluvium and residuum derived from sedimentary rocks, basalt, tuff, and other igneous rocks. Typical landforms are mountain slopes and hillslopes. Ultisols in Oregon generally are of middle to early Pleistocene age.

12.2 Properties and Processes

Ultisols in Oregon contain either an ochric (60% of soil series) or umbric (40%) epipedon over an argillic horizon. The ochric and umbric epipedons average 24 ± 17 cm and 40 ± 16 cm. The argillic horizon averages 99 ± 48 cm. Less than one-quarter (15%) of Ultisols in Oregon are endemic, i.e., are the only soil series in a family.

Nearly three-quarters (72%) of the Ultisols are in the fine particle-size class; 58% are in the mixed; 35% are in the isotic soil mineralogy classes; and 59% are in the active or superactive cation-exchange activity classes. Nearly three-quarters (72%) of Ultisols in Oregon have a mesic soil temperature regime, 23% have an isomesic soil temperature regime, and the remaining 5% have a frigid soil temperature regime. A udic soil moisture regime is present in 71% of Ultisols and a xeric soil moisture regime is in the remaining 29%. Ultisols in Oregon tend moderately deep to very deep and seldom limit plant rooting.

Table 12.1 summarizes the properties of three key Ultisol great groups in Oregon. The soils have an ochric or umbric epipedon (Peavine series) that ranges from 10 to 50 cm thick and an argillic horizon ranging from 32 to 305 cm. The base

T. Thorson et al., *The Soils of Oregon*, World Soils Book Series,
https://doi.org/10.1007/978-3-030-90091-5_12

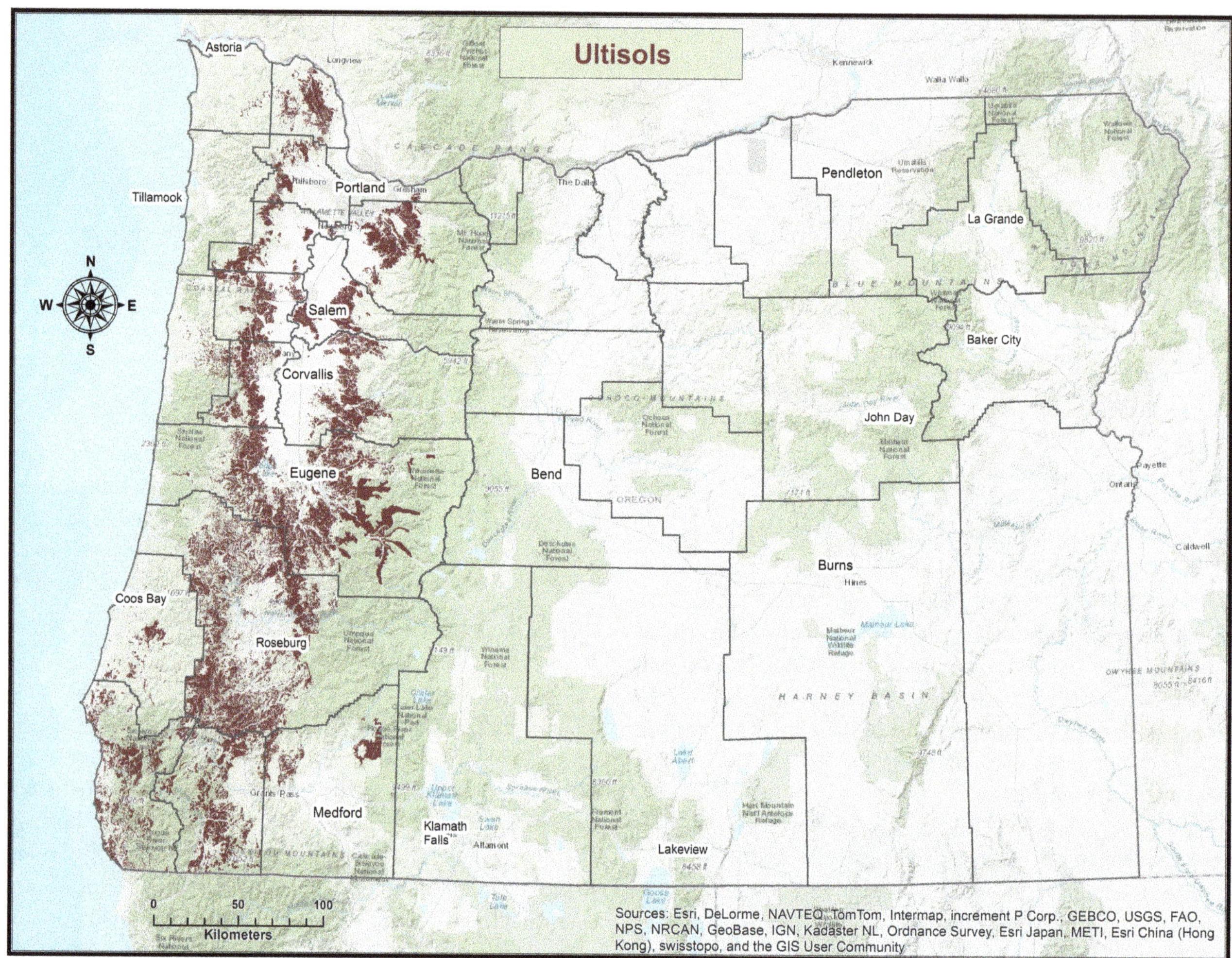

Fig. 12.1 Distribution of Ultisols in Oregon. *Source* Prepared by Whityn Owen

saturation is below 35% throughout the control section. The Humults have an organic carbon concentration above 0.85% in the upper 100 cm, leading to profile quantities (0–100 cm) of 12 kgC/m^2. The dominant soil-forming processes in Oregon Ultisols are argilluviation, humification, and base-cation depletion.

Glassman and Simonson (1985) examined chemical weathering in Haplohumults derived from basalt in western Oregon, including the Bellpine and Honeygrove soil series. Feldspars underwent considerable alteration and smectites, chlorite, gibbsite, and goethite were notable weathering products.

Fig. 12.2 The Burnthill soil series. A fine-loamy, siliceous, superactive, isomesic Typic Palehumults, found on the uplifted Poverty Ridge marine terrace in Curry County (Bockheim et al., 1996). This terrace is the highest and oldest reported along the Oregon coast and may be 500,000 years or older. The soil contains a very dark grayish brown umbric horizon to 28 cm and a brown to reddish brown argillic horizon to 110 cm. *Source* NRCS photo

12.3 Use and Management

Ultisols in Oregon are used primarily for timber production, wildlife, recreation, and watershed protection. Some Ultisol soil series are used for livestock grazing, cultivation, and development. Several Ultisol series, including the Jory, Bellpine, Nekia, and Josephine, are valued for wine grape production.

12.4 Summary

Ultisols are the fifth most abundant soil order in Oregon in terms of land area and the number of soil series. Ultisols occur in 6 of the 17 MLRAs, all of which are west of the Cascade Mountain crest. Suborders ranked by area are Humults, Xerults, Udults, and Aquults. Two great groups account for 80% of the Ultisols on an area basis, including the Haplohumults and the Palehumults. The most extensive Ultisols include the Jory, Peavine, Bellpine, Honeygrove, Nekia, and Josephine soil series. The Jory series is recognized as the State Soil for Oregon.

Ultisols have a mean annual air temperature between 9.5 and 11 °C and receive between 1,100 and 2,000 mm/yr of precipitation. The vegetation on Ultisols is predominantly Douglas-fir forests with bigleaf maple, western hemlock, red alder, and western redcedar. On drier sites, Douglas-fir is accompanied by Oregon white oak, ponderosa pine, sugar pine, and Pacific madrone. Maximum slopes commonly range from 55 to 90%. Parent materials are predominantly colluvium and residuum derived from sedimentary rocks, basalt, tuff, and other igneous rocks. Typical landforms are mountain slopes and hillslopes. Ultisols in Oregon are of middle to early Pleistocene age.

Ultisols in Oregon contain either an ochric or umbric epipedon over an argillic horizon that averages 100 cm in thickness. Less than one-quarter (15%) of Ultisols in Oregon are endemic, i.e., are the only soil series in a family.

Ultisols in Oregon are mainly in the fine particle-size class, the mixed or isotic mineralogy class, the superactive cation exchange activity class, the mesic or isomesic soil temperature regime, and the udic soil moisture regime. Ultisols in Oregon tend to be moderately deep to very deep and seldom limit plant rooting.

Table 12.1 Analytical properties of some Ultisols found in Oregon

Horizon	Depth (cm)	Clay (%)	Silt (%)	Sand (%)	SOC (%)	CEC7 ($cmol_c$/kg)	Base sat (%)	pH H_2O	$CaCO_3$ (%)	1.5 mPa H_2O/clay
Peavine; fine, mixed, active mesic Typic Haplohumults; Yamhill, OR; pedon no. 10N0538										
Oi/Oe	0–2				29.6					
A	2–11	29.2	55.5	15.3	7.3	41.7	54	5.5		0.97
AB	11–22	28.2	56.8	15	6.1	37.6	41	5.5		0.80
Bwc	22–39	26.8	55.2	18	4.1	30.6	19	5.5		0.71
2Btc	39–71	28.6	53.9	17.5	1.3	24.5	28	5.6		0.63
2BC	71–114	33.3	49.5	17.2	0.3	27.2	26	5.4		0.63
2C	114–200	31.2	49.9	18.9	0.2	30.5	32	5.4		0.72
Jory; fine, mixed, active, mesic Xeric Palehumults; Marion, OR; pedon no. 40A0929										
Ap	0–20	51.4	38.6	10.0	3.42	*28.2*	*17*	5.1		0.36
A	20–33	53.4	36.9	9.7	2.71	*26.3*	*23*	5.2		0.43
Bt1	33–69	54.7	35.9	9.4	1.74	*25.8*	*23*	5.2		0.37
Bt21	69–99	58.9	32.8	8.3	0.85	*21.4*	*35*	5.3		0.37
Bt22	99–132	65.5	27.3	7.2	0.33	*20.1*	*31*	5.7		0.36
Bt23	132–185	66.3	26.9	6.8	0.27	*20.3*	*29*	5.6		0.37
Bt31	185–274	64.1	28.8	7.1	0.16	*19.3*	*21*	5.7		0.38
Bt32	274–333	66.1	26.8	7.1	0.19	*19.3*	*22*	5.6		0.38
C	333–361	64.5	26.1	9.4	0.16	*20.6*	*25*	5.4		0.41
Josephine; fine-loamy, mixed, superactive, mesic Typic Haploxerults; Josephine, OR; pedon no. 69C0199										
A1	0–10	22.4	45.8	31.8	4.06	23.5	10	4.7	tr	0.54
Bw1	10–38	27.9	43.1	29.0	1.55	16.2	12	4.9	1	0.48
Bw2	38–53	33.4	38.6	28.0	0.66	13.8	10	4.9	1	0.50
Bt11	53–71	42.3	32.2	25.5	0.50	12.3	14	4.8	1	0.47
Bt12	71–94	62.8	21.2	16.0	0.61	17.0	9	4.9	1	0.43
Bt13	94–108	65.6	19.3	15.2	0.38	18.3	9	4.9		0.43
Bt2	108–132	44.7	26.4	28.9	0.22	15.4	12	4.8	1	0.49
C	132–152	19.4	23.3	57.3	0.08	9.9	12	4.8	1	0.60
R	152									

Bold-face text identifies argillic horizon (Bt)
Italic-face text indicates that CEC7 and Base sat for the Jory soil were determined at a pH of 8.2

References

Bockheim JG, Kelsey HM, Marshall JG III (1996) Soil development, relative dating and correlation of late Quaternary marine terraces in southwestern Oregon. Quat Res 37:60–74

Glassmann JR, Simonson GH (1985) Alteration of basalt in soils of western Oregon. Soil Sci Soc Am J 49:262–273

13 Alfisols

13.1 Distribution

Alfisols comprise 4.1% of the soil area and 5.4% of the soil series of Oregon (Table 6.2). Alfisols occur in 11 of the 17 MLRAs, but are more common in the Willamette Valley and the Siskiyou Mountains and are also common in the east of the Cascade Range in the Blue Mountains (Fig. 13.1). Suborders can be ranked by area: Xeralfs (84%), Aqualfs (12%), Udalfs (3.5%), and Cryalfs (0.5%). Two great groups account for 81% of the Alfisols on an area basis, including the Haploxeralfs (65%) and the Palexeralfs (16%) (Table 6.2).

The most extensive Alfisols include the Dayton, Speaker, Vannoy, Swalesilver, Freezener, and Bateman soil series, each of which occupies an area between 200 and 500 km^2. Photographs of major great groups of Alfisols are given in Chap. 7, including a Haploxeralfs (Dubakella series; Fig. 7.19) and a Palexeralfs (Ruch series; Fig. 7.56). The Dayton soil series (Fig. 13.2), a fine, smectitic, mesic Vertic Albaqualfs, is formed in glaciolacustrine deposits in the Willamette Valley. This soil has an ochric epipedon to 38 cm that includes an albic horizon (23–38 cm) and an argillic horizon to 135 cm. The Dayton soil has pronounced blocky and prismatic structures that develop into vertical cracks during drying periods and masses of brown iron accumulation in the subsoil. Parsons and Balster (1967) reported that the Dayton Planosol in the Willamette Valley was derived from three contrasting glaciolacustrine materials that controlled soil horizonation.

Alfisols in Oregon have a mean annual air temperature ranging between 7.0 and 12 °C and receive between 800 and 1,300 mm/yr of precipitation. The vegetation on Alfisols is predominantly Douglas-fir forests with bigleaf maple, western hemlock, red alder, and western redcedar. On drier sites, Douglas-fir is accompanied by Oregon white oak, California black oak, ponderosa pine, and Pacific madrone. Cryalfs, which occur to a limited extent in the Cascade Mountains, feature subalpine fir, Engelmann spruce, and lodgepole pine. Maximum slopes commonly range from 25 to 75%. Parent materials are predominantly colluvium and residuum derived from sedimentary rocks, basalt, serpentinite, and other igneous rocks. In the Willamette Valley, Alfisols have formed in glaciolacustrine deposits from the Missoula Floods. Typical landforms are mountain slopes, hillslopes, and lake terraces. Alfisols in Oregon commonly are of the late Pleistocene age.

13.2 Properties and Processes

Alfisols in Oregon contain either an ochric (89% of soil series) or umbric (11%) epipedon over an argillic horizon. The ochric and umbric epipedons average 23 ± 12 cm and 28 ± 5.0 cm. An argillic horizon, which occurs in all Alfisols, averages 73 ± 38 cm. A cambic horizon averaging 23 ± 9.3 cm, occurs in 11% of the Alfisols in Oregon. Less than one-quarter (18%) of Alfisols in Oregon are endemic, i.e., are the only soil series in a family.

More than three-quarters (82%) of the Alfisols are in the fine (39%), fine-loamy (31%), or loamy-skeletal (12%) particle-size classes; 56% are in the mixed (56%), 22% in the smectitic, and 17% in the isotic soil mineralogy class; and 35% are in the active or superactive cation exchange activity class. Alfisols in Oregon generally have a mesic (66%) or frigid (25%) soil temperature regime. A xeric soil moisture regime is present in 84% of Alfisols. In Oregon, Alfisols tend to be deep and seldom limit plant rooting.

Table 13.1 provides data for three Alfisol great groups in Oregon. The soils have an ochric epipedon ranging from 10 to 23 cm in thickness and an argillic horizon ranging from 30 to 97 cm. The soils lack a mollic epipedon and the base saturation below 100 cm is more than 35%. The dominant soil-forming processes in Alfisols are argilluviation, base-cation accumulation, and gleization.

T. Thorson et al., *The Soils of Oregon*, World Soils Book Series,
https://doi.org/10.1007/978-3-030-90091-5_13

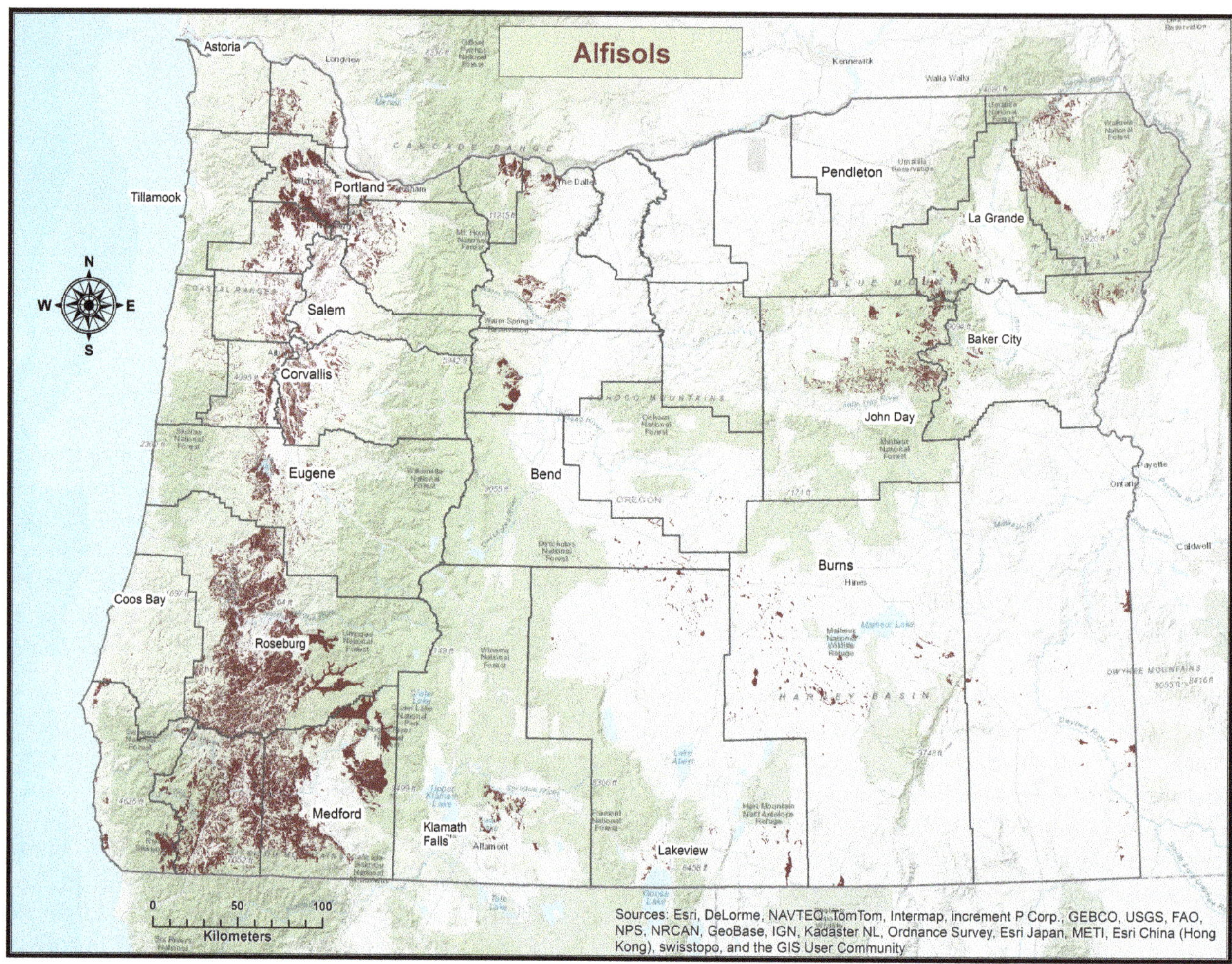

Fig. 13.1 Distribution of Alfisols in Oregon. *Source* Prepared by Whityn Owen

13.3 Use and Management

Alfisols in Oregon are used primarily for agriculture, timber production, recreation, and watershed protection. Some Alfisol soil series are used for livestock grazing, forest products, and wildlife management.

13.4 Summary

Alfisols comprise 4.1% of the soil area and 5.4% of the soil series of Oregon. Alfisols occur in 11 of the 17 MLRAs, but are more common in the Willamette Valley and the Siskiyou Mountains. Predominant suborders are Xeralfs, Aqualfs, with Udalfs, and Cryalfs occupying smaller areas. Two great groups account for 81% of the Alfisols on an area basis, including the Haploxeralfs and the Palexeralfs. The most extensive Alfisols include the Dayton, Speaker, Vannoy, Swalesilver, Freezener, and Bateman soil series.

Alfisols have a mean annual air temperature ranging between 7.0 and 12 °C and receive between 800 and 1,300 mm of annual precipitation. The vegetation on Alfisols is predominantly Douglas-fir forests with bigleaf maple, western hemlock, red alder, and western redcedar. On drier sites, Douglas-fir is accompanied by Oregon white oak, California black oak, ponderosa pine, and Pacific madrone. Maximum slopes commonly range from 25 to 75%. Parent

Fig. 13.2 The Dayton soil series, a fine, smectitic, mesic Vertic Albaqualfs, is formed in glaciolacustrine deposits in the Willamette Valley. This soil has an ochric epipedon to 38 cm that includes an albic horizon (23–38 cm) and an argillic horizon to 135 cm. The Dayton soil has pronounced blocky and prismatic structures that develop into vertical cracks during drying periods and masses of brown iron accumulation in the subsoil. The scale is in inches. *Source* NRCS photo

materials are dominantly colluvium and residuum derived from sedimentary rocks, basalt, serpentinite, and other igneous rocks. In the Willamette Valley, Alfisols have formed in glaciolacustrine deposits from the Missoula Floods. Typical landforms are mountain slopes, hillslopes, and lake terraces. Alfisols in Oregon are of the late Pleistocene age.

Alfisols in Oregon generally have an ochric epipedon over an argillic horizon. Less than one-quarter (18%) of Alfisols in Oregon are endemic, i.e., are the only soil series in a family.

Alfisols are in the fine, fine-loamy, or loamy-skeletal particle-size class, the mixed, smectitic, or isotic soil

Table 13.1 Analytical properties of some Alfisols found in Oregon

Horizon	Depth (cm)	Clay (%)	Silt (%)	Sand (%)	SOC (%)	CEC7 ($cmol_c/kg$)	Base sat (%)	pH H_2O	$CaCO_3$ (%)	EC (dS/m)	Ex. Na (%)	SAR	Tot. salts (%)	1.5 MPa H_2O/clay
Dayton; fine, smectitic, mesic Vertic Albaqualfs; Linn, OR; pedon no. 40A0820														
Ap	0–23	14	81.9	4.1	1.51	*15.3*	*37*	5.1						0.55
E1	23–30	15.4	80.6	4	0.89	*15.5*	*48*	5.2						0.55
E2	30–38	18.1	76.6	5.3	0.55	*16.3*	*59*	5.4						0.50
Bt11	38–56	48.3	48.6	3.1	0.38	*45.0*	*80*	5.2						0.53
Bt12	56–74	43.4	54.2	2.4	0.21	*44.8*	*88*	5.7						0.57
Bt21	74–102	30.6	66.4	3.0	0.13	*39.1*	*92*	6.6						0.67
Bt22	102–135	21.7	75.8	2.5	0.07	*33.4*	*93*	6.8						0.77
C1	135–163	13.9	75.1	11.0	0.05	*24.9*	*91*	7.0						0.83
Vannoy; fine-loamy, mixed, superactive, mesic Mollic Haploxeralfs; Jackson, OR; pedon no. 69C0190														
A1	0–5	13.3	56.8	29.9	3.30	23.9	97	6.6						0.73
A2	5–11	15.2	58.7	26.1	1.58	19.8	87	6.3						0.55
Bw	11–28	17.4	57.8	24.8	0.90	21.4	81	5.4						0.52
Bt11	28–48	27.3	48.6	24.1	0.52	27.9	93	5.6						0.40
Bt12	48–67	27.0	42.4	30.6	0.27	32.4	99	6.0						0.50
Bt21	67–88	23.0	44.9	32.1	0.11	34.7	84	6.0						0.50
Bt22	88–102	14.5	34.4	51.1	0.10	50.8	94	5.9						1.11
C	102–132	16.8	38.8	44.4	0.12	36.7	94	6.0						0.65
Swalesilver; fine, smectitic, frigid Aquic Paleoxeralfs; Harney, OR; pedon no. 96P0346														
A1	0–6	3.3	35.3	61.4	0.21	3.4	100	7.7		0.12	9			0.58
A2	6–10	28.6	42.3	29.1	0.19	12.6	98	8.2		0.08	17			0.35
Bt1	10–28	50.0	35.2	14.8	0.33	34.2	90	8.5		0.2	18			0.52
Bt2	28–40	60.8	29.5	9.7	0.36	43.0	99	8.8		0.29	19	11	tr	0.52
Bk1	40–65	51.1	36.5	12.4	0.22	39.3	100	9.1	1.0	0.5	20	13	tr	0.42
Bk2	65–88	54.3	30.9	14.8	0.18	39.1	100	8.7	tr	0.99	20	16	0.1	0.48
BC1	88–107	53.1	27.4	19.5	0.11	36.4	100	8.4		1.24	20	15	0.1	0.50
BC2	107–137	51.2	26.3	22.5	0.08	37.0	100	7.9		2.57	15	11	0.3	0.53
Cr	137–152	39.8	29.9	30.3	0.06	35.7	100	7.9		2.75	13	10	0.4	0.61

Bold-face text identifies argillic horizon (Bt)
Italic-face text indicates that CEC7 and Base sat for the Jory soil were determined at a pH of 8.2

mineralogy class; the mesic or frigid soil temperature regime; and the xeric or udic soil moisture regime. Alfisols in Oregon tend to be deep and seldom limit plant rooting.

Reference

Parsons RB, Balster CA (1967) Dayton–a depositional Planosol, Willamette Valley, Oregon. Soil Sci Soc Am Proc 31:255–258

14 Entisols, Vertisols, Spodosols, and Histosols

14.1 Distribution

Entisols (3,158 km^2, 2.0%), Vertisols (1,000 km^2, 0.6%), Spodosols (359 km^2, 0.2%), and Histosols (167 km^2, 0.1%) collectively comprise 4,690 km^2, which is 3.0% of the mapped soil area of Oregon (Table 6.2). Entisols occur in all of the MLRAs but are more evident in the Columbia Basin, the Blue Mountain Foothills, the Snake River Plains, and the Malheur High Plateau (Fig. 14.1); Vertisols exist in 11 of the 17 MLRAs. They are more extensive in the Siskiyou Mountains and the Malheur High Plateau, and are also present in the Willamette Valley and the Blue Mountain Foothills (Fig. 14.2); Spodosols occur in the Sitka Spruce Belt and the Cascade Mountains (Fig. 14.3); and Histosols are distributed throughout the state, especially in western Klamath County (Fig. 14.4). The 1:7.5 million-scale "Dominant Soil Orders" map of the United States shows a narrow band of Spodosols along the crest of the Cascade Mountains (Fig. 2.14); however, most of this area has not been mapped. Therefore, the areas for Spodosols in Oregon likely are underestimated.

Entisols suborders can be ranked by area: Psamments (59%), Orthents (32%), Aquents (7%), and Fluvents (2%) (Table 6.2). Two great groups account for 83% of the Entisols on an area basis, including the Torripsamments (53%) and the Torriorthents (30%). The Quincy soil series (500 km^2) is the most extensive Entisol.

Vertisol suborders can be ranked: Aquerts (62%), Xererts (37%), and Uderts (1%) (Table 6.2). Three great groups account for 99% of the Vertisols, including the Endoaquerts (49%), Haploxererts (37%), and the Epiaquerts (13%). The Carney, Bashaw, and Boulder Lake soil series are the most extensive Vertisols, occupying from 100 to 200 km^2 each.

Spodosol suborders can be ranked: Orthods (60%), Cryods (21%), and Aquods (19%) (Table 6.2). Three great groups account for 84% of the Spodosols, including Haplorthods (49%), Haplocryods (21%), and Duraquods (14%). At 100 km^2, the Bullards soil series is the most extensive Spodosol. Spodosols occur in two broad regions in Oregon: on sandy marine terraces along the southern Pacific Coast and in upper montane and subalpine regions of the Cascade Mountains; but much of the high elevation Cascade Mountains are not yet mapped.

Histosol suborders can be ranked: Hemists (87%), Saprists (12%), and Fibrists (1%) (Table 6.2). The Lather soil series, a Limnic Haplohemists, is the most extensive (101 km^2) Histosol in Oregon.

A Torripsamment (Quincy soil series) and a Torriorthent (Garbutt soil series) are discussed in Chap. 7. Additional photographs are provided here of selected Vertisols and Spodosols. The Climax soil series, a very-fine, smectitic, mesic Leptic Haploxererts, is formed in clayey colluvium weathered from basalt, tuff, and volcanic breccia over partially weathered sandstone in the Siskiyou Mountains (Fig. 14.5). The A and Bss horizons are black, contained wedge-shaped aggregates, and display vertical cracks. The Natroy soil series, a very-fine, smectitic, mesic Xeric Endoaquerts, is formed in fine-textured, mixed alluvium on terraces and fans in the southern part of the Willamette Valley and the Umpqua Valley (Fig. 14.6). The A and Bss horizons are very dark gray to dark brown, have a prismatic structure, and vertical cracks. Xererts in the Willamette Valley have formed on geomorphic surfaces of only 550 years and display few changes on older landforms (Parsons et al. 1973; Parsons 1979).

The Netarts soil series, a sandy, isotic, isomesic Entic Haplorthods, is formed in mixed eolian sands on marine terraces and dunes in the Sitka Spruce Belt. This soil has a relatively undecomposed organic O horizon over a black A horizon over a grayish-brown albic E horizon, and over a grayish-brown and reddish-brown Bs spodic horizon to 3 feet (90 cm) (Fig. 14.7). The Joeney soil series, a loamy, isotic, isomesic, ortstein, shallow Typic Duraquods, is formed in medium-textured eolian material overlying stratified marine deposits on the uplifted Silver Butte and Indian Creek marine terraces in Coos and Curry Counties (Bockheim et al. 1996). This soil has a relatively undecomposed organic O horizon over a gray albic E horizon, over a thin,

T. Thorson et al., *The Soils of Oregon*, World Soils Book Series,
https://doi.org/10.1007/978-3-030-90091-5_14

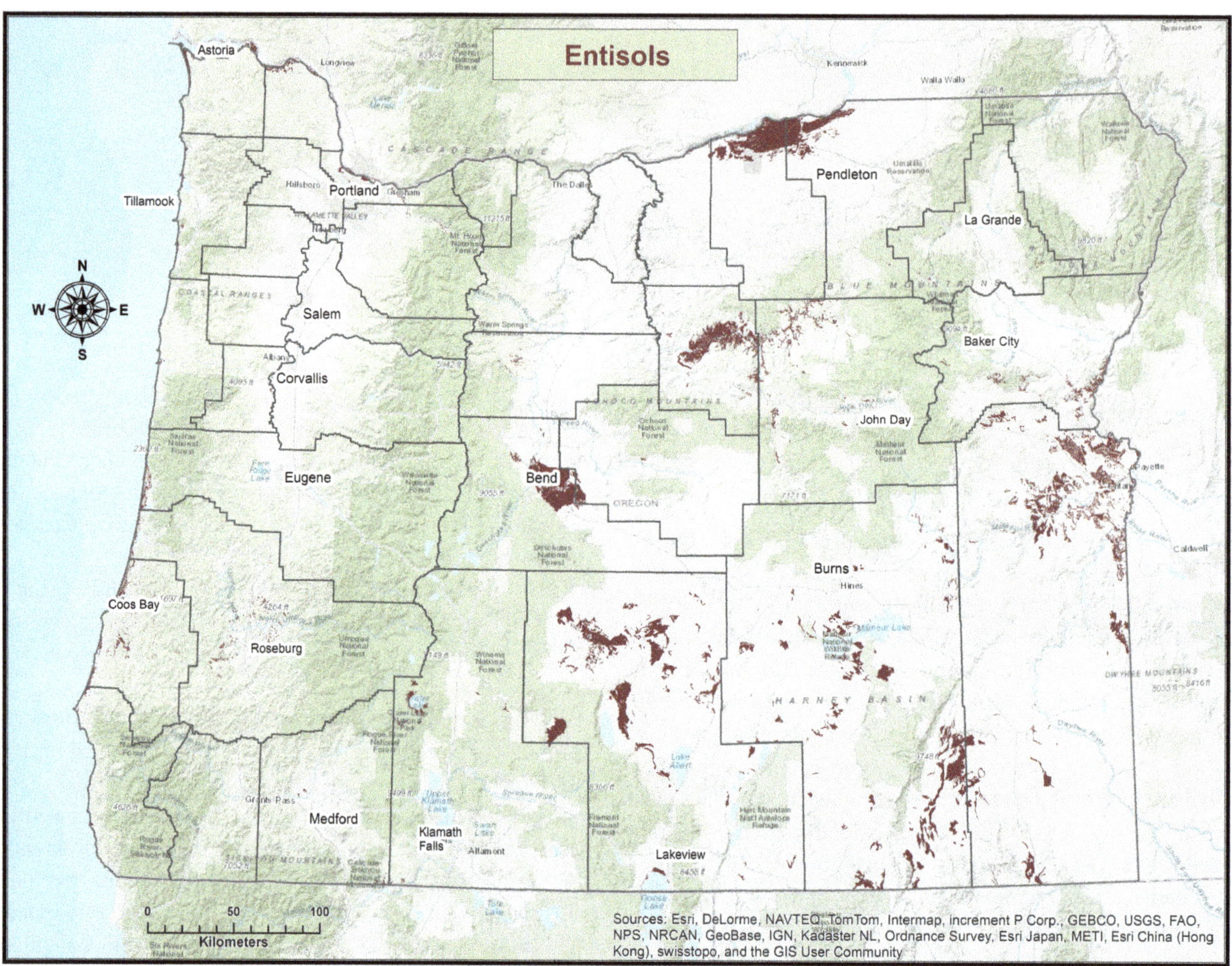

Fig. 14.1 Distribution of Entisols in Oregon. *Source* Prepared by Whityn Owen

dark reddish-brown Bh spodic horizon, and over a reddish yellow and brownish yellow Bs spodic horizon. The lowermost portion of the spodic horizon contains ortstein (Fig. 14.8). Spodosols have been studied along the Oregon coast by Nettleton et al. (1982), Bockheim et al. (1996), and Langley-Turnbaugh and Bockheim (1997, 1998).

Due to their broad geographic distribution, Entisols in Oregon have a mean annual air temperature ranging between 7.0 and 11 °C and receive between 230 and 1,100 mm/yr of precipitation. The vegetation on Entisols is variable but the most common plant communities include sagebrush-bunchgrass, lodgepole pine forest, and shore pine forest. Maximum slopes commonly range from 15 to 60%. Parent materials include eolian sand, alluvium, colluvium, and residuum. Typical landforms are dunes, alluvial plains, and lava plateaus. Entisols in Oregon usually are of Holocene age. Nearly one-third (32%) of Entisols in Oregon are endemic, i.e., they are the only member of the family.

Vertisols have a mean annual air temperature ranging between 7.5 and 11 °C and receive between 250 and 1,000 mm/yr of precipitation. The vegetation on Vertisols is comprised of shrubs and grasses. Maximum slopes commonly range from 0 to 40%. Parent materials are clay-rich alluvium and lacustrine deposits. Vertisols occur on alluvial plains and lake terraces. Nearly one-quarter (22%) of Vertisols are endemic.

Spodosols along the Pacific Coast have a mean annual air temperature (MAAT) of 11 °C and receive from 1,500 to 1,800 mm/yr of precipitation. Spodosols in the Cascade Mountains have a MAAT of 6.0 °C and receive from 2,000 to 3,000 mm/yr of precipitation. Coastal areas with Spodosols feature Sitka spruce, shore pine, grand fir, and Port Orford cedar forests; mountain areas have mountain hemlock, Douglas-fir, Pacific silver fir, and noble fir forests. Maximum slopes for Spodosols in coastal areas are 50% and 75% in the mountains. Although Spodosols along the coast

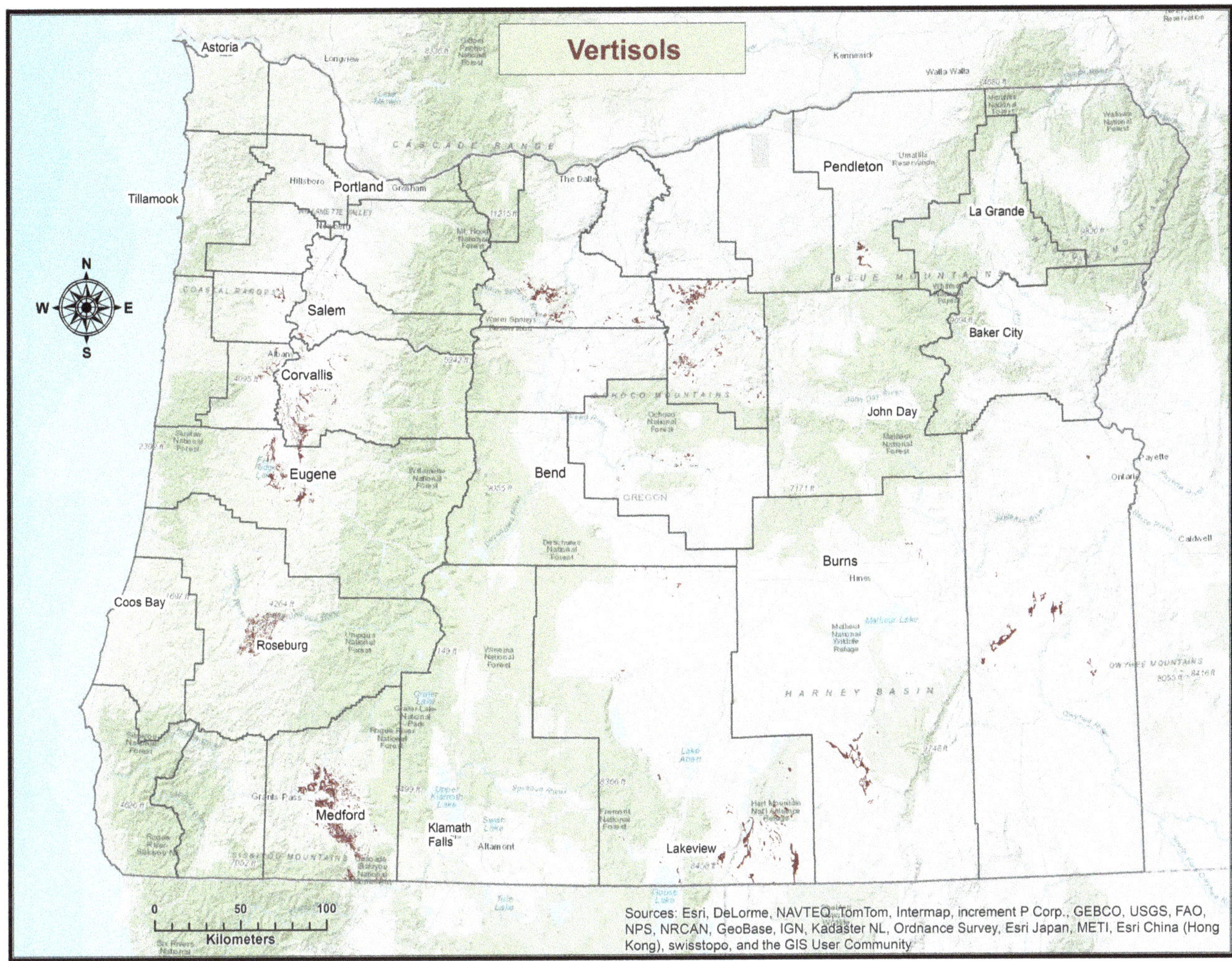

Fig. 14.2 Distribution of Vertisols in Oregon. *Source* Prepared by Whityn Owen

have developed in marine sands and eolian sands on marine terraces, Spodosols in the mountains have formed in colluvium, volcanic ash, and till on mountain slopes and moraines. More than one-half (54%) of the Spodosols in Oregon are endemic.

Histosols occur in a range of environments in Oregon. They generally support sedges, rushes, reeds, and grasses but may have willows, black cottonwood, and other trees. Histosols occur on slopes of less than 3% and are composed of organic materials over a variety of sediments in depressions. One-half (50%) of Histosols are endemic.

14.2 Properties and Processes

Entisols in Oregon contain only an ochric epipedon. Vertisols contain an ochric (67% of soil series), mollic (22%), or umbric epipedon (13%) averaging 18 cm, 72 ± 23 cm, or 66 cm in thickness, respectively. Vertisols either have a cambic horizon (78%) averaging 101 ± 41 cm or lack a diagnostic subsurface horizon (22%). Spodosols have an ochric epipedon averaging 18 cm over an albic horizon averaging 13 ± 9.2 and a spodic horizon averaging

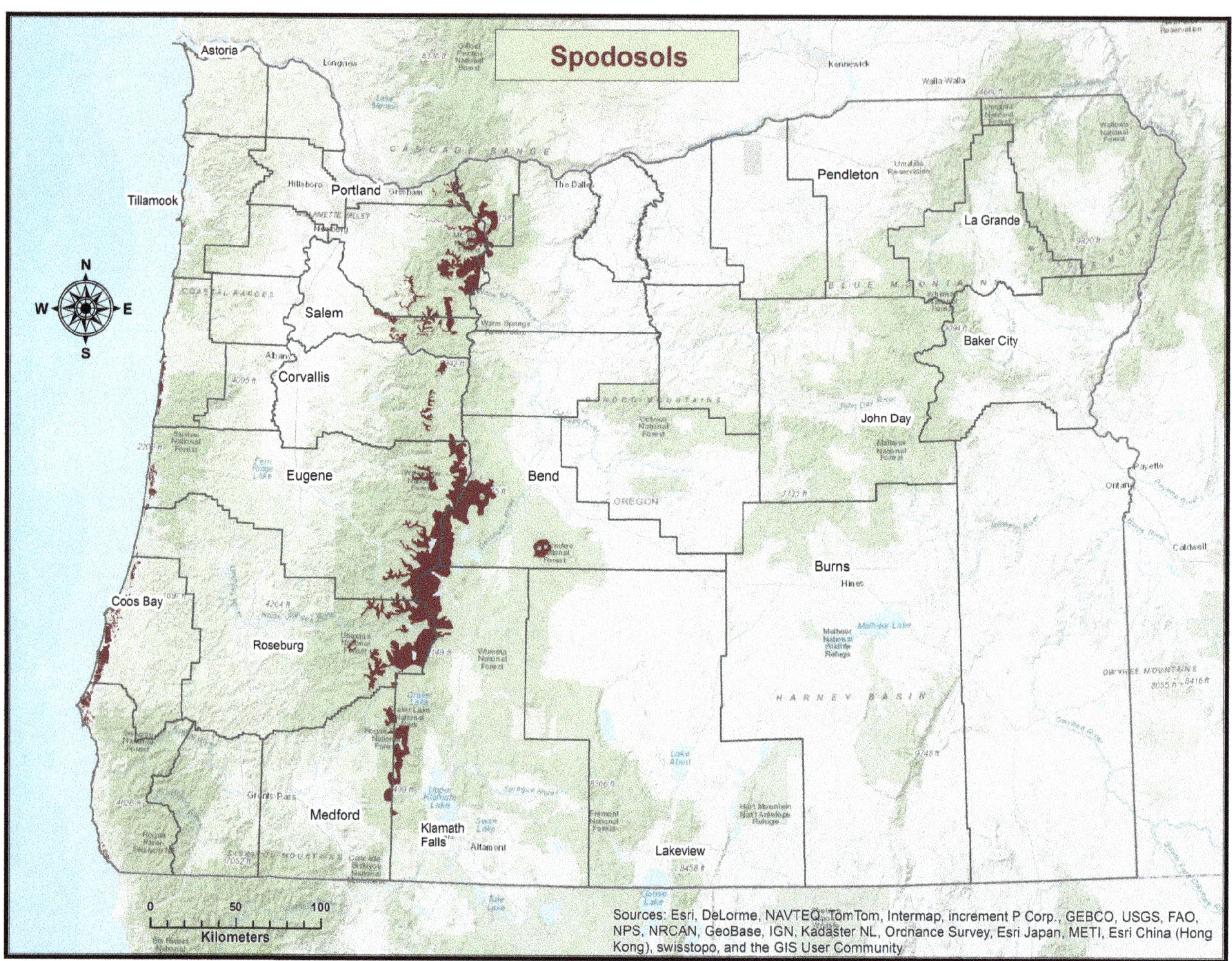

Fig. 14.3 Distribution of Spodosols in Oregon. *Source* Prepared by Whityn Owen

63 ± 29 cm. An ortstein layer, which occurs in 45% of Oregon Spodosols, averages 51 ± 34 cm. Histosols in Oregon contain a histic epipedon that ranges from 97 to 178 cm or more.

Nearly one-half (40%) of the Entisols are in the sandy, 16% are in the ashy-skeletal, and 15% are in the sandy-skeletal particle-size classes; 82% are in the mixed and 16% are in the glassy (soil mineralogy classes); and 33% are in the superactive cation-exchange activity class. Entisols have a mesic (60%), frigid (27%), or isomesic (11%) soil temperature regime and predominantly (57%) have an aridic soil moisture regime. Less than 14% of the Entisols are shallow and limit plant rooting. There are no dominant soil-forming processes in Entisols.

Vertisols are in the very-fine (52%) or fine (48%) particle-size class; the smectitic (100%) mineralogy class; the mesic (74%) or frigid (16%) soil temperature regime; and the aquic (58%) or xeric (37%) soil moisture regime. Vertisol parent materials are typically very deep. The Bashaw and Day soil series are Vertisols that have slickensides (Bss horizons) and abundant clays (>30%). The dominant processes in Vertisols are vertization, gleization, and cambisolization.

Spodosols in Oregon predominantly are in the coarse-loamy (23%), loamy-skeletal (23%), and sandy (23%) particle-size classes; 77% are in the isotic mineralogy class; 69% are in the isomesic and 23% are in the cryic soil temperature regime; and 62% have a udic and 38% have an

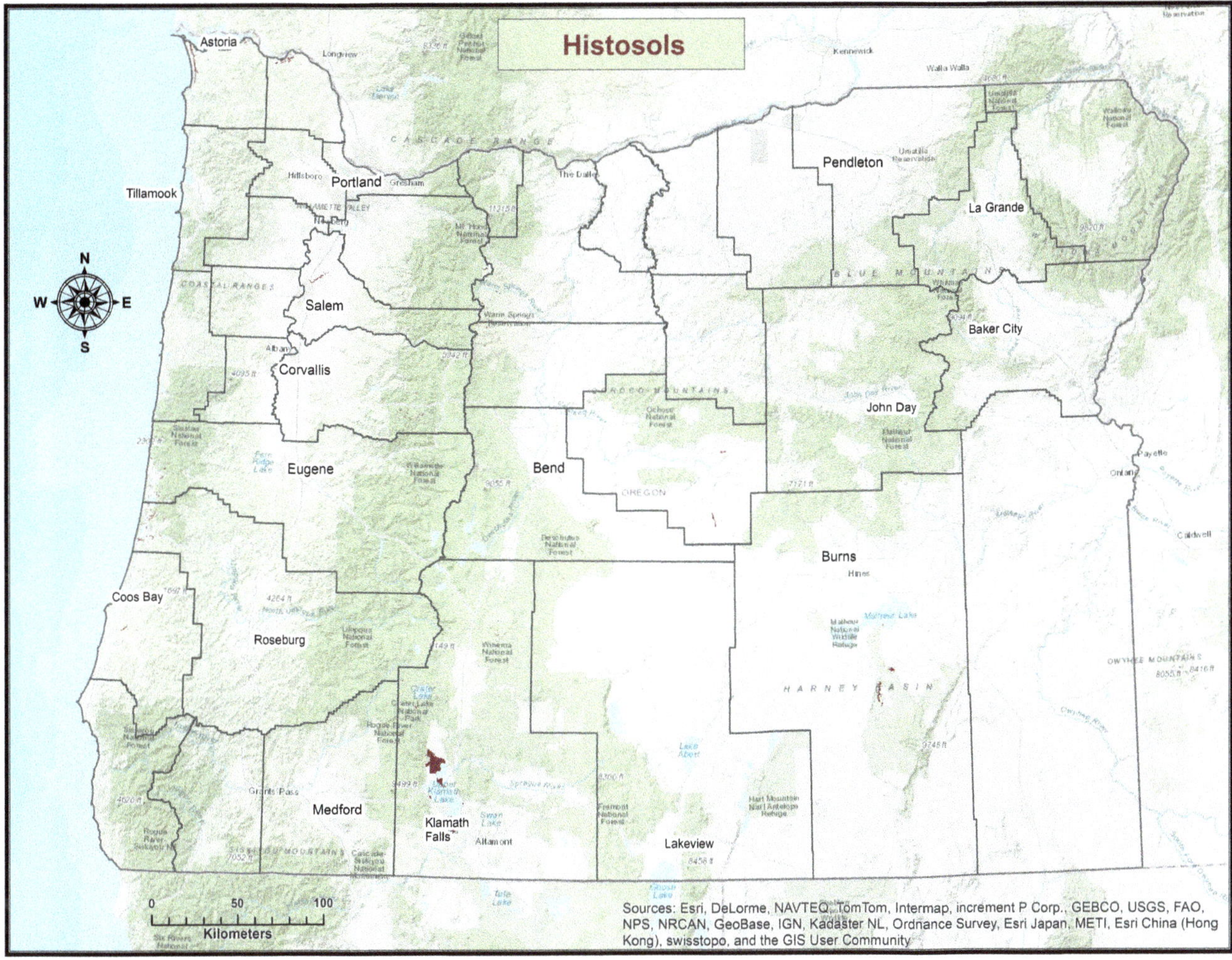

Fig. 14.4 Distribution of Histosols in Oregon. *Source* Prepared by Whityn Owen

aquic soil moisture regime. Spodosol parent materials are deep and usually do not limit plant rooting. The dominant processes in Spodosols are podsolization, gleization, melanization, and possibly silicification.

Two-thirds (67%) of the Histosols are euic, i.e. have a pH value of 4.5 or more, and 33% are dysic. One-third of the Histosols have a cryic soil temperature regime and one-third have an isomesic soil temperature regime. The dominant processes in Histosols are paludization, which is the accumulation of organic materials, and gleization, which is the chemical reduction of iron and other elements due to poor drainage.

Analytical data for five Entisols, Histosols, Spodosols, and Vertisols are given in Table 14.1. The Morehouse soil series is a Vitrandic Torripsamments formed in volcanic ash over lacustrine deposits in Lake County. The Morehouse series contains between 89 and 94% sand, has only an ochric epipedon, and lacks a diagnostic subsurface horizon. The Jimgreen soil series is a Haplosaprist that contains 30 cm of mineral soil over organic materials in the Harney Basin in the Malheur National Wildlife Refuge. The Bandon soil series is a Spodosol with an albic horizon to 13 cm and a spodic horizon to 109 cm that contains ortstein from 76 to 109 cm.

Entisols contain a variety of weakly expressed soil-forming processes; Vertisols are influenced by vertization, gleization, and humification. Spodosols reflect podsolization, base-cation depletion, and humification. Histosols are influenced by paludization and gleization (Chap. 15).

Fig. 14.5 The Climax soil series, a very-fine, smectitic, mesic Leptic Haploxererts, is formed in clayey colluvium weathered from basalt, tuff, and volcanic breccia over partially weathered sandstone in the northern part of the Siskiyou Mountains. The A and Bss horizons are black, contained wedge-shaped aggregates, and display vertical cracking typical of Vertisols. The scale is in feet. *Source* NRCS photo

Fig. 14.6 The Natroy soil series, a very-fine, smectitic, mesic Xeric Endoaquerts, is formed in fine-textured, mixed alluvium on terraces and fans in the southern part of the Willamette Valley and the Umpqua Valley. The A and Bss horizons are very dark gray to dark brown, have a prismatic structure, and vertical cracks common to Vertisols. The scale is in meters. *Source* NRCS photo

Fig. 14.7 The Netarts soil series, a sandy, isotic, isomesic Entic Haplorthods, is formed from mixed eolian sands on marine terraces and dunes in the southern part of the Sitka Spruce Belt. This soil has a relatively undecomposed organic O horizon over a black A horizon, over a grayish-brown E albic horizon, and over a grayish-brown and reddish-brown Bs spodic horizon to 3 ft (90 cm). The scale is in feet. *Source* NRCS photo

Fig. 14.8 The Joeney soil series, a loamy, isotic, isomesic, ortstein, shallow Typic Duraquods, is formed in medium-textured eolian material overlying stratified marine deposits on the uplifted Silver Butte and Indian Creek marine terraces in Curry County (Bockheim et al. 1996). This soil has a relatively undecomposed organic O horizon over a gray E albic horizon over a thin, dark reddish-brown Bh spodic horizon, and over a reddish yellow and brownish yellow Bs spodic horizon. The lowermost portion of the spodic horizon contains ortstein. The photograph shows the upper 75 cm of the soil profile. *Source* NRCS photo

Table 14.1 Analytical properties of some Entisols, Histosols, Spodosols, and Vertisols found in Oregon

	Depth	Clay	Silt	Sand	SOC	CEC7	Base sat	pH	Al_o + $1/2Fe_o$		Ex. Na	1.5 MPa H_2O/
Horizon	(cm)	(%)	(%)	(%)	(%)	(cmol$_c$/ kg)	(%)	H_2O	(%)	ODOE	(%)	clay
Morehouse; ashy, glassy, nonacid, frigid Vitrandic Torripsamments; Lake, OR; pedon no. 79P0429												
A	0–10	4.0	6.6	89.4	0.27	5.7	93	6.7	0.21	0.06		0.88
C1	10–25	3.2	7.4	89.4	0.24	6.3	100	6.9	0.29	0.06	2	1.59
C2	25–49	2.8	7.1	90.1	0.22	6.6	97	7.1	0.18	0.05		1.54
C3	49–75	3.6	4.9	91.5	0.11	6.5	100	7.2	0.23	0.06		1.19
C4	75–111	3.6	2.9	93.5	0.1	7.7	100	7.9	0.19	0.05		1.19
Jimgreen; euic, frigid Hemic Haplosaprists; Harney, OR; pedon no. 98P0561												
2Bw	25–30	43.6	52.2	4.2	6.4	18.2	85.0	6.7			3.0	0.70
3Oa2	81–112				17.6	42.9		6.0				
Bandon; coarse-loamy, isotic, isomesic, ortstein Typic Haplorthods; Coos, OR; pedon no. 74C0147												
E	0–13	1.4	20.6	78.0	1.55	17.5	8	4.4			1	3.00
Bs1	13–31	12.3	22.7	65.0	1.44	16.3	4	5.2			1	0.87
Bs2	31–56	11.1	19.1	69.8	0.79	11.4	4	5.4			1	0.77
Bs3	56–76	9.5	17.8	72.7	1.67	9.7	6	5.3			1	1.42
Bsm	76–109	3.4	11.6	85.0	1.10	7.9	3	5.5				2.94
C	109–140	13.1	14.8	72.1	0.17	6.2	11	5.4			2	0.60
Bashaw; fine, smectitic, mesic Xeric Endoaquerts; Benton, OR; pedon no. 93P0325												
A1	0–11	51.4	47.5	1.1	5.25	60.2	70	4.8			1	0.62
A2	11–22	42.4	54.8	2.8	2.88	50.0	82	5.6			1	0.36
BA	22–39	41.3	54.5	4.2	2.01	45.1	83	5.9			1	0.37
Bw	39–74	51.8	43.7	4.5	1.28	46.6	88	6.2			1	0.44
Bss1	74–114	54.1	42.4	3.5	0.66	48.9	91	6.9			1	0.44
Bss2	114–144	55.5	42.3	2.2	0.32	50.2	92	7.2			1	0.45
Day; very-fine, smectitic Chromic Haploxererts; Grant, OR; pedon no. 98P0221												
A	0–18	59.9	18.5	21.6	10.64	50.3	100	8.1			14	0.35
ABss1	18–51	67.7	15.9	16.4	4.43	60.3	100	8.8			26	0.37
ABss2	51–97	71.7	16.4	11.9	0.25	65.4	100	8.2			36	0.37

Bold-face text identifies histic epipedon (Oa), spodic horizon (Bs, Bsm)

14.3 Use and Management

Entisols are used in Oregon for livestock grazing, irrigated cropland, and wildlife management. Vertisols are used for dryland pasture, irrigated hay and pasture, homesites, wildlife habitat, and livestock grazing. Spodosols are used for timber, pasture, wildlife, recreation, watershed management, and home sites. Histosols are used for wildlife, irrigated pasture, livestock grazing, and specialty crops such as cranberries and (when drained) onions.

14.4 Summary

Entisols, Vertisols, Spodosols, and Histosols collectively comprise 4,198 km^2, which is 3.0% of the mapped soil area of Oregon. Entisols are poorly developed, lack a diagnostic subsurface horizon, and are more extensive in the Columbia Basin, the Malheur High Plateau, and the Sitka Spruce Belt. Vertisols are derived from materials with abundant swelling clays and occur on the Malheur High Plateau, in the Willamette Valley, and the Siskiyou Mountains. Spodosols have an albic and spodic horizon and sometimes ortstein, and occur mainly not only in the Cascade Mountains but also in the Sitka Spruce Belt. Histosols are formed in organic materials and are distributed throughout the state, particularly adjacent to water bodies.

References

Bockheim JG, Kelsey HM, Marshall JG III (1996) Soil development, relative dating and correlation of late Quaternary marine terraces in southwestern Oregon. Quat Res 37:60–74

Langley-Turnbaugh SJ, Bockheim JG (1997) Time-dependent changes in pedogenic processes on marine terraces in coastal Oregon. Soil Sci Soc Am J 61:1428–1440

Langley SJ, Bockheim JG (1998) Mass balance of soil evolution on late Quaternary marine terraces in coastal Oregon. Geoderma 84:265–288

Nettleton WD, Parsons RB, Ness OA, Gelderman FW (1982) Spodosols along the southwest Oregon coast. Soil Sci Soc Am J 46:593–598

Parsons RB (1979) Stratigraphy and land use of the Post-Diamond Hill Paleosol, western Oregon. Geoderma 22:67–70

Parsons RB, Monocharoan L, Knox EG (1973) Geomorphic occurrence of Pelloxererts, Willamette Valley Oregon. Soil Sci Soc Am Proc 37:924–927

15 Soil-Forming Processes in Oregon

15.1 Introduction

Specific soil processes are determined by the soil-forming factors and are expressed in diagnostic horizons, properties, and materials, which are then used to classify soils: soil-forming factors → soil-forming processes → diagnostic horizons, properties, materials → soil taxonomic system. Bockheim and Gennadiyev (2000) identified 17 generalized soil-forming processes, of which 11 are identified in Oregon soils. An additional process, cambisolization, is added here and will be defined forthwith. The dominant soil-forming processes in Oregon are humification, cambisolization, argilluviation, andisolization, and gleization. Vertization, silicification, and calcification occur to a limited extent; solonization, salinization, podzolization, and paludization occur to a very limited extent (Table 15.1).

15.2 Humification

Humification refers to the accumulation of well-humified organic compounds in the upper mineral soil. Soils reflecting humification include the Mollisols, Hum- great groups in Inceptisols, the Hum-suborder in Ultisols, and Humic subgroups in Andisols and Inceptisols. In Oregon, humification is favored by grassland vegetation, base-rich parent materials (in the case of Mollisols), and a time interval of more than 500 years. Common native grassland species that enhance humification include Idaho fescue (*Festuca idahoensis*), Sandberg bluegrass (*Poa secunda*), Indian ricegrass (*Achnatherum hymenoides*), Thurber's needlegrass (*Achnatherum thurberianum*), bluebunch wheatgrass (*Pseudoroegneria spicata*), needle and thread (*Hesperostipa comata*, also known as *Stipa comata*), inland saltgrass (*Distichlis spicata*), and basin wildrye (*Leymus cinereus*). Humification is important in all MLRAs with the exception of the Columbia Basin and the Humboldt Area. Extensive soil great groups reflecting this process include the Argixerolls, Haploxerolls, Humudepts, Palehumults, and Palexerolls.

15.3 Cambisolization

Cambisolization refers to a collection of weak soil-forming processes that leads to the formation of a cambic horizon, which is present largely in Inceptisols, including Dystroxerepts, Dystrudepts, Haplocambids, Haplocryands, Haploxerepts, Haploxerolls, Hapludands, Humudepts, Vitricryands, and soil series in other great groups. Cambic horizons in Oregon feature processes that do not lead to the significant accumulation of clay or salts. A cambic horizon may form in 1,000 to 2,000 years on gravelly parent materials (Gile 1975). Cambisolization is particularly important in all MLRAs. Extensive soil great groups reflecting this process include the Dystroxerepts, Haploxerolls, and Humudepts.

15.4 Argilluviation

Argilluviation (lessivage) refers to the movement and accumulation of clays in the solum. Argilluviation is a dominant process in the Alfisols, Ultisols, and in many Mollisols, particularly in Argixerolls, Haploxeralfs, and Palexerolls, but also in other great groups. This process occurs in Aridisols, especially the Argids suborder and in the Argidurids great group. The evidence for argilluviation in Oregon soils is the presence of argillans (i.e., clay skins) and abrupt increases in the clay content from the eluvial (A horizon) to the Bt horizon. This process is favored by long-duration precipitation, parent materials enriched in carbonate-free clays, stable landscape positions, backslopes rather than eroding shoulders, and a time interval of more than 2,000 years (Bockheim and Hartemink 2013). Argilluviation is especially important in the Argixerolls,

T. Thorson et al., *The Soils of Oregon*, World Soils Book Series,
https://doi.org/10.1007/978-3-030-90091-5_15

Table 15.1 Quantification of soil-forming processes in Oregon

Process	Taxa	Order	Sub-order	Great group	Sub-group	OSDs	Total
Humification	Mollisols; Humults; Hum-, Melan- great groups; Cumulic, Humic, Pachic subgroups	706	48	135	240	0	1129
Cambisolization	Cambids, some Andisols, Inceptisols, Mollisols, Vertisols	0	68	0	0	628	696
Argilluviation	Alfisols, Ultisols; Argids; Argi-, Pale- great groups; Alfic, Argic, Ultic subgroups	159	87	331	58	0	635
Andisolization	Andisols; Andic, Aquandic, Vitrandic, Vitric, Vitritorrandic, Vitrixerandic subgroups	203	0	0	275	0	478
None	Entisols, some Mollisols	62	0	0	0	285	347
Gleization	Aqu- suborders; Aquicambids & Aquisalids great groups; Aquic, Oxyaquic subgroups	0	150	9	78	0	237
Silicification	Durids suborder; Duri- great groups; Duric, Durinodic subgroups	0	58	69	33	0	160
Vertization	Vertisols; Vertic subgroups	23	0	0	46	0	69
Calcification	Calcids; Calci- great groups; Calciargidic, Calcic, Calcidic subgroups	0	7	4	43	0	54
Solonization	Natrids; Natr- great groups; Natric subgroups	0	0	17	2	0	19
Salinization	Salids suborder; Hal- great group; Sodic subgroup	0	2	0	11	0	13
Podzolization	Spodosols; spodic subgroup	13	0	0	0	0	13
Paludization	Histosols; Histic subgroups	10	0	0	3	0	13
	Total	1176	420	565	789	913	3863

Haplohumults, Palehumults, and Palexerolls great groups in the Willamette Valley, Blue Mountain Foothills, Malheur High Plateau, Owyhee High Plateau, Siskiyou Mountains, and Klamath Basin.

15.5 Andisolization

Andisolization results in soils whose fine-earth fraction is dominated by amorphous compounds. Andisols must have andic properties, which include high amounts of acid-oxalate-extractable aluminum and iron, a low bulk density, a high phosphate retention, and in vitric (allophanic) soils an abundance of volcanic glass. This process occurs in Andisols and in Andic, Aquandic, Vitrandic, and related subgroups. Andisolization is especially important in the Sika Spruce Belt, Coast Range, Cascade Mountains, Cascade Mountains—Eastern Slope, and the Blue Mountains. In addition to Andisols, Inceptisols often feature andic properties in the surface layers but lack the thickness requirement for Andisols.

15.6 Gleization

Gleization (hydromorphism) refers to the presence of aquic conditions often evidenced by reductimorphic or redoximorphic features such as mottles and gleying. In Oregon, gleization occurs dominantly in Aqu-suborders of Mollisols, Inceptisols, Entisols, and Alfisols. Gleization in Oregon is favored by depressions in the landscape, proximity to water and by parent materials that restrict drainage by virtue of texture, a layer that restricts moisture movement, or the presence of bedrock. Gleization is more common in soils in the Sitka Spruce Belt, Willamette Valley, basins within the Malheur High Plateau, and the Klamath Basin.

15.7 Silicification

Silicification refers to the secondary accumulation of silica in the form of durinodes or a duripan (Chadwick et al. 1987). This process is dominant in Durids and Durixerolls. Silicification is favored by an aridic and xeric soil moisture regime, parent materials enriched in opaline silica, inputs of loess and volcanic ash, and possibly secondary calcium carbonate ($CaCO_3$) (Chadwick et al. 1987, 1989). Soils strongly influenced by silicification are common on the Malheur High Plateau, which is downwind from where the Mazama ash was deposited 7,700 years ago. Silicification most commonly occurs in soils of the Malheur High Plateau, the Owyhee High Plateau and terraces of the Humboldt Area.

15.8 Vertization

Vertization represents a collection of sub-processes occurring in soils with very high amounts of smectitic clay, which enables soils to undergo shrinking and swelling that leads to cracking on the soil surface, tilted, wedge-shaped aggregates, and slickensides on aggregate faces. In Oregon, vertization is limited to 19 soil series in the Vertisols order and 45 soil series in subgroups of mainly Mollisols that are in fine or very-fine particle-size classes and have a smectitic mineralogy class. Vertization is most common in soils in the Willamette Valley, the Siskiyou Mountains, and basins on the Malheur High Plateau.

15.9 Calcification

Calcification refers to the accumulation of secondary carbonates ($CaCO_3$) in semi-arid and arid soils (Harper 1957). $CaCO_3$ initially fills micropores but over millennia may result in a strongly cemented petrocalcic horizon (Harper 1957; Gile et al. 1965, 1966; Brock and Buck 2005). Petrocalcic horizons are absent in Oregon soils. Calcification occurs primarily in Haploxerolls in the Columbia Basin and the Columbia Plateau because of calcareous loess. Calcification is related to mean annual precipitation, the presence of calcareous parent materials, and dust inputs. Calcification and the accumulation of calcium carbonates are primarily present in the Columbia Basin and occurs primarily in Haplocalcids. The development of calcic horizons in Oregon soils is relatively rare due to the lack of parent materials containing carbonates. However, many of the soils in eastern Oregon having precipitation of 300 mm/yr or less have carbonate accumulation in the cambic or argillic horizon.

15.10 Solonization

Also referred to as alkalization, this process occurs when soils subject to salinization have poor drainage or are drained. The excess soluble salts are leached out, the colloids under the influence of sodium become dispersed, and a strongly alkaline reaction develops. In Oregon, solonization is reflected by the presence of natric horizons in Natrargids and Natridurids. Solonization occurs in soils of arid basins in the Malheur High Platea. Natric horizons may develop in less than 6,600 years (Alexander and Nettleton, 1977).

15.11 Salinization

Nowadays, salinization is often used to describe human-caused increases in soluble salts in soils and surface waters as a result of "desertification." From a soil genesis standpoint, salinization refers to the collection of sub-processes that enable the accumulation of soluble salts of sodium, calcium, magnesium, and potassium as chlorides, sulfates, carbonates, and bicarbonates. In general, these salts are more soluble than gypsum in cold water and may be concentrated in a salic horizon. Salinization is a dominant process in Salids, Halaquepts, and sodic subgroups of Cambids. Salinization in Oregon is favored by depressions in the landscape, seasonally high water tables, and proximity to the edge of playas. The Flagstaff and Icene soil series, Typic Aquisalids, and Halaquepts represented by the Borovall and Reese soil series which occur in basins in the Malheur High Plateau and Humboldt Area.

15.12 Podzolization

Podzolization is a complex collection of processes that includes eluviation of base cations, weathering transformation of iron and aluminum compounds, mobilization of iron and aluminum in surface horizons, and transport of these compounds to the spodic Bs horizon as iron and aluminum complexes with fulvic acids and other complex polyaromatic compounds. Podzolization in Oregon is favored by cool, moist conditions, acid igneous bedrock, and coniferous forest. Only 13 Spodosol soil series that cover 374 km^2 have been identified in Oregon. However, Spodosols are likely to be more extensive in the state, because a large part of the subalpine region in the Cascade Mountains, where they are most apt to occur, has not received detailed mapping (see Figs. 2.4 and 7.1). Spodosols occur on uplifted marine terraces along the south-central Oregon coast. Podzolization occurs in Spodosols in the Sitka Spruce Belt and the Cascade Mountains.

15.13 Paludization

This term pertains primarily to the deep (>40 cm) accumulation of organic matter (histic materials) on the landscape, usually in marshy areas. Most soils featuring paludization are in the Histosol order. There are ten Histosol soil series in Oregon. The Lather soil series, a Limnic Haplohemist, is common on the edges of Upper Klamath Lake. Paludization is more common in Histosols and Humaquepts in the Sitka Spruce Belt, basins within the Malheur High Plateau, and the Klamath Basin.

15.14 Soils with Minimal Soil-Forming Processes

Approximately 13% of the soil series in Oregon lack a diagnostic subsurface horizon and evidence of key soil-forming processes. These soils include nearly all Entisols and many Mollisols, particularly those in the Haploxerolls great groups.

15.15 Summary

The dominant soil-forming processes in Oregon are humification, the accumulation of well-humified materials in the topsoil; cambisolization, the development of B horizons with weak color and structure; argilluviation, the transfer of clay into the subsoil; andisolization, the development of andic soil properties; and gleization, reducing conditions from restricted drainage. The other key processes include vertization, the development of cracking and slickensides in parent materials enriched in smectitic clays; silicification, the plugging of soil pores by secondary silica in the form of opaline materials; and calcification, the accumulation of secondary carbonates. Processes occurring to a limited extent include solonization, the accumulation of sodium salts that leads to the formation of natric horizons; podsolization, the accumulation of iron and aluminum in combination with humic materials which leads to the formation of a spodic horizon; paludization, the accumulation of organic matter under restricted drainage in depressions on the landscape; and salinization, the accumulation of soluble salts in depressions of the landscape.

References

Alexander EB, Nettleton WD (1977) Post-Mazama Natrargids in Dixie Valley Nevada. Soil Sci Soc Am J 41:1210–1212

Bockheim JG, Gennadiyev AN (2000) The role of soil-forming processes in the definition of taxa in Soil Taxonomy and the World Reference Base. Geoderma 95:53–72

Bockheim JG, Hartemink AE (2013) Distribution and classification of soils with clay-enriched horizons in the USA. Geoderma 209–210:153–160

Brock AL, Buck BJ (2005) A new formation process for calcic pendants from Pahranagat Valley, Nevada, USA, and implication for dating Quaternary landforms. Quat Res 63:359–367

Chadwick OA, Hendricks DM, Nettleton WD (1987) Silica in duric soils: I. A depositional model. Soil Sci Soc Am J 51:975–982

Chadwick OA, Hendricks DM, Nettleton WD (1989) Silicification of Holocene soils, in northern Monitor Valley Nevada. Soil Sci Soc Am J 53:158–164

Gile LH (1975) Holocene soils and soil-geomorphic relations in an arid region of southern New Mexico. Quat Res 5:321–360

Gile LH, Peterson FF, Grossman RB (1965) The K horizon: a master soil horizon of carbonate accumulation. Soil Sci 99:74–82

Gile LH, Peterson FF, Grossman RB (1966) Morphological and genetic sequences of carbonate accumulation in desert soils. Soil Sci 101:347–360

Harper WG (1957) Morphology and genesis of Calcisols. Soil Sci Soc Am 21:420–424

16 Benchmark, Endemic, Rare, and Endangered Soils in Oregon

16.1 Introduction

Benchmark soils are those that (i) have a large extent within one or more MLRAs, (ii) hold a key position in the *Soil Taxonomy*, (iii) have a large amount of data, (iv) have special importance to one or more significant land uses, or (v) are of significant ecological importance. About 6.4% of the soil series in the United States have been designated benchmark soils (Table 16.1).

Endemic soils are defined as the only soil in a family (Bockheim 2005). The proportion of soil series identified in the United States that is endemic is not known. Rare soils are those with an area less than 10,000 ha (Ditzler 2003). About 62% of the soil series in the United States occupy less than 10,000 ha (100 km^2).

Endangered soils are those that are endemic and rare. The proportion of soil series in the United States that is endangered is unknown. Moreover, there are certainly endemic soils with an area exceeding 10,000 ha that, depending on land use, have become endangered. A list of all benchmark, endemic, rare, and endangered soil series in Oregon is given in Appendix D.

16.2 Benchmark Soils

Only 5.7% of the soil series in which Oregon is the lead state have been designated as benchmark soils (Table 16.1). This is substantially less than the 6.3% of soils recognized as benchmark soils in the United States.

16.3 Endemic Soils

About 22% of the soil series recognized in Oregon are the only soil in the family and, therefore, may be considered endemic (Table 16.1). This is greater than the values ranging between 14 and 17% for Wisconsin (Bockheim and Hartemink 2017), Nevada (Blackburn et al. 2020), Colorado (Bockheim unpublished), New York State, and the northern New England states of Maine, Vermont, and New Hampshire (Bockheim unpublished).

About three-quarters (73%) of the soil series in Oregon occur only within the confines of the state. Oregon is ranked ninth in the United States in the number of soil series occurring only in the state and tenth in the proportion (percentage) of the total soil series identified only in the state (Bockheim unpublished).

There appear to be several reasons why Oregon has a higher proportion of endemic soils than other states. Deposition of the Mazama ash 7,700 years ago has impacted a large proportion of the state. About 13% of the soils are Andisols and another 23% are in Andic, Vitrandic, Vitritorrandic, Vitrixerandic, and Aquandic subgroups of Mollisols, Inceptisols, Alfisols, Aridisols, and Entisols. A second reason may be related to the fact that 86% of the soil series in Oregon occupy less than 10 km^2 in area. A third reason is that Oregon has a much larger proportion of intergrade soil series (41%) than other states that have been studied (5–26%). Finally, a comparatively large proportion (12%) of the endemic soils in Oregon have an isomesic or isofrigid soil-temperature regime, which is unique to soils along a narrow band of the Pacific Coast from Washington state to central California. Otherwise, the distribution of soil taxa from the order to family level for endemic soils in Oregon is comparable to the overall distribution of all soils.

16.4 Rare Soils

About 78% of the soil series recognized in Oregon occupy less than 100 km^2 (10,000 ha) each and, therefore, may be considered rare (Table 16.1). This is larger than the 62% value for the nation as a whole.

T. Thorson et al., *The Soils of Oregon*, World Soils Book Series,
https://doi.org/10.1007/978-3-030-90091-5_16

Table 16.1 Percentage of benchmark, endemic, rare, endangered, shallow, and lithic soils in Oregon and the USA

Soil class	OR	USA
Benchmark	5.7	6.3
Endemic	22	31
Rare	78	62
Endangered	19	22
Shallow	5	5.6
Lithic	17	5.7

16.5 Endangered Soils

About 19% of the soil series in Oregon are rare and endemic to the state, i.e., are considered "endangered" (Table 16.1). It is not known what proportion of the soil series in the United States is endangered.

16.6 Shallow Soils

Of the more than 1,700 soil series in Oregon, 5.0% are in the shallow family class, due mainly to the presence of a duripan, ortstein, or a paralithic contact that is moderately cemented to less cemented. Another 17% of the soil series are in lithic (bedrock within 50 cm of the surface that is strongly cemented or more) subgroups. However, about 12% of additional soils have a lithic or paralithic contact between 50 and 100 cm of the surface.

16.7 Highly Represented Soil Great Groups

Although there is no soil great group that only occurs in Oregon, there are some that are highly represented in the state. More than 90% of the Vitricryands area occurs in Oregon (Table 16.2). More than one-half of the Humudepts (64%) and Palexerolls (56%) occur in Oregon. Nearly one-half of the Haplohumults (46%), Hapludands (45%), and Palehumults (40%) mapped in the United States occur in Oregon.

16.8 Summary

The uniqueness and high pedodiversity in Oregon are manifested by:

- Mean annual precipitation in Oregon ranges from 175 mm in the Alvord Desert to over 2,400 mm in the Coast Range and Cascades Mountains (Chap. 3);
- Elevations of soils in Oregon range from sea level to over 3,400 m (Chap. 3);
- Sixty-four ecoregions have been identified in Oregon, each bearing a unique collection of soils (Chap. 3);
- Rock types bearing soils in Oregon include igneous rocks ranging from acidic to basic, sedimentary rocks ranging from claystones to sandstones and conglomerates, and metamorphic rocks (Chap. 3);
- Oregon has 17 Major Land Resource Areas, exceeded only by Texas, Alaska, California, Oklahoma, and South Dakota (Chap. 5);

Table 16.2 Great groups that are particularly abundant in Oregon

Taxonomy	Area in OR (km^2)	Area in USA (km^2)	Prop. area in OR (%)
Vitricryands	6518	7200	90.5
Humudepts	10637	16709	63.7
Palexerolls	4514	7988	56.5
Haplohumults	3499	7570	46.2
Hapludands	2572	5700	45.1
Palehumults	2780	6945	40.0
Udivitrands	3202	8437	38.0
Durixerolls	3550	13316	26.7
Haploxerolls	27727	106372	26.1

- Five of the eight epipedons and 10 of the 20 diagnostic subsurface horizons recognized in *Soil Taxonomy* occur in Oregon (Chap. 6);
- Four of the five soil moisture regimes and five of the ten soil temperature regimes identified globally occur in Oregon (Chap. 6);
- Ten of the 12 orders, 40 of the 67 suborders, 112 of 270 great groups, 389 subgroups, 1,080 families, and 1,707 soil series occur in Oregon (Chap. 6);
- 22% of Oregon's soils are endemic and 78% are rare according to the criteria described in this chapter.

References

Blackburn PW; Fisher JB, Dollarhide WE, Merkler DJ, Chiaretti JV, Bockheim JG (2020) The soils of Nevada. Springer Nature Switzerland

Bockheim JG (2005) Soil endemism and its relation to soil formation theory. Geoderma 129:109–124

Bockheim JG, Hartemink AE (2017) The soils of Wisconsin. Springer, NY, p 393

Ditzler C (2003) Endangered soils. National Coop. Soil Surv. Newsletter No. 25, Nov, pp 1–2

17 Land Use in Oregon

17.1 Introduction

The federal government owns 52% of Oregon land, private individuals and entities own 40%, and the remainder is owned by the state, Indigenous Nations, or local government. This chapter examines land use in Oregon from the perspectives of ownership, use, and soil characteristics. It summarizes Oregon's land use policy as well as controversies surrounding those policies. Key natural resource challenges that are especially pertinent to land use and soils are also described: climate change, wetland loss, flooding, landslides, volcanoes, earthquakes, tsunamis, coastal erosion, and wildfire.

17.2 Land Ownership and Management

Oregon is roughly 254,800 km^2 in area, making it the ninth-largest state in the country (World Atlas 2017). Soils have an effect on land ownership, as public entities generally own land that is less agriculturally productive. Land ownership, in turn, affects land management, with soils on public land generally being managed less intensively than on private land (Kelso 1947; Robinson et al. 2019).

17.2.1 Federal

The 52% of Oregon land owned by the federal government includes 15 National Forests, 1 National Grassland, 1 National Park (Crater Lake), and 25 National Wildlife Refuges (Vincent et al. 2020). As Table 17.1 illustrates, the Bureau of Land Management (BLM) and the U.S. Forest Service (USFS) manage most of the federal land in Oregon. BLM is part of the Department of the Interior (DOI) and USFS is part of the United States Department of Agriculture (USDA).

Some federal land is an unusual checkerboard pattern of holdings called the "O&C Lands." These O&C lands date back to a grant of about 15,000 km^2 from the federal government to the Oregon and California Railroad as a development incentive in the 1860s. The grant included odd-numbered sections of public domain land within 20 miles on each side of the proposed railroad line. The railroad was supposed to sell the land to qualified settlers, but by 1916 there were still almost 10,000 km^2 unsold. Congress reclaimed the unsold land, which is mostly forest. It is now managed by the BLM (Bureau of Land Management 2020) and, to a lesser extent, the USFS (Cain 2019). Except for the O&C lands, the BLM mostly manages rangeland in eastern Oregon. The USFS mostly manages federal forests throughout the state. The United States Fish and Wildlife Service (USFWS) and the National Park Service (NPS), both part of DOI, and the Department of Defense (DOD) manage relatively little federal land in Oregon (Fig. 17.1).

There has been long-term, but perhaps increasingly vocal, debate about the management of federal lands in Oregon and other states. In western Oregon, the recent conflict centers on the logging policy of forests managed by the USFS and BLM, especially after the northern spotted owl (*Strix occidentalis caurina*) was designated as a threatened species in 1990. The subsequent curtailment of logging in federal forests where the bird occurs impacted rural communities dependent on logging revenue. This controversy, in simple terms, pits "environmentalists" who want to save old-growth forests and the owls against loggers and their supporters who dislike federal restrictions that affect their livelihoods (Satterfield 2002). The controversy is so strong and the view-

T. Thorson et al., *The Soils of Oregon*, World Soils Book Series,
https://doi.org/10.1007/978-3-030-90091-5_17

Table 17.1 Federal land management in Oregon by the agency

Agency	Area managed (km^2)	Percent of federal lands (%)
Bureau of Land Management	63,692	48.8
U.S. Forest Service	63,510	48.7
U.S. Fish and Wildlife Service	2,328	1.8
National Park Service	794	0.6
Department of Defense[a]	133	0.1
Total	**130,457**	**100**

[a]Does not include land managed by the Army Corps of Engineers
Source Vincent et al. (2020)

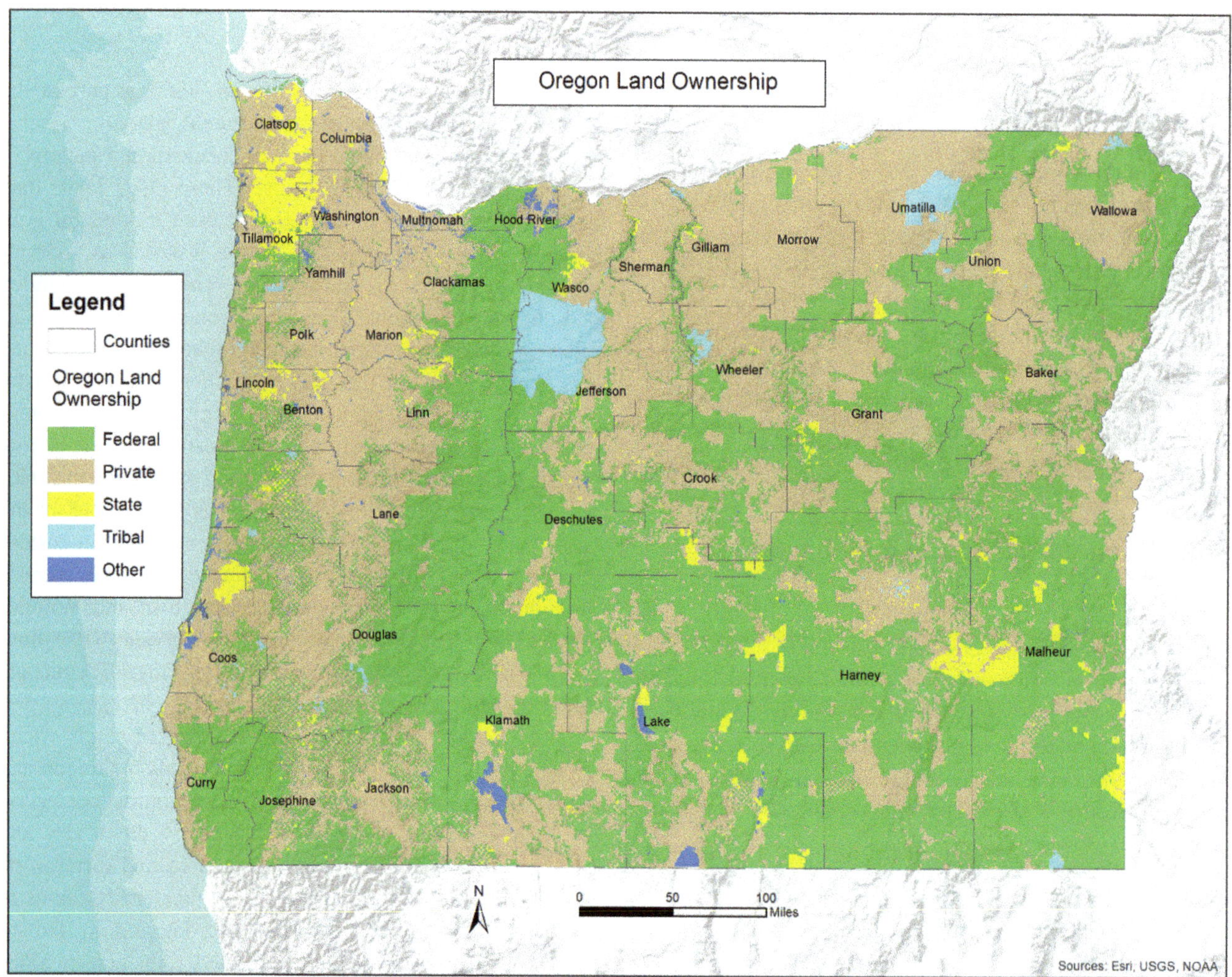

Fig. 17.1 Land ownership in Oregon. "Other" includes water bodies and land owned by local government. *Sources* ESRI, USGS, NOAA, Oregon Department of Forestry (n.d.)

points so entrenched that Oregon Public Broadcasting produced a series called "The Timber Wars" to document the struggle (Oregon Public Broadcasting 2020).

In eastern Oregon, there is an ongoing debate over the administration of grazing permits on federal land.[1] This rangeland debate flared into the 2016 takeover of the federal Malheur National Wildlife Refuge by an armed group of ranching sympathizers from multiple western states. The militants were protesting both the imprisonment of Malheur County ranchers Dwight and Steven Hammond for setting fires on federal land and the larger issue of federal ownership and management of rangeland. The takeover lasted 41 days and resulted in law enforcement officials shooting and killing one of the occupiers and arresting many of the others (Pogue 2018). The refuge takeover is just one example of a movement across the western United States, which is similar to the timber wars described in the preceding paragraph. This dispute primarily involves grazed rangeland and has been termed the "Sagebrush Rebellion," a term that appears to have been coined in the 1970s. Indeed, President Ronald Reagan stated in a campaign speech in 1980, "I happen to be one who cheers and supports the Sagebrush Rebellion. Count me in as a rebel" (Coates 1986, p. 1).

Two agencies within USDA, the Natural Resources Conservation Service (NRCS) and the Farm Service Agency (FSA), do not manage federal lands but instead assist private land managers and Indigenous Nations to implement voluntary conservation projects. NRCS also operates the Web Soil Survey, which provides soil data and information produced by the National Cooperative Soil Survey (Soil Survey Staff 2019). FSA administers the Conservation Reserve Program (CRP), in which private landowners can enroll land to temporarily remove it from production (USDA-Farm Service Agency 2020).

[1] Until the federal Organic Administration Act of 1897 was passed, federal lands in the west were essentially open for grazing without regulation. Under the auspices of that act, grazing regulations on national forestland were implemented in 1910. The 1934 federal Taylor Grazing Act allowed for grazing regulations on other federal land and eventually led to the establishment of the BLM. These regulations require ranchers to obtain a permit from the local BLM or USFS office for grazing livestock on specific parcels ("allotments") of federal land. Although permits are usually authorized for 10 years, the BLM or USFS provide annual specifications for the number and type of livestock and the allowable grazing periods. Ranchers buy permits based on the animal unit months (AUM) allowed via the permit. An AUM is the amount of forage consumed by a 450 kg (1,000 pound) cow with calf for one month (Galbraith and Anderson 1971), (Keyes and Keyes 2015).

17.2.2 Indigenous Peoples[2]

There are nine federally recognized Indigenous Nations in Oregon:

- Burns Paiute Tribe
- Confederated Tribes of Coos, Lower Umpqua, and Siuslaw Indians
- Coquille Tribe
- Cow Creek Band of Umpqua Tribe of Indians
- Confederated Tribes of the Grand Ronde Community of Oregon
- The Klamath Tribes
- Confederated Tribes of Siletz
- Confederated Tribes of the Umatilla Indian Reservation
- Confederated Tribes of the Warm Springs Indian Reservation

Five additional Indigenous Nations maintain interest in Oregon:

- Nez Perce Tribe
- Tolowa Dee-ni' Nation
- Fort McDermitt Paiute and Shoshone Tribes of the Fort McDermitt Indian Reservation, Nevada and Oregon
- Shoshone-Paiute Tribes of the Duck Valley Reservation, Nevada
- Yakama Nation

Indigenous Nations are sovereign and self-governing. Approximately 3,632 km^2 or 1.6% of land in Oregon are Tribal reservations or trust lands.[3] Reservations and trust lands are held by the US government for the use or benefit of Indigenous Nations and are mostly managed as forestland or rangeland, although most reservations also include land developed for housing, tribal government, cultural facilities, tourism, and recreation. Indigenous Nations ceded over 140,000 km^2 of Oregon land to the US government as part of treaties signed in the 1800s. Some Oregon tribes retain "reserved rights" to hunt, fish, gather Indigenous foods, and co-manage cultural and natural resources in the ceded areas. (Oregon Legislative Policy and Research Office 2016).

The original treaties in the Pacific Northwest were written by Isaac Stevens and Joel Palmer on behalf of President Franklin Pierce in 1854 and 1855. Stevens was the Territorial Governor of Washington and Palmer was the

[2] The terminology used here for Indigenous Peoples and Cultures follows, as much as possible, guidance set forth in Gregory Younging's *Elements of Indigenous Style* (2018).

[3] This land is included as part of the 52.3% of Oregon land owned by the federal government because it is held in trust by the Bureau of Indian Affairs.

Superintendent for Indian Affairs in the Oregon Territory. There is widespread agreement that these treaties were unfairly negotiated to benefit Euro-American settlers. For example, Judge George H. Boldt offered a summary of the treaty process in his landmark decision reaffirming Indigenous fishing rights in the *United States v. State of Washington* (1974, p. 330): "The treaties were written in English, a language unknown to most of the tribal representatives, and translated for the Indians by an interpreter in the service of the United States using Chinook Jargon, which was also unknown to some tribal representatives. Having only about three hundred words in its vocabulary, the Jargon was capable of conveying only rudimentary concepts, but not the sophisticated or implied meaning of treaty provisions…."

Subsequent interpretations of the treaties entail a complicated and mercurial set of policies beyond the scope of this volume. Interested readers are directed to tribal websites of the Indigenous Peoples of Oregon, institutions such as The Tamástslikt Cultural Institute (2021), The Museum at Warm Springs (The Confederated Tribes of Warm Springs 2021), or the Chachalu Museum and Cultural Center (Confederated Tribes of Grand Ronde 2021). Another good reference is *The Oregon Historical Quarterly Special Issue: The Isaac I. Stevens and Joel Palmer Treaties—1855–2005* (Oregon Historical Society 2005).

17.2.3 State and Local Government

The state of Oregon owns 7,184 km^2 or 2.8% of land in Oregon. The Oregon Department of Forestry manages about 40% of this state land, the Oregon Department of State Lands manages another 40% (most of which is rangeland), the Oregon Department of Fish and Wildlife and the Parks and Recreation Department each manage a little over 5%, and the balance of state land is managed by other agencies such as the university system and the Department of Transportation (Oregon Department of State Lands 2017a). Multiple Oregon state agencies enforce land use and environmental protection regulations and/or administer permits to manipulate private land. These include the Departments of Environmental Quality, State Lands, Agriculture, Forestry, Geology and Mineral Resources, Land Conservation and Development, and Energy.[4]

City and county governments also administer permit systems for some types of land manipulation, such as zoning, and removal or fill in floodplains. County governments and, in the Portland metropolitan area, Metro Regional Government ("Metro"), develop land-use plans. Metro, county, and city governments also own and manage parks and greenspaces.

Soil and Water Conservation Districts (SWCDs) are special districts that help landowners voluntarily implement conservation projects on their land. Oregon passed legislation in 1939 allowing the formation of SWCDs. By 1972, SWCDs had formed across Oregon, generally along county boundaries (Oregon Department of Agriculture 2016). SWCDs collaborate with other agencies (most notably NRCS and the Oregon Department of Agriculture) in prioritizing land conservation efforts (Oregon Association of Conservation Districts 2018). Oregon State University (OSU) administers the Extension Service, which, among other duties, conducts research and provides educational programming to help Oregon land managers (Oregon State University 2020).

17.2.4 Private

About 40% of Oregon land is privately owned, with the greatest percentage being forestland, followed by rangeland, cropland, pasture, and developed land (NRCS 2018). The National Agricultural Statistics Service (NASS) 2017 Census of Agriculture reported 37,616 Oregon farms,[5] a 6% decrease since 1997. The land in Oregon farms (about 64,600 km^2) decreased by about 10%, and the average farm size (about 170 ha)[6] decreased by about 4% between 1997 and 2017. As Fig. 17.2 illustrates, small (less than 20 ha) farms have increased in number, while larger farms have decreased in number since 1997. However, Figs. 17.3 and 17.4 show that large farms (greater than 202 ha) account for most of Oregon's farmland area and market value of production (NASS 2019, 1997). From a land use perspective, Fig. 17.4 illustrates the economic importance of maintaining relatively large blocks of land available for farming, as will be discussed further in Sect. 17.5.

In 2017, NASS identified 7% of Oregon producers as less than 35 years old, 57% aged 35–64, and 36% over 64 years old. About 97% of Oregon producers were identified as white, about 1% identified as American Indian/Alaskan native, about 1% Asian, 1% more than one race, and less than 0.2% as Black or Native Hawaiian/Pacific Islander. In 2017, 84% of Oregon farms were owned by individuals or families, about 6% were organized as partnerships, 7% as

[4] The Department of Land Conservation and Development provides staffing for the Land Conservation and Development Commission, and the Department of Energy provides staffing for the Oregon Energy Facility Siting Council.

[5] NASS considers farms to be those operations from which $1,000 or more of agricultural products were produced and sold, or normally would have been sold, during the census year. NASS statistics do not include forestry operations.

[6] NASS statistics on farm size do not include rangeland that a farm or ranch leases from government agencies like BLM.

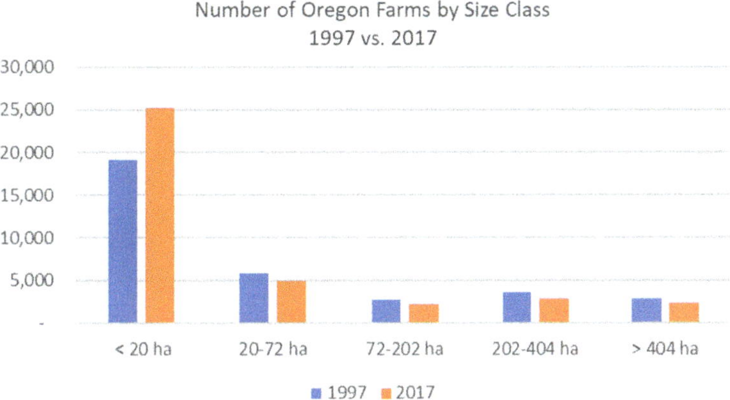

Fig. 17.2 Number of Oregon farms by size class, 1997 versus 2017. Small (less than 20 ha) farms increased in number, while larger farms decreased in number between 1997 and 2017. *Sources* NASS (1997) and NASS (2019)

Fig. 17.3 Land area in Oregon farms by size class, 1997 versus 2017. Although there are many more small farms, most of Oregon's farmland is held in larger operations. *Sources* NASS (1997) and NASS (2019)

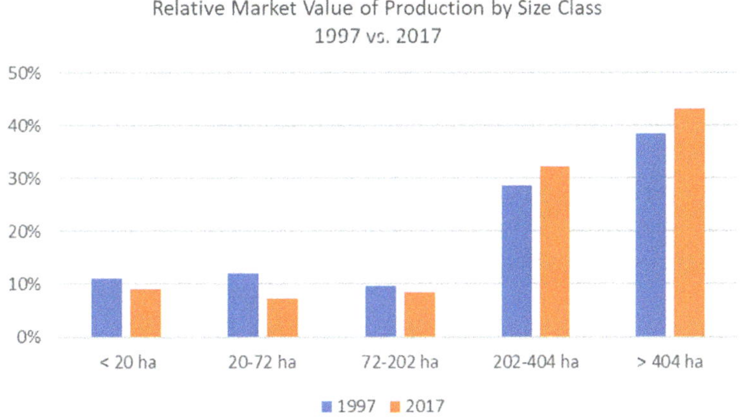

Fig. 17.4 Relative market value of production by size class, 1997 versus 2017. Large Oregon farms produce more economic value of agricultural products than small farms, a relationship that was even more marked in 2017 than in 1997. Percentages for each year add to 100%. *Sources* NASS (1997) and NASS (2019)

corporations, and the remainder as some other legal entity. This breakdown in legal organization has changed little since 1997 (NASS 2019).

However, Horst (2019) found evidence of an increasing trend of corporate farm ownership in Oregon. She analyzed farmland sales exceeding $1,000 in 33 of Oregon's 36 counties over the period 2010–2015, classifying these sales according to the type of purchaser, acreage, price, and location. Horst's results showed a median sale of 8.1 ha, an average sale of 48.2 ha, and a median price of about $11,000 per undeveloped[7] hectare. In terms of the type of purchaser, 71% were individuals, 19% were corporations, and 8% were trusts, limited partnerships, or estates. Corporations, however, purchased 40% of the land area during this time period, indicating they tended to buy larger properties than individuals. Fewer than half of the corporate buyers appeared to be primarily farming interests. More than half were real estate, land development, banks, or investment companies. The Willamette Valley had the most sales, but eastern Oregon had the most land area sold. Horst's data also indicated a growing trend of "amenity owners," who purchase rural Oregon property mainly for lifestyle reasons rather than for farming and forestry livelihoods. She concluded that most Oregon farmland is owned by individual "family farmers," but there appears to be a trend to increasing corporate ownership of farmland in the state (Ibid.).

Private forestland classification systems differ by source and have changed over time, resulting in a somewhat confusing set of definitions. For example, the USFS National Woodland Owner Survey (NWOS) might be thought of as the forestry equivalent to the NASS Census of Agriculture. The NWOS recognizes three types of private forestland ownership: family, corporate, and other. "Other" includes organizations, associations, and clubs (Butler and Butler 2016a). The 2011–2013 Oregon NWOS estimated there to be 120,000 km^2 of Oregon forestland,[8] about 60% of which was federal, 15% family plus other, 21% corporate, and 4% state and local government. NWOS estimated there to be 44,000 Oregon family forests, over 70% of which were less than 20 ha. About 50% of Oregon family forests were owned by people 55–64 years old (Butler and Butler 2016b).

USDA uses classification systems that distinguish between industrial and non-industrial private forestland (NIPF). USDA agencies such as NRCS, USFS, and FSA define NIPF slightly differently, but they generally agree that the term applies to ownerships of smaller land holdings that do not include the operation of a commercial lumber mill. For example, NRCS recently offered the following clarification, but noted that the agency is requesting comments to modify the definition (Federal Register 2020, p. 81,872): "NRCS will identify someone as a nonindustrial private landowner if they: (1) (i) Own fewer than 45,000 acres[9] of forest land in the United States; and (ii) Do not own or operate an industrial mill for the primary processing of raw wood products as determined by NRCS in consultation with the State Technical Committees; or (2) Meet criteria established for a nonindustrial private landowner by NRCS in a State in consultation the State Technical Committee."

The Oregon Department of Forestry (ODF) defines NIPF landowners as "…individuals, partnerships and privately held corporations that have less than 25 percent of their income coming from a primary forest products milling facility, and do not employ a professional forestry staff" (Oregon Department of Forestry 2006, p. 3). However, ODF and the Oregon Forest Resources Institute now mostly use a system classifying private forestland as either "large" (ownerships of 2,023 or more hectares[10]) or "small" (ownerships of less than 2,023 hectares). For the sake of simplicity, this chapter will use ODF's "small" and "large" terminology to describe Oregon forestland.

Defining private forestland classes based solely on the size of the operation is well-adapted to a growing trend in forestland ownership: the use of Real Estate Investment Trusts (REITs) and Timber Investment Management Organizations (TIMOs).[11] Some REITs maintain a "vertically integrated" structure by owning both forestland and mills; other forest REITs do not own mills. No TIMOs own mills. Thus, definitions of industrial forestland that include mill ownership are not as useful as they once were to distinguish between large corporate versus small "family" owned forestland. As of 2020, approximately 40% of private forestland in western Oregon is owned by REITs or managed by TIMOs, and debate exists concerning the social, economic, and environmental effects of this form of ownership (Bliss and Kelly 2008; Rusignola 2019; Schick et al. 2020).

Some private landowners choose to sell an easement on their property, or convey development rights by other means, to a land trust. The land trust then protects the

[7] "Undeveloped" farmland was presumably land that did not include buildings.

[8] NWOS data exclude forestland holdings less than 4.0 ha in size.

[9] Approximately 18,211 hectares.

[10] 2,023 hectares equals 5,000 acres.

[11] REITs are companies that own and manage income-producing real estate. Some REITs specialize in forestland. TIMOs manage forestland owned by institutional investors such as pension funds. Over the past 30 years, many large timber companies converted their forestland to REITs and TIMOS for tax and other economic reasons (Mendell 2016).

property from development. As of 2015, land trusts had protected over 3,100 km^2 in Oregon from development (Land Trust Alliance 2020).

17.3 Land-Use Designations

The NRCS (2014) currently defines nine land-use designations in its National Planning Procedures Handbook:

1. Developed land: land used for non-farm residences, commercial sites, roads, schools, airports, and urban open spaces.
2. Farmstead: land used for farm dwellings, storage of equipment and crops, livestock confinement, and the storage and handling of farm supplies.
3. Cropland: land used for producing and harvesting annual or perennial plants for food, fiber, forage, or energy.
4. Pasture: land planted to forage species that are used primarily for grazing livestock.
5. Rangeland: land predominantly covered with herbaceous species and shrubs managed mostly as a natural ecosystem.
6. Forestland: land predominantly covered with trees that are managed for wood products.
7. Water: sites where the dominant characteristic is open water or permanent ice and snow.
8. Associated agriculture lands: incidental land associated with farms and ranches that are not managed for food or fiber production. Examples include ditches, farm roads, and riparian areas.
9. Other: barren land or land used for extraction of minerals, fossil fuels, gravel, or sand.

The first six land uses above are listed in order from generally most intensive to least intensive human inputs and are further described in Sects. 17.3.1–17.3.5. The management intensities of water, associated agricultural lands, and other lands vary greatly by site and are not addressed in this chapter. NRCS identifies eight "land use modifiers" to provide further specificity of land use: irrigated, wildlife, grazed, drained, organic, water feature, protected, and hayed. Soil characteristics are certainly one of the deciding factors in how humans use land, but land use also affects soil characteristics, sometimes markedly so. For example, many urban areas (developed land) and quarries (other land) have been so altered by land use as to fall under the category of "anthropogenic features" in soil surveys (Schoeneberger et al. 2017).

Land-use classification can be imprecise, and people sometimes use land in ways that seem to be a hybrid between two or more land-use classes. For example, organic vegetable farms in the Willamette Valley sometimes rotate several years of perennial pasture into their fields to break up disease cycles for the vegetable crops and to improve soil health.[12] Considerable irrigated land in the Columbia Basin has been planted to hybrid poplars, which are intensively managed like cropland but are harvested for wood or paper products like forestland. Oak savanna in the Siskiyou-Trinity area may have significant herbaceous and shrub communities that are grazed like rangeland, but have firewood harvested like forestland. Additionally, land use can change, for example when cropland is developed for housing or converted to designated protected areas such as wildlife refuges. Figure 17.5 summarizes Oregon land uses based on the most recent (2011) USGS data (U.S. Geological Survey 2011). The NRCS National Resource Inventory (NRI) provides land-use data for non-federal rural land, but the USGS data are used here because they represent the entire state, including rural and urban, and federal and non-federal (NRCS n.d.).

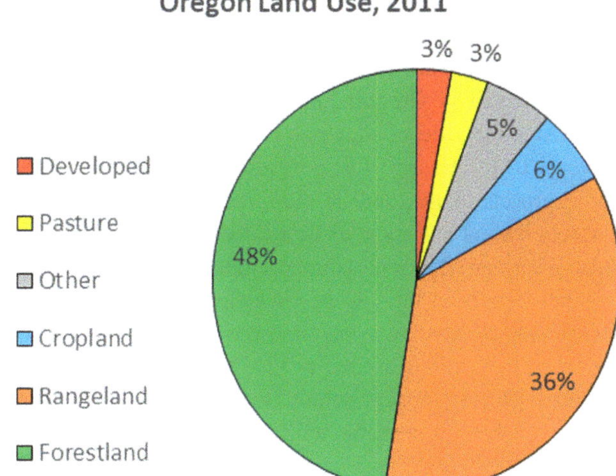

Fig. 17.5 Oregon land use in 2011. Data are compiled from United States Geological Survey data, which are based on 2011 imagery. USGS categorizes vegetative cover according to 12 "class names." For this chart, those USGS vegetative cover types were categorized into the land uses shown above. Developed land, pasture, cropland, and forestland are clear categories in the USGS data. "Rangeland" in this chart includes the USGS shrub and herb, desert and semi-desert, and recently burned grassland and shrubland. "Other" in this chart includes open water, wetlands, bedrock, dunes, mined land, and ice fields. *Source* U.S. Geological Survey (2011)

[12] Karlen et al. (1997, p. 6) defined soil health, also known as soil quality, as "the capacity of a specific kind of soil to function, within natural or managed ecosystem boundaries, to sustain plant and animal productivity, maintain or enhance water and air quality, and support human health and habitation."

17.3.1 Developed Land

In 2011, USGS estimated there to be about 6,900 km^2 of developed land in Oregon, approximately 3% of the total land area in the state, as shown in Fig. 17.5 (U.S. Geological Survey 2011). Except for government buildings, roads, major airports, dams, and the like, most developed land is privately owned and most of Oregon's developed land is in cities. There are 241 incorporated cities in Oregon, ranging in population from the city of Greenhorn in Baker County (2018 population of two full-time residents) to the city of Portland in Multnomah County (2018 population of about 650,000). The ten most populous Oregon cities are Portland, Eugene, Salem, Gresham, Hillsboro, Beaverton, Bend, Medford, Springfield, and Corvallis. Of these, all but Bend and Medford are in the Willamette Valley (Oregon Secretary of State 2020). The 2019 United States Census estimated Oregon's population at 4.2 million, which ranked 27th in the country (U.S. Census Bureau 2020). A population density of 103 people/km^2 makes Oregon the 38th most densely populated state in the country.[13] In the course of development, humans alter soils by excavating, filling, and paving. This alteration has led to special urban soil survey processes that acknowledge the existence of "human-altered and human-transported soils" (Galbraith and Shaw 2017).

17.3.2 Farmsteads

Except for minor exceptions, such as farmsteads connected to Oregon State University farms or those connected to USFWS refuges, farmsteads in Oregon are privately owned. Farmsteads are important components of farming operations. All farms, whether primarily cropping or livestock operations, store materials and equipment on farmsteads. Farmers, of course, typically live on farmsteads. Swine, poultry, and mink usually are housed for most or all of their lives in confinement on farmsteads. The duration and type of confinement for livestock on dairy, beef, and horse farms varies by operation.

Farmsteads where animals are kept and raised in confinement tend to store feed and manure in a small area, resulting in potential risk to air and water quality due to odors, nutrients, pathogens, antibiotics, hormones, organic matter, and heavy metals. EPA and USDA define the term "animal feeding operation" (AFO) as facilities in which livestock are confined and fed for 45 days or more per year in an area where "crops, vegetation forage growth, or post-harvest residues are not sustained in the normal growing season…" (USDA and U.S. EPA 1999, p. 10). Thus, livestock on AFOs is entirely dependent on farmers bringing them feed for 45 or more days per year. AFOs with the greatest risk of contributing to water pollution are termed "concentrated animal feeding operations" (CAFOs) and are subject to permitting under Sect. 402 of the federal Clean Water Act. Risk is generally defined in terms of the number of animals confined. EPA has authorized 43 states, including Oregon, to issue permits for CAFOs. In 2019, ODA reported a total of 508 active CAFO[14] permits in Oregon (Oregon Department of Agriculture 2019).

As with developed land, farmstead soils have often been highly manipulated via excavation, filling, and paving. This is especially true for CAFO operations that have constructed waste storage ponds. However, farmsteads are usually small in size compared to developed land, and soil maps generally do not delineate human-altered and human-transported soils on farmsteads.

17.3.3 Cropland

There are few states that can boast of such cropland diversity as Oregon (Miles 1985). Some Oregon farmers produce crops grown for direct human consumption, such as Marionberries, wheat, pears, and potatoes. Some farmers grow corn for silage and hay, which is harvested, usually stored for some period of time, and then fed to livestock. There are Oregon farms that produce seed crops such as grass, clover, and carrot seed destined for national and international markets. Ornamentals, such as containerized plants, shrubs, cut flowers, and Christmas trees, are economically important Oregon crops marketed across the country. In 2015, the Oregon legislature legalized the cultivation of recreational marijuana, which quickly became an important crop in terms of economics, if not land area.

USGS estimates there to be about 14,900 km^2 of cropland in Oregon, approximately 6% of the total land area in the state, as shown in Fig. 17.5 (U.S. Geological Survey 2011). Almost all cropland in Oregon is privately owned. The amount of dry (not irrigated) cropland has been declining in Oregon, as illustrated in Fig. 17.6 (NRCS 2018). Table 17.2 shows the top 15 Oregon cropland commodities, ranked in terms of 2019 value of production. It does not, however, include livestock production because it is not possible to distinguish the value of livestock produced from feed raised

[13] In other words, 37 states have a higher population density than Oregon.

[14] Federal policy defines "*concentrated* animal feeding operations" as CAFOs, while Oregon uses the same acronym for "*confined* animal feeding operations." Oregon's definition of CAFO is somewhat broader than the federal definition and encompasses a larger number of operations (Hessler, Luk, and McMillan), (Oregon Secretary of State 2009), (USDA and U.S. EPA 1999).

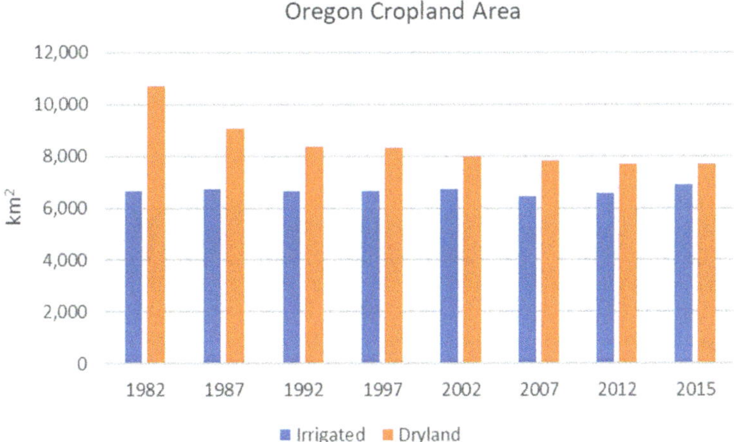

Fig. 17.6 Cropland area in Oregon since 1982. Dry (non-irrigated) cropland has decreased since 1982, but irrigated cropland has remained fairly stable over this period. Data include both cultivated crops (e.g., wheat, corn) and uncultivated crops (e.g., permanent hay, orchard). The decrease in cropland during the 1980s was largely due to the land being enrolled in USDA's Conservation Reserve Program. *Source* USDA-Natural Resources Conservation Service (2018)

Table 17.2 The top 15 Oregon cropland commodities, ranked by the National Agricultural Statistics Service in terms of 2019 value of production. Marijuana and hemp sales are not included in these data

Rank	Crop	2019 Value of Production (millions of dollars)
1	Greenhouse & nursery	$955
2	Hay	$674
3	Grass seed	$517
4	Wheat	$283
5	Wine grapes	$238
6	Potatoes	$199
7	Blueberries	$134
8	Pears	$109
9	Onions	$108
10	Christmas trees	$104
11	Hazelnuts	$84
12	Cherries	$75
13	Hops	$72
14	Corn for grain	$51
15	Mint for oil	$41

Source Oregon Department of Agriculture (2021)

on Oregon cropland versus that grown on rangeland and pasture or imported from outside the state.

Table 17.2 also does not include marijuana because the underlying data were collected by the National Agricultural Statistics Service, a division of USDA, and the federal government currently considers marijuana to be an illegal crop. In 2019, the value of recreational marijuana in Oregon was estimated at $725.8 million and thus this crop ranks high in the total value of production, especially if combined with medical marijuana and hemp (Danko 2020; Mortenson 2016).

In 2018, more hectares of hay were grown than any other crop in Oregon, followed by wheat, and then seed crops (predominantly grass seed). In 2019, Oregon ranked first in the nation in the production of hazelnuts, crimson clover seed, orchardgrass seed, fescue seed, ryegrass seed, red

Fig. 17.7 Beef cattle grazing pasture land near Seneca Oregon, in Grant County. *Source* Photograph by NRCS Oregon

clover seed, sugarbeets for seed, white clover seed, potted florist azaleas, Christmas trees, and rhubarb. Essentially the entire US production of hazelnuts is from Oregon (Oregon Department of Agriculture 2021). Not surprisingly, over 90% of Oregon cropland soils are in Land Capability Classes 1 through 4 (NRCS 2003). See Sect. 17.4.1 for a description of Land Capability Classes.

17.3.4 Pasture and Rangeland

In 2011, USGS estimated Oregon to have 7,500 km^2 of pasture (approximately 3% of the total land area in the state) and 90,400 km^2 of rangeland (approximately 36% of the total land area), as shown in Fig. 17.5 (U.S. Geological Survey 2011). The distinction between pasture and grazed rangeland is based on the intensity of use. NRCS (2014) defines rangeland as being occupied by primarily herbaceous vegetation that is "managed as a natural ecosystem" (p. 600-A.13). Although pasture also is established to herbaceous vegetation, pastures "receive periodic renovation and cultural treatments, such as tillage, fertilization, mowing, weed control, and may be irrigated" (Fig. 17.7) (Ibid., p. 600-A.13). Thus, rangeland is not necessarily grazed, whereas pasture is, by definition, used for grazing. Furthermore, pastures usually receive regular management inputs, but the management of grazed rangelands is generally confined to fencing, livestock water developments, and restricting when and where livestock have access to the land. As in most agricultural enterprises, there are gray areas to these distinctions: some farmers provide few management inputs to their pastures, but some ranchers seed rangeland with desirable forage species or actively control undesirable species such as western juniper (Bedell 1993).

Almost all pasture in Oregon is privately owned and may occur on soils suitable for cropping. Oregon rangeland is a mix of federal (largely managed by the BLM), state (managed by the Oregon Division of State Lands), and private ownership. Rangeland is typified by soils and climate that prevent the economic production of crops. Pasture occurs across Oregon, but rangeland is generally considered to lie east of the Cascade Mountains. Over half (55%) of the Oregon rangeland soils with the highest yield potentials are in the Haploxerolls and Argixerolls great groups (Appendix E).[15]

In 2019, Oregon produced $625 million worth of cattle and calves and $552 million of milk, ranking third and fourth, respectively, in the value of all Oregon commodities. Sheep can be locally important, but 2019 wool sales totaled only $1.4 million. Beef and sheep are raised on both pasture and rangeland and dairy animals are pastured some of the time on many farms. However, livestock are also fed hay and, usually, grain. This makes it difficult to identify the value of pasture and rangeland forage contributing to the value of beef, milk, and wool. Horses, goats, llamas, and alpacas are grazing animals kept on some farms as pets and on other farms for commercial sale, but are not in the top 40 Oregon agricultural commodities for 2019 sales (Oregon Department of Agriculture 2021). Poultry and swine are raised in Oregon but are not typically grazed. Mink are never grazed.

17.3.5 Forestland

USGS (2011) estimated there to be 120,900 km^2 of Oregon forestland, which is approximately 48% of the total land area in the state. About 60% of Oregon forestland is federal, 36% private, and the rest state and local (Butler and Butler 2016b). See Figs. 17.8 and 17.9. Although always an important part of the state's economy, Oregon timber harvests soared following World War II. By the 1970s, the forest industry accounted for about 10% of Oregon's private-sector jobs, 12% of state gross domestic product, and 13% of private-sector wages. Timber was especially important in rural areas, where it provided good paying jobs. A nationwide recession and poor housing market in the early 1980s reduced timber harvests and caused mills to close or to modernize in ways that required less labor. There was also increasing competition with timber from southern states and British Columbia (Lehner 2017).

Then, in the early 1990s, the USFWS listed the northern spotted owl (*Strix occidentalis caurina*) and the marbled murrelet (*Brachyramphus marmoratus*), both residents of Pacific Northwest old-growth forests, as threatened species. These listings led to the protection of federal old-growth forest under the 1994 Northwest Forest Plan developed by the USFS and resulted in substantial reductions in timber harvest on federal land (Spies et al. 2019). Oregon's timber harvest declined from around nine billion board feet[16] per year in the 1960s to around four billion board feet in the 1990s, a level at which it appears to have stabilized (Fig. 17.10), (Gale et al. 2012). Nevertheless, Oregon consistently ranks at or near the top nationally in the production of softwood[17] lumber (Latta et al. 2019).

Oregon's forests are also important culturally, as exemplified by the Douglas-fir silhouette on Oregon license plates and the 23-foot gold-plated bronze statue of a pioneer, complete with an ax, which sits atop the Oregon State Capitol. Oregon forests feature prominently in such literary works as *Sometimes a Great Notion* (Kesey 1964) and *The Overstory* (Powers 2018). Portland is home to the World Forestry Center, a non-profit educational organization founded in 1966 (World Forestry Center 2021). Portland itself is known by the nickname "Stumptown," and the city's soccer teams are called the "Timbers" and the "Thorns" (Miner 2013).

Key Oregon forest products are paper, lumber, exports, plywood and veneer, and energy. In 2016, forest-product sales in Oregon were $8.1 billion, 3.7% of the Oregon Gross Domestic Product. In 2019, 66% of the harvested timber volume was on large private forestland, 13% on federal land, 12% on small private forestland, and 9% on state land (Fig. 17.9; Oregon Forest Resources Institute 2019). Douglas-fir (*Pseudotsuga menziesii*) is the leading species (Fig. 17.11),[18] accounting for almost 70% of the total harvest, followed by western hemlock (*Tsuga heterophylla*) at

[15] This percentage is approximate, based on land area (km^2) as mapped. The percentage is of all rangeland, regardless of whether or not it is grazed by livestock.

[16] A board foot is a unit of volume for timber equal to 12 inches by 12 inches by 1 inch, or 144 in^3. 9.0 billion board feet equals about 21 million m^3 and 4.0 billion board feet equals about 9.4 million m^3. The convention in Oregon forestry is to report yields in board feet rather than m^3 and that convention will be followed in this section in order to allow the use of historical charts such as Fig. 17.10. There are several methods for calculating board feet based on the dimensions of logs. The common method used for Oregon softwood was originally described by J.M. Scribner in 1846 but has since been modified. Data based on variations of the Scribner method are often described in units of "board-feet Scribner."

[17] Softwood lumber is wood produced from coniferous trees such as Douglas-fir, pine, or hemlock.

[18] Douglas-fir is also Oregon's state tree.

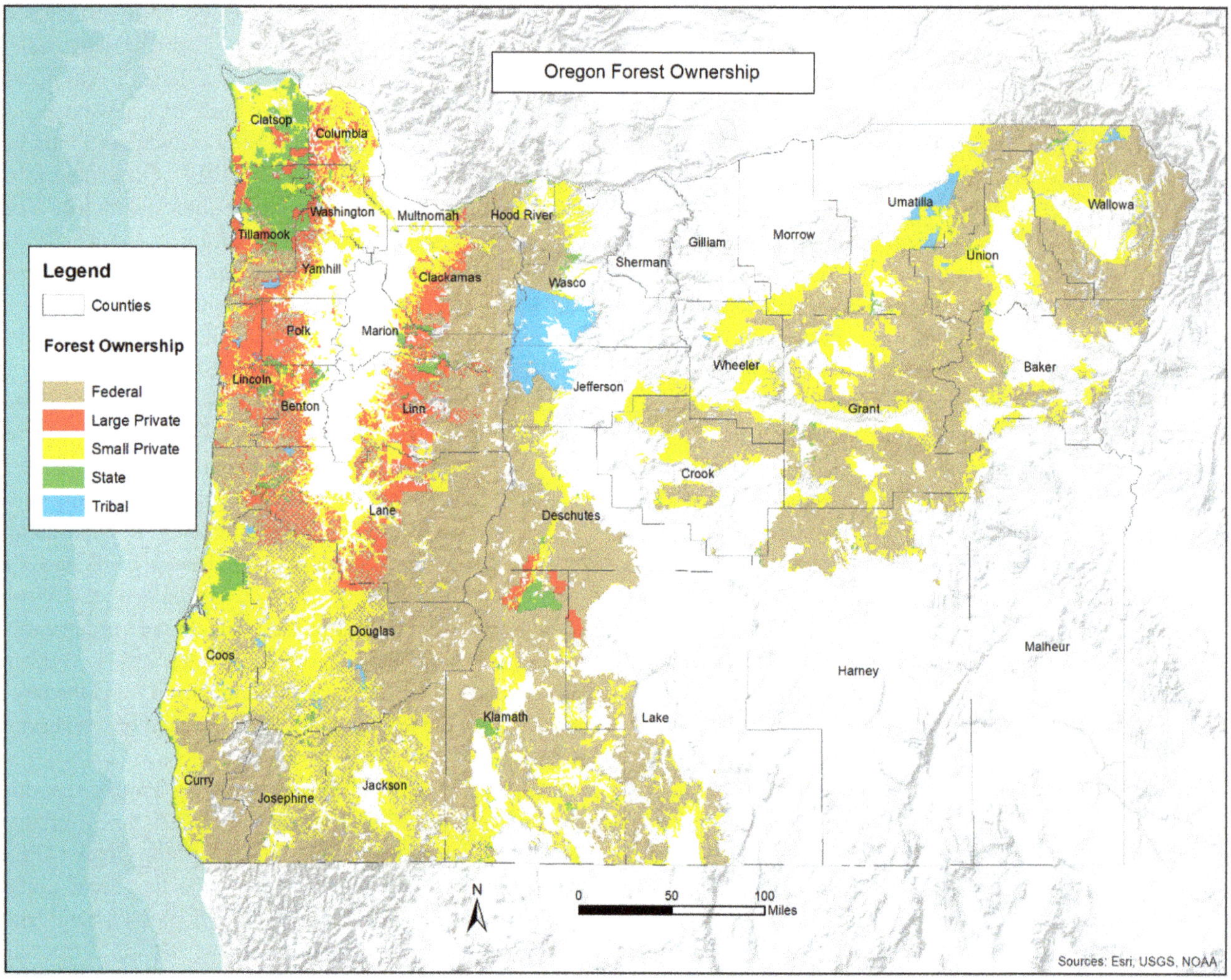

Fig. 17.8 Oregon forest ownership. "Large private" ownerships are operations of 2,023 hectares (5,000 acres) or more. "Small private" ownerships are less than 2,023 hectares. Land areas identified as "Tribal" are generally only partially forested. *Sources* ESRI, USGS, NOAA, INR Portland and Oregon Biodiversity Information Center (2014), Oregon Department of Forestry (n.d.)

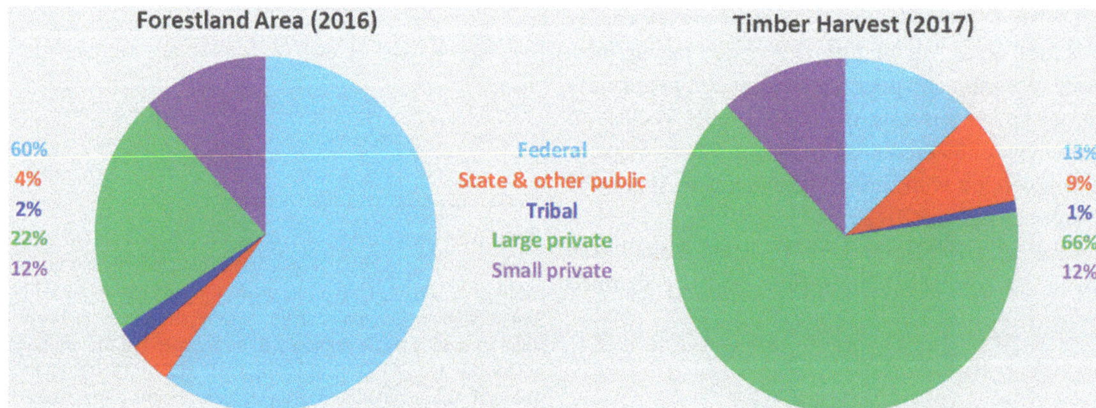

Fig. 17.9 Oregon forestland: area and timber harvest, by ownership. *Source* Oregon Forest Resources Institute (2019)

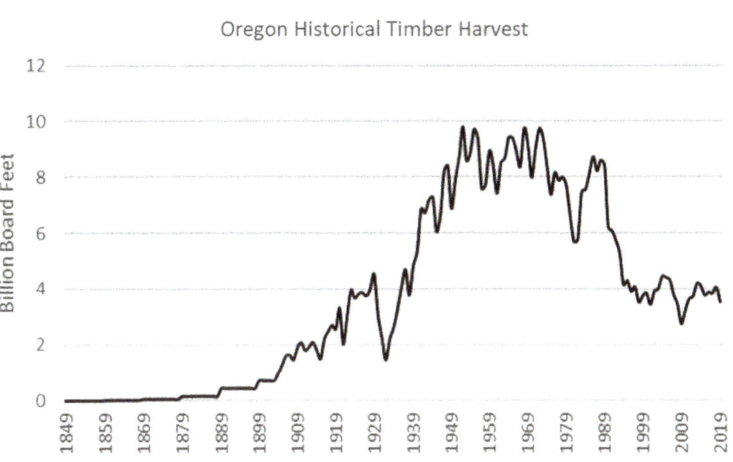

Fig. 17.10 Oregon timber harvest, 1849–2019. *Sources* Andrew and Kutara (2005), Oregon.gov (2021), Simmons et al. (2016)

Fig. 17.11 A Douglas-fir/western hemlock forest in the Coast Range of Tillamook County. This forest was harvested decades ago, as evidenced by the decayed stump visible in the lower right portion of the photo. It has grown back into a diverse stand with multiple understory species. *Source* Photograph by Dean Moberg

Table 17.3 Oregon timber harvest and standing volume by species, 2013

Species	Harvested		Standing[a]	
	Volume (MMBF[b])	Percent of Total[b] (%)	Volume (MMBF[b])	Percent of Total[b] (%)
Douglas-fir	2,953.4	69.5	224,215	58.1
Hemlock	476.9	11.2	29,489	7.6
True firs	340.2	8.0	39,502	10.2
Pines	205.9	4.8	49,524	12.8
Cedar	63.7	1.5	9,451	2.4
Spruce	49.1	1.2	6,293	1.6
Other softwoods	8.3	0.2	8,136	2.1
TOTAL softwoods	**4,097.5**	**96.5**	**366,610**	**94.9**
Red alder	82.3	1.9	10,773	2.8
Other hardwoods[c]	66.9	1.6	8,737	2.3
TOTAL hardwoods	**149.2**	**3.5**	**19,510**	**5.1**
ALL SPECIES	**4,246.7**	**100.0**	**386,120**	**100.0**

[a]Sawlog portion of growing-stock trees with diameter at breast height greater than or equal to 9 inches on non-reserved forestland
[b]MMBF = million board feet Scribner
[c]Other hardwoods include cottonwood/poplar, maple, oak, and others
Source Simmons et al. (2016)

11%, other softwoods (16%), and hardwoods 3.5% in 2013. Red alder (*Alnus rubra*) is the predominant hardwood species harvested (Table 17.3). Over half (54%) of the Oregon forestland soils with the highest forest site indices[19] are in the Humudepts and Argixerolls great groups (Appendix E).

17.4 Soil Survey Management Groups

Soil surveys lump multiple soil map units into categories called "management groups" based on shared characteristics that can help simplify land-use decisions. The advantage of management groups is that the user only needs to be familiar with a small number of groups rather than potentially hundreds or even thousands of individual soil map units. This is especially useful to people and organizations who work with land use planning over large areas. Land capability classes and national inventory groupings are examples of management groups (Dobos et al. 2017).

17.4.1 Land Capability Classes

Perhaps the most widely used management group is the NRCS classification of U.S. soils into eight land capability classes primarily based on limitations to the use of the soils for agricultural[20] purposes and for their risks of soil damage if mismanaged (Klingebiel and Montgomery 1961). The eight classes are described in the NRCS Web Soil Survey (Soil Survey Staff 2019) as[21]:

- *Class 1 soils have few limitations that restrict their use.*
- *Class 2 soils have moderate limitations that reduce the choice of plants or that require moderate conservation practices.*
- *Class 3 soils have severe limitations that reduce the choice of plants or that require special conservation practices, or both.*
- *Class 4 soils have very severe limitations that reduce the choice of plants or that require very careful management, or both.*

[19] A forest site index (SI) is the expected height, in feet, of a tree species at a given age. For example, a soil that historically produces 130-foot-tall Douglas-fir in 100 years would have a SI of 130 for that species at a base age of 100 years. High forest site indices in Appendix E are defined as greater than 130 for Douglas-fir with a base age of 100 years, greater than 120 for ponderosa pine (*Pinus ponderosa*) with a base age of 100 years, and greater than 90 for grand fir (*Abies grandis*) with a base age of 50 years.

[20] The term "agricultural" is used here in a broad sense and includes the production of cultivated crops, pasture, range, forest, and even recreational or habitat purposes.

[21] Land Capability Classes are often given as Roman numerals (I through VIII),; however, here they will be referred to with Arabic numerals (1 through 8), which is the convention used by the Web Soil Survey.

- *Class 5 soils are subject to little or no erosion but have other limitations, impractical to remove, that restrict their use mainly to pasture, rangeland, forestland, or wildlife habitat.*
- *Class 6 soils have severe limitations that make them generally unsuitable for cultivation and that restrict their use mainly to pasture, rangeland, forestland, or wildlife habitat.*
- *Class 7 soils have very severe limitations that make them unsuitable for cultivation and that restrict their use mainly to grazing, forestland, or wildlife habitat.*
- *Class 8 soils and miscellaneous areas have limitations that preclude commercial plant production and that restrict their use to recreational purposes, wildlife habitat, watershed, or esthetic purposes.*

Capability classes 2 through 8 are further defined by sub-class designations, denoted by single letters identifying the main factor limiting their use. The letter "s" indicates the soil is limited by shallow depth or stoniness, "e" indicates erosion is the main limiting factor, "w" indicates excessively wet conditions, and "c" indicates limitations based on very cold or dry conditions. For example, class 2e soils have moderate limitations to use, and the main limitation is erodibility.

The capability class system does not account for intense manipulation of soils such as land leveling with earth moving equipment or the installation of greenhouses. However, capability classes are generally defined separately for irrigated and non-irrigated conditions in Oregon, and the class designations can vary significantly based on irrigation. For example, in Jackson County, Medford silty clay loam with 0–3% slopes is rated capability class 4c when not irrigated, but is rated class 1 when irrigated (Soil Survey Staff 2019).

17.4.2 National Inventory Groupings

There are four national inventory groups that have been referenced in legislation pertaining to land use and management:

- Prime farmland soils have physical, chemical, and climatic characteristics best suited to the production of crops (Fig. 17.12).
- Unique farmland soils are not classified as prime farmland, but have special characteristics that enable the production of specialty crops. An example of a unique Oregon soil is Labish mucky clay, a Willamette Valley soil that is naturally poorly drained and frequently flooded, but has historically been managed to grow high-value root crops like carrots and onions.
- Hydric soils formed under conditions of saturation, flooding, or ponding long enough during the growing season to develop anaerobic conditions near the surface. Hydric soils are one of the criteria used to identify wetlands (Fig. 17.13).
- Highly erodible farmland soils require special management to control water and/or wind erosion within tolerable limits.

17.5 Oregon Land-Use Planning

17.5.1 Zoning

In 1961, Oregon enacted legislation allowing counties to develop Exclusive Farm Use (EFU) zoning districts, thus restricting the land use to farming plus exceptions such as schools, churches, golf courses, parks, and utilities. Although the legislation allowed EFU zones, it did not require them, and it set few criteria for their establishment. Property taxes for farms in EFU are typically much lower than taxes for similar sized parcels in urban areas (Sullivan and Eber 2009).

17.5.2 Oregon Land Use Act of 1973

A growing population and increasing urbanization in the 1960s and 1970s stoked concern among many Oregonians about uncontrolled growth. In 1963, the Oregon legislature passed Senate Bill 10, which required all Oregon cities and counties to implement comprehensive zoning to achieve seven planning goals, including the conservation of prime farmland. Senate Bill 10 proved ineffective due to lack of funding and unclear standards upon which to base the zoning. In 1973, then Oregon Governor Tom McCall gave a speech to the state legislature criticizing "sagebrush subdivisions, coastal condomania, and the ravenous rampages of suburbia" (Ibid., p. 4). That same year, the legislature passed and Governor McCall signed the Oregon Land Use Act (originally Senate Bill 100). The act was popular, but not unanimously so, with 58 of 90 state senators voting in favor. Most of those favorable votes were from senators representing Willamette Valley districts (Abbott 2020). The act was widely seen as one of the country's most restrictive land-use laws and generated nationwide interest, including a series of New York Times editorials (Barringer 2004).

The 1973 Oregon Land Use Act required all Oregon cities and counties to prepare and adopt comprehensive plans consistent with the act and to develop zoning and other ordinances or regulations to implement those plans. The act also created a new state agency, the Department of Land

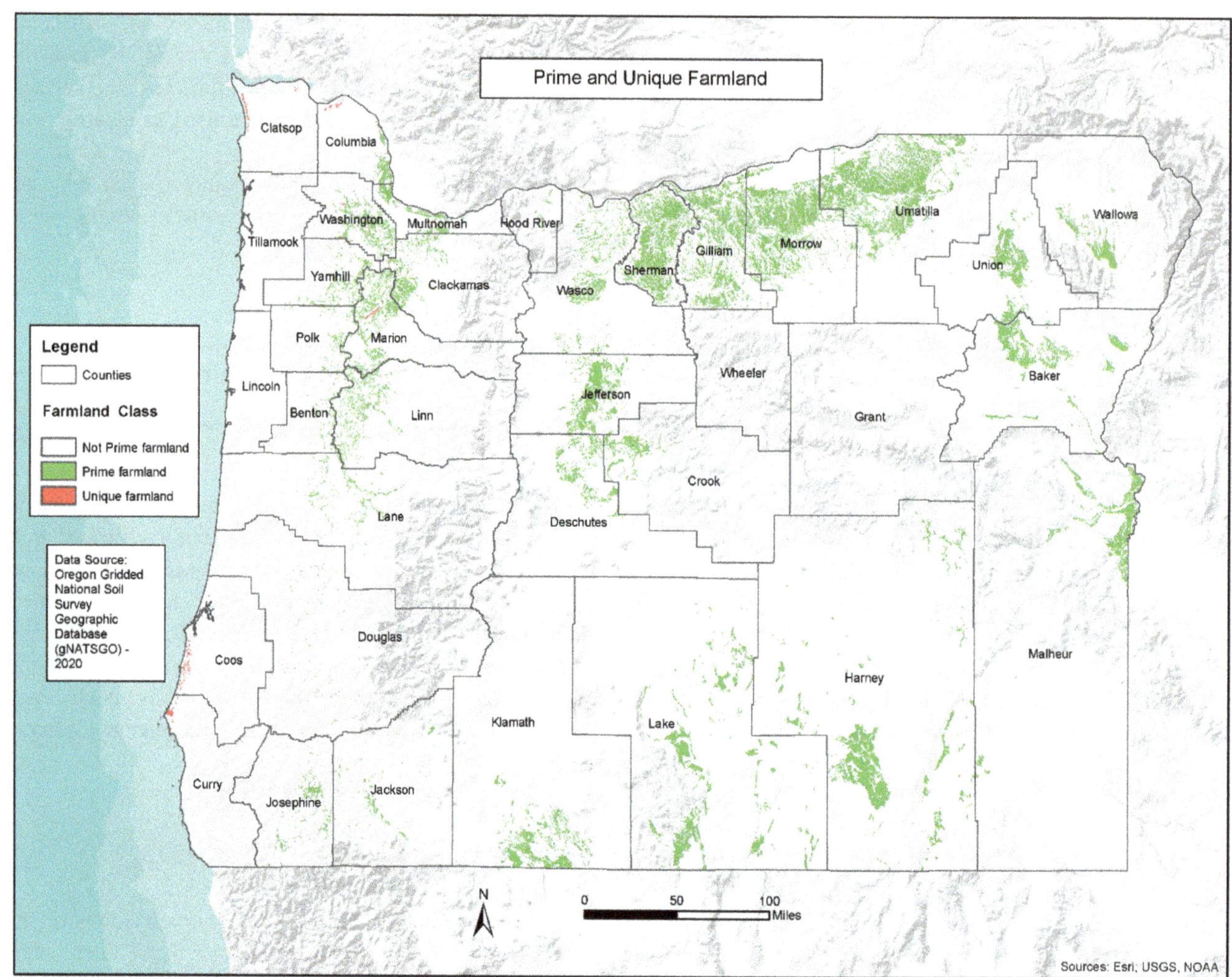

Fig. 17.12 Distribution of Oregon prime and unique farmland. *Source* Soil Survey Staff (2020)

Conservation and Development (DLCD), and established the Land Conservation and Development Commission (LCDC), comprising seven citizen members appointed by the governor. The Commission directs the Department, establishes state-wide planning goals and guidelines, and reviews city and county comprehensive plans to ensure conformance with the state-wide goals and guidelines (Oregon Legislative Assembly 1973).

The LCDC approved 19 state-wide planning goals, including procedural goals such as requiring public involvement in land-use planning, general goals pertaining to broadscale land uses such as agriculture, and goals focused on more narrowly defined uses such as the Willamette River Greenway. Agriculture is addressed under Goal 3, which employs the NRCS land capability classes to help define "agricultural land." Specifically, western Oregon lands composed of predominantly Class 1–4 soils and eastern Oregon lands composed of predominantly Class 1–6 soils, plus additional lands suitable or necessary to permit farming, are defined as "agricultural." These definitions intentionally broadened protections of agricultural lands from just the prime soils specified under Senate Bill 10 because "…many crops like orchards, wine grapes, grass seed, alfalfa, and hay grow very well on lower quality soils" (Sullivan and Eber 2009, p. 8). However, regardless of soil type, Goal 3 states that land within Urban Growth Boundaries (UGB) is not considered agricultural (Oregon Department of Land Conservation and Development 2019b).

Statewide Planning Goal 14 requires each of Oregon's 240 cities to designate a UGB around the city to accommodate urbanization needs consistent with a 20-year population forecast (Fig. 17.14). Land inside a UGB may be developed for housing, employment opportunities, public facilities, schools, parks, and other urban features. In simple

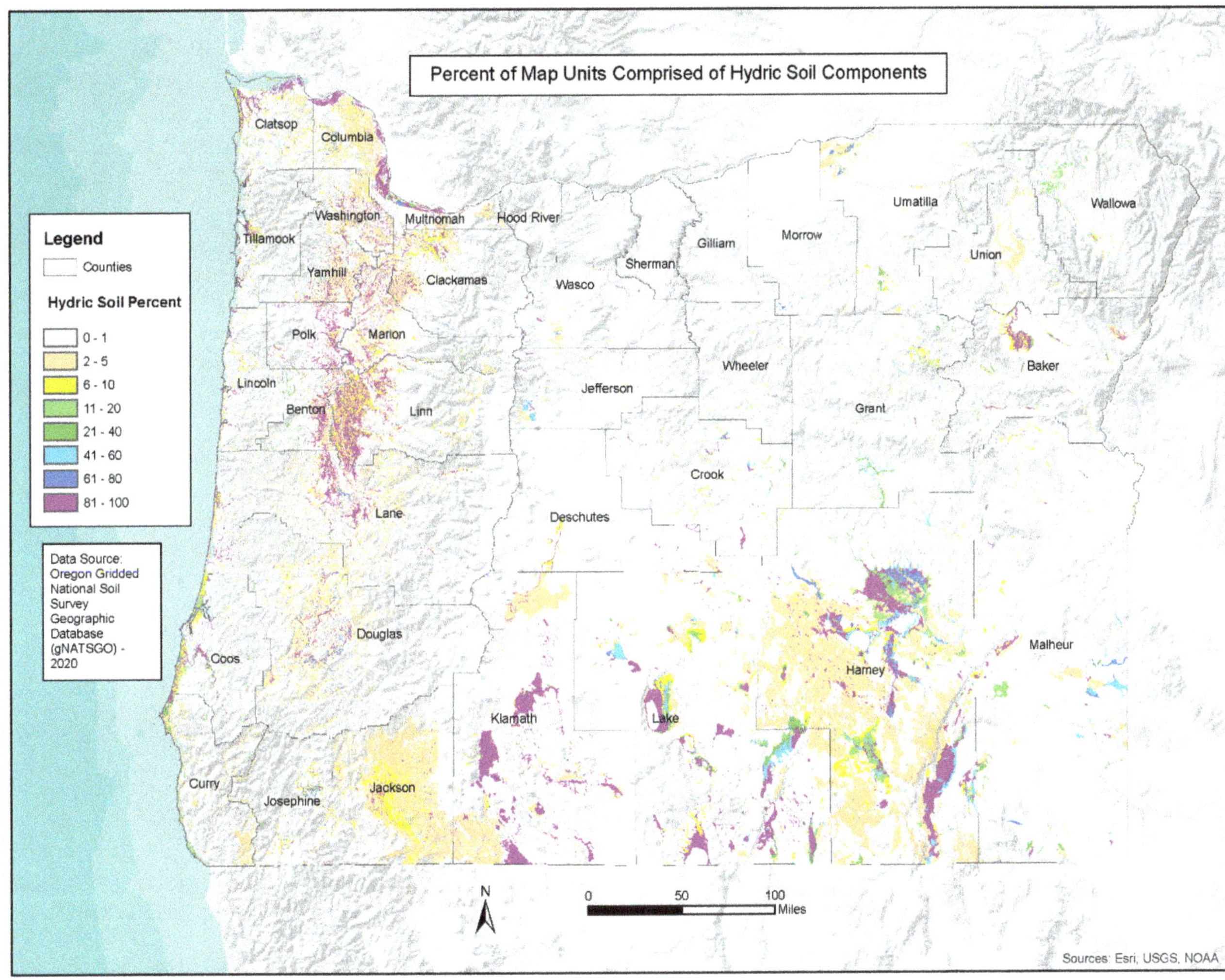

Fig. 17.13 Distribution of Oregon's hydric soils. *Source* Soil Survey Staff (2020)

terms, the 1973 Oregon Land Use Act restricted urban development to confined areas in and adjacent to existing cities and largely preserved rural land for agriculture, forestry, open spaces, and recreation (Ibid.).

17.5.3 Changes to the Oregon Land Use Act: Measures 7, 37, and 49

The Oregon Land Use Act was effective in reducing the development of rural land. Between 1978 and 1992, Oregon lost only 2.5% of its farmland to development, while California lost 11.5% and Washington lost 6.0% (Abbott 2020). However, critics of the act complained that it inhibited economic growth, was excessively regulatory, and restricted private property rights. Those critics put forth three ballot measures attempting to repeal the act in 1976 through 1982, but each of the measures failed (Abbott and Howe 1993).

In 2000, the opponents of the Oregon Land Use Act tried a different approach and introduced Ballot Measure 7, which did not attempt to repeal the act. Instead, Measure 7 was a constitutional amendment requiring the state to compensate landowners when state regulations reduced the value of their property. Some legal scholars interpreted Measure 7 to require compensation for "virtually any type of government regulation" (Sullivan and Eber 2009, p. 20). Measure 7 passed with 53% of the vote but was soon declared unconstitutional by the Oregon Supreme Court.

Fig. 17.14 This aerial photograph illustrates the clear separation of land uses at an Urban Growth Boundary (UGB, yellow line). The residential area in the upper right is inside the UGB and is part of the City of Forest Grove in Washington County. The land to the left and below the UGB is agricultural land. *Data source* Oregon Department of Land Conservation and Development (2021)

Opponents of the Oregon Land Use Act then proposed Measure 37, which was simple legislation rather than a constitutional amendment. Measure 37 required state and local governments to either waive regulations passed since the property was purchased or to compensate landowners for a decrease in property value due to those regulations. This measure was passed in 2004 with 61% of the vote. Faced with billions of dollars in potential compensation costs under Measure 37, the Oregon legislature wrote and referred Measure 49 to the voters in 2007. Measure 49 attempted to clarify aspects of Measure 37 and to reduce its applicability. It allowed some additional home building on farmland outside the UGB, but still prohibited large housing subdivisions and commercial developments (Abbott 2020). Measure 49 passed with 62% of the vote (Oregon Secretary of State 2007).

17.5.4 After Measure 49

Since Measure 49 was passed in 2007, there have been no further measures that seek to broadly modify the 1973 Oregon Land Use Act. However, debate continues between those who wish to protect farm and forest soils from development and those who believe economic development on farm and forest land is more important.

For example, 1000 Friends of Oregon is a non-governmental organization co-founded by Governor McCall in 1974 to champion the Oregon Land Use Act. The organization argues that the original 1973 act allowed only a limited number of non-farm uses, such as schools and utility facilities, in areas zoned as Exclusive Farm Use (EFU). However, the Oregon legislature has regularly added permitted uses to EFU land. Currently, there are approximately

60 such uses, including commercial wedding sites, bed and breakfast ventures, non-farm dwellings, guest ranches, and outdoor mass gatherings such as music festivals. The 1000 Friends of Oregon organization maintains that these permitted uses make farming more difficult through increased traffic on rural roads, complaints and lawsuits filed against farmers engaged in accepted farming practices, land value inflation, land fragmentation, and eventual loss of farm businesses such as equipment dealers. Potential solutions to this erosion of farm viability include improved funding for enforcing land-use laws, clarification of vaguely defined permitted uses,[22] requiring cumulative impact analyses for permitted uses, reducing the number of uses currently permitted, and considering tools outside of land-use policy, such as tax policy (Shackelford 2020).

Representing an opposing viewpoint, the Oregon Property Owners Association (OPOA) is a non-governmental organization that seeks to help property owners use their land. Their work involves a range of activities including helping individual landowners with land-use applications, representing landowners in court, and working to change regulations. OPOA argues that current Oregon land-use laws excessively restrict private property rights in ways that violate the civil rights of property owners, hurt the economy, and damage society. OPOA also contends that the 1973 Oregon Land Use Act concentrates too much land-use planning power in state government, resulting in a system that is slow to change and has "one-size-fits-all" regulations that don't work well in a state as geographically and economically diverse as Oregon (Hunnicutt and Gorman 2010).

An additional source of conflict develops when cities or the Portland metropolitan area consider expanding their UGB to allow urbanization on land previously considered agricultural. As might be expected, deciding how far to extend a UGB and where the new boundaries should be drawn is controversial and, increasingly, litigious. For example, in 2005, the city of Woodburn sought to add 396 hectares to its UGB. After review by the Oregon LCDC, the case was taken up by the Oregon Court of Appeals, where it was still undecided after 10 years (Christensen 2015).

In sum, Oregon's efforts to preserve the best soils for farming and forestry have been uniquely progressive, but not without controversy. Much of the history of this process is detailed in an aptly named paper *The Long and Winding Road: Farmland Protection in Oregon 1961–2009* by Sullivan and Eber (2009).

[22] Examples of permitted uses that 1000 Friends of Oregon contend are poorly defined include home occupations, agri-tourism, and commercial activities in conjunction with farm use.

17.6 Key Natural Resource Challenges Related to Land Use and Soil

This section describes significant natural resource challenges related to land use and soil: climate change, wetland loss, flooding, landslides, volcanoes, earthquakes, tsunamis, coastal erosion, and wildfires. The state of Oregon recognizes the importance of addressing climate change, includes climate resilience or mitigation as an "emerging issue" in Oregon's land-use planning program, and notes the importance of maintaining the working farm and forest lands in sequestering carbon in soils and perennial vegetation (Oregon Department of Land Conservation and Development 2019a). Section 34 of the 1973 Oregon Land Use Act identifies wetland loss as one of the areas requiring "priority consideration" (Oregon Legislative Assembly 1973). Other challenges, as identified in Goal 7 of the Oregon Land Use Act include flooding, landslides, earthquakes, tsunamis, coastal erosion, and wildfires as natural hazards that must be addressed in land-use plans (Oregon Department of Land Conservation and Development 2001).

These resource challenges are addressed in the following sections. However, conservation issues like cropland erosion, invasive species on rangeland, and water quality leaving confined animal feeding operations are more site-specific to individual agricultural operations, aren't generally solved with land-use policy, and are therefore addressed in Chap. 18.

17.6.1 Climate Change

The State of Oregon has prioritized both reducing greenhouse gas (GHG) emissions and adapting to the changing climate. Oregon Governor Kate Brown issued an executive order in March 2020 that set goals for addressing climate change and directed state agencies to develop plans to achieve those goals (Brown 2020). The state provides funding for the Oregon Climate Change Research Institute (OCCRI), which is required to develop regular reports on how climate change affects Oregon (Mote et al. 2019).

The OCCRI notes that the Pacific Northwest is predicted to warm by 2.2–5.0 °C by 2100. Oregon's total annual precipitation is not predicted to significantly change, but winter precipitation will increase, summer precipitation will decrease, and heavy rainfall events will become more frequent, with the potential for increased soil erosion. Precipitation will fall increasingly as rain and what does fall as snow will tend to melt more quickly (Fig. 17.15). This will increase winter flood risk (see Sect. 17.6.3 on flooding) and decrease summer streamflow. Low summer streamflow will present challenges to irrigation and fisheries. The OCCRI

Fig. 17.15 NRCS monitors snowpack via sites like this on Mt. Hood. Snowpack is expected to decrease across Oregon as the climate changes (Mote et al. 2019). *Source* Photograph by NRCS Oregon

projects longer growing seasons for Oregon crops, which may benefit farms that have adequate irrigation water but may also cause unexpected impacts from insect pests and weeds. Forest wildfires have already increased in severity in Oregon, and wildfire risk is expected to increase even more in coming years (see Sect. 17.6.7 on wildfire) (Ibid.).

The global soil and terrestrial vegetation carbon (C) pools combined are roughly four times greater than the atmospheric C pool (Ciais et al. 2013), and considerable research has focused on combating climate change by sequestering atmospheric carbon in healthy soils and perennial vegetation. Lal et al. (2007, p. 943) noted that "Most agricultural soils have lost 30–75% of their antecedent soil organic carbon (SOC) pool or 30 to 40 t C ha^{-1}." This SOC was oxidized, largely as a result of tillage, and the resulting carbon dioxide (CO_2) was emitted to the atmosphere. Lal states that much of this SOC can be restored, and he estimated that practices like conservation tillage, cover crops, manuring, and the like could sequester 4,000 to 8,000 Mt C per year on 1.35 billion hectares of cropland globally, roughly equivalent to 0.3 to 0.6 t C ha^{-1} per year.[23] Lal estimated that the total potential SOC sequestration in the US cropland, grazing land, and forestland equals 144 to 432 Mt C (528 to 1,585 Mt CO_2e) per year.[24] Lal also described the benefits of increased SOC to make agricultural soils more resilient to a changing climate, for example by increasing a soil's water-holding capacity.

Zomer et al. (2017) estimated potential SOC sequestration rates of 0.56–1.15 t C ha^{-1} per year (approximately twice that of Lal). However, Zomer cautioned that soils with greater than 85% sand or greater than 400 t C ha^{-1} (for example, Histosols) are unlikely to sequester additional SOC. Furthermore, Zomer noted that, after 20–40 years of the sequestration rates in his analysis, SOC would approach equilibrium and significant further net sequestration would be unlikely.

The Rodale Institute (2014) is one of several organizations advocating the use of regenerative agriculture to increase SOC in order to improve soil health, make a farm more resilient to a changing climate, and mitigate GHG emissions. The definition of "regenerative agriculture" varies by source, but generally includes enhanced crop rotations, cover crops, use of compost or manure, and integration of livestock production with cropping. Ranganathan et al. (2020) agree that regenerative agriculture can benefit soil health, but question its role in mitigating GHG emissions due to limitations in scientific understanding of SOC sequestration, "faulty carbon accounting" in studies that tout significant SOC sequestration, and barriers to widespread adoption of regenerative practices.

In Oregon, Graves and her colleagues (2020) simulated the potential for 12 Natural Climate Solutions (NCS) to sequester GHG from 2020 to 2050. NCS used in the model included land management changes, ecosystem restoration, and avoidance of ecosystem conversions such as urban development of agricultural and forest land. The simulations were based on SOC sequestration rates similar to those by Zomer, for example 0.35–0.46 t SOC ha^{-1} per year for a combination of no-till plus cover crops. Graves' simulations

[23] The unit "t" denotes a metric tonne, which is 1,000 kg. The unit "Mt" denotes one million metric tonnes. The Intergovernmental Panel on Climate Change (IPCC) developed the Global Warming Potential (GWP) concept to compare the ability of different greenhouse gases to trap heat in the atmosphere relative to CO_2. The GWP is expressed as carbon dioxide equivalents (CO_2e). For example, methane (CH_4) has 25 times the GWP of CO_2 over a 100-year timeframe, and thus 1.0 t CH_4 equals 25 t CO_2e. Nitrous oxide (N_2O) has even higher GWP, and 1 t N_2O equals 298 t CO_2e. Based on the atomic masses of carbon and oxygen, there are approximately 44 units of CO_2 per 12 units of C, a ratio of 3.67:1. Thus, 1.00 t SOC × 3.67 = 3.67 t CO_2e (U.S. EPA 2020).

[24] Put in perspective, the EPA estimated 2018 U.S. greenhouse gas emissions at 6,677 Mt CO_2e per year (U.S. EPA 2020).

included projections of the adoption rates of practices in Oregon, and the results indicated that Oregon NCS could sequester 2.9–9.9 Mt CO_2e per year by 2050, which is roughly 5–18% of Oregon's estimated 1990 GHG production of 56.4 Mt CO_2e per year. Of the simulated NCS, 76–94% of GHG reductions were from forest-based activities, such as deferred timber harvest, riparian restoration, and replanting federal forests after wildfire. Changes in agricultural management, such as increased use of no-till farming, cover crops, and nitrogen management accounted for only 3–15% of modeled GHG reductions.

Unfortunately, climate change itself may reduce the ability of agricultural NCS, such as no-till and crop rotations, to sequester carbon in the soil. Morrow, Huggins, and Reganold (2017) assessed SOC and soil nitrogen (N) at four agricultural research sites in eastern Washington and eastern Oregon. The sites differ in mean annual air temperature (MAAT) and mean annual precipitation (MAP), and each site had long-term trials involving different non-irrigated crop rotations and tillage regimes. The results indicated that SOC was strongly negatively correlated to the "climate ratio" (MAAT/MAP).[25] Furthermore, tillage intensity and crop rotation explained little of the variability in SOC or total soil N. Based on future increases in MAAT/MAP predicted for eastern Washington and Oregon, the authors concluded that SOC could decrease between 12.3% and 21.2% by 2070 for the cropping systems at the eastern Oregon (Pendleton) site.[26]

In a concise review of the current science behind SOC sequestration and climate change, Daniel Kane (2015, p. 3) states "Currently the atmosphere and ocean have too much carbon while soils have lost carbon at an alarming rate due to development, conversion of native grasslands and forests to cropland, and agricultural practices that decrease soil organic matter." Thus, it makes sense to consider ways to increase C sequestration in soils and perennial vegetation, both to help mitigate GHG emissions and to improve the resiliency of agricultural systems to a changing climate. Those efforts, however, surely will be only part of Oregon's plan to address climate change.

[25] This relationship of MAAT/MAP to SOC is widely accepted and was exquisitely documented in Jenny's classic work (1941) comparing grassland SOC to gradients of MAAT and MAP across the North American Great Plains.

[26] The authors modeled two scenarios based on Representative Concentration Pathways (RCP) that are commonly used in climate change research: RCP 4.5 and RCP 8.5 represent future climates that stabilize by 2100 with radiative forcing of 4.5 W/m^2 and 8.5 W/m^2, respectively. Radiative forcing is the difference between global incoming and outgoing radiation, compared to the year 1750, roughly the beginning of the industrial revolution. As GHG concentrations increase, radiative forcing increases, and the planet warms (National Oceanic and Atmospheric Administration 2021).

17.6.2 Wetland Loss

The USFWS, the Army Corps of Engineers, and NRCS each employ a different definition of "wetland." The technical differences between these definitions are not especially important to this volume, but each definition includes the criterion of soil saturation (Tiner 1996). NRCS defines hydric soils as those that "formed under conditions of saturation, flooding, or ponding long enough during the growing season to develop anaerobic conditions in the upper part" (Dobos et al. 2017, p. 460). Historical records of vegetation and flooding lack detail, but hydric soils generally retain distinct characteristics even after a wetland is converted. Thus, the extent of hydric soils is sometimes used to estimate the area of wetlands prior to the Euro-American settlement in Oregon (Christy 2010).

Critical wetland ecosystem functions include providing wildlife habitat, maintaining biodiversity, filtering surface water runoff, controlling flooding, sequestering carbon, and providing sites for human recreation (Lomnicky et al. 2019). Historical wetland losses from draining, filling, diking, and/or vegetation removal vary by region and by type. For example, approximately 57% of Willamette Valley wetlands (all types), 99% of Willamette Valley wet prairies, 91% of Klamath Basin wetlands (all types), and 95% of tidal forested wetlands have been converted to non-wetland land uses—primarily cropland and pasture (Brophy 2019; Christy 2010).

The state of Oregon has set a goal of no net loss of wetlands (Oregon Department of State Lands 2017b). Federal and state regulations, such as Sect. 404 of the federal Clean Water Act, the "swampbuster" provisions of the 1985 federal Food Security Act (and subsequent farm bills), and the Oregon Removal-Fill Act, now provide substantial protection of wetlands. There are also significant efforts to restore wetlands in Oregon, most notably through the NRCS Wetland Reserve Easement program on private lands (Fig. 17.16) and USFWS wildlife refuges on public land. Still, of the estimated 930,000 hectares of wetlands in Oregon prior to the Euro-American settlement, only about 530,000 hectares remain today (Kjelstrom and Williams 1996).

17.6.3 Flooding

NRCS classifies soils according to their frequency and duration of flooding, using the categories of none, rare, occasional, frequent, and very frequent (Dobos et al. 2017). The Federal Emergency Management Agency (FEMA) develops floodplain maps and administers the National Flood Insurance Program (NFIP), in which most Oregon cities and counties participate (Oregon Department of Land

Fig. 17.16 Land enrolled in the USDA Wetland Easement Program. This site in Marion County was restored to a wetland and now provides habitat for the Oregon chub (*Oregonichthys crameri*), a small fish native to the Willamette Valley. The chub and its critical habitat designation were removed from the Endangered Species list, thanks in part to projects like this. *Source* Photograph by NRCS Oregon

Conservation and Development 2020a). Participation in NFIP requires communities to adopt land-use regulations that reduce the risk of property damage by floods, and NFIP regulations require landowners to purchase flood insurance if their home is in a 100-year floodplain and they have a mortgage through a federally regulated or insured lender (Federal Emergency Management Agency, n.d.). Despite the information provided by NRCS and FEMA, floods still regularly cause property damage and sometimes cause loss of life. One reason for this is that NFIP was established by Congress in 1968, and land-use decisions made prior to that did not always include considerations of flood damage.

Although the "Missoula Floods" of the late Pleistocene was not a land-use issue, no review of flooding in Oregon is complete without mentioning these tremendous events that shaped so much of the Pacific Northwest landscape. As explained in Chap. 3, the floods were caused by the periodic rupture of ice dams on what is now the Clark Fork River, releasing cataclysmic floods that swept across eastern Washington and then down the Columbia River gorge. Dozens of Missoula Floods recurred during the period of approximately 15,000– 13,000 years ago (Allen et al. 2009), (Benito and O'Connor 2003). The story of the Missoula Floods was developed through a fascinating series of geological investigations. However, scholarly work has also attempted to gain insights into pre-historic Pacific Northwest floods through the study of the Oral Traditions of Indigenous Peoples (Phillips 2007).

Table 17.4 summarizes significant floods in Oregon since the Euro-American settlement. The "Great Flood" of 1861–2 probably impacted the most land area in Oregon, with one resident claiming that "the whole Willamette Valley was a sheet of water" (Cain 2004, p. 1). The 1894 Portland flood occurred in May and early June. People navigated in canoes and rowboats, and boys fished in downtown streets as the floodwaters inundated 250 Portland city blocks (Fig. 17.17), (Flores and Griffith 2018), (Willingham 2018).

The 1903 Heppner Flood in Morrow County claimed 247 lives, the highest casualties of any recorded Oregon flood. Although most Oregon floods are associated with rain on

Table 17.4 Significant historical Oregon floods

Name	Year	Water Course	Lives Lost
Great Flood	1861–2	Willamette River	Unknown
Portland	1876	Willamette River	Unknown
Portland	1894	Willamette River	Unknown
Heppner	1903	Willow Creek	247
Vanport	1948	Columbia River	15
Christmas	1964	Columbia River	19
Willamette	1996–7	Willamette River	8

Sources Cain (2004), Flores and Griffith (2018), McGregor (2003), Paulsen (1949), Urness (2016), Willingham (2018)

snowpack events, the Heppner Flood was initiated by thunderstorms with heavy rainfall after a protracted drought in the Willow Creek watershed. The Heppner Gazette reported that "...a leaping, foaming wall of water, 40 feet in height, struck Heppner at about 5 o'clock Sunday afternoon, sweeping everything before it and leaving only death and destruction in its wake" (DenOuden 2020, p. 1).

The 1948 Vanport flood is significant for the number of people it left homeless, for the poor land use planning that allowed a city to be built in a floodplain, and for the segregationist real estate policies that helped lead to the creation of Vanport in the first place. In the early 1940s, industrialist Henry Kaiser built shipyards in Portland to satisfy the urgent need for ships in World War II. Kaiser recruited workers from across the United States, but there was insufficient housing available in Portland. Many of the new workers were Black, and racist real estate codes in Portland at the time restricted Black people from living in many parts of the city. In response to the lack of housing for their workers, the Kaiser family purchased land in the Columbia River floodplain, in what is now north Portland, and quickly built the city of Vanport. This small city provided cheap housing for over 40,000 people, about 10,000 of whom were Black. After World War II ended, many workers remained in Vanport. The site was presumably protected from flooding by a railroad embankment that served as a dike, but floodwaters washed out the dike and essentially destroyed Vanport in 1948. Approximately 18,500 were left homeless, a third of whom were Black (McGregor 2003), (Paulsen 1949).

The Willamette Flood of 1996–7 displaced thousands of people from Oregon City to Corvallis, caused eight deaths, and resulted in more than $65 million in damages to public infrastructure. The flood resulted from abnormally high rainfall in the valley, heavy snowfall in the mountains, and unusual warming from a "Pineapple Express," a meteorological phenomenon originating in the waters around the Hawaiian Islands and extending across the Pacific to the U. S. mainland (Urness 2016).

Another type of flooding in Oregon is not represented in Table 17.4 but is highly related to land use and soils. When Euro-Americans settled the Oregon coast, they converted land to pasture and began to raise dairy and beef cattle. The settlers converted upland forest and also built dikes[27] around estuaries to keep high tides from flooding the land. Where dikes crossed streams or drainage-ways, tide gates were installed to allow those water courses to flow downstream. The simplest tide gates consist of a metal culvert installed through a dike, with a metal or wooden flap gate on the downstream end. When tides are high, the gate closes, keeping tides from flowing up through the culvert and flooding the field protected by the dike. When tides recede, water on the upstream side of the dike pushes the gate open and flows out. There are also much larger tide gates installed in streams and rivers, often in conjunction with roadway bridges.

Simple tide gates were relatively inexpensive and allowed farmers to establish pastures in sites that were once estuarine wetlands. However, simple tide gates tend to restrict the movement of aquatic organisms, most notably Oregon coast coho salmon (*Oncorhynchus kisutch*), which is listed as threatened under the Endangered Species Act (ESA). Now, many simple tide gates have reached the end of their design life and are failing, which allows tidal water to flood pastures and make them less productive or, often, no longer usable for livestock production. Due to the ESA listing of coho, and other concerns such as water quality and the presence of Indigenous sites long-since buried under dikes, the costs in both materials and permits to replace a failing tide gate are usually not economical to a farmer.

The proposed land-use solutions involve changing some or all of a given pasture back into a wetland. This could be done by allowing tide gates and dikes to fail over time, purposely removing dikes, or rebuilding the dikes farther back from streams. An alternative to land-use changes

[27] Technically, a dike protects land that would otherwise be flooded most of the time and a levee protects land that is only flooded occasionally. Both systems were built along coastal waterways, but for the sake of simplicity, the term "dike" will be used here for any man-made barrier that protects coastal wetlands from flooding.

Fig. 17.17 Flood of 1894. Men in rowboats navigate near Front Street and Pine Street, Portland. June, 1894. *Source* Photograph courtesy of The Oregonian, from the collection of The Oregon Historical Society

involves employing engineering solutions to replace old tide gates with newer, more sophisticated (and expensive) fish-friendly tide gates. These modern tide gates take various forms, but are often hinged on the side and have mechanisms that keep them open longer and wider, allowing fish better access to swim through them in both directions. Thus, tide gates are an issue that involves flooding, land-use issues, and soil management decisions. Giannico et al. (2018) provide a good overview of tide gate issues and options in Oregon.

In the past, dams have been used as flood control measures throughout Oregon. Approximately 20% of US dams inventoried by the Army Corps of Engineers are used primarily for flood control (Ho et al. 2017). The peak of US dam building was in the middle of the twentieth century and many dams are nearing or have reached their projected lifespans. Regulations such as the 1970 National Environmental Policy Act, considerations for seismic safety, and concerns for economic cost versus benefit make the widespread building of additional flood control dams unlikely (Ibid.).

Land management decisions affect flooding. Farming practices that decrease soil health reduce the infiltration of rain into the soil and increase runoff. Increased impervious surfaces on developed land have the same effect. The net result is "flashy" hydrology: higher winter streamflow with greater risk of flooding, and decreased summer streamflow with negative impacts to aquatic wildlife and irrigation water availability (Carter et al. 1997; Shuster et al. 2007).

In some situations, communities have modified land uses to allow flooding on sites with less economic value in order to protect sites with higher value. One example is in Tillamook County, where flooding at the confluence of five rivers was damaging developed land, including buildings and Highway 101. Oregon Solutions, a third-party mediator, facilitated a planning effort that included dairy farmers, local residents, scientists, and government officials, who met over a period of two years to develop a collaborative solution.

This solution entailed farmers receiving compensation for pasture land that they stopped grazing so that it could be restored to a salt marsh, thereby alleviating the flooding of developed land (Haeffner and Hellman 2020). Thus, as in other land-use issues, the best solutions often involve collaborative efforts between agencies, non-governmental organizations, and local residents affected by a problem.

17.6.4 Landslides

Landslides are the movement of rock, soil, or related debris downslope under the force of gravity. Landslides are particularly common in Oregon because of high rainfall, steep slopes, parent materials with smectitic clays and lithologic discontinuities, and earthquakes (Balster and Parsons 1968; Paeth 1971; Parsons 1978; Swanston 1979). Although landslides occur throughout Oregon, most are triggered by erosion or saturated soil in the Coast Range, the Willamette Valley, and the Columbia River Gorge (Orr and Orr 1999). Oregon landslides cause over $10 million of damage per year on average, but damage can be much greater in years with heavy storms (Wang et al. 2002).

Significant landslides in Oregon are summarized in Table 17.5. In 1450, the massive Bonneville landslide flowed across the Columbia River near what is now the city of Cascade Locks. It created a temporary dam, which allowed Indigenous Peoples to cross the river on foot, and is believed to be the origin of the Klickitat Oral Tradition of the "Bridge of the Gods," when Indigenous Peoples could cross the river without getting their feet wet (Allen et al. 2009), (O'Connor 2018). The river subsequently cut through part of the landslide, creating a set of rapids called the "Cascades of the Columbia," which served as an important Indigenous fishing site for centuries. The construction of Bonneville Dam in 1938 submerged the rapids (Willingham 2019).

During recorded historical times, the deadliest Oregon landslide occurred near Canyonville in January 1974. Heavy rainfall on saturated soil created a catastrophic landslide that killed nine people (Busby 1998). The largest historical Oregon landslide occurred between Gold Beach and Port Orford in 1993. Called the Arizona Inn Slide, it displaced over 3 million cubic meters of material and was believed to have been triggered by an earthquake (Schulz et al. 2012). A landslide in March 2019 occurred between Gold Beach and Brookings in southwestern Oregon; called the Hooskanaden Slide, it displaced Highway 101 and suspended traffic for two weeks (Fig. 17.18), (Alberti et al. 2020).

The landslides described above, and the others listed in Table 17.5, involved the movement of massive amounts of material. However, smaller landslides can also cause significant economic damage. For example, the storm that caused the 1996 Willamette flood resulted in an estimated 168 landslides in Portland, severely damaging almost 40 homes. Most of the damage occurred in the hills west of downtown Portland, where Portland Hills silt is a loess soil prone to landslides (Orr and Orr 1999).

Landslides are so important in Oregon that the USGS and the Oregon Department of Geology and Mineral Industries (DOGAMI) have created factsheets, publications, and maps to help landowners identify their landslide risk (Fig. 17.19). Although landslides are a natural process, they can be triggered by such human land-use decisions as modification of surface or subsurface drainage, construction of buildings, excessive irrigation, clear-cutting forests, and construction of roads (running the spectrum from dirt logging roads to highways). Land-use decisions also determine whether valuable infrastructure is placed on sites prone to landslides in the first place (Ibid.).

Table 17.5 Significant Oregon landslides

Name	Year	County
Bonneville	1450	Hood River
Clatskanie	1933	Columbia
Canyonville	1974	Douglas
Hole-in-the-Wall	1984	Baker
Wilson River	1991	Tillamook
Arizona Inn	1993	Curry
Dodson	1996	Multnomah
Cape Cove	1999	Lane
Mt. Jefferson	2006	Linn
Woodson	2007	Columbia
Clatskanie	2008	Columbia
Hooskanaden	2019	Curry

Sources Alberti et al. (2020), Allen et al. (2009), Busby (1998), Orr and Orr (1999), Schulz et al. (2012)

Fig. 17.18 The Hooskanaden landslide in 2019 between Gold Beach and Brookings, Oregon. *Source* Oregon Department of Transportation (2021)

17.6.5 Volcanoes, Earthquakes, and Tsunamis

Oregon is located in a tectonically active zone where the Juan de Fuca and Pacific Plates are advancing beneath (i.e., subducting) the North American Plate at what is called the "Cascadia Subduction Zone." The edges of these plates may remain locked in place for centuries, creating strain and slowly raising the shoreline. When this strain finally overcomes the frictional forces between the plates, the resulting movement of the plates creates earthquakes and potentially catastrophic tsunamis (Figs. 17.20 and 17.21). The high temperatures and pressures created as the oceanic plates plunge downward forces magma upwards, creating the Cascade Range and its volcanoes (Orr and Orr 1999).

Geologic investigations of buried coastal marshes at seven estuaries along the Oregon coast indicate that earthquakes of magnitude (M) 8.0 or larger have recurred at intervals averaging 400 years for at least the past 2,400 years. Indigenous Oral Traditions and archaeological investigations of Indigenous sites also contribute to current-day knowledge of earthquakes and tsunamis. For example, the Athapaskan name for the area now known as Crook Point in Curry County has been translated as "moving ground place." There is an Indigenous Oral Tradition of elders warning the people to weave long ropes because it is not possible to predict when big tides are coming. This Oral Tradition recounts how the people who disregarded the elders were killed when the ocean rose violently, while those who had prepared sufficiently long ropes were saved. Archeological sites along the Oregon coast clearly show remnants of Indigenous villages that were buried by sediments, and radiocarbon dating of excavated tools and fishing weirs help determine the timing of geologic events (Byram 2007).

Soils have played an important role in establishing the rate of faulting along the Cascadia Subduction Zone and have acted as a stratigraphic marker for buried volcanic ashes. Kelsey and Bockheim (1994) examined coastal landscape evolution as a function of eustasy (a global change in sea level) and surface uplift rate in southern Oregon, and they identified flights of up to seven emergent wave-cut platforms preserved along the interfluves of coastal drainages. The uplifted marine terraces range in elevation from a few meters to 320 m above sea level and range in age from 80,000 to about 500,000 years. Uplift rates range from 0.1 to 0.3 m per 1,000 years (Bockheim et al. 1996).

Soils are also important indicators of risk of damage from earthquakes. Liquefaction of soil occurs when stress, such as shaking from an earthquake, pressurizes the water in soil pores until soil particles are pushed apart, causing the soil to lose its strength and to flow like a liquid. Young soils (e.g., from Holocene parent material) that are poorly drained and composed of similar grain size are most prone to liquefaction (Youd and Idriss 2001). The Oregon Resiliency Plan warns that building infrastructure on liquefiable soils creates a high risk for earthquake damage. Unfortunately, Oregon's Critical Energy Infrastructure hub, where large quantities of liquid fuel are stored, sits atop liquefiable soils along the lower Willamette River (Oregon Seismic Safety Policy Advisory Commission 2013).

Table 17.6 lists tectonic events that have influenced soils in Oregon. The eruption of Mount Mazama in south-central Oregon approximately 7,750 years ago created Crater Lake and deposited ash throughout the central and eastern portions of the state (Fig. 3.17). The Mazama ash is evident in soils throughout Oregon and adjacent states (Bockheim et al. 1969). The eruption of the Newberry crater approximately 1,300 years ago created Paulina and East Lakes, about 40 km south of Bend.

On January 26, 1700, an earthquake of approximately M 9.0 at the Cascadia Subduction Zone created a tsunami that destroyed Japanese coastal villages—an event recorded accurately to the day in Japanese written records. It is likely that the North American side of this tsunami also devastated Indigenous coastal villages, with the probable loss of much human life (Orr and Orr 1999). The last major eruption of Mount Hood, just east of Portland, was in 1781. With this, as well as previous major eruptions, lava reached the surface and debris flows buried forests along the Sandy, Zig-Zag, and White Rivers. The strongest earthquake of modern times occurred at Port Orford—Crescent City in 1873, with M 8.0 (Ibid.).

Several earthquakes and one tsunami significantly affected Oregon in the twentieth century. An earthquake of M 7.0 was recorded in the Milton-Freewater area of northeast Oregon in 1936. A 1964 earthquake (M 9.2) in Prince William Sound, Alaska, created a tsunami that reached Oregon, tragically killing four children sleeping with their parents at Beverly Beach north of Newport. Two earthquakes struck Oregon in 1993, an M 5.6 quake at Scotts Mill, approximately 50 km south of Portland, and an M 6.0 event near Klamath Falls. Two people died as a result of the Klamath Falls earthquake, one of whom was a casualty of a boulder striking their car (Ibid.).

The eruption of Mount St. Helens in 1980 was the most significant volcanic event in the recorded history of the continental United States (Fig. 17.21). The eruption killed 57 people and deposited ash in 11 states. The volcano is located about 85 km northeast of Portland in the Gifford-Pinchot National Forest of Washington. The site has been declared a National Volcanic Monument and the U.S. Forest Service maintains visitor centers and an observatory with interpretive displays of the event (U.S. Forest Service n.d.).

Due to increasing awareness of the dangers of earthquakes and tsunamis, the State of Oregon established the Oregon Seismic Safety Policy Advisory Commission (OSSPAC) in 1991. After that, Oregon began assessing schools and emergency response facilities for earthquake and tsunami risk and providing grants for seismic upgrades. OSSPAC provides a report on its activities every two years and, in 2013, created the Oregon Resilience Plan that classifies land according to damage potential, and specifies measures to reduce risk and improve recovery from Cascadia earthquakes and tsunamis (Oregon Seismic Safety Policy Advisory Commission 2013; Yu et al. 2014). Recommendations in the Resiliency Plan rely heavily on land-use planning to avoid or minimize earthquake and tsunami damage. The plan also extensively describes the importance of mitigating the risk of liquefiable soils. The Oregon Office of Emergency Management provides information on how to prepare for various natural hazards, including earthquakes, tsunamis, and Mount Hood, Mount Jefferson, Three Sisters, Crater Lake, and Newberry volcanoes (Oregon Office of Emergency Management 2020).

17.6.6 Coastal Erosion

Beaches and cliffs along the Oregon coast, like elsewhere on Earth, are in a constant cycle of formation and erosion. The cycle is so slow as to be virtually invisible to the casual observer until a catastrophic event occurs. Twenty thousand years ago, enough of the Earth's water was frozen in glaciers to make the Pacific Ocean about 90 m lower in elevation and Oregon's shoreline extended over 30 km west of its present location (Allen et al. 2009). At that time, rivers meandered through the wide, level plain between the Coast Range and the ocean. There were no headlands[28] and sand migrated freely along the shore, in a net northerly direction.

[28] A headland is a point of land that juts into the ocean and is surrounded by water on three sides. Many headlands on the Oregon coast are named "capes," for example Cape Lookout. Beaches lie between headlands in what are termed "littoral cells" or "pocket beaches." Sand and sediment mostly move within littoral cells rather than from one cell to another.

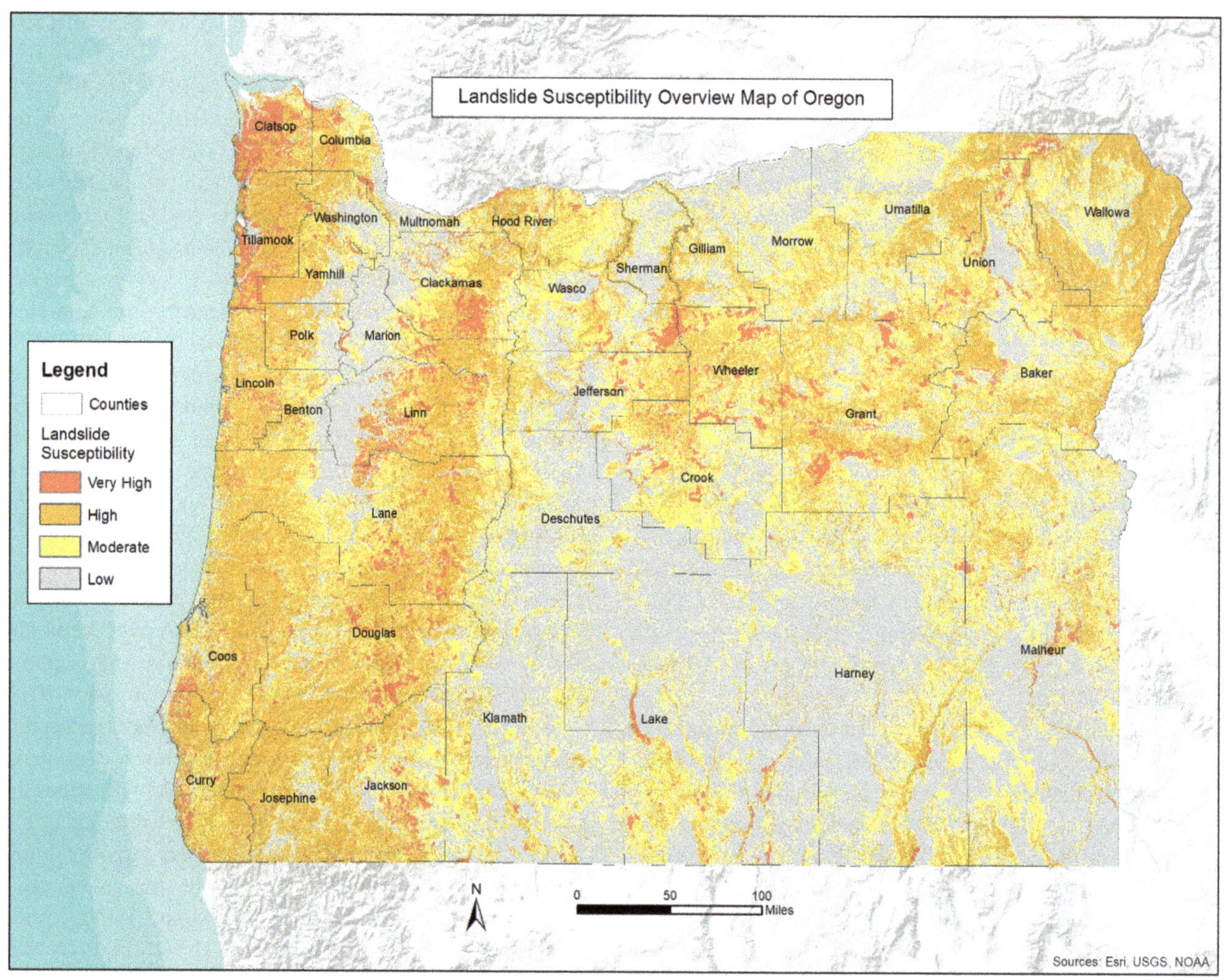

Fig. 17.19 Oregon landslide susceptibility overview map. *Sources* ESRI, USGS, NOAA, Burns et al. (2016), Oregon Department of Geology and Mineral Industries (n.d.)

As the glaciers melted, sea levels rose and gradually inundated the coastal plain until the shoreline approached its current location about 5,000–7,000 years ago. Beaches were then separated into littoral cells and sand migration was confined to much smaller reaches rather than freely occurring along the entire coast. Summer winds from the northwest tend to move sand southward, while winter winds from the southwest tend to move the sand back north. Sand also moves east and west seasonally, as winter waves tend to be larger and move sand into the ocean, where it forms bars. That sand then tends to move back onto the beach with more gentle summer winds. In most littoral cells, though, the long-term net migration of sand is zero. Analysis of sand mineralogy and the shape of sand particles has helped determine the origin and relative age of sand along the Oregon coast (Komar 1992, 1998).

Typical Oregon beach geomorphology can be described as four zones: the beach itself, the first set of dunes ("foredunes"), secondary dunes, and finally inland areas not affected by beach deposits. Beaches have little or no vegetation, and sand moves easily via wave and wind action. As the wind blows sand inland, it forms dunes, which are usually vegetated enough to trap the sand. Over time, soils (e.g., the Waldport soil series) develop in the vegetated dunes and inland areas.

A soils map (Fig. 17.22) based on a 1,500 m transect from the shore through the Gold Beach (Curry County) airport and into the uplands illustrates a typical progression of soils from a beach into the uplands. The soil survey does not classify the beach itself as soil. The foredunes in this area are mapped mostly as Frankport sand, an Entisol with eolian sand parent material at an elevation of approximately 6 m. Much of the city of Gold Beach lies on soils mapped as Ferrelo–Gearhart complex and Bullards–Ferrelo–Hebo complex, with most of these components classified as Inceptisols formed on eolian sands and marine deposits at

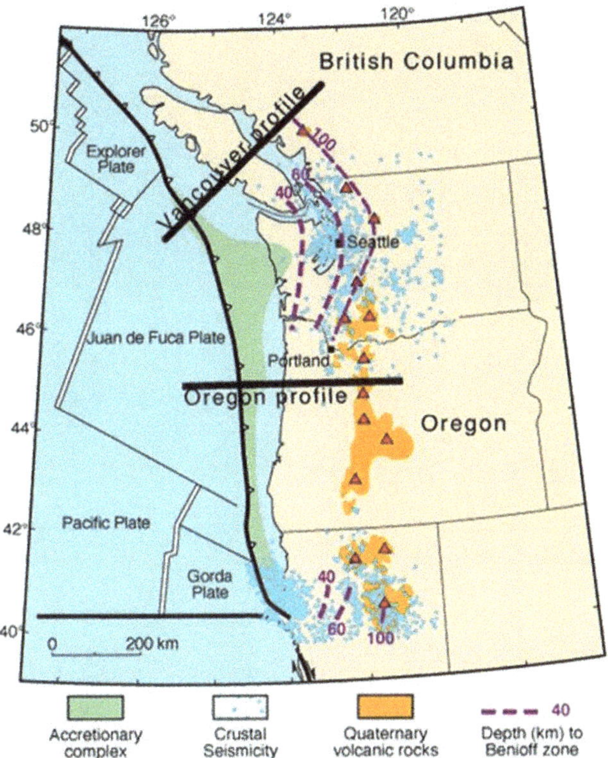

Fig. 17.20 Plate tectonics in the Pacific Northwest. The Explorer, Juan de Fuca, and Pacific-Gorda Plates subduct beneath the North American Plate. Major strato-volcanoes are depicted with red triangles. *Source* U.S. Geological Survey

elevations ranging from about 6 to 30 m. Going further east, one leaves the city of Gold Beach and enters low hills mapped mostly as Cunniff silty clay loam, an Ultisol with mixed alluvium and marine deposit parent material at elevations ranging from about 30 to 75 m. Still further to the east, and higher in elevation, soils are mapped as Millicoma–Whaleshead–Reedsport complex, Rustybutte–Sebastian–Rock outcrop complex, and Bullbulch–Hunterscove complex. These higher elevation soils are a mix of ultisols, inceptisols, and mollisols with colluvial parent materials (Soil Survey Staff 2019).

Beach erosion problems usually involve homes or other infrastructure sitting atop foredunes that are torn away by waves. This erosion tends to occur when storms coincide with high tides, especially during El Niño[29] winters. During these events, large waves may run up the beach and wash part of a foredune away. The parts of a foredune that erode vary considerably depending on the location of rip currents, but the effect in those locations can be significant, with tens of meters of dune eroded away. Thus, the damage is often limited to a small number of houses at a time.

Usually, the runup of storm waves does not overtop foredunes, and the eroded portions of dunes tend to rebuild over a period of 10 years or less. Thus, the cycle of dunes eroding and then reforming occurs over relatively short periods of time. One illustration of this is dune erosion that threatened homes on Siletz Spit (Lincoln County) during the winter of 1972–73. Erosion revealed logs that were sawed (i.e., after the Euro-American settlement), washed ashore, and were covered by dunes prior to housing development. The Oregon coastal process differs from erosion caused by storm surge,[30] which is experienced in areas with hurricanes, like the American Southeast coast (Komar 1992; Komar et al. 1999).

In parts of the world where sand migrates over long distances, jetty[31] construction often disrupts that migration and causes significant long-term beach erosion. This has usually not been a problem on the Oregon coast because sand migration is confined within littoral cells. One exception was the complete destruction of a major tourist development on Bayocean Spit in Tillamook County (Fig. 17.23). In the early twentieth century, the spit was developed with a hotel, heated swimming pool, stores, and homes. However, the construction of a jetty north of the spit disrupted sand migration and robbed the spit of sand. Subsequent erosion tore away parts of the spit, and much of the development, including the hotel and swimming pool, fell into the ocean. The development was finally abandoned after a storm in 1952 (Komar 1992).

Although net long-term sand migration may be zero in Oregon's littoral cells, short-term migration may be sufficient to damage infrastructure. For example, beach erosion at Alsea Spit (Lincoln County) began with the 1982–83 El Niño event. Sand migrated north, relocating the offshore channel leading into Alsea Bay. The subsequent steepening of the beach profile leading into the new channel caused sand to erode for several years, threatening homes built on the foredunes. By 1986, sand had migrated back to this section of beach and the homes were, at least for a time, saved. Interestingly, the deposition of beach sand in unanticipated locations can also cause problems. Storms during

[29] El Niño weather patterns are large-scale events linked to a periodic warming in sea surface temperatures across the central and east-central Equatorial Pacific. During El Niño winters, the ocean surface tends to rise along the Oregon coast, contributing to increased rates of erosion.

[30] Storm surge is the abnormal rise in seawater level during a storm, caused primarily by a storm's winds pushing water onshore. With storm surge, inland areas may be flooded and eroded. This process is contrasted with Oregon beach erosion, where erosion typically is confined to limited stretches of foredunes.

[31] Jetties are engineered structures that extend from the shore into the sea, usually to protect a harbor. Migrating sand tends to deposit as it reaches a jetty. Thus, if sand is migrating northward, it will deposit on the south side of a jetty, building the beach out on that side. On the north side of the jetty, sand will continue migrating but will not be replaced and the beach will tend to erode there.

Fig. 17.21 The eruption of Mount St. Helens, May 18, 1980. *Source* Photograph by U.S. Geological Survey

1978 eroded dunes and threatened homes at Kiwanda Beach (Tillamook County). However, sand deposition at the site began the following year, covered riprap installed to prevent erosion, and then began to inundate the homes (Ibid.).

Cliff[32] erosion along the Oregon coast is highly variable due to two factors. First, the uplift rate of coastal areas varies. Between Newport and Tillamook, the current sea level rise exceeds the tectonic uplift rate, which exposes cliffs in this area to a greater risk of erosion as waves increasingly reach their base. Climate change exacerbates this effect. Second, the stratigraphy of cliffs varies along the coast, with some areas more prone to erosion than others. Depending on these factors of uplift and stratigraphy, homes and other structures built close to the edge of cliffs may face a significant risk of loss.

Perhaps the most well-known Oregon coastal cliff erosion was at an area called "Jump-Off Joe," a bluff composed of

[32] Cliff erosion discussed in this section focuses on those cliffs that erode due primarily to wave action at the foot of the cliff. This is distinguished from landslides described in Sect. 17.6.4, which are triggered by soil saturation and/or earthquakes.

Table 17.6 Tectonic events in and near Oregon

Event	Year	Type	Comments
Glacier Peak	13.1 ka	Volcano	
Mount Mazama	7.7 ka	Volcano	Formed Crater Lake
Newberry Crater	1.3 ka	Volcano	
1700 Tsunami	1700	Tsunami	
Mount Hood	1865	Volcano	
State Line	1936	Earthquake	M 5.6, epicenter Pendleton
Mount St. Helens	1980	Volcano	57 lives lost
Scotts Mill	1993	Earthquake	M 5.6, epicenter Klamath Falls

ka = thousands of years ago
All other dates are current era
M = magnitude
Source Orr and Orr (1999)

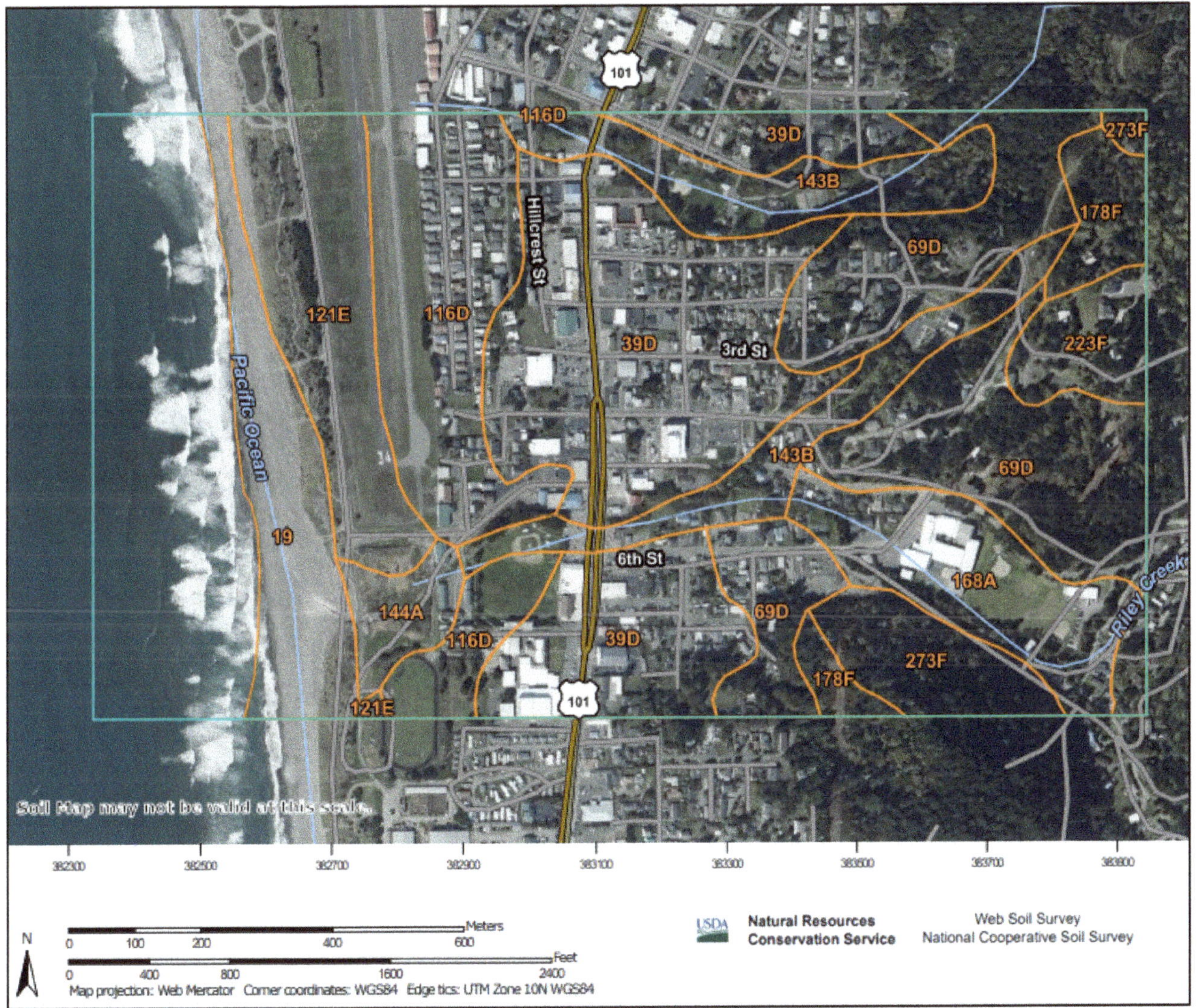

Fig. 17.22 Web Soil Survey map of Gold Beach, Oregon. Key to soil map units: 19 = Beaches, 39D = Bullards-Ferrelo-Hebo complex, 69D = Cunniff silty clay loam, 116D = Ferrelo-Gearhart complex, 121E = Frankport sand, 143B = Hebo silty clay loam, 144A = Heceta fine sand, 168A = Logsden-Euchre complex, 178F = Millicoma-Whiteshead-Reedsport complex, 223F = Rustybutte-Sebastian-Rock outcrop complex, 273F = Whaleshead-Reedport-Millicoma complex. *Source* Soil Survey Staff (2019)

Fig. 17.23 Coastal erosion at Bayocean Spit in Tillamook County, ca 1932. A major tourist attraction was developed on the spit in the early twentieth century. Coastal erosion led to the complete destruction of the hotel and swimming pool. *Source* Photograph courtesy of the Tillamook County Pioneer Museum

concretionary sandstone near Newport (Lincoln County). A 1942 landslide at the site destroyed over a dozen homes. Surprisingly, a condominium development was approved on the same site in 1982. Partway through construction, the shifting ground caused the building foundation to crack and the project was abandoned. Remnants of the concrete foundation remain at the site (Ibid.).

Engineering, vegetative, and land-use solutions have all been employed to reduce the risk of infrastructure damage from beach and cliff erosion on the Oregon coast. Engineering solutions have consisted of riprap and sea walls. Although engineered solutions can protect specific properties, they can also aggravate erosion at nearby sites by interfering with sand migration. Oregon Statewide Planning Goal 18 specifies requirements for beaches and dunes. The implementation requirements for this goal allow permits for engineered structures for beachfront protective structures only where development already existed on January 1, 1977 (Oregon Department of Land Conservation & Development 2019).

Vegetative solutions have included the planting of European beachgrass (*Ammophila arenaria*) on Oregon beach foredunes early in the twentieth century in an effort to stabilize the dunes. The grass proved vigorous on the Oregon coast and stabilized dunes to the extent that it sped ecological succession to woody plant communities. However, this hurt habitat for several at-risk species. For example, western snowy plovers (*Charadrius nivosus nivosus*), listed as threatened under the Endangered Species Act, nest in bare sand. Two plants listed as federal species of concern, Wolf's evening primrose (*Oenothera wolfii*) and pink sand verbena (*Abronia umbellata var. breviflora*) require sandy sites free of heavy competition from other plants. The Oregon Conservation Strategy (Oregon Department of Fish and Wildlife 2016) calls for controlling beachgrass at plover nesting sites and near populations of the sand verbena.

Based on problems associated with both engineered and vegetative solutions, Oregon has moved to land-use decisions as the preferred approach to reducing beach and cliff erosion risk to infrastructure. This process is codified in Oregon Statewide Planning Goal 18, which is administered in conjunction with the Oregon Coastal Management Program (OCMP). Oregon developed the OCMP as its response to the federal Coastal Zone Management Act of 1972.

The OCMP coordinates the efforts of federal, state, and local agencies along the Oregon coast (Oregon Department of Land Conservation and Development 2020b). Goal 18 states, in part, "Local governments and state and federal agencies shall prohibit residential developments and commercial and industrial buildings on beaches, active foredunes, on other foredunes which are conditionally stable and that are subject to ocean undercutting or wave overtopping, and on interdune areas (deflation plains) that are subject to ocean flooding" (Oregon Department of Land Conservation & Development 2019, p. 2).

17.6.7 Wildfires

Fire is a complex topic because it affects multiple land uses, including forest, range, crop, and developed lands. Wildland fire may also be planned and prescribed to achieve ecological objectives, or it may be wildfire[33] that can cause devastating damage and loss of human life. Additionally, the adaptation of different forest types to fire, and thus the potential human responses to wildfire, vary. For example, the Oregon Forest Resources Institute (2020) classifies western Oregon "wet" forests as having a high fire severity with a frequency of every 100 to 450 years, versus eastern Oregon "dry" forests, which have a low to mixed severity and a frequency of every 2 to 50 years. NRCS Ecological Site Descriptions (NRCS 2017), where available, describe the potential response of different rangeland sites to fire and fire suppression. Acknowledging this complexity, the following discussion will focus on wildfires as they affect forestland and rangeland in Oregon, especially in terms of soils and land use. Note, however, that wildfires can also threaten developed land and cropland.

The evidence of wildfires prior to the arrival of Euro-Americans in the Pacific Northwest can be gleaned from geological excavations that reveal layers of charcoal and from burn scars in ancient Douglas-fir and western red cedar trees, which can live for over 1,000 years. Large prehistoric fires in the Pacific Northwest correlate well with periods of warmer, dryer climates. Some of these fires were massive, such as a blaze around the year 1700 that may have burned up to 4 million hectares in western Washington (Henderson et al. 1989). It is coincidental that this wildfire occurred at almost the same time as the 1700 Oregon Coast earthquake and tsunami described previously. However, the timing illustrates how both epic west-side wildfires and devastating earthquakes have been historically infrequent but astonishing in scope (Berger 2019). Also prior to Euro-American arrival, Indigenous Peoples used fire purposely in both western and eastern Oregon to manipulate vegetation, primarily with the goal of maintaining prairies in an early seral stage in order to maximize food sources (Shinn 1977; Walsh et al. 2010).

Climate change has exacerbated the extent and intensity of forest wildfires by creating warmer, dryer late summer conditions (Hartter et al. 2018). Mote et al. (2019) analyzed records from 1895 to 2018 and found that the largest Oregon fires occurred in years that were warmer and dryer than normal during the "fire season" of July through September. Furthermore, models predicted that future (2040–2069) fuels moisture[34] will decrease markedly from historic (1971–2000) levels (Ibid.). In addition to the seasonal effects of climate change, short-term weather conditions (wind speed and direction, temperature, humidity) and topography can determine whether a specific fire will be severe or minor (Sickinger 2020).

Natural disturbances, such as tree mortality from insects, diseases, wind, and avalanches, increase fuel loads and thus affect the risk of wildfire in forests. Management decisions that suppress fires, increase stand densities, allow "ladder fuels,"[35] or decrease plant diversity also increase fuel loads and thus wildfire intensity. In addition to the preceding factors affecting the severity and extent of fires, there are also multiple ways in which fires are started in the first place. Figure 17.24 illustrates that most Oregon wildfires are caused by humans, but lightning causes fires that often burn larger areas. The Oregon Department of Forestry (ODF) provides annual data on causes of fires in forests protected by ODF (primarily state and private lands). These data indicate that the most common human causes of forest fires are debris burning and equipment use. Other causes include recreationists (e.g., campfires), smoking, arson, juveniles, and miscellaneous (Oregon Department of Forestry 2020).

The loss of vegetative cover following fire increases soil erosion during the subsequent rainy season on forest, range, and crop lands (Fig. 17.25). However, extremely hot forest fires can also make soils repellent to water (hydrophobic) from the surface down to depths of 20 cm or more—an effect that can last for five or more years. The decreased

[33] "Wildland fire" occurs in vegetation and natural fuels (as opposed to fires that predominantly burn structures such as houses). Wildland fires may be "prescribed fires" (planned and intentionally ignited to meet management objectives) or "wildfires" (unplanned fires caused by lightning, human accidents, arson, or escaped prescribed fires). Chapter 18 contains additional information on wildland fires in Oregon forests.

[34] Fuels moisture is a measure of the amount of moisture in vegetation. The modeling predicted that, in the future, much of Oregon will have more days of "extreme" low fuels moisture in June through August, indicating a likely longer and more intense fire season.

[35] Ladder fuels allow a fire to climb from the forest floor into the canopy, leading to more severe fires. Ladder fuels include living and dead shrubs, tall herbaceous vegetation, and low tree branches.

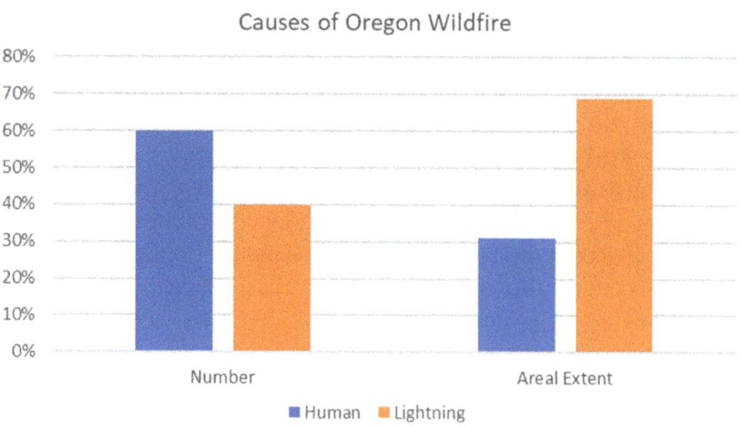

Fig. 17.24 Causes of wildfire in Oregon, 2009–2018. *Source* Northwest Interagency Coordination Center (2019)

ability of soils to absorb rainfall leads to increased runoff and erosion. Hydrophobic soils also experience slower rates of vegetation establishment, further exacerbating the potential for erosion (Dyrness 1976). Soil hydrophobicity, however, is not created in all high-intensity forest wildfires (Parks and Cundy 1989).

Table 17.7 lists historical Oregon wildfires that consumed over 500 km^2. It is reasonable, however, to assume that measurements of nineteenth-century Oregon wildfires may not have been highly accurate. For example, sources disagree about whether the Great Fire of 1845 or the Silverton fire of 1865 was Oregon's largest. Most of the large fires listed in the table were in western Oregon forests, but some also occurred in eastern Oregon rangeland. Ten of these colossal fires occurred in the 150 years between 1845 and 2000. Fourteen burned in just the 20 years between 2001

Fig. 17.25 Aftermath of the 2016 Canyon Creek wildfire in Grant County. This photograph was taken in 2019 after a successful seeding effort by the Grant Soil and Water Conservation District and NRCS. *Source* Photograph by NRCS Oregon

Table 17.7 Oregon wildfires that occurred between Euro-American settlement and 2020 and were greater than 500 km^2

Year	Fire name	km2	Description
1845	The Great Fire	6,000	Lincoln and Tillamook Counties
1853	Yaquina	1,822	Reportedly set by trappers
1853	Nestucca	1,296	Lincoln and Tillamook Counties
1865	Silverton	4,049	Thought by some to be Oregon's largest fire
1868	Coos	1,215	Destroyed most of Port Orford
1902	Columbia	688	Also known as the Yacolt burn, near Mount Hood
1933	Tillamook Burn	972	Washington, Yamhill, Tillamook counties
1936	Bandon	1,162	Killed 13 people
1939	2nd Tillamook Burn	879	Near the Wilson and Salmonberry rivers
1945	3rd Tillamook Burn	700	Burned portions of the two earlier Tillamook burns
2001	Lakeview Complex	726	A cluster of 5 fires near Lakeview
2002	Biscuit	2,024	Largest North American fire in 2002
2007	Egley Complex	568	Threatened the towns of Burns and Hines
2012	Long Draw	2,260	Southeastern Oregon rangeland
2012	Holloway	992	Southeast Oregon rangeland. Additional area in Nevada
2012	Miller Homestead	651	Southeast Oregon rangeland near Frenchglen
2014	Buzzard Complex	1,602	Rangeland southeast of Burns
2017	Chetco Bar	774	Kalmiopsis Wilderness near Brookings
2020	Lionshead	1,215	Detroit Lake area
2020	Beachie Creek	828	Santiam Canyon in Marion County
2020	Holiday Farm	784	Near Springfield
2020	Riverside	702	Near Estacada
2020	Archie Creek	559	Near Glide in Douglas County
2020	Slater	533	Near Cave Junction

Sources KOIN 6 News (2020), Palombo (2019), National Park Service (2018), Burton (2013)

and 2020, providing evidence of the effects of increasingly warmer, dryer conditions.

ODF fights most wildfires on state and private land in Oregon, and the USFS leads the effort in federal forests. Other federal agencies, Indigenous Nations, counties, rural fire departments, and private contractors also fight wildfires. Sometimes, the National Guard and even firefighters from other countries assist in combating extremely large wildfires (U.S. Forest Service n.d.). Local government fire departments focus on fires situated on developed land.

Fighting large wildfires is expensive and becoming more so. For example, federal agency fire suppression[36] efforts averaged about $400 million annually from 1985 through 1994, but have climbed to $1.8 billion annually for 2010 through 2019 (National Interagency Fire Center 2020). Fire suppression consumed 55% of the U.S. Forest Service budget in 2017. Fire suppression is also dangerous to crews. From 1990 to 2006, there were 310 firefighter casualties in the United States. While casualties from burn-overs[37] are significant, aircraft accidents, vehicle accidents, and heart attacks account for more deaths (Mangan 2007).

These casualties, rising costs, and an increased understanding of forest ecology have led to a focus on protection rather than suppression of wildfires. Of course, reducing and reversing climate change would help with this endeavor. Site-specific protection efforts might include reducing fuel loads in forests, as previously mentioned, and allowing some wildfires to burn without suppression or with a modified suppression strategy (U.S. Forest Service 2018). Each of these protection strategies faces its own technical, political, and economic challenges, which are likely to vary across the state.

[36] In the parlance of firefighting, "suppression" includes efforts to extinguish or control blazes already underway. Suppression costs include components like fire fighter salaries, equipment, and fire-retardant chemicals applied by aircraft. "Protection" includes efforts to reduce the risk of wildfires, such as thinning forests, completing prescribed burns, building fire resistant homes, or developing zoning laws to prevent new homes in areas prone to wildfire.

[37] A "burn-over" occurs when flames overtake firefighters so quickly that they cannot escape and likely perish in the fire.

The federal Healthy Forest Restoration Act of 2003 created the Community Wildfire Protection Plan (CWPP) process, by which local communities can work with federal and state agencies to reduce wildfire risk. The CWPP process includes collaboration between local, state, and federal government; prioritizing areas and methods for fuel reduction; as well as recommending methods for homeowners to reduce the ignitability of structures (U.S. Fire Administration 2020). Every Oregon county now has a written CWPP.

Reducing the risk of fire ignition is an important component of educational programs such as the well-known Smokey Bear project (Ad Council 2020). Other avenues to decrease human-caused fires include such regulations as burning permits and allowable "burn days," which are enforced across Oregon (Oregon Department of Environmental Quality 2019; Oregon Department of Forestry 2019).

Still, wildfires in Oregon and other western states continue to cause devastating damage, leading to an interest in more restrictive land-use policies in the Wildland-Urban Interface (WUI). For example, Gorte (2013) outlined possible ways to reduce the number of new homes built in zones with a high risk of wildfire damage. He suggested the relatively simple and direct approach of changing local zoning rules to restrict building in the WUI. He also listed other approaches, which would certainly face political resistance, including states working with insurance companies to authorize higher insurance premiums for homes in the WUI, reducing or eliminating the federal mortgage interest deduction for homes in the WUI, and developing a national wildfire insurance program akin to the National Flood Insurance Program.

In any case, the path forward involves people as much as forests. Hartter et al. (2020) conducted a survey among residents of eastern Oregon's Blue Mountains (Wallowa, Union, and Baker Counties). Respondents to the survey strongly agreed (70%) that wildfires in the region were likely to become more frequent in the future. However, there was less agreement (48%) that climate change is happening and is caused mainly by humans. The authors concluded that a fruitful approach might be to focus on science-based forest management and land-use protections (e.g., thinning, prescribed burning, restricting development in the WUI) without invoking the polarizing issue of climate change. The authors also provide some hopeful evidence that, as people develop social groups that include a diversity of opinions, individual members of those groups may tend to hold less partisan opinions (Ibid.).

17.7 Summary

Oregon is roughly 254,800 km^2 in area and is the ninth-largest state in the country. In 2019, Oregon's population of 4.2 million ranked it 27th in the country. There are 241 cities in Oregon, the largest of which is Portland. Eight of the ten largest Oregon cities are in the Willamette Valley.

The federal government owns 52% of Oregon land, most of which is administered by the United States Forest Service and the Bureau of Land Management. Federal land management in western Oregon has been controversial with respect to logging policy, especially as related to the northern spotted owl (*Strix occidentalis caurina*), and in eastern Oregon in relation to grazing policy on rangeland. There are nine federally recognized Indigenous Nations in Oregon, and five additional Indigenous Nations maintain an interest in Oregon land.

Privately owned land accounts for 40% of Oregon. State and local governments own a minor percentage of land in the state, but exercise significant power in land-use decisions. Combining all land ownership categories, approximately 48% of Oregon land in 2011 was forestland, 36% was rangeland, 6% was cropland, and the remainder was pasture, developed land, or "other" land such as wetlands, dunes, mines, or bedrock. Most of the privately held cropland in Oregon is owned by individuals or families, whereas corporations own more private forestland than do individuals or families.

While land devoted to non-irrigated cropland has decreased somewhat in the last 40 years, due mostly to enrollment of land in the USDA Conservation Reserve Program, irrigated land area has remained fairly constant. Greenhouse and nursery crops accounted for the greatest value of crop production in 2018, followed by hay, grass seed, wheat, and wine grapes. Oregon ranks first in the United States in the production of several types of grass and clover seed, potted florist azaleas, Christmas trees, and rhubarb, and accounts for almost the entire US hazelnut crop. Most of Oregon's crops are grown on soils in Land Capability Classes 1 through 4. Oregon's pasture and rangeland are grazed by a variety of livestock, with beef and dairy cattle being the most economically important.

The state's economy was once strongly dependent on timber, but decreased harvest in federal forests in the 1990s rapidly reduced timber production. Most of Oregon's timber is now harvested on private land rather than in federal or state forests, with Douglas-fir (*Pseudotsuga menziesii*) accounting for almost 70% of the total harvest.

The Natural Resources Conservation Service (NRCS) classifies soils into eight Land Capability Classes, with Class 1 soils having few limitations and Class 8 soils having severe limitations. NRCS also classifies soils by national inventory groups, such as prime farmland, unique farmland, hydric soils, and highly erodible soils. NRCS classifications are referenced in Oregon land-use regulations and planning policies.

Oregon enacted what was probably the most restrictive land-use legislation in the country with the Oregon Land Use

Act of 1973, which restricted development largely to areas in and immediately surrounding cities—inside what is termed the Urban Growth Boundary (UGB) and to previously developed rural lands. The act was controversial, and property rights groups passed Measure 37 in 2004, requiring state or local government to compensate landowners for decreases in property value or to waive any regulations passed after the property was purchased. However, Measure 49, written by the Oregon legislature, referred to the public, and passed in 2007, allows some development on agricultural land but prohibits large housing subdivisions outside the UGB.

Oregon faces significant natural resource challenges related to soils and land use, including climate change, wetland loss, flooding, landslides, volcanoes, earthquakes, tsunamis, coastal erosion, and wildfires. Consequently, a growing tendency in the state is to mitigate risk to human life and infrastructure through land-use policy.

References

Abbott C (2020 February 6) Land use planning. Oregon Encyclopedia, Portland, OR. https://www.oregonencyclopedia.org/articles/land_use_planning/#.X495a-aSmUk. Accessed 14 June 2021

Abbott C, Howe D (1993) The politics of land-use law in Oregon: Senate Bill 100, twenty years after. Oreg Hist Quart 94(1):4–35

Ad Council (2020) Now more than ever we need you to prevent wildfires. https://www.smokeybear.com/en. Accessed 14 June 2021

Alberti S, Senogles A, Kingen K, Booth A, Castro P, DeKoekkoek J, . . . Leshchinsky B (2020) The Hooskanaden Landslide: historic and recent surge behavior of an active earthflow on the Oregon coast. Landslides. https://doi.org/10.1007/s10346-020-01466-8

Allen J, Burns M, Burns S (2009) Cataclysms on the Columbia. Ooligan Press, Portland, OR

Andrew A, Kutara K (2005) Oregon's timber harvests: 1849–2004. Oregon Department of Forestry, Salem, OR. https://www.oregon.gov/ODF/Documents/WorkingForests/oregonstimberharvests.pdf. Accessed June 14 2021

Balster C, Parsons R (1968) Sediment transportation on steep terrain Oregon Coast Range. Northwest Sci 42:62–70

Barringer F (2004 November 26). Property rights law may alter Oregon landscape. New York Times, pp 1, Section A

Bedell TE (1993) Western juniper—Its impact and management in Oregon rangelands. Oregon State University Extension Service, Corvallis, OR

Benito G, O'Connor J (2003) Number and size of last-glacial Missoula floods in the Columbia River valley between the Pasco Basin, Washington, and Portland Oregon. GSA Bulletin 115(5):624–638. https://doi.org/10.1130/0016-7606

Berger K (2019 September 12) Finding hints of our future in the epic blazes of the Northwest's past. Crosscut. https://crosscut.com/2019/09/finding-hints-our-future-epic-blazes-northwests-past. Accessed 14 June 2021

Bliss J, Kelly E (2008) Comparative advantages of small-scale forestry among emerging forest tenures. Small-Scale for 7(1):95–104. https://doi.org/10.1007/s11842-008-9043-5

Bockheim J, Marshall J, Kelsey H (1996) Soil-forming processes and rates on uplifted marine terraces in southwestern Oregon, USA. Geoderma 73(1–5):39–62. https://doi.org/10.1016/0016-7061(96)00017-1

Bockheim J, Schlicte K, Crecelius E, Kummer J, Pongsapich W, Tenbrink N, . . . Gresens R (1969) Compositional variations of the Mazama ash as related to variation in the weathering environment. Northwest Sci 43(4): 162–197

Brophy L (2019) Comparing historical losses of forested, scrub-shrub, and emergent tidal wetlands on the Oregon coast, USA: a paradigm shift for estuary restoration and conservation. Corvallis, OR: Estuary Technical Group, Institute for Applied Ecology

Brown K (2020) Directing state agencies to take actions to reduce and regulate greenhouse gas emissions. Executive Order No. 20–04 of March 10, 2020. https://www.oregon.gov/gov/Documents/executive_orders/eo_20-04.pdf. Accessed 14 June 2021

Bureau of Land Management (2020) O and C lands. https://www.blm.gov/programs/natural-resources/forests-and-woodlands/oc-lands. Accessed 14 June 2021

Burns W, Mickelson K, Madin I (2016) Landslide susceptibility overview map. Oregon Department of Geology and Mineral Industries. https://www.oregongeology.org/pubs/ofr/O-16-02_plate1_lowRes.pdf. Accessed 14 June 2021

Burton L (2013 July 30).The 10 biggest wildfires in recorded U.S. history. Seattle Post-Intelligencer. https://www.seattlepi.com/seattlenews/slideshow/The-10-biggest-wildfires-in-recorded-U-S-history-67296.php. Accessed 14 June 2021

Busby E (1998) The Canyonville landslide of January 16, 1974, Douglas County: Oregon's most deadly landslide. In Burns S (Ed), Environmental, groundwater and engineering geology: Applications from Oregon, pp 391–398. Association of Engineering Geologists, Belmont, CA

Butler B, Butler S (2016a) Supplemental information on National Woodland Owner Survey 2011–2013 two-page summary reports. U.S. Forest Service Northern Research Station, Newton Square, PA. https://doi.org/10.2737/NRS-RN-205

Butler B, Butler S (2016b) Family forest ownership with 10+ acres in Oregon 2011–2013. U.S. Forest Service Northern Research Station, Newtown Square, PA. https://doi.org/10.2737/NRS-RN-233

Byram R (2007) Tectonic history and cultural memory Catastrophe and restoration on the Oregon coast. Oreg Hist Quart 108(2):167–221

Cain A (2004) *The great flood of 1861*. Oregon History Project, Portland, OR. https://www.oregonhistoryproject.org/articles/historical-records/the-great-flood-of-1861/#.X59FkkeSmUk. Accessed 14 June 2021

Cain A (2019 March 14) Oregon and California railroad. Oregon History Project, Portland, OR. https://oregonhistoryproject.org/articles/historical-records/oregon-and-california-railroad/#.X1AVXshKiUk. Accessed 14 June 2021

Carter M, Gregorich E, Anderson D, Doran J, Janzen H, Pierce F (1997) Concepts of soil quality and their significance. In Gregorich E, Carter M (Eds), Soil quality for crop production and ecosystem health, pp 1–19. Elsevier, Amsterdam. https://doi.org/10.1016/S0166-2481(97)80028-1

Christensen N (2015 May 21) Urban growth review: a Q&A on the region's urban growth boundary. Metro News. https://www.oregonmetro.gov/news/urban-growth-review-qa-regions-urban-growth-boundary. Accessed 14 June 2021

Christy J (2010) History of Oregon's wetlands. Oregon Biodiversity Information Center and The Wetlands Conservancy. Oregon Explorer. https://oregonexplorer.info/content/history-oregons-wetlands?topic=4138&ptopic=98. Accessed 14 June 2021

Ciais P, Sabine C, Bala G, Bopp L, Brovkin V, Canadell J, . . . Thornton P (2013) Carbon and Other Biogeochemical Cycles. In Stocker T, Qin D, Plattner G, Tignor M, Allen S, Boschung J, . . . P. Midgley (Eds.), *Climate change 2013:* The physical science basis. Contribution of working group I to the Fifth Assessment Report, pp 465–570. Cambridge, UK: Cambridge University Press. https://www.ipcc.ch/site/assets/uploads/2018/02/WG1AR5_Chapter06_FINAL.pdf. Accessed 14 June 2021

Coates J (1986 March 16) Sagebrush rebellion on hold, group lights other legal fires. Chicago Tribune. https://www.chicagotribune.com/news/ct-xpm-1986-03-16-8601190635-story.html. Accessed 14 June 2021

Confederated Tribes of Grand Ronde (2021) Chachalu Museum and Cultural Center. https://www.grandronde.org/history-culture/culture/chachalu-museum-and-cultural-center/. Accessed 14 June 2021

Danko P (2020 January 3) Oregon cannabis sales soar in 2019, despite vape crisis and depressed prices. Portland Business Journal. https://www.bizjournals.com/portland/news/2020/01/03/oregon-cannabis-sales-soar-in-2019-despite-vape.html. Accessed 14 June 2021

DenOuden B (2020) Heppner flood of 1903. Oregon Encyclopedia, Portland, OR. https://www.oregonencyclopedia.org/articles/heppner_flood/#.X59N_UeSmUI. Accessed 14 June 2021

Dobos R, Seybold C, Chiaretti J, Southard S, Levin M (2017) Interpretations: The impact of soil properties on land use. In Ditzler C, Scheffe K, Monger H (Eds) Soil survey manual. Government Printing Office, Washington, DC

Dyrness C (1976) Effect of wildfire on soil wettability in the high Cascades of Oregon. U.S. Forest Service, Portland, OR

Federal Emergency Management Agency (n.d.) National Flood Insurance Program fact sheet. http://www.houstontx.gov/council/e/nfip/National-Flood-Insurance-Program-Fact-Sheet.pdf#: ~ :text=National%20Flood%20Insurance%20Program%20Fact%20Sheet%20Page%202,the%20insurance%20for%20the%20life%20of%20the%20mortgage. Accessed January 2021

Federal Register (2020 December 17) Guidance for identification of nonindustrial private forest land (NIPF). *85, 243.* https://www.govinfo.gov/content/pkg/FR-2020-12-17/pdf/2020-27703.pdf. Accessed 14 June 2021

Flores T, Griffith S (2018 March 17). Portland flood, 1894. Oregon History Project, Portland, OR. https://www.oregonhistoryproject.org/articles/historical-records/portland-flood-1894/#.X7GOqM6Sk2w Accessed June 14 2021

Galbraith J, Shaw R (2017) Human-altered and human-transported soils. In Ditzler C, Scheffe K, Monger H (Eds) Soil survey manual. Government Printing Office, Washington, DC

Galbraith W, Anderson E (1971 January) Grazing history of the northwest. J Range Manage 24(1): 6–12

Gale C, Keegan III C, Berg E, Daniels JC, Sorenson C, Morgan T, Polzin P (2012) Oregon's forest products industry and timber harvest, 2008: industry trends and impacts of the great recession through 2010. U.S. Forest Service, Portland, OR. https://www.fs.fed.us/pnw/pubs/pnw_gtr868.pdf Accessed June 14 2021

Giannico G, Souder J, Behan J (2018) Ecological effects of tide gate upgrade or removal: a literature review and knowledge synthesis. Institute for Natural Resources, Oregon State University, Corvallis, OR

Gorte R (2013) The rising cost of wildfire protection. Headwater Economics, Bozeman, MT. https://doi.org/10.1371/journal.pone.0230424

Graves R, Hugo R, Holz A, Nielsen-Pincus M, Jones A, Kellogg B, . . . Schindel M (2020) Potential greenhouse gas reductions from Natural Climate Solutions in Oregon, USA. PLoS ONE 15(4). https://doi.org/10.1371/journal

Haeffner M, Hellman D (2020) The social geometry of collaborative flood risk management: a hydrosocial case study of Tillamook County Oregon. Nat Hazards 103:3303–3325. https://doi.org/10.1007/s11069-020-04131-4

Hartter J, Hamilton L, Boag A, Stevens F, Ducey M, Christoffersen N, . . . Palace M (2018 April). Does it matter if people think climate change is human caused? Climate Services 10: 53–62. https://doi.org/10.1016/j.cliser.2017.06.014

Hartter J, Hamilton L, Ducey M, Boag A, Salerno J, Christoffersen N, . . . Stevens F (2020) Finding common ground: agreement on increasing wildfire risk crosses political lines. Environ Res Lett 15 (6): 065002. https://doi.org/10.1088/1748-9326/ab7ace

Henderson J, Peter D, Lesher R, Shaw D (1989) Forested plant associations of the Olympic National Forest. U.S. Forest Service Pacific Northwest Region, Portland, OR

Hessler K, Luk D, McMillan S (n.d.) Report on the Oregon Department of Agriculture's enforcement of the Clean Water Act's NPDES program related to CAFOs. Lewis & Clark Law School. https://law.lclark.edu/live/files/10807-2012-oda-clinic-report Accessed June 14 2021

Ho M, Lall U, Allaire M, Devineni N, Kwon H, Pal I, . . . Wegner D (2017) The future role of dams in the United States of America. Water Resour Res 53(2): 982–998. https://doi.org/10.1002/2016WR019905

Horst M (2019) Changes in farmland ownership in Oregon, USA. Land 8(3):3. https://doi.org/10.3390/land8030039

Hunnicutt D, Gorman K (2010 January 1). Land use vs. property rights. 1859 Oregon's Magazine, 18

INR Portland and Oregon Biodiversity Information Center (2014) Oregon actual vegetation map - 1992. Oregon Spatial Data Library: https://spatialdata.oregonexplorer.info/geoportal/details;id=2509ccf1905340b8ad82c17fb3319c94. Accessed June 14 2021

Jenny H (1941) Factors of soil formation. McGraw-Hill, New York

Kane D (2015) Carbon sequestration potential on agricultural lands: a review of current science and available practices. National Sustainable Agriculture Coalition. https://sustainableagriculture.net/wp-content/uploads/2015/12/Soil_C_review_Kane_Dec_4-final-v4.pdf. Accessed June 14 2021

Karlen D, Mausbach J, Doran J, Cline R, Harris R, Schuman G (1997) Soil quality: a concept, definition, and framework for evaluation. Soil Sci Soc Am J 61(1):4–10. https://doi.org/10.2136/sssaj1997.03615995006100010001x

Kelsey H, Bockheim J (1994) Coastal landscape evolution as a function of eustasy and surface uplift rate, Cascadia margin, southern Oregon. GSA Bull 106(6):840–854. https://doi.org/10.1130/0016-7606(1994)106%3C0840:CLEAAF%3E2.3.CO;2

Kelso MM (1947) Current issues in federal land management in the western United States. J Farm Econ 29(4):1295–1313

Kesey K (1964) Sometimes a great notion. Viking Press, New York

Keyes J, Keyes J (2015) Federal lands grazing permits—Managing rangeland resources. Utah State University Extension. https://digitalcommons.usu.edu/cgi/viewcontent.cgi?article=1697&context=extension_curall. Accessed June 14 2021

Kjelstrom L, Williams J (1996) Oregon wetland resources. In Fretwell J, Williams J, Redman P (Eds) National water summary on wetland resources—Water supply paper 2425, pp. 321–356. U.S. Geological Survey, Washington, DC. https://doi.org/10.3133/wsp2425

Klingebiel A, Montgomery P (1961) Land-capability classification, agricultural handbook 210. Soil Conservation Service, Washington, DC

KOIN 6 News (2020 October 17) Wildfires in Oregon: names, locations, size, containment. https://www.koin.com/news/wildfires/wildfires-in-oregon-names-locations-size-containment/. Accessed June 14 2021

Komar P (1992 January) Ocean processes and hazards along the Oregon coast. Oregon Geol 54(1): 3–19

Komar P (1998) The Pacific Northwest coast—Living with the shores of Oregon and Washington. Duke University Press, London

Komar P, McDougal W, Marra J, Ruggiero P (1999 January) The rational analysis of setback distances: applications to the Oregon coast. Shore & Beach 67(1): 41–49

Lal R, Follett R, Stewart B, Kimble J (2007) Soil carbon sequestration to mitigate climate change and advance food security. Soil Sci 172 (12):943–956. https://doi.org/10.1097/ss.0b013e31815cc498

Land Trust Alliance (2020) National land trust census. https://www.landtrustalliance.org/about/national-land-trust-census. Accessed June 14 2021

Latta G, Watson P, Nadreu T, Kuusela O, Rossi D (2019) 2019 Forest report. Oregon Forest Resources Institute, Portland, OR

Lehner J (2017 October 10) Oregon's timber history, an update. Oregon Office of Economic Analysis. https://oregoneconomicanalysis.com/2017/10/10/oregons-timber-history-an-update/. Accessed June 14 2021

Lomnicky G, Herlihy A, Kaufmann P (2019) Quantifying the extent of human disturbance activities and anthropogenic stressors in wetlands across the conterminous United States: results from the National Wetland Condition Assessment. Environ Monit Assess 191(Suppl 1):324. https://doi.org/10.1007/s10661-019-7314-6

Mangan R (2007) Wildland firefighter fatalities in the United States 1990–2006. National Wildfire Coordinating Group, Missoula, MT

McGregor M (2003) The Vanport flood. Oregon History Project, Portland, OR. https://www.oregonhistoryproject.org/articles/essays/the-vanport-flood/#.X59PY0eSmUk. Accessed June 14 2021

Mendell B (2016) From cigar tax to timberland trusts—A short history of timber REITs and TIMOs. Forest History Today 22(1 & 2):32–36

Miles S (1985) Farming and ranching in Oregon: a picture of diversity. Oregon State University Extension Service, Corvallis, OR

Miner TR (2013 November 8) The state that timber built. Oregon Humanities

Morrow J, Huggins D, Reganold J (2017) Climate change predicted to negatively influence surface soil organic matter of dryland cropping systems in the inland Pacific Northwest, USA. Front Ecol Evol 5:10. https://doi.org/10.3389/fevo.2017.00010

Mortenson E (2016 June 22) Is pot Oregon's top crop? Capital Press. https://www.capitalpress.com/state/oregon/is-pot-oregon-s-top-crop/article_2f43df1c-6ee3-555c-ab5a-d57e2a1c3a87.html. Accessed June 14 2021

Mote P, Abatzoglou J, Dello K, Hegewisch K, Rupp D (2019) Fourth Oregon climate assessment report. Oregon Climate Change Research Institute, Corvallis, OR

NASS (1997) Summary by size of farm, 1997. Accessed December 2020. http://lib-usda-05.serverfarm.cornell.edu/usda/AgCensusImages/1997/01/37/1599/Table-49.pdf

NASS (2019) 2017 Census of agriculture Oregon. https://www.nass.usda.gov/Publications/AgCensus/2017/Full_Report/Volume_1,_Chapter_1_State_Level/Oregon/orv1.pdf. Accessed June 14 2021

National Interagency Fire Center (2020) Federal firefighting costs (suppression only). https://www.nifc.gov/fire-information/statistics/suppression-costs. Accessed June 14 2021

National Oceanic and Atmospheric Administration (2021 January) Climate forcing. Climate.gov. https://www.climate.gov/maps-data/primer/climate-forcing. Accessed June 14 2021

National Park Service (2018) Fire—A curriculum and activity guide to Mammoth Cave National Park. https://www.nps.gov/maca/learn/education/firecurriculum.htm. Accessed 9 Feb 2021

Northwest Interagency Coordination Center (2019) Northwest annual fire report. https://gacc.nifc.gov/nwcc/content/pdfs/archives/2019_NWCC_Annual_Fire_Report_Final.pdf. Accessed February 9 2021

NRCS (2003) 1997 Cropland acreage estimates by land capability class: State of Oregon. https://www.nrcs.usda.gov/Internet/FSE_DOCUMENTS/nrcs142p2_043719.pdf. Accessed February 9 2021

NRCS (2014) National planning procedures handbook (NPPH), 1st edition amendment 6. USDA, Washington DC

NRCS (2017) Part 630–633 national ecological site handbook. USDA, Washington, DC

NRCS (2018) Oregon land use. https://www.nrcs.usda.gov/Internet/NRCS_RCA/reports/nri_or.html. Accessed June 14 2021

NRCS (n.d.) Healthy soil for life. https://www.nrcs.usda.gov/wps/portal/nrcs/main/soils/health/. Accessed June 14 2021

NRCS (n.d.) National resources inventory. https://www.nrcs.usda.gov/wps/portal/nrcs/main/national/technical/nra/nri/. Accessed June 14 2021

O'Connor J (2018) Bridge of the Gods. Oregon Historical Society. Oregon Encyclopedia, Portland, OR. https://www.oregonencyclopedia.org/articles/bridge_of_the_gods/#.X7Hq386Sk2w. Accessed June 14 2021

Oregon Association of Conservation Districts. (2018). *OACD*. https://oacd.org/. Accessed June 14 2021

Oregon Department of Agriculture (2016) Oregon soil and water conservation guidebook. https://www.oregon.gov/ODA/shared/Documents/Publications/NaturalResources/SWCDGuidebook.pdf. Accessed June 14 2021

Oregon Department of Agriculture (2019) Confined animal feeding operation (CAFO) program 2019 annual report. https://www.oregon.gov/ODA/shared/Documents/Publications/NaturalResources/CAFOReport2019.pdf. Accessed June 14 2021

Oregon Department of Agriculture (2021) Oregon agricultural statistics and directory 2021. https://www.oregon.gov/oda/shared/Documents/Publications/Administration/AgStatsDirectory.pdf. Accessed June 14 2021

Oregon Department of Environmental Quality (2019) Oregon outdoor burning guide. https://www.oregon.gov/deq/FilterDocs/OpenBurnEng.pdf. Accessed June 14 2021

Oregon Department of Fish and Wildlife (2016) Oregon conservation strategy. https://oregonconservationstrategy.org/. Accessed June 14 2021

Oregon Department of Forestry (2006) Oregon spatial analysis project. https://www.fs.fed.us/na/sap/products/OR/OR-Methodology.pdf. Accessed June 14 2021

Oregon Department of Forestry. (2019). The forestland burning guide. https://www.oregon.gov/odf/board/Documents/smac/ForestlandBurningGuide_2019update.pdf. Accessed June 14 2021

Oregon Department of Forestry (2020) Information & statistics. https://www.oregon.gov/odf/fire/Pages/firestats.aspx. Accessed June 14 2021

Oregon Department of Forestry (n.d.) Maps & data. https://www.oregon.gov/ODF/AboutODF/Pages/MapsData.aspx. Accessed June 14 2021

Oregon Department of Geology and Mineral Industries (n.d.) DOGAMI open-file report series. https://www.oregongeology.org/pubs/ofr/p-O-16-02.htm. Accessed June 14 2021

Oregon Department of Land Conservation & Development (2019) Oregon's statewide planning goals & guidelines—Goal 18: Beaches and dunes. https://www.oregon.gov/lcd/OP/Documents/goal18.pdf. Accessed June 14 2021

Oregon Department of Land Conservation and Development (2001) Oregon's statewide planning goals and guidelines, Goal 7: Areas subject to natural hazards. https://www.oregon.gov/lcd/OP/Pages/Goal-7.aspx#:~:text=Goal%207%3A%20Areas%20Subject%20to%20Natural%20Disasters%20and,of%20Oregon%27s%20natural%20hazards%20since%20the%20program%20began. Accessed June 14 2021

Oregon Department of Land Conservation and Development (2019a) Combatting climate change is central to DLCD's work. https://www.oregon.gov/lcd/CL/Documents/ClimateChange_Central_to_DLCD_Nov2019.pdf. Accessed June 14 2021

Oregon Department of Land Conservation and Development (2019b) Oregon statewide planning goals and guidelines. https://www.oregon.gov/lcd/Publications/compilation_of_statewide_planning_goals_July2019.pdf. Accessed June 14 2021

Oregon Department of Land Conservation and Development (2020a) National Flood Insurance Program (NFIP) in Oregon. https://www.oregon.gov/lcd/NH/Pages/NFIP.aspx. Accessed June 14 2021

Oregon Department of Land Conservation and Development (2020b) Oregon Coastal Management Program. https://www.oregon.gov/lcd/OCMP/Pages/index.aspx. Accessed June 14 2021

Oregon Department of Land Conservation and Development (2021, May) Maps, data, and tools. https://www.oregon.gov/lcd/About/Pages/Maps-Data-Tools.aspx. Accessed June 14 2021

Oregon Department of State Lands (2017a) Oregon wetland program plan 2017–2021. https://www.oregon.gov/DSL/WW/Documents/oregon_wetland_program_plan.pdf. Accessed June 14 2021

Oregon Department of State Lands (2017b February 27) State of Oregon state land inventory report. https://www.oregon.gov/dsl/Land/Documents/1SLIOwnershipStatewide.pdf. Accessed June 14 2021

Oregon Department of Transportation (2021, January 14) Coastal landslides vs. our roads: Who wins? ODOT Transportation Insights: https://www.oregondot.org/topics/. Accessed June 14 2021

Oregon Forest Resources Institute (2019) Oregon forest facts 2019–2020 edition. https://oregonforests.org/sites/default/files/2019-01/OFRI_2019-20_ForestFacts_WEB.pdf. Accessed June 14 2021

Oregon Forest Resources Institute (2020) Fire in Oregon's forests. https://oregonforests.org/node/96. Accessed June 14 2021

Oregon Historical Society (2005) Oregon historical quarterly special issue: the Isaac I. Stevens and Joel Palmer Treaties—1855–2005 (Vol. 106 (3)). Portland, OR

Oregon Legislative Assembly (1973) Senate Bill 100. https://www.oregon.gov/lcd/OP/Documents/sb100.pdf. Accessed June 14 2021

Oregon Legislative Policy and Research Office (2016) Background brief on tribal governments in Oregon. https://www.oregonlegislature.gov/citizen_engagement/Reports/BB2016TribalGovernmentsinOregon.pdf. Accessed June 14 2021

Oregon Office of Emergency Management (2020) Hazards in Oregon. https://www.oregon.gov/oem/hazardsprep/pages/hazards-in-oregon.aspx. Accessed June 14 2021

Oregon Public Broadcasting (2020) Timber wars. https://www.opb.org/show/timberwars/. Accessed June 14 2021

Oregon Secretary of State (2007) Special election abstract of votes state measure 49. http://records.sos.state.or.us/ORSOSWebDrawer/RecordView/6873595. Accessed June 14 2021

Oregon Secretary of State (2009) Confined animal feeding operation program. Oregon administrative rules database: https://secure.sos.state.or.us/oard/displayDivisionRules.action?selectedDivision=2751. Accessed June 14 2021

Oregon Secretary of State (2020) City populations. Oregon blue book. https://sos.oregon.gov/blue-book/Pages/local/city-population.aspx. Accessed June 14 2021

Oregon Seismic Safety Policy Advisory Commission (2013) The Oregon resiliency plan. https://www.oregon.gov/oem/Documents/Oregon_Resilience_Plan_Executive_Summary.pdf#:~:text=The%20Oregon%20Resilience%20Plan%20maps%20a%20path%20of,steps%20to%20begin%20a%20journey%20along%20that%20path. Accessed June 14 2021

Oregon State University (2020) OSU extension service. https://extension.oregonstate.edu/. Accessed June 14 2021

Oregon.gov (2021) Timber harvest data 1962–2019. https://data.oregon.gov/Natural-Resources/Timber-Harvest-Data-1962-2019/c3sg-dt24?category=Natural-Resources&view_name=Timber-Harvest-Data-1962-2019. Accessed June 14 2021

Orr E, Orr W (1999) The other face of Oregon: geologic processes that shape our state. Or Geol 61(6):130–143

Paeth RK (1971) Factors affecting mass movement of four soils in the western Cascades of Oregon. Soil Sci Soc Am J 35(6):943–947

Palombo L (2019) Oregon's largest wildfires. The Oregonian / Oregon Live: https://projects.oregonlive.com/wildfires/historical.php. Accessed June 14 2021

Parks D, Cundy T (1989) Soil hydraulic characteristics of a small southwest Oregon watershed following high-intensity wildfires. In Berg N (Ed), *Proceedings of the symposium on fire and watershed management* (pp 63–67). Berkeley, CA: U.S. Forest Service

Parsons R (1978) Soil-geomorphology relations in mountains of Oregon, USA. Geoderma 21(1):25–39

Paulsen C (1949) Floods of May-June 1948 in Columbia River Basin. U.S. Geological Survey. https://pubs.usgs.gov/wsp/1080/report.pdf. Accessed June 14 2021

Phillips P (2007) Tsunamis and floods in Coos Bay mythology. Oreg Hist Quart 108(2):181

Pogue J (2018) Chosen country: a rebellion in the West. Henry Holt and Company, New York

Powers R (2018) The overstory. W.W. Norton & Company, New York

Ranganathan J, Waite R, Searchinger T, Zionts J (2020) Regenerative agriculture: good for soil health, but limited potential to mitigate climate change. World Resources Institute. https://www.wri.org/insights/regenerative-agriculture-good-soil-health-limited-potential-mitigate-climate-change. Accessed June 14 2021

Robinson N, Allred B, Naugle D, Jones M (2019) Patterns of rangeland productivity and land ownership: Implications for conservation and management. Ecol Appl 29(3). https://doi.org/10.1002/eap.1862

Rodale Institute (2014) Regenerative organic agriculture and climate change. https://rodaleinstitute.org/wp-content/uploads/rodale-white-paper.pdf. Accessed June 14 2021

Rusignola D (2019) How the new Weyerhaeuser CEO keeps the timber REIT rooted in its values. REIT Magazine. https://www.reit.com/news/reit-magazine/may-june-2019/how-new-weyerhaeuser-ceo-keeps-timber-reit-rooted-its-values. Accessed June 14 2021

Satterfield T (2002) Anatomy of a conflict—Identity, knowledge, and emotion in old-growth forests. UBC Press, Vancouver, BC

Schick T, Davis R, Younes L (2020) Big money bought Oregon's forests. Small timber communities are paying the price. Oregon Public Broadcasting. https://www.opb.org/news/article/oregon-investigation-timber-logging-forests-policy-taxes-spotted-owl/. Accessed June 14 2021

Schoeneberger P, Wysocki D, Busskohl C, Libohova Z (2017) Landscapes, geomorphology, and site description. In Ditzler C, Scheffe K, Monger H (Eds) Soil survey manual. Government Printing Office, Washington, DC

Schulz W, Higgins J, Galloway S (2012) Evidence for earthquake triggering of large landslides in coastal Oregon, USA. Geomorphology 141–142:88–98. https://doi.org/10.1016/j.geomorph.2011.12.026

Shackelford A (2020) Death by 1000 cuts - A 10-point plan to protect Oregon's farmland. 1000 Friends of Oregon, Portland, OR

Shinn D (1977) Man and the land: an ecological history of fire and grazing on eastern Oregon rangelands. Oregon State University, Corvallis, OR

Shuster W, Bonta J, Thurston H, Warnemuende E, Smith D (2007) Impacts of impervious surface on watershed hydrology: A review. Urban Water J. 2(4):263–275. https://doi.org/10.1080/15730620500386529

Sickinger T (2020, September 15) Oregon's historic wildfires: Unusual but not unprecedented. The Oregonian / Oregon Live. https://www.oregonlive.com/news/2020/09/oregons-historic-wildfires-the-unprecedented-was-predictable.html. Accessed June 14 2021

Simmons E, Scudder M, Morgan T, Berg E, Christensen G (2016) Oregon's forest products industry and timber harvest 2013 with trends through 2014. U.S. Forest Service. https://www.fs.fed.us/pnw/pubs/pnw_gtr942.pdf. Accessed June 14 2021

Soil Survey Staff (2019) Web soil survey. https://websoilsurvey.sc.egov.usda.gov/App/HomePage.htm. Accessed June 14 2021

Soil Survey Staff (2020) Gridded National Soil Survey Geographic (gNATSGO) Database for Oregon. https://nrcs.app.box.com/v/soils . Accessed June 14 2021

Spies T, Long J, Charnley S, Hessburg P, Marcot B, Reeves G, . . . Raphael M (2019) Twenty-five years of the Northwest Forest Plan: what have we learned? Front Ecol Environ 17(9): 511–520. https://doi.org/10.1002/fee.2101

Sullivan E, Eber R (2009, November 18) The long and winding road: Farmland protection in Oregon 1961 - 2009. San Joaquin Agricultural Law Review. https://ssrn.com/abstract=3177855. Accessed June 14 2021

Swanston D (1979) Effect of geology on soil mass movement activity in the Pacific Northwest. In Youngberg C (Ed) Proceedings of the Fifth North American Forest Soils Conference (pp 89–116). Ft. Collins, CO: Colorado State University

Tamastslikt Cultural Institute (2021) Visit the Museum at Tamastslikt Cultural Institute. https://www.tamastslikt.org/#. Accessed June 14 2021

The Confederated Tribes of Warm Springs (2021) The Museum at Warm Springs. https://www.museumatwarmsprings.org/. Accessed June 14 2021

Tiner R (1996).Wetland definitions and classifications in the United States. In Fretwell J, Williams J, Redman P (Eds) National water summary on wetland resources - Water supply paper 2425 (pp 27–34). Washington, DC: U.S. Geological Survey. https://doi.org/10.3133/wsp2425

U.S. Census Bureau (2020) *Explore census data*. https://data.census.gov/cedsci/. Accessed June 14 2021

U.S. EPA (2020) Inventory of U.S. greenhouse gas emissions and sinks 1990–2018. https://www.epa.gov/sites/production/files/2020-04/documents/us-ghg-inventory-2020-main-text.pdf. Accessed June 14 2021

Fire Administration US (2020) Creating a community wildfire protection plan. Federal Emergency Management Agency, Washington, DC

U.S. Forest Service (2018) Toward shared stewardship across landscapes: An outcome-based investment strategy. https://www.fs.usda.gov/sites/default/files/toward-shared-stewardship.pdf. Accessed June 14 2021

U.S. Forest Service (n.d.) Mount St. Helens area. https://www.fs.usda.gov/recarea/giffordpinchot/recarea/?recid=34143. Accessed June 14 2021

U.S. Forest Service (n.d.) *Partners*. https://www.fs.usda.gov/managing-land/fire/partners

U.S. Geological Survey (2011) Gap analysis project—Land cover data download. https://www.usgs.gov/core-science-systems/science-analytics-and-synthesis/gap/science/land-cover-data-download?qt-science_center_objects=0#qt-science_center_objects. Accessed June14 2021

United States of America v. State of Washington, 384 F. Supp. 312 (United States District Court, W.D. Washington at Tacoma February 12, 1974)

Urness Z (2016, February 21) Memories of the 1996 flood: "There was so much water, everywhere you looked'. *Statesman Journal*. Retrieved June 14, 2021, from https://www.statesmanjournal.com/story/travel/outdoors/2016/02/21/memories-1996-flood-there-so-much-water-everywhere-you-looked/80297232/

USDA and U.S. EPA (1999) Unified national strategy for animal feeding operations. https://www.epa.gov/sites/production/files/2015-10/documents/finafost.pdf. Accessed 14 June 2021

USDA-Farm Service Agency (2020) Farm Service Agency Oregon. https://www.fsa.usda.gov/state-offices/Oregon/index. Accessed 14 June 2021

Vincent C, Hanson L, Bermejo L (2020) Federal land ownership: Overview and data. Congressional Research Service. https://crsreports.congress.gov/product/pdf/R/R42346. Accessed 14 June 2021

Walsh M, Whitlock C, Bartlein P (2010, November 10) 1200 years of fire and vegetation history in the Willamette Valley, Oregon and Washington, reconstructed using high-resolution macroscopic charcoal and pollen analysis. Palaeogeogr Palaeocl 297(2), 273–289. https://doi.org/10.1016/j.palaeo.2010.08.007

Wang Y, Summers R, Hofmeister R (2002) Landslide loss estimation pilot project in Oregon Open-File Report O-02–05. Oregon Department of Geology and Mineral Industries. https://www.oregongeology.org/pubs/ofr/O-02-05.pdf. Accessed 14 June 2021

Willingham W (2018) Willamette River flood of 1894. Portland, OR: Oregon Encyclopedia. https://www.oregonencyclopedia.org/articles/willamette_flood_1894_/#.X59LzEeSmUI. Accessed 14 June 2021

Willingham W (2019) Cascade Locks. Oregon Encyclopedia, Portland, OR. https://www.oregonencyclopedia.org/articles/cascade_locks/#.X6CSOkeSmUk. Accessed 14 June 2021

World Atlas (2017) U.S. states by size. World Map: https://www.worldatlas.com/aatlas/infopage/usabysiz.htm. Accessed 14 June 2021

World Forestry Center (2021) Our mission. https://www.worldforestry.org/about-us/mission-vision/. Accessed 14 June 2021

Youd T, Idriss I (2001) Liquefaction resistance of soils: Summary report from the 1996 NCEER and 1998 NCEER/NSF workshops on evaluation of liquefaction resistance of soils. J Geotech Geoenviron 127(4):297–313. https://doi.org/10.1061/(ASCE)1090-0241(2001)127:4(297)

Younging G (2018) Elements of indigenous style—A guide for writing by and about indigenous peoples. Brush Education Inc., Edmonton, AB

Yu Q, Wilson J, Wang Y (2014) Overview of the Oregon Resilience Plan for next Cascadia earthquake and tsunami. In: Proceedings of the 10th national conference in earthquake engineering. Anchorage, AK: Earthquake Engineering Research Institute. https://doi.org/10.4231/D3CN6Z08T

Zomer R, Bossio D, Sommer R, Verchot L (2017 November). Global sequestration potential of increased carbon in cropland soils. Sci Rep-UK. 7:15554. https://doi.org/10.1038/s41598-017-15794-8

18 Yields, Soil Conservation, and Production Systems

18.1 Introduction

This chapter examines the relationship of Oregon soils to what are often called "working lands," which include the cropland, pasture, range, and forests managed to produce food, fiber, ornamental, or other plant or animal products for human use (Law Insider Inc. 2020). Pasture and range used for livestock production are lumped here into "grazing" land. This chapter's focus on working lands is not meant to diminish the importance of other land uses such as developed land, farmsteads, lands used for mining, or lands managed for wildlife habitat.

Section 18.2 provides a statewide view of crop, grazing land, and forestry yields. Section 18.3 summarizes the current efforts aimed at the conservation of soil and related resources, touching both on legal requirements and voluntary efforts. Sections 18.4–18.6 focus on the management systems used to produce important commodities from cropland, grazing lands, and forestland. It seems logical to end this chapter with a section on the First Foods of Indigenous Peoples, in essence finishing this view of working lands with what came first.

The state's diverse climate and soils allow the production of a remarkably wide range of commodities. This diversity necessitates simplification and generalization in the following sections, and many working lands with limited areal extent, economic significance, or cultural importance are not addressed here.

18.2 Yields of Oregon Working Lands

18.2.1 Cropland Yields

Irrigated and non-irrigated crop yield data included in Natural Resources Conservation Service (NRCS) soil surveys[1] indicate yields expected in an average year under a high level of management. Web Soil Survey (WSS) reports for Oregon include potential yields for the crops listed in Table 18.1 (Soil Survey Staff 2019). Yield estimates for a given crop are only provided for those soils that commonly are used to produce that crop. Additional crops, not included in Table 18.1, with yield estimates provided in only a few soil surveys include cabbage, sugar beets, red clover seed, bluegrass seed, bentgrass seed, walnuts, and distilled mint. Since some soil surveys were completed prior to 1980 (Fig. 2.4), they are not always useful for crop selection or other planning purposes. For example, wine grapes are grown in multiple Oregon counties, but most soil surveys do not include estimated yields for wine grapes. Also, yields shown in WSS reports can be out of date. In those cases, actual yields are usually greater than those estimated by WSS.

Soil requirements vary considerably between crops. For example, a soil with occasional flooding may have a high yield rating for corn (which is planted after flood season has ended), have a medium rating for grass-legume hay (a perennial crop that often tolerates some winter flooding), but be unsuitable for growing alfalfa hay (a perennial crop that

[1] NRCS tabular soil survey data are maintained in the National Soil Information System (NASIS). Data are available in concise reports for user-defined areas via the Web Soil Survey (WSS). Datasets for entire soil survey areas may also be downloaded from WSS for delivery in a Soil Survey Geographic Database (SSURGO) format. Managers of working lands typically use WSS reports, while research or large-scale planning efforts may require SSURGO data sets. In any case, WSS reports and SSURGO downloads are generated from the same data.

T. Thorson et al., *The Soils of Oregon*, World Soils Book Series,
https://doi.org/10.1007/978-3-030-90091-5_18

Table 18.1 Crops for which yield data commonly are provided in WSS reports, by Oregon survey area

Survey area	Apples, Pears	Barley	Beans, Snap	Cherries	Corn	Cranberries	Hazelnuts	Hay, Alfalfa	Hay, Grass-Legume	Pasture	Peas	Potatoes, Irish	Raspberries	Ryegrass Seed	Straw-berries	Wheat, Winter	Wine Grapes
Alsea area																	
Baker Co. area		x						x	x	x						x	
Benton Co.				x	x					x				x		x	
Clackamas Co. area					x		x			x			x		x	x	
Clatsop Co.									x	x							
Columbia Co.		x						x		x			x		x	x	
Coos Co.						x			x	x							
Crater Lake Natl. Park																	
Curry Co.						x			x	x							
Douglas Co. area								x	x	x						x	x
Gilliam Co.		x						x		x						x	
Grant Co., central								x	x	x						x	
Harney Co. area								x	x	x							
Hood River Co. area	x	x								x							
Jackson Co. area	x				x				x	x						x	
Josephine Co.					x			x	x	x							
Klamath Co., South		x						x		x		x				x	
Lake County, North																	
Lake County, South																	
Lane Co. area			x		x		x			x					x	x	
Lincoln Co. area									x	x					x		
Linn Co. area			x		x					x				x		x	
Malheur Co., northeast					x			x		x		x		x			
Marion Co. area		x	x	x	x		x	x		x					x	x	
Morrow Co. area		x			x			x		x		x				x	
Multnomah Co.					x					x		x	x			x	
Polk Co.		x			x					x						x	
Prineville area		x	x	x				x	x			x				x	

(continued)

Table 18.1 (continued)

Survey area	Apples, Pears	Barley	Beans, Snap	Cherries	Corn	Cranberries	Hazelnuts	Hay, Alfalfa	Hay, Grass-Legume	Pasture	Peas	Potatoes, Irish	Raspberries	Ryegrass Seed	Straw-berries	Wheat, Winter	Wine Grapes
Sherman Co.		x						x								x	
Tillamook Co.									x	x							
Trout Creek-Shaniko area		x						x		x						x	
Umatilla Co. area		x			x			x			x					x	
Union Co. area		x						x		x	x	x					
South Umpqua area																	
Upper Deschutes River area																	
Wallowa Co. area		x						x	x	x						x	
Warm Springs Indian Reservation	x	x		x				x		x						x	
Wasco Co., north									x	x						x	
Washington Co.			x				x		x	x						x	
Yamhill Co.		x	x		x			x		x						x	

usually is killed by winter flooding). Many soils do not currently have ratings for certain crops, but this does not mean those crops cannot be grown on those soils.

Perhaps of even greater importance, crop yields are determined to a great extent by factors that cannot be factored into WSS reports. Weather, pest pressure, tillage, crop rotations, previous management, and varieties are just some factors that play important roles in determining yields. Indeed, Stermitz et al. (1999) found that yield maps generated using soil survey data were neither accurate nor precise and were of little value in predicting site-specific yields on Montana wheat farms. Thus, yield data provided in WSS reports should be used as generalized information rather than accurate predictions. This is true for both cropland yields and the grazing yields described in the next section.

18.2.2 Grazing Land Yields

The WSS provides yield estimates for soils typically used for pasture or grazed rangeland. Because pasture and rangeland are managed quite differently, their yield estimates are provided in two different formats with units that are not interchangeable.

The most informative yield estimates for pastures in WSS are via Forage Suitability Groups (FSG), a type of soil survey management group (see Chap. 17). NRCS (2003) defines FSGs as a group of soils with similar plant species adaptation, productivity, and management needs. FSGs are used as planning tools to help determine which pasture species to seed, which practices to use, and what initial livestock stocking rates to employ. The WSS provides a FSG code for each soil that is typically pastured. Detailed reports for each FSG are located in the NRCS Field Office Technical Guide, Section II (NRCS 2020a).

FSGs are designated with codes, for example, "G001XY007OR." The codes can be deciphered as follows: "G" stands for Forage Suitability Group, the following three digits designate the Major Land Resource Area (MLRA),[2] the next letter designates the MLRA letter code ("X" is used if no code exists), the next letter designates the Land Resource Unit code ("Y" is used if no code exists), the next three digits are a unique identifier for the FSG, and the final two letters designate the state (NRCS 2003).

The FSG report for G001XY007OR describes a group of somewhat poorly drained soils in the Coast Range (MLRA 1). These soils are found on terraces, low hills, and mountains, have a mean annual precipitation of 1,000–3,300 mm, and a mean annual air temperature of 11.1–12.8 °C. They have a seasonal water table that varies from about 30–75 cm beneath the soil surface during winter and early spring. Adapted forage species include tall fescue[3] (*Schedonorus arundinaceus*), perennial ryegrass (*Lolium perenne*), white clover (*Trifolium repens*), and others. When these soils are not irrigated, they support about 5–30 animal unit months[4] (AUM) ha^{-1}, and support 37–40 AUM ha^{-1} when irrigated. With a medium level of management, peak forage production usually occurs in May and June; there is insignificant forage growth in December through February.

The WSS provides rangeland yield estimates via Ecological Site Descriptions (ESD). Ecological rangeland sites are another type of soil survey management group, which designate distinctive kinds of land with specific vegetation production characteristics (NRCS 2003). As with FSG, detailed reports for each ESD are located in the NRCS Field Office Technical Guide, Section II (NRCS 2020b). WSS also provides online links to ESDs. ESD codes have the same format as for FSGs, except rangeland ESD codes begin with the letter "R" for rangeland (NRCS 2003).

An example of an ecological site in the Palouse Prairie of Union County is the Shallow Clayey 17–22 PZ[5] site, which has code R009XY021OR. The ESD describes soils in this site as occurring on table lands and mountain plateaus, typically within the northern portion of the Blue Mountains and at elevations of 900–1,200 m. The mean annual precipitation for this site is 430–560 mm and the mean annual air temperature is 8.3 °C. The soils are shallow over basalt bedrock and areas of rock outcrop may occur. Permeability is slow and the potential for erosion is slight to moderate. The historic (undisturbed) plant community is dominated by Idaho fescue (*Festuca idahoensis*) and bluebunch wheatgrass (*Pseudoroegneria spicata*), with forbs such as common yarrow (*Achillea millefolium*), arrowleaf balsamroot (*Balsamorhiza sagittata*), and other species. Annual plant production[6] varies from 780 to 1,300 kg ha^{-1}. This ecological site is suitable for use by cattle and sheep in summer and fall and provides forage for deer and elk in spring, summer, and fall. The state and transition model[7] for this site predicts that overgrazing will first decrease Idaho fescue while

[2] See Chaps. 4 and 5 for descriptions of Oregon's Major Land Resource Areas.

[3] Tall fescue, an introduced species of grass grown widely throughout North America, is also known as *Festuca arundinacea* and *Lolium arundinaceum*. The scientific name currently recognized by the USDA Plants Database, however, is *Schedonorus arundinaceus* (NRCS 2020b).

[4] One AUM is about 360 kg of dry forage, which is the amount a beef cow with calf will typically consume in one month (NRCS 2003).

[5] Ecological site names typically reference a mean annual precipitation zone (PZ), given in inches.

[6] Estimated annual production provided in ESDs (and shown for rangeland in this chapter) is for total air-dry weight of annual plant growth, regardless of whether the plant is palatable to livestock.

[7] ESDs often include state and transition models that describe the interaction between management and vegetation dynamics for that site (Duniway et al. 2010).

increasing bluebunch wheatgrass populations. Further deterioration will reduce bluebunch wheatgrass, leading to invasion by annual grasses such as cheatgrass (*Bromus tectorum*) and increasing populations of unpalatable forbs like yarrow. Continued overgrazing can lead to the loss of biotic crusts on the soil surface, erosion, and irreversible changes to the soil profile.

Rangeland productivity of Oregon soils is roughly categorized in Appendix F, with ratings of high (greater than 2,240 kg ha^{-1}); medium (1,680–2,240 kg ha^{-1}); and low (less than 1,680 kg ha^{-1}).

18.2.3 Forestry Yields

The WSS provides two metrics of expected timber productivity for soils: a site index number and a volume growth rate. The site index is the estimated height that dominant and codominant trees of a given species will attain at a specified "base age." It is common for the WSS to provide more than one site index for each soil map unit. For example, Josephine gravelly loam, 35–55% south slopes, a Typic Haploxerult in Josephine County's Siskiyou Mountains, has a site index of 27 m for Douglas-fir (*Pseudotsuga menziesii*) with a base age of 50 years and a site index of 35 m for ponderosa pine (*Pinus ponderosa*) with a base age of 100 years.[8] The site index applies to fully stocked, even-aged, unmanaged stands (Soil Survey Staff 2019).

The volume growth rate is the projected maximum annual increase in wood volume for a given tree species on a specified soil. Each tree species has a characteristic growth curve in which wood is produced slowly at first, reaches a maximum at a certain age, and then slows down. Those growth curves are documented in the literature, referenced in the NRCS National Forestry Handbook (NRCS 2004), and used to predict the growth rate for the year of maximum increase in wood volume. Not surprisingly, growth rates are positively correlated to site indices. The annual volume growth rates for the Josephine gravelly loam soil described above are 8.0 m^3 ha^{-1} for Douglas-fir and 9.0 m^3 ha^{-1} for ponderosa pine.

Appendix E categorizes Oregon forest soils as low, medium, or high timber productivity. Medium productivity ratings in Appendix E signify site indices of 30–40 m for Douglas-fir with a base age of 100 years, 27–37 m for ponderosa pine with a base age of 100 years, or 21–27 m for grand fir (*Abies grandis*) with a base age of 50 years. Accordingly, low or high productivity ratings in Appendix E are soils with site indices lower than or higher than those medium values. In general, highly productive forest soils have a mean annual precipitation of 1,150–2,400 mm, a mean annual air temperature of 9–11 °C, are deeper than 100 cm, and are well drained. The parent materials usually are colluvium and residuum derived from sedimentary rocks (siltstone and sandstone), igneous rocks, or basalt. For example, the Orford, Rinearson, Kinney, and Aschoff soil series each occupy more than 300 km^2 are among the most productive forest soils in the state and are all in western Oregon.

Three caveats pertain to site indices and volume growth rates. First, WSS usually doesn't provide site indices or volume growth rates for soil map units used for cropland, even though those soils are sometimes used for timber production. Second, some work has begun to craft ESDs for forestland west of the Cascade crest.[9] These could function in much the same way as ESDs for rangeland in eastern Oregon and may prove especially useful in managing oak savanna systems in western Oregon. However, ESDs for western Oregon are not available in the kind of detail needed for forest management or yield predictions, and thus land managers use site indices and volume growth rates. Third, while site indices and volume growth rates are good metrics for forest soil productivity, actual timber harvest yields are commonly measured in board feet.[10]

18.2.4 Productivity Indices and Yield Modeling

To address the concerns of the accuracy and precision of yield estimates in WSS reports, there have been efforts to develop crop productivity indices, which provide stable comparative values of yields for given crops and soils. These numerical ratings indicate the relative potential yield of soil from 0 (minimum) to 100 (maximum) for a wide diversity of adapted crops. Indices remain constant, unlike average actual yields, which tend to increase over time due to changing farming practices. Thus, indices offer the advantage of keeping soil survey yield data relevant without requiring regular updates to the soil survey database. The disadvantage is that indices do not provide actual yield estimates (for example, bushels of wheat per acre), which are useful to land managers.

Huddleston (1982) first illustrated how yield indices (he used the term "ratings") could be determined for soils in Oregon's Willamette Valley. He calculated "native productivity" indices by adding or subtracting values based on soil taxonomy and ten soil properties. Huddleston's method also

[8] The Web Soil Survey provides site indices in feet and volume growth rates in ft^3 ac^{-1}. Those values have been converted to metric equivalents here.

[9] Throughout this chapter, the term "Cascade crest" refers to the crest of the Cascade Mountain range.

[10] A board foot is a unit of volume for timber equal to 12 inches by 12 inches by one inch, which equals 144 in^3. 1,000 board feet is approximately 2.36 m^3.

provided for the modification of indices based on the management of specific sites. For example, the Woodburn silt loam, 0–3% map unit is an Aquultic Argixeroll. This map unit receives +100 points as a Mollisol, −20 points due to the xeric suborder, and −35 points because of the Aquultic subgroup. Its rating is modified with +15 points because it is moderately well drained and +5 points because its pH tends to be greater than 5.6. Other factors, such as the map unit's mesic temperature regime and silt loam surface texture, provide 0 points in the calculation. Adding the values together, Huddleston calculated 100 − 20 − 35 + 15 + 5 = 65, which is the soil's native productivity rating. If this soil occupies a site that is drained (+8), regularly receives fertilizer and lime (+5), and is irrigated (+16), its rating would be 65 + 8 + 5 + 16 = 94. Huddleston calculated "calibration scores" for 23 soils by averaging yield estimates for ten crops from published soil surveys and converting them to values varying from 0 (lowest) to 100 (highest) yields. Linear correlation between his ratings and the calibration scores was very good ($r = 0.98$). Note that this correlation is between the native productivity ratings and soil survey yield estimates. There does not appear to have been an effort to compare the native productivity ratings to actual yield measurements in the field.

Since Huddleston's 1982 publication, additional work has been pursued in the same vein. NRCS has developed crop productivity indices for hay and small grains, which are now available via WSS for some areas. One reason for this work was to develop accurate relative yield estimates for soils to aid in determining rental rates paid to farmers enrolling land in USDA's Conservation Reserve Program. Unlike Huddleston's ratings, which are based on soil taxonomy, soil survey data, and management, these crop productivity indices are based solely on soil survey data (NRCS 2012).

A similar technique was applied to forestry yield predictions in Idaho and Washington. In this work, a multiple linear regression model was developed to predict Douglas-fir site indices using soil survey data, terrain factors such as elevation), and climate factors (Kimsey 2014). There do not appear to be any published efforts to use crop productivity indices methods for grazed rangeland.

18.3 Conservation of Soil and Related Resources

18.3.1 Background

Many sources provide information about conservation practices[11] used on Oregon working lands to reduce soil erosion, improve soil health, and address other natural resource concerns. For example, NRCS provides information in its online Field Office Technical Guide (NRCS 2020c) and in county long-range strategies (NRCS 2020c). State agencies, soil and water conservation districts, Oregon State University (OSU), non-profit, and for-profit organizations also provide such information. Many conservation practices are implemented voluntarily by landowners and managers; some practices are required by regulation. Note that regulatory information provided below is composed of brief summaries, which are not meant to substitute for legal regulatory language. Also note that rules change over time. Readers are directed to the appropriate agencies listed below for specific questions about regulatory compliance.

18.3.2 State Regulations

State regulations applicable to conservation on Oregon working lands include the Agricultural Water Quality Management Act, Confined Animal Feeding Operation (CAFO[12]) permits, the Oregon Forest Practices Act, and water rights.

The Oregon legislature passed the Agricultural Water Quality Management Act in 1993 to address water pollution originating from farms and ranches. The Oregon Department of Agriculture (ODA) administers this law and, acknowledging the diversity of agriculture across the state, created 38 watershed-based Agricultural Water Quality Management Areas. Each area has its own plan and rules, which are developed, reviewed, and updated collaboratively by ODA and a local advisory committee. The plans outline how farmers and ranchers will protect water quality and how the local Soil and Water Conservation District can provide technical and financial assistance for those efforts. The rules do not dictate specific practices but rather describe conditions that farmers and ranchers must achieve on their operations. Three priorities are common in the rules for every area: allowing riparian vegetation to shade streams and filter contaminated runoff, controlling erosion, and utilizing manure and fertilizer efficiently to keep nutrients out of streams and ditches. ODA can fine landowners who violate area rules (Oregon Department of Agriculture 2019a). Although the plans and rules do not dictate specific practices, and appropriate practices vary across the state, common practices that farmers use to comply with the rules include fences along riparian areas, waste storage facilities

[11] Table 18.2 lists conservation practices commonly used in Oregon.

[12] Federal policy defines "*concentrated* animal feeding operations" as CAFOs, while Oregon uses the same acronym for "*confined* animal feeding operations." Oregon's definition of CAFO is somewhat broader than the federal definition and encompasses a larger number of operations (Oregon Secretary of State 2009; USDA and U.S. EPA 1999; Hessler et al. 2021).

Table 18.2 Common conservation practices on Oregon's working lands. Some practice names and definitions have been shortened due to limitations of space. *Source* NRCS (2020b)

Practice name	NRCS code	Definition
Brush management	314	The management or removal of woody plants
Conservation cover	327	Establishing and maintaining permanent vegetative cover
Contour farming	330	Aligning ridges formed by field operations around the hillslope
Cover crop	340	Grasses, legumes, and forbs planted for seasonal vegetative cover
Crop rotation	328	A planned sequence of crops grown on the same ground over a period of time
Fence	382	A constructed barrier to animals or people
Field border	386	A strip of permanent vegetation at the edge of a field
Field operation emissions reduction	376	Adjusting field operations to reduce particulate matter emissions
Filter strip	393	A strip of herbaceous vegetation that removes contaminants from overland flow
Forest stand improvement	666	Killing selected trees or understory vegetation to achieve desired forest conditions
Forest trails and landings	655	A temporary or infrequently used path or cleared area
Habitat management	644/645	Managing habitat for wildlife
Hedgerow	422	Establishment of dense vegetation in a linear design
Herbaceous weed treatment	315	The removal or control of herbaceous weeds
High tunnel	325	An enclosed structure used to extend the crop growing season
Irrigation water management	449	Controlling the volume, frequency, and application rate of irrigation water
Micro-irrigation	441	A system for application of small quantities of water as drops or miniature spray
Mulching	484	Applying plant residues or other materials produced off site
No-till	329	Managing surface residue year-round in fields without tillage
Nutrient management	590	Managing the amount, source, placement, and timing of plant nutrients
Pest management system	595	A specific combination of pest prevention, avoidance, monitoring, and suppression
Prescribed burning	338	Controlled fire applied to a predetermined area
Prescribed grazing	528	Managing vegetation with grazing animals
Range planting	550	Establishment of adapted self- sustaining vegetation
Reduced till	345	Managing surface residue year-round while limiting soil-disturbance in tilled fields
Roof runoff structure	558	A structure to collect and convey precipitation runoff from a roof
Sprinkler system	442	A system that applies water by means of nozzles operated under pressure
Tailwater recovery	447	A system to store and convey irrigation tailwater and/or rainfall runoff for reuse
Terrace	600	An earth embankment constructed across the field slope
Tree/shrub establishment	612	Establishing woody plants
Tree/shrub pruning	660	The removal of selected lower branches from trees
Tree/shrub site preparation	490	Treatment to improve conditions for establishing trees and/or shrubs

(continued)

Table 18.2 (continued)

Practice name	NRCS code	Definition
Underground outlet	620	An underground conduit to convey surface water to a suitable outlet
Waste storage facility	313	A pond or building to store manure
Watering facility	614	A means of providing drinking water to livestock or wildlife
Wildlife habitat planting	420	Establishing wildlife habitat by planting herbaceous vegetation or shrubs
Windbreak	380	Single or multiple rows of trees or shrubs in linear configurations
Woody residue treatment	384	Burning, chipping, lop/scatter, or removal of woody plant residues

for manure, prescribed grazing, and irrigation water management (Oregon Department of Agriculture 2020a).

As described in Chap. 17, the Federal Clean Water Act requires National Pollutant Discharge Elimination System permits for CAFOs. In Oregon, ODA administers those permits, and the similar Water Pollution Control Facilities permits. For either type of permit, ODA requires CAFOs to submit an animal waste management plan for their operation, maintain records, and allow regular inspections by ODA staff. The management plans and records include soil maps and soil test results for nitrate nitrogen (NO_3-N), phosphorus (P), potassium (K), pH, and other parameters. The plans and records are designed to safeguard both the storage of animal waste and the application of that waste to cropland and pastures (Soil Survey Staff 2019; Oregon Department of Agriculture 2020b; Oregon Secretary of State 2009). Common practices in animal waste management plans include nutrient management, waste storage facilities, and roof runoff structures (i.e., gutters and downspouts).

The Oregon Forest Practices Act was passed in 1971 and was the first such act in the nation. The Act seeks to protect soil, air, fish, wildlife, and water quality on non-federal forestland in the state. The Oregon Board of Forestry, a seven-member citizen board appointed by the governor and confirmed by the state senate, adopts and revises rules under the Act, and the Oregon Department of Forestry administers the rules. Landowners, operators, or their agents must notify the Oregon Department of Forestry at least 15 days prior to beginning forestry operations such as road construction, thinning, harvesting, applying chemicals, and preparing sites for reforestation. The notification includes what will be done, where the operation will take place, and who will do the work.

An important part of the Oregon Forest Practices Act is the designation of mandatory buffers along streams, wetlands, and lakes where forest practices are either restricted or prohibited altogether. Similarly, the Act provides for buffers around sensitive wildlife sites, for example, nesting trees used by bald eagles (*Haliaeetus leucocephalus*), osprey (*Pandion haliaetus*), northern spotted owls (*Strix occidentalis caurina*), or great-blue herons (*Arsea herodias*).

Soil-related concerns addressed by the Oregon Forest Practices Act include compaction, erosion, wetland degradation, hazardous waste spills, and pesticide and fertilizer application. The act prohibits soil disturbance that could significantly impact water quality or vegetation productivity, and it encourages returning slash (tree tops, limbs, and unmarketable wood remaining after harvest) to the soil to recycle nutrients and organic matter. The Act recognizes the linkage between soils and forest site classes,[13] erodibility, and logging road stability. It provides different requirements based on site class, size of harvest, type of harvest, steepness of slope, and proximity to surface water features. Oregon Forest Practices Act rules also designate special considerations for the Tyee core area, a region prone to landslides due to shallow soils overlying relatively impermeable sandstone in parts of Douglas, Lane, and Coos Counties (Cloughesy and Woodward 2018). Common practices used by foresters to ensure compliance with the Act include tree site preparation, tree establishment, forest stand improvement, and forest trails and landings.

Irrigated farms, municipal water systems, and some other significant consumers of water must obtain a water right from the Oregon Water Resources Department (OWRD) to use surface water or groundwater. Like most states west of the Mississippi River, Oregon water laws (first adopted in 1909) are based on the doctrine of prior appropriation, which means the oldest water right on a stream is the last to be denied water in times of low streamflow. In addition to prior

[13] Forest site class is a measure of how well trees grow on a specific soil. Forest site indices provided in Web Soil Survey reports can be converted to forest site classes through the use of tables published by the Oregon Department of Forestry.

appropriation, there are three other fundamental aspects of Oregon water law: (1) water may only be used for beneficial purposes enumerated by the state; (2) irrigation water rights are attached to specific pieces of land and are transferred with the land when it is sold; and (3) water rights not used at least once every 5 years are subject to cancellation by the state.

In order to obtain a water right to irrigate land, applicants follow a three-step process. They first apply to OWRD for a permit that designates the exact place from which water will be diverted (i.e., taken from a stream or pumped from groundwater) and the exact piece of land to be irrigated. If OWRD approves the permit, the applicant must then develop their irrigation system within 5 years and hire a certified water right examiner to determine if the provisions of the permit are being met. In the final step, if permit provisions have been met, examiners file a "claim of beneficial use" detailing their findings, and the OWRD issues a water rights certificate. In addition to designating the point of diversion and land to be irrigated, the certificate states the maximum flow (typically measured in cubic feet per second (cfs) per acre)[14] and often states the maximum annual amount (typically measured in acre-feet per acre) of water that may be applied. For example, a water right providing up to 1/80 cfs per acre and 2.5 acre-feet per acre on 40.0 acres of land allows irrigation up to 0.5 cfs at any one time and up to 100 acre-feet for the irrigation season, if applied to the entire 40.0 acres.

Oregon state agencies charged with protecting fish and wildlife hold "instream water rights" that establish flow levels to remain in stream reaches in order to protect fish, dilute pollutants, or provide for recreation. OWRD has established a program through which irrigators may lease their water rights for instream use, designating that water they won't use for irrigation (typically for a period of 5 years) will be kept in the stream. The Oregon legislature and OWRD have closed or restricted the issuance of new water rights for some surface water and groundwater sources to ensure adequate water supplies for existing water rights, including water reserved for in-stream flow. In most parts of the state, water rights to irrigate from streams during the summer are no longer available. Examples of conservation practices (Table 18.2) associated with irrigation in Oregon include irrigation water management, micro-irrigation, and low energy precision application sprinkler systems (Oregon Water Resources Department 2018; NRCS 2016).

[14] Note: in those parts of this chapter that describe legal processes codified in US customary units (for example, feet and acres), this chapter will use those legally specified units rather than the metric equivalents.

18.3.3 Federal Regulations

Federal laws such as the Clean Water Act (CWA) and the Endangered Species Act (ESA) are critically important for conservation. However, most CWA provisions that affect Oregon working lands have been delegated to state agencies, as for example with the CAFO permit program. ESA provisions are often tied to programs administered by government agencies, such as logging on federal forests or permits administered by the Army Corps of Engineers. With over 50 species of plants and animals listed as threatened or endangered in Oregon (Ballotpedia 2016), each with specific conservation needs, ESA provisions are beyond the scope of this volume. Instead, this section will focus on the conservation compliance provisions of federal farm bills, which have widespread applicability to the management of Oregon cropland soils.

The Food Security Act of 1985 ("1985 farm bill") included highly erodible land and wetland conservation compliance provisions, also known respectively as "sodbuster" and "swampbuster" rules. The rules have been modified every 5–6 years with successive farm bills, but have consistently linked eligibility for a wide range of USDA program benefits to minimum levels of conservation on highly erodible land (HEL) and to avoid the conversion of wetlands to annual crops. Although farmers may elect not to comply with these conservation provisions, some might consider the provisions essentially regulatory if they perceive USDA program benefits as essential to their farming operation (Stubbs 2012).

HEL with respect to water erosion is determined with Universal Soil Loss Equation (USLE)[15] factors. The USLE is an empirically based equation that estimates sheet and rill erosion[16] based on the product of five factors (Wischmeier and Smith 1978):

$$A = R(K_w)(LS)CP$$

where:

A average annual soil loss (tons ac^{-1} yr^{-1}).

R a factor to account for the average energy and intensity of rainfall (unitless).

[15] USLE is no longer widely used but was the standard USDA erosion prediction methodology in 1985 when the conservation compliance provisions were first included in the federal farm bill. Current erosion prediction methods are the Revised Universal Soil Loss Equation (RUSLE2), the Integrated Erosion Tool (IET), and the Wind Erosion Prediction System (WEPS) (NRCS, n.d.).

[16] Sheet erosion is the detachment and movement of soil from the land surface by the shallow sheet flow of runoff. Rill erosion is the detachment and movement of soil by the concentration of runoff in small channels less than 10 cm deep, which are obliterated by tillage, and which typically form in different locations from year to year (Grigar et al. 2020).

K_w a factor to account for the erodibility of soil due to water (tons ac^{-1} yr^{-1}).
LS a factor to account for the length and steepness of a slope (unitless).
C a factor to account for cover and management (unitless).
P a factor to account for supporting practices like terraces or contour farming (unitless).

These factors are available online via the NRCS Field Office Technical Guide (NRCS 2020a). Since HEL determinations apply to the soil regardless of management, the C_w and P factors do not apply in deciding which soils are HEL. The HEL calculation for water erosion is:

$$EI_{water} = RK_w(LS)/T$$

where:

EI_{water} the erodibility index for water. Any soil with EI_{water} greater than or equal to 8.0 is considered HEL, and
R, K_w, and LS are as defined above
T is the soil loss tolerance in tons ac^{-1} yr^{-1}. For the purpose of HEL calculations, each soil's T factor is available in the Field Office Technical Guide. Erosion at rates equal to or less than T are considered tolerable, and T values vary from 1 to 5 tons ac^{-1} yr^{-1}, depending on the thickness of the topsoil.

Wind erosion HEL determinations are based on factors from the Wind Erosion Equation (WEQ) (Woodruff and Siddoway 1965):

$$E = f(IKCLV)$$

where:

E average annual soil loss (tons ac^{-1} yr^{-1}).
f is a non-linear function, originally calculated with nomograms.
I a factor to account for the erodibility of soil due to wind (tons ac^{-1} yr^{-1}).
K a factor to account for the roughness of the soil surface (unitless).
C a factor to account for the erosive potential of wind speed and surface moisture at a specific site (unitless).
L a factor to account for the unsheltered distance across a specific field (unitless).
V a factor to account for vegetative cover on a specific field (unitless).

These factors are available via the Field Office Technical Guide. As with HEL due to water, the management factors (K, L, and V) are not used in calculating HEL due to wind, resulting in a wind erodibility index (EI_{wind}):

$$EL_{\mathrm{wind}} = CI/T$$

As with water, EI_{wind} values greater than or equal to 8.0 are HEL for wind. NRCS has determined HEL status for all Oregon soil map units commonly used for cropland. The USDA Farm Service Agency determines field boundaries for all cropland producers who participate in USDA programs. NRCS determines the erodibility of a field by calculating the area of HEL soils in that field. If HEL area is 33.33% or more of a field, or if the HEL soils equal 50.0 acres or more, then that field is considered HEL. In order to maintain eligibility for most USDA programs, farmers must implement a conservation system that controls erosion to acceptable levels[17] on all of their HEL fields where they plant annual crops[18] (NRCS 2015). Common Oregon practices to comply with the HEL conservation compliance provisions include crop rotation, no-till, reduced till, and terraces.

The wetland conservation compliance rules require USDA program participants to notify USDA prior to manipulating land by such practices as draining, filling, or removing trees for the purpose of making a site capable of producing annual crops. Qualified NRCS employees, after receiving a landowner's permission, visit the site proposed for manipulation and make a determination of USDA program eligibility, which typically includes a certified wetland determination. Wetlands, for the purposes of conservation compliance, exhibit each of three characteristics. First, there must be a predominance of hydric[19] soils. Second, the site must exhibit wetland hydrology.[20] Third, under normal

[17] The criteria for "acceptable levels" of erosion depend on how long the field has produced annual crops, when initial conservation plans were approved, and other factors. NRCS evaluates conservation systems with erosion prediction model results compared to the soil loss tolerance value (T) of the soils in each evaluated field. Farmers must also control gully erosion, which is measured in the field rather than predicted with models, in order to maintain eligibility for USDA programs.

[18] Conservation compliance rules use the term "agricultural commodity" to designate annually planted crops. Conservation compliance rules do not pertain to perennial crops like alfalfa, blueberries, or orchards.

[19] Section 514.4 of the National Food Security Act Manual, 5th edition (NFSAM) defines hydric soils as "soils that, in an undrained condition, are saturated, flooded, or ponded long enough during a growing season to develop an anaerobic condition that supports the growth and regeneration of hydrophytic vegetation" (NRCS 2015). A list of hydric soils can be found in Section II of the NRCS Field Office Technical Guide and on WSS.

[20] Section 514.6 of the NFSAM defines wetland hydrology as "inundation or saturation by surface or groundwater at a frequency and duration sufficient to support a prevalence of hydrophytic vegetation typically adapted for life in saturated soil conditions" (Ibid.).

circumstances, the site must support a prevalence of hydrophytic[21] vegetation.

The certified wetland determination includes delineating the outline of wetlands and identifying what type of exemptions, if any, apply. Each wetland exemption type is identified by an abbreviation and has a unique set of allowable management activities assigned to it. Perhaps the simplest types are wetland (W) or non-wetland (NW). Other types include manipulated wetlands (WX), artificial wetland (AW), prior-converted cropland wetland (PC), farmed wetland (FW), and farmed wetland pasture or hayland (FWP). If the certified wetland determination finds that the site was a wetland and was converted after 1985 for the purpose of making the production of an agricultural commodity possible, the site is labeled as a converted wetland (CW)[22] and the landowner is subject to losing eligibility for most USDA programs. There are several avenues for landowners to maintain eligibility after converting wetlands, including appeals of determinations, NRCS declaring the conversion as having minimal effect, and the landowner mitigating for the lost functions and values of the converted wetland (Ibid.).

18.3.4 Local, State, and Federal Funding for Conservation

Managers of working lands have multiple opportunities to receive financial and technical help to implement conservation practices. NRCS programs include the Environmental Quality Incentives Program, which helps farmers, ranchers, and foresters implement one or more practices on their land with contracts awarded via a competitive application process. Funding in this program usually ranges between 50 and 90% of estimated costs, with higher payments for historically underserved populations, beginning farmers, and organic farmers. In the Agricultural Conservation Easement Program, farmers can sell an easement[23] on some or all of their land in exchange for preserving the property as working lands or for restoring wetlands (NRCS 2021a). The Conservation Reserve Program and Conservation Reserve Enhancement Program are administered by the USDA Farm Service Agency, and entail contracts of 10–15 years in which farmers convert working lands to perennial vegetation (trees, shrubs, or perennial herbaceous plants) in exchange for annual payments and partial funding to establish the required vegetation (USDA-Farm Service Agency n.d.). USDA programs tend to change somewhat with each new federal farm bill.

The Oregon Watershed Enhancement Board provides grants that help protect working lands (Oregon Watershed Enhancement Board n.d.). Often these grants are administered locally by Soil and Water Conservation Districts or watershed councils, some of which have their own funding for conservation practices. There may be potential for managers of working lands to sell carbon credits in markets established to reduce greenhouse gases, but there has not been widespread use of that funding mechanism in Oregon to date.

18.3.5 Practices

This chapter mentions practices that conserve resources, especially soil resources, on working lands. NRCS provides a useful catalog of conservation practices in its Field Office Technical Guide, including a unique code number for each practice, a practice standard, and specifications or implementation requirements to tailor practices to specific situations. Table 18.2 lists NRCS conservation practices commonly used in Oregon.

18.3.6 Soil Health

Karlen et al. (1997, p. 6) defined soil health, also known as soil quality, as “the capacity of a specific kind of soil to function, within natural or managed ecosystem boundaries, to sustain plant and animal productivity, maintain or enhance water and air quality, and support human health and habitation.” Various other definitions exist, but all stress the importance of soil function, the relationship between soil and other aspects of ecosystems such as water and air, and the interaction between soil physical, chemical, and biological components (Fig. 18.1).

The concept of healthy soil function is not new. The ancient Greek philosopher Plato commented on it, Darwin (1881) alluded to it, and agricultural research in the early twentieth century—for example Wallace (1910) and Yoder (1937)—began to explore the processes behind it (Karlen et al. 2019). Warkentin and Fletcher (1977) used the term “soil quality” in the 1970s, and other works on the subject began to be published in the 1990s, by authors such as

[21] Section 514.5 of the NFSAM defines hydrophytic vegetation as a plant growing in water or a substrate that is at least periodically deficient in oxygen during a growing season as a result of excessive water content (Ibid.).

[22] Wetlands converted between 1985 and 1990 are labeled “CW” and wetlands converted after 1990 are labeled with “CW” plus the year of conversion. For example, a wetland converted in 2015 would be labeled “CW+2015”.

[23] Easements convey specific property rights from the seller to the buyer. Easements are usually conveyed legally via an attachment to the deed. Easements established via the Agricultural Conservation Easement Program either legally restrict the landowner from developing the land or allow USDA to restore wetlands on the property.

Fig. 18.1 A healthy soil with stable aggregates and an earthworm. The small, pinkish-white, kidney shaped bumps on the roots of this field pea cover crop are nodules where *Rhizobium* bacteria fix nitrogen in a symbiotic relationship with the plant. *Source* Photograph by NRCS Oregon

Larson and Pierce (1991), Haberern (1992), and Doran, et al. (1994). Soon thereafter, Acton and Gregorich (1995) used the term "soil health." Whichever term is used, the concept has generated much interest in both academia and the public, including scientific papers, USDA and land grant university programs, popular books, and even widely viewed documentaries.

NRCS and OSU describe the principles for improving or maintaining soil health: keep the soil covered, avoid disturbing the soil, keep living roots growing, and maximize plant diversity[24] (Duyck et al. 2015). Keeping the soil covered with plant residues, for example through the use of mulching, reduces evapotranspiration and moderates soil temperatures. Avoiding soil disturbance, for example with no-till, preserves soil structure and habitat for soil biota. Cover crops can keep living roots growing in the soil, which provides a food source for soil organisms and adds organic matter to the soil. Plant diversity can be maximized by practicing long-term crop rotations that employ a range of annual and perennial crops, grasses and legumes, and cool and warm season plants. Prescribed grazing is a way to integrate livestock into working lands, which can help increase nutrient cycling and also allow perennial pasture to be rotated with annual crops.

NRCS, land grant universities, and private soil testing firms are exploring a variety of ways to measure soil health, with the hope that these methods will help guide management decisions on working lands. Just as the concept of soil health involves the interrelationship of soil biology, chemistry, and physics, so too do the analytical methods now being refined. The OSU Soil Health Lab (2021d) offers a variety of soil health analyses, either singly or as a package. For example, soil respiration analysis measures the amount of carbon dioxide generated under controlled conditions to assess soil biological function. Measuring soil pH is a proven and relatively easy method to gain information on one of the chemical properties of soil. Water stable aggregate analysis measures the physical component of soil structure. There are various methods, such as those developed by Cornell University, to integrate soil health analyses into one

[24] Some sources also include integrating livestock into cropland systems, either by rotating pasture with crops or by grazing crop residues after harvest.

or more indices of soil health to help managers make sense of testing results (Moebius-Clune et al. 2016).

Insights into soil health are also available from models and the WSS. The RUSLE2 model, in addition to estimating sheet and rill erosion, also provides a Soil Conditioning Index that integrates estimated erosion, the intensity of field operations, and the balance of soil organic matter as a means of predicting soil health impacts of cropland management systems (Campbell 2000; NRCS n.d.). The WSS provides a variety of reports under the category of soil health, including limitations for siting composting facilities, a fragile soil index that indicates a soil's susceptibility to degradation, a risk rating for organic matter depletion, a susceptibility to compaction rating, and others (Soil Survey Staff 2019).

18.4 Cropland Management Systems

Chapter 17 presented economic data of Oregon crops. Table 18.3 lists Oregon crops with 2017 production of over 90 km^2, and the following subsections discuss soils and common production practices suitable to each of these crops. The USDA Conservation Reserve Program (CRP) is not a crop but is included as a sub-section because it is applied to cropland. Small acreage vegetable production is included as a sub-section because it reflects the increasing trend to small farms raising fresh local produce. There are many Oregon crops that cannot be included here due to limited space. Examples include mint, dry legumes, hops, sugar beets, and other crops, which can be locally important and add diversity to Oregon working lands.

18.4.1 Hay and Haylage

Hay[25] is the most widely grown commodity on Oregon cropland, with about 4,350 km^2 reported in production in 2017. Every Oregon County produces hay, but over 75% of Oregon's hayland area is east of the Cascade crest, where dry spring and summer weather provides excellent growing conditions for this crop (NASS 2019). Hay may be grown, stored, and then fed to livestock on the same farm, or it may be sold to other farms. For example, wet weather conditions on the Oregon coast prevent dairy farmers in Tillamook County from raising high-quality alfalfa hay, so those dairies typically buy hay from eastern Oregon (Tillamook County Soil and Water Conservation District 2010). Hay is also exported overseas, especially to China, Japan, Saudi Arabia, the United Arab Emirates, and Korea; these export markets generally require brokers, quality certifications, and the use of metric units to measure quantities (Oregon Hay Products Inc. 2020; Rankin 2018).

Farmers grow various species for hay, depending on soil, market demand, the availability of irrigation, and climate. Perennial legume hay species include alfalfa (*Medicago sativa*), red clover (*Trifolium pratense*), white clover (Trifolium repens), and big trefoil (*Lotus pedunculatus*). Perennial grass hay species include orchardgrass (*Dactylis glomerata*), Timothy (*Phleum pratense*), tall fescue (*Schedonorus arundinaceus*), Reed canarygrass (*Phalaris arundinacea*), and meadow foxtail (*Alopecurus pratensis*). Annual grains such as oats (*Avena sativa*) and wheat (*Triticum aestivum*) are sometimes harvested as hay prior to grain formation.

Alfalfa, red clover, and orchardgrass require well-drained soils. For example, Deter loam with 0–5% slopes occurs in Lake County, a major producer of Oregon hay; it is rated as "high" for irrigated alfalfa yields,[26] is classified as a well-drained Argixeroll, and is considered prime farmland if irrigated. Timothy, white clover, and big trefoil grow well in somewhat poorly drained soil, while Reed canarygrass[27] and meadow foxtail can grow in very poorly drained soils. Tall fescue is adapted to a wide range of soils, except those that are very poorly drained (Fransen and Chaney 2002). Annual grain hay crops are adapted to a wide range of cropland soils, including floodplains if sown after the winter flood season.

Farmers generally try to keep an established hay field in production for as long as possible in order to avoid the costs of re-establishment, which can include herbicide, tillage, and the loss of production in the seeding year. When perennial hay fields no longer meet the farmer's needs, they are usually tilled and planted to one or more years of annual crops, providing a rotation that helps control weeds, diseases, and

[25] Hay and haylage are grown for livestock feed, predominantly cattle, sheep, and horses. Hay is forage that is cut in the field, allowed to dry to a stable moisture content (typically 15–20% or drier), and then stored in bales of various sizes and shapes (Bohle et al. 2020). Haylage is harvested at a higher moisture content (typically 65–75% moisture) and then stored anaerobically in silos, plastic bales, or plastic tubes, where fermentation forms organic acids that preserve it (Fransen 2020). NASS reports combine hay and haylage into a category called "forage" and subsequently this section will generally use the term "hay" to refer to both methods of storing forage. Data and discussion in this section do not include corn (*Zea mays*) and sorghum (*Sorghum bicolor*) that is made into silage.

[26] Web Soil Survey for Oregon provides crop productivity indices for irrigated alfalfa, irrigated grass, and non-irrigated grass hay. The indices have five categories ranging from low to high and are based on soil physical properties, soil chemical properties, climate, landscape, and (for the non-irrigated grass hay rating) soil water (Soil Survey Staff 2019).

[27] Reed canarygrass forms dense long-lived monocultures in poorly drained soils and can prove difficult to eradicate. Although not classified as a noxious weed, Oregon State University discourages seeding this invasive species at sites where it does not already occur.

Table 18.3 Cropland area for selected Oregon crops. Marijuana data not available for 1997 or 2017, but licenses totaled about 2 km^2 approved for production as of March 2021. *Sources* Jones (2021), Oregon Liquor Control Commission (2021), NASS (2021), NRCS (2017)

Crop	Cropland area (km^2)	
	1997	2017
Hay and haylage	4,496	4,349
Grains	4,378	3,645
CRP[1]	1,954	1,747
Seed crops	2,213	1,707
Vegetables	644	605
Orchards[2]	376	443
Corn	206	367
Christmas trees[3]	275	183
Nurseries[4]	n/a	139
Hemp[5]	n/a	111
Berries[6]	99	109
Grapes[2]	42	97

n/a = not available
[1]CRP is the USDA Conservation Reserve Program
[2]Includes fields being harvested plus those not yet producing
[3]Includes all Christmas trees in production (not just harvested area)
[4]Includes greenhouse plus production in the open
[5]Hemp datum is area registered for production and does not include marijuana. Actual area planted may be less
[6]Berry data for 1997 not available. Value shown in 1997 column is for 2007

insects. Perennial hay species can be seeded in the spring or fall, depending on the species, the local climate, and the soil.

Most, but not all, perennial hay crops in Oregon are irrigated, and estimated average net irrigation[28] requirements for high-yielding alfalfa fields are around 440 mm per year in Harney and Lake Counties and 820 mm per year in Malheur County (Cuenca 1999). Fields being seeded to alfalfa or grass may require applications of lime, P, K, boron (B), and other nutrients based on soil test results. After establishment, alfalfa requires no additional nitrogen (N) fertilizer, but some growers will apply up to 45 kg N ha^{-1} prior to seeding. Established grass hay generally receives 50–225 kg N ha^{-1} each year, depending on yield. Both grass and alfalfa hay often require applications of K and S after the establishment year (Hannaway et al. 2019; Shewmaker and Bohle 2010).

Because most hay crops are relatively long-lived perennial species, they tend to provide excellent soil cover, increase soil organic matter, and have few erosion issues. Probably, the most important conservation practice for hay crops is irrigation water management for alfalfa.

18.4.2 Grain Crops

In 2017, about 3,650 km^2 of wheat, barley, oats, and rye were grown for grain in Oregon. Over 80% of the Oregon land area devoted to these crops was in Umatilla, Morrow, Sherman, Gilliam, and Wasco Counties in the Columbia Plateau and Palouse Prairie. Soil and climate for non-irrigated grain production in these MLRAs vary from those that can support annual cropping (for example, Athena silt loam, a Pachic Haploxeroll with 430–560 mm mean annual precipitation) to those that are usually farmed in a wheat—fallow rotation, where 2 years of precipitation produce one crop (for example, Cantala silt loam, a Typic Haploxeroll with 300–360 mm mean annual precipitation).

Wheat (*Triticum aestivum*) is by far the most common Oregon grain crop, representing over 95% of the 2017 land area in grain production. Early Euro-American settlers brought wheat seed with them to Oregon in the nineteenth century, and much of that wheat was then seeded in the

[28] Net irrigation requirement is the difference between a crop's water requirement and the effective precipitation during the growing season. Net irrigation requirement does not account for additional water required due to irrigation system inefficiencies. However, actual application rates are often lower than "requirements" because sufficient irrigation water is unavailable, water rights are limiting, irrigation systems are inadequate, or farmers delay irrigation while crops utilize soil moisture stored from winter precipitation. Values presented here are based on average growing season weather data for 27 regions across Oregon, and thus pertain to specific crops in specific areas. That is, net irrigation requirements for alfalfa hay in Harney County are different from requirements in Malheur County. Weather data are all from twentieth century records, which may result in estimates that are too low as climates become warmer. In any case, the net irrigation requirement data presented in this chapter is best used to compare relative water needs between different crops or different parts of the state rather than to predict actual amounts of water applied.

Fig. 18.2 The Columbia Plateau in April. Soils are Walla Walla silt loam and Anderly silt loam, which are both Typic Haploxerolls. The winter wheat growing on the left side of this photo was seeded last fall and will be harvested this summer after it forms grain. The golden-colored cropland in the distance was wheat harvested last summer, is now beginning a period of summer fallow, and will probably be seeded to wheat in the coming fall. The excellent stand of perennial bunchgrass to the right may have been enrolled in the Conservation Reserve Program. Wind turbines are a common sight in the Columbia Plateau. *Source* Photograph by NRCS Oregon

Willamette Valley because transportation costs were too high to ship wheat from eastern Oregon to Portland. Railroad access, beginning in the 1880s, provided more economical shipping and resulted in expanded wheat production east of the Cascade crest. Dams and locks built on the Columbia River in the 1950s allowed grain shipments by barge. Most Oregon wheat is soft white winter wheat, which is milled into relatively low protein (8.5–10.5%) flour, suitable for use in crackers, pastries, flatbreads, and noodles. Currently, over 90% of Oregon's wheat crop is exported, primarily to countries in Asia (Schillinger and Papendick 2008; USDA-Economic Research Service 2020a; Weaver 2020).

Production systems for wheat vary by soil and climate. In the Columbia Plateau and Palouse Prairie, crop rotations are limited due to low precipitation rates. Where annual cropping is possible, winter wheat can be rotated with spring barley (*Hordeum vulgare*), spring wheat, peas (*Pisum sativum*), lentils (*Lens culinaris*), chickpeas (*Cicer arietinum*), canola (*Brassica napus*), and condiment mustard (*Brassica nigra or Sinapis alba*). In drier locations (generally less than 300 mm mean annual precipitation), winter wheat is rotated with summer fallow, as in Fig. 18.2. In some locations, a hybrid of these systems is used, with a rotation such as winter wheat—chickpeas—summer fallow (Prakriti et al. 2017). In western Oregon and on irrigated fields, grain crops are grown in annual rotations with a variety of other crops. One reason western Oregon farmers include wheat in their crop rotations is to break up pest cycles. In most parts of Oregon, winter wheat is not irrigated, but the estimated average net annual irrigation requirement for winter wheat grown in the Columbia Basin is 530 mm (Cuenca 1999).

Nutrient management for grains, like production systems in general, varies between eastern and western Oregon. In eastern Oregon, soil tests determine the appropriate levels of N, P, and S to apply. Nitrogen application rates vary according to soil tests for mineral[29] N, expected yield, whether excessive straw remains from the previous crop, and

[29] The term "mineral N" used in this chapter means the total elemental N contained in ammonium (NH_4^+) plus nitrate (NO_3^-).

(in annual cropping systems) whether the previous crop was a legume. With those considerations, N fertilizer rates in non-irrigated eastern Oregon wheat are generally in the 50–125 kg N ha^{-1} range. Potassium and lime applications to non-irrigated eastern Oregon wheat generally are not needed (Lutcher et al. 2007; Wysocki, et al. 2007).

In western Oregon, soil tests determine appropriate levels of P, K, and lime. Unlike eastern Oregon, fall soil tests are not used for mineral N in western Oregon because soil NO_3-N in high precipitation zones is ephemeral; much of the soil NO_3-N that was present in the fall leaches below the root zone before the wheat crop can take it up in the spring. However, a relatively new "N-min" test estimates mineralization of soil organic N for western Oregon winter wheat crops. With the N-min test, soil samples are collected in January, incubated anaerobically in a lab, and then analyzed for the N mineralized by soil microbes. The results indicate how much soil organic N will be converted into mineral N available to the wheat and thus help determine the appropriate amount of N fertilizer to apply in the following months.

N-min test results vary, depending on such factors as whether the previous crop in the rotation was a legume. Minimum N applications to winter wheat in western Oregon are 90 kg N ha^{-1}, but usually N-Min test results indicate higher rates are needed. Fertilizer rates for P, K, calcium (Ca), magnesium (Mg), and lime are also determined by soil test results. Sulfur (S) is usually applied at rates up to 35 kg S ha^{-1} (Hart et al. 2011a). Nitrogen fertilizer application rates for other grains, such as oats and barley, are typically less than for wheat in both eastern and western Oregon.

The most commonly used conservation practices for Oregon grain crops are no-till, reduced till, and nutrient management. In eastern Oregon, contour farming and terraces are sometimes used to control erosion. In western Oregon, crop rotation and underground outlets can be used for erosion control. Grain growers often participate in USDA programs, and grain is often grown on steeper soils, so erosion control practices are often necessary to comply with the HEL conservation compliance provisions of the federal farm bill.

18.4.3 Conservation Reserve Program

The Conservation Reserve Program (CRP) began with the 1985 federal farm bill, the same bill that established conservation compliance provisions. CRP has been included in every subsequent farm bill, although changes have been made over time. One such change is the 1998 addition of a CRP category called the "Conservation Reserve Enhancement Program," through which farmers can establish riparian forest buffers and similar projects. CRP is administered by the USDA Farm Service Agency, with technical help from NRCS. The goals of the program are to conserve soil, protect water quality, and provide habitat on environmentally sensitive land, especially soils prone to erosion.

Land enrolled in CRP must be planted to perennial vegetation, generally grass and/or trees (Fig. 18.2). Farmers enroll land in CRP via 10-to-15-year contracts with USDA, although there are provisions for renewing contracts. Generally, CRP contracts prohibit planting crops, grazing, or haying the land in return for cost-sharing to establish the perennial cover plus an annual rental payment for the land. When a CRP contract ends, the landowner can apply to either renew the contract, return the land to crop production,[30] begin grazing the land, or maintain the land as wildlife habitat (National Sustainable Agriculture Coalition 2019; USDA-Farm Service Agency, 2021).

Almost 90% of Oregon's CRP land is in Umatilla, Morrow, Sherman, Gilliam, and Wasco Counties, primarily in the Columbia Plateau (Environmental Working Group 2021). A typical soil enrolled in CRP in this MLRA is Condon silt loam, a Typic Haploxeroll with a mean annual precipitation of 305–355 mm. Condon soil map units are highly erodible for water erosion and have a soil loss tolerance (T) value of 2 tons ac^{-1} yr^{-1}. When farmed, Condon silt loam is typically in a wheat-fallow rotation.

In 1995, McLeod et al. (1996) conducted a survey in the five counties listed above and received responses from 75% of the CRP enrollees. The median size of respondents' operations was about 800 ha and the median area enrolled in CRP was about 200 ha. Most (78%) of the respondents employed a wheat-fallow rotation on the land prior to CRP enrollment, and their reasons for enrolling in CRP included soil protection (84% of respondents), stable income generation (77%), low crop yields or poor soil (70%), wildlife habitat improvement (62%), and watershed protection (51%). Fewer respondents (12–17%) enrolled in CRP because they wanted to get out of farming or retire. Because wheat-fallow systems only produce a crop once every 2 years, the annual CRP rental payment was a strong incentive to enroll. Most (65%) of the respondents listed a return to wheat-fallow cropping as a potential option when their CRP contracts ended, while around 40% thought leaving the land in grass and managing it for grazing or hay was a likely option.[31]

Conservation practices for CRP may include conservation cover, tree/shrub site preparation, tree/shrub establishment, and herbaceous weed treatment. Habitat management and

[30] If the CRP land is highly erodible, farmers will need to use approved conservation practices to control erosion in order to maintain eligibility for USDA programs under the conservation compliance rules.

[31] The survey responses exceed 100% because farmers could choose more than one option.

associated practices can be employed to improve habitat on CRP land, which, in Oregon, is not fertilized and generally not irrigated.

18.4.4 Seed Crops

In 2017, about 1,710 km^2 of seed crops were grown in Oregon. Over 90% of the Oregon land area devoted to these crops was in Linn, Marion, Yamhill, Polk, Benton, Lane, and Washington Counties, all in the Willamette Valley (Fig. 18.3). Soils used for seed crops in this MLRA can be divided into two broad categories. Moderately well-drained soils (for example Woodburn silt loam, an Aquultic Argixeroll), generally found in the northern Willamette Valley, are suitable for clover and perennial grass seed crops. Poorly drained soils, such as Dayton silt loam, a Typic Albaqualf generally found in the southern Willamette Valley, are used for annual ryegrass seed crops. Jefferson County raised about 68 km^2 of seed crops in 2017 and produces most of the nation's hybrid carrot seed on irrigated soils such as Madras loam, an Aridic Argixeroll (Oregon State University 2021a).

In 2017, about 80% of Oregon seed crop hectares produced ryegrass or fescue seed. Several species of ryegrass are grown for seed, including annual, intermediate, and perennial (respectively *Lolium multiflorum*, *L. hybridum*, and *L. perenne*). The most commonly grown fescue is tall fescue (*Schedonorus arundinaceus*), but "fine fescue" species are also grown for seed (e.g., *Festuca rubra*, *F. brevipila*). Bentgrass (*Agrostis* spp.), Kentucky bluegrass (*Poa pratensis*), and orchardgrass (*Dactylis glomerata*) seed is also grown in Oregon. All of these are cool season grasses. Red clover (*Trifolium pratense*) and crimson clover (*T. incarnatum*) are often grown for seed in rotation with tall fescue and perennial ryegrass on better-drained soils (NASS 2019).

Oregon produces almost two-thirds of the cool season grass seed in the United States. Some Oregon grass seed varieties are sold for lawns, some for forage, and some for use as cover crops. Red clover is mostly marketed for pasture seedings, while crimson clover is used as cover crops. Most of Oregon's grass and clover seed is shipped out of state, with about 15–20% of the production shipped overseas (Oregon Seed Council 2016).

The Willamette Valley's cool, wet winters and warm dry summers are nearly ideal for seed production and grass seed

Fig. 18.3 Most of Oregon's seed crop is produced in Linn, Marion, Yamhill, Polk, Benton, Lane, and Washington Counties in the Willamette Valley. *Source* Photograph by Dean Moberg

production began there in the 1920s (Oregon State University 2021b). In the northern Willamette Valley, grass and clover seed are grown in rotation on non-irrigated farms with wheat and oats. Annual ryegrass is often grown continuously without a rotation (Chastain et al. 2017). In the 1940s, many grass seed growers began burning their fields after harvest as an inexpensive means to remove excess straw, control weeds, and reduce plant pests. Concern over the health effects of smoke arose in the 1960s and temporary burn bans in Willamette Valley grass seed fields were instituted. Then, in 1988, smoke from grass seed burning drifted onto Highway 5 and a multi-car crash resulted in death and injuries, leading to strict permanent burning limits. In 2009, Oregon passed legislation essentially phasing out field burning of grass seed in the Willamette Valley (Giombolini 2018).

There are some interesting connections between seed farming and livestock production. Sheep (*Ovis aries*) are commonly trucked from southern Oregon farms to graze Willamette Valley annual ryegrass fields during late winter and early spring, and then are trucked back south to allow the ryegrass crop to mature and produce seed in the summer (Young et al. 1996). Red clover is typically harvested as forage and sold to dairy farms for ensilage in late May or early June. This delays the bloom so that when the clover grows back the flowers bloom at the same time, allowing for a late summer seed harvest (Oregon Clover Commission 2020). Some farmers bale straw remaining after the grass seed harvest; the straw can be sold for livestock bedding or shipped overseas—primarily to Japan—where it is used as a component of cattle feed. Harvesting straw developed at least in part to prohibitions against burning residue (Hart et al. 2012).

Similar to hay crops, perennial grass and legume seed crops tend to have few erosion problems after establishment. However, fall-planted grass seed crops typically employ a weed control technique in which activated carbon is applied in a 2.5 cm band directly over the drill rows, followed by a broadcast pre-emergence[32] application of herbicide to control weeds. The activated carbon protects the germinating grass seed, but the herbicide effectively controls most other weeds between drill rows (Hulting 2019). This clever method, however, can result in erosion on moderate to steep slopes during the winter immediately after seeding because, in order for the banded carbon to work properly, the soil must be tilled enough to reduce large surface aggregates and surface residue that would otherwise hold soil in place. Still, when averaged over the entire multi-year stand of perennial grass seed, most fields are able to stay at or below the soil loss tolerance level "T."

Due to stored soil moisture from winter precipitation, Willamette Valley grass seed crops do not require irrigation. However, Huettig et al. (2013) and Chastain et al. (2015) found that one application of about 90 mm of water to tall fescue and multiple applications totaling about 170 mm of water to perennial ryegrass can significantly increase seed yields on Woodburn silt loam (an Aquultic Argixeroll). Accordingly, grass seed farms with water rights and available equipment often irrigate these crops in the spring.

Fertilizing Willamette Valley grass seed is similar to fertilizing western Oregon wheat. Soil tests determine appropriate levels of P, K, Mg, Ca, and lime. The N-min test has not been calibrated for grass seed crops, so instead a general N recommendation is about 145 kg N ha^{-1} for perennial ryegrass, 160 kg N ha^{-1} for annual ryegrass, and 170 kg N ha^{-1} for tall fescue, almost all applied in the spring. Sulfur is usually applied at a rate of around 15–45 kg S ha^{-1} in the spring (Anderson et al. 2014; Hart et al. 2011b, 2017). For western Oregon clover seed, soil tests determine appropriate levels of P, K, Mg, Ca, and lime. Being legumes, no N fertilizer is required for clover seed crops, but around 30 kg S ha^{-1} is needed. It's not unusual for soil tests to indicate small amounts of boron (around 2 kg B ha^{-1}) are needed for clover seed crops.

Crop rotation is a common practice for perennial grass and legume crops. Reduced till can be used for annual ryegrass. Underground outlets are a common solution to concentrated flow erosion in Willamette Valley seed fields.

18.4.5 Vegetables

In 2017, about 600 km^2 of vegetable crops[33] were grown in Oregon, with potatoes (*Solanum tuberosum*), onions (*Allium cepa*), sweet corn (*Zea mays*), and peas[34] (*Pisum sativum*) accounting for over 75% of that area. Over 80% of the potato crop area was in Umatilla, Morrow, and Klamath Counties. About 75% of the onion hectares were in Malheur and Morrow Counties. Over 70% of sweet corn hectares were in Marion, Linn, Umatilla, and Morrow Counties, and over 80% of pea production was in the Palouse Prairie of Umatilla County. A wide variety of vegetables are also grown in other counties where soils, climate, and the availability of irrigation water allow (NASS 2019).

[32] "Pre-emergence" means the herbicide is applied prior to crop and weed seedlings emerging from the soil. "Broadcast" means the herbicide is applied to the entire field rather than being banded in strips.

[33] NASS data for vegetables include melons. Although a locally important crop, melons were grown on only about 550 ha in Oregon in 2017.

[34] References to peas in this section pertain solely to green peas eaten as vegetables, rather than dried peas.

Potatoes, onions, and sweet corn are almost always irrigated in Oregon, and typical soils vary by MLRA. For example, a typical soil used for vegetables in the Columbia Basin is Quincy loamy fine sand, a Xeric Torripsamment. A typical soil for onion production in Malheur County is Owyhee silt loam, a Xeric Haplocalcid. In Klamath County, a typical soil for potato production is Laki loam, a Typic Haploxeroll. Chehalis silt loam, an Ultic Haploxeroll, can be used for sweet corn production in Linn and Marion Counties. Although occasionally flooded in winter, Chehalis soils are well drained and thus, when irrigated, are suited to summer crops like sweet corn. Unlike other vegetable crops in Oregon, much of the pea crop is not irrigated, but instead is grown in rotation with winter wheat on Umatilla County soils such as Athena silt loam, a Pachic Haploxeroll that receives about 430 mm mean annual precipitation (Soil Survey Staff 2019).

Nationally, Oregon ranked second among states in the production of onions (Fig. 18.4), fourth in potatoes and peas, and fifth in sweet corn in 2019 (Oregon Department of Agriculture 2021; NASS 2020). About 15% of Oregon potatoes are exported to South America, Mexico, and Asia, and about 75% of Oregon potatoes are processed into chips, soups, and French fries (World Atlas 2019). Oregon onions and sweet corn are marketed both fresh and processed, while almost the entire pea crop is sold frozen or canned (NASS 2019).

Oregon farmers rotate their potato, onion, sweet corn, and pea crops with other vegetables, grains, hay, and seed crops. Vegetables are prone to diseases and insect pests, so crop rotation is an essential part of production (Shock et al. 2021). Irrigation requirements for vegetables vary by crop, MLRA, weather, type of system, and availability of water. Average calculated net requirements are provided in Table 18.4. Irrigation water management is a critical practice for efficient irrigation on these crops, and many growers use soil moisture monitors, daily evapotranspiration predictions, and consultants to help in that effort (Washington State University 2021). In Malheur County, many vegetable growers are converting from flooded furrow to micro-irrigation or sprinkler systems to conserve water and reduce irrigation-induced erosion (NRCS 2018).

Soil test results determine the required fertilizer rates of P, K, Ca, Mg, lime, and any micronutrients[35] for vegetable crops. Optimal N applications for these crops are determined in different ways, which often attempt to account for the mineralization of soil organic N sources (including cover crops) during the growing season. As with other crops, pre-plant testing for mineral N can be useful east of the Cascade Mountains, where soil NO_3-N does not tend to leach below the root zone over winter, but this testing is not used in western Oregon. It is often useful to vary N application rates to match the expected yield. It is sometimes useful to account for N present in irrigation water; this is recommended for eastern Oregon onions (Sullivan et al. 2001). Analyzing potato leaf petioles for NO_3-N during the early growing season can help inform how much additional N fertilizer, if any, is needed for the crop (Lang et al. 1981). Sweet corn growers in western Oregon can use a pre-sidedress nitrate test (PSNT) to help determine how much additional N fertilizer, if any, is needed during the growing season. With the PSNT test, soil is sampled between corn rows to a depth of 30 cm when the corn has five or six leaves. The NO_3-N concentration in the soil provides an index for mineral N that will be available[36] during the remainder of the growing season. As legumes, peas have simple N fertilizer recommendations—they typically do not need any (Kaiser et al. 2016). Vegetable crops generally need small annual applications of S.

Typical conservation practices for vegetable crops include crop rotation, cover crop, reduced till, irrigation water management, nutrient management, and pest management system.

18.4.6 Orchards

In 2017, there were about 440 km^2 of orchards[37] in Oregon, and hazelnuts[38] (*Corylus avellana*), pears (*Pyrus communis*), cherries (*Prunus avium* and *P. cerasus*), and apples (*Malus domestica*) accounted for over 95% of that land. Essentially, all of Oregon's production of hazelnuts is from the Willamette Valley, and Marion County is the largest producer. About 70% of Oregon's pear orchards are in Hood River County and about 25% are in Jackson County. Almost 80% of the state's cherry orchards are in Hood River and Wasco Counties. Most of Oregon's apple orchards are in Umatilla County (41%) and Hood River County (23%), with Willamette Valley counties accounting for much of the remaining production. Oregon produces essentially the entire US hazelnut crop. Nationally, the state ranked second in

[35] Micronutrients are any plant mineral nutrients except N, P, K, Ca, Mg, and S. Examples of micronutrients are zinc (Zn), manganese (Mn), copper (Cu), molybdenum (Mo), and boron (B).

[36] PSNT test results account for soil mineral N present when the test is given, plus estimated additional N that will be mineralized during the growing season. PSNT results are not accurate for all western Oregon soils.

[37] Includes both orchards in production and young orchards not yet producing a crop.

[38] Hazelnuts are also known as "filberts." European hazelnuts (*Corylus avellana*) are grown commercially, and a related species (*C. cornuta* var. *californica*) is native and common in western Oregon (Olsen 2013c).

Fig. 18.4 Oregon is the second leading producer of onions in the United States. These sweet onions were grown in Umatilla County. *Source* Photograph by NRCS Oregon

Table 18.4 Average net irrigation requirements of common vegetable crops. These values are most useful as a comparison between crops and MLRA. Actual irrigation applications vary based on weather, soil moisture stored from winter precipitation, irrigation system efficiency, availability of water, and water rights. *Source* Cuenca (1999)

Crop	Average net irrigation requirements (mm)			
	MLRA			
	2	7	11	21
Onions	381	692	578	n/a[2]
Potatoes[1]	407	645	695	543
Sweet corn	370	484	461	n/a

[1]Some potato data are an average for early, mid-season, late crops
[2]n/a = not available
MLRA 2 = Willamette Valley
MLRA 7 = Columbia Basin
MLRA 11 = Snake River Plains
MLRA 21 = Klamath Basin

pear, third in sweet cherry, and seventh in apple production in 2019 (Oregon Department of Agriculture 2021).

In the Willamette Valley, Nekia silty clay loam, a Xeric Haplohumult located in foothill landscapes, is regularly used for hazelnut orchards, although many hazelnuts are also grown lower in the valley. In the Cascade Mountains—Eastern Slope (Hood River and Wasco Counties), Oak Grove loam, an Ultic Haploxeralf, supports pear, cherry, and apple orchards. In the Siskiyou Mountains MLRA of Jackson County, Carney clay, a Udic Haploxerert, supports pear orchards. Freewater very cobbly loam, a Fluventic Haploxeroll in the Columbia Basin of Umatilla County is commonly planted to apple orchards. As Table 18.5 illustrates, typical orchard soil map units represent four soil orders and vary considerably by texture, slope, precipitation, and land capability class. They all, however, are at least moderately deep, moderately well drained, and are not prone to frequent flooding.

Table 18.5 Representative soil map units for the four main Oregon orchard crops

Map unit	Typical orchard crops	MLRA[1]	Order	MAP[2]	Drainage[3]	Flooding	Depth[4]	LCC[5]
Nekia silty clay loam, 7–12% slopes	Hazelnuts	2	Ultisol	1143	W	None	MD	3e
Carney clay, 1–5% slopes	Pears	5	Vertisol	686	MW	None	MD	3s
Oak Grove loam, 0–8% slopes	Pears, cherries, apples	6	Alfisol	1016	W	None	D	2e
Freewater very cobbly loam, 0–3% slopes	Apples	7	Mollisol	356	SE	Rare	D	4s

Source Soil Survey Staff (2019)
[1]MLRAs:
2 = Willamette Valley
5 = Siskiyou Mountains
6 = Cascade Mountains, Eastern Slope
7 = Columbia Basin
[2]Mean annual precipitation (mm)
[3]MW = moderately well; W = well; SE = somewhat excessively
[4]MD = moderately deep; D = deep
[5]Irrigated land capability class and subclass. e = erosion is main limiting factor; s = shallow depth or stoniness is main limiting factor

Table 18.6 Oregon orchards in 1997 and 2017

Crop	Area in orchards (ac)	
	1997	2017
Hazelnuts	35,023	68,378
Pears	23,534	16,774
Cherries	16,858	15,575
Apples	10,958	5,791
Other	6,439	2,988
Totals	92,812	109,506

Source NASS (2021)

European hazelnuts were planted in Oregon as early as 1858, but the first commercial orchard was established by George Dorris in 1903 on a Lane County farm that is now a living museum (McBee 2016). Interestingly, hazelnut trees flower in winter and are pollinated by wind, but the nuts don't begin to form until June. Nuts are harvested in September and October after they fall to the ground. Roughly half of the Oregon hazelnut harvest is exported overseas (Oregon Aglink n.d.). As shown in Table 18.6, Oregon land planted to hazelnuts almost doubled between 1997 and 2017.

Commercial tree fruit production began without irrigation in Jackson County's Rogue River Valley in 1885. By the 1930s, most of those orchards were devoted to irrigated pears, which are well adapted to the region's warm days, cool nights, and clay soils. Pears grown in this region have long been known for their high quality and were exported as far away as Europe in the early part of the twentieth century. Pear farms in the Rogue River Valley declined from about 4,900 ha in the 1930s to only 1,600 ha in 2017, due in part to expanded pear production in Washington and urban encroachment on orchard land in Jackson County (Oregon State University 2007). Jackson County pears are marketed both fresh and processed, and a facility in Medford produces juice, baby food, and other products with the crop (Tree Top Inc. n.d.). About 20–30% of Oregon's pears are exported, largely to Latin America and Canada (Northwest Horticultural Council 2020a).

Apples were first planted commercially in Hood River County in 1876, and production was soon thriving. However, a devastating freeze in 1919 destroyed many orchards and growers replanted to pears, which are more resistant to cold weather. Hood River County also has sweet cherries—about 830 ha in 2017—but, just to the east, Wasco County had almost 4,200 ha of sweet cherries in 2017. The dry sunny summers in these counties are well suited to cherry production. Most of the Oregon sweet cherry crop is sold for the fresh market, and 30–35% is exported, with the main markets being Canada and Asia (Northwest Horticultural Council 2020b). Some cherry production also occurs in the Willamette Valley, where tart cherries are grown for processing and sweet cherries are sold for the fresh market or processed into maraschino cherries[39] (NASS 2021).

Apples are the main tree fruit crop in the Milton-Freewater area of northeastern Umatilla County, just over the state line and close to Walla Walla, Washington. One challenge apple growers face is the continued development of new varieties and the subsequent changes in consumer preferences. Establishing orchards is expensive, and growers must try to anticipate which variety will remain popular for years after an orchard is planted. Also, retailers complain about the number of apple varieties and limited supermarket shelf space (Eddy 2020). Current trends in Oregon apple production are the growing number of organically certified orchards and an increasing interest in establishing apple orchards for hard cider production (Heinrich 2017; Northwest Horticultural Council 2021).

Prune plums are no longer a mainstay of the Oregon fruit industry despite the fact that this crop grows well in the Willamette Valley. As Charles Dailey (1899) stated in a talk about prune plums at an early meeting of the Oregon Horticultural Society, "We do not irrigate our orchards, there being ample rainfall to stimulate and grow all vegetation in a natural, healthy way, and this without the eternal downpour that is sometimes supposed to occur." Prunes are, of course, dried plums. However, specific varieties of plums, such as the Italian prune plum, are best adapted to drying due to their high sugar and fiber content. There were once prune drying facilities scattered across the Willamette Valley, and in 1929 there were about 22,000 ha in production. However, by 2017 only about 350 ha of plums remained in the state (Rohse 2017; NASS 2021). The loss of the Oregon prune industry was associated with decreased consumer demand and a migration of production to California (Boriss and Brunke 2005).

Farmers hope for many years of production from orchards, and thus there are no standard rotations for these crops. Spacing of trees in orchards varies by species and production system. Hazelnuts have traditionally been planted at a density of about 270 trees ha^{-1}, but some growers are now doubling that density by planting at a spacing of about 3.0 rather than 6.0 m between trees within a row. Fruit tree planting densities vary from the traditional 340 trees ha^{-1} to about 1300 trees ha^{-1} for high density and even 2400 trees ha^{-1} for ultra-high-density orchards. Higher density orchards have greater establishment costs, involving more young trees, more planting labor, and usually a trellis or support system, but they bring orchards into production sooner, have potentially higher yields, eliminate the need for tall ladders during harvest, and may be adaptable to harvest mechanization. Because the trees are shorter and often trained on wires, ultra-high-density orchards can produce superior fruit quality due to more effective application of pesticides and better penetration of sunlight to the fruit (Long & Kaiser 2013; Olsen 2013a; Parker et al. 1998), (Washington State University, n.d.).

Nutrient management decisions for Oregon orchards are usually based on soil test pH for lime need and leaf tissue analyses for other nutrients. Nitrogen application rates can vary from less than 100 to over 200 kg N ha^{-1} (Hart et al. 1997; Olsen 2013b; Righetti et al. 1998). Historically, Oregon hazelnut orchards were not irrigated, but some growers are beginning to use micro-irrigation to reduce mortality during establishment, decrease the time required for an orchard to begin producing a harvestable crop, and increase yields (Fig. 18.5). Pears, cherries, and apples are usually irrigated in Oregon. Irrigation requirements vary by crop, MLRA, weather, type of system, and availability of water. Average calculated net requirements for fruit orchards are around 700–800 mm in the Siskiyou Mountains, Cascade Mountains—Eastern Slope, and Columbia Basin, and 500–600 mm in the Willamette Valley. Irrigation water management is important for fruit orchards, and many growers use micro-irrigation to conserve water. Fruit orchards often have wind machines and/or heaters to reduce the risk of frost damage to blossoms.

Additional conservation practices for Oregon orchard crops include field borders and hedgerows to provide habitat for beneficial insects, windbreaks in sites with high wind,[40] and pest management system to control myriad insect and disease problems. Fruit orchards often use conservation cover and cover crops to establish vegetation between orchard rows for erosion control and soil health. Some hazelnut growers are trying those practices, but they face challenges because farmers use machines to sweep nuts off the soil surface during harvest, which has traditionally required a bare, hard soil surface. New hazelnut harvesting equipment is helping to overcome those challenges. Finally, orchard growers are using practices such as field operations emissions reduction to address air quality problems caused by old orchard frost protection systems (smudge pots) or the burning of pruning debris.

[39] Working from 1925 to 1931, Ernest Wiegand at Oregon State University perfected a method of making maraschino cherries that is still in use today. Royal Ann cherries are light in color and are the common variety used for maraschino cherries in a process that involves brining the cherries in calcium salts, dying them a bright red color, and then preserving them in jars for use in cocktails and desserts (Verzemnieks 2018).

[40] Cherries are especially prone to wind damage.

Fig. 18.5 A young Willamette Valley hazelnut orchard in early spring, just beginning to leaf out. Increasingly, young hazelnut orchards are established with micro-irrigation (tubing and micro-sprayers shown here). This grower planted strawberries between tree rows to provide a cash crop during the years before the orchard begins producing sufficient yields for harvest. *Source* Photograph by NRCS Oregon

18.4.7 Corn for Silage or Grain

Between 1997 and 2017, Oregon's grain corn crop increased from about 113–211 km^2, and the silage corn crop increased from about 93–154 km^2. Much of the state's grain and silage corn is grown on dairy farms where it is used for feed, and high grain prices incentivize dairy farmers to grow corn in lieu of buying it (Mortensen 2019). Crop farms without livestock also grow corn, which is then sold for feed on other operations. In 2017, over 50% of Oregon's grain corn production was in the Columbia Basin of Morrow, Umatilla, and Gilliam Counties, and over 40% was in the Snake River Plains of Malheur County. Oregon silage corn statistics are not reported here—the National Agricultural Statistics

Service does not provide those data for counties with few producers to prevent information about individual farms from being derived.

The soils used for corn production in the Willamette Valley, Columbia Basin, and Snake River Plains are generally the same soils used for vegetables. In the Sitka Spruce Belt of Tillamook County, farmers have begun to grow silage corn on soils such as Quillamook silt loam, a Pachic Melanudand with a mean annual precipitation of over 2,000 mm. With that level of precipitation, silage corn has been grown in Tillamook County without irrigation (Tillamook Headlight Herald (newspaper) 2008), but in other MLRAs, irrigation is almost always needed. Average calculated net irrigation requirements for corn range from around 300 mm in the Willamette Valley to around 700 mm in the Columbia Basin and Snake River Plains.

On livestock farms, corn may be rotated with hay or small grain crops, although it is sometimes grown continuously for several years. On farms without livestock, corn may be rotated with hay, grain, or vegetable crops. Nutrient management recommendations for grain and silage corn are similar to sweet corn in that preplant soil tests determine appropriate rates of P, K, and lime. The PSNT test can be used in western Oregon and a pre-plant NO_3-N test can be used in eastern Oregon, just as with sweet corn. Typical N requirement is around 175–225 kg N ha^{-1}, which is higher than for sweet corn. Where grain and silage corn are grown on livestock farms, special considerations are needed, including manure testing and calculations to estimate nutrient contributions from manure. Significant N is lost from manure due to ammonia volatilization during storage and application, but P does not volatilize. This causes the ratio of available N:P from manure applications to be lower than the N:P ratio required by the corn crop. Thus, manure rates that provide the required N levels for corn (or other crops) invariably lead to over application of P, which can create water quality problems (Sharpley et al. 1996). Another long-term concern on dairies is the use of copper (Cu) or zinc (Zn) footbath solutions to control hoof diseases. These solutions are typically disposed of in manure storage systems and then applied to the land, potentially leading to excessively high levels of Cu and Zn in the soil (Gardner et al. 2000; Hart et al. 2009a).

Thus, nutrient management is an important conservation practice for corn grain and silage, especially where manure is applied. Other common practices include irrigation water management, filter strips, and crop rotation. Some corn growers are trying no-till, but corn is usually not grown on steep slopes in Oregon, so erosion is usually not a major concern.

18.4.8 Christmas Trees

Oregon is the nation's leading grower of Christmas trees and in 2017 produced 31% of the US crop on approximately 180 km^2 of land (Oregon Department of Agriculture 2019b). Much of the crop is grown in the Willamette Valley. A typical soil planted to Christmas trees in Washington County is Laurelwood silt loam, an Ultic Haploxeralf that formed in silty loessial-like materials on hills with a mean annual precipitation of about 1,320 mm (Fig. 18.6). These and similar soils commonly used for Christmas tree production are well drained and often occur at the interface between lower elevation lands used for grain, seed, and hazelnut crops and higher elevations dominated by forests.

Noble fir (*Abies procera*) and Douglas-fir (*Pseudotsuga menziesii*) are the most common Christmas tree species grown in Oregon, and over 90% of the harvest is shipped out of state. The greatest share goes to California, but trees are also shipped across the United States, as well as to Mexico and Asia. Trees generally are pruned each year to achieve a desirable shape and density, and then harvested at 7–9 years of age (Pacific Northwest Christmas Tree Association 2021).

Nutrient management recommendations include testing soil for P, K, pH, Ca, and Mg and then incorporating needed minerals with tillage prior to planting trees. Some growers deeply till the soil prior to planting to encourage better root growth, and most growers plant Christmas trees at densities of about 2,500–4,300 trees ha^{-1}. After trees are established, needle tissue analyses determine the need for additional fertilizer. Trees need more N as they grow in size, and applications of between 100 and 180 kg N ha^{-1} are typical (Hart et al. 2009b). Some growers harvest entire fields over the course of 1 or 2 years, and then remove stumps, till the soil, and apply lime if needed before planting a new crop. Other growers harvest selected trees throughout a field and then replant between stumps, resulting in fields of varying sizes of trees. Christmas trees are rarely irrigated in Oregon (Landgren et al. 2003).

The main conservation practices for Christmas trees address erosion or soil health concerns. Establishing annual cover crops or conservation cover between tree rows helps control erosion and adds organic matter to the soil (Fig. 18.6). Where gully erosion is a concern, growers can install underground outlets to convey runoff downslope through a pipe that outlets safely at the bottom of the field.

Fig. 18.6 Douglas-fir Christmas trees in the Willamette Valley MLRA of Washington County. The soil is Laurelwood silt loam, an Ultic Haploxeralf. This field is managed with the harvest of selected trees, followed by replanting between stumps, and resulting in a stand of multiple-aged trees. The aisles between trees in this field were seeded to a cover of fine-leafed fescue (e.g., *Festuca brevipila*) 24-years before this photograph was taken. This perennial cool-season grass was chosen because it is long-lived and vigorous, controls erosion, improves soil health, and is not highly competitive with the trees. *Source* Photograph by Dean Moberg, courtesy of Logan Family Tree Farm

18.4.9 Nursery Crops

In 2017, there were approximately 139 km^2 of nursery crops[41] in Oregon. Marion, Clackamas, Washington, Yamhill, and Multnomah counties, all in the northern Willamette Valley, account for over 95% of the Oregon cropland area planted to nursery crops. Nursery crops are relatively labor-intensive, and one source estimates that over 22,000 workers are employed in Oregon nurseries (Oregon Aglink n.d.). About 75% of Oregon nursery sales are from plant materials shipped out of state.

Nurseries require relatively level soil that has good drainage and is not flooded. For example, Willamette silt loam, 0–3% slopes, is well suited to nursery production. It is a Pachic, Ultic Argixeroll that occurs on terraces with a mean annual precipitation of about 1,000–1,300 mm and a mean annual air temperature of about 10.0–12.2 °C. A mild climate and the availability of high-quality irrigation water allow a wide variety of nursery crops to be grown, including flower bulbs, cut flowers, bedding plants, potted plants for indoor use, shrubs, and trees.

Nurseries employ various production systems (Oregon State University 2020). Trees and shrubs grown in natural soil[42] may be harvested as bare-root stock or via a balled-and-burlapped (B&B) system. With bare-root production, the plants are typically dug with a machine that leaves most of the soil in the field. Soil remaining on the

[41] Includes trees and shrubs, bedding plants, cut flowers, potted plants, aquatic plants, bulbs, cuttings, seedlings, liners, and plugs. Also includes vegetable transplants sold to other farms. Does not include flower and vegetable seed, sod, edible crops grown in greenhouses, or mushrooms.

[42] The term "natural soil" is used here to describe the soil naturally found onsite, as distinguished from human-made media (sometimes called "potting soil") produced from ingredients such as peat moss, sand, and perlite. The natural soil employed in bare-root and balled-and-burlapped systems is manipulated with tillage and fertilizer, similar to other crops.

Fig. 18.7 This nursery operation in Washington County grows plants in outdoor containers on geotextile fabric (foreground), greenhouses (background), and in natural soil (not pictured). A long bank of solar thermal panels heat water that is stored in an insulated tank for use in heating the greenhouses. *Source* Photograph by Dean Moberg, courtesy of Blooming Nursery

roots may be washed off at the packing facility. B&B trees and shrubs are usually harvested at an older age and larger size than bare-root stock, using a machine that leaves an intact ball of soil held in place around the roots with burlap.

Alternatively, nursery plants may be grown in containers filled with human-made media. A wide range of containers is used, depending on the type and size of the product. The containers, or "pots," are typically plastic and are either placed inside greenhouses or in outdoor production areas on gravel or geotextile fabric (Fig. 18.7). Outdoor containerized products can vary from relatively small bedding plants to trees several meters in height; greenhouse products are limited to smaller plants. One specialized outdoor container system places potted trees inside larger containers buried in the ground. In those "pot in pot" systems, the aisles between the plants are often seeded to perennial grass to reduce mud and to control erosion. Greenhouses vary from simple systems with plastic stretched over hoops, to sophisticated buildings that control temperature and ventilation and may provide artificial lighting.

Some Oregon nurseries produce cut-flowers, which may be grown in fields or greenhouses. Other nurseries produce rhizomes, corms, or bulbs of ornamental plants (e.g., irises and tulips). Finally, some nursery products are gathered in the wild rather than grown on cropland—for example, evergreen boughs for wreaths, or native plant seed and cuttings of native woody species for habitat restoration work.

Nursery crops generally require higher levels of nutrient, pest, and irrigation water management than other crops because markets are competitive and blemished nursery products are difficult to sell. There are too many plant species grown in Oregon nurseries for Oregon State University or other land-grant institutions to provide detailed fertilizer recommendations for each. Thus, nurseries develop nutrient management programs based on somewhat general land-grant recommendations (Owen and Stoven 2011), advice from consultants, and their own observations.

Irrigation water management for bare-root and B&B is similar to non-nursery crops, but managing irrigation in containers is more difficult due to the low water-holding capacity of human-made potting media and the relatively small root mass the containers allow. Also, micro-irrigation systems are difficult to design and manage in nurseries.

Bare-root production systems generally have plants spaced in narrow rows that are planted and dug frequently, which would require frequent replacement of drip tubing. For most container production systems, extensive tubing arrays with one or more emitters per container usually prove to be cost-prohibitive. The agricultural water quality management plans described in Sect. 18.3.2 generally prohibit the runoff of irrigation water into streams in the summer. For container nurseries, this usually requires the use of tailwater recovery or the equivalent to capture and reuse water that runs off container yards.

Oregon nurseries have implemented energy conservation practices, including installing more efficient lighting systems, improving greenhouse insulation and heating, and increasing vehicle fuel efficiency (Hensey 2013), (Oregon Association of Nurseries; Oregon Environmental Council 2011). A life cycle assessment of greenhouse gases found that the atmospheric carbon sequestered by trees planted in urban areas quickly surpasses the amount of greenhouse gases emitted during the nursery production of the trees (Kendall and McPherson 2012). Another conservation benefit offered by Oregon nurseries is the increasing production of native plants for use in both large-scale ecosystem restoration projects as well as small-scale residential habitat plantings (Anderson 2019).

18.4.10 Hemp and Marijuana

The National Agricultural Statistics Service does not currently publish data on hemp or marijuana (*Cannabis sativa*)[43] production, however the Oregon Department of Agriculture reported 11,100 ha of land registered to grow hemp in 2020. Jackson and Josephine Counties had the most land registered for hemp production, followed by Deschutes, Umatilla, and Malheur Counties (Jones 2021). As of March 2021, there were about 214 ha of flowering canopy licensed for marijuana production, with Jackson and Josephine Counties accounting for about 60% of that land area (Oregon Liquor Control Commission 2021).

Cannabis products raised on Oregon farms can be classified based on their concentration of delta-9 tetrahydrocannabinol (THC), a psychoactive compound. Marijuana is defined by the federal government as a cannabis plant with greater than 0.3% THC. Marijuana is considered an illegal Schedule I drug by the federal government but was legalized by the state of Oregon in 2015. Marijuana can be marketed in Oregon as either medical (purchasers must possess a signed recommendation from a medical doctor registered to practice in Oregon) or recreational (purchasers must be at least 21 years old). However, by the close of 2018, almost all retailers licensed to sell medical marijuana had transitioned to recreational licensing (Oregon Liquor Control Commission 2019). Marijuana growers must obtain a license from the Oregon Liquor Control Commission prior to planting a crop (Oregon Legislative Assembly 2019).

Cannabis plants with THC not exceeding 0.3% are classified as hemp,[44] which were widely grown in the United States until the 1930s. This crop can be harvested for fiber used in clothing and paper products, for edible grain, and for edible oil. Henry Ford used hemp fiber for automotive components (Jeliazkov et al. 2019). Much of Oregon's current hemp production is for the purpose of extracting compounds such as cannabidiol or CBD, which are marketed to address human ailments including pain, anxiety, acne, nausea from cancer treatment, and other maladies. The Food and Drug Administration has approved only one medical CBD product, specifically for the treatment of two rare forms of epilepsy (US Food and Drug Administration 2020). Hemp growers must register their grow sites with the Oregon Department of Agriculture each year prior to planting (Oregon Department of Agriculture 2020c).

Marijuana is grown indoors under artificial lighting, in greenhouses, and outdoors. In 2021, the area licensed for outdoor marijuana production totaled roughly six times the licenses for indoor crops. Marijuana grown indoors is typically sold for direct use of flowers, buds, and leaves, and demands a higher price; outdoor crops are more commonly used for extracts or THC concentrates (Oregon Liquor Control Commission 2019). Most hemp is produced outdoors in Oregon.

Growers of indoor cannabis utilize human-made potting soil media. Cannabis grown outdoors is best adapted to deep, loamy, fertile soils with a pH of 6.0–7.5. OSU recommends soil testing prior to planting to determine needs for P, K, Ca, Mg, and lime and then using fertilizer rates recommended for spring wheat. Outdoor cannabis is often planted into plastic row covers, which warm the soil and control weeds, and then irrigated with micro-irrigation (Roseberg et al. 2019). Other conservation practices useful in marijuana and hemp production include cover crops, irrigation water management, nutrient management, and pest management system (Jones et al. 2019).

[43] Botanists debate Cannabis taxonomy, with some believing just one species (*C. sativa*) exists, and others recognizing three species (*C. sativa*, *C. indica*, and *C. ruderalis*). In any case, growers have successfully cross-bred plants, and it may be more useful to identify *Cannabis* plants by their chemical attributes rather than their taxonomic heritage (Jeliazkov et al. 2019). In this chapter, the term "cannabis" will be used generically for any of the species grown commercially in Oregon.

[44] Hemp is also sometimes called "industrial hemp.".

18.4.11 Berries

Oregon was the nation's leading grower of blueberries (*Vaccinium corymbosum*)[45] in 2018, producing 25% of the US crop. The state ranked third for the production of cranberries (*V. macrocarpon*) and fourth for strawberries (*Fragaria × ananassa*) (Oregon Department of Agriculture 2021). Blackberries and raspberries (*Rubus spp.*),[46] as well as other species of berries are also grown in Oregon. Oregon had 109 km^2 of berries in 2017, with blueberries and blackberries accounting for 69% of that area.

Almost all of Oregon's cranberries are grown in the Sitka Spruce Belt of Curry and Coos Counties, on soils such as Nelscott silt loam, a moderately well-drained Typic Durorthod with a mean annual precipitation of about 1800 mm. Over 90% of the blueberry, blackberry, raspberry, and strawberry production happens in the Willamette Valley (NASS 2021) on fertile soils such as Woodburn silt loam, a moderately well drained Aquultic Argixeroll with 1,100 mm mean annual precipitation (Fig. 18.8). Interestingly, the state's most urban county, Multnomah, grew over 4 km^2 of berries in 2017.

The long growing season and abundant rainfall on Oregon's coast are well suited to cranberries and Indigenous Peoples traditionally harvested the native bog cranberry (*Vaccinium oxycoccos*) for food and medicinal purposes. In 1885, Charles McFarlin planted the first commercial field of Oregon cranberries in Coos County, using cuttings he brought from Massachusetts (Miller 2020). In preparation for planting cranberries, growers manipulate soil rather intensively by placing sand on the site and grading the land into tiered fields. Typically, there is a reservoir at the bottom of a set of fields to supply water for irrigation and flooding. Although fields are irrigated during the summer, they are only flooded in the fall to allow the berries—which float—to be easily harvested. Growers often apply additional sand to cranberry beds, which stimulates the growth of this long-lived perennial crop and helps to control insect pests. Fertilizer needs are determined primarily by tissue testing of leaves, and typical N application rates are 45–70 kg N ha^{-1}. The recommended low pH levels of 4.0–5.5 for cranberry beds inhibit soil nitrification and retain most of the soil mineral N in the form of NH_4^+, which cranberries prefer (Hart et al. 2015). Most Oregon cranberries are used for juice or are sweetened and dried.

Several relatives of the highbush blueberry are native to Oregon, most notably huckleberry species (*Vaccinium membranaceum*, *V. parviflorum*, and *V. ovatum*), which are important First Foods and are still gathered in the wild today. Likewise, Oregon's native trailing blackberry or dewberry (*Rubus ursinus*) was used by Indigenous Peoples for food and medicinal purposes (Pojar and MacKinnon 2014). Domesticated blueberries, blackberries, raspberries, and strawberries are perennial crops, although with varying productive lifespans. Strawberry plantings typically remain commercially productive for just 3 to 4 years, while blueberries can remain productive for decades.

Nutrient management recommendations include testing soil for P, K, pH, Ca, and Mg and then incorporating needed minerals with tillage prior to planting berries. Blueberries prefer acidic soils and elemental sulfur application may be required to achieve the recommended pH of 4.5–5.5. Caneberries and strawberries do best at pH 5.5–6.5. After establishment, leaf tissue testing is the best method of determining additional fertilizer applications. Mature blueberries require applications of about 160–190 kg N ha^{-1}. Like cranberries, blueberries require N fertilizers based on NH_4^+ rather than NO_3^-. Caneberries and strawberries require about 45–110 kg N ha^{-1} once established. Actual N rates for berries may be adjusted on the basis of tissue testing results, crop performance, and (for blueberries) whether sawdust mulch has been applied recently, which can immobilize soil mineral N (Dixon et al. 2019; Hart et al. 2006a, b).

Although caneberries and strawberries can be grown in western Oregon without irrigation, essentially all commercial production of these crops, as well as blueberries and cranberries, now use irrigation (Bubl 2014). Average irrigation requirements for cranberries in Oregon are not available. Average total seasonal evapotranspiration in the Willamette Valley is about 860 mm for strawberries, and 950 mm for blueberries and caneberries (Smesrud et al. 2000).[47]

[45] There are other blueberry species cultivated in the United States, but Oregon's commercial production is almost entirely the northern highbush blueberry, *V. corymbosum*.

[46] Growers often use the generic term of "caneberry" to refer to both blackberries and raspberries. Many varieties have been developed through cross-breeding of berry strains introduced from Europe and, for some varieties, the Oregon native *Rubus ursinus* (Anderson & Finn 1995).

[47] Note that seasonal evapotranspiration totals provided here for Willamette Valley berry crops are from the *Western Oregon Irrigation Guides* (Smesrud et al. 2000) and cannot be directly compared with the net average irrigation requirements provided in other sections, which are based on estimates in *Oregon Crop Water Use and Irrigation Requirements* (Cuenca 1999). The Cuenca publication is used in other sections because it provides information on a variety of crops across the state, while the Smesrud, et al. publication only provides data on selected Willamette Valley crops. However, the Cuenca data for berries are quite limited as it lumps all berry species together, and so it is not used in this section. Cuenca data are based on average evapotranspiration (ET) minus effective precipitation, while Smesrud, et al. data are just average ET.

Fig. 18.8 This blueberry field in the Willamette Valley MLRA of Washington County is located on Woodburn silt loam, 0–3% slope. The photograph was taken in April and the berries are in bloom, with beehives placed to facilitate pollination. Micro-irrigation conserves water, with tubing suspended above the ground to reduce damage from mammals and to allow the quick identification of leaks. As is typical of blueberry farms, there is also a sprinkler irrigation system, which may be used for frost protection in the spring or for cooling on hot days when berries are ripening in the summer. The perennial grass conservation cover between rows improves soil health and allows better trafficability in winter. A heavy layer of sawdust mulch in each row conserves soil moisture and moderates soil temperature. *Source* Photograph by Dean Moberg, courtesy of Alfred Dinsdale

Important conservation practices for Oregon berries include irrigation water management, nutrient management, and pest management system. For blueberries and caneberries, micro-irrigation can conserve significant amounts of water. However, most growers using drip irrigation systems also maintain sprinkler systems for frost protection in the spring or to cool plants during excessively hot summer days. Establishing perennial grass between blueberry and blackberry rows via conservation cover helps to reduce erosion, improve soil health and provide a firm surface for equipment during wet weather. Annual cover crops are more commonly used with raspberries because their habit of creeping underground and then sending up shoots between crop rows makes regular tilling desirable. Hedgerows and field borders can provide important habitat for pollinators and beneficial insects, but there is some concern that hedgerow plants that produce fruit can also provide habitat for spotted wing Drosophila (*Drosophila suzukii*), a harmful pest of berries and other soft fruit (Dreves et al. 2009).

18.4.12 Grapes

Oregon ranks fourth nationally in area[48] devoted to grapes, with at least 97 km^2 in 2017. Almost all of Oregon's grape production is for wine, using the European *Vitis vinifera* species,[49] and the most common varietals are pinot noir, pinot gris, and chardonnay. About 60% of Oregon's vineyard land is in the Willamette Valley, with Yamhill County having the greatest share. About 20% of the state's vineyards are in Jackson, Douglas, and Josephine Counties.

Soils are an important component of wine terroir,[50] and thus growers (and oenophiles!) are keenly interested in the soils where vineyards are located. Well-drained soils with moderate, south-facing slopes, and moderate fertility are prized for Oregon vineyards, and the state soil, Jory silty clay loam, is a good example. Jory is a Xeric Palehumult formed in colluvium and residuum derived mostly from basaltic bedrock. It occurs on foothills at elevations of about 75–750 m in the Willamette and Umpqua Valleys (Soil Survey Staff 2019). There are numerous other important soil series for Oregon vineyards (Table 18.7).

Historical records indicate that the first grapes cultivated in the Pacific Northwest probably originated from seeds that a Hudson's Bay Company officer, Aemilius Simpson, brought from Britain in 1826. Those seeds established a few vines of table grapes at Fort Vancouver in what is now the state of Washington. The Jean Mathiot family, who were French immigrants, made the first major planting of wine grapes in Oregon, in the community of Butteville just east of the Champoeg[51] area of Marion County. The Mathiots used grape planting stock that they purchased in California, reportedly trying about 50 different varieties in an effort to determine what was best adapted to the relatively cool Oregon climate. The Mathiots produced wine, which was mostly consumed locally by fellow French immigrants (Stursa 2019).

By the 1870s and 1880s, wine grapes had been established in Josephine, Jackson, and Douglas Counties, using planting stock brought to Oregon from California. By the end of the nineteenth century, wine produced in western Oregon was being marketed within the state and also exported to neighboring states. Some wine was distilled into brandy, and some grapes were sold for making jelly. In 1915, Oregon outlawed the manufacture of all alcoholic beverages, which marked an end to the early period of winemaking in the state, and many wine grape vineyards were converted to other crops (Ibid.)

In 1933, the 21st Amendment ended prohibition in the United States, and the Oregon legislature also repealed the state's prohibition. Wine grape production increased, but only slowly. Dr. Hoya Yang, a Chinese immigrant and OSU professor, planted about 40 varieties of grapes in the 1950s to study which were best adapted to Oregon. In the 1960s, pinot noir grapes began to be recognized as well-adapted to Oregon, with the first commercial plantings in the Umpqua Valley and then, soon after, in the Willamette Valley. The number of Oregon vineyards and land in production increased steadily until around 1999, after which the industry expanded dramatically. The high quality of Oregon wines is now recognized nationally and internationally, and by 2020, there were 18 American Viticultural Areas[52] in Oregon (Oregon Wine Board 2021). For more information on the history of Oregon wine, readers are directed to Scott Stursa's *Oregon Wine—A Deep-Rooted History* (2019).

Wine grapes are long-lived perennials that take several years to come into production. As with other perennial crops, soil testing is most helpful prior to planting the grapes. Soil pH in the range of 6.5–6.8 is considered optimal for grapes. After planting, leaf blade and/or leaf petiole analyses, along with observations of vine vigor, indicate whether fertilizer is needed. In general, the quality of wine grapes decreases with excessive levels of fertility (Skinkis and Schreiner 2011). Willamette Valley vineyards may be

[48] NASS data for 2017 report 97 km^2 total planted area (includes vineyards not yet being harvested). The Southern Oregon University Research Center (2017) reported a total Oregon vineyard area of 123 km^2 for 2016.

[49] The most common grape species cultivated in North America are *Vitis vinifera* and *V. labrusca*. *V. vinifera* is commonly called the "European grape" because the species evolved in the Mediterranean basin and most cultivars of *V. vinifera* originated in Europe. Pinot noir, pinot gris, and chardonnay are all *V. vinifera* cultivars used for wine making. Thompson seedless table grapes are also *V. vinifera*. *V. vinifera* is commonly grown in California, Oregon, and Washington, but is difficult to grow in the eastern United States due to its cold, wet climate. *V. labrusca*, however, evolved in eastern North America and is grown there commercially. Concord is a well-known *V. labrusca* cultivar, is considered a table grape, and is the common ingredient in grape jelly. *V. labrusca* grapes are sometimes used for wine but are considered inferior to *V. vinifera* for that purpose. Both species, as well as hybrids, have been and currently are planted in Oregon, but the great majority of Oregon wine grapes are *V. vinifera*. Table grapes tend to be large, have thin skins, and are often seedless. Wine grapes are smaller, have thick skins, many seeds, and higher sugar content than table grapes.

[50] Terroir is the environment in which wine grapes are grown, including the soil, bedrock geology, slope orientation, elevation, climate, and farming practices (Van Leeuwen & Seguin 2006). Terroir imparts unique characteristics to a wine such that, for example, pinot noir grapes grown on one type of soil may result in wine with noticeably different flavor, aroma, and color from pinot noir grapes grown on a different soil.

[51] Champoeg was the seat of the first provisional government before Oregon became a territory. The towns of Champoeg and Butteville were severely damaged in the 1861 flood (see Chap. 17) and abandoned.

[52] An American Viticultural Area is a legally designated grape-growing region with geography, soil, and climate (i.e., some of the components of terroir) that distinguish it from other regions. The first Oregon viticultural area was established in the Willamette Valley in 1983.

irrigated during establishment but are not generally irrigated after vines begin to be harvested. Vineyards in the Siskiyou Mountains MLRA and in eastern Oregon are typically irrigated with micro-irrigation (Levin 2018).

The most common conservation practices for Oregon vineyards are conservation cover and cover crops, seeded between grape rows to control erosion and improve soil health (Fig. 18.9). Introduced perennial grass species and native perennial herbaceous species provide excellent erosion control and soil health benefits, but they can compete with the grape vines for soil moisture in the summer. Annual cover crops like oats can be killed in early spring, thus preserving soil moisture for the grapes during dry summer months. In order to achieve a balance between too much and too little moisture, while still controlling erosion and improving soil health, some vineyards plant alternating aisles to perennial cover and annual cover (NRCS 2020b).

Oregon vineyards are increasingly certified by one or more organizations attesting to environmentally responsible grape production. These certifications include USDA

Fig. 18.9 A wine grape vineyard in the Willamette Valley. The mixed conservation cover herbaceous seeding controls erosion, improves soil health, and provides habitat for beneficial insects. *Source* Photograph by NRCS Oregon

Table 18.7 Some important soil series for Oregon vineyards

American viticultural area (AVA)	Soil series	Soil taxonomy	Soil depth[1]	Drainage class[2]	MAAT[3] (°C)	MAP[4] (mm)
Willamette valley	Bellpine*	Xeric Haplohumults	MD	WD	11.1	1270
	Carlton*	Aquultic Haploxerolls	VD	MWD	11.1	1145
	Jory*	Xeric Palehumults	VD	WD	11.1	1145
	Laurelwood*	Ultic Haploxeralfs	VD	WD	11.1	1320
	Nekia*	Xeric Haplohumults	MD	WD	11.7	1145
	Steiwer*	Ultic Haploxerolls	MD	WD	11.1	1270
	Willakenzie*	Ultic Haploxeralfs	MD	WD	11.1	1270
	Willamette*	Pachic Ultic Argixerolls	VD	WD	11.1	1345
	Woodburn*	Aquultic Argixerolls	VD	MWD	11.7	1145
Columbia valley	Latourell*	Ultic Haploxeralfs	VD	WD	11.7	1270
	Multnomah*	Humic Dystroxerepts	D	WD	11.7	1270
	Oak Grove*	Ultic Palexeralfs	VD	WD	9.4	1015
	Van Horn*	Ultic Argixerolls	D	WD	10.0	685
	Wind River*	Ultic Haploxerolls	VD	WD	10.0	760
Umpqua valley	Oakland	Ultic Haploxeralfs	MD	WD	11.7	1015
	Rosehaven*	Ultic Haploxeralfs	VD	WD	11.1	1145
	Speaker	Ultic Haploxeralfs	MD	WD	10.0	1145
	Sutherlin	Ultic Haploxeralfs	VD	MWD	11.7	1145
Rogue valley	Josephine*	Typic Haploxerults	D	WD	10.0	1145
	Manita*	Mollic Haploxeralfs	D	WD	10.6	760
	Ruch*	Mollic Palexeralfs	VD	WD	11.1	760
	Speaker	Ultic Haploxeralfs	MD	WD	10.0	1145
Snake valley	Greenleaf*	Xeric Calciargids	VD	WD	10.0	229
	Nyssa*	Xeric Haplodurids	MD	WD	11.1	254

*Prime soil
[1]MD = moderately deep (51–102 cm), D = deep (102–152 cm), VD = very deep (>152 cm)
[2]MWD = moderately well drained, WD = well drained
[3]MAAT = mean annual air temperature
[4]MAP = mean annual precipitation

Organic, Demeter Biodynamic, LIVE, Food Alliance, and Salmon-Safe. As a response to consumer confusion over multiple certification types, the Oregon Wine Board has launched an overarching certification called "Oregon Certified Sustainable Wine" (Buckenmeyer).

Climate change is a concern for Oregon vineyards. As temperatures increase, there is a risk of losing the delicate balance of sugar to acid in grapes that produce excellent pinot noir and other wines for which Oregon is known. Climate change also brings a risk of new disease and insect pests, and the smoke from wildfires can drift down to vineyards and ruin a crop if excessively absorbed by the grapes.

18.4.13 Small Acreage Vegetables and Specialty Crops

The following discussion of small acreage farms is different from, and in ways more difficult than, the previous sections that focused on individual crops or types of crops. However, it's important to discuss smaller farms because they represent a growing trend in agriculture, play a unique role in providing local food, and help connect urban populations to agriculture and thus the soil.

Figure 17.2 illustrated that Oregon operations smaller than 20 ha were the only farm size category that recently increased in number, from about 19,000 farms in 1997 to about 25,000 farms in 2017. Note that these data from the National Agricultural Statistics Service reflect operations with potential or actual gross annual sales of $1,000 or more and so, at least theoretically, don't include rural properties where no crops or livestock are ever raised for sale.

Small acreage farms produce a remarkable variety of commodities, including fresh vegetables, eggs, meat, cheese, honey, cut flowers, herbs, jams, pickles, and other products. The farms usually market these products locally. For example, some farms sell directly to consumers at "fruit stands" on or near their property. Others sell to restaurants, many of which advertise the locally grown aspect of their offerings. As of 2019, there were at least 108 farmers' markets operating in Oregon. Farms pay a fee to sell their produce directly to consumers at these markets, which generally operate during summer months (Oregon Farmers Markets Association 2021). There is a growing trend toward community supported agriculture (CSA), in which consumers subscribe to receive a regular (usually weekly) box of produce from a farm. The first CSAs in the United States began in 1986 and were partly inspired by the biodynamic movement advanced by Austrian philosopher Rudolph Steiner, who stressed the importance of a close linkage between producers and consumers. Many CSAs publish newsletters to inform their members about events on the farm; some CSAs encourage members to help with farm labor (Strochlic and Shelley 2004).

Agencies have responded to the growth in small acreage farms with programs specifically focused on their needs. For example, OSU has a small farm program with faculty and staff who provide educational events, technical publications, webinars, and an online small farm newsletter (Oregon State University 2021c, d). NRCS provides funding for high tunnels and other practices used by small acreage farms (NRCS 2021b). People wishing to learn how to begin farming have multiple opportunities that combine on-farm experiences with more formal educational programs. Examples of these opportunities in Oregon include Rogue Farm Corps, Adelante Mujeres Sustainable Farming Class Series, the Pathways to Farming Program, and the Headwaters Incubator Program (Rogue Farm Corps, n.d.).

Many small farms use organic methods (Fig. 18.10), but some eschew organic certification if their clientele trusts the farm's practices regardless of formal certification status. Common conservation practices on small acreage crop farms include high tunnels, crop rotation, cover crops, and mulching. Small acreage livestock farms often employ prescribed grazing. It's common for small acreage farms to integrate livestock and crop production.

Organic farms using manure and compost as nutrient sources can face the same issue of elevated soil test P levels commonly seen on dairy farms (see the section on corn for silage or grain in this chapter). Those farms benefit from nutrient management and the use of legume cover crops to provide at least some of the crop's N needs, thus reducing the need for manure or compost (Nelson and Janke 2007). OSU provides a Cover Crop Calculator to help farmers estimate N available from cover crops and a variety of organic amendments, including manure and compost (Sullivan et al. 2019).

18.5 Pasture and Grazed Rangeland Management Systems

In 2011, the US Geological Survey[53] estimated Oregon to have roughly 7,500 km^2 of pasture and 90,400 km^2 of rangeland (US Geological Survey 2011). As explained in Chap. 17, the distinction between pasture and rangeland is based on the intensity of use. Rangeland is characterized as primarily herbaceous vegetation that is managed as a natural ecosystem, whereas pasture receives cultural treatments and periodic renovation that may include tillage, fertilization, mowing, weed control, and irrigation NRCS (2014). Rangeland is not necessarily grazed, whereas pasture is by definition used for grazing. Because this chapter focuses on working lands, the aspects of rangeland that relate to grazing management are emphasized here.

Cattle and sheep are the predominant Oregon livestock types. NASS (2019) reported 2017 Oregon totals of approximately 1.2 million cattle and calves (includes dairy and beef) and about 180,000 sheep and lambs. Malheur, Morrow, Harney, Lake, Baker, and Klamath Counties accounted for just over half of Oregon's total cattle inventory; Morrow and Tillamook Counties had about 56% of the dairy cows. Linn, Douglas, Curry, and Lane Counties accounted for about 56% of the recorded sheep and lambs. Horses, goats, llamas, and alpacas are important locally, and for many families personally, but they do not rank in the top 40 Oregon commodities economically and will not be included in this chapter due to limitations on space.

Additionally, it is noted that livestock can and do graze cropland, forestland, and land managed primarily for wildlife. For example, cattle often graze wheat stubble in parts of the Columbia Plateau and sheep graze annual ryegrass and white clover (*Trifolium repens*) seed fields in the southern portion of the Willamette Valley. Cattle and sheep also graze

[53] U.S. Geological Survey (USGS) data are used here for pasture and rangeland because NASS data do not include federal land, much of which is rangeland in eastern Oregon. The USGS 2011 estimate of 7,500 km^2 of pasture in Oregon is within the margin of error of estimated pasture reported in the 2015 NRCS National Resources Inventory (7,000 ± 900 km^2). The values for rangeland used here include the USGS vegetation categories of shrub and herb, desert and semi-desert, and recently burned grassland and shrubland. These USGS data do not indicate what percentage of that land is grazed.

Fig. 18.10 Small farms in Oregon are increasing in number. Many small farms use organic growing techniques and sell their produce to local consumers. Pictured is a crop of garlic on a small, certified organic and biodynamic farm in the Willamette Valley of Benton County. *Source* Photograph by NRCS Oregon

oak savanna or forests of widely spaced pine where sunlight penetration and rainfall are sufficient to produce understory forage for grazing. Examples of this grazing on forestland abound in the Siskiyou Mountains MLRA and in eastern Oregon. Although not common, grazing of cattle or goats can occur on land managed principally for wildlife habitat in order to control weeds, maintain a prairie at an early seral stage, or to provide income to landowners whose primary goal is habitat.

18.5.1 Pasture

Oregon pastureland is overwhelmingly under private ownership and occurs in both western and eastern Oregon. Pastures are typically seeded to one or more of the same cool season grasses used for hay production, including orchardgrass (*Dactylis glomerata*), tall fescue (*Schedonorus arundinaceus*), perennial ryegrass (*Lolium perenne*), Reed canarygrass (*Phalaris arundinacea*), or meadow foxtail (*Alopecurus pratensis*). OSU does not recommend seeding Reed canarygrass due to its invasive nature, but pastures dominated by the species are not uncommon either because it was seeded historically or because it invaded a pasture and crowded out other species. OSU also does not recommend seeding Timothy (*Phleum pratense*) for pasture, because it does not tolerate grazing well. It's common for pastures to contain one or more legume species, such as alfalfa (*Medicago sativa*), red clover (*Trifolium pratense*), white clover (*Trifolium repens*), big trefoil (*Lotus pedunculatus*), or birdsfoot trefoil (*Lotus corniculatus*). Two annual pasture species sometimes employed are subclover (*Trifolium subterraneum*) and annual ryegrass[54] (*Lolium perenne* ssp. *multiflorum*), which can be managed to reseed themselves (Fransen and Chaney 2002). See the preceding section on hay and haylage for adaptations of forage species to soil drainage characteristics.

[54] Annual ryegrass is sometimes called by the common name "Italian ryegrass" and by the Latin name *Lolium multiflorum*. Although some cultivars of this species truly are annual plants, other cultivars are short-lived perennials that can live 4 or more years. There are also cultivars that are hybrids between perennial ryegrass and annual ryegrass (Jung et al. 1996; NRCS 2020c).

The narrow coastal plains and valleys in the Sitka Spruce Belt have limited farmland due to steep soils and wet, cool weather. However, where soils are suitable, pastures are highly productive and grazing can occur year-round, creating a near-ideal environment for dairy cows. Tillamook County is a prime example of this, with many farms providing milk to the Tillamook County Creamery Association, a co-op of about 80 dairy farmers founded in 1909 (Tillamook County Creamery Association 2019).

Grazing in the northern part of the Willamette Valley (Multnomah, Clackamas, Washington, Yamhill, Marion, and Polk Counties) is mostly by beef and dairy cattle. However, these enterprises have declined in recent years, due in part to the high cost of farmland. For example, between 1997 and 2017, milk cow numbers in these six counties declined from 32,284 to 28,290 (NASS 2021). Cattle grazing in the southern Willamette Valley (Linn, Benton, and Lane Counties) has remained relatively stable over these years. Southern Willamette Valley counties also have significant sheep grazing during the winter, when sheep are moved north from the Siskiyou Mountains MLRA to graze on annual ryegrass that will later be harvested for seed.

Grazing lands in the Siskiyou Mountains MLRA include both sheep and cattle production and both dryland and irrigated fields. Sheep numbers expanded in the 1950s when Oregon State University promoted clearing brush and oak from hillsides and then seeding subterranean clover for pasture. After peaking in the 1980s, sheep production in this MLRA decreased and cattle ranching expanded. This change was due partly to predation on sheep by coyotes, bears, and cougars (Reed 2017), and partly to decreased American consumption of lamb and mutton (USDA-Economic Research Service 2020b). The 2017 agriculture census estimated about 85,000 cattle and calves and 28,000 sheep and lambs in Douglas, Jackson, and Josephine Counties combined.

Grazing occurs on a wide range of soils. For example, Locoda silt loam, protected, is a very poorly drained Typic Fluvaquent that occurs in floodplains protected by dikes and drainage systems along the Columbia River in Clatsop and Columbia Counties. The mean annual precipitation is about 1,700 mm and the mean annual air temperature is 11.7 °C. This map unit is part of Forage Suitability Group[55] (FSG) G001XY009OR. For this FSG, common forage species include tall fescue, white clover, birdsfoot trefoil, and Reed canarygrass. Productivity is about 20 animal unit months (AUM) ha^{-1} when not irrigated; the FSG report does not provide data for irrigated conditions. May is typically the month of greatest forage growth, and some growth occurs every month except December through February. Grazing should be restricted when soils are saturated.

A rather different soil used for pasture is Coburg silty clay loam, 0–5% slopes, a moderately well-drained Ultic Argixeroll that occurs on terraces in the Siskiyou Mountains MLRA. The mean annual precipitation is 1,150 mm, and the mean annual air temperature is about 11 °C. This map unit is part of FSG G005XY006OR. Common forage species for this FSG are tall fescue, orchardgrass, perennial ryegrass, red clover, and subterranean clover. This FSG produces about 20–45 AUM ha^{-1} when irrigated and about 5–25 AUM ha^{-1} when not. The greatest forage growth is in May, but insignificant growth occurs in July through September unless irrigated. Whether irrigated or not, insignificant growth occurs in December and January. Grazing should be restricted when the soil is wet.

Pastures in eastern Oregon are often managed in conjunction with grazed rangeland, where beef cattle or sheep are moved between pastures (often located along streams where irrigation water is available) and rangeland, depending on the season and availability of forage. Figure 18.11 illustrates such an example with Wingville silt loam, 0–2% slopes, managed for pasture. This is a somewhat poorly drained Pachic Haploxeroll that occurs on floodplains in the Blue Mountain Foothills of Baker County. It is Land Capability Class 3w, is prime farmland if irrigated and drained, with a mean annual precipitation of about 280 mm and a mean annual air temperature of about 8.9 °C. Web Soil Survey reports do not provide a FSG for this map unit but estimate the irrigated pasture productivity at 36 AUM ha^{-1}.

Nutrient management recommendations include testing soil for P, K, Ca, Mg, and pH and then applying nutrients or lime prior to establishing a new pasture. Generally, a pH of 5.5 or greater is recommended for pastures. After establishment, N fertilizer rates depend on pasture productivity and whether irrigation is used. Nitrogen can be applied between one and three times per year, at a rate of about 55–70 kg N ha^{-1} for each application. OSU has developed a method using heat units calculated from daily air temperatures to schedule late winter/early spring N applications based on the growth curves of pasture grasses (Hart et al. 2000). Irrigation of pasture varies by climate, productivity, and availability of water. Average net annual irrigation requirements range from about 180 mm in northern coastal counties to over 900 mm in the Columbia Basin (Cuenca 1999). Common pasture irrigation systems include sprinkler systems in all parts of the state and flood irrigation in parts of eastern Oregon.

Conservation practices, useful for Oregon pastures, include nutrient management, fences, prescribed grazing, filter strips, and watering facilities. It's often necessary to set aside a "sacrifice area" where livestock can be confined when there is inadequate pasture forage available or the

[55] See the section on grazing land yields in this chapter for an explanation of Forage Suitability Groups and animal unit months.

ground is too wet to support livestock. Livestock can be fed hay in these sacrifice areas, which should be relatively small, have well-drained soils, and be located on flat or gently sloping land away from water sources. Pasture vegetation is invariably damaged due to the concentration of livestock in sacrifice areas, but this is preferable to the destruction of an entire pasture (Fransen and Chaney 2002).

18.5.2 Grazed Rangeland

Although resource professionals may have theoretical disagreements about whether rangeland occurs west of the Cascade crest, this chapter confines the discussion of grazed rangeland to the large portions of eastern Oregon that are grazed but not cropped, tilled, fertilized, or irrigated. Most Oregon rangeland is owned by the Bureau of Land Management and private ranches or farms, but some is owned by the US Forest Service and the state of Oregon. Bureau of Land Management rangelands are mostly in southeast Oregon, while private rangelands are more common in the northeast part of the state. See Fig. 17.1.

Much of the grazed rangeland in Oregon is characterized by plant communities that historically were dominated by bunchgrasses, such as bluebunch wheatgrass (*Pseudoroegneria spicata*), Idaho fescue (*Festuca idahoensis*), Sandberg bluegrass (*Poa secunda*), and various needlegrasses (*Stipa* spp.). Rangeland shrub vegetation in Oregon includes sagebrush (*Artemisia* spp.), juniper (*Juniperus* spp.), mountain mahogany (*Cercocarpus ledifolius*), saltbush (*Atriplex* spp. and others), and greasewood (*Sarcobatus* spp.). A shrub canopy zone often exists with a dominant shrub, an understory, and interspace area consisting of smaller shrubs, bunchgrasses, forbs, and biological soil crusts (lichens, mosses and cyanobacteria at the soil surface).

The early history of grazing on Oregon rangelands is both remarkable and sad. Lewis and Clark noted horses in a Shoshone village, and it is believed that the tribe probably procured those horses around 1700 from a Spanish mission in present-day New Mexico. Horses, elk, deer, and antelope were the main grazing animals present in Oregon's rangelands before the 1830s, when Euro-American settlers began arriving in significant numbers. Settlers brought livestock to the Pacific Northwest via the Oregon Trail and also imported them from California. Early on, most of Oregon's livestock were confined to western Oregon. In the 1860s, however, grazing operations began expanding in eastern Oregon, where bunchgrasses appeared to offer unlimited forage. In the 1870s, cattle, sheep, and horses were being driven from Oregon to Wyoming, and in 1883, railroads began to move Oregon livestock to eastern markets.

Early on, livestock operations experienced staggering setbacks. Several severe winters in the late 1800s killed large numbers of Oregon cattle, a result of the belief that livestock could be left on rangeland with little thought to winter shelter or feeding. Because much of the eastern Oregon rangeland was considered a free resource open to all, there was little incentive for conservation. By the 1880s, overgrazing was causing erosion, and native bunchgrasses were being displaced by cheatgrass (*Bromus tectorum*), an introduced annual grass that produces little forage. Also, decreased bunchgrass populations led to the expansion of native woody species like sagebrush and juniper, neither of which is desirable forage for cattle.

In addition to these conservation problems, ranchers faced problems with their neighbors. Eastern Oregon farming operations were plowing out prairies, seeding crops, and building fences, which led to conflicts between cattle producers and farmers. Around 1900, severe conflicts also erupted between cattle and sheep operations, with violent repercussions such as the establishment of "sheep shooter committees." Finally, between 1906 and 1910, the federal government began to develop grazing laws. Initially, the US Forest Service administered these laws, but the 1934 Taylor Grazing Act transferred the management of much of the federal rangeland in the western United States to what would later become the Bureau of Land Management (Galbraith and Anderson 1971).

In modern times, ranchers obtain permits from the Bureau of Land Management or US Forest Service for grazing livestock on specific parcels ("allotments") of federal land. These agencies usually authorize permits for 10 years but provide annual specifications for the number and type of livestock and the allowable grazing periods. Ranchers buy permits based on the animal unit months allowed via the permit (Keyes and Keyes 2015).

Figure 18.11 illustrates rangeland that lies on hills above an irrigated floodplain in the Blue Mountain Foothills of Baker County. The lower hills in Fig. 18.11 are predominantly Encina gravelly silt loam, 12–35% south slopes. The mean annual precipitation for the Encina series is about 300 mm, and the mean annual air temperature is about 9 °C. Encina is a deep, well-drained Calciargidic Argixeroll formed in loess and lacustrine sediments on south slopes of dissected terraces. This Encina map unit's Ecological Site Description[56] is SR South 9–12 PZ, which has a predominant natural plant community of Wyoming big sagebrush (*Artemesia tridentata*) and bluebunch wheatgrass (*Pseudoroegneria spicata*), and an estimated annual production of 700–1,100 kg ha^{-1}. If over-grazed, bluebunch wheatgrass will decrease, while sagebrush and rabbitbrush (*Ericameria*

[56] See the section on grazing land yields in this chapter for an explanation of Ecological Site Descriptions and the associated annual production estimates.

Fig. 18.11 Irrigated pasture with grazing cattle in the foreground and rangeland with some forestland in the hills. This is a site in the Blue Mountain Foothills MLRA of Baker County. The predominant soil in the pasture is Wingdale silt loam, and the predominant soil in the rangeland of the lower hills is Encina gravelly silt loam. *Source* Photograph by NRCS Oregon

nauseosa) will increase. Continued deterioration will result in annual grasses and shrubs dominating the site, and potentially irreversible changes to the soil profile.

Generally, a conservation rancher's goal for Oregon rangeland is to achieve sustainable income from livestock production while maintaining the historic climax plant community shown in the Ecological Site Description. If a site has degraded to a disturbed state, it is sometimes possible to return it to its historic community; if not, the goal is to keep the site from degrading further. Often, managers have additional goals regarding wildlife habitat. For example, partnerships between agencies and landowners are seeking to restore habitat for the greater sage-grouse (*Centrocercus urophasianus*), an at-risk species considered to be an indicator of the health of sagebrush rangeland ecosystems (Sage-Grouse Conservation Partnership 2015). Common rangeland practices include prescribed grazing, prescribed burning, fences, watering facilities, range planting, wildlife habitat planting, habitat management, brush management, and herbaceous weed treatment.

18.6 Forestry Management Systems

The US Forest Service began a Forest Inventory and Analysis (FIA) program in 1928 to collect information on the nation's forest resources, including public and private forestlands. The FIA defines forests as sites at least 0.4 ha in size and at least 37 m in width that currently have, or formerly had in the past 30 years, at least a 10% canopy of trees. Current FIA data, based on inventories completed in 2006 through 2015, estimated total Oregon forestland at

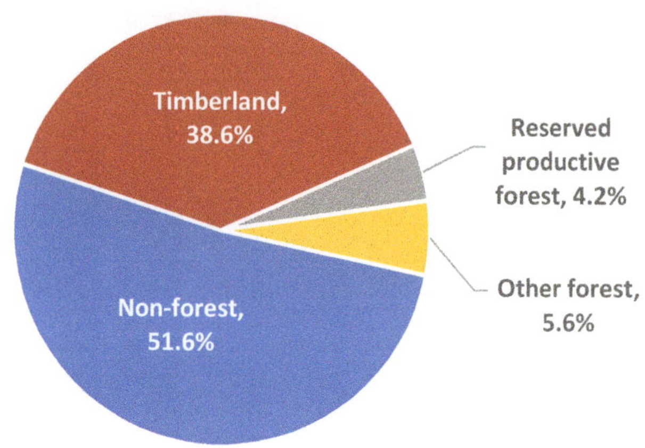

Fig. 18.12 Oregon forest inventory and analysis classes. *Source* Chase et al. (2018)

about 120,000 km^2, which is about 48% of Oregon's total surface area and is essentially the same as 2011 US Geological Survey estimates (US Geological Survey 2011). The FIA defines categories of land as shown in Fig. 18.12: non-forest, timberland, reserved productive forest, and other forest[57] (Chase et al. 2018). Figure 3.8 illustrates the general location of Oregon forest types (Oregon Department of Forestry, n.d.).

According to the current FIA data, conifers, also called "softwoods," accounted for more than 85% of Oregon's forest area, and hardwoods accounted for 11%. Non-stocked areas, which temporarily have less than 10% forest cover due to fire or recent harvest, accounted for 4% of the state's forestland.

Oregon's timber industry expanded in the mid-1800s in response to Euro-American settlement, the rapid growth of Portland, and the California gold rush. Prior to the 1880s, logging operations were usually located near rivers, and logs were moved through the forest by livestock teams, often composed of oxen. The "donkey engine,[58]" a steam-powered device that could pull logs up to a mile through the forest by means of drums or winches and cables, was invented in 1881. It helped to mechanize logging operations and also allowed logging further from rivers and on steep slopes where oxen teams could not function (Kamholz 2018; Tucker 2002). Around the same time, the expansion of railroads allowed timber to be shipped to distant markets (World Forestry Center 2020).

Even after the donkey engine allowed logs to be moved long distances through the forest, the logs were often floated down streams to mills during winter high flow periods. That practice, however, created a dearth of logs for mills during summer months. In 1884, Charles Granholm built a "splash dam" on the north fork of Oregon's Coos River, beginning a practice that soon was used widely throughout western Oregon. Splash dams were wooden structures that dammed a stream or river and created a reservoir where loggers could stockpile logs. When the reservoir was full of water and logs, the dam was opened quickly and a torrent of water carried the logs downstream. Splash dams allowed logs to be moved via water even during the summer months, but the tremendous rush of logs and water scoured streambeds and destroyed habitat for aquatic organisms. Chinook salmon (*Oncorhynchus tshawytscha*), which spawn in the larger streams and rivers downstream from splash dams, were especially impacted. Logging trucks were increasingly used in the 1900s to move logs to mills, but splash dams continued to be used into the 1950s. The scars from this practice are still visible on some Oregon streams (Lichatowich 1999).

Two immigrants, Frederick Weyerhauser and Simon Benson, developed large and profitable lumber businesses in the Pacific Northwest, beginning in the late 1800s (Abbott 2018). The Weyerhauser Company is now one of the largest timber companies in the country. In addition to running a large and profitable lumber company, Simon Benson helped found Benson Polytechnic School, which still operates in Portland. Benson also donated iconic four-lobed brass water fountains to the City of Portland, reportedly to provide drinking water to his workers as an alternative to alcohol. These "Benson Bubblers" continue to function throughout downtown Portland today. By 1938, Oregon was the nation's leading producer of lumber and is still first in softwood lumber production (16% of US totals in 2019) and plywood (28% of US totals in 2019) (Oregon Forest Resources Institute, 2021).

18.6.1 Western Oregon Forests

Douglas-fir (*Pseudotsuga menziesii*) is the most common Oregon forest type, occupying about 44,000 km^2 or 37% of Oregon's forestland area, mostly in western Oregon's Coast Range, the Willamette Valley, and the Cascade Mountains. Western Oregon Douglas-fir forests often have components of red alder (*Alnus rubra*), big leaf maple (*Acer macrophyllum*), western redcedar (*Thuja plicata*), western hemlock (*Tsuga heterophylla*), grand fir (*Abies grandis*), and other

[57] "Timberland" includes areas capable of producing at least 1.40 m^3 ha^{-1} (20 ft^3 ac^{-1}) of industrial wood per year in natural stands. "Reserved" forestlands are areas permanently removed from wood product production through statute or administrative designation (e.g., wilderness areas and national parks). "Other" forestlands are not capable of producing 1.40 m^3 ha^{-1} of industrial wood per year, usually due to soil limitations. "Non-forest" includes all other land uses, such as rangeland, pasture, cropland, and developed land.

[58] Donkey engines were also called "steam donkeys.".

Fig. 18.13 A Douglas-fir forest in the Coast Range of Polk County. Oregon Douglas-fir forests such as this are typically harvested as clearcuts and then replanted within two years. This results in a patchwork of forest stands. Within each stand, trees are generally the same age. At least four stands are visible in this photograph, including a young stand in the foreground, older stands to the left and in the distance, and a medium-aged stand in the mid-ground. *Source* Photograph by Dean Moberg

species. Although these forests can also have diverse understory vegetation, clearcut harvests[59] followed by replanting with just Douglas-fir seedlings decreases diversity of species and age classes in those stands (Fig. 18.13). Red alder is a pioneer species, and while Douglas-fir is a long-lived early to mid-successional species, and western hemlock is the dominant climax species, these forests are commonly logged or destroyed by fire before a climax community is reached. Douglas-fir forests can be harvested with rotations as short as 30 years, whereas old growth Douglas-fir can be hundreds of years old. Douglas-fir wood is strong and is typically available in a wide range of sizes, making it a desirable species for a variety of dimensional wood products used in construction projects (Oregon Forest Resources Institute 2020; Chase, et al. 2018).

A typical soil for Douglas-fir forests is Rinearson silt loam, 30–50% slopes, a deep, well-drained Typic Humudept in the Coast Range with a mean annual precipitation of about 2,000 mm and a mean annual air temperature of 9.4 °C. Its site index[60] is 40 m (50 year base age) with an average estimated annual productivity of 14 m^3 ha^{-1}. Associated trees are big-leaf maple, red alder, and western hemlock. Typical understory plants are vine maple (*Acer circinatum*), low Oregon grape (*Mahonia repens*), salal (*Gaultheria shallon*), western swordfern (*Polystichum munitum*), and red huckleberry (*Vaccinium parvifolium*) (Soil Survey Staff 2019).

[59] A clearcut is a logging method in which an entire timber stand, except required "leave trees," is harvested. The Oregon Forest Practices Act regulates the size of clearcuts, requires forested buffers to remain around streams and lakes, and requires trees to be replanted within 2 years of harvest. Depending on the size of the clearcut, the Act may require some "leave trees" to remain standing after harvest for wildlife habitat. Clearcut harvesting is often used for trees such as Douglas-fir, which are intolerant of shade and thus are more successfully replanted in open sites created by clearcutting (Cloughesy & Woodward 2018). See the section on state regulations earlier in this chapter for more information about the Oregon Forest Practices Act.

[60] See the section on forestry yields earlier in this chapter for a review of the terms "site index" and "annual productivity.".

Spruce/hemlock[61] forests occur along the Oregon coast in the Sitka Spruce Belt. These forests may also contain Douglas-fir, red alder, western redcedar, shore pine (*Pinus contorta* var. *contorta*), and other species. Understory vegetation can be thick, providing excellent wildlife habitat for songbirds, amphibians, and small mammals. Current FIA data indicate that spruce/hemlock forests have the greatest live tree volume per hectare of all Oregon forest types. Spruce wood is light but strong and has been used in airplane frames; it is also highly resonant and commonly used to build pianos, guitars, and violins. Western hemlock is often used to make high-quality paper (Oregon Forest Resources Institute 2020; Chase et al. 2018). A typical soil for spruce/hemlock forests is Templeton silt loam, 30–50% slopes, a deep, well-drained Andic Humudept with a mean annual precipitation of about 2,000 mm and a mean annual air temperature of 9.4 °C. It has a site index of 52 m for Sitka spruce (100-year base age) with average estimated annual productivity of 18 m^3 ha^{-1}. In addition to spruce and hemlock, Templeton silt loam native plant communities often include red alder, western redcedar, Douglas-fir, salal, western swordfern, salmonberry (*Rubus spectabilis*), red huckleberry, and other species (Soil Survey Staff 2019).

Oregon white oak (*Quercus garryana*), also called "Garry oak," occurs mostly in the Willamette Valley and Siskiyou Mountains. It is a slow-growing tree that reaches heights of 25 m and can occur in thick single species stands of stunted individual trees, in mixed conifer/deciduous forests, or in savannas. When Oregon white oak grows in savannas, the widely spaced trees develop a characteristic spreading habit with lower branches growing outward and down to almost touch the ground. Oregon white oaks are resistant to fire damage and thrive in savannas that experience frequent, low-intensity surface fires.[62]

Oak wood historically was used for ship building, railroad ties, and fence posts; it is still used for wine barrels, fire wood, and furniture. Oaks provide excellent habitat and are used by over 200 species of wildlife. Birds and mammals use white oak cavities for nesting or den sites, and many species depend on the acorns for food, especially during the fall and winter when other food sources are scarce (Oregon Forest Resources Institute 2020). A typical soil for oaks is the appropriately named Oakland silt loam, 12–20% slopes, a moderately deep, well-drained Ultic Haploxeralf in the Siskiyou Mountains MLRA with a mean annual precipitation of about 1,000 mm and a mean annual air temperature of 11.7 °C. It has a site index of 23 m (base age of 50 years) with an estimated annual productivity of 6 m^3 ha^{-1} for Douglas-fir. In addition to oaks, native plants adapted to Oakland soils include big-leaf maple, Pacific madrone (*Arbutus menziesii*), beaked hazelnut (*Corylus cornuta*), western fescue (*Festuca occidentalis*), common snowberry (*Symphoricarpos albus*), Pacific poison oak (*Toxicodendron diversilobum*), and other species (Soil Survey Staff 2019).

There are other forest types in western Oregon, such as alpine forests at high elevations in the Cascade Mountains, pine (*Pinus spp.*) and California black oak (*Quercus kelloggii*) forests in southwest Oregon, and redwood forests (*Sequoia sempervirens*) in southwestern Curry County. There are also trees such as the Pacific yew (*Taxus brevifolia*), Bentham's ponderosa pine (*Pinus ponderosa var. benthamiana*),[63] and understory species such as cascara (*Rhamnus purshiana*), which are not covered here due to limitations in space. Readers are referred to *Plants of the Pacific Northwest Coast* (Pojar and MacKinnon 2014) for more information on western Oregon plant species.

Common conservation practices for western Oregon coniferous forests include tree/shrub site preparation and tree/shrub establishment, which are required by the Oregon Forest Practices Act after harvest on most sites. Brush management is often used, especially after a site is replanted, to control non-native weedy species like Himalayan blackberry[64] (*Rubus armeniacus*) and Scotch broom (*Cytisus scoparius*). Forest stand improvement (pre-commercial thinning) is used as stands mature to ensure adequate spacing between trees, and woody residue treatment is usually employed after thinning to ensure residues do not present a wildfire hazard. Forest managers employ these same practices in Oregon white oak forests, and often also use prescribed burning, prescribed grazing, and wildlife habitat planting to restore or maintain native

[61] "Spruce/hemlock" forest refers to a plant community dominated by Sitka spruce (*Picea sitchensis*) and western hemlock (*Tsuga heterophylla*).

[62] Terminology for working lands fire is as follows. "Wildland fire" occurs in vegetation and natural fuels (as opposed to fires that predominantly burn structures such as houses). Wildland fires may be "prescribed fires" (planned and intentionally ignited to meet management objectives) or "wildfires" (unplanned fires caused by lightning, human accidents, arson, or escaped prescribed fires). Fires are further classified by the height at which they occur. "Ground fires" are slow-burning, smoldering fires that burn organic soils, duff, decomposing litter, buried logs, underground stumps, and roots. "Surface fires" burn needles, leaves, fallen branches, herbaceous plants, shrubs, and small trees, with flame lengths typically less than 1 m. "Crown fires" burn through the canopy of a forest. "Catastrophic wildland fires" have significant negative economic, social, and/or ecological effects. Catastrophic fires in forests are usually crown fires and often kill a majority of the trees in a stand (National Park Service 2017), (Wooten). See the Chap. 17 section on wildfires for information about the relationship of land use and soils to wildfire.

[63] Bentham's ponderosa pine is also known as "valley pine" or "Willamette Valley pine" (Oregon Flora 2021a).

[64] Himalayan blackberries are also called "Armenian blackberries," however the NRCS Plants Database (NRCS 2020c) lists the name as "Himalayan blackberry.".

understory plant communities. There are several serious challenges to western Oregon forest management, including the risk of catastrophic wildfire (see Chap. 17), sudden oak death,[65] Swiss needle cast,[66] and laminated root rot.[67] Damage from high winds and ice storms can damage forests, including the loss of branches and tops, or the blow-down of entire trees or groups of trees.

18.6.2 Eastern Oregon Forests

Ponderosa pine (*Pinus ponderosa*) is the second most common Oregon forest type, with about 21,000 km^2, or 17% of the state's forestland area, mostly east of the Cascade crest in the Cascade Mountains, Cascade Mountains—Eastern Slope, Blue Mountains, and Klamath Basin. Ponderosa pine forests in these MLRAs occupy lower elevations (700–1,800 m), soils, and slope aspects that satisfy this species' precipitation and temperature requirements. Ponderosa pine may occur in pure stands or be mixed with species such as lodgepole pine (*Pinus contorta*), quaking aspen (*Populus tremuloides*), or others. Ponderosa pine forests transition through elevation gradients into other forest types, such as Douglas-fir and grand fir (*Abies grandis*) at higher elevations or western juniper (*Juniperus occidentalis*) on south-facing slopes. Where well-adapted, ponderosa pine is the climax forest species, responds well to selective harvesting, and is thus often managed in uneven-aged stands. Ponderosa pines are well adapted to low-intensity surface fires, but when fire suppression allows exuberant understory growth or crowded conditions, these forests become vulnerable to crown fires. Ponderosa pines can live for over 700 years and grow to heights of 60 m. Because the wood has a uniform grain and doesn't shrink and swell much after processing, it is used in construction and millwork (Oregon Forest Resources Institute 2020; NRCS 2006).

A typical soil that supports ponderosa pine in the Cascade Mountains—Eastern Slope is Smiling sandy loam, 0–15% slopes, a deep, well-drained Alfic Vitrixerand with a mean annual precipitation of 760 mm and a mean annual air temperature of 6.7 °C. This map unit has a ponderosa pine site index of 24 m (100 year base age) with an average estimated annual productivity of 4.7 m^3 ha^{-1}. In addition to ponderosa pine, native plants adapted to Smiling soils include Douglas-fir, antelope bitterbrush (*Purshia tridentata*), snowbrush (*Ceanothus velutinus*), Idaho fescue (*Festuca idahoensis*), and other species (Soil Survey Staff 2019).

Lodgepole pine forests occur in the same MLRAs as ponderosa pine forests, but are pioneer species that can colonize sites disturbed by fire or clearcut harvests. After colonization, lodgepole pines may be slowly replaced by other species like ponderosa pine. However, lodgepole pine forests may persist where conditions are less favorable, such as frost pockets or soils that are either too wet or too dry for other species. These trees are adapted to surface fires, which help to open the cones and release seeds. They can form excessively thick stands, which results in stunted growth and susceptibility to crown fires and insect damage. The name "lodgepole" derives from this tree's habit of forming straight and moderately slim trunks, which are often used for log cabins or utility poles. Lodgepole pines can also be used for paper products. A typical lodgepole pine soil is Timbercrater ashy paragravelly loamy sand, a Typic Vitricryand in the Cascade Mountains near Crater Lake. This is a very deep, excessively drained soil that formed in deposits of ash and pumice. The mean annual precipitation is about 1,500 mm, and the mean annual air temperature is 5.0 °C. In addition to lodgepole pine, native plant communities may include mountain hemlock (*Tsuga mertensiana*), grouse huckleberry (*Vaccinium scoparium*), and other species (Soil Survey Staff 2019).

There are other forest types in eastern Oregon, such as sub-alpine forests in the Wallowa Mountains and juniper forests in central and southeast Oregon, and species, such as the western larch (*Latrix occidentalis*) and saskatoon serviceberry (*Amelanchier alnifolia*), that are not covered here due to limitations on space. References such as *Flora of Oregon* (Meyers et al. 2015), the Oregon Flora Project (Oregon Flora 2021b), or institutions such as the High

---

[65] Sudden oak death (SOD) was diagnosed fairly recently in Oregon and has the potential to cause devastating losses to both native ecosystems and the Oregon nursery industry. It is caused by a fungus, *Phytophthora ramorum*, that infects a variety of species, including California black oak (*Quercus kelloggii*), tanoak (*Notholithocarpus densiflorus*), Douglas-fir, and others. Currently, SOD appears to be confined to Curry County in southwest Oregon. Efforts at control include identifying infected plants and then cutting, piling, and burning infected plant material and host plants within a radius of about 90 m (Oregon Department of Forestry 2019; Rizzo & Garbelotto 2003).

[66] Swiss needle cast (SNC) is a disease of Douglas-fir caused by the fungus *Phaeocryptopus gaeumannii*. Although the fungus is native to the Pacific Northwest, the incidence of SNC has increased dramatically since the 1990s. The disease is specific to Douglas-fir and can cause needle loss and reduced growth, but rarely kills a tree. One management strategy is to diversify forests by adding species such as western hemlock when forests are replanted on sites with moderate to high SNC. This does not reduce the incidence of the disease on Douglas-fir, but the other species can compensate for yield losses if Douglas-fir growth is reduced (Mulvey et al. 2013).

[67] Laminated root rot is the most damaging western Oregon forest disease. Douglas-fir, western hemlock, and the true firs are highly susceptible. Management options include planting resistant species such as western red-cedar, thinning to reduce the spread of disease from one tree to another, and removing infected stumps (Shaw et al. 2009).

Fig. 18.14 This ponderosa pine forest in eastern Oregon has been thinned and pruned to improve forest health, enhance wildlife habitat, and reduce the risk of catastrophic wildfire. *Source* Photograph by NRCS Oregon

Desert Museum near Bend can provide more information on eastern Oregon forests and plant species.

Although harvest rotations, yields, and timber products are different from those in western Oregon, forest management practices east of the Cascades are broadly similar. Tree/shrub site preparation and establishment, brush management, forest stand improvement, and woody residue treatment are all used in eastern Oregon forests. Prescribed burning is especially important in many eastern Oregon forests to prevent overcrowding, insect damage, and wildfire risk. Tree/shrub pruning can be beneficial, especially in ponderosa pine forests, to improve the quality of lumber and to decrease "ladder fuels" that enable fires to climb from the forest floor into tree canopies (Fig. 18.14). In addition to wildfire, insects such as pine beetles (*Dendroctonus* spp.), pine engravers (*Ips* spp.), western spruce budworm (*Choristoneura occidentalis*), and others can severely damage eastern Oregon forests. As in western Oregon forests, damage from wind can hurt forests east of the Cascade crest (Oregon Forest Resources Institute 2020; Shaw et al. 2009).

18.7 First Foods of Indigenous Peoples

It is appropriate to end this chapter with a description of what came first for the area now known as Oregon. For Indigenous Peoples, First Foods[68] represent more than nutrition. First Foods are also an integral part of tribal cultures and spirituality based on a reciprocal responsibility to care for the land, non-human beings, air, and water that provide for the people (Donatuto et al. 2020). Information about First Foods is somewhat overwhelming, given the tremendous range of species that Indigenous Peoples used in the past and still use today for food, medicines, dyes, and fibers.

For example, Chap. 17 lists nine federally recognized Indigenous Nations in Oregon and five additional Nations that maintain an interest in Oregon land. Each of these Nations has a unique worldview, maintains a

[68] First foods are foods eaten by Indigenous Peoples beginning in the time prior to Euro-American contact (Karten 2011; Oregon Department of Education).

distinct culture, and holds rights in their own traditional use area. Thus, although there are plant species and land management characteristics common to many Indigenous Nations in the Pacific Northwest, each Nation has its own relationship with First Foods. Given the complexity of this subject, the following section admittedly provides only a limited description of First Foods. More information about First Foods in specific regions of Oregon is available from websites of Indigenous Nations or institutions such as The Tamástslikt Cultural Institute (2021), The Museum at Warm Springs (The Confederated Tribes of Warm Springs 2021), or the Chachalu Museum and Cultural Center (Confederated Tribes of Grand Ronde 2021). Numerous publications provide details about First Foods. For example, the following informative publications were used as references for this section:

- *Aligning environmental management with ecosystem resilience: A First Foods example from the Confederated Tribes of the Umatilla Indian Reservation, Oregon, USA*, (journal article) (Quaempts et al. 2018).
- *Salmon Without Rivers: A History of the Pacific Salmon Crisis*, (book), (Lichatowich 1999).
- *Native American Food Plants: An Ethnobotanical Dictionary*, (book), (Moerman 2010).
- *Native American Medicinal Plants: An Ethnobotanical Dictionary*, (book), (Moerman 2009)
- *Vegetable Food Products of the Foraging Economies of the Pacific Northwest*, (journal article), (Norton et al. 1984).
- *Plants Database,* (website), (NRCS 2020c).
- *Oregon Flora Project*, (website), (Oregon Flora 2021a).
- *Plants of the Pacific Northwest Coast*, (book), (Pojar and MacKinnon 2014)

The serving order of First Foods, such as during feasts and other ceremonial meals, is culturally important. For example, the Confederated Tribes of the Umatilla Indian Reservation serve First Foods in the order in which Oral Histories indicate the foods promised to care for the people: water, fish, big game, roots, berries, and then again water (Quaempts et al. 2018). Due to limitations of space, the following sections only describe a limited number of First Foods, selected because of their applicability to different parts of Oregon and their associations to soils. The sections are generally ordered alphabetically by scientific name, beginning with the section on *Camas*. However, the section on water and salmon is presented at the end of this chapter, and thus of this book, because those First Foods have such special importance to Indigenous Peoples, are emblematic of the Pacific Northwest, and are inextricably linked to soils across the Oregon landscape.

18.7.1 Camas

Small, also known as “common,” camas (*Camassia quamash*), (Fig. 18.15), is an important First Food across Oregon, and great camas (*C. leichtlinii*) is an important First Food in western Oregon. Camas are members of the lily family (Liliaceae) that grow on grassy slopes and meadows at low to middle elevations. Both species are classified as facultative wetland plants.[69] Natroy silty clay is a typical soil for camas beds. It is a very deep, poorly drained Xeric Endoaquert that occurs on terraces and fans with a mean annual precipitation of 1,100 mm and a mean annual air temperature of 11.7 °C in the Willamette Valley.

Camas bulbs were harvested during or just after flowering in the late spring, using specialized digging implements made of wood, antlers, or other materials. Typically, women working in groups related by kinship harvested camas. They placed the bulbs in pits lined with rocks and leaves, covered the pits with soil, and then burned a fire over the pits for 1–3 days. After cooking, the camas bulbs were either eaten immediately or dried for future use, often after being baked into cakes. The fresh bulbs do not keep well, and the long cooking periods were necessary to break down inulin, a carbohydrate in the bulbs that is fairly indigestible and unpalatable. After cooking, the taste has been compared to baked pears. The dried camas or cakes were often used for trading with other tribes and provided valuable nutrients, especially during winter months. Camas provides slightly more calories and protein and significantly more calcium and iron per unit weight than cultivated Irish potatoes (*Solanum tuberosum*). Indigenous Peoples managed camas meadows by periodic burning to maintain an early seral stage. During harvest, small camas bulbs were returned to the soil for future crops. Often, camas plots were owned by families and passed down through generations.

18.7.2 *Lomatium* Species

At least eight species in the genus *Lomatium* are First Foods. These plants are in the Apiaceae (formerly Umbelliferae) family, which includes carrots, celery, and parsley. *Lomatium* species commonly used as First Foods in Oregon are perennial forbs with stout taproots and are also known as “biscuitroots.” *L. nudicaule*, commonly called “bare-stem

[69] Plants are classified according to the probability that they occur in wetlands. There are five categories in this rating system: obligate wetland plants occur almost always in wetlands, facultative wetland plants usually occur in wetlands, facultative plants are equally likely to occur in wetlands and non-wetlands, facultative upland plants usually occur in non-wetlands, and upland plants almost always occur in non-wetlands (Lichvar et al. 2012).

Fig. 18.15 Small, or common, camas (*Camassia quamash*) at The Nature Conservancy's Camassia Natural Area in the Willamette Valley MLRA of Clackamas County. The soil is probably Cascade silt loam, a somewhat poorly drained Humic Fragixerept. *Source* Photograph by Christopher Reidy

biscuitroot" or "bare-stem desert-parsley," occurs widely across both eastern and western Oregon. It is a facultative upland plant in western mountains, valleys, and coastal regions and occurs in dry, open, or sparsely wooded areas at low to middle elevations. The Oregon Flora Project notes a collection of *L. nudicaule* just north of Corvallis on soil mapped as Willamette silt loam, a very deep, well-drained Pachic Ultic Argixeroll in the Willamette Valley. The mean annual precipitation is about 1,100 mm, and the mean annual air temperature is 11.1 °C. Another species, *L. grayi*, also known as "butterfly-bearing[70] biscuitroot" or "Gray's biscuitroot," is also adapted to well-drained sites and grows in rocky areas with open habitat such as sagebrush, juniper, or ponderosa pine communities (Fig. 18.16). The Oregon Flora Project notes a collection of *L. grayi* near the community of Pine Grove on Mutton gravelly loam, a very deep, well-drained Vitrandic Haploxeralf in the Cascade Mountains—Eastern Slope. The mean annual precipitation is about 460 mm, and the mean annual air temperature is 8.3 °C.

The young leaves or sprouts of *L. nudicaule* and *L. grayi* are high in vitamin C, and Indigenous Peoples ate them raw or cooked. The roots were also an important food source, and multiple species of *Lomatium* have similar protein and carbohydrate levels, but higher calcium and iron content than Irish potatoes. *Lomatium* roots were eaten raw or cooked and were preserved by drying for later use. After cooking, Indigenous Peoples mashed the roots or used them

[70] Some populations of *L. grayi* are also known as *L. papilioniferum*, a name based on the Papilionidae family of swallowtail butterflies. It is a host plant for the rare Indra swallowtail butterfly (*Papilio indra*) and is one of two host plants used by the Anise swallowtail (*P. zelicaon*). Because of its habit of beginning growth soon after snowmelt, *L. grayi* is an important plant for early spring insects, which in turn provide a food source for greater sage-grouse (*Centrocercus urophasianus*) chicks.

Fig. 18.16 *Lomatium grayi* (also known as *L. papilioniferum*) in bloom at the Nature Conservancy's Tom McCall Nature Preserve in the Cascade Mountains, Eastern Slope MLRA of Wasco County. *Lomatium columbianum*, not flowering and with lighter green foliage, is also present at this site. *Source* Photograph by Christopher Reidy

in soups or stews that might also include fish. Sometimes the roots were pulverized and baked into cakes. Indigenous Peoples used the seeds of *L. nudicaule* for flavoring and to make tea, or chewed them to treat colds, sore throats, and tuberculosis. Indigenous Peoples managed *Lomatium* and other root crops by leaving most roots unharvested, replanting seeds, or delaying harvest until after seed production.

18.7.3 Oregon White Oak

Oregon white oak (*Quercus garryana*) acorns were a food staple of Indigenous Peoples west of the Cascade crest. This tree is a facultative upland plant that is adapted to a variety of sites, including dry rocky slopes, deep well-drained soils, and even well-drained floodplains. They are common in the Willamette Valley and Siskiyou Mountains. The species also extends somewhat into eastern Oregon, mostly in the Cascade Mountains—Eastern Slope near the Columbia River. A typical soil is described in the section on western Oregon forests in this chapter.

The acorns were harvested after falling from the tree and then roasted or steamed. After the initial cooking, Indigenous Peoples removed the acorn meat from the shells and dried it. Before eating, the meat was soaked in water to remove bitter tannins and then cooked again. Acorns are an energy-rich food due to their high-fat content and have a higher niacin content than pinto beans (USDA-Agricultural Research Service n.d.). Indigenous Peoples used oak to fashion camas digging tools and for firewood. Some tribes used a decoction of the bark as a treatment for tuberculosis. Indigenous Peoples used periodic burning to control brush and small trees in oak savannas, which resulted in large oaks that produced good crops of acorns. Burning understory vegetation also facilitated acorn harvest.

18.7.4 The Huckleberries

Several species of huckleberries, including black (*Vaccinium membranaceum*), red (*V. parvifolium*), and evergreen (*V. ovatum*) are First Foods. Huckleberries are perennial shrubs in the Ericaceae (heath) family, which also includes blueberries and cranberries. Huckleberries are classified as facultative upland plants in western mountains, valleys, and coastal regions.

Black huckleberries (Fig. 18.17) are common in middle to upper elevation coniferous forests in western and north-eastern Oregon. They can grow as understory plants but are especially abundant in forest openings created by fire. A typical black huckleberry soil is Tolo ashy silt loam, a deep or very deep, well-drained Alfic Vitrixerand in the Blue Mountains near the Confederated Tribes of the Umatilla Indians Reservation.

Fig. 18.17 Black huckleberries (*Vaccinium membranaceum*), with both ripe and unripe berries. *Source* Photograph by Christopher Reidy

Red huckleberries are mostly a western Oregon species, occupying moist sites in coniferous forests at low to middle elevations. Red huckleberries are shade tolerant and often grow on old stumps, logs, or soils with abundant decaying organic matter. The Oregon Flora Project notes a collection of red huckleberries at the west end of Bull Run Lake[71] at an elevation of approximately 990 m on Lastance stony fine sandy loam, a deep, well-drained Typic Haplocryod in the Cascade Mountains.

Evergreen huckleberries, also known as "California huckleberries," typically occur in western Oregon, especially in coastal coniferous forests at lower elevations. Evergreen huckleberries can be found in Harris Beach State Park at an elevation of approximately 60 m on soil mapped as Bullards sandy loam, a very deep, well-drained Typic Haplorthod in the Sitka Spruce Belt.

Black and red huckleberries are gathered from summer to fall, while evergreen huckleberries are gathered in fall or even into the winter months. Indigenous Peoples often traveled long distances to harvest huckleberries. Black huckleberries are sweet, juicy, and can produce abundant crops, especially on old burn sites. Red huckleberries are tart. All three species can be eaten raw or used as ingredients in cooking. To preserve huckleberries, Indigenous Peoples dried them like raisins, sometimes using a smoky fire made from the huckleberry branches. In addition to being a good source of calories, huckleberries are fairly high in vitamin C. A decoction of red huckleberry bark was sometimes used as a cold remedy. Indigenous Peoples used periodic burning to manage huckleberry stands and left some berries on huckleberry bushes for use by wildlife.

18.7.5 Water and Salmon

No discussion of First Foods in the Pacific Northwest would be complete without water and salmon. The Columbia River Inter-Tribal Fish Commission (2021) relates the traditional importance of water and salmon as follows: "From a tribal legend, we learn that when the Creator was preparing to bring forth people onto the earth, He called a grand council of all creation. From them, He asked for a gift for these new creatures—a gift to help the people survive, since they would be quite helpless and require much assistance from them all. The very first to come forward was Salmon, who offered his body to feed the people. The second to come forward was Water, who promised to be the home to the salmon. In turn, everyone else gathered at the council gave the coming humans a gift, but it is significant that the very first two were Salmon and Water. In accordance with their sacrifice, these two receive a place of honor at traditional feasts throughout the Columbia Basin."

Pacific salmon and their kin are members of the genus *Oncorhynchus*. North American species in this genus include Chinook (*O. tshawytscha*), coho (*O. kisutch*), chum (*O. keta*), sockeye (*O. nerka*), pink (*O. gorbuscha*), steelhead and rainbow trout (*O. mykiss*), and cutthroat trout (*O. clarkii*). Except for rainbow trout, inland cutthroat trout, and some landlocked populations of *O. nerka* (called "kokanee"), the *Oncorhynchus* are anadromous, hatching and rearing in freshwater and then migrating to the ocean, where they grow for 1–7 years before completing the journey back to spawn in their natal stream. Each *Oncorhynchus* species has characteristic spawning times and habitat, but the relatively high genetic diversity in these fish ensures that not all individuals of any one population are at the same place at the same time. This diversity has made *Oncorhynchus* species robust and highly adaptable to changing conditions, at least until recent times.

Archaeological evidence demonstrates that humans were living in North America at least 15,000 years ago and may have arrived much earlier than that (Hodges 2015). In any case, both humans and salmon were present in Oregon during the latter part of the Wisconsin glaciation period and the Missoula floods. However, during the ice age and its aftermath of warm temperatures, disturbed soils, unstable river channels, and changing vegetation, Pacific salmon populations were modest at best and did not contribute greatly to the diets of Indigenous Peoples in the Pacific Northwest, who mostly hunted mammals as a source of protein. By around 5,000 years ago, Oregon's river channels and sea levels had largely stabilized. At that time, modern climax plant communities were in place, including riparian vegetation and old forests that added large wood to streams, creating habitat and contributing the insects that young salmon need to survive. The climate became cooler and wetter. Salmon populations began to increase, eventually reaching enormous levels and becoming a key component for both First Foods and ecosystem function (Lichatowich 1999).

As salmon populations expanded, Indigenous Peoples adapted to this new food source by developing harvesting technologies and social structures. These technologies included hooks, nets, seines, spears, and weirs. Another key adaptation was the preservation of salmon with smoking and drying, which began about 3,000 years ago. Salmon preservation enabled Indigenous Peoples to utilize this tremendous food source for much of the year, rather than only during the few months that adult salmon are in rivers. As the use of abundant salmon runs increased, Indigenous populations grew and developed social structures with more permanent villages and a way of life sometimes called a "gift economy" (Ibid.).

[71] The highly protected Bull Run watershed is the primary source of drinking water for the city of Portland.

In gift economies, valuable goods are given away, creating a relationship between giver and receiver, with obligations to provide gifts in return. The coastal tribes had well-developed gift economies that included potlatch gatherings in which a visiting tribe received abundant gifts from the host tribe. The visiting tribe was then obligated to host a future potlatch. The concept of gifts also applied to the relationship of Indigenous Peoples to nature, as embodied in the legend referenced above in which Salmon provided the gift of his body as food for the humans. In other words, Indigenous Peoples treated the salmon, water, and other parts of nature as beings who were willing to help humans as long as humans respected and honored them in return. This belief system translated into practices that conserved salmon runs (Ibid.). Quaempts et al. (2018, p. 29) describe this relationship between Indigenous Peoples and First Foods with the concept of "reciprocity" that "acknowledges a moral and practical obligation for humans and biota to care for and sustain one another, and arises from human gratitude and reverence for the contributions and sacrifices made by other biota to sustain human kind."

Indigenous conservation practices included such things as limiting the amount of time weirs were left in a river, which enabled large numbers of salmon to proceed upstream for spawning. Wasting salmon was considered disrespectful and legends warned that such waste would anger the salmon and cause them to reduce their runs in future years. With practices built on the idea of reciprocal gift-giving between humans, water, and salmon, Indigenous Peoples and salmon thrived together for thousands of years. Arguably, activities that Indigenous Peoples did *not* practice were more important for water and salmon conservation than those activities they *did* practice. It is not coincidental that the expansion of Euro-American culture in Oregon coincides with the decimation of salmon populations.

Activities not used in the Pacific Northwest prior to Euro-American contact can be categorized by impacts on hydrology, water quality, habitat, and harvests. Impacts on hydrology include building dams[72] and dikes, irrigation and municipal water withdrawals from rivers, manipulation of stream channels, the decrease in permeable surfaces due to urbanization, degradation of soil health, and the drainage of wetlands. Sediment, mining spoils, pesticide runoff, and industrial pollution have decreased water quality. Habitat has been impaired by the destruction of riparian vegetation, the removal of woody debris from streams, splash dams historically used to move logs down streams, placer mining, reduced frequency of surface fires, the introduction of invasive species, and culverts that interfere with salmon migration. Harvests of salmon were quickly industrialized by Euro-Americans, which allowed the export of salmon to other places. The use of hatcheries to raise salmon smolt reduced the critically important genetic diversity of these species. Finally, climate change looms large as a stressor of salmon (Long 2021). Chinook, coho, chum, sockeye, and steelhead all have one or more populations listed as threatened or endangered on the United States west coast: (National Oceanic and Atmospheric Administration, n.d.).

Three relationships between soil, water, and salmon are of special note. First, some of the most negative impacts to water and *Oncorhynchus* species listed above, and avoided by Indigenous Peoples for millennia, involve soil erosion and subsequent sediment transport to streams. Urbanization, farming, grazing, and logging all increased erosion rates compared to the relatively undisturbed vegetation present at the time of Euro-American contact. Eroded soils that enter streams as sediment can cover salmon redds,[73] smothering the eggs and alevin, resulting in a significant decrease in reproductive success (Koski 1966).

Second, soil plays a role in helping *Oncorhynchus* adults migrate upstream to spawn. Wisby and Hasler (1954) established that salmon rely at least in part on their olfactory sense to complete their remarkable journey from the ocean back to their natal stream. In other words, adult salmon migrate back to the stream where they hatched using the unique smell of the water in that stream. The most important constituents of this smell may be dissolved free amino acids derived from the stream's ecosystem, including vegetation, microbial activity, and soil (Yamamoto et al. 2013). Interestingly, there are parallels between the salmon's extraordinary olfactory sense that guides it back to its natal ecosystem's smell and the human ability to detect unique characteristics of wine derived from terroir.

A third noteworthy connection between soil, water, and salmon, recognized as early as the 1930s but only recently quantified, is the flux of nutrients and energy from the ocean

[72] Some dams, especially in early days of construction, did not include fish ladders to help adult salmon return upstream. Dams also hurt young salmon ("smolts") migrating downstream to the ocean. Some dams destroyed traditional fishing locations, most notably the inundation of Celilo Falls on the Columbia River following the construction of the Dalles Dam in 1957 (Barber 2021), (Fig. 18.18).

[73] A female salmon makes a nest, called a "redd," in the gravelly bottom of the same natal stream where her parents spawned. The female releases eggs, which are immediately fertilized by a male and then settle into the redd. After 1–3 months, the eggs hatch into baby salmon called "alevin," which are about 2–3 cm in length and are still attached to the yolk sac of the egg. After spending about 1 month in the gravels of the redd, the alevin will have absorbed the nutrients in the yolk sac and emerge from the redds as fry. If redds are covered with fine sediments before fry emerge, the dissolved oxygen levels around the eggs or alevin can drop to lethal levels. Fry that successfully emerge from redds will stay in their natal stream for periods ranging up to several years, depending on the species. Fry eventually transform into smolts and migrate into the ocean, where they become adult salmon.

Fig. 18.18 Celilo Falls on the Columbia River, 1954. Indigenous fishers balance on scaffolding and pursue salmon with nets on long poles. The Dalles Dam and reservoir permanently inundated Celilo Falls in 1957, destroying this fishing site that Indigenous Peoples had used for thousands of years. *Source* Photograph by Herb Alden, from the collection of The Oregon Historical Society, and courtesy of The Oregonian

via salmon returning to their natal streams. This flux derives from the fact that most *Oncorhynchus* species are not only anadromous, leaving their natal stream when very small and returning from the ocean as large adults, but are also semelparous (spawning once and then dying soon thereafter). This results in what has been termed a "conveyor belt" of nutrients and energy from the ocean to terrestrial and freshwater ecosystems. To illustrate the potential magnitude of this flux, Wheeler and Kavanaugh (2017) note that annual pre-contact salmon runs to the Columbia River Basin equaled around 10.5 million kg of carbon, 3 million kg of nitrogen, and significant amounts of other nutrients in the form of salmon biomass, over 95% of which was derived from marine sources consumed by the adult salmon.

Bears and other predators capture live salmon, which they generally consume on land adjacent to the stream. Salmon that escape predation die after spawning and their bodies remain in the water or (if floods occur) in the riparian area. Salmon eggs are also a concentrated source of nutrients. Whatever the fate of an individual salmon, the entire salmon run provides significant nutrients to a wide range of organisms and trophic levels, including mammals, birds, fish, invertebrates, and both aquatic and terrestrial vegetation (Schindler et al. 2003). Helfield and Naiman (2001) found that about 22–24% of the foliar N of trees and shrubs along salmon spawning streams in southeast Alaska was derived from spawning salmon.[74] They also found that Sitka spruce (*Picea sitchensis*) in riparian areas along spawning streams grew significantly faster than trees in the same region that grew along streams not used for spawning. Wheeler and Kavanaugh (*op. cit.*) found that salmon carcasses significantly increased such soil health parameters as soil N, soil dissolved organic carbon, and soil respiration in Idaho riparian forests. Sadly, the conspicuous decrease in salmon populations reduces this conveyance of nutrients and may hurt the terrestrial and freshwater ecosystems that, in turn, provide habitat for young salmon and many other species (Gresh et al. 2011).

Returning to the legend of Salmon providing its body as a gift to feed people, it is clear that this gift also benefits many other organisms and the soil itself. The soil, in combination with other ecosystem factors, provides the smell that guides salmon home. However, soil erosion that results in sediment suffocating salmon eggs and alevin can cause great harm to salmon populations. Whether one thinks of this system as part of a reciprocal gift economy or describes it in scientific terms, if the ecological interrelationships are not understood and honored, then poor management decisions may cause disruptions from which recovery is difficult or impossible.

Space limitations do not allow descriptions of many other First Foods, including shellfish, halibut, elk, deer, waterfowl, mushrooms, nettle, salmonberry, soapberry, thimbleberry, wild coastal strawberry, elderberry, oval-leaved blueberry, cranberry, wild trailing blackberry, salal, Indian plum, fiddleheads, hazelnuts, evergreen tree tips, and others. Interested readers are directed to the references shown earlier in this section for more information. Also, it is important to note that many descriptions above are written in the past tense, indicating that First Foods *were* consumed by Indigenous Peoples. In fact, Oregon tribes continue to use First Foods both as part of their diets and for ceremonial purposes. Finally, notable work currently underway by the Confederated Tribes of the Umatilla Indian Reservation (2021) seeks to link the past, present, and future by developing a climate adaptation plan that relies heavily on First Foods (Quaempts et al. 2018).

[74] This calculation is based on the isotopic ratio of ^{15}N to ^{14}N being higher in marine systems than in freshwater or terrestrial ecosystems. Because an adult salmon's mass is almost entirely derived from marine food sources, analyses of foliage isotopic N ratios reveal the relative amount of marine N transformed from salmon biomass to soil mineral N and utilized by plants.

18.8 Summary

"Working lands" include the cropland, pasture, range, and forests managed to produce food, fiber, ornamental, or other plant or animal products for human use. Oregon's diverse climate and soils allow the production of a remarkably wide range of commodities. Different commodities require different yield metrics, but soil characteristics are an important determinant for the potential productivity of most working land systems. Soil survey data, available via Web Soil Survey reports for user-defined areas or SSURGO database downloads for entire soil survey areas, provide estimates of productivity for cropland, grazing land, and forestland. Productivity indices address the issue of maximum yields tending to increase over time, especially for cropland, and are now available for some crops.

Many practices are available to conserve soil and related resources, and the Natural Resources Conservation Service (NRCS) provides a system that names, defines, and provides standards and specifications for these practices. Oregon regulations pertaining to soil and water conservation include the Agricultural Water Quality Management Act, Confined Animal Feeding Operation permits, the Oregon Forest Practices Act, and water rights certification. The conservation compliance provisions of federal farm bills link eligibility for most USDA program benefits to controlling erosion on highly erodible land and to avoiding the conversion of wetlands to annual cropland. A variety of local, state, and federal programs exist to provide funding and technical assistance to managers wishing to implement conservation practices on working lands, but most managers also implement conservation practices without government assistance. Interest in practices to improve soil health, also known as soil quality, is increasing.

This chapter describes Oregon cropland management systems for hay and haylage, grains, seed crops, the Conservation Reserve Program, vegetables, orchards, corn, Christmas trees, nurseries, hemp and marijuana, berries, and wine grapes. Oregon crops are consumed within state, exported to other states, or sold internationally.

Oregon State University provides nutrient management recommendations based on crop type, soil test results, plant tissue test results, and location. When irrigation is used, water requirements vary by crop and location. The net irrigation requirement data presented in this chapter are best used to compare relative water needs between different crops or different parts of the state rather than to predict actual amounts of water applied.

Pasture land occurs across the state, while grazed rangeland occurs throughout eastern Oregon. The main Oregon livestock species are beef cattle, dairy cattle, and sheep. Production systems include various combinations of grazing and feeding in confinement. Commonly pastured soils are categorized by Forage Suitability Groups, which describe recommended forage species and potential yields. Common rangeland soils are categorized by Ecological Site Descriptions, which list species likely to occur on the site, potential annual productivity, and state and transition models regarding the interaction between management and vegetation dynamics.

In western Oregon, Douglas-fir and spruce/hemlock are the main forest types, and commercial timber productivity is very high. Oregon white oak is not as important commercially but provides important wildlife habitat, especially when it occurs in oak savanna systems. In eastern Oregon, ponderosa pine and lodgepole pine are the main forest types. For soils that commonly support forests, the Web Soil Survey provides site index and annual productivity reports to indicate potential timber production.

Indigenous Peoples have eaten First Foods since long before Euro-American contact. These plants and animals are integral parts of Indigenous culture and spirituality based on a reciprocal responsibility to care for the land and the non-human beings that provide food, medicine, dyes, and fiber for the people. This chapter briefly described camas, *Lomatium* species, huckleberries, Oregon white oak, water, and salmon, which are just some of the First Foods in the Pacific Northwest. The relationship between Indigenous Peoples, the land, and First Foods allowed their cultures to thrive sustainably for thousands of years.

References

Abbott C (2018) Simon Benson (1851–1942). Oregon Encyclopedia, Portland, OR. https://www.oregonencyclopedia.org/articles/benson_simon_1851_1942_/. Accessed 14 June 2021

Acton D, Gregorich L (eds) (1995) The health of our soils: toward a sustainable agriculture in Canada. Centre for Land and Biological Resources Research, Research Branch, Agriculture and Agri-Food Canada, Ottawa, ON

Anderson A (2019) Native Plant Production and Marketing Digger 15 (3):33–36

Anderson A, Finn C (1995) Variation in reproductive traits of western trailing blackberry (Rubus ursinus) in the Pacific Northwest. HortScience 30(4):833. https://doi.org/10.21273/HORTSCI.30.4.833F

Anderson N, Chastain T, Hart J, Young III W, Christensen N (2014) Tall fescue grown for seed—a nutrient management guide for western Oregon. Oregon State University, Corvallis, OR. https://catalog.extension.oregonstate.edu/sites/catalog/files/project/pdf/em9099.pdf. Accessed 14 June 2021

Ballotpedia (2016) Endangered species in Oregon https://ballotpedia.org/Endangered_species_in_Oregon#:~:text=As%20of%20July%202016%2C%20Oregon,Endangered%20Species%20Act%20(ESA). Accessed 14 June 2021

Barber K (2021) Celilo falls. Oregon Encyclopedia, Portland, OR. https://www.oregonencyclopedia.org/articles/celilo_falls/. Accessed 14 June 2021

Bohle M, Shewmaker G, Norberg S (2020) Hay. Oregon State University, Corvallis, OR. https://forages.oregonstate.edu/oregon/topics/harvest/hay#:~:text=Dry%20the%20hay%20to%2015,evenly%2C%20and%20completely%20as%20possible. Accessed 14 June 2021

Boriss H, Brunke H (2005) Commodity profile: Dried plums (prunes). University of California Agricultural Marketing Resource Center, Davis, CA. https://aic.ucdavis.edu/wp-content/uploads/2019/01/agmr-profile-Prunes-2005.pdf. Accessed 14 June 2021

Bubl C (2014) Crops that don't require irrigation (and big equipment). Oregon State University, Corvallis, OR. https://smallfarms.oregonstate.edu/crops-don%E2%80%99t-require-irrigation-and-big-equipment. Accessed 14 June 2021

Buckenmeyer H (n.d.) Oregon certified sustainable wine. Oregon State Bar, Tigard, OR. https://sustainablefuture.osbar.org/section-newsletter/20102summer5buckenmeyer/. Accessed 14 June 2021

Campbell S (2000) Soil conditioning index for cropland management systems. NRCS, Portland, OR. https://www.nrcs.usda.gov/Internet/FSE_DOCUMENTS/nrcs142p2_040629.pdf. Accessed 14 June 2021

Chase J, Fried J, Jovan S, Mercer K, Gray A, Bell D, Morgan T et al (2018) Oregon's forest resources, 2006–2015: ten-year forest inventory and analysis report. In: Palmer M, Kuegler O, Christensen G (eds). U.S. Forest Service Pacific Northwest Research Station, Portland, OR. https://doi.org/10.2737/PNW-GTR-971

Chastain T, Garbacik C, Young WI (2017) Tillage and establishment system effects on annual ryegrass seed crops. Field Crop Res 209:144–150. https://doi.org/10.1016/j.fcr.2017.04.017

Chastain T, King C, Garbacik C, Young WI (2015) Irrigation frequency and seasonal timing effects on perennial ryegrass (Lolium perenne L.) seed production. Field Crop Res 180:126–134. https://doi.org/10.1016/j.fcr.2015.05.021

Cloughesy M, Woodward J (2018) Oregon's forest protection laws—an illustrated manual, 3rd edn. Oregon Forest Resources Institute, Portland, OR

Columbia River Inter-Tribal Fish Commission (2021) We are all salmon people. https://www.critfc.org/salmon-culture/we-are-all-salmon-people/. Accessed 14 June 2021

Confederated tribes of grand ronde (2021) Chachalu Museum and Cultural Center. https://www.grandronde.org/history-culture/culture/chachalu-museum-and-cultural-center/. Accessed 14 June 2021

Confederated Tribes of the Umatilla Indian Reservation (2021) Climate adaptation plan—chapter one: CTUIR First Foods and Indigenous food system. https://ctuir.org/departments/natural-resources/climate-adaptation/ctuir-climate-adaptation-plan-drafts-for-comment/cap-chapter-one-ctuir-first-foods-and-indigenous-food-system/. Accessed 14 June 2021

Cuenca R (1999) Oregon crop water use and irrigation requirements. Oregon State University, Corvallis, OR. https://catalog.extension.oregonstate.edu/sites/catalog/files/project/pdf/em8530.pdf. Accessed 14 June 2021

Dailey C (1899) The prune in Oregon. Pacific Rural Press 58(7). https://cdnc.ucr.edu/?a=d&d=PRP18990812.2.9.1&e=———en–20–1–txt-txIN———1. Accessed 14 June 2021

Darwin C (1881) The formation of vegetable mold through the action of worms, with observations on their habits. John Murray, London

Dixon E, Strik B, Fernandez-Salvador J, Devetter L (2019) Strawberry nutrient management guide for Oregon and Washington. Oregon State University, Corvallis, OR. https://catalog.extension.oregonstate.edu/em9234/html. Accessed 14 June 2021

Donatuto J, Campbell L, LeCompte J, Rohlman D, Tadlock S (2020) The story of 13 moons: developing an environmental health and sustainability curriculum founded on indigenous first foods and technologies. Sustainability 12(21):8913. https://doi.org/10.3390/su12218913

Doran J, Coleman D, Bezdicek D, Stewart B (eds) (1994) Defining soil quality for a sustainable environment. Soil Science Society of America (SSSA), Madison, WI

Dreves A, Walton V, Fisher G (2009) A new pest attacking healthy ripening fruit in Oregon—Spotted wing Drosophila: Drosophila suzukii (Matsumura). Oregon State University, Corvallis, OR. https://extension.oregonstate.edu/pests-weeds-diseases/insects/new-pest-attacking-healthy-ripening-fruit-oregon-spotted-wing. Accessed 14 June 2021

Duniway M, Bestelmeyer B, Tugel A (2010) Soil processes and properties that distinguish ecological sites and states. Rangelands 32:9–15. https://doi.org/10.2111/RANGELANDS-D-10-00090.1

Duyck G, Tuck B, Kerr S, Olson S, Hammond E (2015) Soil health principles. Oregon State University, Corvallis, OR. https://catalog.extension.oregonstate.edu/ec1647. Accessed 14 June 2021

Eddy D (2020) Apple variety selection tougher than ever for growers. Meister Media Worldwide—Growing Produce, Willoughby, OH. https://www.growingproduce.com/fruits/apple-variety-selection-tougher-than-ever-for-growers/. Accessed 14 June 2021

Environmental Working Group (2021) Conservation database—Oregon. https://conservation.ewg.org/crp.php?fips=41000®ionname=Oregon. Accessed 14 June 2021

Fransen S (2020) Silage. Oregon State University, Corvallis, OR. https://forages.oregonstate.edu/oregon/topics/harvest/silage. Accessed 14 June 2021

Fransen S, Chaney M (2002) Pasture and hayland renovation for western Washington and Oregon. Washington State University, Pullman, WA. https://s3.wp.wsu.edu/uploads/sites/2079/2015/06/Pasture-and-Hayland-Renovation-for-Western-Washington-and-Oregon-WSU.pdf. Accessed 14 June 2021

Galbraith W, Anderson E (1971) Grazing history of the northwest. J Range Manage 24(1):6–12

Gardner E, Hall L, Pumphrey F (2000) Field corn eastern Oregon—East of Cascades fertilizer guide. Oregon State University, Corvallis, OR. https://ir.library.oregonstate.edu/downloads/6q182k869. Accessed 14 June 2021

Giombolini K (2018) Grass seed industry. Oregon Historical Society, Portland, OR. https://www.oregonencyclopedia.org/articles/grass_seed_industry/. Accessed 14 June 2021

Gresh T, Lichatowich J, Schoonmaker P (2011) An estimation of historic and current levels of salmon production in the northeast Pacific ecosystem: evidence of a nutrient deficit in freshwater systems of the Pacific Northwest. Fisheries 25(1):15–21. https://doi.org/10.1577/1548-8446(2000)025%3c0015:AEOHAC%3e2.0.CO;2

Grigar J, Dulker S, Flanagan D (2020) Understanding soil erosion by water to improve soil conservation. Crops Soils 53(3):47–55. https://doi.org/10.1002/crso.20030

Haberern J (1992) Coming full circle—the new emphasis on soil quality. Am J Altern Agric 7(1–2):3–4. https://doi.org/10.1017/S0889189300004355

Hannaway D, Bohle M, Miles D, Lin Y, Randow B (2019) Alfalfa soil fertility and fertilizer requirements. Oregon State University, Corvallis, OR. https://extension.oregonstate.edu/crop-production/pastures-forages/alfalfa-soil-fertility-fertilization-requirements. Accessed 14 June 2021

Hart J, Anderson N, Chastain T, Flowers M, Ocamb C, Melbye M, Young WI (2017) Perennial ryegrass grown for seed (Wester Oregon) nutrient management guide. Oregon State University, Corvallis, OR. https://catalog.extension.oregonstate.edu/sites/catalog/files/project/pdf/em9086.pdf. Accessed 14 June 2021

Hart J, Anderson N, Hulting A, Chastain T, Melbye M, Young WI, Silberstein T (2012) Postharvest residue management for grass seed production in western Oregon. Oregon State University, Corvallis, OR. https://catalog.extension.oregonstate.edu/sites/catalog/files/project/pdf/em9051.pdf. Accessed 14 June 2021

Hart J, Flowers M, Anderson N, Roseburg R, Christensen N, Melbye M (2011) Soft white winter wheat (western Oregon) nutrient management guide. Oregon State University, Corvallis, OR. https://catalog.extension.oregonstate.edu/sites/catalog/files/project/pdf/em8963.pdf. Accessed 14 June 2021

Hart J, Landgren C, Fletcher R, Bondi M, Withrow-Robinson B, Chastagner G (2009) Christmas tree nutrient management guide—western Oregon and Washington. Oregon State University, Corvallis, OR. https://catalog.extension.oregonstate.edu/sites/catalog/files/project/pdf/em8856.pdf. Accessed 14 June 2021

Hart J, Melbye M, Young WI, Silberstein T (2011) Annual ryegrass grown for seed (western Oregon) nutrient management guide. Oregon State University, Corvallis, OR. https://catalog.extension.oregonstate.edu/sites/catalog/files/project/pdf/em8854.pdf. Accessed 14 June 2021

Hart J, Pirelli G, Cannon L, Fransen S (2000) Pastures fertilizer guide western Oregon and western Washington. Oregon State University, Corvallis, OR. https://smallfarms.oregonstate.edu/sites/agscid7/files/fg63-e.pdf. Accessed 14 June 2021

Hart J, Righetti T, Stebbins B, Lombard P, Burkhart D, Van Buskirk P (1997) Pears nutrient management guide. Oregon State University, Corvallis, OR. https://agsci.oregonstate.edu/sites/agscid7/files/horticulture/attachments/fg59-e.pdf. Accessed 14 June 2021

Hart J, Strik B, Rempel H (2006) Caneberries nutrient management guide. Oregon State University, Corvallis, OR. https://catalog.extension.oregonstate.edu/sites/catalog/files/project/pdf/em8903.pdf. Accessed 14 June 2021

Hart J, Strik B, DeMoranville C, Davenport J, Roper T (2015) Cranberries—a nutrient management guide for south coastal Oregon. Oregon State University, Corvallis, OR. https://catalog.extension.oregonstate.edu/sites/catalog/files/project/pdf/em8672.pdf. Accessed 14 June 2021

Hart J, Strik B, White L, Yang W (2006) Nutrient management for blueberries in Oregon. Oregon State University, Corvallis, OR. http://www.ucanr.org/sites/nm/files/76680.pdf. Accessed 14 June 2021

Hart J, Sullivan D, Gamroth M, Downing T, Peters A (2009) Silage corn nutrient management guide. Oregon State University, Corvallis, OR. https://catalog.extension.oregonstate.edu/sites/catalog/files/project/pdf/em8978.pdf. Accessed 14 June 2021

Heinrich A (2017) Cider taking root. Oregon State University, Corvallis, OR. https://archive.progress.oregonstate.edu/summer-2017/cider-taking-root. Accessed 14 June 2021

Helfield J, Naiman R (2001) Effects of salmon-derived nitrogen on riparian forest growth and implications for stream productivity. Ecology 82(9):2403–2409. https://doi.org/10.1890/0012-9658(2001)082%5B2403:EOSDNO%5D2.0.CO;2

Hensey A (2013) The profitable side of green growing. Portl Bus J. Portland, OR. https://www.bizjournals.com/portland/blog/sbo/2013/04/the-profitable-side-of-green-growing.html. Accessed 14 June 2021

Hessler K, Luk D, McMillan S (n.d.) Report on the Oregon Department of Agriculture's enforcement of the Clean Water Act's NPDES program related to CAFOs. Lewis & Clark Law School. https://law.lclark.edu/live/files/10807-2012-oda-clinic-report. Accessed 14 June 2021

Hodges G (2015) Natl Geogr 227(1):125–137

Huddleston, J.H. (1982) Agricultural productivity ratings for soils of the Willamette Valley. Oregon State University Extension Service, Corvallis, OR.

Huettig K, Chastain T, Garbacik C, Young WI (2013) Spring irrigation of tall fescue for seed production. Field Crop Res 144:297–304. https://doi.org/10.1016/j.fcr.2013.01.023

Hulting A (2019) Grass seed crops. In: Peachey E (ed) Pacific Northwest weed management handbook. Oregon State University, Corvallis, OR, pp D1–D38

Jeliazkov V, Noller J, Angima S, Rondon S, Roseberg R, Summers S, Sikora V et al (2019) What is industrial hemp? Oregon State University, Corvallis, OR. https://catalog.extension.oregonstate.edu/sites/catalog/files/project/pdf/em9240.pdf. Accessed 14 June 2021

Jones G (2021) Map of registered outdoor hemp acreage—November 2020. Oregon State University, Corvallis, OR. https://extension.oregonstate.edu/crop-production/hemp/map-registered-outdoor-hemp-acreage-november-2020. Accessed 14 June 2021

Jones G, Jeliazkov V, Roseberg R, Angima S (2019) Basics of fall cover cropping for hemp in Oregon. Oregon State University, Corvallis, OR.https://catalog.extension.oregonstate.edu/sites/catalog/files/project/pdf/em9240.pdf. Accessed 14 June 2021

Jung G, Van Wijk A, Hunt W, Watson C (1996) Ryegrasses. Cool-season forage grasses. American Society of Agronomy, Crop Science Society of America, Soil Science Society of America, Madison, WI, pp 605–641

Kaiser C, Horneck D, Koenig T, Porter L, Brewer L (2016) Green pea nutrient management inland northwest—east of the Cascades. Oregon State University, Corvallis, OR. https://catalog.extension.oregonstate.edu/sites/catalog/files/project/pdf/em9140.pdf. Accessed 14 June 2021

Kamholz E (2018) Donkey engine. Oregon Encyclopedia, Portland, OR. https://www.oregonencyclopedia.org/articles/donkey_engine/#.Xv4lx2pKh0t. Accessed 14 June 2021

Karlen D, Mausbach J, Doran J, Cline R, Harris R, Schuman G (1997) Soil quality: a concept, definition, and framework for evaluation. Soil Sci Soc Am J 61:4–10. https://doi.org/10.2136/sssaj1997.03615995006100010001x

Karlen D, Veum K, Sudduth K, Obrycki J, Nunes M (2019) Soil health assessment: past accomplishments, current activities, and future opportunities. Soil till Res 195:1–10. https://doi.org/10.1016/j.still.2019.104365

Karten R (2011) Tribal government day honors native first foods. Smoke Signals. https://www.smokesignals.org/articles/2011/05/16/tribal-government-day-honors-native-first-foods/. Accessed 14 June 2021

Kendall A, McPherson E (2012) A life cycle greenhouse gas inventory of a tree production system. Int J Life Cycle Assess 17(4):444–452. https://doi.org/10.1007/s11367-011-0339-x

Keyes J, Keyes J (2015) Federal lands grazing permits—managing rangeland resources. Utah State University Extension, Logan, UT. https://digitalcommons.usu.edu/cgi/viewcontent.cgi?article=1697&context=extension_curall. Accessed 14 June 2021

Kimsey M (2014) Geo-spatial Douglas-fir site index modeling for northern Idaho and northeast Washington. Technical Report 063014. Intermountain Forest Tree Nutrition Cooperative, Moscow, ID. College of Natural Resources, University of Idaho

Koski K (1966) The survival of coho salmon (Oncorhynchus kisutch) from egg deposition to emergence in three Oregon coastal streams. Oregon State University, Corvallis, OR

Landgren C, Fletcher R, Bondi M, Barney D, Mahoney R (2003) Growing Christmas trees in the Pacific Northwest. Oregon State University, Corvallis, OR. https://catalog.extension.oregonstate.edu/sites/catalog/files/project/pdf/pnw6.pdf. Accessed 14 June 2021

Lang N, Stevens R, Thornton R, Pan W, Victory S (1981) Nutrient management guide: Central Washington irrigated potatoes. Washington State University, Pullman, WA. http://potatoes.wsu.edu/wp-content/uploads/2014/11/nutrient-central-wa.pdf. Accessed 14 June 2021

Larson W, Pierce F (1991) Conservation and enhancement of soil quality. Evaluation for sustainable land management in the developing world, vol 2, pp 175–203. International Board for Soil Research and Management, Bangkok, Thailand

Law Insider Inc (2020) Working lands definition. https://www.lawinsider.com/dictionary/working-lands. Accessed 14 June 2021

Levin A (2018) Management of grapevine water status under irrigated and non-irrigated conditions. Oregon State University, Corvallis, OR. https://extension.oregonstate.edu/water/irrigation/management-grapevine-water-status-under-irrigated-non-irrigated-conditions. Accessed 14 June 2021

Lichatowich (1999) Salmon without rivers: a history of the pacific salmon crisis. Island Press, Washington, DC

Lichvar R, Melvin N, Butterwick M, Kirchner W (2012) National wetland plant list indicator rating definitions. U.S. Army Corps of Engineers, Washington, DC

Long L, Kaiser C (2013) Sweet cherry orchard establishment in the Pacific Northwest. Oregon State University, Corvallis, OR. https://catalog.extension.oregonstate.edu/sites/catalog/files/project/pdf/pnw642.pdf. Accessed 14 June 2021

Long P (2021) Salmon in the Pacific Northwest. History Link, Seattle, WA. https://www.historylink.org/File/10443. Accessed 14 June 2021

Lutcher L, Horneck D, Wysocki D, Hart J, Petrie S, Christensen N (2007) Winter wheat in summer-fallow systems (low precipitation zone) fertilizer guide. Oregon State University, Corvallis, OR. https://catalog.extension.oregonstate.edu/sites/catalog/files/project/pdf/fg80.pdf. Accessed 14 June 2021

McBee B (2016) Oregon's first commercial filbert farm. 1859—Oregon's Magazine. https://1859oregonmagazine.com/think-oregon/art-culture/dorris-ranch-filbert-hazelnut-farm/. Accessed 14 June 2021

McLeod D, Miller S, Perry G (1996) North central Oregon conservation reserve program survey: a summary of results. Oregon State University, Corvallis, OR

Meyers S, Jaster T, Mitchell K, Hardison L (eds) (2015) Flora of Oregon, Volume 1: Pteridophytes, gymnosperms, and monocots. Oregon State University, Corvallis, OR

Miller G (2020) Cranberry industry. Oregon Encyclopedia, Portland, OR. https://www.oregonencyclopedia.org/articles/cranberry_industry/#:~:text=Nearly%203%2C000%20acres%20of%20cranberries,pounds%20of%20berries%20each%20year. Accessed 14 June 2021

Moebius-Clune B, Moebius-Clune D, Gugino B, Idowu O, Schindelbeck R, Ristow A, Abawi G et al (2016) Comprehensive assessment of soil health—The Cornell framework manual, edition 3.1. Cornell University, Geneva, NY

Moerman D (2009) Native American medicinal plants: an ethnobotanical dictionary. Timber Press, Portland, OR

Moerman D (2010) Native American food plants: an ethnobotanical dictionary. Timber Press, Portland, OR

Mortensen E (2019) Oregon corn crop leaps in acreage and value, Midwest drought may send it even higher. The Oregonian (newspaper). https://www.oregonlive.com/business/2012/08/oregon_corn_crop_leaps_in_acre_1.html. Accessed 14 June 2021

Mulvey R, Shaw D, Filip G, Chastagner G (2013) Swiss needle cast. U. S. Forest Service, Pacific Northwest Region, Portland, OR

NASS (2019) 2017 Census of agriculture Oregon. USDA, Washington, DC. https://www.nass.usda.gov/Publications/AgCensus/2017/Full_

Report/Volume_1,_Chapter_1_State_Level/Oregon/orv1.pdf. Accessed 14 June 2021
NASS (2020) Vegetables 2019 summary. USDA, Washington, DC. https://www.nass.usda.gov/Publications/Todays_Reports/reports/vegean20.pdf. Accessed 14 June 2021
NASS (2021) Quick stats. https://quickstats.nass.usda.gov/. Accessed 14 June 2021
National Oceanic and Atmospheric Administration (n.d.) Species directory. https://www.fisheries.noaa.gov/species-directory/threatened-endangered?title=&species_category=any&species_status=any®ions=1000001126&items_per_page=all&sort=. Accessed Mar 2021
National Park Service (2017) Wildland fire spread and suppression. https://www.nps.gov/articles/wildland-fire-spread-and-suppression.htm. Accessed 14 June 2021
National Sustainable Agriculture Coalition (2019) Conservation reserve program. https://www.fsa.usda.gov/programs-and-services/conservation-programs/conservation-reserve-program/. Accessed 14 June 2021
Nelson N, Janke R (2007) Phosphorus sources and management in organic production systems. HortTechnology 17(4):442–454. https://doi.org/10.21273/HORTTECH.17.4.442
Northwest Horticultural Council (2020a) Pacific Northwest pears. https://nwhort.org/industry-facts/pear-fact-sheet/
Northwest Horticultural Council (2020b) Pacific Northwest sweet cherries. https://nwhort.org/industry-facts/cherry-fact-sheet/. Accessed 14 June 2021
Northwest Horticultural Council (2021) Organics. https://nwhort.org/organics/. Accessed 14 June 2021
Norton H, Hunn E, Martinsen C, Keely P (1984) Vegetable food products of the foraging economies of the Pacific Northwest. Ecol Food Nutr 14(3):219–228. https://doi.org/10.1080/03670244.1984.9990789
NRCS (2003) National range and pasture handbook, Revision 1. USDA, Washington, DC
NRCS (2004) National forestry handbook. USDA, Washington, DC
NRCS (2006) Land resource regions and major land resource areas of the United States, the Caribbean, and the Pacific Basin. USDA, Washington, DC
NRCS (2012) User guide for the National Commodity Crop Productivity Index (NCCPI) version 2.0. USDA, Washington, DC
NRCS (2014) National planning procedures handbook (NPPH), 1st edition amendment 6. USDA, Washington DC
NRCS (2015) National food security act manual, 5th edn. USDA, Washington, DC
NRCS (2016) Chapter 11—Sprinkler irrigation. In: National engineering handbook part 623—Irrigation. USDA, Washington, DC
NRCS (2017) National Resources Inventory results. https://www.nrcs.usda.gov/wps/portal/nrcs/main/national/technical/nra/nri/results/. Accessed 14 June 2021
NRCS (2018) NRCS natural resources long range strategy in Malheur County, Oregon. USDA, Ontario, OR
NRCS (2020a) Field office technical guide (FOTG). https://www.nrcs.usda.gov/wps/portal/nrcs/main/national/technical/fotg/. Accessed 14 June 2021
NRCS (2020b) Oregon's strategic approach to conservation. https://www.nrcs.usda.gov/wps/portal/nrcs/detailfull/or/technical/cp/?cid=stelprdb1262209. Accessed 14 June 2021
NRCS (2020c) The plants database. http://plants.usda.gov. Accessed 14 June 2021
NRCS (2021a) EQIP high tunnel system initiative. https://www.nrcs.usda.gov/wps/portal/nrcs/detail/or/programs/financial/eqip/?cid=nrcseprd398832. Accessed 14 June 2021
NRCS (2021b) Natural resources conservation service Oregon programs. https://www.nrcs.usda.gov/wps/portal/nrcs/main/or/programs/. Accessed 14 June 2021
NRCS (n.d.) Erosion. https://www.nrcs.usda.gov/wps/portal/nrcs/main/national/landuse/crops/erosion/. Accessed 14 June 2021
NRCS (n.d.) Soil conditioning index. https://www.nrcs.usda.gov/wps/portal/nrcs/detail/null/?cid=nrcs142p2_008548. Accessed 14 June 2021
Olsen J (2013a) Growing hazelnuts in the Pacific Northwest—Introduction. Oregon State University, Corvallis, OR. https://catalog.extension.oregonstate.edu/em9072/html#: ~ :text=In%201858%2C%20Sam%20Strickland%2C%20an,present%2Dday%20Albany%2C%20Oregon. Accessed 14 June 2021
Olsen J (2013b) Growing hazelnuts in the Pacific Northwest—Orchard design. Oregon State University, Corvallis, OR. https://catalog.extension.oregonstate.edu/sites/catalog/files/project/pdf/em9077.pdf . Accessed 14 June 2021
Olsen J (2013c) Growing hazelnuts in the Pacific Northwest: Orchard nutrition. Oregon State University, Corvallis, OR. https://catalog.extension.oregonstate.edu/em9080. Accessed 14 June 2021
Oregon Aglink (n.d.) Oregon agriculture production. https://oregonfresh.net/education/oregon-agriculture-production/. Accessed 14 June 2021
Oregon Association of Nurseries; Oregon Environmental Council (2011) Best management practices for climate friendly nurseries, Version 2.0. Wilsonville, OR. http://www.climatefriendlynurseries.org/resources/best_management_practices_for_climate_friendly_nurseries.pdf. Accessed 14 June 2021
Oregon Clover Commission (2020) Red clover. https://www.oregonclover.org/pages/redclover.html. Accessed 14 June 2021
Oregon Department of Agriculture (2019a) Oregon agriculture facts & figures. Salem, OR. https://www.nass.usda.gov/Statistics_by_State/Oregon/Publications/facts_and_figures/facts_and_figures.pdf. Accessed 14 June 2021
Oregon Department of Agriculture (2019b) Water quality & agriculture—it's your responsibility. Salem, OR. https://www.oregon.gov/oda/shared/Documents/Publications/NaturalResources/WaterQualityandAgriculture2Pages.pdf. Accessed 14 June 2021
Oregon Department of Agriculture (2020a) Agricultural water quality plans. https://www.oregon.gov/oda/programs/NaturalResources/AgWQ/Pages/AgWQPlans.aspx. Accessed 14 June 2021
Oregon Department of Agriculture (2020b) Confined animal feeding operations (CAFO). https://www.oregon.gov/oda/programs/NaturalResources/Pages/CAFO.aspx. Accessed 14 June 2021
Oregon Department of Agriculture (2020c) Hemp growers. https://www.oregon.gov/oda/programs/Hemp/Pages/HempGrowers.aspx. Accessed 14 June 2021
Oregon Department of Agriculture (2021d) Oregon agricultural statistics and directory 2021. Salem, OR. https://www.oregon.gov/oda/shared/Documents/Publications/Administration/AgStatsDirectory.pdf. Accessed 14 June 2021
Oregon Department of Education (n.d.) Native nutrition. Salem, OR. https://www.oregon.gov/ode/students-and-family/equity/NativeAmericanEducation/Documents/SB13%20Curriculum/ODE_G8_Science_Native%20Nutrition.pdf. Accessed 14 June 2021
Oregon Department of Forestry (2019) Sudden oak death (Phytophthera ramorum). Salem, OR. https://www.oregon.gov/ODF/Documents/ForestBenefits/SOD.pdf. Accessed 14 June 2021
Oregon Department of Forestry (n.d.) About Oregon's forests. https://www.oregon.gov/odf/forestbenefits/pages/aboutforests.aspx. Accessed 14 June 2021
Oregon Farmers Markets Association (2021) Beginning vendor support network. https://www.oregonfarmersmarkets.org/beginning-vendor-support-network. Accessed 14 June 2021

Oregon Flora (2021a) Pinus ponderosa var. benthamiana (Harw.) Vasey. https://oregonflora.org/taxa/index.php?taxon=14103. Accessed 14 June 2021

Oregon Flora (2021b) Welcome to Oregon Flora, your comprehensive guide to the vascular plants of Oregon. https://oregonflora.org/. Accessed 14 June 2021

Oregon Forest Resources Institute (2020) Know your forests. https://oregonforests.org/node/86. Accessed 14 June 2021

Oregon Forest Resources Institute (2021) Oregon forest facts 2021–2022 edition. Portland, OR. https://oregonforests.org/sites/default/files/2021-01/OFRI_2021ForestFacts_WEB3.pdf. Accessed 14 June 2021

Oregon Hay Products, Inc. (2020) Welcome to Oregon Hay Products, Inc. worldwide shipping. https://oregonhayproducts.com/. Accessed 14 June 2021

Oregon Legislative Assembly (2019) ORS 475B - Cannabis regulation. https://www.oregonlegislature.gov/bills_laws/ors/ors475B.html. Accessed 14 June 2021

Oregon Liquor Control Commission (2019) 2019 Recreational marijuana supply and demand legislative report. Portland, OR. https://www.oregon.gov/olcc/marijuana/Documents/Bulletins/2019%20Supply%20and%20Demand%20Legislative%20Report%20FINAL%20for%20Publication(PDFA).pdf. Accessed 14 June 2021

Oregon Liquor Control Commission, p. c. (2021, March 8)

Oregon Secretary of State (2009) Confined animal feeding operation program. https://secure.sos.state.or.us/oard/displayDivisionRules.action?selectedDivision=2751. Accessed 14 June 2021

Oregon Seed Council (2016) FAQs. http://www.oregonseedcouncil.org/faqs/. Accessed 14 June 2021

Oregon State University (2007) How the Rogue Valley became famous for pears. https://extension.oregonstate.edu/crop-production/fruit-trees/how-rogue-valley-became-famous-pears. Accessed 14 June 2021

Oregon State University (2020) Production systems. https://horticulture.oregonstate.edu/department-horticulture/nursery/production-systems. Accessed 14 June 2021

Oregon State University (2021a) Carrot seed. https://agsci.oregonstate.edu/coarec/carrot-seed. Accessed 14 June 2021

Oregon State University (2021b) Grass seed. https://valleyfieldcrops.oregonstate.edu/grass-seed. Accessed 14 June 2021

Oregon State University (2021c) OSU small farms. https://extension.oregonstate.edu/smallfarms/small-farms-technical-reports. Accessed 14 June 2021

Oregon State University (2021d) Soil health lab. https://cropandsoil.oregonstate.edu/shl/soil-health-osu. Accessed 14 June 2021

Oregon Water Resources Department (2018) Water rights in Oregon, An introduction to Oregon's water laws. Salem, OR. https://www.oregon.gov/owrd/WRDPublications1/aquabook.pdf. Accessed 14 June 2021

Oregon Watershed Enhancement Board (n.d.) Grant programs. https://www.oregon.gov/oweb/grants/Pages/grant-programs.aspx. Accessed 14 June 2021

Oregon Wine Board (2021) Explore Oregon's American Viticultural Areas (AVAs). https://www.oregonwine.org/discover-oregon-wine/place/. Accessed 14 June 2021

Owen J, Stoven H (2011) Feeding bare-root trees. Digger 7(1):41–45. https://agsci.oregonstate.edu/sites/agscid7/files/horticulture/osu-nursery-greenhouse-and-christmas-trees/DiggerJAN2011_41-45.pdf. Accessed 14 June 2021

Pacific Northwest Christmas Tree Association (2021) Facts at a glance. http://www.pnwcta.org/news-events/facts-at-a-glance/. Accessed 14 June 2021

Parker M, Unrath C, Safley C, Lockwood D (1998) High density apple orchard management. North Carolina State University, Raleigh, NC. https://content.ces.ncsu.edu/high-density-apple-orchard-management#: ~ :text=Another%20advantage%20is%20the%20potential,higher%20density%20orchards%20as%20well. Accessed 14 June 2021

Pojar J, MacKinnon A (2014) Plants of the Pacific Northwest Coast. Lone Pine Publishing, Vancouver BC

Prakriti B, Machado S, Ghimire R, Yorgey G, Wysocki D (2017) Conservation tillage systems. In: Yorgey G, Kruger C (eds) Advances in dryland farming in the inland Pacific Northwest. Washington State University, Pullman, WA, pp 99–124

Quaempts E, Jones K, O'Daniel T (2018) Aligning environmental management with ecosystem resilience: a First Foods example from the Confederated Tribes of the Umatilla Indian Reservation, Oregon USA. Ecol Soc 23(2):29. https://doi.org/10.5751/ES-10080-230229

Rankin M (2018) Hay exports hit record high. Hay & Forage Grower. https://www.hayandforage.com/article-1791-Hay-exports-hit-record-high.html. Accessed 14 June 2021

Reed C (2017) Ag series: Douglas County agriculture has adapted through the years. The News-Review (Roseburg Oregon). https://www.nrtoday.com/business/ag-series-douglas-county-agriculture-has-adapted-through-the-years/article_ad397ea1-4ba7-5b9e-a92b-647582dd4101.html. Accessed 14 June 2021

Righetti T, Wilder K, Stebbins R, Burkhart D, Hart J (1998) Apples nutrient management guide. Oregon State University, Corvallis, OR. https://catalog.extension.oregonstate.edu/sites/catalog/files/project/pdf/em8712.pdf. Accessed 14 June 2021

Rizzo D, Garbelotto M (2003) Sudden oak death: endangering California and Oregon forest ecosystems. Front Ecol Environ 1(4):197–204. https://doi.org/10.1890/1540-9295(2003)001[0197:SODECA]2.0.CO;2

Rogue Farm Corps (n.d.) We are the next generation of farmers and ranchers. https://www.roguefarmcorps.org/. Accessed 14 June 2021

Rohse E (2017) Prunes were once a treasure of local ag. Yamhill County News-Register (newspaper). https://newsregister.com/article?articleTitle=rohse-prunes-were-once-a-treasure-of-local-ag–1497385046–26407–rohse. Accessed 14 June 2021

Roseberg R, Angima S, Jeliazkov V (2019) Soil, seedbed preparation and seeding for hemp in Oregon. Oregon State University, Corvallis, OR. https://catalog.extension.oregonstate.edu/sites/catalog/files/project/pdf/em9239.pdf. Accessed 14 June 2021

Sage-Grouse Conservation Partnership (2015) The Oregon Sage-Grouse Action Plan. Governor's Natural Resources Office, Salem, OR

Schillinger W, Papendick R (2008) Then and now: 125 years of dryland wheat in the inland Pacific Northwest. Agron J 100(53):166–182. https://doi.org/10.2134/agronj2007.0027c

Schindler D, Scheuerell M, Moore J, Gende S, Francis T, Palen W (2003) Pacific salmon and the ecology of coastal ecosystems. Front Ecol Environ 1(1):31–37. https://doi.org/10.1890/1540-9295(2003)001[0031:PSATEO]2.0.CO;2

Sharpley A, Daniel T, Sims J, Pote D (1996) Determining environmentally sound soil phosphorus levels. J Soil Water Conserv 51(2):160–166

Shaw D, Oester P, Filip G (2009) Managing insects and diseases of Oregon conifers. Oregon State University, Corvallis, OR. https://catalog.extension.oregonstate.edu/em8980. Accessed 14 June 2021

Shewmaker G, Bohle M (2010) Pasture and grazing management in the Northwest. University of Idaho, Moscow ID. https://www.extension.uidaho.edu/publishing/pdf/PNW/PNW0614.pdf. Accessed 14 June 2021

Shock C, Cheatham N, Harden J, Mahony A, Shock B (2021) Sustainable onion production. Oregon State University, Corvallis, OR. https://agsci.oregonstate.edu/mes/sustainable-onion-production/crop-rotation. Accessed 14 June 2021

Skinkis P, Schreiner R (2011) Grapevine nutrition. Oregon State University, Corvallis, OR. https://catalog.extension.oregonstate.edu/sites/catalog/files/project/media/em9024/index.html#screen/00-128-6385103003164-10250014588-91-874-5925-37-71. Accessed 14 June 2021

Smesrud J, Hess M, Selker J (2000) Western Oregon irrigation guides. Oregon State University, Corvallis, OR

Soil Survey Staff (2019) Web soil survey. https://websoilsurvey.sc.egov.usda.gov/App/HomePage.htm. Accessed 14 June 2021

Southern Oregon University Research Center (2017) 2016 Oregon vineyard and winery census report. Ashland, OR. https://inside.sou.edu/assets/sou-announcements/2016_Oregon_Vineyard_and_Winery_Production_Final_Report_08.28.17.pdf. Accessed 14 June 2021

Stermitz R, Nielsen G, Long D (1999) Testing quality of soil survey geographic (SSURGO) database for precision farming. In: Proceedings of the fourth international conference on precision agriculture, pp. 319–326. ASA-CSSA-SSSA, Madison WI

Strochlic R, Shelley C (2004) Community supported agriculture in California, Oregon, and Washington: challenges and opportunities. California Institute for Rural Studies, Davis, CA. https://citeseerx.ist.psu.edu/viewdoc/download?doi=10.1.1.208.9442&rep=rep1&type=pdf. Accessed 14 June 2021

Stubbs M (2012) Conservation compliance and U.S. farm policy. Congressional Research Service, Washington, DC. https://fas.org/sgp/crs/misc/R42459.pdf. Accessed 14 June 2021

Stursa S (2019) Oregon wine, a deep-rooted history. American Palate, A Division of The History Press, Charleston, SC

Sullivan D, Andrews N, Sullivan C, Brewer J (2019) OSU organic fertilizer & cover crop calculator: predicting plant-available nitrogen. Oregon State University, Corvallis, OR. https://catalog.extension.oregonstate.edu/em9235/html. Accessed 14 June 2021

Sullivan D, Brown B, Shock C, Horneck D, Stevens R, Pelter G, Feiber E (2001) Nutrient management for onions in the Pacific Northwest. Oregon State University, Corvallis, OR. https://catalog.extension.oregonstate.edu/sites/catalog/files/project/pdf/pnw546.pdf. Accessed 14 June 2021

Tamastslikt Cultural Institute (2021) Visit the Museum at Tamastslikt Cultural Institute. https://www.tamastslikt.org/#. Accessed 14 June 2021

The Confederated Tribes of Warm Springs (2021) The museum at warm springs. https://www.museumatwarmsprings.org/. Accessed 14 June 2021

Tillamook County Creamery Association (2019) Tillamook. https://www.tillamook.com/. Accessed 14 June 2021

Tillamook County Soil & Water Conservation District (2010) Business plan. Tillamook, OR

Tillamook Headlight Herald (newspaper) (2008, September 30) Field of corn creates farm community buzz. https://www.tillamookheadlightherald.com/news/field-of-corn-creates-farm-community-buzz/article_1e490b0e-0748-56f5-907c-f95a3433e948.html. Accessed 14 June 2021

Tree Top, Inc. (n.d.) Our locations. https://www.treetop.com/our-locations/. Accessed 14 June 2021

Tucker K (2002) Coos Bay Lumber Company steam donkeys. The Oregon History Project, Portland, OR. https://www.oregonhistoryproject.org/articles/historical-records/coos-bay-lumber-company-steam-donkeys/. Accessed 14 June 2021

U.S. Food and Drug Administration (2020) What you need to know (and what we're working to find out) about products containing cannabis or cannabis-derived compounds, including CBD. https://www.fda.gov/consumers/consumer-updates/what-you-need-know-and-what-were-working-find-out-about-products-containing-cannabis-or-cannabis. Accessed 14 June 2021

U.S. Geological Survey (2011) Gap analysis project—land cover data download. https://www.usgs.gov/core-science-systems/science-analytics-and-synthesis/gap/science/land-cover-data-download?qt-science_center_objects=0#qt-science_center_objects. Accessed 14 June 2021

USDA and U.S. EPA (1999) Unified national strategy for animal feeding operations. Washington, DC. https://www.epa.gov/sites/production/files/2015-10/documents/finafost.pdf. Accessed 14 June 2021

USDA-Agricultural Research Service (n.d.) FoodData Central. https://fdc.nal.usda.gov/index.html. Accessed 7 Mar 2021

USDA-Economic Research Service (2020a) Sheep, lamb, mutton sector at a glance. https://www.ers.usda.gov/topics/animal-products/sheep-lamb-mutton/sector-at-a-glance/. Accessed 14 June 2021

USDA-Economic Research Service (2020b) Wheat sector at a glance. https://www.ers.usda.gov/topics/crops/wheat/wheat-sector-at-a-glance/#classes. Accessed 14 June 2021

USDA-Farm Service Agency (2021) Conservation reserve program. https://www.fsa.usda.gov/programs-and-services/conservation-programs/conservation-reserve-program/. Accessed 14 June 2021

USDA-Farm Service Agency (n.d.) Farm Service Agency Oregon state programs. https://www.fsa.usda.gov/state-offices/Oregon/programs/index. Accessed 14 June 2021

Van Leeuwen C, Seguin G (2006) The concept of terroir in viticulture. J Wine Res 17(1):1–10. https://doi.org/10.1080/09571260600633135

Verzemnieks I (2018) Maraschino cherries. Oregon Encyclopedia, Portland, OR. https://www.oregonencyclopedia.org/articles/maraschino_cherries/. Accessed 14 June 2021

Wallace H (1910) Relation between livestock farming and the fertility of the land. Iowa State University, Ames, IA. https://doi.org/10.31274/rtd-180813-7404

Warkentin B, Fletcher H (1977) Soil quality for intensive agriculture. In: Proceedings of the international seminar on soil environment and fertilizer management in intensive agriculture, pp 594–598. Society of Science of Soil and Manure—National Institute of Agricultural Science, Tokyo

Washington State University (2021) Irrigation in the Pacific Northwest. http://irrigation.wsu.edu/Content/Resources/Irrigation-Scheduling-Aids-Tools.php. Accessed 14 June 2021

Washington State University (n.d.) Orchard establishment. http://treefruit.wsu.edu/orchard-management/orchard-establishment/. Accessed 14 June 2021

Weaver M (2020) World demand grows for soft white wheat. Capital Press (newspaper). https://www.capitalpress.com/ag_sectors/grains/world-demand-grows-for-soft-white-wheat/article_ca62f2b4-feb3-11ea-be3b-0f00dc95a5d7.html. Accessed 14 June 2021

Wheeler T, Kavanaugh K (2017) Soil biogeochemical responses to the deposition of anadromous fish carcasses in inland riparian forests of the Pacific Northwest USA. Can J Res 47:1506–1516. https://doi.org/10.1139/cjfr-2017-0194

Wisby W, Hasler A (1954) Effect of olfactory occlusion on migrating silver salmon (Oncorhynchus kisutch). J Fish Res Bd Can 11 (4):472–478. https://doi.org/10.1139/f54-031

Wischmeier WH, Smith DD (1978) Predicting rainfall erosion losses. USDA, Washington, DC

Woodruff N, Siddoway F (1965) A wind erosion equation. Soil Sci Soc Am J 29(5):602–608. https://doi.org/10.2136/sssaj1965.03615995002900050035x

Wooten G (n.d.) Fire and fuels management: definitions, ambiguous terminology and references. National Park Service, Twisp, WA. https://www.nps.gov/olym/learn/management/upload/fire-wildfire-definitions-2.pdf. Accessed 14 June 2021

World Atlas (2019) The top 10 potato producing states in the US. https://www.worldatlas.com/articles/the-top-10-potato-producing-states-in-the-us.html. Accessed 14 June 2021

World Forestry Center (2020) Tracing Oregon's timber culture. https://www.worldforestry.org/tracing-oregons-timber-culture/. Accessed 14 June 2021

Wysocki D, Horneck D, Lutcher L, Hart J, Petrie S, Corp M (2007) Winter wheat in continuous cropping systems (intermediate precipitation zone). Oregon State University, Corvallis, OR. https://catalog.extension.oregonstate.edu/sites/catalog/files/project/pdf/fg83.pdf. Accessed 14 June 2021

Yamamoto Y, Shibata H, Ueda H (2013) Olfactory homing of chum salmon to stable compositions of natal stream water. Zool Sci 30 (8):607–612. https://doi.org/10.2108/zsj.30.607

Yoder R (1937) The significance of soil structure in relation to the tilth problem. Soil Sci Soc Am Proc 2:21–22

Young WI, Chilcote D, Youngberg H (1996) Annual ryegrass seed yield response to grazing during early stem elongation. Agron J 88 (2):211–215. https://doi.org/10.2134/agronj1996.00021962008800020015x

19 Summary

- Oregon is roughly 254,800 km^2 in area and is the ninth largest state in the country. Portland is the largest city in the state, and eight of the ten largest Oregon cities are in the Willamette Valley.
- The federal government owns 52% of Oregon land, while 40% of the land is in private ownership. There are nine federally recognized Indigenous Nations in Oregon, and five additional Indigenous Nations maintain an interest in Oregon land.
- Soil mapping began in Oregon in 1903 in Baker City and Salem in 1903; the first countywide soil survey was of Yamhill County in 1917.
- A general soil map of Oregon was published in 1973, followed by revisions in 1975 and 1986. The latter map has remained in use to the present date.
- Nearly two-thirds (64%) of the soils mapped to date were between 1975 and 1990.
- Soil survey data, available online via Web Soil Survey reports or SSURGO database downloads, are freely available to the public.
- The Jory soil series, a fine, mixed, active, mesic Xeric Palehumults, is the official state soil of Oregon.
- 32% of Oregon soils have not been mapped, a proportion exceeded only by Alaska and Idaho.
- Due to broad latitudinal and elevation differences, Oregon has an exceptionally diverse climate, which exerts a major influence on soil formation. The mean annual air temperature in Oregon ranges from 0 °C in the Wallowa and Blue Mountains of northeastern Oregon to 12 °C in south-central Oregon. The mean annual precipitation ranges from 180 mm in southeastern Oregon to over 3,500 mm at higher elevations in the Coast Range.
- The dominant vegetation type in Oregon is temperate shrublands, followed by lodgepole-pine-dominated forest, Douglas-fir-dominated forest, mixed coniferous forests, grasslands, subalpine forests, maritime Sitka spruce-western hemlock forests, and ponderosa pine-dominated forests.
- There are strong climate, vegetation, and soil gradients in the Cascade and Coast Ranges and the Blue and Klamath Mountains.
- Bedrock is exposed throughout the state, with basalt being dominant in the Columbia Basin, Blue Mountains, Lava Plains, Basin and Range, Owyhee Uplands, and East Cascades physiographic provinces. Sedimentary and volcanic rocks are predominant in the West Cascades. The Willamette Valley contains alluvial and lacustrine sediments of Quaternary age. The Coast Range mainly is composed of sandstone and siltstone rocks. The Klamath Mountains contain intrusive granitic rocks and fragments of accreted terranes.
- Colluvium/residuum is the dominant surficial deposit in Oregon, followed by alluvium, colluvium, volcanic ash complex, lacustrine, ash/colluvium, and loess.
- Oregon is divided into 17 Major Land Resource Areas, the most of any state except the considerably larger Texas, California, Alaska, and two other states. The Malheur High Plateau is the most extensive MLRA, followed by the Cascade Mountains, Blue Mountain Foothills, and Blue Mountains.
- A mollic epipedon is present in 43% of the soil series, followed by the ochric (42%), and umbric (14%) epipedons. The cambic horizon is the dominant diagnostic subsurface horizon (41%), followed by the argillic (38%), duripan (7.5%), and calcic (2.6%) horizon. The histic epipedon is the thickest of the epipedons at 106 ± 61 cm, followed by the mollic at 49 ± 26 cm, melanic at 47 ± 5.9 cm, umbric at 44 ± 21 cm, and ochric epipedon at about 18 cm. The spodic and salic horizons are the thickest subsurface horizons at 63 cm, followed by the spodic (62 cm), argillic (58 cm), and calic and cambic horizons (50 cm).
- Oregon has soil series representative of 10 orders, 40 suborders, 112 great groups, 389 subgroups, 1,080 families, and 1,707 soil series.

T. Thorson et al., *The Soils of Oregon*, World Soils Book Series,
https://doi.org/10.1007/978-3-030-90091-5_19

- Mollisols are the dominant order in Oregon (42%), followed by Aridisols (17%), Inceptisols (16%), Andisols (13%), Ultisols (5.0%), and Alfisols (4.1%). Xerolls are the most extensive suborder (39%), followed by Udepts (9.8%), Durids (6.7%), Argids (6.3%), and Cryands (5.2%). The dominant great group is Haploxerolls (18%), followed by Argixerolls (16%), Humudepts (6.7%), and Haplargids and Argidurids (4.3% each).
- The dominant soil-forming processes in Oregon are humification, cambisolization, argilluviation, andisolization, gleization, and silicification.
- Only 5.7% of the Oregon's soil series are benchmark soils; 78% are rare (i.e., contain less than 10,000 acres (40 km^2), 22% are endemic (i.e., are the single example at the family level), and 19% are endangered (rare and endemic).
- Oregon adopted what was probably the most restrictive land-use legislation in the United States with the Oregon Land Use Act of 1973, which restricted development largely to areas in and immediately surrounding cities. This law, subsequent state land use legislation, and policies regarding logging and grazing on federal land engender vigorous debate.
- Key issues in Oregon that affect, and/or are affected by, soils and land use are climate change, wetland loss, flooding, landslides, volcanoes, earthquakes, tsunamis, coastal erosion, and wildfires. Although engineered and vegetative solutions to these issues continue to be important, there is a growing tendency in the state to mitigate risk to human life and infrastructure through land-use policy.
- Working lands in Oregon are used primarily for forest products, livestock grazing, agricultural crops, and wildlife management. Many products from Oregon working lands are exported to other states or countries.
- Approximately 48% of Oregon land is classified as forestland. The state's timber production peaked in the 1960s, however, Oregon is still consistently the leading, or one of the leading, states in the production of softwood lumber. Douglas-fir is the predominant timber species. In addition to the role of forest products in the Oregon economy, forests have historically played an important role in Oregon's culture.
- Livestock grazing consists primarily of cattle (beef and dairy) and sheep. Livestock graze on rangeland and pasture. Rangeland occurs primarily in eastern Oregon and occupies about 36% of Oregon. Pasture occurs throughout the state and usually is irrigated in eastern Oregon.
- Greenhouse and nursery crops, in recent years, account for the greatest value of Oregon crops, but hay accounts for the greatest land area of crop production. Oregon ranks first in the nation in the production of several crops, including grass and clover seed, Christmas trees, and hazelnuts.
- Working lands managers in Oregon use a variety of practices to conserve soil and related resources. A variety of local, state, and federal programs are available to provide funding and/or technical support for planning and implementing practices.
- State regulations pertaining to working lands include the Oregon Forest Practices Act, the Agricultural Water Quality Management Act, Confined Animal Feeding Operations permits, water rights certification, and others.
- Indigenous Peoples have eaten First Foods since long before Euro-American contact in the land now known as "Oregon." These plants and animals are integral parts of Indigenous culture and spirituality based on a reciprocal responsibility to care for the land and the non-human beings that provide food, medicine, dyes, and fiber for the people. The relationship between Indigenous Peoples, the land, and First Foods have allowed their cultures to thrive sustainably for thousands of years.

Appendix A
Soil-Forming Factors for Soil Series in Oregon with an Area of 50 km^2 or More

T. Thorson et al., *The Soils of Oregon*, World Soils Book Series,
https://doi.org/10.1007/978-3-030-90091-5

Series name	Area (km^2)	Subgroup	MAAT (°C)	MAP (mm)	Elev. lower (m)	Elev. upper (m)	Max slope (%)	Vegetatiom	Parent material	Landform
Abegg	78.3	Ultic Haploxeralfs	11.9	1223	241	504	30	California black oak, Douglas-fir, Pacific madrone, ponderosa pine, sugar pine, California fescue, Idaho fescue	Alluvium and colluvium from igneous, metamorphic and sedimentary rock	Stream terraces, fans
Abert	144.8	Sodic Xeric Haplocambids	7.0	229	1304	1469	2	Basin wildrye, basin big sagebrush, rabbitbrush, inland saltgrass	Volcanic ash over lacustrine deposits	Lakebeds
Abiqua	57.2	Cumulic Ultic Haploxerolls	11.3	1294	82	232	5	Douglas-fir, Oregon ash, Oregon white oak, snowberry, wild rose	Alluvium	Fans, stream terraces, flood plains
Absaquil	54.0	Typic Haplohumults	10.0	1862	71	762	60	Douglas-fir, grand fir, Port Orford cedar, tanoak, western hemlock, western redcedar	Colluvium and residuum derived from sandstone and siltstone	Mountains
Acker	217.4	Typic Palexerults	9.4	1509	543	1114	60	Douglas-fir, incense cedar, Pacific madrone, Oregongrape, salal	Colluvium and residuum derived from metamorphic rock	Mountains
Actem	749.1	Xeric Argidurids	6.8	267	1433	1674	20	Bluebunch wheatgrass, Thurber's needlegrass, Wyoming big sagebrush, Indian ricegrass	Colluvium derived from igneous rock	Hills, plateaus
Ada	84.9	Typic Argixerolls	7.0	280	1302	1504	40	Idaho fescue, bluebunch wheatgrass, mountain big sagebrush	Colluvium derived from igneous and metamorphic rock	Hills, plateaus
Adkins	107.5	Xeric Haplocalcids	12.0	229	96	309	25	Needle-and-thread, bluebunch wheatgrass, Sandberg bluegrass	Eolian sands	Hills
Agate	50.3	Typic Durixerepts	12.0	610	335	564	15	Bluebunch wheatgrass, lemmon's needlegrass, ceanothus	Alluvium derived from igneous, metamorphic and sedimentary rock	Fan terraces
Agency	135.8	Aridic Haploxerolls	9.3	251	663	975	15	Bluebunch wheatgrass, Idaho fescue, antelope bitterbrush, mountain big sagebrush	Loess over residuum derived from volcaniclastic sediments of the Deschutes Formation	Plateaus
Albee	123.6	Vitrandic Haploxerolls	6.2	570	1099	1565	60	Idaho fescue, bluebunch wheatgrass, balsamroot, lupine, prairie junegrass	Loess and volcanic ash mixed with colluvium derived from basalt	Plateaus, mountains
Alding	63.1	Lithic Argixerolls	6.0	509	1138	1785	70	Idaho fescue, curl-leaf mountainmahogany, ponderosa pine, bluebunch wheatgrass, western juniper	Colluvium derived from metavolcanics	Hills
Aloha	236.3	Aquic Haploxerepts	12.0	1298	41	98	8	Douglas-fir, Oregon ash, Oregon white oak, Oregongrape	Glaciolacustrine deposits	Glaciolacustrine terraces
Alspaugh	130.5	Typic Paleudults	10.0	1842	244	549	50	Douglas-fir, red alder, salal, western swordfern	Colluvium derived from andesite and tuff	Hills
Alstony	82.6	Alic Hapludands	9.0	1595	91	481	90	Douglas-fir, red alder, salal, vine maple	Colluvium derived from igneous rock	Mountains

(continued)

Series name	Area (km^2)	Subgroup	MAAT (°C)	MAP (mm)	Elev. lower (m)	Elev. upper (m)	Max slope (%)	Vegetatiom	Parent material	Landform
Althouse	59.8	Typic Dystroxerepts	6.0	2459	947	1676	90	Douglas-fir, golden chinquapin, sugar pine, tanoak, white fir	Colluvium and residuum derived from metasedimentary and metavolcanic rock	Mountains
Alvodest	263.2	Sodic Aquicambids	8.2	223	1250	1412	3	Black greasewood, inland saltgrass, basin wildrye	Lacustrine deposits	Lakebeds
Amity	434.3	Argiaquic Xeric Argialbolls	11.5	1122	48	110	3	Douglas-fir, Oregon white oak, Oregon ash, wild rose	Glaciolacustrine deposits	Glaciolacustrine terraces
Analulu	264.8	Vitrandic Haploxerepts	4.9	701	1355	1647	90	Douglas-fir, common snowberry, ponderosa pine, elk sedge, pinegrass, western larch	Colluvium derived from metasedimentary rocks or metavolcanic rocks	Mountains
Anatone	1,588.3	Lithic Haploxerolls	6.0	578	1199	1595	90	Curl-leaf mountainmahogany, Idaho fescue, bluebunch wheatgrass, sagebrush, Sandberg bluegrass	Residuum and colluvium derived from volcanic rocks	Mountains
Anawalt	1,279.2	Lithic Xeric Haplargids	6.6	280	1527	1833	50	Bluebunch wheatgrass, low sagebrush, Sandberg bluegrass	Colluvium and residuum from tuff and basalt	Mountains, plateaus
Anderly	185.0	Typic Haploxerolls	10.8	320	254	648	35	Bluebunch wheatgrass, Sandberg bluegrass, arrowleaf balsamroot, western yarrow	Loess over fractured basalt	Plateaus, hills
Angelbasin	78.7	Andic Dystrocryepts	1.9	995	1938	2220	90	Subalpine fir, Engelmann spruce, grouse huckleberry, heartleaf arnica, western larch	Volcanic ash over glacial till derived from granite	Mountains
Angelpeak	127.0	Typic Vitricryands	2.9	811	1650	1864	90	Subalpine fir, Engelmann spruce, grouse huckleberry, heartleaf arnica, western larch	Volcanic ash over colluvium derived from argillite	Mountains
Apt	148.8	Typic Haplohumults	9.8	2318	111	656	50	Bigleaf maple, Douglas-fir, red alder, western hemlock, salal	Residuum and colluvium derived from sedimentary rock	Mountains
Arcia	92.5	Vitrandic Argixerolls	6.0	343	1602	1945	60	Idaho fescue, bluebunch wheatgrass, mountain big sagebrush	Colluvium derived from igneous and sedimentary rock	Hills, mountains
Ascar	120.8	Typic Fulvudands	9.4	2385	64	515	90	Douglas-fir, red alder, Sitka spruce, western hemlock, salal	Colluvium derived from basalt	Mountains
Aschoff	310.7	Andic Humudepts	9.6	1788	175	694	95	Bigleaf maple, Douglas-fir, red alder, western hemlock, salal	Colluvium derived from andesite and basalt	Mountains
Astoria	145.8	Andic Humudepts	9.2	2466	98	683	90	Douglas-fir, red alder, western hemlock, western redcedar, salal	Colluvium and residuum from sedimentary rock	Mountains
Ateron	1,287.7	Lithic Argixerolls	5.9	374	1197	1744	90	Bluebunch wheatgrass, Sandberg bluegrass, Idaho fescue, mountain big sagebrush	Colluvium and residuum derived from volcanic rock	Hills, mountains
Athena	210.3	Pachic Haploxerolls	10.0	445	457	701	12	Idaho fescue, bluebunch wheatgrass	Loess	Hills
Atlow	357.0	Lithic Xeric Haplargids	7.9	229	1306	1591	50	Indian ricegrass, Wyoming big sagebrush, spiny hopsage, Thurber's needlegrass	Residuum and colluvium derived from volcanic rock	Mountains, hills

(continued)

Series name	Area (km²)	Subgroup	MAAT (°C)	MAP (mm)	Elev. lower (m)	Elev. upper (m)	Max slope (%)	Vegetatiom	Parent material	Landform
Atring	479.8	Typic Dystroxerepts	10.4	1646	271	933	90	Douglas-fir, Oregongrape, Pacific madrone, Oregon white oak	Colluvium derived from sandstone and siltstone	Mountains
Ausmus	235.1	Aquic Natrargids	6.8	229	1229	1321	2	Basin wildrye, black greasewood, inland saltgrass	Lacustrine deposits	Lakebeds
Ayres	86.5	Argiduridic Durixerolls	10.0	252	945	1163	20	Bluebunch wheatgrass, low sagebrush, Sandberg bluegrass, Idaho fescue	Colluvium derived from rhyolite	Fans
Bacona	212.5	Typic Palehumults	9.0	1715	122	488	30	Douglas-fir, red alder, western hemlock, western redcedar, salal	Colluvium derived from siltstone, shale, and sandstone	Mountains
Baconcamp	372.9	Pachic Haplocryolls	5.0	625	1859	2541	80	Mountain snowberry, mountain big sagebrush, Idaho fescue, rough fescue, tufted hairgrass	Residuum and colluvium derived from volcanic rocks	Mountains
Bakeoven	1,224.0	Aridic Lithic Haploxerolls	9.5	289	648	1050	60	Sandberg bluegrass, stiff sagebrush	Colluvium and residuum derived from basalt	Plateaus
Baker	77.9	Haploduridic Durixerolls	9.0	267	656	930	7	Bluebunch wheatgrass, Thurber's needlegrass, needle-and-thread, Wyoming big sagebrush	Alluvium	Stream terraces
Barkshanty	69.7	Typic Palehumults	10.0	2794	61	914	40	Douglas-fir, Port Orford cedar, red alder, tanoak, western hemlock, western redcedar	Residuum and colluvium	Mountains
Bashaw	171.8	Xeric Endoaquerts	11.7	1217	58	139	12	Oregon ash, sedges, rushes	Alluvium	Flood plains, terraces
Bateman	204.5	Ultic Palexeralfs	12.0	1202	76	792	60	Douglas-fir, incense cedar, Oregon white oak, Pacific madrone	Colluvium and residuum derived from sandstone and siltstone	Mountains, hills
Bearpawmeadow	58.3	Andic Haplocryepts	3.1	1033	1911	2117	90	Subalpine fir, grouse huckleberry, Engelmann spruce, heartleaf arnica, western larch, lodgepole pine	Colluvium derived from andesite or basalt	Mountains
Beden	106.4	Aridic Lithic Argixerolls	6.7	268	1255	1477	50	Bluebunch wheatgrass, Idaho fescue, Sandberg bluegrass, Wyoming big sagebrush, western juniper	Residuum derived from basalt	Hills plateaus
Beekman	646.8	Typic Dystroxerepts	10.3	1514	228	1064	90	Douglas-fir, ponderosa pine, Pacific madrone, tanoak	Colluvium derived from metasedimentary and metavolcanic rock	Mountains
Bellpine	769.0	Xeric Haplohumults	11.4	1241	99	476	75	Bigleaf maple, Douglas-fir, Oregon white oak, snowberry	Colluvium and residuum derived from sandstone and siltstone	Hills
Bennettcreek	107.3	Vitrandic Haploxeralfs	4.8	586	1389	1582	60	Pinegrass, ponderosa pine, grand fir, elk sedge, Douglas-fir, western larch, heartleaf arnica	Volcanic ash over colluvium derived from andesite	Mountains, plateaus
Beoska	55.9	Durinodic Natrargids	8.0	212	1240	1462	15	Shadscale, bud sagebrush, bottlebrush squirreltail, Indian ricegrass	Alluvium	Fans

(continued)

Series name	Area (km^2)	Subgroup	MAAT (°C)	MAP (mm)	Elev. lower (m)	Elev. upper (m)	Max slope (%)	Vegetatiom	Parent material	Landform
Berdugo	114.6	Xeric Paleargids	8.0	229	1423	1635	5	Wyoming big sagebrush, bottlebrush squirreltail, Sandberg bluegrass, Thurber's needlegrass, Indian ricegrass	Lacustrine deposits	Lake terraces
Bingville	68.7	Pachic Palexerolls	6.1	349	1118	1796	75	Idaho fescue, bluebunch wheatgrass, low sagebrush, Sandberg bluegrass	Colluvium or residuum derived from basalt	Plateaus
Blachly	509.7	Humic Dystrudepts	9.9	2361	142	810	75	Bigleaf maple, Douglas-fir, red alder, western hemlock	Colluvium derived from igneous and sedimentary rock	Mountains
Bluecanyon	66.9	Lithic Haploxerolls	5.1	674	1360	1742	90	Bluebunch wheatgrass, Idaho fescue, Sandberg bluegrass, mountain big sagebrush	Colluvium derived from argillite and other metasedimentary or metavolcanic rocks	Mountains
Bly	60.8	Vitrandic Argixerolls	6.7	467	1278	1448	35	Ponderosa pine, antelope bitterbrush, Idaho fescue	Alluvium	Stream terraces, fans
Boardtree	112.2	Alfic Vitrixerands	6.0	503	1014	1539	70	Douglas-fir, common snowberry, ponderosa pine, elk sedge, pinegrass	Volcanic ash over colluvium derived from basalt	Mountains
Bocker	1,472.8	Lithic Haploxerolls	6.0	559	1227	1583	90	Bluebunch wheatgrass, Sandberg bluegrass, low sagebrush, stiff sagebrush, Idaho fescue	Colluvium and residuum derived from igneous rocks	Mountains, plateaus, hills
Bodell	83.9	Lithic Haploxerolls	9.1	578	552	1038	75	Idaho fescue, bluebunch wheatgrass, Sandberg bluegrass	Colluvium derived from basalt	Mountains
Bohannon	2,044.0	Andic Humudepts	10.0	2277	74	846	100	Douglas-fir, red alder, western hemlock, western redcedar	Colluvium derived from sandstone and siltstone	Mountains
Boilout	52.7	Vitrixerandic Argidurids	7.0	228	1323	1497	10	Indian ricegrass, Thurber's needlegrass, bluebunch wheatgrass, basin big sagebrush	Residuum derived from basalt or tuff	Plateaus, hills
Bolobin	66.3	Vitrandic Argixerolls	5.9	456	1160	1424	60	Pinegrass, ponderosa pine, Douglas-fir, elk sedge, heartleaf arnica	Colluvium derived from basalt	Plateaus, mountains
Bombadil	62.2	Lithic xeric Haplargids	8.5	230	1250	2000	8	Thurber's needlegrass, Wyoming big sagebrush	Colluvium derived from volcanic rock	Hills
Bonnick	96.7	Vitritorrandic Haploxerolls	6.7	253	1317	1395	15	Mountain big sagebrush, needle-and-thread, Indian ricegrass	Volcanic ash over lacustrine deposits	Lake terraces
Booth	593.3	Vertic Palexerolls	7.4	401	1407	1894	65	Idaho fescue, low sagebrush, bluebunch wheatgrass	Colluvium derived from basalt and tuff	Hills, mountains
Bordengulch	71.1	Andic Haplocryepts	2.8	880	1723	1969	90	Engelmann spruce, lodgepole pine, western larch, grouse huckleberry	Colluvium and residuum derived from argillite	Mountains
Bornstedt	67.7	Typic Palexerults	11.0	1435	122	198	30	Douglas-fir, western redcedar, western swordfern	Alluvium	Hills, terraces
Borobey	215.1	Vitritorrandic Haploxerolls	6.6	275	1293	1473	15	Mountain big sagebrush, Idaho fescue, Thurber's needlegrass	Volcanic ash over alluvium	Lake terraces
Boulder Lake	81.4	Xeric Epiaquerts	6.8	343	1418	1574	2	Sandberg bluegrass, beardless wildrye, silver sagebrush	Lacustrine deposits	Lakebeds, plateaus

(continued)

Series name	Area (km^2)	Subgroup	MAAT (°C)	MAP (mm)	Elev. lower (m)	Elev. upper (m)	Max slope (%)	Vegetatiom	Parent material	Landform
Bouldrock	55.9	Humic Haploxerepts	6.0	457	1219	1890	80	Bluebunch wheatgrass, mountain big sagebrush, Thurber's needlegrass, arrowleaf balsamroot	Colluvium and residuum derived from quartz diorite	Mountains
Brace	1,570.2	Xeric Argidurids	7.2	265	1363	1636	20	Bluebunch wheatgrass, Thurber's needlegrass, Wyoming big sagebrush	Colluvium and residuum derived from tuff and basalt	Plateaus
Brader	50.4	Typic Haploxerepts	11.3	776	305	866	60	Oregon white oak, Idaho fescue	Colluvium and residuum derived from sandstone and siltstone	Hills
Braun	167.6	Dystric Eutrudepts	9.1	1964	143	610	90	Bigleaf maple, Douglas-fir western hemlock, western redcedar	Colluvium derived from siltstone	Mountains
Bravo	188.7	Humic Dystrudepts	10.0	2794	305	914	90	Douglas-fir, red alder, tanoak, western hemlock, Pacific madrone	Colluvium and residuum derived from metavolcanics and metasedimentary rock	Mountains
Bridgecreek	75.5	Typic Palexerolls	6.0	517	1006	1311	35	Idaho fescue, bluebunch wheatgrass	Residuum derived from tuffaceous material	Plateaus
Brisbois	66.5	Xeric Haplargids	9.5	254	687	1055	85	Bluebunch wheatgrass, shadscale, Sandberg bluegrass	Colluvium and residuum derived from tuff	Hills
Broyles	97.3	Durinodic Haplocambids	8.0	257	1133	1303	8	Shadscale, bud sagebrush, Indian ricegrass	Volcanic ash over alluvium	Fans
Btree	116.5	Alfic Udivitrands	5.6	633	1322	1577	90	Douglas-fir, Engelmann spruce, grand fir, western larch, big huckleberry	Volcanic ash over colluvium and residuum derived from tuff	Mountains
Buckcreek	79.6	Pachic ultic Haploxerolls	6.5	635	610	1219	70	Mallow ninebark, common snowberry, Idaho fescue	Colluvium derived from basalt	Hills
Bucketlake	120.8	Typic Vitricryands	2.7	943	1682	1938	90	Engelmann spruce, big huckleberry, subalpine fir, western larch, heartleaf arnica, elk sedge	Volcanic ash over glacial till	Mountains
Bull Run	78.1	Eutric Fulvudands	10.7	1962	183	564	80	Bigleaf maple, Douglas-fir, red alder, western hemlock, western redcedar	Silty material mixed with volcanic ash	Mountains
Bullards	104.1	Typic Haplorthods	11.4	1720	13	160	60	Douglas-fir, shore pine, Sitka spruce, western hemlock, western redcedar	Eolian sands	Marine terraces
Bullump	103.8	Pachic Argixerolls	7.5	412	1593	1933	70	Bluebunch wheatgrass, mountain big sagebrush, Idaho fescue, antelope bitterbrush, common snowberry	Colluvium and residuum derived from rhyolite, tuff, and basalt	Mountains
Burgerbutte	61.2	Lithic Humicryepts	1.9	1069	1906	2132	90	Mountain big sagebrush, mountain snowberry, Idaho fescue	Colluvium derived from basalt	Mountains, plateaus
Burke	62.6	Xeric Haplodurids	11.0	229	198	396	30	Bluebunch wheatgrass, Sandberg bluegrass, basin big sagebrush	Loess over glaciolacustrine deposits	Lake terraces
Bybee	57.4	Typic Haploxerolls	6.0	1016	1097	1676	35	Douglas-fir, ponderosa pine, white fir, common snowberry	Colluvium derived from andesite or tuff	Hills
Calderwood	155.7	Lithic Xeric Haplocambids	8.0	236	1312	1598	50	Thurber's needlegrass, bluebunch wheatgrass, Indian ricegrass, Wyoming big sagebrush, bottlebrush squirreltail, spiny hopsage	Colluvium derived from volcanic rock	Hills

(continued)

Series name	Area (km^2)	Subgroup	MAAT (°C)	MAP (mm)	Elev. lower (m)	Elev. upper (m)	Max slope (%)	Vegetatiom	Parent material	Landform
Calimus	156.9	Pachic Haploxerolls	8.6	330	1257	1401	35	Bluebunch wheatgrass, basin wildrye, antelope bitterbrush, Idaho fescue, Wyoming big sagebrush	Alluvium derived from tuff, and basalt	Fans, lake terraces
Camas	111.2	Fluventic Haploxerolls	11.4	1084	128	453	3	Oregon ash, Oregon white oak, red alder, wildrose, willow	Alluvium	Flood plains
Campcreek	56.1	Vertic Palexerolls	5.5	430	1329	1520	60	Bluebunch wheatgrass, Idaho fescue, mountain big sagebrush	Alluvium	Stream terraces
Canest	353.2	Lithic Argixerolls	6.6	404	1224	1618	60	Bluebunch wheatgrass, Sandberg bluegrass, stiff sagebrush, low sagebrush	Colluvium and residuum derived from igneous rock	Plateaus
Cantala	165.2	Typic Haploxerolls	10.0	311	451	975	35	Idaho fescue, bluebunch wheatgrass	Loess over alluvium	Plateaus
Caris	207.9	Typic Haploxerepts	10.0	826	305	1219	80	California black oak, Douglas-fir, Pacific madrone, ponderosa pine, sugar pine, Idaho fescue	Colluvium derived from metavolcanic and or metasedimentary rock	Mountains
Carney	235.1	Udic Haploxererts	10.4	672	473	1202	35	Idaho fescue, bluebunch wheatgrass	Colluvium derived from tuff and breccia	Hills
Carryback	1,188.9	Vertic Palexerolls	6.6	335	1449	1728	50	Idaho fescue, low sagebrush, bluebunch wheatgrass, Sandberg bluegrass	Colluvium and residuum derived from tuff and basalt	Plateaus, hills
Carvix	151.0	Aridic Haploxerolls	6.8	318	1328	1533	8	Basin wildrye, basin big sagebrush, bluebunch wheatgrass, Idaho fescue	Alluvium	Stream terraces
Cascade	178.7	Humic Fragixerepts	11.1	1397	78	383	60	Bigleaf maple, Douglas-fir, red alder, western redcedar, salal, western swordfern	Loess	Hills
Cassiday	183.2	Humic Dystrudepts	10.0	2794	61	914	90	Douglas-fir, red alder, tanoak, western hemlock, Pacific madrone	Colluvium derived from metavolcanics and/or metasedimentary rock	Mountains
Castlecrest	265.6	Typic Vitricryands	4.3	1463	1647	2066	80	Grouse huckleberry, Shasta red fir, mountain hemlock, lodgepole pine	Ash and pumice	Mountains
Caterl	253.0	Alic Hapludands	6.8	2876	546	887	90	Douglas-fir, red alder, western hemlock, salal, huckleberry	Colluvium derived from igneous rock	Mountains
Catherine	89.2	Cumulic Endoaquolls	9.2	419	616	981	3	Tufted hairgrass, sedge, rush	Alluvium	Flood plains
Catlow	171.6	Durinodic Xeric Haplocambids	7.8	229	1328	1487	20	Indian ricegraass, Thurber's needlegrass, bluebunch wheatgrass, Wyoming big sagebrush	Alluvium	Lake terraces
Cazadero	82.0	Rhodic Paleudults	11.0	1733	183	379	60	Douglas-fir, red alder, western hemlock, western redcedar, salal	Alluvium	Hills
Chapman	58.1	Cumulic Ultic Haploxerolls	11.6	1270	62	287	3	Bigleaf maple, black cottonwood, Douglas-fir, grand fir, Oregon white oak	Alluvium	Flood plains, stream terraces
Chehalis	238.5	Cumulic Ultic Haploxerolls	11.7	1202	37	203	3	Bigleaf maple, Douglas-fir, grand fir, red alder, western redcedar, western swordfern	Alluvium	Flood plains

(continued)

Series name	Area (km^2)	Subgroup	MAAT (°C)	MAP (mm)	Elev. lower (m)	Elev. upper (m)	Max slope (%)	Vegetatiom	Parent material	Landform
Chen	111.8	Aridic Lithic Argixerolls	5.4	332	1768	2092	50	Idaho fescue, low sagebrush, bluebunch wheatgrass, Sandberg bluegrass	Colluvium and residuum derived from volcanic rock	Mountains, hills
Cherrycreek	71.2	Vitrandic Haploxerolls	5.9	533	990	1509	90	Idaho fescue, bluebunch wheatgrass, lupine, common snowberry, mallow ninebark	Colluvium and residuum derived from basalt	Plateaus
Chilcott	93.3	Abruptic Xeric Argidurids	9.3	268	945	1057	25	Wyoming big sagebrush	Alluvium	Stream terraces
Choptie	64.5	Lithic Haploxerolls	6.4	386	1227	1653	40	Idaho fescue, antelope bitterbrush, bluebunch wheatgrass, mountain big sagebrush	Residuum derived from tuff and breccia	Hills
Clackamas	82.5	Typic Argiaquolls	12.0	1191	50	186	3	Oregon ash, wildrose, sedges, rushes	Alluvium	Stream terraces
Clamp	134.2	Lithic Haplocryolls	5.0	536	1746	2321	70	Idaho fescue, bluebunch wheatgrass, low sagebrush, Sandberg bluegrass, Thurber's needlegrass	Residuum and colluvium derived from basalt or andesite	Mountains
Cloquato	159.7	Cumulic Ultic Haploxerolls	11.6	1232	39	175	3	Bigleaf maple, Douglas-fir, red alder, western redcedar, western swordfern	Alluvium	Flood plains
Clovercreek	54.4	Lithic Argixerolls	9.0	350	776	1016	90	Bluebunch wheatgrass, Sandberg bluegrass, Idaho fescue, mountain big sagebrush	Colluvium derived from greenstone	Hills
Clovkamp	50.0	Vitritorrandic Haploxerolls	10.0	272	892	1219	25	Idaho fescue, needle-and-thread, antelope bitterbrush, mountain big sagebrush	Volcanic ash over alluvium	Plateaus
Coburg	178.8	Oxyaquic Argixerolls	11.8	1188	51	181	5	Douglas-fir, Oregon white oak	Alluvium	Stream terraces
Coglin	100.1	Xeric Paleargids	7.2	244	1494	1664	15	Black sagebrush, bluebunch wheatgrass, Sandberg bluegrass	Colluvium and residuum derived from tuff and basalt	Plateaus
Colbar	76.8	Xeric Haplargids	9.0	230	1213	1537	50	Thurber's needlegrass, Wyoming big sagebrush	Colluvium derived from volcanic rock	Hills
Colestine	180.2	Typic Dystroxerepts	9.5	1693	237	1263	90	Douglas-fir, incense cedar, Pacific madrone, tanoak, sugar pine	Colluvium and residuum derived from metavolcanics or metasedimentary rock	Mountains
Collier	411.0	Xeric Vitricryands	5.2	889	1293	1747	80	Ponderosa pine, Idaho fescue, antelope bitterbrush	Ash and pumice	Ash flows
Concord	123.0	Typic Endoaqualfs	11.8	1118	47	113	3	Oregon ash, sedges, rushes	Glaciolacustrine deposits	Glaciolacustrine terraces
Condon	1,170.5	Typic Haploxerolls	9.6	310	468	898	40	Idaho fescue, bluebunch wheatgrass	Loess over fractured basalt	Plateaus
Connleyhills	122.0	Vitritorrandic Argixerolls	7.3	306	1337	1596	60	Mountain big sagebrush, western juniper, bluebunch wheatgrass, Idaho fescue	Residuum derived from basalt	Hills
Conser	93.2	Vertic Argiaquolls	11.5	1196	38	290	3	Oregon ash, sedges, rushes	Alluvium	Stream terraces
Coquille	108.4	Fluvaquentic Endoaquepts	10.3	2088	1	5	1	Tufted hairgrass, tussocks, saltgrass	Alluvium	Flood plains
Cornelius	101.3	Mollic Fragixeralfs	12.0	1246	106	235	60	Bigleaf maple, Douglas-fir, western redcedar	Loess	Hills

(continued)

Series name	Area (km^2)	Subgroup	MAAT (°C)	MAP (mm)	Elev. lower (m)	Elev. upper (m)	Max slope (%)	Vegetatiom	Parent material	Landform
Cornutt	79.5	Ultic Haploxeralfs	10.2	1372	281	1268	60	Incense cedar, Douglas-fir, Pacific madrone	Residuum and colluvium derived from serpentine	Mountains
Corral	112.9	Xeric Haplargids	7.7	246	1498	1639	15	Indian ricegrass, Thurber's needlegrass, bluebunch wheatgrass, Wyoming big sagebrush	Colluvium and residuum derived from tuff	Plateaus
Cove	83.1	Vertic Endoaquolls	11.6	1268	78	252	8	Sedges, rushes, willows	Alluvium	Flood plains
Cowsly	269.7	Xerertic Argialbolls	6.5	606	1062	1377	30	Douglas-fir, ponderosa pine, western larch, common snowberry, pinegrass, elk sedge	Colluvium and residuum derived from basalt	Plateaus
Coyata	54.6	Humic Dystroxerepts	9.0	1143	634	1211	80	Douglas-fir, incense cedar, ponderosa pine, white fir, Pacific madrone	Colluvium and residuum derived from igneous rocks	Hills, plateaus
Coztur	117.4	Lithic Xeric Haplargids	7.0	278	1366	1676	20	Thurber's needlegrass, bluebunch wheatgrass, mountain big sagebrush, Indian ricegrass	Residuum and colluvium derived from basalt or tuff	Hills, mountains
Crackedground	60.1	Vitritorrandic Haploxerolls	7.0	266	1332	1395	15	Idaho fescue, mountain big sagebrush, Thurber's needlegrass	Colluvium and residuum derived from volcanic rock	Plateaus
Crackercreek	99.6	Alfic Vitrixerands	5.6	599	1383	1659	75	Big huckleberry, grand fir, ponderosa pine, western larch, pinegrass, Douglas-fir	Volcanic ash over colluvium derived from basalt	Mountains
Crowcamp	108.5	Vertic Palexerolls	7.0	254	1226	1342	2	Nevada bluegrass, creeping wildrye, silver sagebrush	Lacustrine deposits	Lakebeds
Cruiser	51.5	Typic Haplocryands	6.0	2328	762	1229	70	Douglas-fir, noble fir, mountain hemlock, western redcedar, common beargrass	Colluvium derived from basic igneous rock	Mountains
Crump	70.6	Histic Humaquepts	9.0	227	1289	1472	1	Cattails, rushes, sedges	Decomposed organic layer over lacustrine deposits	Lakebeds
Cullius	79.3	Aridic Lithic Argixerolls	10.0	254	762	914	15	Bluebunch wheatgrass, basin big sagebrush, Sandberg bluegrass, Idaho fescue, antelope bitterbrush	Loess over colluvium and residuum from sediments of the Deschutes Formation	Plateaus
Cumley	136.0	Oxyaquic Palehumults	9.9	1702	244	597	20	Bigleaf maple, Douglas-fir, red alder, western redcedar, western hemlock	Colluvium and glacial till from basic igneous rock	Mountains
Curant	203.5	Calcic Pachic Haploxerolls	8.0	305	640	1061	75	Idaho fescue, bluebunch wheatgrass, Sandberg bluegrass	Loess	Hills
Curtin	55.5	Aquic Haploxererts	12.0	1080	122	610	20	Oregon white oak, wild rose, sedges	Alluvium	Fans
Dacker	50.1	Xeric Argidurids	8.6	184	1550	1767	15	Bluebunch wheatgrass, Thurber's needlegrass, Wyoming big sagebrush	Alluvium	Fans
Damewood	180.1	Andic Humudepts	10.0	2286	61	671	90	Bigleaf maple, Douglas-fir, red alder, western hemlock	Colluvium derived from sandstone	Mountains
Davey	76.9	Xeric Haplocambids	7.8	222	1284	1417	45	Indian ricegrass, Wyoming big sagebrush, spiny hopsage	Alluvium	Fans
Day	53.4	Chromic Haploxererts	10.0	264	557	1051	70	Bluebunch wheatgrass, shadscale, giant wildrye, basin big sagebrush	Residuum derived from tuff in the John Day Formation	Fans, hills

(continued)

Series name	Area (km²)	Subgroup	MAAT (°C)	MAP (mm)	Elev. lower (m)	Elev. upper (m)	Max slope (%)	Vegetatiom	Parent material	Landform
Dayton	417.5	Vertic Albaqualfs	11.5	1132	52	116	3	Oregon ash, wild rose, Oregon white oak	Glaciolacustrine deposits	Glaciolacustrine terraces
Deadline	107.7	Humic Dystrudepts	10.0	2794	61	914	90	Canyon live oak, Douglas-fir, Pacific madrone, tanoak, Douglas-fir	Colluvium and residuum derived from phyllite or schist	Mountains
Defenbaugh	58.0	Typic Haplocambids	8.0	210	1204	1387	4	Black greasewood, bud sagebrush, shadscale, spiny hopsage, bottlebrush squirreltail	Alluvium	Fans
Dehlinger	58.9	Pachic Haploxerolls	9.2	356	1263	1770	70	Bluebunch wheatgrass, Wyoming big sagebrush, Thurber's needlegrass	Colluvium and residuum derived from basalt, tuff, and andesite	Mountains
Deppy	308.4	Argidic Argidurids	8.1	217	1302	1498	50	Shadscale, bud sagebrush, Indian ricegrass, bottlebrush squirreltail	Alluvium	Lake terraces, fans
Deschutes	245.1	Vitritorrandic Haploxerolls	9.8	253	844	1161	30	Idaho fescue, needle-and-thread, antelope bitterbrush, mountain big sagebrush	Volcanic ash over residuum derived from basalt	Plateaus
Deseed	64.5	Xeric Haplargids	7.0	279	1501	1728	50	Bluebunch wheatgrass, Sandberg bluegrass, Wyoming big sagebrush	Colluvium and residuum derived from tuff and basalt	Plateaus, hills
Deskamp	172.3	Vitritorrandic Haploxerolls	9.5	261	931	1191	15	Idaho fescue, needle-and-thread, antelope bitterbrush, mountain big sagebrush	Volcanic ash over basalt	Plateaus
Dester	364.8	Vitritorrandic Argixerolls	6.7	279	1264	1437	15	Idaho fescue, mountain big sagebrush, Thurber's needlegrass	Colluvium and residuum derived from basalt	Plateaus
Devnot	65.5	Lithic Argixerolls	8.5	406	671	1219	35	Sandberg bluegrass, stiff sagebrush	Colluvium and residuum derived from basalt	Hills
Dewar	117.0	Xeric Argidurids	9.0	230	1403	1976	15	Wyoming big sagebrush, Thurber's needlegrass, Sandberg bluegrass, Indian ricegrass	Alluvium	Fans
Digger	786.1	Dystric Eutrudepts	9.7	2303	101	811	90	Bigleaf maple, Douglas-fir, western hemlock, salal, western swordfern	Colluvium and residuum derived from sandstone and siltstone	Mountains
Divers	50.9	Typic Haplocryands	5.8	2280	867	1382	90	Douglas-fir, grand fir, noble fir, western white pine	Colluvium and residuum derived from basalt or andesite	Mountains
Dixon	61.7	Xeric Haplocambids	8.0	216	1257	1356	15	Indian ricegrass, basin wildrye, basin big sagebrush, black greasewood	Alluvium	Fans
Dixonville	266.1	Pachic Ultic Argixerolls	11.5	1245	88	404	60	Douglas-fir, grand fir, Oregon white oak, Pacific madrone, common snowberry	Colluvium and residuum derived from basalt	Hills
Doyn	248.4	Aridic Lithic Haploxerolls	6.1	335	1285	1763	30	Stiff sagebrush, Thurber's needlegrass, Sandberg bluegrass	Residuum and colluvium derived from igneous rock	Plateaus, hills
Drews	97.7	Pachic Argixerolls	8.0	392	1421	1548	30	Idaho fescue, antelope bitterbrush, bluebunch wheatgrass, mountain big sagebrush	Alluvium	Lake terraces

(continued)

Series name	Area (km^2)	Subgroup	MAAT (°C)	MAP (mm)	Elev. lower (m)	Elev. upper (m)	Max slope (%)	Vegetatiom	Parent material	Landform
Drewsey	258.0	Xeric Haplocambids	9.5	252	781	998	40	Bluebunch wheatgrass, Thurber's needlegrass, Wyoming big sagebrush	Eolian sands and colluvium derived from tuffaceous rock	Hills
Drinkwater	74.5	Xeric Haplocambids	9.4	253	612	1095 ara>	90	Thurber's needlegrass, bluebunch wheatgrass, basin big sagebrush, basin wildrye, antelope bitterbrush	Colluvium and residuum derived from basalt	Hills, fans
Droval	113.1	Sodic Aquicambids	8.0	216	1238	1341	3	Basin wildrye, black greasewood, spiny hopsage, basin big sagebrush	Lacustrine deposits	Lakebeds
Dubakella	189.6	Mollic Haploxeralfs	9.9	1228	332	1182	75	Incense cedar, Jeffrey pine, ponderosa pine, California fescue	Colluvium and residuum derived from serpentine and peridotite	Mountains
Duff	71.9	Pachic Haplocryolls	5.0	542	1865	2463	80	Idaho fescue, mountain brome, slender wheatgrass	Residuum and colluvium derived from basalt or andesite	Mountains
Dufur	61.9	Calcic Haploxerolls	9.4	331	244	549	40	Bluebunch wheatgrass, Sandberg bluegrass, arrowleaf balsamroot	Loess over sedimentary rock	Hills
Dumont	179.2	Typic Palexerults	9.1	1273	538	1191	60	Douglas-fir, western hemlock, white fir, Oregongrape, Pacific madrone	Colluvium and residuum derived from metavolcanics and metasedimentary rock	Hills, mountains
Dunres	111.5	Vitrandic Durixerolls	7.0	302	1386	1547	20	Idaho fescue, antelope bitterbrush, bluebunch wheatgrass, mountain big sagebrush	Colluvium and residuum derived from basalt	Plateaus
Dunstan	52.4	Vitrandic Haploxeralfs	4.6	590	1339	1552	90	Douglas-fir, common snowberry, ponderosa pine, elk sedge, pinegrass, western larch	Colluvium derived from basalt or breccia	Mountains
Dupee	167.3	Aquultic Haploxeralfs	11.2	1200	70	274	30	Oregon white oak, Douglas-fir	Colluvium and residuum derived from sandstone and siltstone	Hills
Dupratt	126.4	Vitrandic Argixerolls	5.9	426	1133	1697	65	Ponderosa pine, common snowberry, elk sedge	Colluvium and residuum derived from basalt	Mountains
Durkee	223.2	Calcic Argixerolls	6.0	356	1097	1524	60	Idaho fescue, bluebunch wheatgrass, mountain big sagebrush	Colluvium derived from argillite and rhyolite	Hills
Edemaps	132.1	Argiduridic Durixerolls	6.8	321	1360	1562	20	Idaho fescue, Thurber's needlegrass, bluebunch wheatgrass, basin big sagebrush	Residuum derived from volcanic rock	Plateaus, hills
Eglirim	51.5	Aridic Argixerolls	7.8	299	1219	1448	55	Bluebunch wheatgrass, Idaho fescue, Wyoming big sagebrush	Colluvium derived from tuff and basalt	Fans
Egyptcreek	111.0	Vitrandic Haploxerolls	5.1	426	1247	1589	70	Ponderosa pine, western juniper, Idaho fescue, antelope bitterbrush, bluebunch wheatgrass, mountain big sagebrush	Residuum and colluvium derived from basalt or tuff	Hills
Eightlar	62.6	Typic Dystroxerepts	10.0	1918	297	1027	90	Incense cedar, Jeffrey pine, ponderosa pine, manzanita	Colluvium and residuum derived from peridotite or serpentinite	Mountains, fans
Eilertsen	59.6	Ultic Hapludalfs	10.6	1919	27	219	7	Bigleaf maple, Douglas-fir, grand fir, red alder, western hemlock, western redcedar	Alluvium	Stream terraces

(continued)

Series name	Area (km^2)	Subgroup	MAAT (°C)	MAP (mm)	Elev. lower (m)	Elev. upper (m)	Max slope (%)	Vegetatiom	Parent material	Landform
Elijah	284.6	Xeric Argidurids	9.9	268	830	973	20	Thurber's needlegrass, bluebunch wheatgrass, Wyoming big sagebrush	Alluvium	Stream terraces
Encina	148.4	Calciargidic Argixerolls	8.8	281	835	1174	60	Idaho fescue, Wyoming big sagebrush, bluebunch wheatgrass, Sandberg bluegrass	Loess over lacustrine deposits	Stream terraces
Enko	506.4	Durinodic Xeric Haplocambids	7.8	229	1370	1548	15	Basin wildrye, basin big sagebrush, Indian ricegrass, needle-and-thread	Alluvium	Fans
Era	105.9	Vitritorrandic Haploxerolls	10.0	262	853	1204	70	Idaho fescue, bluebunch wheatgrass, mountain big sagebrush, Sandberg bluegrass, antelope bitterbrush	Colluvium derived from rocks from the Deschutes Formation	Hills
Erakatak	494.8	Vitrandic Argixerolls	6.0	355	1455	1891	80	Bluebunch wheatgrass, Idaho fescue, Thurber's needlegrass, western juniper, mountain big sagebrush	Colluvium and residuum derived from volcanic rock	Mountains
Etelka	191.9	Oxyaquic Dystrudepts	10.3	2240	97	596	70	Bigleaf maple, Douglas-fir, grand fir, Port Orford cedar, tanoak	Colluvium and residuum derived from sedimentary rock	Mountains
Exfo	201.6	Lithic Torriorthents	9.0	324	695	1244	90	Bluebunch wheatgrass, curl-leaf mountainmahogany, Idaho fescue, purple sage, antelope bitterbrush	Residuum and colluvium derived from basalt	Mountains
Farva	218.0	Typic Haploxerepts	6.0	1080	1097	1859	70	Douglas-fir, white fir, common snowberry	Colluvium derived from andesite	Mountains
Felcher	693.6	Xeric Haplocambids	7.1	278	1406	1855	70	Bluebunch wheatgrass, Thurber's needlegrass, Wyoming big sagebrush	Colluvium and residuum derived from volcanic rock	Mountains
Fendall	86.2	Andic Humudepts	10.0	2170	11	183	75	Douglas-fir, red alder, Sitka spruce, western hemlock, western redcedar	Colluvium and residuum derived from sedimentary rock	Hills, mountains
Fernhaven	77.9	Typic Paleudults	10.0	1842	61	914	75	Bigleaf maple, Douglas-fir, red alder, western hemlock, western redcedar, salal	Colluvium and residuum derived from sandstone and siltstone	Mountains
Fernwood	71.1	Andic Humudepts	6.0	2032	549	914	90	Douglas-fir, red alder, western hemlock, western redcedar, salal, bigleaf maple	Colluvium and residuum derived from andesite and basalt	Mountains
Fertaline	84.5	Abruptic Xeric Argidurids	6.8	274	1486	1614	15	Bluebunch wheatgrass, low sagebrush, Sandberg bluegrass, Thurber's needlegrass	Colluvium and residuum from tuff and basalt	Plateaus
Fitzwater	186.6	Aridic Haploxerolls	6.9	332	1626	1957	70	Bluebunch wheatgrass, Idaho fescue, Wyoming big sagebrush, low sagebrush, Thurber's needlegrass	Colluvium and residuum derived from basalt or tuff	Plateaus
Fivebeaver	548.3	Lithic Utic Haploxerolls	5.8	619	1298	1572	90	Douglas-fir, common snowberry, ponderosa pine, elk sedge, pinegrass	Colluvium and residuum derived from basalt	Mountains
Fivebit	160.5	Lithic Ultic Haploxerolls	6.1	557	1200	1616	90	Western juniper, curl-leaf mountainmahogany, Idaho fescue, Sandberg bluegrass, bluebunch wheatgrass	Colluvium and residuum derived from igneous rocks	Mountains

(continued)

Series name	Area (km^2)	Subgroup	MAAT (°C)	MAP (mm)	Elev. lower (m)	Elev. upper (m)	Max slope (%)	Vegetatiom	Parent material	Landform
Flagstaff	100.8	Typic Aquisalids	7.0	229	1310	1331	1	Black greasewood, bud sagebrush, basin wildrye, inland saltgrass, shadscale	Lacustrine deposits	Lakebeds
Floke	246.9	Abruptic Xeric Argidurids	7.0	280	1484	1829	15	Bluebunch wheatgrass, low sagebrush, Sandberg bluegrass, Thurber's needlegrass	Colluvium and residuum derived from tuff and basalt	Plateaus
Fopiano	121.2	Vitrandic Argixerolls	6.0	339	1153	1548	65	Idaho fescue, low sagebrush, bluebunch wheatgrass	Colluvium derived from tuff	Hills
Fordney	124.2	Torripsammentic Haploxerolls	8.3	359	1276	1371	20	Bluebunch wheatgrass, antelope bitterbrush, Thurber's needlegrass, basin big sagebrush	Alluvium	Stream terraces
Formader	154.3	Alic Hapludands	10.0	2371	91	610	80	Bigleaf maple, Douglas-fir, western hemlock, red alder, western redcedar	Colluvium and residuum derived from basalt	Mountains
Fort Rock	97.9	Vitritorrandic Haploxerolls	6.9	252	1316	1392	8	Basin wildrye, mountain big sagebrush, needle-and-thread, Indian ricegrass	Volcanic ash over lacustrine deposits	Lake terraces
Fourwheel	168.6	Xeric Paleargids	7.0	280	1331	1859	40	Bluebunch wheatgrass, Sandberg bluegrass, Wyoming big sagebrush, Thurber's needlegrass	Residuum and colluvium derived from basalt, andesite, rhyolite or tuff	Hills, mountains
Freezener	264.5	Ultic Haploxeralfs	9.0	1016	457	1219	60	Douglas-fir, ponderosa pine, sugar pine, white fir, Pacific madrone	Colluvium and residuum derived from basalt	Mountains
Fremkle	86.9	Lithic Vitrixerands	6.6	394	955	1333	30	Ponderosa pine, western juniper, Idaho fescue, bluebunch wheatgrass, antelope bitterbrush, Sandberg bluegrass	Volcanic ash over tuff or basalt	Hills
Freznik	344.2	Xeric Paleargids	6.8	276	1626	1860	15	Bluebunch wheatgrass, low sagebrush, Sandberg bluegrass, Thurber's needlegrass	Colluvium and residuum from tuff and basalt	Plateaus
Fritsland	192.2	Humic Dystrudepts	10.0	2794	305	914	60	Douglas-fir, tanoak, salal, western swordfern	Colluvium and residuum derived from metavolcanics and metasedimentary rock	Mountains
Frohman	171.8	Xeric Haplodurids	10.6	259	697	848	40	Sandberg bluegrass, bluebunch wheatgrass, Wyoming big sagebrush	Lacustrine deposits	Lake terraces
Gaib	169.2	Lithic Ultic Argixerolls	6.0	497	1106	1683	75	Ponderosa pine, western juniper, Idaho fescue, antelope bitterbrush, bluebunch wheatgrass, mountain big sagebrush	Residuum and colluvium derived from basalt or tuff	Hills, mountains
Gardone	127.6	Vitritorrandic Haploxerolls	6.7	279	1250	1480	50	Idaho fescue, mountain big sagebrush	Volcanic ash	Plateaus
Geisercreek	63.3	Alfic Udivitrands	5.6	598	1349	1554	60	Douglas-fir, grand fir, lodgepole pine, western larch, big huckleberry, Engelmann spruce	Volcanic ash over residuum and colluvium derived from tuff	Mountains
Geppert	97.0	Typic Dystroxerepts	9.0	1016	457	1219	70	Douglas-fir, ponderosa pine, white fir, Pacific madrone	Colluvium derived from igneous rock	Mountains
Getaway	194.7	Vitrandic Argixerolls	6.1	645	1087	1485	90	Douglas-fir, pinegrass, common snowberry, ponderosa pine, elk sedge	Colluvium derived from basalt or andesite	Mountains

(continued)

Series name	Area (km^2)	Subgroup	MAAT (°C)	MAP (mm)	Elev. lower (m)	Elev. upper (m)	Max slope (%)	Vegetatiom	Parent material	Landform
Ginsberg	79.6	Alic Hapludands	9.4	2540	61	671	60	Bigleaf maple, Douglas-fir, red alder, western hemlock, salal, western swordfern	Colluvium and residuum derived from tuffaceous sedimentary rock	Mountains
Ginser	63.4	Pachic Haploxerolls	6.9	363	1152	1433	70	Idaho fescue, bluebunch wheatgrass, Wyoming big sagebrush	Colluvium and residuum derived from tuff	Hills
Glencabin	57.6	Vitrandic Haploxerolls	6.2	310	1403	1681	65	Bluebunch wheatgrass, Idaho fescue, Thurber's needlegrass, western juniper, mountain big sagebrush	Colluvium and residuum derived from rhyolite, basalt, or tuff	Hills
Goble	261.3	Andic Fragiudepts	9.2	1635	113	440	60	Douglas-fir, western hemlock, western red cedar, red alder, bigleaf maple	Loess	Mountains
Goldrun	87.5	Xeric Torripsamments	8.0	211	1267	1345	15	Indian ricegrass, basin big sagebrush, basin wildrye, spiny hopsage, needle-and-thread	Eolian deposits	Dunes
Goodin	51.2	Ultic Haploxeralfs	11.1	1274	65	272	60	Douglas-fir, Oregon white oak, bigleaf maple, western swordfern	Colluvium and residuum derived from sandstone and siltstone	Hills
Gooding	169.5	Vertic Paleargids	6.9	308	1348	1575	60	Bluebunch wheatgrass, Thurber's needlegrass, Wyoming big sagebrush, Sandberg bluegrass	Loess	Fans
Goodtack	596.5	Vitritorrandic Durixerolls	6.7	285	1351	1479	20	Bluebunch wheatgrass, Thurber's needlegrass, basin big sagebrush, Idaho fescue, mountain big sagebrush	Colluvium and residuum derived from volcanic rock	Plateaus, hills
Gooserock	62.6	Vitritorrandic Haploxerolls	9.4	254	608	1148	90	Idaho fescue, bluebunch wheatgrass, basin wildrye, basin big sagebrush	Colluvium and residuum derived from tuff	Hills
Gosney	235.4	Lithic Torripsamments	9.7	256	931	1191	15	Bluebunch wheatgrass, antelope bitterbrush, Idaho fescue, western juniper	Volcanic ash over basalt	Plateaus
Gradon	81.1	Argiduridic Durixerolls	7.0	292	1326	1533	8	Idaho fescue, Thurber's needlegrass, bluebunch wheatgrass, Wyoming big sagebrush	Alluvium	Fans
Gravecreek	57.7	Typic Dystroxerepts	9.3	1598	618	1248	90	Douglas-fir, incense cedar, Jeffrey pine, sugar pine, common beargrass	Colluvium and residuum derived from serpentine and peridotite	Hills
Greenmountain	54.8	Vitritorrandic Durixerolls	7.0	274	1367	1482	15	Idaho fescue, mountain big sagebrush, bluebunch wheatgrass, western juniper	Residuum derived from basalt	Plateaus, hills
Greystoke	66.3	Pachic ultic Argixerolls	7.0	635	999	1490	75	Douglas-fir, incense cedar, ponderosa pine, sugar pine, common snowberry	Colluvium and residuum derived from andesite and basalt	Mountains
Gribble	59.9	Haplic Durixerolls	8.3	335	988	1213	20	Bluebunch wheatgrass, Idaho fescue, Sandberg bluegrass	Colluvium and residuum derived from basalt or rhyolite	Hills
Grousehill	194.3	Oxyaquic Duricryands	4.4	1439	1560	2057	50	Shasta red fir, mountain hemlock, white fir, incense cedar	Volcanic ash over glacial till	Moraines

(continued)

Series name	Area (km^2)	Subgroup	MAAT (°C)	MAP (mm)	Elev. lower (m)	Elev. upper (m)	Max slope (%)	Vegetatiom	Parent material	Landform
Grouslous	50.2	Lithic Dystrudepts	10.0	2794	61	914	90	Canyon live oak, Douglas-fir, Pacific madrone, tanoak, western swordfern	Colluvium and residuum derived from metasedimentary or metavolcanic rock	Mountains
Gumble	173.1	Xeric Haplargids	8.4	270	1050	1267	40	Bluebunch wheatgrass, Thurber's needlegrass, Wyoming big sagebrush, basin wildrye	Residuum and colluvium derived from sedimentary rock	Hills
Gurdane	189.9	Pachic Argixerolls	8.0	508	518	1372	45	Idaho fescue, bluebunch wheatgrass, Sandberg bluegrass, prairie junegrass	Loess over residuum derived from basalt	Hills
Gustin	69.0	Aquic Palehumults	9.0	1651	244	1012	45	Douglas-fir, grand fir, western hemlock, western redcedar, salal, western swordfern	Colluvium and residuum derived from tuff	Mountains
Gutridge	216.6	Typic Udivitrands	4.0	718	1431	1671	90	Grand fir, engelman spruce, Douglas-fir, ponderosa pine, western larch, big huckleberry	Volcanic ash over colluvium derived from sandstone	Mountains
Gwin	649.0	Lithic Argixerolls	8.4	492	591	1336	90	Bluebunch wheatgrass, Idaho fescue, Sandberg bluegrass	Colluvium and residuum derived from basalt	Hills, plateaus, mountains
Gwinly	990.0	Lithic Argixerolls	8.6	451	787	1291	90	Idaho fescue, bluebunch wheatgrass, mountain big sagebrush, Sandberg bluegrass	Colluvium and residuum derived from basalt	Plateaus, hills, mountains
Haar	278.8	Xeric Torriorthents	9.8	275	776	1018	60	Bluebunch wheatgrass, Wyoming big sagebrush, Indian ricegrass	Colluvium and residuum derived from sedimentary rock	Hills
Hack	55.9	Calcic Argixerolls	9.0	385	590	1178	40	Basin wildrye, bluebunch wheatgrass, Idaho fescue, big sagebrush	Alluvium	Fans, stream terraces
Hackwood	91.8	Pachic Haplocryolls	5.2	559	1812	2426	50	Quaking aspen, mountain brome, tall bluegrass	Loess over colluvium derived from mixed rocks	Mountains
Hall Ranch	107.6	Vitrandic Haploxerolls	6.5	616	1021	1542	65	Douglas-fir, ponderosa pine, pinegrass, elk sedge	Colluvium and residuum derived from andesite and rhyolite	Mountains
Hankins	361.7	Vertic Palexerolls	6.0	500	1108	1508	70	Douglas-fir, ponderosa pine, elk sedge, pinegrass, Idaho fescue	Residuum and colluvium derived from rhyolite, tuff, and basalt	Mountains, hills
Hanning	76.6	Pachic Argixerolls	9.5	280	772	1078	55	Idaho fescue, bluebunch wheatgrass, Sandberg bluegrass	Loess	Hills
Harcany	119.4	Pachic Haplocryolls	6.0	376	1897	2235	70	Idaho fescue, mountain big sagebrush, common snowberry, mountain brome	Colluvium and residuum derived from tuff and basalt	Mountains
Hardtrigger	57.6	Xeric Haplargids	8.4	238	1197	1547	50	Bluebunch wheatgrass, Thurber's needlegrass, Wyoming big sagebrush	Alluvium	Fans
Harl	126.2	Typic Udivitrands	5.3	761	1375	1854	90	Grand fir, big huckleberry, western larch, twinflower, Douglas-fir	Volcanic ash over colluvium derived from basalt	Mountains, plateaus
Harlow	394.4	Lithic Argixerolls	6.2	540	1128	1555	90	Idaho fescue, bluebunch wheatgrass	Colluvium and residuum derived from basalt	Mountains, hills

(continued)

Series name	Area (km^2)	Subgroup	MAAT (°C)	MAP (mm)	Elev. lower (m)	Elev. upper (m)	Max slope (%)	Vegetatiom	Parent material	Landform
Harriman	61.6	Pachic Argixerolls	8.5	287	1269	1352	35	Bluebunch wheatgrass, basin wildrye, antelope bitterbrush, Idaho fescue, mountain big sagebrush	Lacustrine deposits	Lake terraces
Harrington	206.3	Typic Humudepts	9.3	1810	239	888	90	Douglas-fir, grand fir, western hemlock, western redcedar, salal, western swordfern	Colluvium and residuum derived from basalt or andesite	Mountains
Harslow	189.9	Alic Hapludands	9.7	2381	86	627	90	Bigleaf maple, Douglas-fir, red alder, western hemlock, salal, western swordfern	Colluvium and residuum derived from basalt	Mountains
Hart	119.3	Duric Palexerolls	7.0	356	1646	1905	50	Idaho fescue, low sagebrush, bluebunch wheatgrass, black sagebrush, Sandberg bluegrass	Colluvium and residuum derived from basalt and tuff	Plateaus
Hayespring	73.5	Vitritorrandic Durixerolls	7.0	279	1409	1511	20	Bluebunch wheatgrass, antelope bitterbrush, mountain big sagebrush, Idaho fescue	Colluvium and residuum derived from basalt	Plateaus
Hazelair	175.3	Vertic Haploxerolls	11.6	1237	86	270	35	Oregon white oak, baldhip rose, Douglas-fir, california hazel	Colluvium and residuum derived from sandstone and siltstone	Hills
Hazelcamp	52.2	Typic Haplohumults	10.0	2794	305	914	30	Douglas-fir, Pacific madrone, tanoak, western hemlock, salal, western swordfern	Colluvium and residuum derived from metasedimentary or metavolcanic rock	Mountains
Hehe	115.5	Vitrandic Argixerolls	8.7	483	721	1026	65	Incense cedar, ponderosa pine, Idaho fescue, antelope bitterbrush	Colluvium and residuum derived from andesite and basalt	Mountains
Helphenstein	85.7	Sodic Aquicambids	7.5	233	1292	1353	5	Black greasewood, basin wildrye, inland saltgrass	Lacustrine deposits	Lakebeds
Helvetia	76.7	Ultic Argixerolls	12.0	1143	75	152	30	Douglas-fir, bigleaf maple, Oregon white oak, western swordfern	Glaciolacustrine deposits	Glaciolacustrine terraces
Hembre	205.4	Andic Humudepts	9.6	2353	148	704	90	Bigleaf maple, Douglas-fir, red alder, western hemlock, western redcedar	Colluvium and residuum derived from basalt	Mountains
Hemcross	402.9	Alic Hapludands	9.7	2345	102	612	90	Bigleaf maple, Douglas-fir, red alder, western hemlock, western redcedar	Colluvium and residuum derived from basalt	Mountains
Henkle	158.9	Lithic Vitrixerands	7.0	396	1199	1499	65	Idaho fescue, antelope bitterbrush, ponderosa pine, mountain big sagebrush	Volcanic ash over colluvium derived from basalt	Hills
Henley	63.0	Aquic Haplodurids	8.2	387	1245	1284	2	Inland saltgrass, black greasewood	Alluvium	Stream terraces
Henline	86.0	Typic Humicryepts	6.0	2270	857	1490	90	Douglas-fir, noble fir, mountain hemlock, Pacific silver fir, salal	Colluvium and residuum from basalt	Mountains
Hermiston	67.1	Cumulic Haploxerolls	10.4	318	239	650	3	Giant wildrye, bluebunch wheatgrass	Alluvium	Flood plains, stream terraces
Hesslan	93.7	Typic Haploxerolls	8.2	432	152	1067	70	Bluebunch wheatgrass, Oregon white oak, Sandberg bluegrass, Idaho fescue, antelope bitterbrush	Colluvium and residuum derived from sandstone	Hills

(continued)

Series name	Area (km^2)	Subgroup	MAAT (°C)	MAP (mm)	Elev. lower (m)	Elev. upper (m)	Max slope (%)	Vegetatiom	Parent material	Landform
Highcamp	86.4	Typic Haplocryands	6.0	2159	853	1463	90	Douglas-fir, noble fir, western hemlock, common beargrass, western swordfern	Colluvium and residuum derived from andesite	Mountains
Holcomb	83.6	Xeric Argialbolls	11.8	1118	61	143	3	Oregon white oak, baldhip rose, Oregon ash	Glaciolacustrine deposits	Glaciolacustrine terraces
Holland	66.5	Ultic Haploxeralfs	11.0	889	244	1219	35	Douglas-fir, Pacific madrone, ponderosa pine, sugar pine	Colluvium and residuum derived from granite	Mountains
Hondu	77.2	Andic Haploxerepts	4.6	741	1423	1709	90	Douglas-fir, common snowberry, ponderosa pine, elk sedge, pinegrass, western larch	Colluvium and residuum derived from metasedimentary or metavolcanic rocks	Mountains
Honeygrove	696.2	Typic Palehumults	10.3	1893	152	650	75	Bigleaf maple, Douglas-fir, red alder, western hemlock, western redcedar, salal, western swordfern	Residuum and colluvium derived from sandstone, siltstone, basalt	Mountains
Hot Lake	80.7	Aquic Haploxerands	9.0	419	792	853	2	Tufted hairgrass, sedge, rush	Volcanic ash over diatomaceous sediment	Lakebeds
Houstake	71.2	Vitritorrandic Haploxerolls	9.7	257	823	1140	8	Idaho fescue, needle-and-thread, antelope bitterbrush, mountain big sagebrush	Volcanic ash over residuum derived from basalt	Plateaus
Howash	162.9	Humic Udivitrands	4.8	1492	1036	1600	65	Douglas-fir, grand fir, Pacific silver fir, western larch	Volcanic ash mixed with colluvium from andesite	Mountains
Humarel	107.0	Vitrandic Argixerolls	5.4	498	1233	1465	90	Douglas-fir, elk sedge, ponderosa pine, pinegrass, common snowberry	Colluvium and residuum derived from breccia	Mountains
Hummington	119.6	Typic Haplocryands	6.5	2153	1020	1478	90	Douglas-fir, noble fir, Pacific silver fir, mountain hemlock, big huckleberry	Colluvium and residuum derived from basalt	Mountains
Hurwal	85.5	Vitrandic Argixerolls	6.4	446	1078	1506	60	Idaho fescue, bluebunch wheatgrass, common snowberry, chokecherry	Loess over basalt	Hills, mountains
Hutchley	80.8	Lithic Argixerolls	6.0	375	1077	1815	75	Idaho fescue, antelope bitterbrush, bluebunch wheatgrass	Colluvium and residuum derived from basalt	Hills
Icene	95.3	Typic Aquisalids	8.6	231	1313	1435	1	Black greasewood, bud sagebrush, shadscale, spiny hopsage, Indian ricegrass	Lacustrine deposits	Lakebeds
Igert	505.2	Durinodic Xeric Haplargids	8.0	290	1450	1553	15	Bluebunch wheatgrass, Wyoming big sagebrush, Sandberg bluegrass	Colluvium and residuum derived from volcanic rock	Hills
Illahee	59.2	Typic Humudepts	6.0	1761	877	1276	90	Douglas-fir, western hemlock, western white pine, salal, western swordfern	Colluvium and residuum derived from volcanic rock	Mountains
Immig	223.3	Typic Argixerolls	8.2	418	852	1277	70	Bluebunch wheatgrass, Idaho fescue, mountain big sagebrush, Sandberg bluegrass	Colluvium and residuum derived from basalt	Hills
Imnaha	173.7	Vitrandic Argixerolls	6.2	542	1066	1564	90	Bluebunch wheatgrass, Idaho fescue, Sandberg bluegrass, arrowleaf balsamroot	Colluvium and residuum derived from basalt	Plateaus
Inkler	66.7	Andic Haploxerepts	5.3	910	1490	1819	90	Douglas-fir, ponderosa pine, western larch, common snowberry, pinegrass, elk sedge	Colluvium and residuum derived from rhyolite and andesite	Mountains

(continued)

Series name	Area (km^2)	Subgroup	MAAT (°C)	MAP (mm)	Elev. lower (m)	Elev. upper (m)	Max slope (%)	Vegetatiom	Parent material	Landform
Ironside	78.6	Vitrandic Haploxerolls	4.8	711	1380	1740	90	Ponderosa pine, elk sedge, Idaho fescue, bluebunch wheatgrass, Douglas-fir, mountain big sagebrush	Colluvium and residuum derived from argillite	Mountains
Jacksplace	82.6	Vitritorrandic Argixerolls	6.9	273	1363	1508	20	Idaho fescue, mountain big sagebrush, western juniper, Thurber's needlegrass	Residuum derived from basalt	Plateaus
Jayar	150.1	Typic Dystroxerepts	6.0	1976	1019	1643	90	Douglas-fir, white fir, Oregongrape, oceanspray, Pacific serviceberry	Colluvium derived from metamorphic rock	Mountains
Jett	112.8	Cumulic Haploxerolls	8.2	300	919	1219	4	Giant wildrye, big sagebrush	Alluvium	Flood plains
Jojo	56.2	Typic Vitricryands	2.2	2371	1300	1910	70	Lodgepole pine, mountain hemlock, subalpine fir, Douglas-fir, big huckleberry, common beargrass	Colluvium and residuum derived from ash flow and andesite	Mountains
Jory	872.2	Xeric Palehumults	11.3	1249	91	409	90	Bigleaf maple, Douglas-fir, grand fir, Oregon white oak	Colluvium derived from basalt	Hills
Josephine	520.9	Typic Haploxerults	9.7	1288	267	1088	75	Douglas-fir, incense cedar, ponderosa pine, California black oak, Pacific madrone	Colluvium and residuum derived from metavolcanics and metasedimentary rock	Mountains
Kahler	113.0	Vitrandic Haploxerolls	6.2	644	947	1546	90	Pinegrass, ponderosa pine, Douglas-fir, elk sedge	Colluvium and residuum derived from basalt	Mountains
Kamela	249.7	Vitrandic Haploxerepts	5.8	745	1373	1684	90	Douglas-fir, grand fir, ponderosa pine, western larch, pinegrass, elk sedge	Colluvium and residuum derived from basalt	Mountains
Kanid	249.7	Typic Dystroxerepts	10.2	1803	296	993	90	Douglas-fir, Pacific madrone, sugar pine, tanoak, salal	Colluvium and residuum derived from metasedimentary rock	Mountains
Kaskela	63.4	Typic Haploxererts	9.0	356	750	1292	50	Antelope bitterbrush, Idaho fescue, bluebunch wheatgrass	Colluvium and residuum derived from sedimentary rock	Hills
Keating	60.9	Typic Argixerolls	8.9	360	914	1180	35	Idaho fescue, bluebunch wheatgrass, mountain big sagebrush	Colluvium and residuum derived from greenstone	Hills
Keel	156.5	Typic Haplocryands	6.0	2114	906	1372	75	Douglas-fir, noble fir, Pacific silver fir, western hemlock, common beargrass	Colluvium and residuum derived from igneous rock	Mountains
Kerrfield	74.9	Durinodic Xeric Haplocambids	8.0	235	1280	1463	20	Indian ricegrass, Thurber's needlegrass, Sandberg bluegrass, Wyoming big sagebrush, spiny hopsage	Colluvium and residuum derived from tuffaceous sandstone, siltstone, basalt	Hills
Ketchly	51.1	Vitrandic Haploxeralfs	6.4	724	610	1097	65	Douglas-fir, ponderosa pine, Oregon white oak, elk sedge	Loess over colluvium derived from andesite	Hills
Kettenbach	83.0	Pachic Argixerolls	8.8	498	563	1194	90	Arrowleaf balsamroot, buckwheat, common snowberry, silky lupine	Colluvium and residuum derived from basalt	Hills
Kewake	131.4	Vitrandic Torripsamments	8.0	230	1300	1409	45	Indian ricegrass, basin big sagebrush, basin wildrye, black greasewood, fourwing saltbush, needle-and-thread	Eolian sands	Dunes on lakebeds
Kilchis	121.7	Lithic Humudepts	9.6	2289	212	830	100	Douglas-fir, red alder, western hemlock, bigleaf maple, salal, western swordfern	Colluvium and residuum derived from basalt	Mountains

(continued)

Series name	Area (km^2)	Subgroup	MAAT (°C)	MAP (mm)	Elev. lower (m)	Elev. upper (m)	Max slope (%)	Vegetatiom	Parent material	Landform
Kimberly	75.1	Torrifluventic Haploxerolls	10.3	259	320	588	3	Basin wildrye, needle-and-thread, thickspike wheatgrass	Alluvium	Flood plains
Kingbolt	124.6	Typic Vitrixerands	4.4	700	1441	1688	60	Big huckleberry, grand fir, subalpine fir, western larch, pinegrass, Douglas-fir	Volcanic ash over colluvium and residuum derived from sandstone	Mountains
Kinney	787.1	Andic Humudepts	9.2	1798	259	819	90	Bigleaf maple, Douglas-fir, grand fir, western hemlock, salal, western swordfern	Colluvium derived from tuff breccia	Mountains
Kinton	60.6	Typic Fragixerepts	12.0	1222	95	198	60	Douglas-fir, bigleaf maple, western redcedar, western swordfern	Loess	Hills
Kinzel	75.1	Typic Fulvicryands	5.9	2476	902	1353	90	Douglas-fir, noble fir, mountain hemlock, Pacific silver fir, salal, western swordfern	Colluvium derived from andesite and basalt mixed with glacial till	Mountains
Kiona	108.5	Xeric Haplocambids	8.5	237	955	1180	70	Indian ricegrass, Thurber's needlegrass, Wyoming big sagebrush, bluebunch wheatgrass, needle-and-thread	Colluvium derived from basalt	Hills
Kirk	107.5	Typic Cryaquands	6.7	533	1263	1360	1	Tufted hairgrass, northern mannagrass, Nebraska sedge	Alluvium derived from ash and pumice	Flood plains, lakebeds
Kirkendall	54.1	Oxyaquic Humudepts	10.9	1968	15	242	3	Bigleaf maple, Douglas-fir, red alder, western hemlock, salal, western swordfern	Alluvium	Flood plains
Kishwalk	191.6	Pachic Argixerolls	8.9	347	717	1201	80	Bluebunch wheatgrass, Idaho fescue, basin big sagebrush, antelope bitterbrush	Colluvium and residuum derived from basalt	Mountains
Klamath	95.9	Cumulic Cryaquolls	6.5	483	1268	1420	1	Tufted hairgrass, Baltic rush, northern mannagrass	Alluvium	Flood plains
Klicker	1,199.8	Vitrandic Argixerolls	6.0	609	1199	1589	90	Douglas-fir, elk sedge, ponderosa pine, bluebunch wheatgrass, pinegrass, common snowberry, Idaho fescue	Colluvium and residuum derived from basalt	Mountains
Klickitat	1,018.7	Typic Humudepts	9.5	2167	202	905	90	Douglas-fir, western hemlock, western redcedar, salal, western swordfern	Colluvium and residuum derived from basalt	Mountains
Klickson	185.4	Vitrandic Argixerolls	6.0	591	1157	1487	90	Douglas-fir, common snowberry, ponderosa pine, elk sedge, pinegrass, western larch	Colluvium and residuum derived from basalt	Mountains
Klistan	383.4	Alic Hapludands	9.5	2366	93	624	90	Douglas-fir, red alder, western hemlock, salal, western swordfern	Colluvium and residuum derived from basalt	Mountains
Klootchie	494.9	Typic Fulvudands	9.4	2421	15	549	90	Douglas-fir, red alder, Sitka spruce, western hemlock, salal, western swordfern	Colluvium and residuum derived from basalt	Mountain
Koehler	95.3	Xeric Haplodurids	11.5	210	127	234	12	Needle-and-thread, Indian ricegrass, antelope bitterbrush, big sagebrush	Eolian sands	Terraces
Krackle	59.7	Xeric Haplocryolls	5.0	538	1873	2340	65	Bluebunch wheatgrass, mountain big sagebrush, Idaho fescue, sheep fescue	Residuum and colluvium derived from basalt or andesite	Mountains

(continued)

Series name	Area (km^2)	Subgroup	MAAT (°C)	MAP (mm)	Elev. lower (m)	Elev. upper (m)	Max slope (%)	Vegetatiom	Parent material	Landform
Kunaton	56.7	Abruptic Xeric Argidurids	8.0	280	1220	1392	40	Wyoming big sagebrush, Sandberg bluegrass, Thurber's needlegrass, bluebunch wheatgrass	Loess over silty alluvium derived from basalt	Plateaus
Kunceider	146.9	Aridic Lithic Haploxerolls	7.0	270	1324	1452	15	Idaho fescue, antelope bitterbrush, needle-and-thread, mountain big sagebrush	Volcanic ash over colluvium and residuum derived from basalt	Plateaus
Kutcher	68.1	Alfic Udivitrands	4.8	1492	1074	1494	30	Douglas-fir, noble fir, Pacific silver fir, western hemlock, common snowberry	Volcanic ash over colluvium derived from andesite or basalt	Mountains
La Grande	66.8	Pachic Haploxerolls	10.0	407	671	1097	2	Giant wildrye, Nebraska sedge, Baltic rush	Alluvium	Fans, stream terraces
Lackeyshole	110.1	Typic Vitricryands	2.8	1098	1796	2061	90	Subalpine fir, grouse huckleberry, Douglas-fir, Engelmann spruce, western larch, lodgepole pine	Volcanic ash over colluvium and residuum derived from andesite or basalt	Mountains
Laderly	167.7	Alic Hapludands	6.7	2854	539	906	90	Douglas-fir, noble fir, Pacific silver fir, western hemlock, salal, western swordfern	Colluvium derived from basalt	Mountains
Laidlaw	53.1	Humic Vitrixerands	7.0	359	1286	1476	40	Ponderosa pine, mountain big sagebrush, Idaho fescue, antelope bitterbrush, western juniper	Volcanic ash over alluvium	Plateaus
Lakeview	95.3	Cumulic Haploxerolls	8.3	380	1331	1510	2	Basin wildrye, basin big sagebrush	Alluvium	Flood plains
Laki	81.7	Typic Haploxerolls	8.3	351	1242	1305	2	Basin wildrye, inland saltgrass, black greasewood	Alluvium	Stream terraces
Lambring	349.0	Pachic Haploxerolls	6.1	358	1466	1831	70	Idaho fescue, antelope bitterbrush, bluebunch wheatgrass, mountain big sagebrush	Residuum and colluvium derived from basalt or tuff	Mountains, hills
Lamonta	93.6	Abruptic Argiduridic Durixerolls	9.4	270	969	1085	15	Bluebunch wheatgrass, Idaho fescue, antelope bitterbrush, big sagebrush	Colluvium and residuum derived from sedimentary rocks	Plateaus
Lapine	2,511.2	Xeric Vitricryands	5.2	686	1372	1733	70	Lodgepole pine, ponderosa pine, white fir, Douglas-fir, antelope bitterbrush, western needlegrass	Volcanic ash and pumice	Plateaus
Larabee	370.3	Vitrandic Argixerolls	5.9	565	1235	1562	90	Pinegrass, Douglas-fir, ponderosa pine, grand fir, western larch, common snowberry, elk sedge	Colluvium and residuum derived from basalt	Plateaus, mountains
Larmine	58.9	Lithic Haploxerepts	11.6	1207	76	792	90	Bigleaf maple, Douglas-fir, Oregon white oak, western swordfern	Colluvium and residuum derived from sandstone and siltstone	Mountains
Lasere	86.4	Typic Palexerolls	8.0	407	1463	1657	50	Idaho fescue, low sagebrush, bluebunch wheatgrass, mountain big sagebrush	Lacustrine deposits over colluvium derived from tuff and basalt	Hills
Lastcall	70.2	Vitritorrandic Argixerolls	7.0	279	1381	1496	15	Idaho fescue, low sagebrush, Sandberg bluegrass, western juniper, Thurber's needlegrass	Colluvium and residuum derived from basalt	Plateaus

(continued)

Series name	Area (km^2)	Subgroup	MAAT (°C)	MAP (mm)	Elev. lower (m)	Elev. upper (m)	Max slope (%)	Vegetatiom	Parent material	Landform
Lather	101.0	Limnic Haplohemists	7.4	514	1262	1282	1	Bullrush, tules	Moderately decomposed organic material	Lakebeds
Latourell	61.6	Ultic Haploxeralfs	12.0	1254	17	118	30	Douglas-fir, Oregon white oak, bigleaf maple, western swordfern	Glaciolacustrine deposits	Glaciolacustrine terraces
Laurelwood	155.4	Ultic Haploxeralfs	11.6	1355	101	445	60	Douglas-fir, bigleaf maple, Oregon white oak	Loess	Hills
Lawen	131.0	Calciargidic Argixerolls	7.0	229	1219	1372	5	Thurber's needlegrass, basin big sagebrush, basin wildrye, Indian ricegrass	Alluvium	Lake terraces
Legler	122.7	Xeric Haplocambids	8.9	257	865	1056	20	Basin wildrye, bluebunch wheatgrass, Idaho fescue	Alluvium	Stream terraces
Lettia	140.6	Ultic Haploxeralfs	9.1	1224	364	1069	70	Douglas-fir, Pacific madrone, ponderosa pine, sugar pine, salal	Colluvium and residuum derived from granodiorite	Mountains
Lickskillet	1,935.0	Aridic lithic Haploxerolls	9.5	303	628	1161	90	Bluebunch wheatgrass, Wyoming big sagebrush, Thurber's needlegrass, Sandberg bluegrass	Colluvium and residuum derived from basalt and rhyolite	Hills, mountains
Limberjim	994.8	Alfic Udivitrands	5.9	676	1342	1611	90	Douglas-fir, grand fir, lodgepole pine, western larch, twinflower, big huckleberry	Volcanic ash over colluvium and residuum derived from basalt or breccia	Mountains
Lithgow	179.0	Xeric Haplargids	8.0	259	1209	1454	60	Bluebunch wheatgrass, Sandberg bluegrass, Wyoming big sagebrush	Colluvium and residuum derived from tuff and rhyolite	Hills
Llaorock	100.9	Vitric Haplocryands	4.4	1487	1676	2096	80	Shasta red fir, mountain hemlock, lodgepole pine	Volcanic ash mixed with colluvium derived from andesite	Mountains
Lobert	76.7	Vitrandic Haploxerolls	7.1	439	1295	1400	35	Ponderosa pine, Idaho fescue, antelope bitterbrush	Eolian and lacustrine deposits	Stream terraces
Lonely	336.3	Xeric Haplocambids	7.0	280	1342	1821	30	Thurber's needlegrass, bluebunch wheatgrass, Wyoming big sagebrush, Indian ricegrass	Residuum and colluvium derived from basalt, andesite, rhyolite, or tuff	Hills
Longbranch	66.5	Pachic Argixerolls	6.0	358	1211	1683	65	Basin wildrye, Idaho fescue, mountain big sagebrush	Colluvium derived from greenstone	Hills
Lookingglass	130.9	Xerertic Argialbolls	8.2	538	712	1097	30	Douglas-fir, ponderosa pine, common snowberry, pinegrass, elk sedge	Loess over colluvium and residuum derived from basalt	Plateaus, hills
Lookout	217.3	Abruptic Xeric Argidurids	7.9	273	967	1229	20	Idaho fescue, Wyoming big sagebrush, bluebunch wheatgrass, Sandberg bluegrass	Colluvium and residuum derived from basalt and tuff	Hills
Lorella	479.0	Lithic Argixerolls	8.4	375	1373	1660	70	Bluebunch wheatgrass, antelope bitterbrush, Idaho fescue, western juniper	Colluvium and residuum derived from tuff and basalt	Hills
Lostbasin	132.5	Typic Haploxerepts	6.0	356	1203	1609	80	Bluebunch wheatgrass, Idaho fescue, mountain big sagebrush, Sandberg bluegrass	Colluvium and residuum derived from graywacke and schist	Hills
Loupence	58.2	Cumulic Haploxerolls	8.0	274	1067	1174	3	Basin wildrye, bluebunch wheatgrass, Sandberg bluegrass, basin big sagebrush	Alluvium	Terrace
Mackatie	102.5	Alfic Udivitrands	4.8	1492	1074	1494	30	Douglas-fir, grand fir, western hemlock, western larch, common snowberry, pinegrass	Volcanic ash over colluvium and residuum derived from andesite	Mountains

(continued)

Series name	Area (km^2)	Subgroup	MAAT (°C)	MAP (mm)	Elev. lower (m)	Elev. upper (m)	Max slope (%)	Vegetatiom	Parent material	Landform
Madeline	69.9	Aridic Lithic Argixerolls	6.6	308	1337	1817	60	Bluebunch wheatgrass, Thurber's needlegrass, Idaho fescue, Wyoming big sagebrush	Residuum and colluvium derived from basalt, andesite, rhyolite or tuff	Plateaus, hills, mountains
Madras	301.4	Aridic Argixerolls	9.5	250	625	909	40	Bluebunch wheatgrass, mountain big sagebrush, Idaho fescue, antelope bitterbrush	Colluvium and residuum derived from sediments from the Deschutes Formation	Plateaus
Mahogee	54.9	Lithic Argicryolls	5.0	432	1937	2243	50	Curl-leaf mountainmahogany, bluebunch wheatgrass, Idaho fescue, mountain big sagebrush	Colluvium and residuum derived from rhyolite	Mountains
Mahoon	152.2	Aridic Palexerolls	8.1	272	1085	1316	40	Bluebunch wheatgrass, Thurber's needlegrass, Wyoming big sagebrush, Sandberg bluegrass	Residuum and colluvium derived from tuffaceous sedimentary rock and diatomaceous earth	Hills
Maklak	150.5	Xeric Vitricryands	4.4	1179	1355	1760	10	Lodgepole pine, ponderosa pine, white fir, Douglas-fir, antelope bitterbrush, western needlegrass	Volcanic ash and pumice	Plateaus
Malabon	167.1	Pachic Ultic Argixerolls	11.9	1189	65	224	3	Douglas-fir, Oregon white oak, bigleaf maple, baldhip rose	Alluvium	Stream terraces
Malin	58.1	Fluvaquentic Endoaquolls	8.5	356	1334	1366	1	Inland saltgrass, Nutall's alkaligrass	Alluvium and lacustrine deposits	Flood plains, lakebeds
Mallory	58.3	Pachic Argixerolls	8.8	484	740	1157	90	Idaho fescue, bluebunch wheatgrass	Colluvium and residuum derived from basalt	Hills
Manita	122.9	Mollic Haploxeralfs	10.5	776	285	1085	50	Douglas-fir, Pacific madrone, ponderosa pine, California fescue, California black oak	Colluvium and residuum derived from siltstone	Fans
Marack	129.8	Calciargidic Argixerolls	6.0	267	1158	1341	35	Idaho fescue, mountain big sagebrush, bluebunch wheatgrass, prairie junegrass	Lacustrine deposits	Lake terraces
Marblepoint	69.7	Andic Haplocryepts	2.1	1025	1836	2120	90	Grouse huckleberry, subalpine fir, Engelmann spruce, Douglas-fir, lodgepole pine, elk sedge	Volcanic ash over glacial till	Mountains
Maset	230.0	Alfic Vitrixerands	6.0	462	1298	1544	60	Ponderosa pine, antelope bitterbrush, Idaho fescue, curl-leaf mountainmahogany	Volcanic ash over residuum and colluvium derived from tuffaceous sandstone	Hills
Maupin	97.0	Haploduridic Durixerolls	9.5	276	537	927	12	Bluebunch wheatgrass, buckwheat, Sandberg bluegrass, Idaho fescue	Loess	Plateaus
Mayger	71.9	Aquic Palehumults	9.5	1651	152	366	30	Bigleaf maple, Douglas-fir, red alder, western hemlock, western redcedar, salal	Residuum and colluvium derived from shale	Mountains
McAlpin	138.4	Aquic Cumulic Haploxerolls	11.1	1309	82	249	6	Douglas-fir, grand fir, Oregon ash, baldhip rose	Alluvium	Fans, stream terraces
McBee	124.5	Aquic Cumulic Haploxerolls	11.5	1221	27	232	3	Black cottonwood, Douglas-fir, Oregon ash, wild rose	Alluvium	Flood plains, stream terraces
McCartycreek	80.5	Vitrandic Haploxerolls	5.9	881	1225	1603	90	Idaho fescue, mountain big sagebrush, antelope bitterbrush, prairie junegrass, bluebunch wheatgrass	Colluvium derived from andesite or basalt	Mountains, plateaus

(continued)

Series name	Area (km^2)	Subgroup	MAAT (°C)	MAP (mm)	Elev. lower (m)	Elev. upper (m)	Max slope (%)	Vegetatiom	Parent material	Landform
McConnel	265.4	Xeric Haplocambids	8.1	228	1333	1508	50	Indian ricegrass, Thurber's needlegrass, bluebunch wheatgrass, Wyoming big sagebrush	Alluvium	Lake terraces
McCully	292.5	Typic Humudepts	10.1	1795	269	659	70	Douglas-fir, western hemlock, salal, western swordfern, bigleaf maple	Colluvium and residuum derived from igneous rock	Mountains
McDuff	166.9	Typic Haplohumults	9.7	2190	153	758	75	Bigleaf maple, Douglas-fir, red alder, western hemlock, salal, western swordfern	Colluvium and residuum derived from sedimentary rocks	Mountains
McKay	66.0	Calcic Argixerolls	9.2	406	453	739	25	Bluebunch wheatgrass, Idaho fescue	Loess over alluvium	Fans
McMullin	285.7	Lithic Ultic Haploxerolls	9.5	810	396	1192	75	Idaho fescue, wedgeleaf ceanothus, pine bluegrass, Oregon white oak, California black oak	Colluvium and residuum derived from metasedimentary rock	Hills
McNull	323.2	Ultic Argixerolls	9.0	802	495	1219	60	Douglas-fir, ponderosa pine, California fescue, Idaho fescue, Pacific madrone	Colluvium and residuum derived from tuff or breccia	Hills
McWillar	89.8	Alfic Vitrixerands	4.1	697	1433	1657	90	Big huckleberry, grand fir, subalpine fir, heartleaf arnica, western larch, pinegrass, Douglas-fir	Volcanic ash over colluvium derived from metasedimentary rocks	Mountains
Meadowridge	123.4	Vitritorrandic Argixerolls	9.7	257	796	1164	80	Bluebunch wheatgrass, Thurber's needlegrass, basin big sagebrush	Loess over colluvium and residuum derived from tuffaceous sedimentary rock	Hills
Meda	76.2	Typic Humudepts	10.9	1949	43	260	20	Bigleaf maple, Douglas-fir, red alder, western hemlock, western redcedar, salal	Alluvium	Fans, stream terraces
Medco	298.1	Ultic Haploxerolls	9.0	777	537	1219	50	Douglas-fir, ponderosa pine, Oregon white oak, Idaho fescue	Colluvium and residuum derived from andesite, tuff, or breccia	Hills
Medford	64.9	Pachic Argixerolls	11.5	813	274	854	15	Oregon white oak, California black oak, common snowberry, Idaho fescue, Pacific madrone	Alluvium	Fans, stream terraces
Melbourne	129.3	Ultic Palexeralfs	11.2	1276	72	297	60	Douglas-fir, red alder, western red cedar, bigfleaf maple, salal, western swordfern	Colluvium derived from siltstone or sandstone	Hills
Melby	137.1	Humic Dystrudepts	9.6	1947	152	630	90	Bigleaf maple, Douglas-fir, red alder, western hemlock, salal, western swordfern	Colluvium and residuum derived from siltstone	Mountains
Melhorn	88.8	Vitrandic Argixerolls	5.7	599	1139	1438	90	Douglas-fir, pinegrass, common snowberry, grand fir, ponderosa pine	Loess over colluvium derived from basalt	Plateaus, mountains
Menbo	59.8	Vitrandic Argixerolls	6.8	355	1348	1588	65	Idaho fescue, antelope bitterbrush, bluebunch wheatgrass, mountain big sagebrush	Colluvium and residuum derived from basalt	Hills
Merlin	936.0	Lithic Argixerolls	5.6	371	1295	1728	60	Idaho fescue, bluebunch wheatgrass, Sandberg bluegrass, antelope bitterbrush, low sagebrush	Colluvium and residuum derived from tuff and basalt	Hills, plateaus
Mesman	158.7	Xeric Natrargids	8.6	229	1302	1441	15	Basin big sagebrush, Indian ricegrass, spiny hopsage, black greasewood, basin wildrye	Lacustrine deposits	Lake terraces

(continued)

Series name	Area (km^2)	Subgroup	MAAT (°C)	MAP (mm)	Elev. lower (m)	Elev. upper (m)	Max slope (%)	Vegetatiom	Parent material	Landform
Middlebox	96.1	Vitrandic Torriorthents	7.0	280	1433	1829	40	Bluebunch wheatgrass, Thurber's needlegrass, Wyoming big sagebrush	Residuum and colluvium derived from tuffaceous rock	Hills
Mikkalo	386.7	Calcidic Haploxerolls	10.3	269	296	669	40	Bluebunch wheatgrass, big sagebrush	Loess over basalt	Hills
Milbury	265.1	Typic Humudepts	9.6	2625	81	725	90	Douglas-fir, red alder, western hemlock, salal, western swordfern	Colluvium derived from sandstone	Mountains
Millicoma	190.5	Andic Humudepts	10.6	2105	40	328	90	Douglas-fir, grand fir, red alder, Sitka spruce, tanoak, western hemlock	Colluvium and residuum derived from sandstone	Mountains
Minveno	193.9	Xeric Haplodurids	7.8	230	1050	1310	8	Thurber's needlegrass, Indian ricegrass, bluebunch wheatgrass, Wyoming big sagebrush, Sandberg bluegrass	Loess over residuum derived from basalt	Plateaus
Moe	63.2	Andic Humudepts	7.0	2032	853	1097	75	Douglas-fir, western hemlock, western redcedar, Pacific madrone, western swordfern	Colluvium derived from tuffaceous rock and breccia	Mountains
Moonbeam	475.1	Vitritorrandic Durixerolls	7.0	288	1399	1559	20	Idaho fescue, bluebunch wheatgrass, low sagebrush, Sandberg bluegrass, Thurber's needlegrass	Colluvium and residuum derived from basalt	Plateaus
Morehouse	231.3	Vitrandic Torripsamments	7.0	233	1328	1392	35	Indian ricegrass, needle-and-thread, basin big sagebrush, basin wildrye	Volcanic ash over lacustrine deposits	Dunes on lakebeds
Morfitt	51.2	Xeric Haplargids	8.8	262	1056	1168	15	Creeping wildrye, basin big sagebrush, basin wildrye	Alluvium	Fans
Morganhills	82.3	Vitrandic Torriorthents	7.0	280	1219	1524	35	Thurber's needlegrass, Indian ricegrass, Sandberg bluegrass, Wyoming big sagebrush	Residuum and colluvium derived from tuffaceous sandstone	Hills
Morrow	707.3	Calcic Argixerolls	9.0	332	450	1035	40	Idaho fescue, bluebunch wheatgrass	Loess over basalt	Plateaus
Mountemily	226.9	Typic Vitricryands	3.2	881	1646	1879	90	Engelmann spruce, big huckleberry, subalpine fir, western larch, lodgepole pine, pinegrass	Volcanic ash over colluvium and residuum derived from andesite or basalt	Mountains
Mountireland	54.1	Alfic Vitricryands	3.1	756	1616	1788	30	Grand fir, grouse huckleberry, Engelmann spruce, western larch, Douglas-fir, subalpine fir	Volcanic ash over colluvium and residuum derived from andesite or basalt	Mountains
Mudlakebasin	70.4	Typic Vitricryands	2.3	1181	1843	2088	90	Grouse huckleberry, lodgepole pine, subalpine fir, Engelmann spruce	Volcanic ash over glacial till	Mountains
Mudpot	62.0	Chromic Endoaquerts	8.0	242	1524	1798	2	Foxtail barley, curly dock	Lacustrine deposits	Lakebeds
Multnomah	61.8	Humic Dystroxerepts	12.0	1234	42	107	60	Bigleaf maple, Douglas-fir, Oregon white oak, western redcedar	Glaciolacustrine deposits	Glaciolacustrine terraces
Muni	435.0	Haploxeralfic Argidurids	8.0	245	1232	1535	15	Thurber's needlegrass, Indian ricegrass, bluebunch wheatgrass, Wyoming big sagebrush, spiny hopsage	Alluvium	Fans
Murtip	201.5	Alic Hapludands	6.7	2867	546	905	90	Douglas-fir, noble fir, Pacific silver fir, western hemlock, salal, western swordfern	Colluvium derived from basalt	Mountains

(continued)

Series name	Area (km^2)	Subgroup	MAAT (°C)	MAP (mm)	Elev. lower (m)	Elev. upper (m)	Max slope (%)	Vegetatiom	Parent material	Landform
Mutton	51.1	Vitrandic Haploxeralfs	8.0	457	691	1128	80	Douglas-fir, ponderosa pine, Oregon white oak, common snowberry	Residuum and colluvium derived from tuff	Mountains
Nailkeg	61.9	Typic Dystrudepts	10.0	2794	61	914	90	Douglas-fir, Port Orford cedar, red alder, tanoak, western hemlock, salal, western swordfern	Colluvium and residuum derived from schist or phyllite	Mountains
Nansene	109.9	Pachic Haploxerolls	10.0	321	196	627	70	Idaho fescue, bluebunch wheatgrass, Sandberg bluegrass	Loess over basalt	Hills
Natroy	61.6	Xeric Endoaquerts	12.0	1222	106	228	2	Sedges, rushes, willows, Oregon ash	Alluvium	Fans, stream terraces
Necanicum	356.3	Typic Fulvudands	9.3	2430	25	543	90	Douglas-fir, red alder, Sitka spruce, western hemlock, western redcedar, salal	Residuum and colluvium derived from basalt	Mountains
Nehalem	57.5	Fluventic Humudepts	10.3	2131	3	70	3	Douglas-fir, red alder, Sitka spruce, western hemlock, western redcedar, salal, western swordfern	Alluvium	Flood plains
Nekia	602.9	Xeric Haplohumults	11.8	1245	102	388	60	Bigleaf maple, Douglas-fir, grand fir, Oregon white oak	Residuum and colluvium derived from basalt	Hills
Nekoma	54.9	Fluventic Humudepts	11.0	1968	24	206	3	Douglas-fir, red alder, western hemlock, salal, western swordfern	Alluvium	Flood plains
Nestucca	50.1	Fluvaquentic Humaquepts	10.8	2018	4	83	8	Red alder, Sitka spruce, western hemlock, western redcedar, skunk cabbage	Alluvium	Flood plains
Nevador	1,111.9	Durinodic Xeric Haplargids	7.8	227	1193	1400	15	Thurber's needlegrass, bluebunch wheatgrass, Indian ricegrass, Wyoming big sagebrush	Alluvium	Fans
Newberg	270.5	Fluventic Haploxerolls	11.5	1164	83	390	3	Bigleaf maple, black cottonwood, Douglas-fir, Oregon ash, red alder	Alluvium	Flood plains
Ninemile	1,807.0	Aridic Lithic Argixerolls	6.4	327	1470	1816	60	Idaho fescue, bluebunch wheatgrass, low sagebrush, Sandberg bluegrass, Thurber's needlegrass	Colluvium and residuum derived from basalt	Plateaus
Ninetysix	80.5	Calcic Haploxerolls	8.3	358	719	1304	90	Bluebunch wheatgrass, basin big sagebrush, Sandberg bluegrass	Colluvium derived from basalt	Fans
Nonpareil	87.4	Typic Dystroxerepts	11.9	1085	105	708	90	Oregon white oak, Pacific madrone, wildrose, grasses	Colluvium and residuum derived from sandstone and siltstone	Hills
Norad	144.1	Xeric Haplargids	8.0	229	1387	1448	2	Winterfat, Indian ricegrass, Nuttall's saltbush, bud sagebrush	Lacustrine deposits	Lake terraces
Norling	126.3	Ultic Haploxeralfs	9.2	1362	617	1152	60	Douglas-fir, incense cedar, Pacific madrone, salal, western swordfern	Colluvium and residuum derived from metavolcanics and metasedimentary rock	Mountains
Nuss	214.4	Lithic Haploxerolls	7.2	411	1389	1785	70	Idaho fescue, curl-leaf mountainmahogany, mountain big sagebrush, bluebunch wheatgrass, western juniper	Colluvium and residuum derived from tuff and basalt	Hills

(continued)

Series name	Area (km^2)	Subgroup	MAAT (°C)	MAP (mm)	Elev. lower (m)	Elev. upper (m)	Max slope (%)	Vegetatiom	Parent material	Landform
Nyssa	117.7	Xeric Haplodurids	11.0	256	664	799	20	Beardless wheatgrass, Sandberg bluegrass, bud sagebrush, Wyoming big sagebrush	Lacustrine deposits	Lake terraces
Oakland	147.3	Ultic Haploxeralfs	12.0	1080	91	715	60	Douglas-fir, Oregon white oak, Pacific madrone, California black oak, grasses	Colluvium and residuum derived from sandstone and siltstone	Hills
Oatman	133.9	Typic Haplocryands	6.0	1078	1521	1924	90	Shasta red fir, western white pine, white fir, mountain brome	Glacial till	Mountains
Observation	429.0	Typic Argixerolls	7.3	367	1253	1747	50	Curl-leaf mountainmahogany, Idaho fescue, antelope bitterbrush, bluebunch wheatgrass, mountain big sagebrush	Residuum and colluvium derived from basalt, andesite, rhyolite or tuff	Hills
Offenbacher	84.8	Typic Haploxerepts	10.0	826	305	1219	80	California black oak, Douglas-fir, Pacific madrone, ponderosa pine, California fescue	Colluvium derived from metavolcanics or metasedimentary rock	Mountains
Old Camp	51.6	Lithic xeric Haplargids	7.8	230	1369	1642	50	Thurber's needlegrass, bluebunch wheatgrass, Indian ricegrass, Wyoming big sagebrush	Residuum and colluvium derived from basalt	Hills
Olex	96.8	Calcidic Haploxerolls	11.1	254	160	358	40	Bluebunch wheatgrass, needle-and-thread, Sandberg bluegrass	Loess over alluvium	Plateaus
Olot	342.0	Typic Vitrixerands	5.9	668	1354	1618	65	Douglas-fir, common snowberry, ponderosa pine, elk sedge, pinegrass, western larch	Volcanic ash over colluvium and residuum derived from basalt	Mountains
Olyic	227.3	Typic Haplohumults	9.7	1693	152	610	90	Bigleaf maple, Douglas-fir, red alder, western hemlock, salal, western swordfern	Residuum and colluvium derived from basalt	Mountains
Opie	62.7	Cumulic Endoaquolls	7.2	243	1219	1372	5	Alkali sacaton, inland saltgrass, Sandberg bluegrass, alkali cordgrass	Lacustrine deposits	Lakebeds
Oreneva	122.2	Xeric Haplocambids	6.9	272	1502	1766	60	Thurber's needlegrass, bluebunch wheatgrass, Wyoming big sagebrush, Sandberg bluegrass	Colluvium and residuum derived from tuff or basalt	Plateaus
Orford	304.3	Typic Palehumults	9.8	1876	163	895	60	Douglas-fir, grand fir, western hemlock, western redcedar, salal, western swordfern	Residuum and colluvium derived from metasedimentary or metavolcanic rock	Mountains
Orovada	143.6	Durinodic Xeric Haplocambids	8.2	230	1126	1329	15	Basin wildrye, Indian ricegrass, basin big sagebrush, bluebunch wheatgrass, Thurber's needlegrass	Loess over alluvium	Fans
Outerkirk	184.0	Durinodic Haplocalcids	8.0	207	1210	1381	15	Spiny hopsage, Indian ricegrass	Alluvium and colluvium	Fans
Owsel	65.6	Durinodic Xeric Haplargids	10.0	260	730	925	15	Wyoming big sagebrush, Thurber's needlegrass, bluebunch wheatgrass	Loess and alluvium	Stream terraces

(continued)

Series name	Area (km^2)	Subgroup	MAAT (°C)	MAP (mm)	Elev. lower (m)	Elev. upper (m)	Max slope (%)	Vegetatiom	Parent material	Landform
Owyhee	76.9	Xeric Haplocalcids	11.0	242	643	767	20	Wyoming big sagebrush, bluebunch wheatgrass, Sandberg bluegrass, basin wildrye	Lacustrine deposits	Lake terraces
Oxwall	116.0	Palexerollic Durixerolls	8.3	406	1021	1366	50	Bluebunch wheatgrass, Idaho fescue	Alluvium	Outwash terraces
Ozamis	184.0	Fluvaquentic Endoaquolls	8.6	240	1275	1424	1	Alkali sacaton, alkali bluegrass, inland saltgrass	Lacustrine deposits	Lakebeds
Palouse	122.4	Pachic Ultic Haploxerolls	10.0	534	644	1032	45	Idaho fescue, bluebunch wheatgrass	Loess	Hills
Panther	59.5	Vertic Epiaquolls	11.5	1262	79	397	25	Oregon white oak, wildrose, sedges, rushes	Colluvium and residuum derived from basalt and siltstone	Hills
Parsnip	97.1	Lithic Argixerolls	6.1	498	1179	1502	30	Bluebunch wheatgrass, Idaho fescue	Loess over colluvium and residuum derived from basalt	Plateaus
Pearlwise	60.2	Pachic Haploxerolls	5.7	345	1685	1921	65	Idaho fescue, Thurber's needlegrass, bluebunch wheatgrass, mountain big sagebrush	Colluvium and residuum derived from tuff and basalt	Mountains
Pearsoll	232.3	Lithic Dystroxerepts	10.1	1875	215	974	90	Incense cedar, Jeffrey pine, sheep fescue, ceanothus	Colluvium and residuum derived from serpentine and peridotite	Mountains
Peavine	851.0	Typic Haplohumults	10.8	1762	115	502	75	Bigleaf maple, Douglas-fir, grand fir, red alder, western hemlock, western redcedar	Colluvium and residuum derived from basalt, sandstone, and siltstone	Mountains
Pengra	106.2	Vertic Epiaquolls	11.7	1216	95	361	30	Douglas-fir, Oregon ash, Oregon white oak, wildrose, grasses	Alluvium	Fans, hills
Perdin	87.9	Ultic Haploxeralfs	6.0	2350	975	1333	90	Jeffrey pine, incense cedar, knobcone pine, manzanita, common beargrass	Colluvium and residuum derived from serpentine and peridotite	Mountains
Pernty	346.8	Aridic Lithic Argixerolls	6.1	334	1424	1815	70	Bluebunch wheatgrass, Idaho fescue, Thurber's needlegrass, western juniper, mountain big sagebrush	Colluvium and residuum derived from tuff, basalt, or rhyolite	Mountains
Philomath	232.8	Vertic Haploxerolls	11.8	1232	108	454	70	Oregon white oak, baldhip rose, grasses	Colluvium and residuum derived from basalt	Hills
Piersonte	55.7	Vitrandic Haploxerolls	5.4	555	1250	1799	80	Douglas-fir, ponderosa pine, pinegrass, elk sedge	Colluvium derived from schist or shale	Mountains
Piline	73.7	Xeric Epiaquerts	8.0	278	1260	1294	2	Nevada bluegrass, creeping wildrye, silver sagebrush, mat muhly	Alluvium	Lakebeds
Pilot Rock	142.0	Haploxerollic Durixerolls	11.0	356	335	640	40	Bluebunch wheatgrass, Idaho fescue, Sandberg bluegrass	Loess over alluvium	Fan terraces
Pinehurst	147.9	Pachic Ultic Argixerolls	6.5	878	1083	1579	35	Douglas-fir, ponderosa pine, white fir, common snowberry	Colluvium derived from andesite, breccia, tuff	Mountains

(continued)

Series name	Area (km^2)	Subgroup	MAAT (°C)	MAP (mm)	Elev. lower (m)	Elev. upper (m)	Max slope (%)	Vegetatiom	Parent material	Landform
Pinhead	141.4	Vitric Haplocryands	2.6	2254	1196	1764	65	Douglas-fir, noble fir, Pacific silver fir, mountain hemlock, subalpine fir, common beargrass	Colluvium and residuum derived from andesite	Mountains
Pipp	91.6	Humic Vitrixerands	6.4	811	758	1236	65	Douglas-fir, grand fir, ponderosa pine, western larch, Idaho fescue	Volcanic ash mixed with colluvium and residuum derived from andesite	Mountains
Poall	322.3	Xeric Paleargids	9.8	270	779	1084	60	Sandberg bluegrass, Idaho fescue, bluebunch wheatgrass	Loess over lacustrine sediments	Hills
Pokegema	317.5	Humic Haploxerands	6.6	649	1243	1647	35	Douglas-fir, ponderosa pine, white fir, common snowberry, western fescue	Volcanic deposits over colluvium derived from andesite	Plateaus
Pollard	172.8	Typic Palexerults	10.5	1459	264	928	60	Douglas-fir, ponderosa pine, sugar pine, California black oak, Pacific madrone	Colluvium derived from metavolcanics and metasedimentary rock	Mountains
Poujade	290.6	Durinodic Xeric Natrargids	7.0	229	1227	1349	5	Basin wildrye, basin big sagebrush, black greasewood	Lacustrine deposits	Lake terraces
Powder	132.0	Cumulic Haploxerolls	10.0	254	581	820	3	Basin wildrye	Alluvium	Flood plains
Prag	162.4	Pachic Palexerolls	6.2	359	1144	1434	70	Idaho fescue, bluebunch wheatgrass, big sagebrush, western juniper	Residuum and colluvium derived from rhyolite, tuff, and basalt	Hills
Preacher	1,469.1	Andic Humudepts	9.9	2231	72	802	90	Bigleaf maple, Douglas-fir, red alder, western hemlock, salal, western swordfern	Colluvium and residuum derived from sandstone	Mountains
Prill	241.1	Pachic Palexerolls	9.2	353	814	1244	60	Bluebunch wheatgrass, antelope bitterbrush, Idaho fescue, Oregon white oak	Colluvium and residuum derived from tuff	Hills
Prouty	67.7	Andic Dystrocryepts	3.5	912	1906	2180	90	Subalpine fir, Engelmann spruce, grouse huckleberry, grand fir, common beargrass	Colluvium and residuum derived from granite	Mountains
Quatama	85.2	Aquultic Haploxeralfs	12.0	1143	30	102	30	Douglas-fir, western redcedar, Oregon white oak, Oregon ash	Glaciolacustrine deposits	Glaciolacustrine terraces
Quincy	545.2	Xeric Torripsamments	11.6	226	215	398	40	Needle-and-thread, antelope bitterbrush, Indian ricegrass, big sagebrush	Eolian sands	Dunes
Quirk	110.2	Vitrandic Palexerolls	5.7	497	1228	1475	60	Ponderosa pine, Douglas-fir, common snowberry, Idaho fescue, pinegrass	Loess over colluvium and residuum derived from tuff and basalt	Plateaus
Rabbithills	198.4	Xereptic Haplodurids	7.7	221	1332	1439	15	Thurber's needlegrass, bluebunch wheatgrass, Indian ricegrass, Wyoming big sagebrush	Alluvium and lacustrine deposits	Lake terraces
Ratto	186.5	Xeric Argidurids	7.3	258	1481	1710	15	Bluebunch wheatgrass, Sandberg bluegrass, Wyoming big sagebrush	Colluvium and residuum derived from tuff, basalt, and rhyolite	Plateaus

(continued)

Series name	Area (km^2)	Subgroup	MAAT (°C)	MAP (mm)	Elev. lower (m)	Elev. upper (m)	Max slope (%)	Vegetatiom	Parent material	Landform
Raz	2,277.7	Xeric Haplodurids	7.0	270	1367	1667	20	Thurber's needlegrass, bluebunch wheatgrass, Indian ricegrass, Wyoming big sagebrush	Colluvium and residuum derived from basalt or tuff	Plateaus
Reallis	240.4	Durinodic Xeric Haplocambids	7.0	260	1363	1587	8	Thurber's needlegrass, bluebunch wheatgrass, Wyoming big sagebrush, basin big sagebrush, Sandberg bluegrass	Alluvium	Lake terraces, fans
Redcliff	143.2	Aridic Haploxerolls	9.3	268	752	1280	75	Idaho fescue, bluebunch wheatgrass, Sandberg bluegrass, Wyoming big sagebrush	Residuum and colluvium derived from basalt	Hills
Redmond	99.5	Vitritorrandic Haploxerolls	9.9	265	881	1126	15	Idaho fescue, bluebunch wheatgrass, antelope bitterbrush, mountain big sagebrush	Residuum derived from basalt	Plateaus
Reedsport	254.1	Andic Humudepts	10.0	2261	35	398	90	Bigleaf maple, Douglas-fir, grand fir, Port Orford cedar, Sitka spruce, western hemlock, western redcedar	Colluvium and residuum derived from sandstone	Mountains
Reese	165.6	Duric Halaquepts	8.6	231	1292	1423	2	Alkali sacaton, inland saltgrass, basin wildrye, black greasewood	Lacustrine and alluvial deposits	Lakebeds
Reluctan	342.7	Aridic Argixerolls	6.7	296	1357	1648	30	Idaho fescue, Thurber's needlegrass, bluebunch wheatgrass, mountain big sagebrush	Colluvium and residuum derived from basalt, tuff, or rhyolite	Hills, plateaus
Remote	233.7	Typic Dystrudepts	9.7	2486	137	782	90	Douglas-fir, grand fir, bigleaf maple, red alder, salal, western swordfern	Colluvium and residuum derived from sandstone and siltstone	Mountains
Rhea	122.7	Calcic Haploxerolls	9.7	319	488	962	50	Idaho fescue, bluebunch wheatgrass, Sandberg bluegrass	Loess	Hills
Riddleranch	114.6	Aridic Haploxerolls	6.9	275	1389	1724	70	Bluebunch wheatgrass, Thurber's needlegrass, basin big sagebrush, Idaho fescue, Wyoming big sagebrush	Residuum and colluvium derived from basalt and tuff	Mountains
Rinconflat	119.5	Xeric Haplocambids	7.0	280	1368	1665	10	Thurber's needlegrass, bluebunch wheatgrass, Sandberg bluegrass, Wyoming big sagebrush	Alluvium	Fans
Rinearson	402.1	Typic Humudepts	9.9	2003	124	399	90	Bigleaf maple, Douglas-fir, red alder, western hemlock, salal, western swordfern	Colluvium and residuum derived from siltstone	Mountains
Rio King	54.5	Aridic Haploxerolls	8.0	216	1240	1351	6	Basin wildrye, basin big sagebrush	Alluvium	Stream terraces
Risley	199.0	Xeric Haplargids	8.1	271	1045	1289	40	Bluebunch wheatgrass, big sagebrush, Sandberg bluegrass	Residuum and colluvium derived from sandstone, shale, andesite	Hills
Ritner	291.6	Humic Haploxerepts	11.2	1267	104	531	90	Bigleaf maple, Douglas-fir, Oregon white oak, western swordfern	Colluvium derived from basalt	Hills
Ritzville	1,132.6	Calcidic Haploxerolls	10.4	265	285	686	50	Bluebunch wheatgrass, Sandberg bluegrass, Wyoming big sagebrush	Loess	Plateaus

(continued)

Series name	Area (km^2)	Subgroup	MAAT (°C)	MAP (mm)	Elev. lower (m)	Elev. upper (m)	Max slope (%)	Vegetatiom	Parent material	Landform
Robson	665.6	Lithic Xeric Haplargids	6.2	274	1444	1787	35	Bluebunch wheatgrass, Sandberg bluegrass, Wyoming big sagebrush, Idaho fescue, low sagebrush	Residuum and colluvium derived from basalt, andesite, rhyolite, or tuff	Hills, mountains
Roca	103.3	Xeric Haplargids	6.3	260	1290	1581	40	Bluebunch wheatgrass, Thurber's needlegrass, Wyoming big sagebrush	Residuum and colluvium derived from basalt, andesite, rhyolite, or tuff	Hills, mountains
Rockly	774.3	Lithic Haploxerolls	8.7	447	744	1296	90	Bluebunch wheatgrass, Sandberg bluegrass, stiff sagebrush	Residuum and colluvium derived from basalt	Mountains, hills
Rogger	68.1	Ultic Haploxerolls	8.0	703	1702	2012	60	Ponderosa pine, white fir, common snowberry, Wheeler's bluegrass	Colluvium and residuum derived from tuff, basalt, and andesite	Plateaus
Rogue	50.6	Typic Dystroxerepts	6.0	1633	1045	1633	80	Douglas-fir, incense cedar, white fir, bush chinquapin, greenleaf manzanita	Colluvium and residuum derived from granite	Mountains
Roloff	52.1	Aridic Haploxerolls	11.1	243	104	280	20	Bluebunch wheatgrass, Sandberg bluegrass, giant wildrye, big sagebrush	Loess over basalt	Hills
Roostercomb	178.3	Typic Argixerolls	6.0	414	1197	1625	65	Bluebunch wheatgrass, Idaho fescue, mountain big sagebrush, antelope bitterbrush	Colluvium and residuum derived from greenstone	Hills
Rosehaven	107.5	Ultic Haploxeralfs	11.8	1207	84	741	90	Douglas-fir, grand fir, incense cedar, Oregon white oak, Pacific madrone	Colluvium and residuum derived from sandstone and siltstone	Mountains
Royst	259.2	Vitrandic Argixerolls	7.1	429	1399	1741	70	Ponderosa pine, Idaho fescue, antelope bitterbrush, bluebunch wheatgrass, curl-leaf mountainmahogany, western juniper	Colluvium and residuum derived from basalt and rhyolite	Hills
Ruch	86.0	Mollic Palexeralfs	11.7	1049	250	653	20	Douglas-fir, incense cedar, ponderosa pine, California black oak, Pacific madrone	Alluvium	Fans, stream terraces
Ruckles	629.2	Aridic Lithic Argixerolls	9.5	262	672	1067	80	Bluebunch wheatgrass, Idaho fescue, Wyoming big sagebrush, western juniper, Sandberg bluegrass	Colluvium and residuum derived from basalt and tuff	Hills, mountains
Ruclick	410.8	Aridic Argixerolls	9.1	272	805	1219	70	Idaho fescue, Wyoming big sagebrush, bluebunch wheatgrass, Sandberg bluegrass	Colluvium and residuum derived from basalt and tuff	Mountains
Sagehen	78.7	Lithic Xeric Haplocambids	7.0	280	1543	1964	70	Bluebunch wheatgrass, low sagebrush, Sandberg bluegrass	Residuum and colluvium derived from basalt, andesite, rhyolite, or tuff	Mountains
Sagehill	254.9	Xeric Haplocalcids	10.6	220	268	462	35	Needle-and-thread, bluebunch wheatgrass, Sandberg bluegrass, Wyoming big sagebrush	Loess over lacustrine sediments	Strath terraces
Salander	153.1	Typic Fulvudands	10.6	2091	14	374	90	Douglas-fir, Sitka spruce, western hemlock, western redcedar, salal, western swordfern	Colluvium and residuum derived from igneous rock	Mountains
Salem	108.0	Pachic Ultic Argixerolls	11.4	1241	52	223	12	Bigleaf maple, Douglas-fir, Oregon white oak, baldhip rose	Alluvium	Stream terraces

(continued)

Series name	Area (km^2)	Subgroup	MAAT (°C)	MAP (mm)	Elev. lower (m)	Elev. upper (m)	Max slope (%)	Vegetatiom	Parent material	Landform
Salhouse	66.6	Vitrandic Torripsamments	7.0	229	1309	1332	20	Indian ricegrass, basin big sagebrush, black greasewood, inland saltgrass, spiny hopsage	Volcanic ash over lacustrine deposits	Dunes on lakebeds
Salkum	120.4	Xeric Palehumults	12.0	1257	114	257	20	Bigleaf maple, Douglas-fir, grand fir, red alder, western redcedar, salal, western swordfern	Alluvium	Stream terraces
Santiam	67.7	Aquultic Haploxeralfs	11.6	1244	85	126	30	Douglas-fir, Oregon white oak, western hazel, creambush oceanspray	Glaciolacustrine deposits	Glaciolacustrine terraces
Saum	158.4	Ultic Palexeralfs	11.4	1199	87	344	75	Douglas-fir, grand fir, Oregon white oak, common snowberry, western swordfern	Colluvium and residuum derived from basalt	Hills
Sauvie	100.2	Fluvaquentic Endoaquolls	12.0	1206	3	6	2	Red alder, Oregon white oak, willows, black cottonwood, tussocks	Alluvium	Flood plains
Scaponia	173.5	Humic Dystrudepts	9.1	1964	143	610	90	Douglas-fir, western hemlock, western redcedar, red alder, salal, western swordfern	Residuum and colluvium derived from siltstone	Mountains
Searles	117.8	Aridic Argixerolls	9.9	257	887	1177	80	Idaho fescue, antelope bitterbrush, bluebunch wheatgrass, Wyoming big sagebrush, western juniper	Colluvium and residuum derived from basalt and rhyolite	Hills
Segundo	58.8	Typic Haploxerepts	6.0	584	1219	1829	75	Douglas-fir, ponderosa pine, elk sedge, pinegrass, Idaho fescue	Colluvium derived from andesite and rhyolite	Mountains
Seharney	122.7	Xereptic Haplodurids	7.0	227	1338	1514	20	Thurber's needlegrass, bluebunch wheatgrass, Indian ricegrass, Wyoming big sagebrush	Colluvium and residuum derived from andesite or basalt	Hills
Senra	171.7	Vitritorrandic Durixerolls	7.0	271	1405	1525	20	Idaho fescue, bluebunch wheatgrass, antelope bitterbrush, low sagebrush	Volcanic ash over residuum derived from basalt or breccia	Plateaus
Shanahan	332.3	Xeric Vitricryands	5.9	523	1348	1624	45	Lodgepole pine, ponderosa pine, antelope bitterbrush, western needlegrass	Volcanic ash over colluvium and residuum derived from basalt	Plateaus
Shano	207.0	Xeric Haplocambids	11.0	229	198	457	40	Bluebunch wheatgrass, Wyoming big sagebrush, Sandberg bluegrass	Loess	Strath terraces
Sharesnout	242.5	Typic Argixerolls	5.0	398	1780	2059	60	Idaho fescue, bluebunch wheatgrass, low sagebrush, antelope bitterbrush	Residuum and colluvium derived from tuff	Mountains
Sharpshooter	61.0	Ultic Haploxerolls	9.0	1207	305	914	90	Douglas-fir, incense cedar, Pacific madrone, ponderosa pine, salal, western swordfern	Colluvium and residuum derived from schist	Mountains
Shefflein	56.8	Mollic Haploxeralfs	10.0	826	305	1219	35	California black oak, Douglas-fir, Pacific madrone, ponderosa pine, Idaho fescue	Colluvium and residuum derived from granite	Mountains
Shukash	602.5	Xeric Vitricryands	6.0	574	1418	1803	65	Lodgepole pine, ponderosa pine, antelope bitterbrush, western needlegrass	Volcanic ash over colluvium and residuum derived from basalt	Plateaus

(continued)

Series name	Area (km^2)	Subgroup	MAAT (°C)	MAP (mm)	Elev. lower (m)	Elev. upper (m)	Max slope (%)	Vegetatiom	Parent material	Landform
Sidlake	120.9	Xeric Haplargids	8.0	240	1140	1455	15	Bluebunch wheatgrass, Wyoming big sagebrush, Thurber's needlegrass	Loess over residuum and colluvium derived from basalt and rhyolite	Plateaus
Silverash	63.8	Aquandic Palexeralfs	6.5	262	1339	1480	2	Nevada bluegrass, creeping wildrye, silver sagebrush, mat muhly, sedges	Alluvium	Depressions on plateaus
Silvies	87.3	Vertic Cryaquolls	5.8	484	1137	1569	15	Tufted hairgrass, Baltic rush, sedge	Alluvium	Flood plains
Simas	771.9	Vertic Palexerolls	9.6	259	572	987	80	Bluebunch wheatgrass, Wyoming big sagebrush, Sandberg bluegrass, Thurber's needlegrass	Colluvium and residuum derived from tuff	Hills
Simnasho	110.7	Alfic Vitrixerands	6.5	800	792	1158	40	Douglas-fir, ponderosa pine, grand fir, common snowberry, Idaho fescue	Volcanic ash over colluvium and residuum derived from andesite	Mountains
Sinker	100.0	Pachic Haploxerolls	5.4	387	1276	1734	80	Idaho fescue, bluebunch wheatgrass, mountain big sagebrush, prairie junegrass	Colluvium derived from schist and graywacke	Hills
Siskiyou	94.1	Typic Dystroxerepts	10.6	1102	304	981	90	Douglas-fir, ponderosa pine, sugar pine, California black oak, common snowberry, Pacific madrone	Colluvium and residuum derived from granite	Mountains
Skedaddle	184.6	Lithic Xeric Torriorthents	8.0	232	1295	1573	70	Wyoming big sagebrush, bluebunch wheatgrass, Thurber's needlegrass	Residuum and colluvium derived from basalt, andesite, rhyolite, or tuff	Mountains
Skidoosprings	104.2	Duric Halaquepts	7.0	231	1244	1301	3	Basin wildrye, black greasewood, inland saltgrass	Lacustrine deposits	Lake terraces
Skipanon	215.2	Andic Humudepts	9.4	2413	15	457	60	Douglas-fir, red alder, Sitka spruce, western hemlock, western redcedar, salal, western swordfern	Residuum and colluvium derived from sandstone and siltstone	Mountains, hills
Skookumhouse	62.0	Typic Haplohumults	10.0	2794	305	914	30	Douglas-fir, red alder, western hemlock, Pacific madrone, salal, western swordfern	Colluvium and residuum derived from sandstone	Mountains
Skullgulch	57.8	Pachic Palexerolls	6.6	364	1219	1520	60	Bluebunch wheatgrass, Idaho fescue, mountain big sagebrush, prairie junegrass	Alluvium	Stream terraces, fans
Skunkfarm	71.7	Typic Endoaquolls	7.0	229	1252	1274	2	Creeping wildrye	Lacustrine deposits	Lake terraces
Skyline	99.6	Typic Haploxerolls	8.5	407	213	945	65	Bluebunch wheatgrass, Idaho fescue, Oregon white oak, ponderosa pine	Residuum and colluvium derived from sandstone	Hills
Slickrock	380.0	Alic Hapludands	10.0	2256	42	604	75	Douglas-fir, western hemlock, western redcedar, bigleaf maple, salal, western swordfern	Colluvium and residuum derived from sandstone	Mountains
Smiling	236.2	Alfic Vitrixerands	6.4	777	957	1246	70	Douglas-fir, ponderosa pine, antelope bitterbrush, Idaho fescue, greenleaf manzanita	Volcanic ash over colluvium and residuum derived from andesite	Mountains
Snaker	62.4	Lithic Xeric Torriorthents	8.8	272	690	1237	80	Bluebunch wheatgrass, Sandberg bluegrass	Colluvium and residuum derived from schist	Hills

(continued)

Series name	Area (km^2)	Subgroup	MAAT (°C)	MAP (mm)	Elev. lower (m)	Elev. upper (m)	Max slope (%)	Vegetatiom	Parent material	Landform
Snell	335.6	Pachic Argixerolls	6.3	494	1092	1570	90	Idaho fescue, bluebunch wheatgrass, Sandberg bluegrass, big sagebrush	Colluvium and residuum derived from basalt	Mountains, hills
Snellby	50.6	Aridic Argixerolls	6.9	266	1100	1258	80	Idaho fescue, bluebunch wheatgrass, mountain big sagebrush	Colluvium and residuum derived from basalt	Hills
Snowmore	1,379.9	Xeric Argidurids	7.2	248	1140	1518	15	Thurber's needlegrass, bluebunch wheatgrass, Wyoming big sagebrush, Indian ricegrass	Colluvium and residuum derived from basalt and rhyolite	Hills, mountains
Sorf	117.9	Vertic Paleargids	9.9	255	508	878	60	Bluebunch wheatgrass, Sandberg bluegrass, basin big sagebrush	Colluvium and residuum derived from tuff	Hills
Soughe	86.5	Lithic Xeric Haplargids	8.0	250	1210	1676	30	Thurber's needlegrass, bluebunch wheatgrass, Indian ricegrass, Wyoming big sagebrush, spiny hopsage	Residuum and colluvium derived from andesite and basalt	Hills
Spangenburg	470.2	Xeric Paleargids	7.9	242	1367	1425	2	Thurber's needlegrass, Indian ricegrass, needle-and-thread, Wyoming big sagebrush, Sandberg bluegrass	Lacustrine deposits	Lake terraces
Speaker	469.4	Ultic Haploxeralfs	10.3	1216	229	984	75	Douglas-fir, ponderosa pine, Pacific madrone, California black oak, incense cedar, Oregongrape	Colluvium and residuum derived from metavolcanics, or metasedimentary rock	Mountains
Stampede	74.3	Vertic Durixerolls	7.0	292	1306	1425	5	Thurber's needlegrass, bluebunch wheatgrass, basin big sagebrush	Loess over alluvium	Fans
Starkey	98.9	Typic Argixerolls	8.7	391	782	1234	50	Bluebunch wheatgrass, Idaho fescue, Sandberg bluegrass	Colluvium and residuum derived from tuff	Hills, mountains
Statz	71.4	Vitritorrandic Durixerolls	8.6	271	838	1219	30	Bluebunch wheatgrass, antelope bitterbrush, Idaho fescue, mountain big sagebrush	Volcanic ash over basalt	Hills, plateaus
Steiger	537.5	Xeric Vitricryands	5.8	579	1364	1585	65	Lodgepole pine, ponderosa pine, western needlegrass, antelope bitterbrush	Volcanic ash and pumice	Plateaus
Steiwer	74.5	Ultic Haploxerolls	11.4	1238	76	200	60	Douglas-fir, Oregon white oak, wild rose	Residuum and colluvium derived from sandstone and siltstone	Hills
Stookmoor	212.8	Vitritorrandic Haploxerolls	6.9	280	1311	1463	50	Idaho fescue, mountain big sagebrush, bluebunch wheatgrass, Thurber's needlegrass	Volcanic ash over residuum from basalt	Plateaus
Straight	81.2	Typic Dystroxerepts	9.0	1143	610	1219	70	Douglas-fir, sugar pine, white fir, Oregongrape	Residuum and colluvium derived from breccia and andesite	Mountains
Stukel	259.8	Aridic Lithic Haploxerolls	9.8	254	873	1174	30	Bluebunch wheatgrass, antelope bitterbrush, Idaho fescue, mountain big sagebrush, western juniper	Volcanic ash over residuum derived from basalt	Plateaus
Suckerflat	160.9	Aridic Lithic Haploxerolls	6.2	242	1331	1469	40	Bluebunch wheatgrass, Thurber's needlegrass, basin big sagebrush	Volcanic ash over colluvium derived from basalt or breccia	Plateaus

(continued)

Series name	Area (km^2)	Subgroup	MAAT (°C)	MAP (mm)	Elev. lower (m)	Elev. upper (m)	Max slope (%)	Vegetatiom	Parent material	Landform
Sutherlin	112.8	Ultic Haploxeralfs	12.0	1080	91	610	50	Douglas-fir, Oregon white oak, Pacific madrone, common snowberry	Colluvium and residuum derived from sandstone and siltstone	Hills
Svensen	119.0	Andic Humudepts	10.0	2146	9	292	90	Douglas-fir, grand fir, red alder, Sitka spruce, western hemlock, western swordfern, salal	Colluvium and residuum derived from sandstone	Mountains
Swaler	90.6	Xeric Paleargids	6.8	280	1326	1478	3	Idaho fescue, mountain big sagebrush, Thurber's needlegrass, western needlegrass	Lacustrine deposits	Lake terraces on plateaus
Swalesilver	204.5	Aquic Palexeralfs	6.9	264	1368	1571	2	Nevada bluegrass, creeping wildrye, silver sagebrush, mat muhly	Lacustrine deposits	Depressions on plateaus
Syrupcreek	787.3	Alfic Udivitrands	5.9	698	1336	1618	90	Big huckleberry, lodgepole pine, grand fir, western larch, Douglas-fir, Engelmann spruce, pinegrass	Volcanic ash over colluvium and residuum derived from basalt and andesite	Mountains
Tallowbox	107.4	Typic Haploxerepts	10.0	826	305	1219	70	California black oak, Douglas-fir, Pacific madrone, ponderosa pine, Idaho fescue, whiteleaf manzanita	Colluvium and residuum derived from granite	Mountains
Tamara	144.8	Alfic Udivitrands	5.8	645	1269	1568	70	Douglas-fir, grand fir, ponderosa pine, western larch, grouse huckleberry, Engelmann spruce, lodgepole pine, subalpine fir	Volcanic ash over colluvium and residuum derived from basalt	Mountains
Tamarackcanyon	65.0	Vitrandic Haploxeralfs	6.2	609	1134	1712	90	Douglas-fir, ponderosa pine, western larch, pinegrass, common snowberry, elk sedge	Residuum and colluvium derived from basalt	Mountains
Tatouche	78.3	Typic Argixerolls	6.0	1016	1097	1676	65	Douglas-fir, white fir, Oregongrape, incense cedar	Colluvium and residuum derived from andesite, tuff, breccia	Mountains
Taunton	60.0	Xeric Haplodurids	10.3	217	200	309	12	Needle-and-thread, bluebunch wheatgrass, Sandberg bluegrass, Wyoming big sagebrush	Eolian sands over alluvium	Strath terraces
Teewee	63.3	Vitrandic Argixerolls	8.4	453	762	988	75	Ponderosa pine, Idaho fescue, antelope bitterbrush, greenleaf manzanita	Residuum and colluvium derived from basalt or andesite	Mountains, hills
Teguro	239.8	Lithic Argixerolls	5.3	357	1230	1525	30	Idaho fescue, antelope bitterbrush, mountain big sagebrush, bluebunch wheatgrass	Residuum and colluvium derived from basalt, andesite, rhyolite, or tuff	Hills, plateaus
Templeton	666.4	Andic Humudepts	10.2	2174	16	410	90	Douglas-fir, red alder, Sitka spruce, western hemlock, western redcedar, salal, western swordfern	Colluvium and residuum derived from sandstone and siltstone	Mountains, hills
Tenmile	131.5	Xeric Haplargids	8.2	280	901	1230	55	Bluebunch wheatgrass, Thurber's needlegrass, Wyoming big sagebrush, Sandberg bluegrass	Alluvium	Stream terraces

(continued)

Series name	Area (km^2)	Subgroup	MAAT (°C)	MAP (mm)	Elev. lower (m)	Elev. upper (m)	Max slope (%)	Vegetatiom	Parent material	Landform
Thenarrows	63.5	Typic Halaquepts	7.0	229	1247	1251	2	Alkali sacaton, inland saltgrass, alkali cordgrass	Lacustrine deposits	Lake terraces
Thornlake	156.2	Sodic Xeric Haplocambids	7.0	229	1308	1329	5	Black greasewood, basin big sagebrush, basin wildrye, inland saltgrass	Lacustrine deposits	Lakebeds
Threebuck	76.0	Alfic Vitrixerands	6.3	550	1019	1565	90	Douglas-fir, common snowberry, ponderosa pine, elk sedge, pinegrass, western larch	Volcanic ash over colluvium derived from tuff or basalt	Mountains
Timbercrater	102.0	Typic Vitricryands	4.3	1389	1616	2050	80	Mountain hemlock, Shasta red fir, lodgepole pine	Volcanic ash and pumice	Mountains
Tolany	64.9	Alic Hapludands	6.4	1796	457	785	60	Douglas-fir, western hemlock, western redcedar, red alder, salal, western swordfern	Colluvium	Mountains
Tolke	118.3	Alic Hapludands	9.1	2191	225	652	60	Bigleaf maple, Douglas-fir, red alder, western hemlock, salal, western swordfern	Colluvium and residuum derived from tuffaceous sedimentary rock	Mountains
Tolo	607.5	Alfic Vitrixerands	6.2	581	1166	1463	65	Douglas-fir, grand fir, lodgepole pine, ponderosa pine, western larch, common snowberry, elk sedge	Volcanic ash over colluvium derived from basalt	Mountains, plateaus
Tolovana	220.1	Typic Fulvudands	9.4	2413	15	549	85	Douglas-fir, red alder, Sitka spruce, western hemlock, western redcedar, salal, western swordfern	Colluvium and residuum derived from sandstone and siltstone	Mountains
Tonor	76.7	Sodic Xeric Haplocambids	7.0	229	1306	1348	3	Inland saltgrass, basin big sagebrush, basin wildrye, black greasewood	Lacustrine deposits	Lakebeds
Top	83.9	Vertic Argixerolls	6.1	531	990	1587	75	Douglas-fir, ponderosa pine, elk sedge, white fir, pinegrass	Residuum and colluvium derived from basalt	Mountains
Troutmeadows	290.6	Typic Vitricryands	3.2	961	1692	1944	90	Engelmann spruce, lodgepole pine, subalpine fir, grand fir, grouse huckleberry, western larch	Volcanic ash over colluvium and residuum derived from basalt	Mountains
Tub	629.1	Vertic Argixerolls	9.0	333	825	1191	70	Idaho fescue, bluebunch wheatgrass, big sagebrush	Colluvium derived from basalt and tuff	Hills
Tulana	107.5	Aquandic Humaquepts	8.2	371	1261	1290	1	Cattails, rushes	Lacustrine deposits	Lakebeds
Tumtum	312.0	Typic Argidurids	8.0	225	1309	1566	15	Thurber's needlegrass, bluebunch wheatgrass, Indian ricegrass, Wyoming big sagebrush, spiny hopsage	Alluvium	Lake terraces
Turpin	202.8	Sodic Xeric Haplocambids	8.1	237	1320	1378	15	Black greasewood, bud sagebrush, shadscale, spiny hopsage	Lacustrine deposits	Lake terraces
Tutni	83.4	Typic Cryaquands	5.8	629	1263	1402	3	Lodgepole pine, antelope bitterbrush, ross sedge, western needlegrass	Volcanic ash and pumice over alluvium	Plateaus
Tweener	93.5	Lithic Argixerolls	6.0	356	1219	2012	20	Bluebunch wheatgrass, Sandberg bluegrass, mountain big sagebrush, Idaho fescue	Colluvium and residuum derived from basalt, andesite, and tuff	Mountains

(continued)

Series name	Area (km²)	Subgroup	MAAT (°C)	MAP (mm)	Elev. lower (m)	Elev. upper (m)	Max slope (%)	Vegetatiom	Parent material	Landform
Twelvemile	54.0	Typic Vitrixerands	7.9	730	1676	2092	60	Ponderosa pine, white fir, common snowberry, heartleaf arnica, Wheeler's bluegrass	Volcanic ash over colluvium and residuum derived from rhyolite	Mountains
Ukiah	84.4	Vertic Argixerolls	8.8	467	884	1419	50	Bluebunch wheatgrass, Idaho fescue, Sandberg bluegrass	Colluvium and residuum derived from tuff or basalt	Hills
Umapine	151.4	Typic Halaquepts	9.9	260	575	845	3	Basin wildrye, inland saltgrass, black greasewood, basin big sagebrush	Alluvium	Flood plains
Umatilla	139.6	Vitrandic Haploxerolls	6.2	768	780	1485	70	Douglas-fir, grand fir, ponderosa pine, elk sedge, pinegrass	Loess over colluvium from basalt	Mountains
Umpcoos	645.7	Lithic Eutrudepts	9.7	2409	77	817	99	Douglas-fir, Pacific madrone, tanoak, western hemlock, salal, western swordfern	Colluvium and residuum derived from sandstone or siltstone	Mountains
Unionpeak	145.5	Typic Duricryands	4.4	1460	1557	2064	35	Mountain hemlock, Shasta red fir, lodgepole pine	Ash, pumice, and cinders	Mountains
Valby	464.0	Calcic Haploxerolls	9.7	327	475	1029	35	Bluebunch wheatgrass, Sandberg bluegrass, Idaho fescue	Loess over basalt	Hills, plateaus
Valsetz	155.7	Alic Haplocryands	6.0	3070	805	1242	90	Douglas-fir, noble fir, Pacific silver fir, mountain hemlock, salal, western swordfern	Colluvium and residuum and residuum derived from basalt	Mountains
Vandamine	68.5	Andic Haplocryepts	2.7	803	1662	1917	90	Douglas-fir, western larch, subalpine fir, big huckleberry, grouse huckleberry, lodgepole pine	Volcanic ash over colluvium and residuum derived from argillite	Mountains
Vannoy	425.2	Mollic Haploxeralfs	10.2	760	300	1219	75	Douglas-fir, ponderosa pine, California black oak, Idaho fescue, Pacific madrone, common snowberry	Colluvium and residuum derived from metavolcanics and metasedimentary rock	Hills
Vanwyper	61.7	Xeric Haplargids	7.6	258	1614	1850	70	Bluebunch wheatgrass, Thurber's needlegrass, Wyoming big sagebrush, Sandberg bluegrass	Colluvium and residuum derived from andesite	Mountains
Venator	135.2	Lithic Haploxerolls	8.8	350	855	1320	80	Bluebunch wheatgrass, Thurber's needlegrass, Idaho fescue, Sandberg bluegrass	Colluvium and residuum derived from shale, sandstone, limestone, and conglomerate	Hills
Veneta	59.6	Ultic Haploxeralfs	12.0	1101	50	1406	20	Douglas-fir, Oregon white oak, Pacific madrone, ponderosa pine	Alluvium	Stream terraces
Vergas	249.0	Durinodic Xeric Haplargids	6.7	279	1271	1446	8	Basin wildrye, basin big sagebrush, bluebunch wheatgrass, Thurber's needlegrass	Alluvium	Lake terraces
Vermisa	542.3	Lithic Dystroxerepts	10.5	1663	232	1011	99	Canyon live oak, Douglas-fir, Pacific madrone, tanoak, deerbrush ceanothus	Colluvium and residuum derived from graywacke	Mountains
Vernonia	90.3	Ultic Hapludalfs	9.0	1524	106	580	30	Bigleaf maple, Douglas-fir, red alder, western hemlock, western redcedar	Colluvium and residuum derived from siltstone and shale	Mountains

(continued)

Series name	Area (km^2)	Subgroup	MAAT (°C)	MAP (mm)	Elev. lower (m)	Elev. upper (m)	Max slope (%)	Vegetatiom	Parent material	Landform
Virtue	225.7	Xeric Argidurids	9.6	263	755	982	20	Bluebunch wheatgrass, Thurber's needlegrass, Wyoming big sagebrush, Sandberg bluegrass	Lacustrine deposits	Lake terraces
Vitale	270.2	Typic Argixerolls	5.3	359	1304	1879	60	Bluebunch wheatgrass, antelope bitterbrush, Idaho fescue, mountain big sagebrush	Residuum and colluvium derived from basalt, andesite, rhyolite or tuff	Hills, mountains
Voltage	60.9	Xeric Haplocalcids	7.0	267	1226	1342	2	Basin wildrye, basin big sagebrush	Lacustrine deposits	Lake terraces
Voorhies	92.4	Mollic Haploxeralfs	10.0	762	305	1219	55	Douglas-fir, ponderosa pine, California black oak, Idaho fescue, Pacific madrone, common snowberry	Colluvium and residuum derived from metavolcanics and metasedimentary rock	Hills
Waha	504.4	Pachic Argixerolls	9.6	532	424	1079	70	Idaho fescue, bluebunch wheatgrass, common snowberry, ponderosa pine	Colluvium and residuum derived from basalt	Hills
Wahstal	68.1	Palexerollic Durixerolls	6.1	398	1255	1585	15	Bluebunch wheatgrass, Sandberg bluegrass, onespike oatgrass, stiff sagebrush	Alluvium	Stream terraces
Waldo	160.8	Fluvaquentic Vertic Endoaquolls	11.6	1270	56	254	3	Black cottonwood, Oregon ash, red alder	Alluvium	Flood plains
Waldport	91.9	Typic Udipsamments	11.0	1954	2	72	90	Douglas-fir, shore pine, Sitka spruce, western hemlock	Eolian sands	Coastal dunes
Walla Walla	1,269.0	Typic Haploxerolls	10.5	335	205	693	50	Bluebunch wheatgrass, Sandberg bluegrass	Loess over basalt	Plateaus, hills
Wallowa	83.3	Vitrandic Haploxerolls	8.0	432	1088	1353	30	Bluebunch wheatgrass, Sandberg bluegrass, Idaho fescue	Loess over basalt	Plateaus
Wamic	247.2	Vitrandic Haploxerepts	8.3	432	305	1097	70	Oregon white oak, ponderosa pine, Idaho fescue, bluebunch wheatgrass, antelope bitterbrush	Loess over colluvium derived from basalt or andesite	Hills
Wanoga	331.8	Humic Vitrixerands	6.6	387	1239	1512	65	Mountain big sagebrush, ponderosa pine, antelope bitterbrush, Idaho fescue, western juniper	Volcanic ash over basalt, rhyolite, or tuff	Hills
Wapato	164.1	Fluvaquentic Endoaquolls	11.7	1219	48	256	3	Oregon ash, black cottonwood, wildrose, red alder	Alluvium	Flood plains
Warden	280.3	Xeric Haplocambids	10.1	207	163	370	40	Needle-and-thread, bluebunch wheatgrass, Sandberg bluegrass, big sagebrush	Loess over lacustrine deposits	Strath terraces
Watama	261.3	Pachic Haploxerolls	9.1	402	709	1167	70	Bluebunch wheatgrass, Idaho fescue, Sandberg bluegrass	Loess over colluvium and residuum derived from basalt	Plateaus
Waterbury	170.1	Lithic Argixerolls	9.3	383	691	1128	80	Bluebunch wheatgrass, Idaho fescue, low sagebrush, Sandberg bluegrass	Colluvium and residuum derived from basalt or tuff	Plateaus, hills
Wato	54.5	Typic Haploxerolls	10.5	324	76	534	35	Idaho fescue, bluebunch wheatgrass, Sandberg bluegrass	Loess	Plateaus
Wegert	190.3	Vitritorrandic Haploxerolls	6.8	259	1335	1463	15	Needle-and-thread, Indian ricegrass, mountain big sagebrush, basin wildrye	Volcanic ash over residuum derived from basalt	Plateaus

(continued)

Series name	Area (km^2)	Subgroup	MAAT (°C)	MAP (mm)	Elev. lower (m)	Elev. upper (m)	Max slope (%)	Vegetatiom	Parent material	Landform
Weglike	81.0	Vitritorrandic Haploxerolls	6.5	239	1324	1407	6	Basin big sagebrush, basin wildrye	Volcanic ash over residuum derived tuff	Plateaus
Welch	61.5	Cumulic Endoaquolls	6.5	373	1605	2006	10	Tufted hairgrass, sedges, creeping wildrye, Nevada bluegrass	Alluvium	Flood plains
Westbutte	642.1	Pachic Haploxerolls	5.8	351	1440	1839	70	Bluebunch wheatgrass, Idaho fescue, mountain big sagebrush, Sandberg bluegrass, western juniper	Residuum and colluvium derived from basalt, andesite, rhyolite, or tuff	Hills, mountains
Whetstone	53.6	Typic Haplocryods	6.0	2032	914	1219	75	Douglas-fir, noble fir, Pacific silver fir, mountain hemlock, Pacific rhododendron, common beargrass	Colluvium and residuum derived from tuff and basalt	Mountains
Whobrey	95.8	Aquertic Eutrudepts	9.8	2540	126	701	60	Douglas-fir, grand fir, western redcedar, western hemlock, salal, western swordfern	Colluvium and residuum derived from metasedimentary rock	Mountains
Widowspring	98.5	Cumulic Haploxerolls	6.8	278	1301	1432	5	Basin wildrye, basin big sagebrush	Alluvium	Stream terraces
Wieland	59.2	Durinodic Xeric Haplargids	9.0	250	1740	2100	15	Bluebunch wheatgrass, Thurber's needlegrass, Wyoming big sagebrush	Alluvium	Fans
Wilhoit	75.9	Andic Humudepts	6.0	2032	549	914	60	Douglas-fir, red alder, western hemlock, salal, western swordfern	Colluvium and residuum derived from andesite, tuff, breccia	Mountains
Willakenzie	157.3	Ultic Haploxeralfs	11.4	1184	72	267	75	Douglas-fir, grand fir, Oregon white oak, common snowberry	Colluvium and residuum derived from sandstone or siltstone	Hills
Willamette	155.0	Pachic Ultic Argixerolls	11.5	1135	43	101	20	Douglas-fir, Oregon white oak, bigleaf maple, common snowberry	Glaciolacustrine deposits	Glaciolacustrine terraces
Willis	148.7	Haploduridic Durixerolls	10.1	271	305	695	40	Bluebunch wheatgrass, Sandberg bluegrass, big sagebrush	Loess over cemented alluvium	Fan terraces
Willowdale	62.7	Cumulic Haploxerolls	10.5	255	686	957	3	Basin wildrye, bluebunch wheatgrass	Alluvium	Flood plains
Winchester	58.0	Xeric Torripsamments	11.9	238	130	251	12	Needle-and-thread, antelope bitterbrush, Indian ricegrass	Eolian sands	Dunes on outwash plains
Windego	70.4	Alfic Vitrixerands	6.5	764	989	1251	70	Douglas-fir, ponderosa pine, antelope bitterbrush, Idaho fescue, greenleaf manzanita	Volcanic ash over colluvium derived from basalt	Mountain slopes
Windygap	313.0	Xeric Haplohumults	12.0	1207	162	790	60	Douglas-fir, grand fir, Pacific madrone, ponderosa pine, salal, western swordfern	Colluvium and residuum derived from sandstone and siltstone	Mountains
Wingville	78.3	Pachic Haploxerolls	9.5	292	671	1097	2	Tufted hairgrass, sedges, Baltic rush	Alluvium	Fans, stream terraces
Winterim	117.8	Pachic Argixerolls	7.7	647	1543	1876	60	Ponderosa pine, western juniper, antelope bitterbrush	Colluvium and residuum derived from tuff and basalt	Plateaus

(continued)

Series name	Area (km^2)	Subgroup	MAAT (°C)	MAP (mm)	Elev. lower (m)	Elev. upper (m)	Max slope (%)	Vegetatiom	Parent material	Landform
Witzel	113.2	Lithic Ultic Haploxerolls	11.0	1279	88	479	75	Douglas-fir, Oregon white oak, baldhip rose, common snowberry	Colluvium and residuum derived from basalt	Hills
Wolfpeak	56.3	Ultic Palexeralfs	10.1	1034	283	897	60	Douglas-fir, incense cedar, sugar pine, Pacific madrone, salal	Colluvium and residuum derived from granite	Hills
Woodburn	910.6	Aquultic Argixerolls	11.5	1126	43	104	55	Douglas-fir, Oregon white oak, baldhip rose, salal	Glaciolacustrine deposits	Glaciolacustrine terraces
Woodchopper	77.0	Pachic Ultic Argixerolls	8.0	762	1676	2195	40	Ponderosa pine, white fir, common snowberry, Wheeler's bluegrass	Colluvium and residuum from tuff, andesite, and basalt	Plateaus
Woodcock	576.3	Alfic Humic Haploxerands	6.6	623	1322	1665	60	Douglas-fir, ponderosa pine, white fir, common snowberry, western fescue	Colluvium from glacial outwash	Plateaus
Woodseye	77.5	Humic Lithic Dystroxerepts	6.0	1641	1013	2352	90	Canyon live oak, Douglas-fir, golden chinquapin, sugar pine, tanoak, white fir	Colluvium derived from metavolcanics and metasedimentary rock	Mountains
Wrentham	333.6	Pachic Haploxerolls	9.0	335	440	1116	70	Idaho fescue, bluebunch wheatgrass, Sandberg bluegrass	Loess over colluvium and residuum derived from basalt	Hills
Xanadu	72.8	Typic Palehumults	10.0	1842	61	914	60	Bigleaf maple, Douglas-fir, grand fir, red alder, western hemlock, western redcedar, salal, western swordfern	Colluvium and residuum derived from sandstone and siltstone	Mountains
Yancy	52.3	Palexerollic Durixerolls	6.9	422	1302	1480	8	Idaho fescue, low sagebrush, Sandberg bluegrass, bluebunch wheatgrass	Alluvium	Fan terraces
Yankeewell	169.6	Xeric Natridurids	6.8	277	1401	1782	20	Wyoming big sagebrush, bluebunch wheatgrass, shadscale	Colluvium and residuum derived from basalt	Hills, plateaus
Yawhee	144.1	Alfic Udivitrands	6.0	572	1431	2040	40	Ponderosa pine, white fir, sugar pine, incense cedar	Volcanic ash mixed with colluvium	Mountains
Yawkey	84.2	Vertic Palexerolls	5.9	483	1116	1512	70	Douglas-fir, elk sedge, ponderosa pine, pinegrass, Idaho fescue	Colluvium and residuum derived from tuff	Mountains
Yellowstone	81.1	Lithic Haplocryands	6.2	3006	816	1283	90	Douglas-fir, noble fir, Pacific silver fir, western hemlock, western white pine	Colluvium and residuum derived from basalt	Mountains
Zevadez	200.1	Durinodic Xeric Haplargids	8.0	270	1210	1600	15	Thurber's needlegrass, bluebunch wheatgrass, Wyoming big sagebrush, Indian ricegrass	Alluvium	Fans
Zing	64.1	Aquultic Haploxeralfs	9.3	1201	305	920	45	Douglas-fir, grand fir, Pacific madrone, ponderosa pine, sugar pine, salal	Colluvium and residuum derived from granodiorite	Mountains
Zygore	221.7	Andic Humudepts	6.4	2139	568	962	90	Douglas-fir, red alder, western hemlock, grand fir, salal, western swordfern	Colluvium and glacial till from basalt and andesite	Mountains

Appendix B
Thicknesses (cm) of Diagnostic Horizons in Soil Series with an Area of 50 km^2 and Greater in Oregon

T. Thorson et al., *The Soils of Oregon*, World Soils Book Series,
https://doi.org/10.1007/978-3-030-90091-5

Series name	Area (km^2)	Subgroup	Mollic	Ochric	Umbric	Histic	Melanic	Cambic	Argillic	Duripan	Calcic	Albic	Natric	Spodic	Fragipan	Ortstein	Salic	Glossic	None	Andic 1*	Andic 2*
Abegg	78.4	Ultic Haploxeralfs		44					100												
Abert	144.9	Sodic Xeric Haplocambids		18				28													
Abiqua	57.3	Cumulic Ultic Haploxerolls	53					84													
Absaquil	54.0	Typic Haplohumults		10					89												
Acker	217.6	Typic Palexerults		20					**109**												
Actem	749.7	Xeric Argidurids		5					33	13											
Ada	84.9	Typic Argixerolls	33						**119**												
Adkins	107.6	Xeric Haplocalcids		18				43			**91**										
Agate	50.3	Typic Durixerepts		15				33		12											
Agency	135.9	Aridic Haploxerolls	41					20													
Albee	123.7	Vitrandic Haploxerolls	48					23													
Alding	63.1	Lithic Argixerolls	41						27												
Aloha	236.5	Aquic Haploxerepts		20				114													
Alspaugh	130.6	Typic Paleudults			36				61												
Alstony	82.6	Alic Hapludands		12				46												X	
Althouse	59.9	Typic Dystroxerepts		18				71													
Alvodest	263.4	Sodic Aquicambids		18				92													
Amity	434.7	Argiaquic Xeric Argialbolls	41						33			15									
Analulu	265.0	Vitrandic Haploxerepts		18				48													
Anatone	1589.5	Lithic Haploxerolls	43																X		
Anawalt	1280.1	Lithic Xeric Haplargids		15					33												
Anderly	185.2	Typic Haploxerolls	46					30													
Angelbasin	78.8	Andic Dystrocryepts		18				51													
Angelpeak	127.1	Typic Vitricryands		30				38													X
Apt	148.9	Typic Haplohumults			46				140												
Arcia	92.6	Vitrandic Argixerolls	53						63												
Ascar	120.9	Typic Fulvudands			30			58												X	
Aschoff	311.0	Andic Humudepts			31			56													
Astoria	145.9	Andic Humudepts			30			122													
Ateron	1288.7	Lithic Argixerolls	30						12												
Athena	210.4	Pachic Haploxerolls	66																X		
Atlow	357.3	Lithic Xeric Haplargids		8					30												
Atring	480.2	Typic Dystroxerepts		23				27													
Ausmus	235.3	Aquic Natrargids		5							67		18								
Ayres	86.6	Argiduridic Durixerolls	20						18	114											
Bacona	212.6	Typic Palehumults		7.5					**122**												
Baconcamp	373.2	Pachic Haplocryolls	89																X		
Bakeoven	1225.0	Aridic Lithic Haploxerolls	18																X		
Baker	77.9	Haploduridic Durixerolls	28					51		20											
Barkshanty	69.7	Typic Palehumults		33					135												
Bashaw	171.9	Xeric Endoaquerts	36					114													
Bateman	204.6	Ultic Palexeralfs		18					142												
Bearpawmeadow	58.3	Andic Haplocryepts		7				23													
Beden	106.5	Aridic Lithic Argixerolls	28						18												
Beekman	647.3	Typic Dystroxerepts		36				27													
Bellpine	769.6	Xeric Haplohumults		15					41												
Bennettcreek	107.4	Vitrandic Haploxeralfs		7				18	33												
Beoska	55.9	Durinodic Natrargids		33							96		28								
Berdugo	114.7	Xeric Paleargids		18					27												
Bingville	68.8	Pachic Palexerolls	53						38												

(continued)

Series name	Area (km^2)	Subgroup	Mollic	Ochric	Umbric	Histic	Melanic	Cambic	Argillic	Duripan	Calcic	Albic	Natric	Spodic	Fragipan	Ortstein	Salic	Glossic	None	Andic 1*	Andic 2*
Blachly	510.1	Humic Dystrudepts		38				124													
Bluecanyon	67.0	Lithic Haploxerolls	33																X		
Bly	60.8	Vitrandic Argixerolls	71						79	**33**											
Boardtree	112.2	Alfic Vitrixerands		18					71												X
Bocker	1473.9	Lithic Haploxerolls	25																X		
Bodell	83.9	Lithic Haploxerolls	46																X		
Bohannon	2045.6	Andic Humudepts			27			15													
Boilout	52.8	Vitrixerandic Argidurids		15					13	**116**											
Bolobin	66.3	Vitrandic Argixerolls	74						54												
Bombadil	62.3	Lithic Xeric Haplargids		5					20												
Bonnick	96.7	Vitritorrandic Haploxerolls	25																		X
Booth	593.7	Vertic Palexerolls	30						20												
Bordengulch	71.1	Andic Haplocryepts		10				81													
Bornstedt	67.7	Typic Palexerults		20					**157**												
Borobey	215.2	Vitritorrandic Haploxerolls	38																X		
Boulder lake	81.4	Xeric Epiaquerts		18				36													
Bouldrock	56.0	Humic Haploxerepts		15				36													
Brace	1571.4	Xeric Argidurids		25					33	8											
Brader	50.5	Typic Haploxerepts		15				18													
Braun	167.7	Dystric Eutrudepts		15				53													
Bravo	188.9	Humic Dystrudepts		18				56													
Bridgecreek	75.6	Typic Palexerolls	40						40												
Brisbois	66.6	Xeric Haplargids		7.5					15												
Broyles	97.4	Durinodic Haplocambids		18				20													
Btree	116.6	Alfic Udivitrands		5				33	68												X
Buckcreek	79.6	Pachic Ultic Haploxerolls	91																X		
Bucketlake	120.9	Typic Vitricryands		13				144													X
Bull run	78.1	Eutric Fulvudands			33			58												X	
Bullards	104.2	Typic Haplorthods		18										86							
Bullump	103.9	Pachic Argixerolls	58						79												
Burgerbutte	61.2	Lithic Humicryepts			25			20													
Burke	62.6	Xeric Haplodurids		10						**97**											
Bybee	57.5	Typic Haploxerolls	25					71													
Calderwood	155.8	Lithic Xeric Haplocambids		18				38													
Calimus	157.0	Pachic Haploxerolls	56					66													
Camas	111.3	Fluventic Haploxerolls	25																X		
Campcreek	56.1	Vertic Palexerolls	38						**127**												
Canest	353.5	Aridic Lithic Argixerolls	23						10												
Cantala	165.4	Typic Haploxerolls	45					90													
Caris	208.1	Typic Haploxerepts		30				48													
Carney	235.3	Udic Haploxererts	30																X		
Carryback	1189.8	Vertic Palexerolls	28						43												
Carvix	151.1	Aridic Haploxerolls	48					104													
Cascade	178.8	Humic Fragixerepts			41			27							84						
Cassiday	183.3	Humic Dystrudepts		18				46													
Castlecrest	265.8	Typic Vitricryands		48															X		X
Caterl	253.1	Alic Hapludands			41														X	X	
Catherine	89.2	Cumulic Endoaquolls	102																X		
Catlow	171.7	Durinodic Xeric Haplocambids		23				35													

(continued)

Series name	Area (km^2)	Subgroup	Mollic	Ochric	Umbric	Histic	Melanic	Cambic	Argillic	Duripan	Calcic	Albic	Natric	Spodic	Fragipan	Ortstein	Salic	Glossic	None	Andic 1*	Andic 2*
Cazadero	82.0	Rhodic Paleudults			30				74												
Chapman	58.1	Cumulic Ultic Haploxerolls	58					48													
Chehalis	238.7	Cumulic Ultic Haploxerolls	70					80													
Chen	111.9	Aridic Lithic Argixerolls	33						13												
Cherrycreek	71.3	Vitrandic Haploxerolls	71					64													
Chilcott	93.3	Abruptic Xeric Argidurids		20					56	43	15										
Choptie	64.6	Lithic Haploxerolls	41																X		
Clackamas	82.5	Typic Argiaquolls	61						23												
Clamp	134.3	Lithic Haplocryolls	30																X		
Cloquato	159.8	Cumulic Ultic Haploxerolls	102																X		
Clovercreek	54.5	Lithic Argixerolls	25						25												
Clovkamp	50.0	Vitritorrandic Haploxerolls	30																X		
Coburg	178.9	Oxyaquic Argixerolls	71						89												
Coglin	100.1	Xeric Paleargids		18					12												
Colbar	76.9	Xeric Haplargids		23					18												
Colestine	180.3	Typic Dystroxerepts		18				56													
Collier	411.4	Xeric Vitricryands		10															X		X
Concord	123.1	Typic Endoaqualfs		38					35			23									
Condon	1171.4	Typic Haploxerolls	18					33													
Connleyhills	122.1	Vitritorrandic Argixerolls	38						53												
Conser	93.3	Vertic Argiaquolls	69						68												
Coquille	108.5	Fluvaquentic Endoaquepts		15				25													
Cornelius	101.4	Mollic Fragixeralfs		18					15						36						
Cornutt	79.6	Ultic Haploxeralfs		29					76												
Corral	113.0	Xeric Haplargids		10					20												
Cove	83.2	Vertic Endoaquolls	41					84													
Cowsly	269.9	Xerertic Argialbolls	38						104			10									
Coyata	54.6	Humic Dystroxerepts		36				30													
Coztur	117.5	Lithic Xeric Haplargids		28					15												
Crackedground	60.2	Vitritorrandic Haploxerolls	33					64													
Crackercreek	99.7	Alfic Vitrixerands		18					91												X
Crowcamp	108.6	Vertic Palexerolls	58						69												
Cruiser	51.5	Typic Haplocryands		36				79												X	
Crump	70.6	Histic Humaquepts				20													X		
Cullius	79.4	Aridic Lithic Argixerolls	23						28												
Cumley	136.1	Oxyaquic Palehumults			25				79												
Curant	203.6	Calcic Pachic Haploxerolls	84																X		
Curtin	55.5	Aquic Haploxererts	152																X		
Dacker	50.1	Xeric Argidurids		18					46	**53**											
Damewood	180.3	Andic Humudepts			74														X		
Davey	77.0	Xeric Haplocambids		18				23													
Day	53.5	Chromic Haploxererts		8				70													
Dayton	417.9	Vertic Albaqualfs		38					97			15									
Deadline	107.8	Humic Dystrudepts		18				124													
Defenbaugh	58.1	Typic Haplocambids		12				61													
Dehlinger	58.9	Pachic Haploxerolls	69					**66**													
Deppy	308.6	Argidic Argidurids		15					23	15											
Deschutes	245.3	Vitritorrandic Haploxerolls	25					36													
Deseed	64.6	Xeric Haplargids		10					56												

(continued)

Series name	Area (km^2)	Subgroup	Mollic	Ochric	Umbric	Histic	Melanic	Cambic	Argillic	Duripan	Calcic	Albic	Natric	Spodic	Fragipan	Ortstein	Salic	Glossic	None	Andic 1*	Andic 2*
Deskamp	172.4	Vitritorrandic Haploxerolls	25																X		
Dester	365.1	Vitritorrandic Argixerolls	43						43												
Devnot	65.6	Lithic Argixerolls	48						38												
Dewar	117.1	Xeric Argidurids		13					5	109											
Digger	786.8	Dystric Eutrudepts		38				56													
Divers	51.0	Typic Haplocryands		33				84												X	
Dixon	61.7	Xeric Haplocambids		5				41													
Dixonville	266.3	Pachic Ultic Argixerolls	86						56												
Doyn	248.6	Aridic Lithic Haploxerolls	20																X		
Drews	97.8	Pachic Argixerolls	69						41												
Drewsey	258.2	Xeric Haplocambids		7.5				74													
Drinkwater	74.6	Xeric Haplocambids		7.5				71													
Droval	113.2	Sodic Aquicambids		18				71													
Dubakella	187.7	Mollic Haploxeralfs		28					25												
Duff	72.0	Pachic Haplocryolls	69																X		
Dufur	62.0	Calcic Haploxerolls	30					75													
Dumont	179.4	Typic Palexerults		28					218												
Dunres	111.6	Vitrandic Durixerolls	20						38	94											
Dunstan	52.4	Vitrandic Haploxeralfs		10				15	63			13									
Dupee	167.4	Aquultic Haploxeralfs		61					69												
Dupratt	126.5	Vitrandic Argixerolls	53						15												
Durkee	223.4	Calcic Argixerolls	36						30		12										
Edemaps	132.2	Argiduridic Durixerolls	25						46	12											
Eglirim	51.5	Aridic Argixerolls	41						53												
Egyptcreek	111.1	Vitrandic Haploxerolls	20					41													
Eightlar	62.7	Typic Dystroxerepts		25				86													
Eilertsen	59.7	Ultic Hapludalfs			43				51												
Elijah	284.8	Xeric Argidurids		18					36	27											
Encina	148.5	Calciargidic Argixerolls	30						12												
Enko	506.7	Durinodic Xeric Haplocambids		18				36													
Era	106.0	Vitritorrandic Haploxerolls	41					53													
Erakatak	495.1	Vitrandic Argixerolls	16						18												
Etelka	192.0	Oxyaquic Dystrudepts		18				119													
Exfo	201.7	Lithic Torriorthents		18															X		
Farva	218.2	Typic Haploxerepts		31				38													
Felcher	694.2	Xeric Haplocambids		25				31													
Fendall	86.3	Andic Humudepts			33			20													
Fernhaven	77.9	Typic Paleudults		10					107												
Fernwood	71.1	Andic Humudepts			25			38													
Fertaline	84.6	Abruptic Xeric Argidurids		20					31	15		7									
Fitzwater	186.8	Aridic Haploxerolls	25					23													
Fivebeaver	548.7	Lithic Ultic Haploxerolls	35																X		
Fivebit	160.7	Lithic Ultic Haploxerolls	46																X		
Flagstaff	100.9	Typic Aquisalids		18				20				10					79				
Floke	247.1	Abruptic Xeric Argidurids		20					10	12											
Fopiano	121.3	Vitrandic Argixerolls	38						18												
Fordney	124.3	Torripsammentic Haploxerolls	25																X		
Formader	154.4	Alic Hapludands			48			38												X	
Fort rock	97.9	Vitritorrandic Haploxerolls	46																X		
Fourwheel	168.7	Xeric Paleargids		18					38												

(continued)

Series name	Area (km^2)	Subgroup	Mollic	Ochric	Umbric	Histic	Melanic	Cambic	Argillic	Duripan	Calcic	Albic	Natric	Spodic	Fragipan	Ortstein	Salic	Glossic	None	Andic 1*	Andic 2*
Freezener	264.7	Ultic Haploxeralfs		25					118												
Fremkle	86.9	Lithic Vitrixerands	25																X		X
Freznik	344.5	Xeric Paleargids		8					83												
Fritsland	192.3	Humic Dystrudepts		18				79													
Frohman	172.0	Xeric Haplodurids		18						15											
Gaib	169.4	Lithic Ultic Argixerolls	23						23												
Gardone	127.7	Vitritorrandic Haploxerolls	25																X		
Geisercreek	63.3	Alfic Udivitrands		7.5				30	53												X
Geppert	97.0	Typic Dystroxerepts		33				43													
Getaway	194.8	Vitrandic Argixerolls	76						109												
Ginsberg	79.6	Alic Hapludands			46			112												X	
Ginser	63.5	Pachic Haploxerolls	64					20													
Glencabin	57.7	Vitrandic Haploxerolls	64																X		
Goble	261.5	Andic Fragiudepts			36			57							**33**						
Goldrun	87.5	Xeric Torripsamments		18															X		
Goodin	51.2	Ultic Haploxeralfs		23					51												
Gooding	169.6	Vertic Paleargids		18					64		38										
Goodtack	597.0	Vitritorrandic Durixerolls	18						30	69											
Gooserock	62.6	Vitritorrandic Haploxerolls	86																X		
Gosney	235.6	Lithic Torripsamments		20															X		
Gradon	81.1	Argiduridic Durixerolls	25						56	59											
Gravecreek	57.8	Typic Dystroxerepts		18				56													
Greenmountain	54.9	Vitritorrandic Durixerolls	33						28	13											
Greystoke	66.4	Pachic Ultic Argixerolls	58						48												
Gribble	59.9	Haplic Durixerolls	25						69	15											
Grousehill	194.4	Oxyaquic Duricryands		18				74		**43**										X	
Grouslous	50.2	Lithic Dystrudepts		10				30													
Gumble	173.2	Xeric Haplargids		10					36												
Gurdane	190.0	Pachic Argixerolls	76					27	25												
Gustin	69.0	Aquic Palehumults		18					**135**												
Gutridge	216.7	Typic Udivitrands		18				40													X
Gwin	649.5	Lithic Argixerolls	43						30												
Gwinly	990.8	Lithic Argixerolls	41						23												
Haar	279.0	Xeric Torriorthents		15															X		
Hack	55.9	Calcic Argixerolls	36						41												
Hackwood	91.9	Pachic Haplocryolls	53																X		
Hall ranch	107.7	Vitrandic Haploxerolls	43					36													
Hankins	362.0	Vertic Palexerolls	69						112												
Hanning	76.6	Pachic Argixerolls	53						71												
Harcany	119.5	Pachic Haplocryolls	183																X		
Hardtrigger	57.7	Xeric Haplargids		10					41												
Harl	126.3	Typic Udivitrands		5				56													X
Harlow	394.7	Lithic Argixerolls	36						20												
Harriman	61.7	Pachic Argixerolls	107						61												
Harrington	206.5	Typic Humudepts			30			41													
Harslow	190.1	Alic Hapludands			27			15												X	
Hart	119.4	Duric Palexerolls	23						25												
Hayespring	73.6	Vitritorrandic Durixerolls	43						36	51											
Hazelair	175.4	Vertic Haploxerolls	18					18													
Hazelcamp	52.2	Typic Haplohumults			30				61												
Hehe	115.6	Vitrandic Argixerolls	27						69												

(continued)

Series name	Area (km²)	Subgroup	Mollic	Ochric	Umbric	Histic	Melanic	Cambic	Argillic	Duripan	Calcic	Albic	Natric	Spodic	Fragipan	Ortstein	Salic	Glossic	None	Andic 1*	Andic 2*
Helphenstein	85.8	Sodic Aquicambids		18				23													
Helvetia	76.8	Ultic Argixerolls	25						97												
Hembre	205.5	Andic Humudepts			30			51													
Hemcross	403.3	Alic Hapludands			48			74												X	
Henkle	159.0	Lithic Vitrixerands	30																X		X
Henley	63.1	Aquic Haplodurids		27						**61**											
Henline	86.1	Typic Humicryepts			25														X		
Hermiston	67.1	Cumulic Haploxerolls	60																X		
Hesslan	93.8	Typic Haploxerolls	46					36													
Highcamp	86.5	Typic Haplocryands		43				56												X	
Holcomb	83.7	Typic Argialbolls	46						66			15									
Holland	66.6	Ultic Haploxeralfs		23					134												
Hondu	77.2	Andic Haploxerepts		4				18													
Honeygrove	696.8	Typic Palehumults		10					147												
Hot lake	80.7	Aquic Haploxerands	36					12												X	
Houstake	71.2	Vitritorrandic Haploxerolls	38																X		
Howash	163.0	Humic Udivitrands			36			86													X
Humarel	107.1	Vitrandic Argixerolls	30						46												
Hummington	119.7	Typic Haplocryands			79														X	X	
Hurwal	85.5	Vitrandic Argixerolls	68						74												
Hutchley	80.9	Lithic Argixerolls	38						15												
Icene	95.4	Typic Aquisalids		18													46				
Igert	505.6	Durinodic Xeric Haplargids		46					5												
Illahee	59.2	Typic Humudepts			38			81													
Immig	223.5	Typic Argixerolls	30						36												
Imnaha	173.9	Vitrandic Argixerolls	61						18												
Inkler	66.7	Andic Haploxerepts		10				43													
Ironside	78.7	Vitrandic Haploxerolls	33					38													
Jacksplace	82.7	Vitritorrandic Argixerolls	30						27												
Jayar	150.2	Typic Dystroxerepts		7.5				71													
Jett	112.9	Cumulic Haploxerolls	130																X		
Jojo	56.2	Typic Vitricryands		18															X	X	
Jory	872.9	Xeric Palehumults		48					**205**												
Josephine	521.4	Typic Haploxerults		7.5					84												
Kahler	113.1	Vitrandic Haploxerolls	56																X		
Kamela	249.9	Vitrandic Haploxerepts		15				48													
Kanid	249.9	Typic Dystroxerepts		50				105													
Kaskela	63.4	Typic Haploxererts	48					30													
Keating	60.9	Typic Argixerolls	30						25												
Keel	156.6	Typic Haplocryands			38			23												X	
Kerrfield	75.0	Durinodic Xeric Haplocambids		7.5				38													
Ketchly	51.1	Vitrandic Haploxeralfs			28				64												
Kettenbach	83.1	Pachic Argixerolls	53						38												
Kewake	131.5	Vitrandic Torripsamments		17															X		
Kilchis	121.8	Lithic Humudepts			20														X		
Kimberly	75.2	Torrifluventic Haploxerolls	25																X		
Kingbolt	124.6	Typic Vitrixerands		10				75													X
Kinney	787.7	Andic Humudepts			38			64													
Kinton	60.6	Typic Fragixerepts		25				36						**76**							

(continued)

Series name	Area (km^2)	Subgroup	Mollic	Ochric	Umbric	Histic	Melanic	Cambic	Argillic	Duripan	Calcic	Albic	Natric	Spodic	Fragipan	Ortstein	Salic	Glossic	None	Andic 1*	Andic 2*
Kinzel	75.2	Typic Fulvicryands			33			63												X	
Kiona	108.6	Xeric Haplocambids		10				41													
Kirk	107.6	Typic Cryaquands	23																X		X
Kirkendall	54.1	Oxyaquic Humudepts			41			25													
Kishwalk	191.7	Pachic Argixerolls	79						60												
Klamath	96.0	Cumulic Cryaquolls	79					20													
Klicker	1200.7	Vitrandic Argixerolls	45						46												
Klickitat	1019.5	Typic Humudepts			38			36													
Klickson	185.5	Vitrandic Argixerolls	37						97												
Klistan	383.7	Alic Hapludands			52			84												X	
Klootchie	495.3	Typic Fulvudands			30			33												X	
Koehler	95.4	Xeric Haplodurids		10						**1**											
Krackle	59.7	Xeric Haplocryolls	38																X		
Kunaton	56.8	Abruptic Xeric Argidurids		10					23	30											
Kunceider	147.0	Aridic Lithic Haploxerolls	23																X		
Kutcher	68.2	Alfic Udivitrands		46					51												X
La grande	66.9	Pachic Haploxerolls	53					58													
Lackeyshole	110.2	Typic Vitricryands		12				25													X
Laderly	167.8	Alic Hapludands			41			53												X	
Laidlaw	53.2	Humic Vitrixerands	38					76													X
Lakeview	95.4	Cumulic Haploxerolls	51																X		
Laki	81.8	Typic Haploxerolls	48					33													
Lambring	349.2	Pachic Haploxerolls	102																X		
Lamonta	93.7	Abruptic Argiduridic Durixerolls	23						36	**1**											
Lapine	2513.1	Xeric Vitricryands		63															X		X
Larabee	370.6	Vitrandic Argixerolls	25						23												
Larmine	59.0	Lithic Haploxerepts		7.5				41													
Lasere	86.5	Typic Palexerolls	25						33												
Lastcall	70.2	Vitritorrandic Argixerolls	33						46												
Lather	101.1	Limnic Haplohemists				**178**													X		
Latourell	61.6	Ultic Haploxeralfs		41					53												
Laurelwood	155.5	Ultic Haploxeralfs			27				74												
Lawen	131.1	Calciargidic Argixerolls	25						38												
Legler	122.8	Xeric Haplocambids		18				132													
Lettia	140.7	Ultic Haploxeralfs		31					117												
Lickskillet	1936.4	Aridic Lithic Haploxerolls	23					25													
Limberjim	995.5	Alfic Udivitrands		20				51	53												X
Lithgow	179.2	Xeric Haplargids		7.5					36												
Llaorock	100.9	Vitric Haplocryands		36				119												X	
Lobert	76.8	Pachic Haploxerolls	104																X		
Lonely	336.5	Xeric Haplocambids		10				51													
Longbranch	66.5	Pachic Argixerolls	75						38												
Lookingglass	131.0	Xerertic Argialbolls	27						43			25									
Lookout	217.4	Abruptic Xeric Argidurids		23					30	53											
Lorella	479.4	Lithic Argixerolls	25						23												
Lostbasin	132.6	Typic Haploxerepts		33				38													
Loupence	58.3	Cumulic Haploxerolls	71					99													
Mackatie	102.6	Alfic Udivitrands		18					58												X
Madeline	69.9	Aridic Lithic Argixerolls	36						33												

(continued)

Series name	Area (km^2)	Subgroup	Mollic	Ochric	Umbric	Histic	Melanic	Cambic	Argillic	Duripan	Calcic	Albic	Natric	Spodic	Fragipan	Ortstein	Salic	Glossic	None	Andic 1*	Andic 2*
Madras	301.6	Aridic Argixerolls	25						33												
Mahogee	55.0	Lithic Argicryolls	43						38												
Mahoon	152.3	Aridic Palexerolls	23						56												
Maklak	150.6	Xeric Vitricryands		68															X		X
Malabon	167.3	Pachic Ultic Argixerolls	74						76												
Malin	58.2	Fluvaquentic Endoaquolls	36																X		
Mallory	58.3	Pachic Argixerolls	64						41												
Manita	123.0	Mollic Haploxeralfs		20					127												
Marack	129.9	Calciargidic Argixerolls	30						33		76										
Marblepoint	69.7	Andic Haplocryepts		10				45													
Maset	230.2	Alfic Vitrixerands		53					20												X
Maupin	97.1	Haploduridic Durixerolls	25					38		15											
Mayger	72.0	Aquic Palehumults		28					69												
Mcalpin	138.5	Aquic Cumulic Haploxerolls	58					46													
Mcbee	124.6	Aquic Cumulic Haploxerolls	89																X		
Mccartycreek	80.5	Vitrandic Haploxerolls	23					26													
Mcconnel	265.6	Xeric Haplocambids		18				18													
Mccully	292.7	Typic Humudepts			25			119													
Mcduff	167.0	Typic Haplohumults			66				48												
Mckay	66.0	Calcic Argixerolls	27						27												
Mcmullin	286.0	Lithic Ultic Haploxerolls	18																X		
Mcnull	323.4	Ultic Argixerolls	30						66												
Mcwillar	89.8	Alfic Vitrixerands		5				33	41												X
Meadowridge	123.5	Vitritorrandic Argixerolls	38						20												
Meda	76.3	Typic Humudepts			25			56													
Medco	298.3	Ultic Haploxerolls	30					46													
Medford	64.9	Pachic Argixerolls	89						**150**												
Melbourne	129.4	Ultic Palexeralfs		20					**175**												
Melby	137.2	Humic Dystrudepts		18				84													
Melhorn	88.9	Vitrandic Argixerolls	48						**135**												
Menbo	59.9	Vitrandic Argixerolls	66						46												
Merlin	936.7	Lithic Argixerolls	30						15												
Mesman	158.8	Xeric Natrargids		23									28								
Middlebox	96.2	Vitrandic Torriorthents		18															X		
Mikkalo	387.0	Calcidic Haploxerolls	41					28													
Milbury	265.3	Typic Humudepts			46			46													
Millicoma	190.7	Andic Humudepts			46			43													
Minveno	194.0	Xeric Haplodurids		10						18	15										
Moe	63.2	Andic Humudepts			38			117													
Moonbeam	475.4	Vitritorrandic Durixerolls	20						26	23											
Morehouse	231.5	Vitrandic Torripsamments		13															X		
Morfitt	51.2	Xeric Haplargids		12					74												
Morganhills	82.3	Vitrandic Torriorthents		20															X		
Morrow	707.9	Calcic Argixerolls	36						25		30										
Mountemily	227.1	Typic Vitricryands		15				36													X
Mountireland	54.2	Alfic Vitricryands						40	73												X
Mudlakebasin	70.4	Typic Vitricryands		56				89													X
Mudpot	62.0	Chromic Endoaquerts		36				23													
Multnomah	61.9	Humic Dystroxerepts		20				43													

(continued)

Series name	Area (km^2)	Subgroup	Mollic	Ochric	Umbric	Histic	Melanic	Cambic	Argillic	Duripan	Calcic	Albic	Natric	Spodic	Fragipan	Ortstein	Salic	Glossic	None	Andic 1*	Andic 2*
Muni	435.4	Haploxeralfic Argidurids		20					26	15											
Murtip	201.7	Alic Hapludands			27			94												X	
Mutton	51.1	Vitrandic Haploxeralfs		20				69	**41**												
Nailkeg	62.0	Typic Dystrudepts		15				53													
Nansene	110.0	Pachic Haploxerolls	94																X		
Natroy	61.6	Xeric Endoaquerts	66																X		
Necanicum	356.6	Typic Fulvudands			30			58												X	
Nehalem	57.6	Fluventic Humudepts			41			112													
Nekia	603.4	Xeric Haplohumults			61				46												
Nekoma	54.9	Fluventic Humudepts			28			23													
Nestucca	50.1	Fluvaquentic Humaquepts			36			69													
Nevador	1112.8	Durinodic Xeric Haplargids		15					46												
Newberg	270.7	Fluventic Haploxerolls	48																X		
Ninemile	1808.4	Aridic Lithic Argixerolls	20						31												
Ninetysix	80.5	Calcic Haploxerolls	36					61													
Nonpareil	87.4	Typic Dystroxerepts		18				33													
Norad	144.2	Xeric Haplargids		5					51												
Norling	126.4	Ultic Haploxeralfs		25					23												
Nuss	214.5	Lithic Haploxerolls	38																X		
Nyssa	117.8	Xeric Haplodurids		33				30		69											
Oakland	147.4	Ultic Haploxeralfs		25					56												
Oatman	134.0	Typic Haplocryands		18				66												X	
Observation	429.3	Typic Argixerolls	30						48												
Offenbacher	84.9	Typic Haploxerepts		13				76													
Old camp	51.6	Lithic Xeric Haplargids		5					31												
Olex	96.8	Calcidic Haploxerolls	30					30													
Olot	342.3	Typic Vitrixerands		18				61													X
Olyic	227.5	Typic Haplohumults			33				56												
Opie	62.7	Cumulic Endoaquolls	66																X		
Oreneva	122.3	Xeric Haplocambids		18				48													
Orford	304.5	Typic Palehumults		61					91												
Orovada	143.7	Durinodic Xeric Haplocambids		18				31													
Outerkirk	184.1	Durinodic Haplocalcids		18							16										
Owsel	65.7	Durinodic Xeric Haplargids		7.5					43												
Owyhee	76.9	Xeric Haplocalcids		28				28			56										
Oxwall	116.1	Palexerollic Durixerolls	33						13	17											
Ozamis	184.2	Fluvaquentic Endoaquolls	25																X		
Palouse	122.5	Pachic Ultic Haploxerolls	61					**91**													
Panther	59.6	Vertic Epiaquolls	36					56													
Parsnip	97.2	Lithic Argixerolls	23						10												
Pearlwise	60.2	Pachic Haploxerolls	64					25													
Pearsoll	232.5	Lithic Dystroxerepts		12				23													
Peavine	851.7	Typic Haplohumults			25				66												
Pengra	106.3	Vertic Epiaquolls	33					20													
Perdin	88.0	Ultic Haploxeralfs		18					43												
Pernty	347.1	Aridic Lithic Argixerolls	20						28												
Philomath	233.0	Vertic Haploxerolls	48																X		
Piersonte	55.7	Vitrandic Haploxerolls	69					86													
Piline	73.7	Xeric Epiaquerts		18				117													

(continued)

Series name	Area (km²)	Subgroup	Mollic	Ochric	Umbric	Histic	Melanic	Cambic	Argillic	Duripan	Calcic	Albic	Natric	Spodic	Fragipan	Ortstein	Salic	Glossic	None	Andic 1*	Andic 2*
Pilot rock	142.2	Haploxerollic Durixerolls	51					25		46											
Pinehurst	148.1	Pachic Ultic Argixerolls	53						51												
Pinhead	141.5	Vitric Haplocryands		18				30												X	
Pipp	91.7	Humic Vitrixerands	30					99													X
Poall	322.6	Xeric Paleargids		23					41												
Pokegema	317.8	Humic Haploxerands	64					66												X	
Pollard	172.9	Typic Palexerults		18					109												
Poujade	290.8	Durinodic Xeric Natrargids		15									18								
Powder	132.1	Cumulic Haploxerolls	69																X		
Prag	162.5	Pachic Palexerolls	56						66												
Preacher	1470.2	Andic Humudepts			36			74													
Prill	241.2	Pachic Palexerolls	51						74												
Prouty	67.8	Andic Dystrocryepts		20				25													
Quatama	85.3	Aquultic Haploxeralfs		18					86												
Quincy	545.6	Xeric Torripsamments		18															X		
Quirk	110.3	Vitrandic Palexerolls	30						41												
Rabbithills	198.5	Xereptic Haplodurids		13						26											
Ratto	186.6	Xeric Argidurids		15					18	10											
Raz	2279.5	Xeric Haplodurids		18						28											
Reallis	240.6	Durinodic Xeric Haplocambids		18				40													
Redcliff	143.3	Aridic Haploxerolls	36					43													
Redmond	99.6	Vitritorrandic Haploxerolls	33					51													
Reedsport	254.3	Andic Humudepts			81														X		
Reese	165.7	Duric Halaquepts		18															X		
Reluctan	343.0	Aridic Argixerolls	23						46												
Remote	233.8	Typic Dystrudepts		12				102													
Rhea	122.8	Calcic Haploxerolls	35					48													
Riddleranch	114.7	Aridic Haploxerolls	38					41													
Rinconflat	119.6	Xeric Haplocambids		18				64													
Rinearson	402.4	Typic Humudepts			38			48													
Rio king	54.5	Aridic Haploxerolls	51					20													
Risley	199.2	Xeric Haplargids		15					31												
Ritner	291.8	Humic Haploxerepts		38				58													
Ritzville	1133.4	Calcidic Haploxerolls	46					45													
Robson	666.1	Lithic Xeric Haplargids		25					23												
Roca	103.4	Xeric Haplargids		20					41												
Rockly	774.9	Lithic Haploxerolls	23																X		
Rogger	68.1	Ultic Haploxerolls	30					61													
Rogue	50.6	Typic Dystroxerepts		15				71													
Roloff	52.1	Aridic Haploxerolls	20					20													
Roostercomb	178.5	Typic Argixerolls	30						61												
Rosehaven	107.5	Ultic Haploxeralfs		30					**130**												
Royst	259.4	Pachic Argixerolls	61						48												
Ruch	86.0	Mollic Palexeralfs		18					160												
Ruckles	629.7	Aridic Lithic Argixerolls	20						15												
Ruclick	411.1	Aridic Argixerolls	30						48												
Sagehen	78.7	Lithic Xeric Haplocambids		18				23													
Sagehill	255.1	Xeric Haplocalcids		18				28			102										
Salander	153.2	Typic Fulvudands			102			38												X	

(continued)

Series name	Area (km^2)	Subgroup	Mollic	Ochric	Umbric	Histic	Melanic	Cambic	Argillic	Duripan	Calcic	Albic	Natric	Spodic	Fragipan	Ortstein	Salic	Glossic	None	Andic 1*	Andic 2*
Salem	108.1	Pachic Ultic Argixerolls	76						53												
Salhouse	66.6	Vitrandic Torripsamments		18															X		
Salkum	120.5	Xeric Palehumults		36					74												
Santiam	67.7	Aquultic Haploxeralfs		33					43												
Saum	158.5	Ultic Palexeralfs			33			33	84												
Sauvie	100.3	Fluvaquentic Endoaquolls	46					52													
Scaponia	173.6	Humic Dystrudepts		18				64													
Searles	117.9	Aridic Argixerolls	46						44												
Segundo	58.8	Typic Haploxerepts		12				41													
Seharney	122.8	Xereptic Haplodurids		18				30		18											
Senra	171.9	Vitritorrandic Durixerolls	25						23	33											
Shanahan	332.6	Xeric Vitricryands		10															X		X
Shano	207.2	Xeric Haplocambids		18				28													
Sharesnout	242.7	Typic Argixerolls	38						31												
Sharpshooter	61.1	Ultic Haploxerolls	25					89													
Shefflein	56.8	Mollic Haploxeralfs		10					117												
Shukash	603.0	Xeric Vitricryands		12															X		X
Sidlake	120.9	Xeric Haplargids		20					41												
Silverash	63.9	Aquandic Palexeralfs		20					33												
Silvies	87.4	Vertic Cryaquolls	71																X		
Simas	772.5	Vertic Palexerolls	36						41												
Simnasho	110.8	Alfic Vitrixerands		18				20	23												X
Sinker	100.1	Pachic Haploxerolls	58					23													
Siskiyou	94.2	Typic Dystroxerepts		18				38													
Skedaddle	184.8	Lithic Xeric Torriorthents		13															X		
Skidoosprings	104.3	Duric Halaquepts		27				30		20											
Skipanon	215.3	Andic Humudepts			48			43													
Skookumhouse	62.1	Typic Haplohumults			28				104												
Skullgulch	57.8	Pachic Palexerolls	61						**91**												
Skunkfarm	71.7	Typic Endoaquolls	33																X		
Skyline	99.7	Typic Haploxerolls	36																X		
Slickrock	380.3	Alic Hapludands			122			102												X	
Smiling	236.4	Alfic Vitrixerands		43					119												X
Snaker	62.5	Lithic Xeric Torriorthents		8															X		
Snell	335.9	Pachic Argixerolls	61						38												
Snellby	50.6	Aridic Argixerolls	25						20												
Snowmore	1381.0	Xeric Argidurids		10					23	15											
Sorf	118.0	Vertic Paleargids		12					18												
Soughe	86.6	Lithic Xeric Haplargids		10					26												
Spangenburg	470.6	Xeric Paleargids		5					81												
Speaker	469.8	Ultic Haploxeralfs		35					43												
Stampede	74.3	Vertic Durixerolls	24						41	13											
Starkey	99.0	Typic Argixerolls	30						15												
Statz	71.4	Vitritorrandic Durixerolls	23							27											
Steiger	537.9	Xeric Vitricryands		53															X		X
Steiwer	74.5	Ultic Haploxerolls	48					20													
Stookmoor	213.0	Vitritorrandic Haploxerolls	36																X		
Straight	81.3	Typic Dystroxerepts		18				51													
Stukel	260.0	Aridic Lithic Haploxerolls	18																X		

(continued)

Series name	Area (km^2)	Subgroup	Mollic	Ochric	Umbric	Histic	Melanic	Cambic	Argillic	Duripan	Calcic	Albic	Natric	Spodic	Fragipan	Ortstein	Salic	Glossic	None	Andic 1*	Andic 2*
Suckerflat	161.0	Aridic Lithic Haploxerolls	33																X		
Sutherlin	112.9	Ultic Haploxeralfs		40					36												
Svensen	119.1	Andic Humudepts			43			53													
Swaler	90.7	Xeric Paleargids		18					41												
Swalesilver	204.7	Aquic Palexeralfs		15					43												
Syrupcreek	787.9	Alfic Udivitrands		10				41	18												X
Tallowbox	107.4	Typic Haploxerepts		33				43													
Tamara	144.9	Alfic Udivitrands		18				41	81												X
Tamarackcanyon	65.1	Vitrandic Haploxeralfs		10				23	64												
Tatouche	78.3	Typic Argixerolls	28						124												
Taunton	60.0	Xeric Haplodurids		13				33		**1**	15										
Teewee	63.4	Vitrandic Argixerolls	51						106												
Teguro	240.0	Lithic Argixerolls	25						26												
Templeton	666.9	Andic Humudepts			41			66													
Tenmile	131.6	Xeric Haplargids		25					69												
Thenarrows	63.6	Typic Halaquepts		18															X		
Thornlake	156.3	Sodic Xeric Haplocambids		18				46													
Threebuck	76.0	Alfic Vitrixerands		10				84	84												X
Timbercrater	102.1	Typic Vitricryands		7.5															X		X
Tolany	65.0	Alic Hapludands		43				**109**												X	
Tolke	118.4	Alic Hapludands		25				**130**												X	
Tolo	608.0	Alfic Vitrixerands		2.5				84	79												X
Tolovana	220.3	Typic Fulvudands			81														X	X	
Tonor	76.7	Sodic Xeric Haplocambids		18				20													
Top	83.9	Vertic Argixerolls	53						**117**												
Troutmeadows	290.8	Typic Vitricryands		5				33													X
Tub	629.6	Vertic Argixerolls	74						43												
Tulana	107.6	Aquandic Humaquepts	58																X		
Tumtum	312.2	Typic Argidurids		5					25	34											
Turpin	203.0	Sodic Xeric Haplocambids		18				38													
Tutni	83.4	Typic Cryaquands		18															X		X
Tweener	93.6	Lithic Argixerolls	20						7												
Twelvemile	54.0	Typic Vitrixerands		27				66													X
Ukiah	84.4	Vertic Argixerolls	71						66												
Umapine	151.5	Typic Halaquepts		23															X		
Umatilla	139.7	Vitrandic Haploxerolls	70					75													
Umpcoos	646.2	Lithic Eutrudepts		7.5				33													
Unionpeak	145.6	Typic Duricryands		20						38											X
Valby	464.4	Calcic Haploxerolls	36					28													
Valsetz	155.8	Alic Haplocryands		10				51												X	
Vandamine	68.5	Andic Haplocryepts		12				99													
Vannoy	425.5	Mollic Haploxeralfs		12					68												
Vanwyper	61.8	Xeric Haplargids		20					79												
Venator	135.4	Lithic Haploxerolls	25					15													
Veneta	59.7	Ultic Haploxeralfs		36					84												
Vergas	249.1	Durinodic Xeric Haplargids		7.5					38												
Vermisa	542.7	Lithic Dystroxerepts		10				33													
Vernonia	90.4	Ultic Hapludalfs		55					79												
Virtue	225.9	Xeric Argidurids		38					25	33											

(continued)

Series name	Area (km^2)	Subgroup	Mollic	Ochric	Umbric	Histic	Melanic	Cambic	Argillic	Duripan	Calcic	Albic	Natric	Spodic	Fragipan	Ortstein	Salic	Glossic	None	Andic 1*	Andic 2*
Vitale	270.4	Typic Argixerolls	38						43												
Voltage	60.9	Xeric Haplocalcids		10							86										
Voorhies	92.5	Mollic Haploxeralfs		23					61												
Waha	504.8	Pachic Argixerolls	71						38												
Wahstal	68.2	Palexerollic Durixerolls	30						15	12											
Waldo	160.9	Fluvaquentic Vertic Endoaquolls	38					56													
Waldport	91.9	Typic Udipsamments		13															X		
Walla walla	1270.0	Typic Haploxerolls	46					68													
Wallowa	83.3	Vitrandic Haploxerolls	56					18													
Wamic	247.4	Vitrandic Haploxerepts		18				53													
Wanoga	332.1	Humic Vitrixerands	30					30													X
Wapato	164.2	Fluvaquentic Endoaquolls	41					110													
Warden	280.5	Xeric Haplocambids		18				33													
Watama	261.5	Pachic Haploxerolls	61					25													
Waterbury	170.2	Lithic Argixerolls	43						20												
Wato	54.5	Typic Haploxerolls	33					**119**													
Wegert	190.5	Vitritorrandic Haploxerolls	51					18													
Weglike	81.0	Vitritorrandic Haploxerolls	30					28													
Welch	61.5	Cumulic Endoaquolls	71																X		
Westbutte	642.6	Pachic Haploxerolls	61																X		
Whetstone	53.7	Typic Haplocryods		12								2.5		46							
Whobrey	95.9	Aquertic Eutrudepts		18				23													
Widowspring	98.6	Cumulic Haploxerolls	109																X		
Wieland	59.3	Durinodic Xeric Haplargids		5					127												
Wilhoit	75.9	Andic Humudepts			84			25													
Willakenzie	157.4	Ultic Haploxeralfs		27					53												
Willamette	155.1	Pachic Ultic Argixerolls	61						51												
Willis	148.8	Haploduridic Durixerolls	20					16		74											
Willowdale	62.8	Cumulic Haploxerolls	102																X		
Winchester	58.1	Xeric Torripsamments		20															X		
Windego	70.4	Alfic Vitrixerands		50					**104**												X
Windygap	313.2	Xeric Haplohumults		18					122												
Wingville	78.4	Pachic Haploxerolls	84																X		
Winterim	117.9	Pachic Argixerolls	53						84												
Witzel	113.3	Lithic Ultic Haploxerolls	43																X		
Wolfpeak	56.3	Ultic Palexeralfs		18					**124**												
Woodburn	911.3	Aquultic Argixerolls	43						94												
Woodchopper	77.0	Pachic Ultic Argixerolls	64						117												
Woodcock	576.8	Alfic humic Haploxerands	51						48											X	
Woodseye	77.5	Humic Lithic Dystroxerepts			36														X		
Wrentham	333.9	Pachic Haploxerolls	81																X		
Xanadu	72.8	Typic Palehumults			20				**132**												
Yancy	52.4	Palexerollic Durixerolls	30						30	**117**											
Yankeewell	169.7	Xeric Natridurids		15						36		7	13								
Yawhee	144.3	Alfic Udivitrands	33						58												X
Yawkey	84.3	Vertic Palexerolls	76						89												
Yellowstone	81.2	Lithic Haplocryands			30														X	X	
Zevadez	200.2	Durinodic Xeric Haplargids		13					28												
Zing	64.1	Aquultic Haploxeralfs		19					**135**												
Zygore	221.9	Andic Humudepts			48																

*Andic 1 represents soils with andic properties that develop in humid climates and have abundant soil organic carbon but without the influence of volcanic glass
Andic 2 represents soil with andic properties that are influenced by volcanic glass

Appendix C
Area and Taxonomy of Soil Series in Oregon

T. Thorson et al., *The Soils of Oregon*, World Soils Book Series,
https://doi.org/10.1007/978-3-030-90091-5

Series name	Area (km^2)	Order	Suborder	Great group	Subgroup	Particle-size class	Mineralogy class	CEC activity class	Reaction class	Soil temp. regime	Soil moisture regime	Other family
Abegg	78.4	Alfisols	Xeralfs	Haploxeralfs	Ultic Haploxeralfs	Loamy-skeletal	Mixed	Superactive		Mesic	Xeric	
Abert	144.9	Aridisols	Cambids	Haplocambids	Sodic Xeric Haplocambids	Ashy	Glassy			Frigid	Aridic	
Abin	6.7	Mollisols	Xerolls	Haploxerolls	Cumulic Haploxerolls	Fine	Mixed	Superactive		Mesic	Xeric	
Abiqua	57.3	Mollisols	Xerolls	Haploxerolls	Cumulic Ultic Haploxerolls	Fine	Mixed	Superactive		Mesic	Xeric	
Absaquil	54.0	Ultisols	Humults	Haplohumults	Typic Haplohumults	Fine	Mixed	Active		Mesic	Udic	
Acanod	6.6	Inceptisols	Udepts	Humudepts	Oxyaquic Humudepts	Fine	Isotic			Mesic	Udic	
Acker	217.6	Ultisols	Xerults	Palexerults	Typic Palexerults	Fine-loamy	Mixed	Superactive		Mesic	Xeric	
Actem	749.7	Aridisols	Durids	Argidurids	Xeric Argidurids	Clayey	Smectitic			Frigid	Aridic	Shallow
Ada	84.9	Mollisols	Xerolls	Argixerolls	Typic Argixerolls	Clayey-skeletal	Smectitic			Mesic	Xeric	
Adieux	0.0	Mollisols	Xerolls	Argixerolls	Pachic Argixerolls	Fine-loamy	Mixed	Superactive		Mesic	Xeric	
Adkins	107.6	Aridisols	Calcids	Haplocalcids	Xeric Haplocalcids	Coarse-loamy	Mixed	Superactive		Mesic	Aridic	
Agate	50.3	Inceptisols	Xerepts	Durixerepts	Typic Durixerepts	Fine-loamy	Mixed	Superactive		Mesic	Xeric	
Agency	135.9	Mollisols	Xerolls	Haploxerolls	Aridic Haploxerolls	Fine-loamy	Mixed	Superactive		Mesic	Aridic	
Agness	2.0	Inceptisols	Udepts	Humudepts	Pachic Humudepts	Fine-loamy	Mixed	Active		Mesic	Udic	
Ahtanum	2.8	Mollisols	Aquolls	Duraquolls	Typic Duraquolls	Coarse-silty	Mixed	Superactive	Calcareous	Mesic	Aquic	
Akerite	18.4	Andisols	Xerands	Vitrixerands	Aquic Vitrixerands	Ashy over loamy	Glassy over isotic			Frigid	Xeric	
Albee	123.7	Mollisols	Xerolls	Haploxerolls	Vitrandic Haploxerolls	Fine-loamy	Mixed	Superactive		Frigid	Xeric	
Alcot	0.4	Andisols	Xerands	Vitrixerands	Typic Vitrixerands	Ashy-pumiceous	Amorphic			Mesic	Xeric	
Alding	63.1	Mollisols	Xerolls	Argixerolls	Lithic Argixerolls	Clayey	Smectitic			Frigid	Xeric	
Algoma	29.4	Mollisols	Aquolls	Endoaquolls	Aquandic Endoaquolls	Ashy over sandy or sandy-skeletal	Glassy over mixed		Calcareous	Mesic	Aquic	
Alicel	39.6	Mollisols	Xerolls	Haploxerolls	Pachic Haploxerolls	Fine-loamy	Mixed	Superactive		Mesic	Xeric	
Alley	2.1	Aridisols	Argids	Haplargids	Durinodic Xeric Haplargids	Fine-loamy	Mixed	Superactive		Mesic	Aridic	
Allingham	18.8	Andisols	Xerands	Vitrixerands	Alfic Vitrixerands	Ashy over medial-skeletal	Glassy over amorphic			Frigid	Xeric	
Almota	1.2	Mollisols	Xerolls	Haploxerolls	Calcic Haploxerolls	Fine-loamy	Mixed	Superactive		Mesic	Xeric	
Aloha	236.5	Inceptisols	Xerepts	Haploxerepts	Aquic Haploxerepts	Fine-silty	Mixed	Superactive		Mesic	Xeric	
Als	8.1	Entisols	Psamments	Torripsamments	Typic Torripsamments	Sandy	Mixed			Mesic	Aridic	
Alsea	6.0	Mollisols	Xerolls	Haploxerolls	Cumulic Ultic Haploxerolls	Fine-loamy	Mixed	Superactive		Mesic	Xeric	
Alspaugh	130.6	Ultisols	Udults	Paleudults	Typic Paleudults	Fine	Mixed	Active		Mesic	Udic	
Alstony	82.6	Andisols	Udands	Hapludands	Alic Hapludands	Medial-skeletal	Ferrihydritic			Mesic	Udic	
Althouse	59.9	Inceptisols	Xerepts	Dystroxerepts	Typic Dystroxerepts	Loamy-skeletal	Mixed	Superactive		Frigid	Xeric	
Alvodest	263.4	Aridisols	Cambids	Aquicambids	Sodic Aquicambids	Fine	Smectitic			Mesic	Aridic	
Alyan	27.8	Mollisols	Xerolls	Argixerolls	Aridic Argixerolls	Fine	Smectitic			Frigid	Aridic	

(continued)

Series name	Area (km^2)	Order	Suborder	Great group	Subgroup	Particle-size class	Mineralogy class	CEC activity class	Reaction class	Soil temp. regime	Soil moisture regime	Other family
Amity	434.7	Mollisols	Albolls	Argialbolls	Argiaquic Xeric Argialbolls	Fine-silty	Mixed	Superactive		Mesic	Aquic	
Analulu	265.0	Inceptisols	Xerepts	Haploxerepts	Vitrandic Haploxerepts	Loamy-skeletal	Isotic			Frigid	Xeric	
Anatone	1589.5	Mollisols	Xerolls	Haploxerolls	Lithic Haploxerolls	Loamy-skeletal	Mixed	Superactive		Frigid	Xeric	
Anawalt	1280.1	Aridisols	Argids	Haplargids	Lithic Xeric Haplargids	Clayey	Smectitic			Frigid	Aridic	
Anderly	185.2	Mollisols	Xerolls	Haploxerolls	Typic Haploxerolls	Coarse-silty	Mixed	Superactive		Mesic	Xeric	
Anders	4.9	Mollisols	Xerolls	Haploxerolls	Typic Haploxerolls	Coarse-loamy	Mixed	Superactive		Mesic	Xeric	
Angelbasin	78.8	Inceptisols	Cryepts	Dystrocryepts	Andic Dystrocryepts	Loamy-skeletal	Isotic			Cryic	Udic	
Angelpeak	127.1	Andisols	Cryands	Vitricryands	Typic Vitricryands	Ashy over loamy-skeletal	Amorphic over isotic			Cryic	Udic	
Anniecreek	3.4	Mollisols	Cryolls	Haplocryolls	Vitrandic Haplocryolls	Ashy	Glassy			Cryic	Udic	
Antelopepeak	7.9	Andisols	Xerands	Vitrixerands	Typic Vitrixerands	Ashy over loamy	Amorphic over isotic			Frigid	Xeric	
Antoken	33.3	Mollisols	Xerolls	Palexerolls	Aridic Palexerolls	Clayey-skeletal	Smectitic			Mesic	Xeric	
Anunde	32.4	Andisols	Udands	Hapludands	Alic Hapludands	Medial	Ferrihydritic			Mesic	Udic	
Applegate	13.1	Mollisols	Xerolls	Argixerolls	Ultic Argixerolls	Fine	Smectitic			Mesic	Xeric	
Apt	148.9	Ultisols	Humults	Haplohumults	Typic Haplohumults	Fine	Isotic			Mesic	Udic	
Arbidge	8.0	Aridisols	Durids	Argidurids	Xeric Argidurids	Fine-loamy	Mixed	Superactive		Mesic	Aridic	
Arcia	92.6	Mollisols	Xerolls	Argixerolls	Vitrandic Argixerolls	Fine	Smectitic			Frigid	Xeric	
Arness	8.6	Mollisols	Xerolls	Durixerolls	Argiduridic Durixerolls	Loamy	Mixed	Superactive		Frigid	Aridic	Shallow
Ascar	120.9	Andisols	Udands	Fulvudands	Typic Fulvudands	Medial-skeletal	Ferrihydritic			Isomesic	Udic	
Aschoff	311.0	Inceptisols	Udepts	Humudepts	Andic Humudepts	Loamy-skeletal	Isotic			Mesic	Udic	
Aspenlake	3.5	Mollisols	Cryolls	Duricryolls	Typic Duricryolls	Coarse-loamy	Mixed	Active		Cryic	Udic	
Astoria	145.9	Inceptisols	Udepts	Humudepts	Andic Humudepts	Fine	Isotic			Mesic	Udic	
Ateron	1288.7	Mollisols	Xerolls	Argixerolls	Lithic Argixerolls	Clayey-skeletal	Smectitic			Frigid	Xeric	
Athena	210.4	Mollisols	Xerolls	Haploxerolls	Pachic Haploxerolls	Fine-silty	Mixed	Superactive		Mesic	Xeric	
Atlow	357.3	Aridisols	Argids	Haplargids	Lithic Xeric Haplargids	Loamy-skeletal	Mixed	Superactive		Mesic	Aridic	
Atring	480.2	Inceptisols	Xerepts	Dystroxerepts	Typic Dystroxerepts	Loamy-skeletal	Mixed	Superactive		Mesic	Xeric	
Ausmus	235.3	Aridisols	Argids	Natrargids	Aquic Natrargids	Fine-silty	Mixed	Superactive		Frigid	Aridic	
Aval	9.5	Aridisols	Cambids	Haplocambids	Lithic Xeric Haplocambids	Ashy	Glassy			Frigid	Aridic	
Averlande	40.3	Ultisols	Udults	Hapludults	Lithic Hapludults	Loamy-skeletal	Isotic			Mesic	Udic	
Awbrig	45.7	Alfisols	Aqualfs	Albaqualfs	Vertic Albaqualfs	Fine	Smectitic			Mesic	Aquic	
Axford	4.2	Mollisols	Xerolls	Argixerolls	Calciargidic Argixerolls	Fine-loamy	Mixed	Superactive		Mesic	Aridic	
Ayres	86.6	Mollisols	Xerolls	Durixerolls	Argiduridic Durixerolls	Loamy-skeletal	Mixed	Superactive		Mesic	Aridic	Shallow

(continued)

Series name	Area (km^2)	Order	Suborder	Great group	Subgroup	Particle-size class	Mineralogy class	CEC activity class	Reaction class	Soil temp. regime	Soil moisture regime	Other family
Ayresbutte	40.0	Mollisols	Xerolls	Durixerolls	Vitritorrandic Durixerolls	Loamy-skeletal	Mixed	Superactive		Mesic	Aridic	
Babbington	0.8	Mollisols	Xerolls	Argixerolls	Calciargidic Argixerolls	Fine-loamy	Mixed	Superactive		Mesic	Aridic	
Bacona	212.6	Ultisols	Humults	Palehumults	Typic Palehumults	Fine-silty	Mixed	Active		Mesic	Udic	
Baconcamp	373.2	Mollisols	Cryolls	Haplocryolls	Pachic Haplocryolls	Loamy-skeletal	Mixed	Superactive		Cryic	Xeric	
Bagness	0.8	Inceptisols	Udepts	Humudepts	Cumulic Humudepts	Fine-loamy	Mixed	Superactive		Isomesic	Udic	
Bakeoven	1225.0	Mollisols	Xerolls	Haploxerolls	Aridic Lithic Haploxerolls	Loamy-skeletal	Mixed	Superactive		Mesic	Aridic	
Baker	77.9	Mollisols	Xerolls	Durixerolls	Haploduridic Durixerolls	Coarse-loamy	Mixed	Superactive		Mesic	Xeric	
Bald	42.7	Mollisols	Xerolls	Haploxerolls	Ultic Haploxerolls	Loamy-skeletal	Mixed	Superactive		Mesic	Xeric	
Balder	18.7	Mollisols	Xerolls	Haploxerolls	Vitrandic Haploxerolls	Loamy	Mixed	Superactive		Mesic	Xeric	Shallow
Baldock	46.4	Mollisols	Aquolls	Calciaquolls	Typic Calciaquolls	Fine-loamy	Mixed	Superactive		Mesic	Aquic	
Baldridge	17.9	Mollisols	Xerolls	Haploxerolls	Pachic Haploxerolls	Loamy-skeletal	Mixed	Superactive		Frigid	Xeric	
Balloontree	3.6	Andisols	Cryands	Vitricryands	Aquic Vitricryands	Ashy over loamy	Amorphic over isotic			Cryic	Udic	
Balm	12.8	Mollisols	Xerolls	Haploxerolls	Fluvaquentic Haploxerolls	Coarse-loamy over sandy or sandy-skeletal	Mixed	Superactive		Mesic	Xeric	
Bandarrow	17.8	Mollisols	Aquolls	Cryaquolls	Typic Cryaquolls	Coarse-loamy	Mixed	Superactive		Cryic	Aquic	
Bandon	46.2	Spodosols	Orthods	Haplorthods	Typic Haplorthods	Coarse-loamy	Isotic			Isomesic	Udic	Ortstein
Banning	23.7	Mollisols	Xerolls	Argixerolls	Pachic Argixerolls	Fine-loamy	Mixed	Superactive		Mesic	Xeric	
Barbermill	16.5	Mollisols	Xerolls	Argixerolls	Aridic Argixerolls	Clayey	Smectitic			Mesic	Aridic	Shallow
Barhiskey	7.6	Inceptisols	Xerepts	Dystroxerepts	Vitrandic Dystroxerepts	Sandy	Mixed			Mesic	Xeric	
Barkley	4.1	Mollisols	Xerolls	Argixerolls	Duric Argixerolls	Fine-loamy	Mixed	Superactive		Frigid	Xeric	
Barkshanty	69.7	Ultisols	Humults	Palehumults	Typic Palehumults	Loamy-skeletal	Mixed	Active		Mesic	Udic	
Barnard	43.1	Mollisols	Xerolls	Durixerolls	Argiduridic Durixerolls	Fine	Smectitic			Mesic	Aridic	
Barneycreek	4.9	Mollisols	Xerolls	Argixerolls	Vitrandic Argixerolls	Loamy-skeletal	Isotic			Frigid	Xeric	
Barron	24.0	Inceptisols	Xerepts	Haploxerepts	Typic Haploxerepts	Coarse-loamy	Mixed	Superactive		Mesic	Xeric	
Bashaw	171.9	Vertisols	Aquerts	Endoaquerts	Xeric Endoaquerts	Very-fine	Smectitic			Mesic	Aquic	
Bata	5.7	Alfisols	Cryalfs	Glossocryalfs	Andic Glossocryalfs	Loamy-skeletal	Mixed	Superactive		Cryic	Udic	
Bateman	204.6	Alfisols	Xeralfs	Palexeralfs	Ultic Palexeralfs	Fine	Mixed	Active		Mesic	Xeric	
Bayside	0.2	Entisols	Aquents	Fluvaquents	Aeric Fluvaquents	Fine	Mixed	Superactive	Nonacid	Isomesic	Aquic	
Beal	36.7	Alfisols	Xeralfs	Haploxeralfs	Ultic Haploxeralfs	Fine-loamy	Mixed	Superactive		Mesic	Xeric	
Bearcamp	19.1	Inceptisols	Xerepts	Humixerepts	Typic Humixerepts	Loamy-skeletal	Mixed	Superactive		Frigid	Xeric	
Bearpawmeadow	58.3	Inceptisols	Cryepts	Haplocryepts	Andic Haplocryepts	Loamy-skeletal	Isotic			Cryic	Udic	
Bearspring	1.7	Mollisols	Xerolls	Haploxerolls	Vitrandic Haploxerolls	Loamy-skeletal	Isotic			Frigid	Xeric	

(continued)

Series name	Area (km^2)	Order	Suborder	Great group	Subgroup	Particle-size class	Mineralogy class	CEC activity class	Reaction class	Soil temp. regime	Soil moisture regime	Other family
Beden	106.5	Mollisols	Xerolls	Argixerolls	Aridic Lithic Argixerolls	Loamy	Mixed	Superactive		Frigid	Aridic	
Bedner	7.0	Mollisols	Xerolls	Durixerolls	Haplic Durixerolls	Fine	Smectitic			Mesic	Xeric	
Beekman	647.3	Inceptisols	Xerepts	Dystroxerepts	Typic Dystroxerepts	Loamy-skeletal	Mixed	Superactive		Mesic	Xeric	
Beeman	3.0	Aridisols	Cambids	Haplocambids	Xerertic Haplocambids	Fine	Smectitic			Mesic	Aridic	
Beetville	0.3	Mollisols	Xerolls	Haploxerolls	Torrifluventic Haploxerolls	Coarse-loamy	Mixed	Superactive		Mesic	Aridic	
Bellpine	769.6	Ultisols	Humults	Haplohumults	Xeric Haplohumults	Fine	Mixed	Active		Mesic	Xeric	
Belrick	35.9	Andisols	Cryands	Vitricryands	Humic Vitricryands	Ashy	Amorphic			Cryic	Udic	
Benderly	11.9	Mollisols	Xerolls	Haploxerolls	Entic Haploxerolls	Sandy-skeletal	Mixed			Mesic	Xeric	
Bennettcreek	107.4	Alfisols	Xeralfs	Haploxeralfs	Vitrandic Haploxeralfs	Loamy-skeletal	Isotic			Frigid	Xeric	
Bensley	24.1	Inceptisols	Cryepts	Dystrocryepts	Typic Dystrocryepts	Loamy-skeletal	Isotic			Cryic	Udic	
Bentilla	5.4	Ultisols	Humults	Palehumults	Typic Palehumults	Fine	Mixed	Superactive		Isomesic	Udic	
Beoska	55.9	Aridisols	Argids	Natrargids	Durinodic Natrargids	Fine-loamy	Mixed	Superactive		Mesic	Aridic	
Berdugo	114.7	Aridisols	Argids	Paleargids	Xeric Paleargids	Fine	Smectitic			Mesic	Aridic	
Bergsvik	4.5	Histosols	Hemists	Haplohemists	Terric Haplohemists	Sandy or sandy-skeletal	Mixed		Dysic	Isomesic	Aquic	
Bickford	1.0	Mollisols	Aquolls	Epiaquolls	Typic Epiaquolls	Fine-silty over clayey	Mixed over smectitic	Superactive		Mesic	Aquic	
Bicondoa	5.8	Mollisols	Aquolls	Endoaquolls	Fluvaquentic Vertic Endoaquolls	Fine	Smectitic		Calcareous	Frigid	Aquic	
Bigbouldercreek	0.0	Andisols	Vitrands	Udivitrands	Typic Udivitrands	Ashy	Amorphic			Frigid	Udic	
Bigcow	47.2	Inceptisols	Xerepts	Haploxerepts	Andic Haploxerepts	Loamy-skeletal	Isotic			Frigid	Xeric	
Bigdutch	3.1	Inceptisols	Udepts	Dystrudepts	Humic Dystrudepts	Loamy-skeletal	Isotic			Frigid	Udic	
Bigelk	35.4	Mollisols	Xerolls	Haploxerolls	Vitrandic Haploxerolls	Loamy-skeletal	Isotic			Frigid	Xeric	
Bigelow	12.4	Inceptisols	Cryepts	Humicryepts	Typic Humicryepts	Loamy-skeletal	Isotic			Cryic	Udic	
Bigfrog	11.9	Aridisols	Durids	Argidurids	Xeric Argidurids	Loamy	Mixed	Superactive		Mesic	Aridic	Shallow
Bigriver	0.7	Entisols	Fluvents	Udifluvents	Typic Udifluvents	Coarse-loamy	Mixed	Superactive	Nonacid	Isomesic	Udic	
Bindle	34.8	Inceptisols	Xerepts	Humixerepts	Vitrandic Humixerepts	Loamy-skeletal	Isotic			Frigid	Xeric	
Bingville	68.8	Mollisols	Xerolls	Palexerolls	Pachic Palexerolls	Clayey-skeletal	Smectitic			Frigid	Xeric	
Bins	47.6	Inceptisols	Xerepts	Humixerepts	Vitrandic Humixerepts	Fine-loamy	Isotic			Frigid	Xeric	
Bittercreek	1.8	Mollisols	Aquolls	Endoaquolls	Aquandic Endoaquolls	Coarse-loamy over sandy or sandy-skeletal	Isotic over mixed			Frigid	Aquic	
Blachly	510.1	Inceptisols	Udepts	Dystrudepts	Humic Dystrudepts	Fine	Isotic			Mesic	Udic	
Blackgulch	17.3	Mollisols	Xerolls	Haploxerolls	Lithic Ultic Haploxerolls	Loamy-skeletal	Magnesic			Frigid	Xeric	
Blackhills	2.5	Mollisols	Xerolls	Haploxerolls	Aridic Lithic Haploxerolls	Ashy-skeletal	Glassy			Mesic	Aridic	

(continued)

Series name	Area (km^2)	Order	Suborder	Great group	Subgroup	Particle-size class	Mineralogy class	CEC activity class	Reaction class	Soil temp. regime	Soil moisture regime	Other family
Blacklock	21.5	Spodosols	Aquods	Duraquods	Typic Duraquods	Sandy	Mixed			Isomesic	Aquic	Ortstein, shallow
Blalock	14.0	Mollisols	Xerolls	Durixerolls	Haploduridic Durixerolls	Loamy	Mixed	Superactive		Mesic	Xeric	Shallow
Blayden	44.8	Mollisols	Xerolls	Durixerolls	Argiduridic Durixerolls	Loamy	Mixed	Superactive		Frigid	Aridic	Shallow
Bler	12.6	Mollisols	Xerolls	Palexerolls	Vitrandic Palexerolls	Clayey-skeletal	Smectitic			Frigid	Xeric	
Blizzard	17.6	Mollisols	Cryolls	Argicryolls	Lithic Argicryolls	Clayey	Smectitic			Cryic	Xeric	
Blodgett	3.3	Inceptisols	Udepts	Dystrudepts	Typic Dystrudepts	Loamy-skeletal	Isotic			Frigid	Udic	Shallow
Bluecanyon	67.0	Mollisols	Xerolls	Haploxerolls	Lithic Haploxerolls	Loamy-skeletal	Mixed	Superactive		Frigid	Xeric	
Bluesters	9.4	Andisols	Xerands	Vitrixerands	Humic Vitrixerands	Ashy over pumiceous or cindery	Glassy			Frigid	Xeric	
Bly	60.8	Mollisols	Xerolls	Argixerolls	Vitrandic Argixerolls	Fine-loamy	Isotic			Frigid	Xeric	
Boardflower	8.3	Alfisols	Xeralfs	Haploxeralfs	Vitrandic Haploxeralfs	Fine	Smectitic			Mesic	Xeric	
Boardtree	112.2	Andisols	Xerands	Vitrixerands	Alfic Vitrixerands	Ashy over clayey	Glassy over smectitic			Frigid	Xeric	
Bobbitt	16.5	Mollisols	Xerolls	Argixerolls	Vitrandic Argixerolls	Loamy-skeletal	Isotic			Mesic	Xeric	
Bobsgarden	25.6	Inceptisols	Udepts	Dystrudepts	Humic Dystrudepts	Loamy-skeletal	Isotic			Frigid	Udic	
Bocker	1473.9	Mollisols	Xerolls	Haploxerolls	Lithic Haploxerolls	Loamy-skeletal	Mixed	Superactive		Frigid	Xeric	
Bodale	2.2	Mollisols	Cryolls	Haplocryolls	Cumulic Haplocryolls	Coarse-loamy	Mixed	Superactive		Cryic	Udic	
Bodell	83.9	Mollisols	Xerolls	Haploxerolls	Lithic Haploxerolls	Loamy-skeletal	Mixed	Superactive		Mesic	Xeric	
Bogus	6.5	Mollisols	Xerolls	Argixerolls	Pachic Ultic Argixerolls	Fine	Smectitic			Mesic	Xeric	
Bohannon	2045.6	Inceptisols	Udepts	Humudepts	Andic Humudepts	Fine-loamy	Isotic			Mesic	Udic	
Boiler	5.8	Mollisols	Xerolls	Palexerolls	Ultic Palexerolls	Clayey-skeletal	Smectitic			Frigid	Xeric	
Boilout	52.8	Aridisols	Durids	Argidurids	Vitrixerandic Argidurids	Ashy	Glassy			Mesic	Aridic	Shallow
Bolobin	66.3	Mollisols	Xerolls	Argixerolls	Vitrandic Argixerolls	Fine-loamy	Isotic			Frigid	Xeric	
Bolony	16.7	Mollisols	Xerolls	Argixerolls	Vitrandic Argixerolls	Fine-loamy	Isotic			Frigid	Xeric	
Bombadil	62.3	Aridisols	Argids	Haplargids	Lithic Xeric Haplargids	Loamy	Mixed	Superactive		Mesic	Aridic	
Bonnick	96.7	Mollisols	Xerolls	Haploxerolls	Vitritorrandic Haploxerolls	Ashy	Glassy			Frigid	Aridic	
Booten	39.1	Mollisols	Xerolls	Argixerolls	Vitrandic Argixerolls	Ashy	Glassy			Mesic	Xeric	
Booth	593.7	Mollisols	Xerolls	Palexerolls	Vertic Palexerolls	Fine	Smectitic			Frigid	Xeric	
Boravall	36.5	Inceptisols	Aquepts	Halaquepts	Aeric Halaquepts	Fine	Smectitic		Calcareous	Mesic	Aquic	
Bordengulch	71.1	Inceptisols	Cryepts	Haplocryepts	Andic Haplocryepts	Loamy-skeletal	Isotic			Cryic	Udic	
Borges	5.6	Inceptisols	Aquepts	Humaquepts	Typic Humaquepts	Fine	Mixed	Superactive	Nonacid	Mesic	Aquic	
Bornstedt	67.7	Ultisols	Xerults	Palexerults	Typic Palexerults	Fine-silty	Mixed	Active		Mesic	Xeric	
Borobey	215.2	Mollisols	Xerolls	Haploxerolls	Vitritorrandic Haploxerolls	Ashy	Glassy			Frigid	Aridic	

(continued)

Series name	Area (km^2)	Order	Suborder	Great group	Subgroup	Particle-size class	Mineralogy class	CEC activity class	Reaction class	Soil temp. regime	Soil moisture regime	Other family
Bosland	44.1	Inceptisols	Udepts	Humudepts	Typic Humudepts	Fine-loamy	Isotic			Isomesic	Udic	
Bott	2.0	Andisols	Cryands	Vitricryands	Alfic Vitricryands	Ashy over loamy-skeletal	Amorphic over isotic			Cryic	Udic	
Boulder Lake	81.4	Vertisols	Aquerts	Epiaquerts	Xeric Epiaquerts	Fine	Smectitic			Frigid	Aquic	
Bouldrock	56.0	Inceptisols	Xerepts	Haploxerepts	Humic Haploxerepts	Coarse-loamy	Mixed	Superactive		Frigid	Xeric	
Bowlus	24.5	Mollisols	Xerolls	Haploxerolls	Pachic Ultic Haploxerolls	Fine-silty	Mixed	Superactive		Frigid	Xeric	
Boyce	31.0	Mollisols	Aquolls	Endoaquolls	Cumulic Endoaquolls	Fine-silty over sandy or sandy-skeletal	Mixed	Superactive		Mesic	Aquic	
Brabble	42.6	Aridisols	Durids	Haplodurids	Xeric Haplodurids	Fine-loamy	Mixed	Superactive		Mesic	Aridic	
Brace	1571.4	Aridisols	Durids	Argidurids	Xeric Argidurids	Fine-loamy	Mixed	Superactive		Frigid	Aridic	
Brader	50.5	Inceptisols	Xerepts	Haploxerepts	Typic Haploxerepts	Loamy	Mixed	Superactive		Mesic	Xeric	Shallow
Bragton	8.9	Histosols	Hemists	Haplohemists	Terric Haplohemists	Loamy	Mixed		Euic	Isomesic	Aquic	
Brallier	12.6	Histosols	Hemists	Haplohemists	Typic Haplohemists				Dysic	Isomesic	Aquic	
Brand	10.4	Inceptisols	Aquepts	Endoaquepts	Typic Endoaquepts	Fine	Mixed	Superactive	Acid	Mesic	Aquic	
Brandypeak	28.6	Inceptisols	Xerepts	Humixerepts	Typic Humixerepts	Loamy-skeletal	Mixed	Superactive		Frigid	Xeric	
Brannan	8.2	Andisols	Xerands	Vitrixerands	Typic Vitrixerands	Ashy over loamy-skeletal	Glassy over isotic			Frigid	Xeric	
Braun	167.7	Inceptisols	Udepts	Eutrudepts	Dystric Eutrudepts	Fine-loamy	Isotic			Mesic	Udic	
Bravo	188.9	Inceptisols	Udepts	Dystrudepts	Humic Dystrudepts	Fine-loamy	Isotic			Mesic	Udic	
Breadloaf	0.6	Vertisols	Xererts	Haploxererts	Leptic Haploxererts	Fine	Smectitic			Mesic	Xeric	
Bregar	40.4	Aridisols	Argids	Haplargids	Lithic Xeric Haplargids	Loamy-skeletal	Mixed	Superactive		Frigid	Aridic	
Brenner	35.7	Inceptisols	Aquepts	Humaquepts	Fluvaquentic Humaquepts	Fine-silty	Mixed	Superactive	Acid	Isomesic	Aquic	
Brezniak	24.7	Mollisols	Xerolls	Argixerolls	Aridic Lithic Argixerolls	Clayey	Smectitic			Mesic	Aridic	
Bridgecreek	75.6	Mollisols	Xerolls	Palexerolls	Typic Palexerolls	Fine	Smectitic			Frigid	Xeric	
Bridgewater	1.7	Mollisols	Xerolls	Haploxerolls	Cumulic Haploxerolls	Loamy-skeletal	Mixed	Superactive		Mesic	Xeric	
Bridgewell	27.7	Mollisols	Aquolls	Endoaquolls	Aquandic Endoaquolls	Ashy	Glassy		Calcareous	Frigid	Aquic	
Briedwell	44.0	Mollisols	Xerolls	Haploxerolls	Ultic Haploxerolls	Loamy-skeletal	Mixed	Superactive		Mesic	Xeric	
Brightwood	14.3	Inceptisols	Udepts	Humudepts	Typic Humudepts	Loamy-skeletal	Isotic			Mesic	Udic	
Brisbois	66.6	Aridisols	Argids	Haplargids	Xeric Haplargids	Clayey	Smectitic			Mesic	Aridic	Shallow
Broadycreek	4.3	Mollisols	Cryolls	Haplocryolls	Aquic Cumulic Haplocryolls	Coarse-loamy	Mixed	Superactive		Cryic	Udic	
Brock	10.6	Aridisols	Durids	Argidurids	Xeric Argidurids	Loamy-skeletal	Mixed	Superactive		Mesic	Aridic	Shallow
Brockman	28.8	Inceptisols	Xerepts	Haploxerepts	Vertic Haploxerepts	Fine	Magnesic			Mesic	Xeric	
Brownlee	18.8	Mollisols	Xerolls	Argixerolls	Ultic Argixerolls	Fine-loamy	Mixed	Superactive		Mesic	Xeric	
Brownscombe	27.3	Mollisols	Xerolls	Argixerolls	Aridic Argixerolls	Fine	Smectitic			Mesic	Aridic	
Broyles	97.4	Aridisols	Cambids	Haplocambids	Durinodic Haplocambids	Ashy over loamy	Glassy over mixed	Superactive		Mesic	Aridic	

(continued)

Series name	Area (km^2)	Order	Suborder	Great group	Subgroup	Particle-size class	Mineralogy class	CEC activity class	Reaction class	Soil temp. regime	Soil moisture regime	Other family
Bruncan	40.5	Aridisols	Durids	Argidurids	Xeric Argidurids	Loamy	Mixed	Superactive		Mesic	Aridic	Shallow
Brunzell	3.5	Mollisols	Xerolls	Haploxerolls	Typic Haploxerolls	Loamy-skeletal	Mixed	Superactive		Frigid	Xeric	
Btree	116.6	Andisols	Vitrands	Udivitrands	Alfic Udivitrands	Ashy over clayey-skeletal	Amorphic over smectitic			Frigid	Udic	
Bubus	13.7	Entisols	Orthents	Torriorthents	Duric Torriorthents	Coarse-loamy	Mixed	Superactive	Calcareous	Mesic	Aridic	
Buckbert	7.2	Mollisols	Xerolls	Haploxerolls	Vitritorrandic Haploxerolls	Fine-loamy	Mixed	Superactive		Mesic	Aridic	
Buckcreek	79.6	Mollisols	Xerolls	Haploxerolls	Pachic Ultic Haploxerolls	Loamy-skeletal	Mixed	Superactive		Frigid	Xeric	
Bucketlake	120.9	Andisols	Cryands	Vitricryands	Typic Vitricryands	Ashy over loamy-skeletal	Amorphic over isotic			Cryic	Udic	
Buckeye	13.9	Mollisols	Xerolls	Argixerolls	Pachic Ultic Argixerolls	Fine-loamy	Mixed	Superactive		Mesic	Xeric	
Bucklake	26.0	Mollisols	Xerolls	Argixerolls	Aridic Argixerolls	Fine	Smectitic			Mesic	Aridic	
Buckshot	13.7	Ultisols	Udults	Paleudults	Typic Paleudults	Fine-loamy	Mixed	Active		Mesic	Udic	
Buckwilder	22.6	Mollisols	Cryolls	Argicryolls	Vertic Argicryolls	Very-fine	Smectitic			Cryic	Xeric	
Budlewis	2.4	Mollisols	Xerolls	Durixerolls	Typic Durixerolls	Fine	Smectitic			Frigid	Xeric	
Buffaran	21.3	Aridisols	Durids	Argidurids	Xeric Argidurids	Clayey	Smectitic			Mesic	Aridic	Shallow
Buford	8.5	Mollisols	Xerolls	Haploxerolls	Vitrandic Haploxerolls	Fine-loamy	Isotic			Frigid	Xeric	
Bulgar	26.1	Andisols	Vitrands	Udivitrands	Typic Udivitrands	Ashy over loamy-skeletal	Amorphic over isotic			Frigid	Udic	
Bull Run	78.1	Andisols	Udands	Fulvudands	Eutric Fulvudands	Medial	Amorphic			Mesic	Udic	
Bullards	104.2	Spodosols	Orthods	Haplorthods	Typic Haplorthods	Coarse-loamy	Isotic			Isomesic	Udic	
Bullgulch	21.3	Ultisols	Humults	Haplohumults	Typic Haplohumults	Fine	Isotic			Isomesic	Udic	
Bullroar	22.4	Andisols	Vitrands	Udivitrands	Typic Udivitrands	Ashy over loamy-skeletal	Amorphic over isotic			Frigid	Udic	
Bullump	103.9	Mollisols	Xerolls	Argixerolls	Pachic Argixerolls	Loamy-skeletal	Mixed	Superactive		Frigid	Xeric	
Bullvaro	35.0	Mollisols	Xerolls	Argixerolls	Pachic Argixerolls	Loamy-skeletal	Mixed	Superactive		Frigid	Xeric	
Bully	14.5	Entisols	Fluvents	Torrifluvents	Xeric Torrifluvents	Coarse-silty	Mixed	Superactive	Nonacid	Mesic	Aridic	
Bunchpoint	14.3	Mollisols	Xerolls	Haploxerolls	Vitrandic Haploxerolls	Coarse-loamy	Isotic			Frigid	Xeric	
Bunyard	2.4	Aridisols	Argids	Natrargids	Durinodic Natrargids	Ashy	Glassy			Frigid	Aridic	
Burbank	46.5	Entisols	Orthents	Torriorthents	Xeric Torriorthents	Sandy-skeletal	Mixed			Mesic	Aridic	
Burgerbutte	61.2	Inceptisols	Cryepts	Humicryepts	Lithic Humicryepts	Loamy-skeletal	Isotic			Cryic	Xeric	
Burke	62.6	Aridisols	Durids	Haplodurids	Xeric Haplodurids	Coarse-silty	Mixed	Superactive		Mesic	Aridic	
Burkemont	21.4	Inceptisols	Aquepts	Halaquepts	Typic Halaquepts	Fine	Smectitic		Calcareous	Mesic	Aquic	
Burlington	8.0	Mollisols	Xerolls	Haploxerolls	Entic Ultic Haploxerolls	Sandy	Mixed			Mesic	Xeric	
Burningman	0.1	Mollisols	Xerolls	Argixerolls	Lithic Argixerolls	Ashy over clayey	Glassy over smectitic			Frigid	Xeric	
Burnthill	12.4	Ultisols	Humults	Palehumults	Typic Palehumults	Fine-loamy	Siliceous	Superactive		Isomesic	Udic	

(continued)

Series name	Area (km^2)	Order	Suborder	Great group	Subgroup	Particle-size class	Mineralogy class	CEC activity class	Reaction class	Soil temp. regime	Soil moisture regime	Other family
Burntriver	25.4	Aridisols	Cambids	Haplocambids	Xeric Haplocambids	Fine-loamy	Mixed	Superactive		Frigid	Aridic	
Burntwoods	6.4	Andisols	Udands	Fulvudands	Typic Fulvudands	Medial-skeletal over loamy-skeletal	Mixed over isotic			Frigid	Udic	
Burrita	2.1	Aridisols	Argids	Haplargids	Lithic Xeric Haplargids	Clayey-skeletal	Smectitic			Mesic	Aridic	
Bybee	57.5	Mollisols	Xerolls	Haploxerolls	Typic Haploxerolls	Fine	Smectitic			Frigid	Xeric	
Bycracky	3.2	Histosols	Fibrists	Cryofibrists	Terric Cryofibrists	Loamy	Isotic		Euic	Cryic	Aquic	
Cabell	4.6	Alfisols	Cryalfs	Haplocryalfs	Andic Haplocryalfs	Fine-loamy	Isotic			Cryic	Udic	
Cabincreek	3.5	Mollisols	Xerolls	Haploxerolls	Vitrandic Haploxerolls	Coarse-loamy	Mixed	Superactive		Mesic	Xeric	
Cabinspring	9.9	Mollisols	Xerolls	Argixerolls	Vitritorrandic Argixerolls	Ashy-skeletal	Glassy			Frigid	Xeric	
Calder	12.5	Alfisols	Xeralfs	Durixeralfs	Abruptic Haplic Durixeralfs	Clayey	Smectitic			Mesic	Xeric	Shallow
Calderwood	155.8	Aridisols	Cambids	Haplocambids	Lithic Xeric Haplocambids	Loamy-skeletal	Mixed	Superactive		Mesic	Aridic	
Calfranch	27.6	Inceptisols	Udepts	Dystrudepts	Typic Dystrudepts	Loamy-skeletal	Mixed	Active		Isomesic	Udic	
Calimus	157.0	Mollisols	Xerolls	Haploxerolls	Pachic Haploxerolls	Fine-loamy	Mixed	Superactive		Mesic	Xeric	
Camas	111.3	Mollisols	Xerolls	Haploxerolls	Fluventic Haploxerolls	Sandy-skeletal	Mixed			Mesic	Xeric	
Camaspatch	0.1	Mollisols	Xerolls	Argixerolls	Lithic Argixerolls	Clayey-skeletal	Smectitic			Mesic	Xeric	
Campcreek	56.1	Mollisols	Xerolls	Palexerolls	Vertic Palexerolls	Fine	Smectitic			Frigid	Xeric	
Campfour	14.6	Mollisols	Xerolls	Argixerolls	Pachic Ultic Argixerolls	Fine-loamy	Mixed	Superactive		Mesic	Xeric	
Camptank	1.0	Aridisols	Argids	Paleargids	Xeric Paleargids	Clayey over loamy-skeletal	Smectitic over mixed	Superactive		Frigid	Aridic	
Canderly	10.4	Mollisols	Xerolls	Haploxerolls	Ultic Haploxerolls	Coarse-loamy	Mixed	Superactive		Mesic	Xeric	
Canest	353.5	Mollisols	Xerolls	Argixerolls	Aridic Lithic Argixerolls	Clayey-skeletal	Smectitic			Frigid	Aridic	
Cant	11.9	Aridisols	Argids	Paleargids	Xeric Paleargids	Clayey-skeletal	Smectitic			Mesic	Aridic	
Cantala	165.4	Mollisols	Xerolls	Haploxerolls	Typic Haploxerolls	Fine-silty	Mixed	Superactive		Mesic	Xeric	
Capeblanco	18.0	Inceptisols	Udepts	Dystrudepts	Typic Dystrudepts	Loamy-skeletal	Mixed	Active		Isomesic	Udic	
Caphealy	28.1	Mollisols	Xerolls	Haploxerolls	Vitritorrandic Haploxerolls	Coarse-loamy	Mixed	Superactive		Mesic	Aridic	
Capona	45.4	Mollisols	Xerolls	Haploxerolls	Aridic Haploxerolls	Fine-loamy	Mixed	Superactive		Mesic	Aridic	
Caris	208.1	Inceptisols	Xerepts	Haploxerepts	Typic Haploxerepts	Loamy-skeletal	Mixed	Superactive		Mesic	Xeric	
Carlton	39.8	Mollisols	Xerolls	Haploxerolls	Aquultic Haploxerolls	Fine-silty	Mixed	Superactive		Mesic	Xeric	
Carney	235.3	Vertisols	Xererts	Haploxererts	Udic Haploxererts	Fine	Smectitic			Mesic	Xeric	
Carpenterville	2.6	Mollisols	Udolls	Argiudolls	Aquic Argiudolls	Clayey-skeletal	Mixed	Superactive		Mesic	Udic	
Carryback	1189.8	Mollisols	Xerolls	Palexerolls	Vertic Palexerolls	Fine	Smectitic			Frigid	Xeric	
Carvix	151.1	Mollisols	Xerolls	Haploxerolls	Aridic Haploxerolls	Fine-loamy	Mixed	Superactive		Frigid	Aridic	
Cascade	178.8	Inceptisols	Xerepts	Fragixerepts	Humic Fragixerepts	Fine-silty	Mixed	Superactive		Mesic	Xeric	

(continued)

Series name	Area (km^2)	Order	Suborder	Great group	Subgroup	Particle-size class	Mineralogy class	CEC activity class	Reaction class	Soil temp. regime	Soil moisture regime	Other family
Cashner	2.7	Spodosols	Aquods	Duraquods	Typic Duraquods	Coarse-loamy	Siliceous	Superactive		Isomesic	Aquic	Ortstein
Cassiday	183.3	Inceptisols	Udepts	Dystrudepts	Humic Dystrudepts	Loamy-skeletal	Isotic			Mesic	Udic	
Castlecrest	265.8	Andisols	Cryands	Vitricryands	Typic Vitricryands	Ashy	Amorphic			Cryic	Udic	
Catchell	36.7	Aridisols	Durids	Argidurids	Abruptic Xeric Argidurids	Fine	Smectitic			Mesic	Aridic	
Caterl	253.1	Andisols	Udands	Hapludands	Alic Hapludands	Medial-skeletal	Ferrihydritic			Frigid	Udic	
Catherine	89.2	Mollisols	Aquolls	Endoaquolls	Cumulic Endoaquolls	Fine-silty	Mixed	Superactive		Mesic	Aquic	
Catlow	171.7	Aridisols	Cambids	Haplocambids	Durinodic Xeric Haplocambids	Loamy-skeletal	Mixed	Superactive		Mesic	Aridic	
Catnapp	3.2	Aridisols	Argids	Natrargids	Xeric Natrargids	Fine	Smectitic			Frigid	Aridic	
Cazadero	82.0	Ultisols	Udults	Paleudults	Rhodic Paleudults	Fine	Mixed	Active		Mesic	Udic	
Cedarcamp	22.5	Inceptisols	Udepts	Eutrudepts	Dystric Eutrudepts	Loamy-skeletal	Magnesic			Frigid	Udic	
Cedargrove	14.3	Alfisols	Xeralfs	Haploxeralfs	Ultic Haploxeralfs	Fine	Mixed	Superactive		Mesic	Xeric	
Cencove	9.3	Entisols	Orthents	Torriorthents	Xeric Torriorthents	Coarse-loamy over sandy or sandy-skeletal	Mixed	Superactive	Calcareous	Mesic	Aridic	
Central Point	33.8	Mollisols	Xerolls	Haploxerolls	Pachic Haploxerolls	Coarse-loamy	Mixed	Superactive		Mesic	Xeric	
Chamate	15.7	Inceptisols	Udepts	Dystrudepts	Typic Dystrudepts	Loamy-skeletal	Isotic			Mesic	Udic	
Chambeam	38.4	Mollisols	Xerolls	Haploxerolls	Pachic Haploxerolls	Loamy-skeletal	Mixed	Superactive		Frigid	Xeric	
Chancelakes	7.0	Vertisols	Aquerts	Epiaquerts	Xeric Epiaquerts	Fine	Smectitic			Frigid	Aquic	
Chapman	58.1	Mollisols	Xerolls	Haploxerolls	Cumulic Ultic Haploxerolls	Fine-loamy	Mixed	Superactive		Mesic	Xeric	
Chard	0.4	Mollisols	Xerolls	Haploxerolls	Calcic Haploxerolls	Coarse-loamy	Mixed	Superactive		Mesic	Xeric	
Chehalem	49.5	Mollisols	Aquolls	Endoaquolls	Cumulic Vertic Endoaquolls	Fine	Smectitic			Mesic	Aquic	
Chehalis	238.7	Mollisols	Xerolls	Haploxerolls	Cumulic Ultic Haploxerolls	Fine-silty	Mixed	Superactive		Mesic	Xeric	
Chehulpum	42.8	Mollisols	Xerolls	Haploxerolls	Ultic Haploxerolls	Loamy	Mixed	Superactive		Mesic	Xeric	Shallow
Chen	111.9	Mollisols	Xerolls	Argixerolls	Aridic Lithic Argixerolls	Clayey-skeletal	Smectitic			Frigid	Aridic	
Chenoweth	14.8	Mollisols	Xerolls	Haploxerolls	Typic Haploxerolls	Coarse-loamy	Mixed	Superactive		Mesic	Xeric	
Cherry Spring	14.5	Aridisols	Durids	Argidurids	Haploxeralfic Argidurids	Fine-loamy	Mixed	Superactive		Mesic	Aridic	
Cherrycreek	71.3	Mollisols	Xerolls	Haploxerolls	Vitrandic Haploxerolls	Loamy-skeletal	Isotic			Frigid	Xeric	
Cherryhill	30.8	Alfisols	Xeralfs	Haploxeralfs	Ultic Haploxeralfs	Fine-loamy	Mixed	Superactive		Mesic	Xeric	
Chesebro	18.5	Mollisols	Xerolls	Argixerolls	Vitrandic Argixerolls	Ashy-skeletal	Glassy			Frigid	Xeric	
Chesnimnus	12.5	Mollisols	Xerolls	Argixerolls	Calcic Argixerolls	Fine-loamy	Mixed	Superactive		Frigid	Xeric	
Chetco	14.1	Inceptisols	Aquepts	Humaquepts	Fluvaquentic Humaquepts	Fine	Mixed	Superactive	Nonacid	Isomesic	Aquic	
Cheval	9.2	Mollisols	Xerolls	Haploxerolls	Aquic Cumulic Haploxerolls	Coarse-loamy over sandy or sandy-skeletal	Mixed	Superactive		Frigid	Xeric	

(continued)

Series name	Area (km^2)	Order	Suborder	Great group	Subgroup	Particle-size class	Mineralogy class	CEC activity class	Reaction class	Soil temp. regime	Soil moisture regime	Other family
Chewaucan	13.8	Mollisols	Xerolls	Argixerolls	Argiduridic Argixerolls	Fine	Smectitic			Mesic	Xeric	
Chilcott	93.3	Aridisols	Durids	Argidurids	Abruptic Xeric Argidurids	Fine	Smectitic			Mesic	Aridic	
Chiloquin	3.4	Inceptisols	Xerepts	Haploxerepts	Oxyaquic Vitrandic Haploxerepts	Fine-loamy	Mixed	Superactive		Frigid	Xeric	
Chimneyrock	23.4	Alfisols	Xeralfs	Haploxeralfs	Ultic Haploxeralfs	Loamy-skeletal	Mixed	Superactive		Mesic	Xeric	
Chinarise	23.4	Mollisols	Xerolls	Haploxerolls	Vitrandic Haploxerolls	Ashy	Glassy			Frigid	Xeric	
Chintimini	8.0	Inceptisols	Udepts	Dystrudepts	Andic Dystrudepts	Loamy-skeletal	Isotic			Frigid	Udic	
Chismore	16.9	Ultisols	Humults	Palehumults	Aquic Palehumults	Fine	Isotic			Mesic	Udic	
Chitwood	22.2	Inceptisols	Udepts	Humudepts	Aquandic Humudepts	Fine	Isotic			Isomesic	Udic	
Chock	23.7	Andisols	Aquands	Cryaquands	Typic Cryaquands	Ashy	Glassy		Nonacid	Cryic	Aquic	
Chocktoot	26.3	Mollisols	Cryolls	Argicryolls	Vitrandic Argicryolls	Loamy-skeletal	Isotic			Cryic	Xeric	
Choptie	64.6	Mollisols	Xerolls	Haploxerolls	Lithic Haploxerolls	Loamy	Mixed	Superactive		Frigid	Xeric	
Chug	24.9	Mollisols	Xerolls	Haploxerolls	Vitrandic Haploxerolls	Fine-loamy	Mixed	Superactive		Frigid	Xeric	
Cinderfall	3.4	Mollisols	Xerolls	Haploxerolls	Vitritorrandic Haploxerolls	Ashy-skeletal	Glassy			Frigid	Aridic	
Circle	14.2	Andisols	Xerands	Vitrixerands	Alfic Vitrixerands	Ashy	Glassy			Frigid	Xeric	
Clackamas	82.5	Mollisols	Aquolls	Argiaquolls	Typic Argiaquolls	Fine-loamy	Mixed	Superactive		Mesic	Aquic	
Clamp	134.3	Mollisols	Cryolls	Haplocryolls	Lithic Haplocryolls	Loamy-skeletal	Mixed	Superactive		Cryic	Xeric	
Clarkscreek	34.0	Inceptisols	Cryepts	Humicryepts	Haploxerandic Humicryepts	Loamy-skeletal	Isotic			Cryic	Xeric	
Clatsop	29.6	Inceptisols	Aquepts	Humaquepts	Histic Humaquepts	Fine-silty	Mixed	Superactive	Nonacid	Isomesic	Aquic	
Clawson	18.2	Inceptisols	Aquepts	Endoaquepts	Typic Endoaquepts	Coarse-loamy	Mixed	Superactive	Nonacid	Mesic	Aquic	
Clearline	42.7	Mollisols	Xerolls	Haploxerolls	Vitrandic Haploxerolls	Ashy-skeletal	Glassy			Frigid	Xeric	
Cleavage	29.2	Mollisols	Xerolls	Argixerolls	Aridic Lithic Argixerolls	Loamy-skeletal	Mixed	Superactive		Frigid	Aridic	
Cleet	35.3	Aridisols	Durids	Argidurids	Xeric Argidurids	Loamy-skeletal	Mixed	Superactive		Mesic	Aridic	Shallow
Cleetwood	28.8	Entisols	Psamments	Cryopsamments	Vitrandic Cryopsamments	Ashy	Glassy		Nonacid	Cryic	Udic	
Clevescove	8.3	Inceptisols	Udepts	Humudepts	Andic Humudepts	Coarse-loamy	Isotic			Isomesic	Udic	
Cleymor	0.5	Mollisols	Xerolls	Argixerolls	Vertic Argixerolls	Fine	Smectitic			Frigid	Xeric	
Climax	17.0	Vertisols	Xererts	Haploxererts	Leptic Haploxererts	Very-fine	Smectitic			Mesic	Xeric	
Clinefalls	7.3	Andisols	Torrands	Vitritorrands	Typic Vitritorrands	Ashy over sandy or sandy-skeletal	Glassy over mixed			Mesic	Aridic	
Cloquato	159.8	Mollisols	Xerolls	Haploxerolls	Cumulic Ultic Haploxerolls	Coarse-silty	Mixed	Superactive		Mesic	Xeric	
Clovercreek	54.5	Mollisols	Xerolls	Argixerolls	Lithic Argixerolls	Loamy-skeletal	Mixed	Superactive		Mesic	Xeric	
Cloverland	22.6	Mollisols	Albolls	Argialbolls	Xeric Argialbolls	Fine-silty	Mixed	Superactive		Frigid	Xeric	

(continued)

Series name	Area (km^2)	Order	Suborder	Great group	Subgroup	Particle-size class	Mineralogy class	CEC activity class	Reaction class	Soil temp. regime	Soil moisture regime	Other family
Clovkamp	50.0	Mollisols	Xerolls	Haploxerolls	Vitritorrandic Haploxerolls	Ashy over sandy or sandy-skeletal	Glassy over mixed			Mesic	Aridic	
Clurde	22.6	Aridisols	Cambids	Haplocambids	Durinodic Xeric Haplocambids	Fine-loamy	Mixed	Superactive		Mesic	Aridic	
Coburg	178.9	Mollisols	Xerolls	Argixerolls	Oxyaquic Argixerolls	Fine	Mixed	Superactive		Mesic	Xeric	
Coglin	100.1	Aridisols	Argids	Paleargids	Xeric Paleargids	Fine	Smectitic			Frigid	Aridic	
Coker	40.6	Vertisols	Aquerts	Endoaquerts	Xeric Endoaquerts	Very-fine	Smectitic			Mesic	Aquic	
Colbar	76.9	Aridisols	Argids	Haplargids	Xeric Haplargids	Fine-loamy	Mixed	Superactive		Mesic	Aridic	
Coleman	15.0	Mollisols	Xerolls	Palexerolls	Typic Palexerolls	Fine	Smectitic			Mesic	Xeric	
Colepoint	16.9	Inceptisols	Udepts	Humudepts	Typic Humudepts	Fine-loamy	Isotic			Mesic	Udic	
Colestine	180.3	Inceptisols	Xerepts	Dystroxerepts	Typic Dystroxerepts	Fine-loamy	Mixed	Superactive		Mesic	Xeric	
Collegecreek	7.1	Andisols	Xerands	Vitrixerands	Typic Vitrixerands	Ashy over loamy	Glassy over mixed	Superactive		Mesic	Xeric	
Collier	411.4	Andisols	Cryands	Vitricryands	Xeric Vitricryands	Ashy	Glassy			Cryic	Xeric	
Concord	123.1	Alfisols	Aqualfs	Endoaqualfs	Typic Endoaqualfs	Fine	Smectitic			Mesic	Aquic	
Condon	1171.4	Mollisols	Xerolls	Haploxerolls	Typic Haploxerolls	Fine-silty	Mixed	Superactive		Mesic	Xeric	
Condorbridge	6.3	Inceptisols	Udepts	Humudepts	Andic Humudepts	Fine-loamy	Isotic			Isomesic	Udic	
Conley	47.5	Mollisols	Albolls	Argialbolls	Xerertic Argialbolls	Fine	Smectitic			Mesic	Xeric	
Connleyhills	122.1	Mollisols	Xerolls	Argixerolls	Vitritorrandic Argixerolls	Clayey-skeletal	Smectitic			Frigid	Xeric	
Conser	93.3	Mollisols	Aquolls	Argiaquolls	Vertic Argiaquolls	Fine	Mixed	Superactive		Mesic	Aquic	
Cookcreek	27.8	Mollisols	Xerolls	Haploxerolls	Vitrandic Haploxerolls	Loamy-skeletal	Mixed	Superactive		Mesic	Xeric	
Cooperdraw	15.5	Aridisols	Durids	Argidurids	Xeric Argidurids	Loamy-skeletal	Mixed	Superactive		Frigid	Aridic	
Cooperopolis	7.8	Mollisols	Xerolls	Argixerolls	Aridic Lithic Argixerolls	Loamy	Mixed	Superactive		Mesic	Aridic	
Copperfield	34.0	Mollisols	Xerolls	Argixerolls	Pachic Argixerolls	Loamy-skeletal	Mixed	Superactive		Mesic	Xeric	
Copsey	5.1	Mollisols	Aquolls	Endoaquolls	Vertic Endoaquolls	Fine	Magnesic			Mesic	Aquic	
Coquille	108.5	Inceptisols	Aquepts	Endoaquepts	Fluvaquentic Endoaquepts	Fine-silty	Mixed	Superactive	Nonacid	Isomesic	Aquic	
Cornelius	101.4	Alfisols	Xeralfs	Fragixeralfs	Mollic Fragixeralfs	Fine-silty	Mixed	Superactive		Mesic	Xeric	
Cornutt	79.6	Alfisols	Xeralfs	Haploxeralfs	Ultic Haploxeralfs	Fine	Mixed	Superactive		Mesic	Xeric	
Corral	113.0	Aridisols	Argids	Haplargids	Xeric Haplargids	Loamy	Mixed	Superactive		Mesic	Aridic	Shallow
Cotant	10.1	Mollisols	Xerolls	Argixerolls	Aridic Argixerolls	Clayey	Smectitic			Frigid	Aridic	Shallow
Cotay	3.5	Alfisols	Xeralfs	Haploxeralfs	Vitrandic Haploxeralfs	Clayey-skeletal	Smectitic			Frigid	Xeric	
Cottrell	47.8	Ultisols	Humults	Haplohumults	Aquic Haplohumults	Fine	Mixed	Active		Mesic	Udic	
Cougarrock	38.6	Alfisols	Xeralfs	Haploxeralfs	Vitrandic Haploxeralfs	Clayey-skeletal	Smectitic			Frigid	Xeric	
Coughanour	29.3	Mollisols	Xerolls	Durixerolls	Argiduridic Durixerolls	Fine-silty	Mixed	Superactive		Mesic	Aridic	

(continued)

Series name	Area (km^2)	Order	Suborder	Great group	Subgroup	Particle-size class	Mineralogy class	CEC activity class	Reaction class	Soil temp. regime	Soil moisture regime	Other family
Court	9.6	Mollisols	Xerolls	Haploxerolls	Vitritorrandic Haploxerolls	Coarse-loamy over sandy or sandy-skeletal	Mixed	Superactive		Mesic	Aridic	
Courtney	9.9	Mollisols	Aquolls	Argiaquolls	Abruptic Argiaquolls	Fine	Smectitic			Mesic	Aquic	
Courtrock	3.4	Mollisols	Xerolls	Haploxerolls	Calcidic Haploxerolls	Coarse-loamy	Mixed	Superactive		Mesic	Aridic	
Cove	83.2	Mollisols	Aquolls	Endoaquolls	Vertic Endoaquolls	Fine	Smectitic			Mesic	Aquic	
Cowsly	269.9	Mollisols	Albolls	Argialbolls	Xerertic Argialbolls	Fine	Smectitic			Frigid	Xeric	
Coyata	54.6	Inceptisols	Xerepts	Dystroxerepts	Humic Dystroxerepts	Loamy-skeletal	Mixed	Superactive		Mesic	Xeric	
Coyotebluff	11.8	Andisols	Xerands	Vitrixerands	Humic Vitrixerands	Ashy-skeletal over loamy-skeletal	Glassy over isotic			Frigid	Xeric	
Coztur	117.5	Aridisols	Argids	Haplargids	Lithic Xeric Haplargids	Loamy	Mixed	Superactive		Frigid	Aridic	
Crabtree	19.6	Inceptisols	Cryepts	Dystrocryepts	Aquic Dystrocryepts	Loamy-skeletal	Isotic			Cryic	Udic	
Crackedground	60.2	Mollisols	Xerolls	Haploxerolls	Vitritorrandic Haploxerolls	Ashy-skeletal	Glassy			Frigid	Aridic	
Crackercreek	99.7	Andisols	Xerands	Vitrixerands	Alfic Vitrixerands	Ashy over loamy-skeletal	Glassy over mixed	Superactive		Frigid	Xeric	
Crackler	18.9	Andisols	Xerands	Vitrixerands	Typic Vitrixerands	Ashy over loamy-skeletal	Glassy over isotic			Frigid	Xeric	
Crannler	23.1	Inceptisols	Cryepts	Humicryepts	Typic Humicryepts	Loamy-skeletal	Isotic			Cryic	Udic	
Crater Lake	4.5	Andisols	Xerands	Vitrixerands	Typic Vitrixerands	Ashy	Amorphic			Mesic	Xeric	
Crawfish	9.8	Inceptisols	Cryepts	Humicryepts	Lithic Humicryepts	Loamy-skeletal	Isotic			Cryic	Udic	
Cressler	13.8	Mollisols	Aquolls	Endoaquolls	Fluvaquentic Endoaquolls	Fine	Smectitic			Frigid	Aquic	
Crims	7.8	Histosols	Hemists	Haplohemists	Terric Haplohemists	Loamy	Mixed		Dysic	Mesic	Aquic	
Crofland	1.5	Ultisols	Humults	Haplohumults	Aquic Haplohumults	Fine	Mixed	Superactive		Isomesic	Udic	
Crooked	7.3	Aridisols	Cambids	Aquicambids	Sodic Aquicambids	Coarse-loamy	Mixed	Superactive		Mesic	Aridic	
Croquib	3.0	Andisols	Aquands	Epiaquands	Alic Epiaquands	Medial over loamy-skeletal	Ferrihydritic over isotic		Acid	Isomesic	Aquic	
Crowcamp	108.6	Mollisols	Xerolls	Palexerolls	Vertic Palexerolls	Fine	Smectitic			Frigid	Xeric	
Cruiser	51.5	Andisols	Cryands	Haplocryands	Typic Haplocryands	Medial	Amorphic			Cryic	Udic	
Crume	7.3	Mollisols	Xerolls	Haploxerolls	Typic Haploxerolls	Fine-loamy	Mixed	Superactive		Frigid	Xeric	
Crump	70.6	Inceptisols	Aquepts	Humaquepts	Histic Humaquepts	Fine-silty	Mixed	Superactive	Nonacid	Mesic	Aquic	
Crutch	7.6	Spodosols	Orthods	Haplorthods	Aquentic Haplorthods	Sandy-skeletal	Mixed			Mesic	Udic	
Crutchfield	16.0	Inceptisols	Udepts	Humudepts	Typic Humudepts	Fine-loamy	Isotic			Mesic	Udic	
Culbertson	22.7	Inceptisols	Xerepts	Humixerepts	Typic Humixerepts	Fine-loamy	Mixed	Superactive		Mesic	Xeric	
Cullius	79.4	Mollisols	Xerolls	Argixerolls	Aridic Lithic Argixerolls	Clayey	Smectitic			Mesic	Aridic	
Cumley	136.1	Ultisols	Humults	Palehumults	Oxyaquic Palehumults	Fine	Isotic			Mesic	Udic	
Cunniff	26.5	Ultisols	Humults	Palehumults	Typic Palehumults	Fine	Mixed	Superactive		Isomesic	Udic	
Cupola	18.7	Andisols	Vitrands	Udivitrands	Typic Udivitrands	Ashy-skeletal	Amorphic			Mesic	Udic	
Curant	203.6	Mollisols	Xerolls	Haploxerolls	Calcic Pachic Haploxerolls	Fine-silty	Mixed	Superactive		Mesic	Xeric	

(continued)

Series name	Area (km^2)	Order	Suborder	Great group	Subgroup	Particle-size class	Mineralogy class	CEC activity class	Reaction class	Soil temp. regime	Soil moisture regime	Other family
Curtin	55.5	Vertisols	Xererts	Haploxererts	Aquic Haploxererts	Very-fine	Smectitic			Mesic	Xeric	
Dabney	5.1	Entisols	Psamments	Udipsamments	Typic Udipsamments	Sandy	Mixed			Mesic	Udic	
Dacker	50.1	Aridisols	Durids	Argidurids	Xeric Argidurids	Fine-loamy	Mixed	Superactive		Mesic	Aridic	
Dahl	9.7	Vertisols	Xererts	Haploxererts	Entic Haploxererts	Fine	Smectitic			Mesic	Xeric	
Damewood	180.3	Inceptisols	Udepts	Humudepts	Andic Humudepts	Loamy-skeletal	Isotic			Mesic	Udic	
Damore	10.6	Mollisols	Xerolls	Haploxerolls	Fluvaquentic Haploxerolls	Fine	Smectitic			Frigid	Xeric	
Darby	7.4	Mollisols	Xerolls	Argixerolls	Pachic Ultic Argixerolls	Fine	Mixed	Superactive		Mesic	Xeric	
Dardry	17.8	Mollisols	Xerolls	Haploxerolls	Cumulic Ultic Haploxerolls	Loamy-skeletal	Mixed	Superactive		Frigid	Xeric	
Darkcanyon	26.1	Aridisols	Cambids	Haplocambids	Xeric Haplocambids	Loamy-skeletal	Mixed	Superactive		Mesic	Aridic	
Darow	20.8	Mollisols	Xerolls	Argixerolls	Vertic Argixerolls	Fine	Smectitic			Mesic	Xeric	
Davey	77.0	Aridisols	Cambids	Haplocambids	Xeric Haplocambids	Sandy	Mixed			Mesic	Aridic	
Daxty	13.7	Inceptisols	Xerepts	Haploxerepts	Vitrandic Haploxerepts	Loamy-skeletal	Isotic			Frigid	Xeric	
Day	53.5	Vertisols	Xererts	Haploxererts	Chromic Haploxererts	Very-fine	Smectitic			Mesic	Xeric	
Dayton	417.9	Alfisols	Aqualfs	Albaqualfs	Vertic Albaqualfs	Fine	Smectitic			Mesic	Aquic	
Dayville	29.2	Mollisols	Aquolls	Endoaquolls	Cumulic Endoaquolls	Fine-silty over sandy or sandy-skeletal	Mixed	Superactive		Mesic	Aquic	
Deadline	107.8	Inceptisols	Udepts	Dystrudepts	Humic Dystrudepts	Loamy-skeletal	Mixed	Active		Mesic	Udic	
Deardorf	36.2	Andisols	Vitrands	Udivitrands	Typic Udivitrands	Ashy over loamy-skeletal	Amorphic over isotic			Frigid	Udic	
Debenger	38.1	Inceptisols	Xerepts	Haploxerepts	Typic Haploxerepts	Fine-loamy	Mixed	Superactive		Mesic	Xeric	
Deck	25.9	Mollisols	Xerolls	Haploxerolls	Vitrandic Haploxerolls	Loamy-skeletal	Isotic			Frigid	Xeric	
Dee	13.5	Andisols	Xerands	Vitrixerands	Aquic Vitrixerands	Ashy	Amorphic			Mesic	Xeric	
Defenbaugh	58.1	Aridisols	Cambids	Haplocambids	Typic Haplocambids	Fine-loamy	Mixed	Superactive		Mesic	Aridic	
Degarmo	22.3	Mollisols	Aquolls	Endoaquolls	Cumulic Endoaquolls	Fine-loamy over sandy or sandy-skeletal	Mixed	Superactive		Frigid	Aquic	
Degner	11.4	Mollisols	Xerolls	Argixerolls	Calcic Argixerolls	Clayey-skeletal	Smectitic			Mesic	Xeric	
Dehill	0.3	Mollisols	Xerolls	Haploxerolls	Pachic Haploxerolls	Coarse-loamy	Mixed	Superactive		Mesic	Xeric	
Dehlinger	58.9	Mollisols	Xerolls	Haploxerolls	Pachic Haploxerolls	Loamy-skeletal	Mixed	Superactive		Mesic	Xeric	
Delena	22.0	Inceptisols	Aquepts	Fragiaquepts	Humic Fragiaquepts	Fine-silty	Mixed	Superactive		Mesic	Aquic	
Demasters	0.0	Mollisols	Xerolls	Argixerolls	Pachic Ultic Argixerolls	Fine-loamy	Mixed	Superactive		Frigid	Xeric	
Dement	43.5	Inceptisols	Udepts	Dystrudepts	Humic Dystrudepts	Fine	Isotic			Mesic	Udic	
Dentdraw	2.7	Mollisols	Aquolls	Endoaquolls	Fluvaquentic Endoaquolls	Fine-loamy	Mixed	Superactive	Calcareous	Frigid	Aquic	
Depoe	11.6	Spodosols	Aquods	Duraquods	Typic Duraquods	Loamy	Isotic			Isomesic	Aquic	Ortstein, shallow
Deppy	308.6	Aridisols	Durids	Argidurids	Argidic Argidurids	Loamy	Mixed	Superactive		Mesic	Aridic	Shallow

(continued)

Series name	Area (km^2)	Order	Suborder	Great group	Subgroup	Particle-size class	Mineralogy class	CEC activity class	Reaction class	Soil temp. regime	Soil moisture regime	Other family
Derallo	11.0	Mollisols	Xerolls	Argixerolls	Vitritorrandic Argixerolls	Ashy-skeletal	Glassy			Frigid	Xeric	
Derapter	8.8	Mollisols	Xerolls	Argixerolls	Calciargidic Argixerolls	Loamy-skeletal	Mixed	Superactive		Frigid	Aridic	
Derringer	23.5	Mollisols	Xerolls	Argixerolls	Typic Argixerolls	Clayey-skeletal	Smectitic			Frigid	Xeric	
Deschutes	245.3	Mollisols	Xerolls	Haploxerolls	Vitritorrandic Haploxerolls	Coarse-loamy	Mixed	Superactive		Mesic	Aridic	
Deseed	64.6	Aridisols	Argids	Haplargids	Xeric Haplargids	Fine	Smectitic			Frigid	Aridic	
Deshler	22.2	Mollisols	Xerolls	Argixerolls	Pachic Argixerolls	Fine	Smectitic			Mesic	Xeric	
Deskamp	172.4	Mollisols	Xerolls	Haploxerolls	Vitritorrandic Haploxerolls	Ashy	Glassy			Mesic	Aridic	
Desolation	4.1	Inceptisols	Udepts	Dystrudepts	Humic Dystrudepts	Fine	Isotic			Mesic	Udic	
Desons	5.1	Ultisols	Humults	Palehumults	Typic Palehumults	Fine	Mixed	Active		Isomesic	Udic	
Dester	365.1	Mollisols	Xerolls	Argixerolls	Vitritorrandic Argixerolls	Fine-loamy	Mixed	Superactive		Frigid	Xeric	
Deter	39.5	Mollisols	Xerolls	Argixerolls	Pachic Argixerolls	Fine	Smectitic			Mesic	Xeric	
Devada	5.5	Mollisols	Xerolls	Argixerolls	Aridic Lithic Argixerolls	Clayey	Smectitic			Mesic	Aridic	
Deven	0.9	Mollisols	Xerolls	Argixerolls	Lithic Argixerolls	Clayey	Smectitic			Mesic	Xeric	
Devnot	65.6	Mollisols	Xerolls	Argixerolls	Lithic Argixerolls	Clayey	Smectitic			Mesic	Xeric	
Devoy	3.8	Mollisols	Cryolls	Argicryolls	Xeric Argicryolls	Clayey-skeletal	Smectitic			Cryic	Xeric	
Dewar	117.1	Aridisols	Durids	Argidurids	Xeric Argidurids	Loamy	Mixed	Superactive		Mesic	Aridic	Shallow
Diablopeak	47.8	Aridisols	Argids	Natrargids	Lithic Natrargids	Clayey	Smectitic			Frigid	Aridic	
Diaz	39.5	Aridisols	Argids	Haplargids	Xeric Haplargids	Fine	Smectitic			Mesic	Aridic	
Dicecreek	7.2	Inceptisols	Xerepts	Dystroxerepts	Lithic Dystroxerepts	Loamy	Mixed	Superactive		Mesic	Xeric	
Dickerson	33.2	Entisols	Orthents	Xerorthents	Lithic Xerorthents	Loamy	Mixed	Superactive	Nonacid	Mesic	Xeric	
Dickle	25.4	Mollisols	Cryolls	Haplocryolls	Lithic Haplocryolls	Loamy	Mixed	Superactive		Cryic	Xeric	
Diffin	6.0	Mollisols	Xerolls	Palexerolls	Vertic Palexerolls	Clayey-skeletal	Smectitic			Frigid	Xeric	
Digger	786.8	Inceptisols	Udepts	Eutrudepts	Dystric Eutrudepts	Loamy-skeletal	Isotic			Mesic	Udic	
Digit	3.4	Andisols	Cryands	Vitricryands	Typic Vitricryands	Ashy over loamy	Amorphic over isotic			Cryic	Udic	
Dilman	24.0	Mollisols	Aquolls	Cryaquolls	Aquandic Cryaquolls	Loamy over ashy or ashy-pumiceous	Mixed over glassy	Superactive		Cryic	Aquic	
Divers	51.0	Andisols	Cryands	Haplocryands	Typic Haplocryands	Medial-skeletal	Amorphic			Cryic	Udic	
Dixiejett	23.3	Mollisols	Xerolls	Argixerolls	Typic Argixerolls	Loamy-skeletal	Mixed	Superactive		Mesic	Xeric	
Dixon	61.7	Aridisols	Cambids	Haplocambids	Xeric Haplocambids	Fine-loamy over sandy or sandy-skeletal	Mixed	Superactive		Mesic	Aridic	
Dixonville	266.3	Mollisols	Xerolls	Argixerolls	Pachic Ultic Argixerolls	Fine	Mixed	Superactive		Mesic	Xeric	
Diyou	0.5	Mollisols	Xerolls	Haploxerolls	Fluvaquentic Haploxerolls	Fine-loamy	Mixed	Superactive		Mesic	Xeric	
Dobbins	11.2	Inceptisols	Udepts	Dystrudepts	Humic Dystrudepts	Clayey-skeletal	Isotic			Mesic	Udic	
Dodes	18.4	Mollisols	Xerolls	Argixerolls	Aridic Argixerolls	Fine-loamy	Mixed	Superactive		Mesic	Aridic	

(continued)

Series name	Area (km^2)	Order	Suborder	Great group	Subgroup	Particle-size class	Mineralogy class	CEC activity class	Reaction class	Soil temp. regime	Soil moisture regime	Other family
Dogmountain	24.3	Aridisols	Durids	Haplodurids	Vitrixerandic Haplodurids	Ashy-skeletal	Glassy			Frigid	Aridic	
Dogtown	37.9	Mollisols	Xerolls	Haploxerolls	Vitrandic Haploxerolls	Loamy-skeletal	Mixed	Superactive		Frigid	Xeric	
Dollarlake	3.4	Inceptisols	Cryepts	Humicryepts	Vitrixerandic Humicryepts	Loamy-skeletal	Isotic			Cryic	Xeric	
Dompier	2.0	Alfisols	Xeralfs	Fragixeralfs	Aquic Fragixeralfs	Fine-silty	Mixed	Superactive		Mesic	Xeric	
Donegan	20.1	Inceptisols	Xerepts	Dystroxerepts	Humic Dystroxerepts	Loamy-skeletal	Mixed	Superactive		Frigid	Xeric	
Donnybrook	23.5	Mollisols	Xerolls	Haploxerolls	Calcidic Haploxerolls	Loamy	Mixed	Superactive		Mesic	Aridic	Shallow
Doublecreek	15.3	Mollisols	Xerolls	Haploxerolls	Vitrandic Haploxerolls	Coarse-loamy	Mixed	Superactive		Mesic	Xeric	
Doubleo	20.4	Mollisols	Aquolls	Endoaquolls	Fluvaquentic Vertic Endoaquolls	Clayey over loamy	Smectitic over mixed	Superactive	Calcareous	Frigid	Aquic	
Douthit	29.4	Andisols	Cryands	Vitricryands	Typic Vitricryands	Ashy-skeletal	Amorphic			Cryic	Udic	
Dowde	37.7	Inceptisols	Udepts	Dystrudepts	Humic Dystrudepts	Fine-silty	Isotic			Mesic	Udic	
Downards	6.7	Mollisols	Udolls	Hapludolls	Andic Hapludolls	Loamy-skeletal	Isotic			Frigid	Udic	
Downeygulch	30.5	Inceptisols	Xerepts	Haploxerepts	Vitrandic Haploxerepts	Coarse-loamy	Isotic			Frigid	Xeric	
Doyn	248.6	Mollisols	Xerolls	Haploxerolls	Aridic Lithic Haploxerolls	Loamy	Mixed	Superactive		Frigid	Aridic	
Drakesflat	49.4	Mollisols	Xerolls	Argixerolls	Calciargidic Argixerolls	Fine	Smectitic			Frigid	Aridic	
Drakespeak	7.8	Inceptisols	Cryepts	Humicryepts	Xeric Humicryepts	Loamy-skeletal	Mixed	Superactive		Cryic	Xeric	
Drews	97.8	Mollisols	Xerolls	Argixerolls	Pachic Argixerolls	Fine-loamy	Mixed	Superactive		Mesic	Xeric	
Drewsey	258.2	Aridisols	Cambids	Haplocambids	Xeric Haplocambids	Coarse-loamy	Mixed	Superactive		Mesic	Aridic	
Drewsgap	9.4	Mollisols	Xerolls	Durixerolls	Typic Durixerolls	Fine-loamy	Mixed	Superactive		Mesic	Xeric	
Drinkwater	74.6	Aridisols	Cambids	Haplocambids	Xeric Haplocambids	Loamy-skeletal	Mixed	Superactive		Mesic	Aridic	
Droval	113.2	Aridisols	Cambids	Aquicambids	Sodic Aquicambids	Fine	Smectitic			Mesic	Aridic	
Drybed	19.6	Mollisols	Xerolls	Argixerolls	Calciargidic Argixerolls	Fine-loamy	Mixed	Superactive		Mesic	Aridic	
Dryck	27.3	Mollisols	Xerolls	Haploxerolls	Cumulic Haploxerolls	Coarse-loamy over sandy or sandy-skeletal	Mixed	Superactive		Mesic	Xeric	
Dryhollow	30.1	Mollisols	Xerolls	Haploxerolls	Vitritorrandic Haploxerolls	Ashy	Glassy			Mesic	Aridic	
Duart	45.7	Mollisols	Xerolls	Haploxerolls	Typic Haploxerolls	Coarse-loamy	Mixed	Superactive		Mesic	Xeric	
Dubakella	187.7	Alfisols	Xeralfs	Haploxeralfs	Mollic Haploxeralfs	Clayey-skeletal	Magnesic			Mesic	Xeric	
Duckclub	46.6	Aridisols	Cambids	Aquicambids	Sodic Aquicambids	Coarse-loamy	Mixed	Superactive		Frigid	Aridic	
Ducklake	48.1	Andisols	Cryands	Vitricryands	Typic Vitricryands	Ashy over loamy-skeletal	Amorphic over isotic			Cryic	Udic	
Duff	72.0	Mollisols	Cryolls	Haplocryolls	Pachic Haplocryolls	Fine-loamy	Mixed	Superactive		Cryic	Xeric	
Dufur	62.0	Mollisols	Xerolls	Haploxerolls	Calcic Haploxerolls	Coarse-loamy	Mixed	Superactive		Mesic	Xeric	
Dulandy	42.7	Inceptisols	Udepts	Dystrudepts	Humic Dystrudepts	Loamy-skeletal	Isotic			Isomesic	Udic	

(continued)

Series name	Area (km^2)	Order	Suborder	Great group	Subgroup	Particle-size class	Mineralogy class	CEC activity class	Reaction class	Soil temp. regime	Soil moisture regime	Other family
Dumont	179.4	Ultisols	Xerults	Palexerults	Typic Palexerults	Fine	Kaolinitic			Mesic	Xeric	
Dunnlake	0.0	Mollisols	Xerolls	Argixerolls	Aridic Lithic Argixerolls	Clayey	Smectitic			Mesic	Aridic	
Dunres	111.6	Mollisols	Xerolls	Durixerolls	Vitrandic Durixerolls	Clayey	Smectitic			Frigid	Xeric	Shallow
Dunstan	52.4	Alfisols	Xeralfs	Haploxeralfs	Vitrandic Haploxeralfs	Clayey-skeletal	Smectitic			Frigid	Xeric	
Dupee	167.4	Alfisols	Xeralfs	Haploxeralfs	Aquultic Haploxeralfs	Fine	Mixed	Superactive		Mesic	Xeric	
Dupratt	126.5	Mollisols	Xerolls	Argixerolls	Vitrandic Argixerolls	Clayey-skeletal	Smectitic			Frigid	Xeric	
Durkee	223.4	Mollisols	Xerolls	Argixerolls	Calcic Argixerolls	Fine	Smectitic			Frigid	Xeric	
Dyarock	1.4	Andisols	Cryands	Vitricryands	Oxyaquic Vitricryands	Ashy	Amorphic			Cryic	Udic	
Ead	26.0	Inceptisols	Udepts	Humudepts	Pachic Humudepts	Fine	Isotic			Mesic	Udic	
Eaglespring	11.0	Mollisols	Xerolls	Haploxerolls	Vitrandic Haploxerolls	Loamy-skeletal	Mixed	Superactive		Mesic	Xeric	
Eastlakesbasin	7.9	Inceptisols	Cryepts	Humicryepts	Lithic Humicryepts	Loamy-skeletal	Isotic			Cryic	Udic	
Eastpine	17.7	Mollisols	Xerolls	Haploxerolls	Vitrandic Haploxerolls	Loamy-skeletal	Isotic			Frigid	Xeric	
Ecola	42.2	Inceptisols	Udepts	Humudepts	Andic Humudepts	Fine-silty	Isotic			Isomesic	Udic	
Edemaps	132.2	Mollisols	Xerolls	Durixerolls	Argiduridic Durixerolls	Fine	Smectitic			Frigid	Aridic	
Edenbower	27.3	Mollisols	Xerolls	Haploxerolls	Lithic Ultic Haploxerolls	Clayey	Smectitic			Mesic	Xeric	
Edson	22.0	Ultisols	Humults	Palehumults	Typic Palehumults	Fine	Mixed	Active		Mesic	Udic	
Eggleson	5.6	Mollisols	Xerolls	Haploxerolls	Oxyaquic Haploxerolls	Sandy-skeletal	Mixed			Frigid	Xeric	
Eglirim	51.5	Mollisols	Xerolls	Argixerolls	Aridic Argixerolls	Clayey-skeletal	Smectitic			Mesic	Aridic	
Egyptcreek	111.1	Mollisols	Xerolls	Haploxerolls	Vitrandic Haploxerolls	Loamy-skeletal	Isotic			Frigid	Xeric	
Eightlar	62.7	Inceptisols	Xerepts	Dystroxerepts	Typic Dystroxerepts	Clayey-skeletal	Magnesic			Mesic	Xeric	
Eilertsen	59.7	Alfisols	Udalfs	Hapludalfs	Ultic Hapludalfs	Fine-silty	Isotic			Mesic	Udic	
Ekoms	4.1	Ultisols	Humults	Haplohumults	Typic Haplohumults	Fine-loamy	Isotic			Isomesic	Udic	
Elbowcreek	31.7	Andisols	Vitrands	Udivitrands	Alfic Udivitrands	Ashy over loamy	Amorphic over isotic			Frigid	Udic	
Elijah	284.8	Aridisols	Durids	Argidurids	Xeric Argidurids	Fine-silty	Mixed	Superactive		Mesic	Aridic	
Elkhorncrest	47.5	Inceptisols	Cryepts	Dystrocryepts	Lithic Dystrocryepts	Loamy-skeletal	Isotic			Cryic	Udic	
Ellisforde	36.4	Mollisols	Xerolls	Haploxerolls	Calcidic Haploxerolls	Coarse-silty	Mixed	Superactive		Mesic	Aridic	
Ellum	5.9	Aridisols	Durids	Haplodurids	Xereptic Haplodurids	Loamy-skeletal	Mixed	Superactive		Mesic	Aridic	
Elsie	14.2	Ultisols	Humults	Haplohumults	Typic Haplohumults	Fine-silty	Isotic			Mesic	Udic	
Embal	29.7	Mollisols	Xerolls	Haploxerolls	Vitritorrandic Haploxerolls	Ashy	Glassy			Frigid	Aridic	
Emerson	1.9	Entisols	Orthents	Torriorthents	Xeric Torriorthents	Coarse-loamy over sandy or sandy-skeletal	Mixed	Superactive	Nonacid	Mesic	Aridic	

(continued)

Series name	Area (km^2)	Order	Suborder	Great group	Subgroup	Particle-size class	Mineralogy class	CEC activity class	Reaction class	Soil temp. regime	Soil moisture regime	Other family
Emily	35.4	Mollisols	Xerolls	Haploxerolls	Vitrandic Haploxerolls	Loamy-skeletal	Isotic			Mesic	Xeric	
Encina	148.5	Mollisols	Xerolls	Argixerolls	Calciargidic Argixerolls	Fine	Smectitic			Mesic	Aridic	
Endcreek	22.5	Andisols	Xerands	Vitrixerands	Typic Vitrixerands	Ashy over loamy-skeletal	Amorphic over isotic			Frigid	Xeric	
Endersby	36.8	Mollisols	Xerolls	Haploxerolls	Cumulic Haploxerolls	Coarse-loamy	Mixed	Superactive		Mesic	Xeric	
Enko	506.7	Aridisols	Cambids	Haplocambids	Durinodic Xeric Haplocambids	Coarse-loamy	Mixed	Superactive		Mesic	Aridic	
Era	106.0	Mollisols	Xerolls	Haploxerolls	Vitritorrandic Haploxerolls	Coarse-loamy	Mixed	Superactive		Mesic	Aridic	
Erakatak	495.1	Mollisols	Xerolls	Argixerolls	Vitrandic Argixerolls	Clayey-skeletal	Smectitic			Frigid	Xeric	
Ermabell	1.8	Andisols	Xerands	Vitrixerands	Humic Vitrixerands	Ashy	Glassy			Frigid	Xeric	
Escondia	18.8	Andisols	Xerands	Vitrixerands	Alfic Vitrixerands	Ashy-skeletal over loamy-skeletal	Glassy over isotic			Frigid	Xeric	
Esquatzel	13.4	Mollisols	Xerolls	Haploxerolls	Torrifluventic Haploxerolls	Coarse-silty	Mixed	Superactive		Mesic	Aridic	
Etelka	192.0	Inceptisols	Udepts	Dystrudepts	Oxyaquic Dystrudepts	Fine	Isotic			Mesic	Udic	
Ettersburg	1.2	Inceptisols	Udepts	Dystrudepts	Humic Dystrudepts	Fine-loamy	Mixed	Superactive		Isomesic	Udic	
Euchrand	13.0	Inceptisols	Udepts	Dystrudepts	Lithic Dystrudepts	Loamy-skeletal	Isotic			Frigid	Udic	
Euchre	2.0	Andisols	Aquands	Endoaquands	Alic Endoaquands	Medial over loamy	Ferrihydritic over isotic		Acid	Isomesic	Aquic	
Evans	38.8	Mollisols	Xerolls	Haploxerolls	Cumulic Haploxerolls	Coarse-loamy	Mixed	Superactive		Mesic	Xeric	
Evick	6.2	Entisols	Psamments	Xeropsamments	Lithic Xeropsamments	Sandy	Mixed			Mesic	Xeric	
Exfo	201.7	Entisols	Orthents	Torriorthents	Lithic Torriorthents	Sandy-skeletal	Mixed			Mesic	Aridic	
Fairylawn	24.4	Alfisols	Xeralfs	Durixeralfs	Abruptic Durixeralfs	Fine	Smectitic			Mesic	Xeric	
Faloma	8.9	Mollisols	Aquolls	Endoaquolls	Fluvaquentic Endoaquolls	Coarse-silty over sandy or sandy-skeletal	Mixed	Superactive		Mesic	Aquic	
Fantz	30.5	Mollisols	Xerolls	Haploxerolls	Pachic Ultic Haploxerolls	Loamy-skeletal	Mixed	Superactive		Mesic	Xeric	
Farmell	44.4	Aridisols	Argids	Haplargids	Xeric Haplargids	Fine	Smectitic			Mesic	Aridic	
Farva	218.2	Inceptisols	Xerepts	Haploxerepts	Typic Haploxerepts	Loamy-skeletal	Mixed	Superactive		Frigid	Xeric	
Fawceter	34.6	Andisols	Udands	Fulvudands	Pachic Fulvudands	Medial-skeletal	Ferrihydritic			Isofrigid	Udic	
Fawnspring	10.1	Mollisols	Xerolls	Palexerolls	Vitrandic Palexerolls	Fine	Smectitic			Mesic	Xeric	
Feaginranch	0.7	Mollisols	Aquolls	Cryaquolls	Aquandic Cryaquolls	Fine	Isotic			Cryic	Aquic	
Felcher	694.2	Aridisols	Cambids	Haplocambids	Xeric Haplocambids	Loamy-skeletal	Mixed	Superactive		Mesic	Aridic	
Feltham	14.6	Entisols	Orthents	Torriorthents	Xeric Torriorthents	Sandy	Mixed			Mesic	Aridic	
Fendall	86.3	Inceptisols	Udepts	Humudepts	Andic Humudepts	Fine	Isotic			Isomesic	Udic	
Ferguson	10.4	Andisols	Vitrands	Udivitrands	Typic Udivitrands	Ashy over loamy-skeletal	Amorphic over isotic			Frigid	Udic	

(continued)

Series name	Area (km^2)	Order	Suborder	Great group	Subgroup	Particle-size class	Mineralogy class	CEC activity class	Reaction class	Soil temp. regime	Soil moisture regime	Other family
Fernhaven	77.9	Ultisols	Udults	Paleudults	Typic Paleudults	Fine-loamy	Mixed	Superactive		Mesic	Udic	
Fernwood	71.1	Inceptisols	Udepts	Humudepts	Andic Humudepts	Loamy-skeletal	Isotic			Frigid	Udic	
Ferrelo	23.5	Inceptisols	Udepts	Humudepts	Typic Humudepts	Coarse-loamy	Isotic			Isomesic	Udic	
Fertaline	84.6	Aridisols	Durids	Argidurids	Abruptic Xeric Argidurids	Fine	Smectitic			Frigid	Aridic	
Final	3.0	Aridisols	Argids	Natrargids	Xerertic Natrargids	Fine	Smectitic			Mesic	Aridic	
Fireball	2.9	Aridisols	Argids	Haplargids	Typic Haplargids	Loamy-skeletal	Mixed	Superactive		Mesic	Aridic	
Firelake	20.7	Entisols	Orthents	Torriorthents	Lithic Xeric Torriorthents	Loamy	Mixed	Superactive	Nonacid	Mesic	Aridic	
Fitzwater	186.8	Mollisols	Xerolls	Haploxerolls	Aridic Haploxerolls	Loamy-skeletal	Mixed	Superactive		Frigid	Aridic	
Fivebeaver	548.7	Mollisols	Xerolls	Haploxerolls	Lithic Ultic Haploxerolls	Loamy-skeletal	Isotic			Frigid	Xeric	
Fivebit	160.7	Mollisols	Xerolls	Haploxerolls	Lithic Ultic Haploxerolls	Loamy-skeletal	Mixed	Superactive		Frigid	Xeric	
Fiverivers	5.7	Inceptisols	Udepts	Dystrudepts	Andic Dystrudepts	Fine-loamy	Isotic			Frigid	Udic	
Flagstaff	100.9	Aridisols	Salids	Aquisalids	Typic Aquisalids	Ashy	Glassy			Frigid	Aridic	
Flane	36.8	Inceptisols	Udepts	Dystrudepts	Humic Dystrudepts	Clayey-skeletal	Isotic			Frigid	Udic	
Flank	9.9	Entisols	Orthents	Torriorthents	Lithic Xeric Torriorthents	Ashy-skeletal	Glassy		Nonacid	Frigid	Aridic	
Flarm	1.1	Alfisols	Xeralfs	Palexeralfs	Ultic Palexeralfs	Fine	Mixed	Superactive		Frigid	Xeric	
Floke	247.1	Aridisols	Durids	Argidurids	Abruptic Xeric Argidurids	Clayey	Smectitic			Frigid	Aridic	Shallow
Floras	48.8	Inceptisols	Udepts	Dystrudepts	Humic Dystrudepts	Fine	Isotic			Isomesic	Udic	
Flowerpot	9.6	Inceptisols	Udepts	Humudepts	Aquandic Humudepts	Fine	Isotic			Isomesic	Udic	
Flybow	0.9	Entisols	Orthents	Xerorthents	Lithic Xerorthents	Loamy-skeletal	Mixed	Superactive	Nonacid	Mesic	Xeric	
Flycatcher	9.6	Inceptisols	Udepts	Eutrudepts	Lithic Eutrudepts	Loamy-skeletal	Magnesic			Frigid	Udic	
Flycreek	41.5	Andisols	Vitrands	Udivitrands	Alfic Udivitrands	Ashy over clayey	Amorphic over smectitic			Frigid	Udic	
Flyvalley	3.3	Andisols	Vitrands	Udivitrands	Lithic Udivitrands	Ashy	Amorphic			Frigid	Udic	
Foehlin	41.2	Mollisols	Xerolls	Argixerolls	Typic Argixerolls	Fine-loamy	Mixed	Superactive		Mesic	Xeric	
Foleylake	11.3	Aridisols	Durids	Argidurids	Abruptic Xeric Argidurids	Fine	Smectitic			Frigid	Aridic	
Fopiano	121.3	Mollisols	Xerolls	Argixerolls	Vitrandic Argixerolls	Clayey	Smectitic			Frigid	Xeric	Shallow
Fordice	4.8	Mollisols	Xerolls	Argixerolls	Ultic argixerolls	Loamy-skeletal	Mixed	Superactive		Mesic	Xeric	
Fordney	124.3	Mollisols	Xerolls	Haploxerolls	Torripsammentic Haploxerolls		Mixed			Mesic	Aridic	
Formader	154.4	Andisols	Udands	Hapludands	Alic Hapludands	Medial over loamy	Ferrihydritic over isotic			Mesic	Udic	
Forshey	5.6	Mollisols	Xerolls	Argixerolls	Vitrandic Argixerolls	Fine-loamy	Isotic			Frigid	Xeric	
Fort Rock	97.9	Mollisols	Xerolls	Haploxerolls	Vitritorrandic Haploxerolls	Ashy over sandy or sandy-skeletal	Glassy over mixed			Frigid	Aridic	
Fossilake	4.2	Inceptisols	Aquepts	Halaquepts	Aquandic Halaquepts	Ashy	Glassy		Calcareous	Frigid	Aquic	

(continued)

Series name	Area (km^2)	Order	Suborder	Great group	Subgroup	Particle-size class	Mineralogy class	CEC activity class	Reaction class	Soil temp. regime	Soil moisture regime	Other family
Fourbeaver	34.9	Mollisols	Xerolls	Argixerolls	Lithic Ultic Argixerolls	Clayey-skeletal	Smectitic			Frigid	Xeric	
Fourthcreek	0.6	Andisols	Xerands	Vitrixerands	Typic Vitrixerands	Ashy over loamy	Amorphic over isotic			Frigid	Xeric	
Fourwheel	168.7	Aridisols	Argids	Paleargids	Xeric Paleargids	Fine	Smectitic			Frigid	Aridic	
Frailey	47.6	Inceptisols	Xerepts	Haploxerepts	Typic Haploxerepts	Coarse-loamy	Mixed	Superactive		Mesic	Xeric	
Frankport	5.7	Entisols	Psamments	Udipsamments	Typic Udipsamments	Sandy	Mixed			Isomesic	Udic	
Freels	13.9	Mollisols	Xerolls	Haploxerolls	Cumulic Haploxerolls	Coarse-loamy	Mixed	Superactive		Frigid	Xeric	
Freewater	26.3	Mollisols	Xerolls	Haploxerolls	Fluventic Haploxerolls	Sandy-skeletal	Mixed			Mesic	Xeric	
Freezener	264.7	Alfisols	Xeralfs	Haploxeralfs	Ultic Haploxeralfs	Fine	Mixed	Superactive		Mesic	Xeric	
Fremkle	86.9	Andisols	Xerands	Vitrixerands	Lithic Vitrixerands	Ashy	Glassy			Frigid	Xeric	
Freznik	344.5	Aridisols	Argids	Paleargids	Xeric Paleargids	Fine	Smectitic			Frigid	Aridic	
Fritsland	192.3	Inceptisols	Udepts	Dystrudepts	Humic Dystrudepts	Fine-loamy	Isotic			Mesic	Udic	
Frizzelcreek	1.3	Inceptisols	Cryepts	Humicryepts	Andic Humicryepts	Coarse-loamy	Isotic			Cryic	Xeric	
Frohman	172.0	Aridisols	Durids	Haplodurids	Xeric Haplodurids	Loamy	Mixed	Superactive		Mesic	Aridic	Shallow
Fruitcreek	18.7	Inceptisols	Cryepts	Humicryepts	Vitrixerandic Humicryepts	Loamy-skeletal	Isotic			Cryic	Xeric	
Fryrear	10.7	Andisols	Xerands	Vitrixerands	Humic Vitrixerands	Ashy-skeletal	Glassy			Frigid	Xeric	
Fuego	6.1	Mollisols	Xerolls	Haploxerolls	Typic Haploxerolls	Loamy-skeletal	Mixed	Superactive		Frigid	Xeric	
Gacey	19.5	Mollisols	Xerolls	Durixerolls	Argiduridic Durixerolls	Clayey-skeletal	Smectitic			Mesic	Aridic	Shallow
Gaib	169.4	Mollisols	Xerolls	Argixerolls	Lithic Ultic Argixerolls	Loamy-skeletal	Mixed	Superactive		Frigid	Xeric	
Gamelake	6.1	Inceptisols	Udepts	Humudepts	Typic Humudepts	Loamy-skeletal	Isotic			Frigid	Udic	
Gamgee	24.5	Aridisols	Argids	Natrargids	Haplic Natrargids	Fine-loamy	Mixed	Superactive		Mesic	Aridic	
Gance	49.7	Aridisols	Argids	Haplargids	Durinodic Xeric Haplargids	Clayey-skeletal	Smectitic			Mesic	Aridic	
Gap	38.3	Andisols	Cryands	Vitricryands	Xeric Vitricryands	Ashy over loamy	Amorphic over isotic			Cryic	Xeric	
Gapcot	4.6	Inceptisols	Xerepts	Haploxerepts	Typic Haploxerepts	Loamy	Mixed	Superactive		Mesic	Xeric	Shallow
Garbutt	24.0	Entisols	Orthents	Torriorthents	Typic Torriorthents	Coarse-silty	Mixed	Superactive	Calcareous	Mesic	Aridic	
Gardiner	7.2	Entisols	Psamments	Udipsamments	Typic Udipsamments	Sandy	Mixed			Mesic	Udic	
Gardone	127.7	Mollisols	Xerolls	Haploxerolls	Vitritorrandic Haploxerolls	Ashy	Glassy			Frigid	Aridic	
Gauldy	7.2	Inceptisols	Udepts	Humudepts	Fluventic Humudepts	Coarse-loamy over sandy or sandy-skeletal	Mixed	Superactive		Isomesic	Udic	
Gearhart	17.6	Inceptisols	Udepts	Humudepts	Psammentic Humudepts		Mixed			Isomesic	Udic	
Geebarc	1.0	Mollisols	Cryolls	Haplocryolls	Oxyaquic Haplocryolls	Sandy-skeletal	Mixed			Cryic	Udic	
Geisel	13.0	Inceptisols	Udepts	Humudepts	Typic Humudepts	Fine	Isotic			Isomesic	Udic	
Geisercreek	63.3	Andisols	Vitrands	Udivitrands	Alfic Udivitrands	Ashy over clayey	Amorphic over smectitic			Frigid	Udic	

(continued)

Series name	Area (km^2)	Order	Suborder	Great group	Subgroup	Particle-size class	Mineralogy class	CEC activity class	Reaction class	Soil temp. regime	Soil moisture regime	Other family
Gelderman	22.8	Ultisols	Humults	Haplohumults	Xeric Haplohumults	Fine	Mixed	Active		Mesic	Xeric	
Gellatly	29.4	Mollisols	Xerolls	Argixerolls	Pachic Argixerolls	Fine	Mixed	Superactive		Mesic	Xeric	
Gelsinger	4.0	Mollisols	Xerolls	Argixerolls	Calcic Pachic Argixerolls	Fine	Smectitic			Mesic	Xeric	
Gem	6.9	Mollisols	Xerolls	Argixerolls	Calcic Argixerolls	Fine	Smectitic			Mesic	Xeric	
Genaw	7.8	Aridisols	Argids	Haplargids	Xeric Haplargids	Loamy	Mixed	Superactive		Mesic	Aridic	Shallow
Geppert	97.0	Inceptisols	Xerepts	Dystroxerepts	Typic Dystroxerepts	Loamy-skeletal	Mixed	Superactive		Mesic	Xeric	
Getaway	194.8	Mollisols	Xerolls	Argixerolls	Vitrandic Argixerolls	Loamy-skeletal	Isotic			Frigid	Xeric	
Gilispie	27.5	Mollisols	Cryolls	Argicryolls	Lithic Argicryolls	Loamy	Mixed	Superactive		Cryic	Xeric	
Ginger	3.1	Andisols	Aquands	Melanaquands	Typic Melanaquands	Medial over clayey	Ferrihydritic over isotic		Nonacid	Isomesic	Aquic	
Ginsberg	79.6	Andisols	Udands	Hapludands	Alic Hapludands	Medial over clayey	Ferrihydritic over isotic			Mesic	Udic	
Ginser	63.5	Mollisols	Xerolls	Haploxerolls	Pachic Haploxerolls	Loamy-skeletal	Mixed	Superactive		Frigid	Xeric	
Giranch	5.4	Mollisols	Xerolls	Durixerolls	Vitritorrandic Durixerolls	Ashy-skeletal	Glassy			Frigid	Aridic	
Giveout	6.6	Andisols	Udands	Hapludands	Alic Hapludands	Medial	Ferrihydritic			Frigid	Udic	
Glasgow	1.3	Aridisols	Argids	Haplargids	Xeric Haplargids	Fine	Smectitic			Mesic	Aridic	
Glassbutte	5.4	Mollisols	Xerolls	Argixerolls	Vitritorrandic Argixerolls	Ashy-skeletal	Glassy			Frigid	Xeric	
Glaze	17.2	Andisols	Cryands	Vitricryands	Xeric Vitricryands	Ashy over loamy-skeletal	Glassy over isotic			Cryic	Xeric	
Glencabin	57.7	Mollisols	Xerolls	Haploxerolls	Vitrandic Haploxerolls	Ashy-skeletal	Glassy			Frigid	Xeric	
Gleneden	9.8	Inceptisols	Udepts	Humudepts	Aquic Humudepts	Fine	Isotic			Isomesic	Udic	
Glide	0.9	Andisols	Xerands	Vitrixerands	Humic Vitrixerands	Ashy	Glassy			Mesic	Xeric	
Glohm	31.3	Inceptisols	Udepts	Fragiudepts	Typic Fragiudepts	Fine-silty	Isotic			Mesic	Udic	
Goble	261.5	Inceptisols	Udepts	Fragiudepts	Andic Fragiudepts	Fine-silty	Isotic			Mesic	Udic	
Gochea	13.9	Mollisols	Xerolls	Argixerolls	Argiduridic Argixerolls	Fine-loamy	Mixed	Superactive		Frigid	Xeric	
Goldbeach	1.3	Inceptisols	Udepts	Humudepts	Lithic Humudepts	Loamy-skeletal	Mixed	Active		Mesic	Udic	
Goldrun	87.5	Entisols	Psamments	Torripsamments	Xeric Torripsamments	Sandy	Mixed			Mesic	Aridic	
Golfer	12.7	Mollisols	Xerolls	Haploxerolls	Vitrandic Haploxerolls	Loamy-skeletal	Isotic			Frigid	Xeric	
Goodin	51.2	Alfisols	Xeralfs	Haploxeralfs	Ultic Haploxeralfs	Fine	Mixed	Superactive		Mesic	Xeric	
Gooding	169.6	Aridisols	Argids	Paleargids	Vertic Paleargids	Fine	Smectitic			Mesic	Aridic	
Goodlow	14.0	Inceptisols	Cryepts	Humicryepts	Andic Humicryepts	Loamy-skeletal	Isotic			Cryic	Udic	
Goodrich	21.3	Mollisols	Xerolls	Haploxerolls	Entic Haploxerolls	Coarse-loamy	Mixed	Superactive		Mesic	Xeric	
Goodtack	597.0	Mollisols	Xerolls	Durixerolls	Vitritorrandic Durixerolls	Ashy	Glassy			Frigid	Aridic	Shallow
Goodwin	26.2	Inceptisols	Xerepts	Dystroxerepts	Humic Dystroxerepts	Loamy-skeletal	Mixed	Superactive		Frigid	Xeric	
Goolaway	35.0	Inceptisols	Xerepts	Dystroxerepts	Typic Dystroxerepts	Fine-silty	Mixed	Superactive		Mesic	Xeric	

(continued)

Series name	Area (km^2)	Order	Suborder	Great group	Subgroup	Particle-size class	Mineralogy class	CEC activity class	Reaction class	Soil temp. regime	Soil moisture regime	Other family
Goose Creek	20.7	Mollisols	Xerolls	Haploxerolls	Cumulic Haploxerolls	Fine-loamy	Mixed	Superactive		Mesic	Xeric	
Goose Lake	16.7	Mollisols	Albolls	Argialbolls	Aquandic Argialbolls	Fine	Smectitic			Mesic	Aquic	
Gooserock	62.6	Mollisols	Xerolls	Haploxerolls	Vitritorrandic Haploxerolls	Loamy-skeletal	Mixed	Superactive		Mesic	Aridic	
Gorhamgulch	32.2	Andisols	Vitrands	Udivitrands	Typic Udivitrands	Ashy over loamy	Amorphic over isotic			Frigid	Udic	
Gosney	235.6	Entisols	Psamments	Torripsamments	Lithic Torripsamments	Ashy	Glassy		Nonacid	Mesic	Aridic	
Gradon	81.1	Mollisols	Xerolls	Durixerolls	Argiduridic Durixerolls	Fine-loamy	Mixed	Superactive		Frigid	Aridic	
Granitemountain	27.5	Inceptisols	Cryepts	Humicryepts	Vitrixerandic Humicryepts	Loamy-skeletal	Isotic			Cryic	Xeric	
Grassmountain	4.9	Inceptisols	Udepts	Dystrudepts	Andic Dystrudepts	Fine-loamy	Isotic			Frigid	Udic	
Grassyknob	11.1	Inceptisols	Udepts	Humudepts	Andic Humudepts	Fine-loamy	Isotic			Isomesic	Udic	
Gravden	12.4	Aridisols	Durids	Haplodurids	Xeric Haplodurids	Loamy-skeletal	Mixed	Superactive		Mesic	Aridic	Shallow
Gravecreek	57.8	Inceptisols	Xerepts	Dystroxerepts	Typic Dystroxerepts	Loamy-skeletal	Magnesic			Mesic	Xeric	
Greengulch	16.1	Alfisols	Xeralfs	Haploxeralfs	Ultic Haploxeralfs	Fine	Mixed	Superactive		Mesic	Xeric	
Greenleaf	24.9	Aridisols	Argids	Calciargids	Xeric Calciargids	Fine-silty	Mixed	Superactive		Mesic	Aridic	
Greenmountain	54.9	Mollisols	Xerolls	Durixerolls	Vitritorrandic Durixerolls	Ashy	Glassy			Frigid	Aridic	
Greenscombe	45.6	Mollisols	Xerolls	Haploxerolls	Typic Haploxerolls	Fine-loamy	Mixed	Superactive		Mesic	Xeric	
Greggo	30.0	Inceptisols	Udepts	Eutrudepts	Lithic Eutrudepts	Loamy-skeletal	Magnesic			Mesic	Udic	
Gregory	13.5	Mollisols	Aquolls	Argiaquolls	Typic Argiaquolls	Fine	Smectitic			Mesic	Aquic	
Grell	12.5	Mollisols	Xerolls	Haploxerolls	Lithic Haploxerolls	Loamy-skeletal	Magnesic			Mesic	Xeric	
Grenet	13.4	Andisols	Xerands	Vitrixerands	Typic Vitrixerands	Ashy-skeletal	Glassy			Frigid	Xeric	
Greystoke	66.4	Mollisols	Xerolls	Argixerolls	Pachic Ultic Argixerolls	Loamy-skeletal	Mixed	Superactive		Frigid	Xeric	
Gribble	59.9	Mollisols	Xerolls	Durixerolls	Haplic Durixerolls	Clayey-skeletal	Smectitic			Mesic	Xeric	
Grindbrook	43.0	Inceptisols	Udepts	Humudepts	Oxyaquic Humudepts	Fine-silty	Isotic			Isomesic	Udic	
Grousehill	194.4	Andisols	Cryands	Duricryands	Oxyaquic Duricryands	Medial-skeletal	Amorphic			Cryic	Udic	
Grouslous	50.2	Inceptisols	Udepts	Dystrudepts	Lithic Dystrudepts	Loamy-skeletal	Isotic			Mesic	Udic	
Grubcreek	19.0	Mollisols	Xerolls	Haploxerolls	Vitrandic Haploxerolls	Loamy-skeletal	Isotic			Frigid	Xeric	
Guano	40.0	Aridisols	Argids	Haplargids	Xeric Haplargids	Loamy	Mixed	Superactive		Frigid	Aridic	Shallow
Guerin	4.2	Inceptisols	Udepts	Humudepts	Lithic Humudepts	Loamy-skeletal	Isotic			Isomesic	Udic	
Gulliford	12.2	Entisols	Orthents	Udorthents	Oxyaquic Udorthents	Sandy-skeletal	Mixed			Frigid	Udic	
Gumble	173.2	Aridisols	Argids	Haplargids	Xeric Haplargids	Clayey	Smectitic			Mesic	Aridic	Shallow
Gurdane	190.0	Mollisols	Xerolls	Argixerolls	Pachic Argixerolls	Clayey-skeletal	Smectitic			Mesic	Xeric	
Gurlidawg	2.4	Andisols	Cryands	Vitricryands	Xeric Vitricryands	Ashy-skeletal	Glassy			Cryic	Xeric	
Gustin	69.0	Ultisols	Humults	Palehumults	Aquic Palehumults	Fine	Mixed	Superactive		Mesic	Udic	
Gutridge	216.7	Andisols	Vitrands	Udivitrands	Typic Udivitrands	Ashy over loamy-skeletal	Amorphic over isotic			Frigid	Udic	

(continued)

Series name	Area (km^2)	Order	Suborder	Great group	Subgroup	Particle-size class	Mineralogy class	CEC activity class	Reaction class	Soil temp. regime	Soil moisture regime	Other family
Gwin	649.5	Mollisols	Xerolls	Argixerolls	Lithic Argixerolls	Loamy-skeletal	Mixed	Superactive		Mesic	Xeric	
Gwinly	990.8	Mollisols	Xerolls	Argixerolls	Lithic Argixerolls	Clayey-skeletal	Smectitic			Mesic	Xeric	
Haar	279.0	Entisols	Orthents	Torriorthents	Xeric Torriorthents	Loamy	Mixed	Superactive	Nonacid	Mesic	Aridic	Shallow
Habenome	5.2	Mollisols	Aquolls	Cryaquolls	Typic Cryaquolls	Coarse-loamy	Mixed	Superactive		Cryic	Aquic	
Hack	55.9	Mollisols	Xerolls	Argixerolls	Calcic Argixerolls	Fine-loamy	Mixed	Superactive		Mesic	Xeric	
Hackwood	91.9	Mollisols	Cryolls	Haplocryolls	Pachic Haplocryolls	Fine-loamy	Mixed	Superactive		Cryic	Xeric	
Haflinger	7.1	Inceptisols	Udepts	Humudepts	Entic Humudepts	Sandy-skeletal	Mixed			Mesic	Udic	
Hager	36.7	Aridisols	Durids	Argidurids	Xeric Argidurids	Fine-loamy	Mixed	Superactive		Mesic	Aridic	
Haines	35.6	Inceptisols	Aquepts	Endoaquepts	Typic Endoaquepts	Coarse-silty	Mixed	Superactive	Calcareous	Mesic	Aquic	
Halfway	3.1	Vertisols	Xererts	Haploxererts	Typic Haploxererts	Very-fine	Smectitic			Mesic	Xeric	
Hall Ranch	107.7	Mollisols	Xerolls	Haploxerolls	Vitrandic Haploxerolls	Fine-loamy	Isotic			Frigid	Xeric	
Hallihan	14.7	Andisols	Cryands	Vitricryands	Xeric Vitricryands	Ashy-skeletal	Glassy			Cryic	Xeric	
Hammersley	3.1	Mollisols	Cryolls	Argicryolls	Pachic Argicryolls	Fine	Mixed	Superactive		Cryic	Xeric	
Hankins	362.0	Mollisols	Xerolls	Palexerolls	Vertic Palexerolls	Fine	Smectitic			Frigid	Xeric	
Hanning	76.6	Mollisols	Xerolls	Argixerolls	Pachic Argixerolls	Fine-silty	Mixed	Superactive		Mesic	Xeric	
Hapgood	39.8	Mollisols	Cryolls	Haplocryolls	Pachic Haplocryolls	Loamy-skeletal	Mixed	Superactive		Cryic	Xeric	
Happus	2.4	Inceptisols	Xerepts	Haploxerepts	Vitrandic Haploxerepts	Ashy-pumiceous	Glassy			Mesic	Xeric	
Harana	21.2	Mollisols	Xerolls	Haploxerolls	Cumulic Haploxerolls	Fine-silty	Mixed	Superactive		Mesic	Xeric	
Harcany	119.5	Mollisols	Cryolls	Haplocryolls	Pachic Haplocryolls	Loamy-skeletal	Mixed	Superactive		Cryic	Xeric	
Hardscrabble	30.9	Alfisols	Xeralfs	Palexeralfs	Aquic Palexeralfs	Fine	Mixed	Superactive		Mesic	Xeric	
Hardtrigger	57.7	Aridisols	Argids	Haplargids	Xeric Haplargids	Fine-loamy	Mixed	Superactive		Mesic	Aridic	
Harl	126.3	Andisols	Vitrands	Udivitrands	Typic Udivitrands	Ashy-skeletal over loamy-skeletal	Amorphic over isotic			Frigid	Udic	
Harlow	394.7	Mollisols	Xerolls	Argixerolls	Lithic Argixerolls	Clayey-skeletal	Smectitic			Frigid	Xeric	
Harriman	61.7	Mollisols	Xerolls	Argixerolls	Pachic Argixerolls	Fine-loamy	Mixed	Superactive		Mesic	Xeric	
Harrington	206.5	Inceptisols	Udepts	Humudepts	Typic Humudepts	Loamy-skeletal	Isotic			Mesic	Udic	
Harslow	190.1	Andisols	Udands	Hapludands	Alic Hapludands	Medial-skeletal	Ferrihydritic			Mesic	Udic	
Hart	119.4	Mollisols	Xerolls	Palexerolls	Duric Palexerolls	Fine	Smectitic			Frigid	Xeric	
Hart Camp	39.4	Mollisols	Xerolls	Argixerolls	Aridic Argixerolls	Loamy	Mixed	Superactive		Frigid	Aridic	Shallow
Hasshollow	19.5	Andisols	Cryands	Vitricryands	Alfic Vitricryands	Ashy-skeletal over loamy-skeletal	Amorphic over isotic			Cryic	Udic	
Hayespring	73.6	Mollisols	Xerolls	Durixerolls	Vitritorrandic Durixerolls	Fine-loamy	Isotic			Frigid	Aridic	
Haynap	13.9	Andisols	Cryands	Vitricryands	Humic Vitricryands	Ashy-skeletal	Glassy over amorphic			Cryic	Udic	
Haystack	36.9	Mollisols	Xerolls	Haploxerolls	Calcidic Haploxerolls	Loamy-skeletal	Mixed	Superactive		Mesic	Aridic	
Hazelair	175.4	Mollisols	Xerolls	Haploxerolls	Vertic Haploxerolls	Very-fine	Smectitic			Mesic	Xeric	
Hazelcamp	52.2	Ultisols	Humults	Haplohumults	Typic Haplohumults	Fine	Mixed	Active		Mesic	Udic	
Hebo	25.2	Inceptisols	Aquepts	Humaquepts	Typic Humaquepts	Fine	Isotic		Acid	Isomesic	Aquic	
Heceta	45.1	Entisols	Aquents	Psammaquents	Typic Psammaquents		Mixed			Isomesic	Aquic	
Hehe	115.6	Mollisols	Xerolls	Argixerolls	Vitrandic Argixerolls	Loamy-skeletal	Mixed	Superactive		Mesic	Xeric	

(continued)

Series name	Area (km^2)	Order	Suborder	Great group	Subgroup	Particle-size class	Mineralogy class	CEC activity class	Reaction class	Soil temp. regime	Soil moisture regime	Other family
Helmick	30.8	Inceptisols	Xerepts	Haploxerepts	Vertic Haploxerepts	Very-fine	Mixed	Superactive		Mesic	Xeric	
Helphenstein	85.8	Aridisols	Cambids	Aquicambids	Sodic Aquicambids	Fine-loamy	Mixed	Superactive		Mesic	Aridic	
Helter	10.1	Andisols	Cryands	Vitricryands	Xeric Vitricryands	Ashy over loamy	Glassy over isotic			Cryic	Xeric	
Helvetia	76.8	Mollisols	Xerolls	Argixerolls	Ultic Argixerolls	Fine	Mixed	Superactive		Mesic	Xeric	
Hembre	205.5	Inceptisols	Udepts	Humudepts	Andic Humudepts	Fine-loamy	Isotic			Mesic	Udic	
Hemcross	403.3	Andisols	Udands	Hapludands	Alic Hapludands	Medial	Ferrihydritic			Mesic	Udic	
Henkle	159.0	Andisols	Xerands	Vitrixerands	Lithic Vitrixerands	Ashy-skeletal	Glassy			Frigid	Xeric	
Henley	63.1	Aridisols	Durids	Haplodurids	Aquic Haplodurids	Coarse-loamy	Mixed	Superactive		Mesic	Aridic	
Henline	86.1	Inceptisols	Cryepts	Humicryepts	Typic Humicryepts	Loamy-skeletal	Isotic			Cryic	Udic	
Heppsie	38.6	Mollisols	Xerolls	Haploxerolls	Vertic Haploxerolls	Fine	Smectitic			Mesic	Xeric	
Hermiston	67.1	Mollisols	Xerolls	Haploxerolls	Cumulic Haploxerolls	Coarse-silty	Mixed	Superactive		Mesic	Xeric	
Hershal	19.0	Mollisols	Aquolls	Endoaquolls	Cumulic Endoaquolls	Coarse-silty over sandy or sandy-skeletal	Mixed	Superactive		Mesic	Aquic	
Hesslan	93.8	Mollisols	Xerolls	Haploxerolls	Typic Haploxerolls	Coarse-loamy	Mixed	Superactive		Mesic	Xeric	
Hezel	12.3	Entisols	Orthents	Torriorthents	Xeric Torriorthents	Sandy over loamy	Mixed	Superactive	Nonacid	Mesic	Aridic	
Hibbard	38.3	Mollisols	Xerolls	Durixerolls	Palexerollic Durixerolls	Fine	Smectitic			Mesic	Xeric	
Highcamp	86.5	Andisols	Cryands	Haplocryands	Typic Haplocryands	Medial-skeletal	Amorphic			Cryic	Udic	
Highhorn	10.9	Inceptisols	Xerepts	Haploxerepts	Vitrandic Haploxerepts	Loamy-skeletal	Isotic			Frigid	Xeric	
Hillsboro	15.2	Mollisols	Xerolls	Argixerolls	Ultic Argixerolls	Fine-silty	Mixed	Superactive		Mesic	Xeric	
Hilltish	24.4	Inceptisols	Xerepts	Dystroxerepts	Typic Dystroxerepts	Loamy-skeletal	Mixed	Superactive		Mesic	Xeric	
Hinton	3.6	Entisols	Orthents	Torriorthents	Duric Torriorthents	Coarse-loamy over sandy or sandy-skeletal	Mixed	Superactive	Nonacid	Mesic	Aridic	
Hobit	10.4	Andisols	Cryands	Fulvicryands	Typic Fulvicryands	Medial	Amorphic			Cryic	Udic	
Hoffer	18.8	Andisols	Cryands	Vitricryands	Typic Vitricryands	Ashy over loamy	Amorphic over isotic			Cryic	Udic	
Hogranch	23.8	Mollisols	Xerolls	Haploxerolls	Vitrandic Haploxerolls	Loamy-skeletal	Mixed	Superactive		Frigid	Xeric	
Holcomb	83.7	Mollisols	Albolls	Argialbolls	Typic Argialbolls	Fine	Smectitic			Mesic	Aquic	
Holderman	14.1	Andisols	Cryands	Haplocryands	Typic Haplocryands	Medial-skeletal	Amorphic			Cryic	Udic	
Holland	66.6	Alfisols	Xeralfs	Haploxeralfs	Ultic Haploxeralfs	Fine-loamy	Mixed	Semiactive		Mesic	Xeric	
Holmzie	43.4	Mollisols	Xerolls	Argixerolls	Aridic Argixerolls	Fine	Smectitic			Mesic	Aridic	
Homefield	6.5	Mollisols	Aquolls	Endoaquolls	Cumulic Endoaquolls	Fine-silty	Mixed	Superactive	Calcareous	Frigid	Aquic	
Hondu	77.2	Inceptisols	Xerepts	Haploxerepts	Andic Haploxerepts	Loamy-skeletal	Isotic			Frigid	Xeric	
Honeygrove	696.8	Ultisols	Humults	Palehumults	Typic Palehumults	Fine	Mixed	Active		Mesic	Udic	
Honeymooncan	19.6	Andisols	Vitrands	Udivitrands	Alfic Udivitrands	Ashy over loamy-skeletal	Amorphic over isotic			Frigid	Udic	
Hood	14.7	Alfisols	Xeralfs	Haploxeralfs	Ultic Haploxeralfs	Fine-loamy	Mixed	Superactive		Mesic	Xeric	
Hooly	18.8	Andisols	Aquands	Endoaquands	Typic Endoaquands	Medial over loamy	Mixed over siliceous	Superactive	Nonacid	Mesic	Aquic	

(continued)

Series name	Area (km^2)	Order	Suborder	Great group	Subgroup	Particle-size class	Mineralogy class	CEC activity class	Reaction class	Soil temp. regime	Soil moisture regime	Other family
Hoopal	11.8	Mollisols	Aquolls	Duraquolls	Typic Duraquolls	Coarse-loamy	Mixed	Superactive	Calcareous	Mesic	Aquic	
Hooskanaden	24.6	Alfisols	Udalfs	Hapludalfs	Andic Hapludalfs	Fine	Mixed	Superactive		Isomesic	Udic	
Horeb	28.4	Inceptisols	Udepts	Humudepts	Oxyaquic Humudepts	Fine-loamy	Isotic			Mesic	Udic	
Horning	17.6	Entisols	Orthents	Torriorthents	Vitrandic Torriorthents	Ashy	Glassy		Calcareous	Frigid	Aridic	
Horseprairie	10.0	Inceptisols	Udepts	Humudepts	Andic Humudepts	Fine-loamy	Isotic			Isomesic	Udic	
Hosley	16.3	Mollisols	Aquolls	Duraquolls	Natric Duraquolls	Fine-loamy	Mixed	Superactive		Mesic	Aquic	
Hot Lake	80.7	Andisols	Xerands	Haploxerands	Aquic Haploxerands	Medial over loamy	Mixed over siliceous	Superactive		Mesic	Xeric	
Housefield	45.9	Mollisols	Aquolls	Endoaquolls	Cumulic Endoaquolls	Fine-silty	Mixed	Superactive		Frigid	Aquic	
Houstake	71.2	Mollisols	Xerolls	Haploxerolls	Vitritorrandic Haploxerolls	Coarse-loamy	Mixed	Superactive		Mesic	Aridic	
Houstenader	3.0	Mollisols	Udolls	Argiudolls	Aquic Argiudolls	Fine-loamy	Mixed	Superactive		Mesic	Udic	
Howash	163.0	Andisols	Vitrands	Udivitrands	Humic Udivitrands	Ashy-skeletal	Glassy			Frigid	Udic	
Howcan	4.0	Mollisols	Xerolls	Argixerolls	Typic Argixerolls	Loamy-skeletal	Mixed	Superactive		Frigid	Xeric	
Howmeadows	15.4	Alfisols	Aqualfs	Epiaqualfs	Vertic Epiaqualfs	Fine	Smectitic			Frigid	Aquic	
Hoxie	7.4	Inceptisols	Aquepts	Cryaquepts	Humic Cryaquepts	Fine-silty	Mixed	Superactive	Nonacid	Cryic	Aquic	
Huberly	13.5	Inceptisols	Aquepts	Fragiaquepts	Typic Fragiaquepts	Fine-silty	Mixed	Superactive		Mesic	Aquic	
Hudspeth	12.6	Mollisols	Xerolls	Palexerolls	Ultic Palexerolls	Clayey-skeletal	Smectitic			Frigid	Xeric	
Huffling	0.5	Ultisols	Aquults	Umbraquults	Typic Umbraquults	Fine	Mixed	Superactive		Isomesic	Aquic	
Hukill	30.3	Alfisols	Xeralfs	Rhodoxeralfs	Typic Rhodoxeralfs	Fine	Halloysitic			Mesic	Xeric	
Hullt	21.5	Inceptisols	Xerepts	Humixerepts	Typic Humixerepts	Fine-loamy	Mixed	Superactive		Mesic	Xeric	
Humarel	107.1	Mollisols	Xerolls	Argixerolls	Vitrandic Argixerolls	Clayey-skeletal	Smectitic			Frigid	Xeric	
Hummington	119.7	Andisols	Cryands	Haplocryands	Typic Haplocryands	Medial-skeletal	Amorphic			Cryic	Udic	
Hunewill	28.2	Aridisols	Argids	Haplargids	Xeric Haplargids	Loamy-skeletal	Mixed	Superactive		Mesic	Aridic	
Hunsaker	38.4	Alfisols	Udalfs	Hapludalfs	Andic Hapludalfs	Fine-loamy	Isotic			Frigid	Udic	
Hunterscove	13.7	Ultisols	Humults	Haplohumults	Typic Haplohumults	Fine	Isotic			Isomesic	Udic	
Huntley	2.2	Mollisols	Udolls	Hapludolls	Lithic Hapludolls	Loamy	Mixed	Superactive		Mesic	Udic	
Huntrock	6.4	Inceptisols	Xerepts	Haploxerepts	Vitrandic Haploxerepts	Loamy-skeletal	Isotic			Frigid	Xeric	
Hurryback	1.8	Mollisols	Xerolls	Argixerolls	Pachic Argixerolls	Fine-loamy	Mixed	Superactive		Frigid	Xeric	
Hurwal	85.5	Mollisols	Xerolls	Argixerolls	Vitrandic Argixerolls	Fine-silty	Mixed	Superactive		Frigid	Xeric	
Hutchinson	17.0	Mollisols	Xerolls	Durixerolls	Typic Durixerolls	Fine	Smectitic			Mesic	Xeric	
Hutchley	80.9	Mollisols	Xerolls	Argixerolls	Lithic Argixerolls	Loamy-skeletal	Mixed	Superactive		Frigid	Xeric	
Hutson	18.9	Andisols	Cryands	Vitricryands	Typic Vitricryands	Ashy	Amorphic			Cryic	Udic	
Hyall	30.1	Mollisols	Xerolls	Argixerolls	Torrertic Argixerolls	Clayey-skeletal	Smectitic			Mesic	Xeric	
Icedee	9.6	Andisols	Cryands	Vitricryands	Alfic Vitricryands	Ashy over loamy	Amorphic over isotic			Cryic	Udic	
Icene	95.4	Aridisols	Salids	Aquisalids	Typic Aquisalids	Fine-loamy	Mixed	Superactive		Mesic	Aridic	
Igert	505.6	Aridisols	Argids	Haplargids	Durinodic Xeric Haplargids	Fine-loamy	Mixed	Superactive		Mesic	Aridic	
Illahee	59.2	Inceptisols	Udepts	Humudepts	Typic Humudepts	Loamy-skeletal	Isotic			Frigid	Udic	

(continued)

Series name	Area (km^2)	Order	Suborder	Great group	Subgroup	Particle-size class	Mineralogy class	CEC activity class	Reaction class	Soil temp. regime	Soil moisture regime	Other family
Imbler	49.2	Mollisols	Xerolls	Haploxerolls	Pachic Haploxerolls	Coarse-loamy	Mixed	Superactive		Mesic	Xeric	
Immiant	2.6	Mollisols	Xerolls	Argixerolls	Typic Argixerolls	Fine-loamy	Mixed	Superactive		Mesic	Xeric	
Immig	223.5	Mollisols	Xerolls	Argixerolls	Typic Argixerolls	Clayey-skeletal	Smectitic			Mesic	Xeric	
Imnaha	173.9	Mollisols	Xerolls	Argixerolls	Vitrandic Argixerolls	Loamy-skeletal	Mixed	Superactive		Frigid	Xeric	
Inkler	66.7	Inceptisols	Xerepts	Haploxerepts	Andic Haploxerepts	Loamy-skeletal	Isotic			Frigid	Xeric	
Ipsoot	10.3	Andisols	Cryands	Vitricryands	Xeric Vitricryands	Ashy over pumiceous or cindery	Glassy			Cryic	Xeric	
Iris	3.9	Mollisols	Xerolls	Haploxerolls	Duridic Haploxerolls	Fine-silty	Mixed	Superactive		Mesic	Xeric	
Irma	7.8	Inceptisols	Udepts	Dystrudepts	Humic Dystrudepts	Fine-loamy	Mixed	Active		Mesic	Udic	
Ironside	78.7	Mollisols	Xerolls	Haploxerolls	Vitrandic Haploxerolls	Loamy-skeletal	Mixed	Superactive		Frigid	Xeric	
Irrigon	32.0	Aridisols	Cambids	Haplocambids	Xeric Haplocambids	Coarse-loamy	Mixed	Superactive		Mesic	Aridic	
Itca	16.1	Mollisols	Xerolls	Argixerolls	Lithic Argixerolls	Clayey-skeletal	Smectitic			Frigid	Xeric	
Izee	28.7	Mollisols	Xerolls	Haploxerolls	Vitrandic Haploxerolls	Fine-loamy	Mixed	Superactive		Frigid	Xeric	
Jacksplace	82.7	Mollisols	Xerolls	Argixerolls	Vitritorrandic Argixerolls	Ashy-skeletal	Glassy			Frigid	Xeric	
James Canyon	0.5	Mollisols	Aquolls	Endoaquolls	Cumulic Endoaquolls	Fine-loamy	Mixed	Superactive		Mesic	Aquic	
Jayar	150.2	Inceptisols	Xerepts	Dystroxerepts	Typic Dystroxerepts	Loamy-skeletal	Mixed	Superactive		Frigid	Xeric	
Jenny	0.0	Vertisols	Xererts	Calcixererts	Entic Calcixererts	Fine	Smectitic			Mesic	Xeric	
Jerome	6.2	Inceptisols	Aquepts	Epiaquepts	Typic Epiaquepts	Coarse-loamy over clayey	Mixed over smectitic	Superactive	Nonacid	Mesic	Aquic	
Jesse Camp	8.4	Aridisols	Cambids	Haplocambids	Xeric Haplocambids	Fine-silty	Mixed	Superactive		Frigid	Aridic	
Jett	112.9	Mollisols	Xerolls	Haploxerolls	Cumulic Haploxerolls	Fine-silty	Mixed	Superactive		Mesic	Xeric	
Jimbo	34.2	Inceptisols	Udepts	Humudepts	Andic Humudepts	Coarse-loamy	Isotic			Mesic	Udic	
Jimgreen	13.3	Histosols	Saprists	Haplosaprists	Hemic Haplosaprists				Euic	Frigid	Aquic	
Joeney	5.9	Spodosols	Aquods	Duraquods	Typic Duraquods	Loamy	Isotic			Isomesic	Aquic	Ortstein, shallow
Jojo	56.2	Andisols	Cryands	Vitricryands	Typic Vitricryands	Medial-skeletal	Amorphic			Cryic	Udic	
Jorn	14.3	Mollisols	Xerolls	Argixerolls	Vitrandic Argixerolls	Fine	Smectitic			Mesic	Xeric	
Jory	872.9	Ultisols	Humults	Palehumults	Xeric Palehumults	Fine	Mixed	Active		Mesic	Udic	
Josephine	521.4	Ultisols	Xerults	Haploxerults	Typic Haploxerults	Fine-loamy	Mixed	Superactive		Mesic	Xeric	
JOSSET	3.4	Mollisols	Xerolls	Haploxerolls	Cumulic Haploxerolls	Coarse-loamy over sandy or sandy-skeletal	Mixed	Superactive		Frigid	Xeric	
Jumpoff	22.0	Alfisols	Xeralfs	Haploxeralfs	Ultic Haploxeralfs	Fine	Mixed	Superactive		Mesic	Xeric	
Kahler	113.1	Mollisols	Xerolls	Haploxerolls	Vitrandic Haploxerolls	Fine-loamy	Isotic			Frigid	Xeric	
Kahneeta	5.4	Mollisols	Aquolls	Duraquolls	Argic Duraquolls	Clayey-skeletal	Smectitic			Mesic	Aquic	
Kamela	249.9	Inceptisols	Xerepts	Haploxerepts	Vitrandic Haploxerepts	Loamy-skeletal	Isotic			Frigid	Xeric	
Kanid	249.9	Inceptisols	Xerepts	Dystroxerepts	Typic Dystroxerepts	Loamy-skeletal	Mixed	Superactive		Mesic	Xeric	
Kanlee	5.7	Mollisols	Xerolls	Argixerolls	Typic Argixerolls	Fine-loamy	Mixed	Superactive		Frigid	Xeric	
Kanutchan	6.1	Vertisols	Aquerts	Endoaquerts	Xeric Endoaquerts	Fine	Smectitic			Frigid	Aquic	

(continued)

Series name	Area (km^2)	Order	Suborder	Great group	Subgroup	Particle-size class	Mineralogy class	CEC activity class	Reaction class	Soil temp. regime	Soil moisture regime	Other family
Kaskela	63.4	Vertisols	Xererts	Haploxererts	Typic Haploxererts	Very-fine	Smectitic			Mesic	Xeric	
Keating	60.9	Mollisols	Xerolls	Argixerolls	Typic Argixerolls	Fine	Smectitic			Mesic	Xeric	
Kecko	13.8	Aridisols	Calcids	Haplocalcids	Xeric Haplocalcids	Coarse-loamy	Mixed	Superactive		Mesic	Aridic	
Keel	156.6	Andisols	Cryands	Haplocryands	Typic Haplocryands	Medial	Amorphic			Cryic	Udic	
Kegler	43.2	Mollisols	Xerolls	Durixerolls	Haploduridic Durixerolls	Fine-loamy	Mixed	Superactive		Frigid	Xeric	
Kenusky	8.1	Ultisols	Aquults	Paleaquults	Umbric Paleaquults	Fine	Isotic			Mesic	Aquic	
Kerby	21.7	Inceptisols	Xerepts	Haploxerepts	Typic Haploxerepts	Fine-loamy	Mixed	Superactive		Mesic	Xeric	
Kerrfield	75.0	Aridisols	Cambids	Haplocambids	Durinodic Xeric Haplocambids	Coarse-loamy	Mixed	Superactive		Mesic	Aridic	
Ketchly	51.1	Alfisols	Xeralfs	Haploxeralfs	Vitrandic Haploxeralfs	Fine-loamy	Isotic			Frigid	Xeric	
Kettenbach	83.1	Mollisols	Xerolls	Argixerolls	Pachic Argixerolls	Loamy-skeletal	Mixed	Superactive		Mesic	Xeric	
Kettlecreek	9.9	Mollisols	Xerolls	Haploxerolls	Vitrandic Haploxerolls	Loamy-skeletal	Isotic			Frigid	Xeric	
Kewake	131.5	Entisols	Psamments	Torripsamments	Vitrandic Torripsamments	Sandy	Mixed			Mesic	Aridic	
Kiesel	4.6	Aridisols	Argids	Natrargids	Xeric Natrargids	Fine	Smectitic			Mesic	Aridic	
Kilchis	121.8	Inceptisols	Udepts	Humudepts	Lithic Humudepts	Loamy-skeletal	Isotic			Mesic	Udic	
Killam	10.6	Andisols	Udands	Fulvudands	Pachic Fulvudands	Medial-skeletal	Ferrihydritic			Isofrigid	Udic	
Killet	4.7	Inceptisols	Xerepts	Dystroxerepts	Humic Dystroxerepts	Fine-loamy	Mixed	Superactive		Frigid	Xeric	
Kilmerque	33.1	Mollisols	Xerolls	Haploxerolls	Vitrandic Haploxerolls	Coarse-loamy	Isotic			Frigid	Xeric	
Kilowan	34.3	Inceptisols	Udepts	Dystrudepts	Humic Dystrudepts	Fine	Isotic			Mesic	Udic	
Kimberly	75.2	Mollisols	Xerolls	Haploxerolls	Torrifluventic Haploxerolls	Coarse-loamy	Mixed	Superactive		Mesic	Aridic	
Kingbolt	124.6	Andisols	Xerands	Vitrixerands	Typic Vitrixerands	Ashy over loamy-skeletal	Amorphic over isotic			Frigid	Xeric	
Kingsriver	3.0	Mollisols	Aquolls	Endoaquolls	Cumulic Endoaquolls	Coarse-loamy	Mixed	Superactive		Mesic	Aquic	
Kinney	787.7	Inceptisols	Udepts	Humudepts	Andic Humudepts	Fine-loamy	Isotic			Mesic	Udic	
Kinton	60.6	Inceptisols	Xerepts	Fragixerepts	Typic Fragixerepts	Fine-silty	Mixed	Superactive		Mesic	Xeric	
Kinzel	75.2	Andisols	Cryands	Fulvicryands	Typic Fulvicryands	Medial-skeletal	Amorphic			Cryic	Udic	
Kiona	108.6	Aridisols	Cambids	Haplocambids	Xeric Haplocambids	Loamy-skeletal	Mixed	Superactive		Mesic	Aridic	
Kirk	107.6	Andisols	Aquands	Cryaquands	Typic Cryaquands	Ashy-pumiceous	Glassy		Nonacid	Cryic	Aquic	
Kirkendall	54.1	Inceptisols	Udepts	Humudepts	Oxyaquic Humudepts	Fine-silty	Mixed	Superactive		Mesic	Udic	
Kishwalk	191.7	Mollisols	Xerolls	Argixerolls	Pachic Argixerolls	Clayey-skeletal	Smectitic			Mesic	Xeric	
Kittleson	6.5	Andisols	Cryands	Vitricryands	Xeric Vitricryands	Ashy	Glassy			Cryic	Xeric	
Klamath	96.0	Mollisols	Aquolls	Cryaquolls	Cumulic Cryaquolls	Fine-silty	Mixed	Superactive		Cryic	Aquic	
Klicker	1200.7	Mollisols	Xerolls	Argixerolls	Vitrandic Argixerolls	Loamy-skeletal	Isotic			Frigid	Xeric	
Klickitat	1019.5	Inceptisols	Udepts	Humudepts	Typic Humudepts	Loamy-skeletal	Isotic			Mesic	Udic	
Klickson	185.5	Mollisols	Xerolls	Argixerolls	Vitrandic Argixerolls	Loamy-skeletal	Isotic			Frigid	Xeric	
Klistan	383.7	Andisols	Udands	Hapludands	Alic Hapludands	Medial-skeletal	Ferrihydritic			Mesic	Udic	

(continued)

Series name	Area (km^2)	Order	Suborder	Great group	Subgroup	Particle-size class	Mineralogy class	CEC activity class	Reaction class	Soil temp. regime	Soil moisture regime	Other family
Klooqueh	4.9	Ultisols	Humults	Palehumults	Typic Palehumults	Fine	Mixed	Superactive		Isomesic	Udic	
Klootchie	495.3	Andisols	Udands	Fulvudands	Typic Fulvudands	Medial	Ferrihydritic			Isomesic	Udic	
Knapke	17.0	Mollisols	Xerolls	Haploxerolls	Entic Ultic Haploxerolls	Loamy-skeletal	Mixed	Superactive		Mesic	Xeric	
Knappa	33.5	Inceptisols	Udepts	Humudepts	Andic Humudepts	Fine-silty	Isotic			Isomesic	Udic	
Koehler	95.4	Aridisols	Durids	Haplodurids	Xeric Haplodurids	Sandy	Mixed			Mesic	Aridic	
Kosh	3.5	Mollisols	Xerolls	Haploxerolls	Lithic Ultic Haploxerolls	Sandy-skeletal	Mixed			Frigid	Xeric	
Krackle	59.7	Mollisols	Cryolls	Haplocryolls	Xeric Haplocryolls	Loamy-skeletal	Mixed	Superactive		Cryic	Xeric	
Krebs	24.1	Mollisols	Xerolls	Argixerolls	Calciargidic Argixerolls	Fine	Smectitic			Mesic	Aridic	
Kubli	12.5	Mollisols	Xerolls	Haploxerolls	Aquic Haploxerolls	Fine-loamy over clayey	Mixed	Superactive		Mesic	Xeric	
Kuck	0.4	Mollisols	Xerolls	Argixerolls	Vertic Argixerolls	Fine	Smectitic			Mesic	Xeric	
Kuckup	1.4	Andisols	Cryands	Haplocryands	Vitric Haplocryands	Medial over pumiceous or cindery	Amorphic over glassy			Cryic	Udic	
Kuhl	10.0	Mollisols	Xerolls	Haploxerolls	Lithic Haploxerolls	Loamy	Mixed	Superactive		Mesic	Xeric	
Kunaton	56.8	Aridisols	Durids	Argidurids	Abruptic Xeric Argidurids	Clayey	Smectitic			Mesic	Aridic	Shallow
Kunceider	147.0	Mollisols	Xerolls	Haploxerolls	Aridic Lithic Haploxerolls	Ashy-skeletal	Glassy			Frigid	Aridic	
Kusu	18.0	Andisols	Xerands	Vitrixerands	Typic Vitrixerands	Ashy-skeletal	Glassy			Frigid	Xeric	
Kutcher	68.2	Andisols	Vitrands	Udivitrands	Alfic Udivitrands	Ashy-skeletal over loamy-skeletal	Amorphic over isotic			Frigid	Udic	
Kweo	8.5	Andisols	Cryands	Vitricryands	Xeric Vitricryands	Ashy over pumiceous or cindery	Glassy			Cryic	Xeric	
La Grande	66.9	Mollisols	Xerolls	Haploxerolls	Pachic Haploxerolls	Fine-silty	Mixed	Superactive		Mesic	Xeric	
Labish	9.9	Inceptisols	Aquepts	Humaquepts	Cumulic Humaquepts	Fine	Smectitic		Acid	Mesic	Aquic	
Labuck	11.6	Inceptisols	Xerepts	Haploxerepts	Vitrandic Haploxerepts	Coarse-loamy	Isotic			Frigid	Xeric	
Lackeyshole	110.2	Andisols	Cryands	Vitricryands	Typic Vitricryands	Ashy over loamy-skeletal	Amorphic over isotic			Cryic	Udic	
Lacy	1.1	Mollisols	Xerolls	Argixerolls	Lithic Ultic Argixerolls	Loamy-skeletal	Mixed	Superactive		Mesic	Xeric	
Ladd	23.7	Mollisols	Xerolls	Argixerolls	Typic Argixerolls	Fine-loamy	Mixed	Superactive		Mesic	Xeric	
Laderly	167.8	Andisols	Udands	Hapludands	Alic Hapludands	Medial-skeletal	Ferrihydritic			Frigid	Udic	
Ladycomb	14.5	Mollisols	Xerolls	Haploxerolls	Aridic Lithic Haploxerolls	Loamy	Mixed	Superactive		Mesic	Aridic	
LaFollette	9.1	Mollisols	Xerolls	Haploxerolls	Vitritorrandic Haploxerolls	Coarse-loamy	Mixed	Superactive		Mesic	Aridic	
Laidlaw	53.2	Andisols	Xerands	Vitrixerands	Humic Vitrixerands	Ashy	Glassy			Frigid	Xeric	
Lakefork	31.4	Andisols	Vitrands	Udivitrands	Typic Udivitrands	Ashy-skeletal over loamy-skeletal	Amorphic over isotic			Frigid	Udic	
Lakeview	95.4	Mollisols	Xerolls	Haploxerolls	Cumulic Haploxerolls	Fine-loamy	Mixed	Superactive		Mesic	Xeric	
Laki	81.8	Mollisols	Xerolls	Haploxerolls	Typic Haploxerolls	Fine-loamy	Mixed	Superactive		Mesic	Xeric	

(continued)

Series name	Area (km^2)	Order	Suborder	Great group	Subgroup	Particle-size class	Mineralogy class	CEC activity class	Reaction class	Soil temp. regime	Soil moisture regime	Other family
Lalos	2.7	Aridisols	Cambids	Haplocambids	Sodic Xeric Haplocambids	Coarse-silty	Mixed	Superactive		Mesic	Aridic	
Lamath	0.7	Inceptisols	Aquepts	Endoaquepts	Aquandic Endoaquepts	Ashy over sandy or sandy-skeletal	Glassy over mixed		Calcareous	Mesic	Aquic	
Lambranch	2.7	Aridisols	Argids	Haplargids	Xeric Haplargids	Clayey-skeletal	Smectitic			Mesic	Aridic	
Lambring	349.2	Mollisols	Xerolls	Haploxerolls	Pachic Haploxerolls	Loamy-skeletal	Mixed	Superactive		Frigid	Xeric	
Lamonta	93.7	Mollisols	Xerolls	Durixerolls	Abruptic Argiduridic Durixerolls	Fine	Smectitic			Mesic	Xeric	
Lamulita	47.1	Mollisols	Xerolls	Argixerolls	Vitrandic Argixerolls	Clayey-skeletal	Smectitic			Frigid	Xeric	
Landermeyer	11.3	Mollisols	Xerolls	Haploxerolls	Vitritorrandic Haploxerolls	Fine-loamy	Mixed	Superactive		Mesic	Aridic	
Langellain	35.1	Alfisols	Xeralfs	Haploxeralfs	Ultic Haploxeralfs	Fine-loamy over clayey	Mixed	Superactive		Mesic	Xeric	
Langlois	29.1	Entisols	Aquents	Fluvaquents	Typic Fluvaquents	Fine	Mixed	Superactive	Nonacid	Isomesic	Aquic	
Langrell	37.8	Mollisols	Xerolls	Haploxerolls	Pachic Haploxerolls	Loamy-skeletal	Mixed	Superactive		Mesic	Xeric	
Langslet	13.5	Aridisols	Cambids	Aquicambids	Xeric Aquicambids	Fine	Smectitic			Frigid	Aridic	
Lapham	32.0	Mollisols	Xerolls	Haploxerolls	Vitritorrandic Haploxerolls	Ashy over loamy-skeletal	Glassy over isotic			Frigid	Aridic	
Lapine	2513.1	Andisols	Cryands	Vitricryands	Xeric Vitricryands	Ashy-pumiceous	Glassy			Cryic	Xeric	
Larabee	370.6	Mollisols	Xerolls	Argixerolls	Vitrandic Argixerolls	Loamy-skeletal	Isotic			Frigid	Xeric	
Larmine	59.0	Inceptisols	Xerepts	Haploxerepts	Lithic Haploxerepts	Loamy-skeletal	Mixed	Superactive		Mesic	Xeric	
Lasere	86.5	Mollisols	Xerolls	Palexerolls	Typic Palexerolls	Fine	Smectitic			Mesic	Xeric	
Lassen	0.6	Vertisols	Xererts	Haploxererts	Leptic Haploxererts	Fine	Smectitic			Mesic	Xeric	
Lastance	23.9	Spodosols	Cryods	Haplocryods	Typic Haplocryods	Loamy-skeletal	Isotic			Cryic	Udic	
Lastcall	70.2	Mollisols	Xerolls	Argixerolls	Vitritorrandic Argixerolls	Ashy	Glassy			Frigid	Xeric	
Lather	101.1	Histosols	Hemists	Haplohemists	Limnic Haplohemists		Diatomaceous		Euic	Frigid	Aquic	
Latourell	61.6	Alfisols	Xeralfs	Haploxeralfs	Ultic Haploxeralfs	Fine-loamy	Mixed	Superactive		Mesic	Xeric	
Laufer	17.9	Mollisols	Xerolls	Argixerolls	Lithic Argixerolls	Clayey-skeletal	Smectitic			Mesic	Xeric	
Laurelwood	155.5	Alfisols	Xeralfs	Haploxeralfs	Ultic Haploxeralfs	Fine-silty	Mixed	Active		Mesic	Xeric	
Lavey	17.2	Mollisols	Xerolls	Argixerolls	Calciargidic Argixerolls	Fine	Smectitic			Mesic	Aridic	
Lawen	131.1	Mollisols	Xerolls	Argixerolls	Calciargidic Argixerolls	Coarse-loamy	Mixed	Superactive		Frigid	Aridic	
Lawyer	8.2	Mollisols	Xerolls	Argixerolls	Pachic Ultic Argixerolls	Loamy-skeletal	Mixed	Superactive		Mesic	Xeric	
Laycock	16.6	Mollisols	Xerolls	Haploxerolls	Lithic Ultic Haploxerolls	Loamy-skeletal	Mixed	Superactive		Frigid	Xeric	
Leathers	39.5	Aridisols	Cambids	Haplocambids	Sodic Xeric Haplocambids	Coarse-loamy	Mixed	Superactive		Mesic	Aridic	
Lebam	46.2	Andisols	Udands	Fulvudands	Typic Fulvudands	Medial over clayey	Ferrihydritic over isotic			Isomesic	Udic	
Leemorris	32.3	Mollisols	Cryolls	Argicryolls	Pachic Argicryolls	Fine	Smectitic			Cryic	Xeric	
Leepcreek	2.0	Mollisols	Xerolls	Argixerolls	Vertic Argixerolls	Fine	Smectitic			Mesic	Xeric	

(continued)

Series name	Area (km^2)	Order	Suborder	Great group	Subgroup	Particle-size class	Mineralogy class	CEC activity class	Reaction class	Soil temp. regime	Soil moisture regime	Other family
Leespeak	1.1	Inceptisols	Cryepts	Humicryepts	Vitrixerandic Humicryepts	Loamy-skeletal	Isotic			Cryic	Xeric	
Leevan	33.3	Mollisols	Xerolls	Argixerolls	Typic Argixerolls	Clayey-skeletal	Smectitic			Frigid	Xeric	
Legler	122.8	Aridisols	Cambids	Haplocambids	Xeric Haplocambids	Fine-loamy	Mixed	Superactive		Mesic	Aridic	
Lemoncreek	8.9	Inceptisols	Xerepts	Haploxerepts	Vitrandic Haploxerepts	Loamy-skeletal	Isotic			Frigid	Xeric	
Lemonex	33.1	Mollisols	Xerolls	Argixerolls	Vitrandic Argixerolls	Fine	Magnesic			Frigid	Xeric	
Lempira	33.8	Andisols	Udands	Hapludands	Typic Hapludands	Medial	Amorphic			Frigid	Udic	
Leopold	2.6	Inceptisols	Udepts	Dystrudepts	Andic Dystrudepts	Fine-loamy	Isotic			Frigid	Udic	
Lequieu	0.0	Entisols	Orthents	Torriorthents	Lithic Xeric Torriorthents	Loamy-skeletal	Mixed	Superactive	Nonacid	Mesic	Aridic	
Lerrow	12.3	Mollisols	Xerolls	Argixerolls	Vitritorrandic Argixerolls	Fine	Smectitic			Frigid	Xeric	
Lettia	140.7	Alfisols	Xeralfs	Haploxeralfs	Ultic Haploxeralfs	Fine-loamy	Mixed	Superactive		Mesic	Xeric	
Lickskillet	1936.4	Mollisols	Xerolls	Haploxerolls	Aridic Lithic Haploxerolls	Loamy-skeletal	Mixed	Superactive		Mesic	Aridic	
Limberjim	995.5	Andisols	Vitrands	Udivitrands	Alfic Udivitrands	Ashy over loamy-skeletal	Amorphic over isotic			Frigid	Udic	
Limpy	0.4	Inceptisols	Udepts	Humudepts	Lithic Humudepts	Loamy-skeletal	Isotic			Frigid	Udic	
Linecreek	36.2	Mollisols	Xerolls	Haploxerolls	Vitrandic Haploxerolls	Ashy-skeletal	Glassy			Frigid	Xeric	
Linkletter	47.7	Aridisols	Argids	Petroargids	Duric xeric Petroargids	Fine	Smectitic			Mesic	Aridic	
Linksterly	38.0	Andisols	Cryands	Vitricryands	Humic Vitricryands	Ashy	Amorphic			Cryic	Udic	
Linslaw	23.5	Alfisols	Xeralfs	Haploxeralfs	Aquultic Haploxeralfs	Fine	Mixed	Superactive		Mesic	Xeric	
Lint	38.0	Andisols	Udands	Fulvudands	Typic Fulvudands	Medial	Ferrihydritic			Isomesic	Udic	
Linville	0.0	Mollisols	Xerolls	Haploxerolls	Pachic Haploxerolls	Fine-loamy	Mixed	Superactive		Mesic	Xeric	
Lithgow	179.2	Aridisols	Argids	Haplargids	Xeric Haplargids	Loamy-skeletal	Mixed	Superactive		Mesic	Aridic	
Littlefawn	13.2	Alfisols	Xeralfs	Haploxeralfs	Vitrandic Haploxeralfs	Fine	Smectitic			Mesic	Xeric	
Littlesand	20.2	Inceptisols	Xerepts	Dystroxerepts	Typic Dystroxerepts	Fine-loamy	Mixed	Superactive		Mesic	Xeric	
Lizard	15.2	Mollisols	Xerolls	Palexerolls	Vitrandic Palexerolls	Clayey-skeletal	Smectitic			Frigid	Xeric	
Llaorock	100.9	Andisols	Cryands	Haplocryands	Vitric Haplocryands	Medial-skeletal	Amorphic			Cryic	Udic	
Lobert	76.8	Mollisols	Xerolls	Haploxerolls	Pachic Haploxerolls	Coarse-loamy	Mixed	Superactive		Frigid	Xeric	
Locane	42.6	Aridisols	Argids	Haplargids	Lithic Xeric Haplargids	Clayey-skeletal	Smectitic			Frigid	Aridic	
Locoda	33.7	Entisols	Aquents	Fluvaquents	Typic Fluvaquents	Fine-silty	Mixed	Superactive	Acid	Mesic	Aquic	
Locolake	39.5	Aridisols	Durids	Natridurids	Typic Natridurids	Loamy	Mixed	Superactive		Frigid	Aridic	Shallow
Loeb	15.3	Ultisols	Humults	Haplohumults	Typic Haplohumults	Fine	Isotic			Isomesic	Udic	
Lofftus	7.8	Aridisols	Durids	Haplodurids	Aquicambidic Haplodurids	Ashy	Glassy			Mesic	Aridic	
Logdell	18.6	Mollisols	Xerolls	Haploxerolls	Lithic Ultic Haploxerolls	Loamy-skeletal	Mixed	Superactive		Frigid	Xeric	

(continued)

Series name	Area (km^2)	Order	Suborder	Great group	Subgroup	Particle-size class	Mineralogy class	CEC activity class	Reaction class	Soil temp. regime	Soil moisture regime	Other family
Logsden	21.4	Inceptisols	Udepts	Humudepts	Typic Humudepts	Fine-silty	Isotic			Isomesic	Udic	
Logsprings	5.3	Alfisols	Xeralfs	Haploxeralfs	Vitrandic Haploxeralfs	Fine-loamy over clayey	Mixed over smectitic	Superactive		Mesic	Xeric	
Lolak	45.5	Inceptisols	Aquepts	Halaquepts	Vertic Halaquepts	Fine	Smectitic		Calcareous	Frigid	Aquic	
Lonely	336.5	Aridisols	Cambids	Haplocambids	Xeric Haplocambids	Fine-loamy	Mixed	Superactive		Frigid	Aridic	
Loneranch	21.2	Inceptisols	Udepts	Humudepts	Aquandic Humudepts	Fine-loamy	Isotic			Isomesic	Udic	
Loneridge	17.1	Alfisols	Xeralfs	Palexeralfs	Vertic Palexeralfs	Clayey-skeletal	Smectitic			Frigid	Xeric	
Longbranch	66.5	Mollisols	Xerolls	Argixerolls	Pachic Argixerolls	Clayey-skeletal	Smectitic			Frigid	Xeric	
Longcreek	21.1	Mollisols	Xerolls	Argixerolls	Aridic Lithic Argixerolls	Clayey-skeletal	Smectitic			Mesic	Aridic	
Longjohn	9.5	Andisols	Cryands	Vitricryands	Humic Vitricryands	Ashy-skeletal	Glassy			Cryic	Udic	
Lookingglass	131.0	Mollisols	Albolls	Argialbolls	Xerertic Argialbolls	Fine	Smectitic			Mesic	Xeric	
Lookout	217.4	Aridisols	Durids	Argidurids	Abruptic Xeric Argidurids	Fine	Smectitic			Mesic	Aridic	
Loomis	3.0	Aridisols	Argids	Haplargids	Lithic Xeric Haplargids	Clayey-skeletal	Smectitic			Mesic	Aridic	
LORELLA	479.4	Mollisols	Xerolls	Argixerolls	Lithic Argixerolls	Clayey-skeletal	Smectitic			Mesic	Xeric	
Lostbasin	132.6	Inceptisols	Xerepts	Haploxerepts	Typic Haploxerepts	Loamy-skeletal	Mixed	Superactive		Frigid	Xeric	
Lostforest	9.3	Aridisols	Cambids	Haplocambids	Vitrixerandic Haplocambids	Ashy	Glassy			Frigid	Aridic	
Lostine	10.0	Mollisols	Xerolls	Haploxerolls	Pachic Haploxerolls	Coarse-silty	Mixed	Superactive		Frigid	Xeric	
Loupence	58.3	Mollisols	Xerolls	Haploxerolls	Cumulic Haploxerolls	Fine-silty	Mixed	Superactive		Mesic	Xeric	
Loveboldt	40.4	Entisols	Fluvents	Torrifluvents	Typic Torrifluvents	Sandy	Mixed			Mesic	Aridic	
Lovline	49.3	Mollisols	Xerolls	Haploxerolls	Aridic Haploxerolls	Fine-loamy	Mixed	Superactive		Mesic	Aridic	
Lowerbluff	34.5	Andisols	Xerands	Vitrixerands	Lithic Vitrixerands	Ashy	Mixed			Frigid	Xeric	
Luckiamute	25.1	Inceptisols	Cryepts	Dystrocryepts	Lithic Dystrocryepts	Loamy-skeletal	Isotic			Cryic	Udic	
Ludi	16.9	Mollisols	Xerolls	Haploxerolls	Vitrandic Haploxerolls	Ashy-skeletal over fragmental or cindery	Glassy over mixed			Frigid	Xeric	
Lundgren	40.8	Andisols	Xerands	Vitrixerands	Humic Vitrixerands	Ashy	Glassy			Frigid	Xeric	
Lurnick	12.8	Inceptisols	Cryepts	Dystrocryepts	Andic Dystrocryepts	Clayey-skeletal	Isotic			Cryic	Udic	
Lyeflat	27.5	Aridisols	Cambids	Haplocambids	Lithic Haplocambids	Loamy	Mixed	Superactive		Mesic	Aridic	
MacDunn	29.7	Inceptisols	Xerepts	Haploxerepts	Humic Haploxerepts	Clayey-skeletal	Mixed	Superactive		Mesic	Xeric	
Mackatie	102.6	Andisols	Vitrands	Udivitrands	Alfic Udivitrands	Ashy over loamy	Amorphic over isotic			Frigid	Udic	
Mackey	14.1	Aridisols	Calcids	Haplocalcids	Xeric Haplocalcids	Loamy-skeletal	Mixed	Superactive		Mesic	Aridic	
Macklyn	10.8	Ultisols	Humults	Haplohumults	Typic Haplohumults	Fine	Isotic			Isomesic	Udic	
Macyflet	4.4	Aridisols	Argids	Paleargids	Vertic Paleargids	Very-fine	Smectitic			Frigid	Aridic	
Madeline	69.9	Mollisols	Xerolls	Argixerolls	Aridic Lithic Argixerolls	Clayey	Smectitic			Frigid	Aridic	
Madras	301.6	Mollisols	Xerolls	Argixerolls	Aridic Argixerolls	Fine-loamy	Mixed	Superactive		Mesic	Aridic	
Mahogee	55.0	Mollisols	Cryolls	Argicryolls	Lithic Argicryolls	Loamy	Mixed	Superactive		Cryic	Xeric	
Mahoon	152.3	Mollisols	Xerolls	Palexerolls	Aridic Palexerolls	Fine	Smectitic			Mesic	Xeric	

(continued)

Series name	Area (km^2)	Order	Suborder	Great group	Subgroup	Particle-size class	Mineralogy class	CEC activity class	Reaction class	Soil temp. regime	Soil moisture regime	Other family
Maklak	150.6	Andisols	Cryands	Vitricryands	Xeric Vitricryands	Ashy-pumiceous	Glassy			Cryic	Xeric	
Malabon	167.3	Mollisols	Xerolls	Argixerolls	Pachic Ultic Argixerolls	Fine	Mixed	Superactive		Mesic	Xeric	
Malheur	2.8	Aridisols	Durids	Natridurids	Xeric Natridurids	Fine-silty	Mixed	Superactive		Mesic	Aridic	
Malin	58.2	Mollisols	Aquolls	Endoaquolls	Fluvaquentic Endoaquolls	Fine	Smectitic		Calcareous	Mesic	Aquic	
Mallory	58.3	Mollisols	Xerolls	Argixerolls	Pachic Argixerolls	Clayey-skeletal	Smectitic			Mesic	Xeric	
Manita	123.0	Alfisols	Xeralfs	Haploxeralfs	Mollic Haploxeralfs	Fine	Mixed	Superactive		Mesic	Xeric	
Manlywham	4.5	Inceptisols	Aquepts	Endoaquepts	Humic Endoaquepts	Coarse-loamy over sandy or sandy-skeletal	Isotic		Acid	Frigid	Aquic	
Marack	129.9	Mollisols	Xerolls	Argixerolls	Calciargidic Argixerolls	Fine	Smectitic			Frigid	Aridic	
Marblepoint	69.7	Inceptisols	Cryepts	Haplocryepts	Andic Haplocryepts	Loamy-skeletal	Isotic			Cryic	Udic	
Marcola	8.4	Mollisols	Xerolls	Argixerolls	Pachic Ultic Argixerolls	Clayey-skeletal	Mixed	Superactive		Mesic	Xeric	
Mariel	7.8	Histosols	Hemists	Cryohemists	Typic Cryohemists				Euic	Cryic	Aquic	
Marty	7.4	Andisols	Udands	Hapludands	Alic Hapludands	Medial	Ferrihydritic			Frigid	Udic	
Mary	0.0	Alfisols	Xeralfs	Haploxeralfs	Mollic Haploxeralfs	Fine-loamy	Mixed	Superactive		Mesic	Xeric	
Maryspeak	0.6	Inceptisols	Cryepts	Dystrocryepts	Andic Dystrocryepts	Sandy-skeletal	Isotic			Cryic	Udic	
Mascamp	0.7	Mollisols	Xerolls	Argixerolls	Aridic Lithic Argixerolls	Loamy-skeletal	Mixed	Superactive		Frigid	Aridic	
Maset	230.2	Andisols	Xerands	Vitrixerands	Alfic Vitrixerands	Ashy over loamy-skeletal	Glassy over isotic			Frigid	Xeric	
Matheny	0.1	Mollisols	Xerolls	Argixerolls	Calcic Pachic Argixerolls	Loamy-skeletal	Mixed	Superactive		Mesic	Xeric	
Matterhorn	2.8	Mollisols	Xerolls	Calcixerolls	Typic Calcixerolls	Sandy-skeletal	Mixed			Frigid	Xeric	
Maupin	97.1	Mollisols	Xerolls	Durixerolls	Haploduridic Durixerolls	Fine-loamy	Mixed	Superactive		Mesic	Xeric	
Mayger	72.0	Ultisols	Humults	Palehumults	Aquic Palehumults	Fine	Isotic			Mesic	Udic	
McAlpin	138.5	Mollisols	Xerolls	Haploxerolls	Aquic Cumulic Haploxerolls	Fine	Mixed	Superactive		Mesic	Xeric	
Mcbain	20.8	Aridisols	Calcids	Haplocalcids	Sodic Xeric Haplocalcids	Fine-loamy	Mixed	Superactive		Frigid	Aridic	
McBee	124.6	Mollisols	Xerolls	Haploxerolls	Aquic Cumulic Haploxerolls	Fine-silty	Mixed	Superactive		Mesic	Xeric	
McCalpinemeadow	2.1	Andisols	Cryands	Vitricryands	Typic Vitricryands	Ashy over loamy-skeletal	Amorphic over isotic			Cryic	Udic	
McCartycreek	80.5	Mollisols	Xerolls	Haploxerolls	Vitrandic Haploxerolls	Loamy-skeletal	Mixed	Superactive		Frigid	Xeric	
McCoin	19.6	Mollisols	Xerolls	Haploxerolls	Aridic Haploxerolls	Loamy	Mixed	Superactive		Mesic	Aridic	Shallow
McComas	3.8	Alfisols	Xeralfs	Palexeralfs	Aquic Palexeralfs	Clayey-skeletal	Smectitic			Mesic	Xeric	
McConnel	265.6	Aridisols	Cambids	Haplocambids	Xeric Haplocambids	Sandy-skeletal	Mixed			Mesic	Aridic	
McCully	292.7	Inceptisols	Udepts	Humudepts	Typic Humudepts	Fine	Isotic			Mesic	Udic	
McCurdy	6.2	Ultisols	Humults	Palehumults	Oxyaquic Palehumults	Fine	Isotic			Mesic	Udic	
McDuff	167.0	Ultisols	Humults	Haplohumults	Typic Haplohumults	Fine	Isotic			Mesic	Udic	

(continued)

Series name	Area (km^2)	Order	Suborder	Great group	Subgroup	Particle-size class	Mineralogy class	CEC activity class	Reaction class	Soil temp. regime	Soil moisture regime	Other family
McEwen	13.9	Alfisols	Xeralfs	Haploxeralfs	Vitrandic Haploxeralfs	Fine-loamy	Isotic			Frigid	Xeric	
McGarr	36.7	Mollisols	Xerolls	Haploxerolls	Vitrandic Haploxerolls	Fine-loamy	Isotic			Frigid	Xeric	
McGinnis	3.1	Ultisols	Xerults	Haploxerults	Typic Haploxerults	Clayey-skeletal	Mixed	Active		Mesic	Xeric	
McIvey	21.4	Mollisols	Xerolls	Argixerolls	Typic Argixerolls	Clayey-skeletal	Smectitic			Frigid	Xeric	
McKay	66.0	Mollisols	Xerolls	Argixerolls	Calcic Argixerolls	Fine-silty	Mixed	Superactive		Mesic	Xeric	
McLoughlin	41.3	Aridisols	Cambids	Haplocambids	Xeric Haplocambids	Fine-silty	Mixed	Superactive		Mesic	Aridic	
McMeen	38.6	Mollisols	Xerolls	Durixerolls	Haplic Haploxerollic Durixerolls	Fine-loamy	Mixed	Superactive		Mesic	Xeric	
McMille	8.9	Inceptisols	Udepts	Humudepts	Andic Humudepts	Fine-silty	Isotic			Frigid	Udic	
McMullin	286.0	Mollisols	Xerolls	Haploxerolls	Lithic Ultic Haploxerolls	Loamy	Mixed	Superactive		Mesic	Xeric	
McMurdie	38.6	Mollisols	Xerolls	Argixerolls	Calcic Pachic Argixerolls	Fine	Smectitic			Mesic	Xeric	
McNab	13.1	Alfisols	Xeralfs	Palexeralfs	Aquic Palexeralfs	Fine	Mixed	Superactive		Mesic	Xeric	
McNamee	47.9	Alfisols	Cryalfs	Haplocryalfs	Andic Haplocryalfs	Loamy-skeletal	Isotic			Cryic	Udic	
McNull	323.4	Mollisols	Xerolls	Argixerolls	Ultic Argixerolls	Fine	Smectitic			Mesic	Xeric	
McNulty	16.2	Inceptisols	Udepts	Dystrudepts	Fluventic Dystrudepts	Coarse-loamy	Mixed	Superactive		Mesic	Udic	
McNye	22.2	Aridisols	Cambids	Haplocambids	Xeric Haplocambids	Sandy-skeletal	Mixed			Mesic	Aridic	
McWillar	89.8	Andisols	Xerands	Vitrixerands	Alfic Vitrixerands	Ashy over loamy-skeletal	Amorphic over isotic			Frigid	Xeric	
Meadowridge	123.5	Mollisols	Xerolls	Argixerolls	Vitritorrandic Argixerolls	Fine-loamy	Mixed	Superactive		Mesic	Xeric	
Meda	76.3	Inceptisols	Udepts	Humudepts	Typic Humudepts	Fine-loamy	Isotic			Mesic	Udic	
Medco	298.3	Mollisols	Xerolls	Haploxerolls	Ultic Haploxerolls	Fine	Smectitic			Mesic	Xeric	
Medford	64.9	Mollisols	Xerolls	Argixerolls	Pachic Argixerolls	Fine	Smectitic			Mesic	Xeric	
Melbourne	129.4	Alfisols	Xeralfs	Palexeralfs	Ultic Palexeralfs	Fine	Mixed	Superactive		Mesic	Xeric	
Melby	137.2	Inceptisols	Udepts	Dystrudepts	Humic Dystrudepts	Fine	Isotic			Mesic	Udic	
Meld	23.1	Mollisols	Xerolls	Durixerolls	Vitritorrandic Durixerolls	Ashy	Glassy			Frigid	Aridic	
Melhorn	88.9	Mollisols	Xerolls	Argixerolls	Vitrandic Argixerolls	Fine-loamy	Isotic			Frigid	Xeric	
Melloe	19.4	Mollisols	Aquolls	Cryaquolls	Typic Cryaquolls	Loamy-skeletal	Mixed	Superactive		Cryic	Aquic	
Mellowmoon	22.1	Inceptisols	Udepts	Humudepts	Typic Humudepts	Fine-loamy	Isotic			Frigid	Udic	
Memaloose	3.3	Inceptisols	Udepts	Dystrudepts	Andic Dystrudepts	Fine-loamy	Isotic			Frigid	Udic	
Menbo	59.9	Mollisols	Xerolls	Argixerolls	Vitrandic Argixerolls	Clayey-skeletal	Smectitic			Frigid	Xeric	
Merlin	936.7	Mollisols	Xerolls	Argixerolls	Lithic Argixerolls	Clayey	Smectitic			Frigid	Xeric	
Mershon	21.2	Inceptisols	Udepts	Humudepts	Aquic Humudepts	Fine-silty	Mixed	Superactive		Mesic	Udic	
Mesman	158.8	Aridisols	Argids	Natrargids	Xeric Natrargids	Fine-loamy	Mixed	Superactive		Mesic	Aridic	
Metolius	19.0	Aridisols	Cambids	Haplocambids	Vitrixerandic Haplocambids	Coarse-loamy	Mixed	Superactive		Mesic	Aridic	
Middlebox	96.2	Entisols	Orthents	Torriorthents	Vitrandic Torriorthents	Ashy-skeletal	Glassy		Nonacid	Frigid	Aridic	

(continued)

Series name	Area (km^2)	Order	Suborder	Great group	Subgroup	Particle-size class	Mineralogy class	CEC activity class	Reaction class	Soil temp. regime	Soil moisture regime	Other family
Mikkalo	387.0	Mollisols	Xerolls	Haploxerolls	Calcidic Haploxerolls	Coarse-silty	Mixed	Superactive		Mesic	Aridic	
Milbury	265.3	Inceptisols	Udepts	Humudepts	Typic Humudepts	Loamy-skeletal	Isotic			Mesic	Udic	
Milcan	45.7	Mollisols	Xerolls	Durixerolls	Vitritorrandic Durixerolls	Ashy	Glassy			Frigid	Aridic	
Milldam	24.4	Mollisols	Xerolls	Durixerolls	Typic Durixerolls	Fine-loamy	Mixed	Superactive		Mesic	Xeric	
Millenium	6.6	Mollisols	Xerolls	Argixerolls	Vitritorrandic Argixerolls	Ashy	Glassy			Frigid	Xeric	
Millerflat	5.1	Mollisols	Xerolls	Argixerolls	Vitrandic Argixerolls	Loamy-skeletal	Isotic			Frigid	Xeric	
Millicoma	190.7	Inceptisols	Udepts	Humudepts	Andic Humudepts	Loamy-skeletal	Isotic			Isomesic	Udic	
Minam	39.6	Mollisols	Xerolls	Haploxerolls	Vitrandic Haploxerolls	Fine-loamy	Isotic			Frigid	Xeric	
Minkwell	8.4	Andisols	Cryands	Vitricryands	Alfic Vitricryands	Ashy over medial	Amorphic over mixed			Cryic	Udic	
Minniece	8.2	Alfisols	Aqualfs	Endoaqualfs	Vertic Endoaqualfs	Fine	Smectitic			Mesic	Aquic	
Minveno	194.0	Aridisols	Durids	Haplodurids	Xeric Haplodurids	Loamy	Mixed	Superactive		Mesic	Aridic	Shallow
Mippon	3.8	Mollisols	Xerolls	Haploxerolls	Fluventic Haploxerolls	Sandy-skeletal	Mixed			Frigid	Xeric	
Mislatnah	31.6	Inceptisols	Udepts	Eutrudepts	Dystric Eutrudepts	Loamy-skeletal	Magnesic			Mesic	Udic	
Moag	7.6	Inceptisols	Aquepts	Endoaquepts	Vertic Endoaquepts	Fine	Mixed	Superactive	Nonacid	Mesic	Aquic	
Modoc	46.2	Mollisols	Xerolls	Durixerolls	Vitritorrandic Durixerolls	Fine-loamy	Mixed	Superactive		Mesic	Aridic	
Moe	63.2	Inceptisols	Udepts	Humudepts	Andic Humudepts	Fine	Isotic			Frigid	Udic	
Molalla	29.2	Inceptisols	Udepts	Humudepts	Typic Humudepts	Fine-loamy	Isotic			Mesic	Udic	
Mondovi	15.7	Mollisols	Xerolls	Haploxerolls	Cumulic Haploxerolls	Coarse-silty	Mixed	Superactive		Mesic	Xeric	
Monroe	4.3	Mollisols	Xerolls	Haploxerolls	Cumulic Haploxerolls	Fine-loamy	Mixed	Superactive		Mesic	Xeric	
Monumentrock	29.2	Inceptisols	Cryepts	Haplocryepts	Andic Haplocryepts	Loamy-skeletal	Isotic			Cryic	Udic	
Moodybasin	13.1	Inceptisols	Cryepts	Humicryepts	Oxyaquic Humicryepts	Loamy-skeletal	Isotic			Cryic	Udic	
Moonbeam	475.4	Mollisols	Xerolls	Durixerolls	Vitritorrandic Durixerolls	Clayey	Smectitic			Frigid	Aridic	Shallow
Moonstone	2.4	Mollisols	Xerolls	Haploxerolls	Pachic Ultic Haploxerolls	Coarse-loamy	Mixed	Superactive		Frigid	Xeric	
Morehouse	231.5	Entisols	Psamments	Torripsamments	Vitrandic Torripsamments	Ashy	Glassy		Nonacid	Frigid	Aridic	
Morfitt	51.2	Aridisols	Argids	Haplargids	Xeric Haplargids	Fine-loamy	Mixed	Superactive		Mesic	Aridic	
Morganhills	82.3	Entisols	Orthents	Torriorthents	Vitrandic Torriorthents	Ashy	Glassy		Nonacid	Frigid	Aridic	Shallow
Morningstar	14.2	Mollisols	Xerolls	Argixerolls	Ultic Argixerolls	Loamy-skeletal	Mixed	Superactive		Frigid	Xeric	
Morrow	707.9	Mollisols	Xerolls	Argixerolls	Calcic Argixerolls	Fine-silty	Mixed	Superactive		Mesic	Xeric	
Mosscreek	29.1	Andisols	Udands	Fulvudands	Pachic Fulvudands	Medial	Ferrihydritic			Isofrigid	Udic	
MOUND	49.6	Mollisols	Xerolls	Argixerolls	Pachic Ultic Argixerolls	Clayey-skeletal	Smectitic			Frigid	Xeric	
Mountemily	227.1	Andisols	Cryands	Vitricryands	Typic Vitricryands	Ashy over loamy-skeletal	Amorphic over isotic			Cryic	Udic	

(continued)

Series name	Area (km^2)	Order	Suborder	Great group	Subgroup	Particle-size class	Mineralogy class	CEC activity class	Reaction class	Soil temp. regime	Soil moisture regime	Other family
Mountireland	54.2	Andisols	Cryands	Vitricryands	Alfic Vitricryands	Ashy over loamy	Amorphic over isotic			Cryic	Udic	
Mowako	7.5	Mollisols	Xerolls	Haploxerolls	Ultic Haploxerolls	Loamy-skeletal	Mixed	Superactive		Mesic	Xeric	
Muddycreek	16.7	Inceptisols	Cryepts	Humicryepts	Andic Humicryepts	Loamy-skeletal	Isotic			Cryic	Udic	
Mudlakebasin	70.4	Andisols	Cryands	Vitricryands	Typic Vitricryands	Ashy over loamy-skeletal	Amorphic over isotic			Cryic	Udic	
Mudpot	62.0	Vertisols	Aquerts	Endoaquerts	Chromic Endoaquerts	Fine	Smectitic			Frigid	Aquic	
Mues	8.9	Andisols	Udands	Fulvudands	Aquic Fulvudands	Medial over loamy-skeletal	Ferrihydritic over isotic			Isomesic	Udic	
Mugwump	7.6	Mollisols	Udolls	Hapludolls	Cumulic Hapludolls	Loamy-skeletal	Mixed	Superactive		Frigid	Udic	
Mulkey	6.0	Andisols	Cryands	Fulvicryands	Pachic Fulvicryands	Medial	Ferrihydritic			Cryic	Udic	
Multnomah	61.9	Inceptisols	Xerepts	Dystroxerepts	Humic Dystroxerepts	Coarse-loamy over sandy or sandy-skeletal	Mixed	Superactive		Mesic	Xeric	
Multorpor	4.2	Entisols	Orthents	Udorthents	Typic Udorthents	Sandy-skeletal	Mixed			Mesic	Udic	
Muni	435.4	Aridisols	Durids	Argidurids	Haploxeralfic Argidurids	Loamy	Mixed	Superactive		Mesic	Aridic	Shallow
Munsoncreek	42.3	Inceptisols	Udepts	Humudepts	Andic Humudepts	Fine	Isotic			Isomesic	Udic	
Murlose	14.4	Mollisols	Xerolls	Durixerolls	Vitrandic Durixerolls	Ashy	Glassy			Frigid	Xeric	Shallow
Murnen	22.9	Andisols	Udands	Fulvudands	Typic Fulvudands	Medial	Ferrihydritic			Frigid	Udic	
Murtip	201.7	Andisols	Udands	Hapludands	Alic Hapludands	Medial	Ferrihydritic			Frigid	Udic	
Musty	11.0	Inceptisols	Xerepts	Dystroxerepts	Humic Dystroxerepts	Loamy-skeletal	Mixed	Superactive		Mesic	Xeric	
Mutt	0.7	Inceptisols	Udepts	Humudepts	Andic Humudepts	Fine-silty	Isotic			Frigid	Udic	
Mutton	51.1	Alfisols	Xeralfs	Haploxeralfs	Vitrandic Haploxeralfs	Ashy-skeletal	Glassy			Mesic	Xeric	
Nagle	21.1	Mollisols	Xerolls	Argixerolls	Pachic Argixerolls	Fine-loamy	Mixed	Superactive		Frigid	Xeric	
Nailkeg	62.0	Inceptisols	Udepts	Dystrudepts	Typic Dystrudepts	Loamy-skeletal	Mixed	Active		Mesic	Udic	
Nansene	110.0	Mollisols	Xerolls	Haploxerolls	Pachic Haploxerolls	Coarse-silty	Mixed	Superactive		Mesic	Xeric	
Natal	6.6	Alfisols	Aqualfs	Endoaqualfs	Umbric Endoaqualfs	Fine	Mixed	Superactive		Mesic	Aquic	
Natroy	61.6	Vertisols	Aquerts	Endoaquerts	Xeric Endoaquerts	Very-fine	Smectitic			Mesic	Aquic	
Necanicum	356.6	Andisols	Udands	Fulvudands	Typic Fulvudands	Medial-skeletal	Ferrihydritic			Isomesic	Udic	
Needhill	9.9	Mollisols	Xerolls	Argixerolls	Vitrandic Argixerolls	Loamy-skeletal	Mixed	Superactive		Frigid	Xeric	
Needle Peak	0.2	Entisols	Orthents	Torriorthents	Oxyaquic Torriorthents	Fine-silty	Mixed	Superactive	Calcareous	Mesic	Aridic	
Nehalem	57.6	Inceptisols	Udepts	Humudepts	Fluventic Humudepts	Fine-silty	Mixed	Superactive		Isomesic	Udic	
Nekia	603.4	Ultisols	Humults	Haplohumults	Xeric Haplohumults	Fine	Mixed	Active		Mesic	Xeric	
Nekoma	54.9	Inceptisols	Udepts	Humudepts	Fluventic Humudepts	Coarse-loamy	Mixed	Superactive		Mesic	Udic	
Nelscott	40.2	Spodosols	Orthods	Durorthods	Typic Durorthods	Fine-loamy over sandy or sandy-skeletal	Isotic over mixed			Isomesic	Udic	Ortstein
Neotsu	38.5	Andisols	Udands	Fulvudands	Typic Fulvudands	Medial	Ferrihydritic			Isomesic	Udic	
Neskowin	43.4	Andisols	Udands	Fulvudands	Typic Fulvudands	Medial	Ferrihydritic			Isomesic	Udic	
Nestucca	50.1	Inceptisols	Aquepts	Humaquepts	Fluvaquentic Humaquepts	Fine-silty	Mixed	Superactive	Acid	Isomesic	Aquic	
Netarts	22.3	Spodosols	Orthods	Haplorthods	Entic Haplorthods	Sandy	Isotic			Isomesic	Udic	

(continued)

Series name	Area (km^2)	Order	Suborder	Great group	Subgroup	Particle-size class	Mineralogy class	CEC activity class	Reaction class	Soil temp. regime	Soil moisture regime	Other family
Nevador	1112.8	Aridisols	Argids	Haplargids	Durinodic Xeric Haplargids	Fine-loamy	Mixed	Superactive		Mesic	Aridic	
Newanna	19.2	Andisols	Cryands	Fulvicryands	Typic Fulvicryands	Medial-skeletal	Ferrihydritic			Cryic	Udic	
Newberg	270.7	Mollisols	Xerolls	Haploxerolls	Fluventic Haploxerolls	Coarse-loamy	Mixed	Superactive		Mesic	Xeric	
Ninemile	1808.4	Mollisols	Xerolls	Argixerolls	Aridic Lithic Argixerolls	Clayey	Smectitic			Frigid	Aridic	
Ninetysix	80.5	Mollisols	Xerolls	Haploxerolls	Calcic Haploxerolls	Loamy-skeletal	Mixed	Superactive		Mesic	Xeric	
Noidee	41.1	Aridisols	Argids	Natrargids	Lithic Xeric Natrargids	Clayey	Smectitic			Frigid	Aridic	
Noname	24.2	Inceptisols	Cryepts	Haplocryepts	Lithic Haplocryepts	Loamy	Mixed	Superactive		Cryic	Xeric	
Nonpareil	87.4	Inceptisols	Xerepts	Dystroxerepts	Typic Dystroxerepts	Loamy	Mixed	Superactive		Mesic	Xeric	Shallow
Norad	144.2	Aridisols	Argids	Haplargids	Xeric Haplargids	Fine-silty	Mixed	Superactive		Mesic	Aridic	
Norcross	42.2	Mollisols	Xerolls	Durixerolls	Vitrandic Durixerolls	Clayey	Smectitic			Frigid	Xeric	Shallow
Norling	126.4	Alfisols	Xeralfs	Haploxeralfs	Ultic Haploxeralfs	Fine-loamy	Mixed	Superactive		Mesic	Xeric	
North Powder	35.0	Aridisols	Cambids	Haplocambids	Xeric Haplocambids	Fine-loamy	Mixed	Superactive		Mesic	Aridic	
Northrup	3.6	Ultisols	Humults	Haplohumults	Oxyaquic Haplohumults	Fine-silty	Isotic			Mesic	Udic	
Noti	15.2	Inceptisols	Aquepts	Humaquepts	Typic Humaquepts	Coarse-loamy over sandy or sandy-skeletal	Mixed	Superactive	Acid	Mesic	Aquic	
Notus	4.7	Entisols	Fluvents	Xerofluvents	Aquic Xerofluvents	Sandy-skeletal	Mixed			Mesic	Xeric	
Nuss	214.5	Mollisols	Xerolls	Haploxerolls	Lithic Haploxerolls	Loamy	Mixed	Superactive		Frigid	Xeric	
Nyssa	117.8	Aridisols	Durids	Haplodurids	Xeric Haplodurids	Coarse-silty	Mixed	Superactive		Mesic	Aridic	
Oak Grove	42.8	Alfisols	Xeralfs	Palexeralfs	Ultic Palexeralfs	Fine	Mixed	Superactive		Mesic	Xeric	
Oakland	147.4	Alfisols	Xeralfs	Haploxeralfs	Ultic Haploxeralfs	Fine	Mixed	Superactive		Mesic	Xeric	
Oatman	134.0	Andisols	Cryands	Haplocryands	Typic Haplocryands	Medial-skeletal	Amorphic			Cryic	Udic	
Oatmanflat	14.8	Mollisols	Xerolls	Haploxerolls	Vitritorrandic Haploxerolls	Ashy	Glassy			Frigid	Aridic	
Observation	429.3	Mollisols	Xerolls	Argixerolls	Typic Argixerolls	Fine	Smectitic			Frigid	Xeric	
Ochoco	28.2	Aridisols	Durids	Argidurids	Vitrixerandic Argidurids	Fine-loamy	Mixed	Superactive		Mesic	Aridic	
Offenbacher	84.9	Inceptisols	Xerepts	Haploxerepts	Typic Haploxerepts	Fine-loamy	Mixed	Superactive		Mesic	Xeric	
Olac	42.1	Aridisols	Argids	Haplargids	Lithic Xeric Haplargids	Loamy-skeletal	Mixed	Superactive		Mesic	Aridic	
Olallie	4.5	Mollisols	Aquolls	Endoaquolls	Cumulic Endoaquolls	Loamy-skeletal	Mixed	Superactive		Mesic	Aquic	
Olaton	3.8	Mollisols	Xerolls	Haploxerolls	Pachic Ultic Haploxerolls	Coarse-loamy	Mixed	Superactive		Mesic	Xeric	
Old Camp	51.6	Aridisols	Argids	Haplargids	Lithic Xeric Haplargids	Loamy-skeletal	Mixed	Superactive		Mesic	Aridic	
Oldblue	7.4	Inceptisols	Udepts	Humudepts	Andic Humudepts	Fine-loamy	Isotic			Frigid	Udic	
Oldsferry	10.8	Mollisols	Xerolls	Haploxerolls	Typic Haploxerolls	Loamy-skeletal	Mixed	Superactive		Mesic	Xeric	
Olex	96.8	Mollisols	Xerolls	Haploxerolls	Calcidic Haploxerolls	Loamy-skeletal	Mixed	Superactive		Mesic	Aridic	
Oliphant	41.1	Mollisols	Xerolls	Haploxerolls	Calcic Pachic Haploxerolls	Coarse-silty	Mixed	Superactive		Mesic	Xeric	

(continued)

Series name	Area (km^2)	Order	Suborder	Great group	Subgroup	Particle-size class	Mineralogy class	CEC activity class	Reaction class	Soil temp. regime	Soil moisture regime	Other family
Olot	342.3	Andisols	Xerands	Vitrixerands	Typic Vitrixerands	Ashy over loamy-skeletal	Glassy over isotic			Frigid	Xeric	
Olyic	227.5	Ultisols	Humults	Haplohumults	Typic Haplohumults	Fine-loamy	Mixed	Active		Mesic	Udic	
Omahaling	3.9	Mollisols	Xerolls	Haploxerolls	Fluvaquentic Haploxerolls	Coarse-loamy	Mixed	Superactive		Frigid	Xeric	
Oneonta	26.6	Andisols	Cryands	Haplocryands	Typic Haplocryands	Medial	Amorphic			Cryic	Udic	
Ontko	48.1	Mollisols	Aquolls	Cryaquolls	Aquandic Cryaquolls	Fine-loamy	Mixed	Superactive		Cryic	Aquic	
Onyx	24.7	Mollisols	Xerolls	Haploxerolls	Cumulic Haploxerolls	Coarse-silty	Mixed	Superactive		Mesic	Xeric	
Opie	62.7	Mollisols	Aquolls	Endoaquolls	Cumulic Endoaquolls	Fine-silty	Mixed	Superactive	Calcareous	Frigid	Aquic	
Oreanna	8.8	Aridisols	Cambids	Haplocambids	Typic Haplocambids	Fine-loamy over sandy or sandy-skeletal	Mixed	Superactive		Mesic	Aridic	
Oreneva	122.3	Aridisols	Cambids	Haplocambids	Xeric Haplocambids	Loamy-skeletal	Mixed	Superactive		Frigid	Aridic	
Orford	304.5	Ultisols	Humults	Palehumults	Typic Palehumults	Fine	Isotic			Mesic	Udic	
Orhood	0.1	Mollisols	Xerolls	Argixerolls	Aridic Lithic Argixerolls	Loamy-skeletal	Mixed	Superactive		Mesic	Aridic	
Ornea	2.6	Aridisols	Argids	Haplargids	Typic Haplargids	Fine-loamy over sandy or sandy-skeletal	Mixed	Superactive		Mesic	Aridic	
Orovada	143.7	Aridisols	Cambids	Haplocambids	Durinodic Xeric Haplocambids	Coarse-loamy	Mixed	Superactive		Mesic	Aridic	
Osoll	12.0	Aridisols	Durids	Haplodurids	Typic Haplodurids	Loamy-skeletal	Mixed	Superactive		Mesic	Aridic	Shallow
Otoole	3.9	Aridisols	Durids	Haplodurids	Xeric Haplodurids	Loamy	Mixed	Superactive		Mesic	Aridic	Shallow
Otwin	9.9	Andisols	Cryands	Haplocryands	Typic Haplocryands	Medial-skeletal	Amorphic			Cryic	Udic	
Outerkirk	184.1	Aridisols	Calcids	Haplocalcids	Durinodic Haplocalcids	Coarse-loamy	Mixed	Superactive		Mesic	Aridic	
Overallflat	26.5	Aridisols	Argids	Paleargids	Aquic Paleargids	Ashy	Glassy			Frigid	Aridic	
Owsel	65.7	Aridisols	Argids	Haplargids	Durinodic Xeric Haplargids	Fine-silty	Mixed	Superactive		Mesic	Aridic	
Owyhee	76.9	Aridisols	Cambids	Haplocambids	Xeric Haplocambids	Coarse-silty	Mixed	Superactive		Mesic	Aridic	
Oxbow	9.6	Mollisols	Xerolls	Durixerolls	Palexerollic Durixerolls	Fine	Smectitic			Mesic	Xeric	
Oxley	8.9	Mollisols	Aquolls	Argiaquolls	Typic Argiaquolls	Loamy-skeletal	Mixed	Superactive		Mesic	Aquic	
Oxman	20.5	Aridisols	Cambids	Haplocambids	Xeric Haplocambids	Fine-loamy	Mixed	Superactive		Mesic	Aridic	
Oxwall	116.1	Mollisols	Xerolls	Durixerolls	Palexerollic Durixerolls	Clayey	Smectitic			Mesic	Xeric	Shallow
Ozamis	184.2	Mollisols	Aquolls	Endoaquolls	Fluvaquentic Endoaquolls	Fine-loamy	Mixed	Superactive		Mesic	Aquic	
Packard	8.5	Mollisols	Xerolls	Haploxerolls	Pachic Haploxerolls	Loamy-skeletal	Mixed	Superactive		Mesic	Xeric	
Padigan	11.1	Vertisols	Aquerts	Endoaquerts	Xeric Endoaquerts	Very-fine	Smectitic			Mesic	Aquic	
Pait	42.2	Mollisols	Xerolls	Haploxerolls	Aridic Haploxerolls	Loamy-skeletal	Mixed	Superactive		Mesic	Aridic	
Palouse	122.5	Mollisols	Xerolls	Haploxerolls	Pachic Ultic Haploxerolls	Fine-silty	Mixed	Superactive		Mesic	Xeric	
Panther	59.6	Mollisols	Aquolls	Epiaquolls	Vertic Epiaquolls	Very-fine	Smectitic			Mesic	Aquic	
Paragon	9.7	Mollisols	Xerolls	Argixerolls	Pachic Ultic Argixerolls	Fine-loamy	Mixed	Superactive		Mesic	Xeric	

(continued)

Series name	Area (km^2)	Order	Suborder	Great group	Subgroup	Particle-size class	Mineralogy class	CEC activity class	Reaction class	Soil temp. regime	Soil moisture regime	Other family
Parkdale	34.3	Andisols	Xerands	Vitrixerands	Humic Vitrixerands	Ashy	Amorphic			Mesic	Xeric	
Parrego	47.1	Alfisols	Xeralfs	Haploxeralfs	Ultic Haploxeralfs	Fine-loamy	Isotic			Frigid	Xeric	
Parsnip	97.2	Mollisols	Xerolls	Argixerolls	Lithic Argixerolls	Loamy	Mixed	Superactive		Frigid	Xeric	
Patit Creek	3.2	Mollisols	Xerolls	Haploxerolls	Cumulic Haploxerolls	Coarse-loamy	Mixed	Superactive		Mesic	Xeric	
Patron	0.3	Mollisols	Xerolls	Palexerolls	Vitrandic Palexerolls	Fine	Smectitic			Mesic	Xeric	
Paulina	36.0	Mollisols	Aquolls	Endoaquolls	Aquandic Endoaquolls	Ashy	Glassy			Frigid	Aquic	
Paynepeak	0.9	Mollisols	Cryolls	Argicryolls	Vitrandic Argicryolls	Ashy-skeletal	Glassy			Cryic	Xeric	
Peahke	5.2	Mollisols	Xerolls	Haploxerolls	Vitrandic Haploxerolls	Loamy-skeletal	Isotic			Frigid	Xeric	
Pearlwise	60.2	Mollisols	Xerolls	Haploxerolls	Pachic Haploxerolls	Fine-loamy	Mixed	Superactive		Frigid	Xeric	
Pearsoll	232.5	Inceptisols	Xerepts	Dystroxerepts	Lithic Dystroxerepts	Clayey-skeletal	Magnesic			Mesic	Xeric	
Peasley	0.6	Vertisols	Xererts	Durixererts	Haplic Durixererts	Fine	Smectitic			Mesic	Xeric	
Peavine	851.7	Ultisols	Humults	Haplohumults	Typic Haplohumults	Fine	Mixed	Active		Mesic	Udic	
Pedigo	23.0	Mollisols	Xerolls	Haploxerolls	Cumulic Haploxerolls	Coarse-silty	Mixed	Superactive		Mesic	Xeric	
Peel	3.3	Mollisols	Xerolls	Argixerolls	Vertic Argixerolls	Fine	Magnesic			Mesic	Xeric	
Pelton	2.8	Mollisols	Xerolls	Haploxerolls	Torrifluventic Haploxerolls	Loamy-skeletal	Mixed	Superactive		Mesic	Aridic	
Pengra	106.3	Mollisols	Aquolls	Epiaquolls	Vertic Epiaquolls	Fine-silty over clayey	Mixed over smectitic	Superactive		Mesic	Aquic	
Perdin	88.0	Alfisols	Xeralfs	Haploxeralfs	Ultic Haploxeralfs	Fine	Magnesic			Frigid	Xeric	
Perla	38.0	Mollisols	Xerolls	Argixerolls	Aridic Argixerolls	Fine	Smectitic			Mesic	Aridic	
Pernog	7.9	Mollisols	Xerolls	Argixerolls	Lithic Argixerolls	Loamy-skeletal	Mixed	Superactive		Frigid	Xeric	
Pernty	347.1	Mollisols	Xerolls	Argixerolls	Aridic Lithic Argixerolls	Loamy-skeletal	Mixed	Superactive		Frigid	Aridic	
Pervina	33.7	Ultisols	Udults	Hapludults	Typic Hapludults	Fine	Mixed	Active		Mesic	Udic	
Philomath	233.0	Mollisols	Xerolls	Haploxerolls	Vertic Haploxerolls	Clayey	Smectitic			Mesic	Xeric	Shallow
Phoenix	3.8	Vertisols	Aquerts	Epiaquerts	Xeric Epiaquerts	Very-fine	Smectitic			Mesic	Aquic	
Phys	31.3	Mollisols	Xerolls	Argixerolls	Typic Argixerolls	Loamy-skeletal	Mixed	Superactive		Mesic	Xeric	
Picturerock	5.4	Mollisols	Xerolls	Haploxerolls	Vitritorrandic Haploxerolls	Ashy	Glassy			Frigid	Aridic	
Piersonte	55.7	Mollisols	Xerolls	Haploxerolls	Vitrandic Haploxerolls	Loamy-skeletal	Isotic			Frigid	Xeric	
Pilchuck	19.8	Entisols	Psamments	Xeropsamments	Dystric Xeropsamments	Sandy	Mixed			Mesic	Xeric	
Piline	73.7	Vertisols	Aquerts	Epiaquerts	Xeric Epiaquerts	Fine	Smectitic			Mesic	Aquic	
Pilot Rock	142.2	Mollisols	Xerolls	Durixerolls	Haploxerollic Durixerolls	Coarse-silty	Mixed	Superactive		Mesic	Xeric	
Pinehurst	148.1	Mollisols	Xerolls	Argixerolls	Pachic Ultic Argixerolls	Fine-loamy	Mixed	Superactive		Frigid	Xeric	
Pineval	15.2	Aridisols	Argids	Haplargids	Durinodic Xeric Haplargids	Loamy-skeletal	Mixed	Superactive		Mesic	Aridic	
Pinhead	141.5	Andisols	Cryands	Haplocryands	Vitric Haplocryands	Medial-skeletal	Amorphic			Cryic	Udic	
Pipp	91.7	Andisols	Xerands	Vitrixerands	Humic Vitrixerands	Ashy-skeletal	Glassy			Frigid	Xeric	

(continued)

Series name	Area (km^2)	Order	Suborder	Great group	Subgroup	Particle-size class	Mineralogy class	CEC activity class	Reaction class	Soil temp. regime	Soil moisture regime	Other family
Pistolriver	0.6	Inceptisols	Aquepts	Humaquepts	Typic Humaquepts	Coarse-loamy over sandy or sandy-skeletal	Mixed	Superactive	Nonacid	Isomesic	Aquic	
Pit	31.7	Vertisols	Aquerts	Endoaquerts	Xeric Endoaquerts	Fine	Smectitic			Mesic	Aquic	
Pitcheranch	7.1	Inceptisols	Aquepts	Endoaquepts	Aquandic Endoaquepts	Ashy	Glassy		Nonacid	Frigid	Aquic	
Piumpsha	21.7	Andisols	Cryands	Vitricryands	Alfic Vitricryands	Medial	Amorphic			Cryic	Udic	
Plainview	20.7	Andisols	Torrands	Vitritorrands	Typic Vitritorrands	Ashy over loamy-skeletal	Glassy over mixed	Superactive		Mesic	Aridic	
Plush	17.4	Aridisols	Argids	Haplargids	Xeric Haplargids	Loamy-skeletal	Mixed	Superactive		Mesic	Aridic	
Poall	322.6	Aridisols	Argids	Paleargids	Xeric Paleargids	Fine	Smectitic			Mesic	Aridic	
Pocan	1.4	Aridisols	Cambids	Haplocambids	Xeric Haplocambids	Fine-loamy	Mixed	Superactive		Mesic	Aridic	
Poden	3.7	Mollisols	Xerolls	Haploxerolls	Cumulic Haploxerolls	Fine-silty over sandy or sandy-skeletal	Mixed	Superactive		Mesic	Xeric	
Podus	1.6	Inceptisols	Xerepts	Durixerepts	Typic Durixerepts	Sandy	Mixed			Mesic	Xeric	Shallow
Poe	24.3	Inceptisols	Xerepts	Durixerepts	Typic Durixerepts	Sandy	Mixed			Mesic	Xeric	
Pokegema	317.8	Andisols	Xerands	Haploxerands	Humic Haploxerands	Medial	Amorphic			Frigid	Xeric	
Polander	40.6	Andisols	Xerands	Vitrixerands	Typic Vitrixerands	Ashy	Glassy			Frigid	Xeric	
Pollard	172.9	Ultisols	Xerults	Palexerults	Typic Palexerults	Fine	Kaolinitic			Mesic	Xeric	
Polly	21.7	Mollisols	Xerolls	Argixerolls	Calciargidic Argixerolls	Fine-loamy	Mixed	Superactive		Mesic	Aridic	
Pomerening	20.5	Entisols	Orthents	Torriorthents	Vitrandic Torriorthents	Ashy	Glassy		Nonacid	Frigid	Aridic	
Ponina	46.3	Alfisols	Xeralfs	Durixeralfs	Abruptic Durixeralfs	Clayey	Smectitic			Frigid	Xeric	Shallow
Poorjug	24.7	Aridisols	Cambids	Haplocambids	Lithic Xeric Haplocambids	Loamy-skeletal	Mixed	Superactive		Mesic	Aridic	
Porterfield	39.5	Entisols	Orthents	Torriorthents	Vitrandic Torriorthents	Ashy	Glassy		Nonacid	Mesic	Aridic	Shallow
Potamus	3.5	Mollisols	Xerolls	Haploxerolls	Typic Haploxerolls	Loamy-skeletal	Mixed	Superactive		Frigid	Xeric	
Poujade	290.8	Aridisols	Argids	Natrargids	Durinodic Xeric Natrargids	Fine-loamy	Mixed	Superactive		Frigid	Aridic	
Powder	132.1	Mollisols	Xerolls	Haploxerolls	Cumulic Haploxerolls	Coarse-silty	Mixed	Superactive		Mesic	Xeric	
Powell	49.9	Inceptisols	Xerepts	Fragixerepts	Humic Fragixerepts	Fine-silty	Mixed	Superactive		Mesic	Xeric	
Power	8.6	Aridisols	Argids	Calciargids	Xeric Calciargids	Fine-silty	Mixed	Superactive		Mesic	Aridic	
Powval	43.6	Mollisols	Xerolls	Haploxerolls	Pachic Haploxerolls	Coarse-silty	Mixed	Superactive		Mesic	Xeric	
Powwatka	49.3	Mollisols	Xerolls	Argixerolls	Vitrandic Argixerolls	Fine-silty	Mixed	Superactive		Frigid	Xeric	
Prag	162.5	Mollisols	Xerolls	Palexerolls	Pachic Palexerolls	Fine	Smectitic			Frigid	Xeric	
Prairie	35.7	Andisols	Cryands	Vitricryands	Xeric Vitricryands	Ashy over loamy	Amorphic over isotic			Cryic	Xeric	
Preacher	1470.2	Inceptisols	Udepts	Humudepts	Andic Humudepts	Fine-loamy	Isotic			Mesic	Udic	
Price	38.2	Inceptisols	Xerepts	Haploxerepts	Humic Haploxerepts	Fine	Mixed	Superactive		Mesic	Xeric	
Prill	241.2	Mollisols	Xerolls	Palexerolls	Pachic Palexerolls	Fine	Smectitic			Mesic	Xeric	
Prineville	23.4	Aridisols	Cambids	Haplocambids	Durinodic Xeric Haplocambids	Coarse-loamy	Mixed	Superactive		Mesic	Aridic	
Pritchard	1.9	Mollisols	Xerolls	Palexerolls	Typic Palexerolls	Fine	Smectitic			Mesic	Xeric	

(continued)

Series name	Area (km^2)	Order	Suborder	Great group	Subgroup	Particle-size class	Mineralogy class	CEC activity class	Reaction class	Soil temp. regime	Soil moisture regime	Other family
Prosser	34.0	Aridisols	Cambids	Haplocambids	Xeric Haplocambids	Coarse-loamy	Mixed	Superactive		Mesic	Aridic	
Prouty	67.8	Inceptisols	Cryepts	Dystrocryepts	Andic Dystrocryepts	Loamy-skeletal	Isotic			Cryic	Udic	
Provig	10.7	Mollisols	Xerolls	Argixerolls	Typic Argixerolls	Clayey-skeletal	Smectitic			Mesic	Xeric	
Puderbaugh	1.2	Mollisols	Xerolls	Argixerolls	Pachic Argixerolls	Loamy-skeletal	Mixed	Superactive		Frigid	Xeric	
Puderbaughridge	2.2	Mollisols	Xerolls	Haploxerolls	Aridic Lithic Haploxerolls	Loamy-skeletal	Mixed	Superactive		Mesic	Aridic	
Puls	17.8	Aridisols	Durids	Argidurids	Abruptic Xeric Argidurids	Clayey	Smectitic			Mesic	Aridic	Shallow
Purple	2.2	Inceptisols	Aquepts	Cryaquepts	Histic Cryaquepts	Coarse-loamy over sandy or sandy-skeletal	Isotic		Acid	Cryic	Aquic	
Puzzlebark	2.3	Aridisols	Durids	Haplodurids	Vitrixerandic Haplodurids	Ashy	Glassy			Frigid	Aridic	Shallow
Puzzlecreek	36.8	Inceptisols	Cryepts	Humicryepts	Haploxerandic Humicryepts	Loamy-skeletal	Isotic			Cryic	Xeric	
Pyburn	10.3	Ultisols	Aquults	Umbraquults	Typic Umbraquults	Fine	Isotic			Mesic	Aquic	
Pyrady	9.0	Ultisols	Humults	Palehumults	Oxyaquic Palehumults	Fine	Isotic			Frigid	Udic	
Pyropatti	0.4	Mollisols	Cryolls	Argicryolls	Vitrandic Argicryolls	Ashy-skeletal	Glassy			Cryic	Xeric	
Quafeno	15.4	Mollisols	Xerolls	Haploxerolls	Aquultic Haploxerolls	Coarse-loamy	Mixed	Superactive		Mesic	Xeric	
Quailprairie	5.8	Inceptisols	Udepts	Humudepts	Pachic Humudepts	Fine-loamy	Isotic			Mesic	Udic	
Quartzville	33.0	Inceptisols	Udepts	Humudepts	Andic Humudepts	Fine	Isotic			Mesic	Udic	
Quatama	85.3	Alfisols	Xeralfs	Haploxeralfs	Aquultic Haploxeralfs	Fine-loamy	Mixed	Superactive		Mesic	Xeric	
Quillamook	14.5	Andisols	Udands	Melanudands	Pachic Melanudands	Medial	Ferrihydritic			Isomesic	Udic	
Quincy	545.6	Entisols	Psamments	Torripsamments	Xeric Torripsamments	Sandy	Mixed			Mesic	Aridic	
Quinton	14.3	Entisols	Psamments	Torripsamments	Xeric Torripsamments	Sandy	Mixed			Mesic	Aridic	
Quirk	110.3	Mollisols	Xerolls	Palexerolls	Vitrandic Palexerolls	Fine	Smectitic			Frigid	Xeric	
Quosatana	12.6	Inceptisols	Aquepts	Humaquepts	Fluvaquentic Humaquepts	Fine-silty	Mixed	Superactive	Nonacid	Mesic	Aquic	
Rabbitcreek	2.8	Aridisols	Cambids	Haplocambids	Typic Haplocambids	Fine-loamy	Mixed	Superactive		Mesic	Aridic	
Rabbithills	198.5	Aridisols	Durids	Haplodurids	Xereptic Haplodurids	Loamy	Mixed	Superactive		Mesic	Aridic	Shallow
Racing	13.4	Inceptisols	Aquepts	Cryaquepts	Aquandic Cryaquepts	Fine-loamy	Mixed	Superactive	Nonacid	Cryic	Aquic	
Rafton	49.6	Inceptisols	Aquepts	Endoaquepts	Fluvaquentic Endoaquepts	Fine-silty	Mixed	Superactive	Nonacid	Mesic	Aquic	
Rail	7.8	Vertisols	Aquerts	Endoaquerts	Xeric Endoaquerts	Fine	Smectitic			Mesic	Aquic	
Rainey	0.5	Mollisols	Xerolls	Haploxerolls	Entic Haploxerolls	Coarse-loamy	Mixed	Superactive		Mesic	Xeric	
Ramo	48.6	Mollisols	Xerolls	Argixerolls	Typic Argixerolls	Fine	Smectitic			Mesic	Xeric	
Randcore	8.0	Entisols	Orthents	Xerorthents	Lithic Xerorthents	Loamy-skeletal	Mixed	Superactive	Nonacid	Mesic	Xeric	
Rastus	35.6	Mollisols	Xerolls	Durixerolls	Palexerollic Durixerolls	Fine	Smectitic			Frigid	Xeric	
Ratsnest	0.1	Aridisols	Argids	Haplargids	Typic Haplargids	Fine	Smectitic			Mesic	Aridic	
Ratto	186.6	Aridisols	Durids	Argidurids	Xeric Argidurids	Clayey	Smectitic			Frigid	Aridic	Shallow
Raz	2279.5	Aridisols	Durids	Haplodurids	Xeric Haplodurids	Loamy	Mixed	Superactive		Frigid	Aridic	Shallow
Raztack	1.4	Alfisols	Xeralfs	Palexeralfs	Vitrandic Palexeralfs	Fine	Smectitic			Frigid	Xeric	

(continued)

Series name	Area (km^2)	Order	Suborder	Great group	Subgroup	Particle-size class	Mineralogy class	CEC activity class	Reaction class	Soil temp. regime	Soil moisture regime	Other family
Reallis	240.6	Aridisols	Cambids	Haplocambids	Durinodic Xeric Haplocambids	Coarse-loamy	Mixed	Superactive		Frigid	Aridic	
Reavis	14.8	Mollisols	Xerolls	Haploxerolls	Calcic Haploxerolls	Fine-loamy	Mixed	Superactive		Frigid	Xeric	
Redbell	22.1	Mollisols	Xerolls	Argixerolls	Aquultic Argixerolls	Fine	Mixed	Superactive		Mesic	Xeric	
Redcanyon	6.2	Mollisols	Xerolls	Haploxerolls	Calcidic Haploxerolls	Loamy-skeletal	Mixed	Superactive		Mesic	Aridic	
Redcliff	143.3	Mollisols	Xerolls	Haploxerolls	Aridic Haploxerolls	Loamy-skeletal	Mixed	Superactive		Mesic	Aridic	
Redcone	5.4	Andisols	Cryands	Duricryands	Typic Duricryands	Ashy-skeletal	Amorphic			Cryic	Udic	
Redflat	14.2	Inceptisols	Udepts	Eutrudepts	Dystric Eutrudepts	Fine-loamy	Magnesic			Mesic	Udic	
Redmond	99.6	Mollisols	Xerolls	Haploxerolls	Vitritorrandic Haploxerolls	Fine-loamy	Mixed	Superactive		Mesic	Aridic	
Redmount	21.6	Mollisols	Xerolls	Haploxerolls	Pachic Haploxerolls	Coarse-loamy	Mixed	Superactive		Frigid	Xeric	
Redslide	11.4	Mollisols	Xerolls	Haploxerolls	Vitritorrandic Haploxerolls	Loamy-skeletal	Mixed	Superactive		Mesic	Aridic	
Reedsport	254.3	Inceptisols	Udepts	Humudepts	Andic Humudepts	Fine-loamy	Isotic			Isomesic	Udic	
Reese	165.7	Inceptisols	Aquepts	Halaquepts	Duric Halaquepts	Fine-loamy	Mixed	Superactive	Calcareous	Mesic	Aquic	
Reinecke	7.3	Andisols	Xerands	Vitrixerands	Typic Vitrixerands	Ashy over loamy	Amorphic over isotic			Mesic	Xeric	
Reinhart	6.4	Inceptisols	Udepts	Humudepts	Lithic Humudepts	Loamy-skeletal	Mixed	Superactive		Isomesic	Udic	
Reluctan	343.0	Mollisols	Xerolls	Argixerolls	Aridic Argixerolls	Fine-loamy	Mixed	Superactive		Frigid	Aridic	
Remote	233.8	Inceptisols	Udepts	Dystrudepts	Typic Dystrudepts	Loamy-skeletal	Isotic			Mesic	Udic	
Reston	12.5	Mollisols	Xerolls	Haploxerolls	Lithic Ultic Haploxerolls	Loamy	Mixed	Superactive		Mesic	Xeric	
Reuter	22.5	Mollisols	Xerolls	Haploxerolls	Vitritorrandic Haploxerolls	Loamy	Mixed	Superactive		Mesic	Aridic	Shallow
Reywat	16.9	Mollisols	Xerolls	Argixerolls	Aridic Lithic Argixerolls	Loamy-skeletal	Mixed	Superactive		Mesic	Aridic	
Rhea	122.8	Mollisols	Xerolls	Haploxerolls	Calcic Haploxerolls	Fine-silty	Mixed	Superactive		Mesic	Xeric	
Ricco	6.3	Mollisols	Aquolls	Endoaquolls	Fluvaquentic Vertic Endoaquolls	Fine	Smectitic			Mesic	Aquic	
Riceton	1.1	Mollisols	Xerolls	Haploxerolls	Pachic Ultic Haploxerolls	Coarse-loamy	Mixed	Superactive		Frigid	Xeric	
Rickreall	20.3	Ultisols	Humults	Haplohumults	Xeric Haplohumults	Clayey	Mixed	Active		Mesic	Xeric	Shallow
Riddleranch	114.7	Mollisols	Xerolls	Haploxerolls	Aridic Haploxerolls	Loamy-skeletal	Mixed	Superactive		Frigid	Aridic	
Ridenbaugh	24.3	Aridisols	Durids	Argidurids	Abruptic Xeric Argidurids	Clayey	Smectitic			Mesic	Aridic	Shallow
Ridley	43.2	Mollisols	Xerolls	Palexerolls	Pachic Palexerolls	Fine	Smectitic			Mesic	Xeric	
Rilea	45.6	Inceptisols	Udepts	Dystrudepts	Typic Dystrudepts	Loamy-skeletal	Isotic			Frigid	Udic	
Rinconflat	119.6	Aridisols	Cambids	Haplocambids	Xeric Haplocambids	Loamy-skeletal	Mixed	Superactive		Frigid	Aridic	
Rinearson	402.4	Inceptisols	Udepts	Humudepts	Typic Humudepts	Fine-silty	Isotic			Mesic	Udic	
Rio King	54.5	Mollisols	Xerolls	Haploxerolls	Aridic Haploxerolls	Coarse-loamy	Mixed	Superactive		Mesic	Aridic	
Risley	199.2	Aridisols	Argids	Haplargids	Xeric Haplargids	Fine	Smectitic			Mesic	Aridic	
Ritner	291.8	Inceptisols	Xerepts	Haploxerepts	Humic Haploxerepts	Clayey-skeletal	Mixed	Superactive		Mesic	Xeric	
Ritzville	1133.4	Mollisols	Xerolls	Haploxerolls	Calcidic Haploxerolls	Coarse-silty	Mixed	Superactive		Mesic	Aridic	
Roanhide	4.2	Mollisols	Xerolls	Haploxerolls	Ultic Haploxerolls	Coarse-loamy	Mixed	Superactive		Frigid	Xeric	

(continued)

Series name	Area (km^2)	Order	Suborder	Great group	Subgroup	Particle-size class	Mineralogy class	CEC activity class	Reaction class	Soil temp. regime	Soil moisture regime	Other family
Robinette	29.6	Mollisols	Xerolls	Argixerolls	Pachic Argixerolls	Fine-loamy	Mixed	Superactive		Mesic	Xeric	
Robson	666.1	Aridisols	Argids	Haplargids	Lithic Xeric Haplargids	Clayey-skeletal	Smectitic			Frigid	Aridic	
Roca	103.4	Aridisols	Argids	Haplargids	Xeric Haplargids	Clayey-skeletal	Smectitic			Frigid	Aridic	
Rocconda	7.5	Aridisols	Argids	Haplargids	Lithic Xeric Haplargids	Clayey-skeletal	Smectitic			Mesic	Aridic	
Rockford	9.5	Mollisols	Xerolls	Haploxerolls	Ultic Haploxerolls	Loamy-skeletal	Mixed	Superactive		Mesic	Xeric	
Rockly	774.9	Mollisols	Xerolls	Haploxerolls	Lithic Haploxerolls	Loamy-skeletal	Mixed	Superactive		Mesic	Xeric	
Rogerson	37.8	Aridisols	Durids	Argidurids	Abruptic Xeric Argidurids	Clayey	Smectitic			Mesic	Aridic	Shallow
Rogger	68.1	Mollisols	Xerolls	Haploxerolls	Ultic Haploxerolls	Loamy-skeletal	Mixed	Superactive		Frigid	Xeric	
Rogue	50.6	Inceptisols	Xerepts	Dystroxerepts	Typic Dystroxerepts	Coarse-loamy	Mixed	Superactive		Frigid	Xeric	
Roloff	52.1	Mollisols	Xerolls	Haploxerolls	Aridic Haploxerolls	Coarse-loamy	Mixed	Superactive		Mesic	Aridic	
Romanose	7.1	Andisols	Udands	Hapludands	Lithic Hapludands	Medial-skeletal	Ferrihydritic			Frigid	Udic	
Rondowa	19.9	Mollisols	Xerolls	Haploxerolls	Pachic Haploxerolls	Loamy-skeletal	Mixed	Superactive		Frigid	Xeric	
Roostercomb	178.5	Mollisols	Xerolls	Argixerolls	Typic Argixerolls	Clayey-skeletal	Smectitic			Frigid	Xeric	
Roschene	44.1	Mollisols	Xerolls	Haploxerolls	Cumulic Haploxerolls	Fine-loamy	Mixed	Superactive		Frigid	Xeric	
Roseburg	22.8	Mollisols	Xerolls	Argixerolls	Pachic Ultic Argixerolls	Fine-loamy	Mixed	Superactive		Mesic	Xeric	
Rosehaven	107.5	Alfisols	Xeralfs	Haploxeralfs	Ultic Haploxeralfs	Fine-loamy	Mixed	Superactive		Mesic	Xeric	
Rouen	13.4	Andisols	Xerands	Vitrixerands	Typic Vitrixerands	Ashy over loamy-skeletal	Glassy over isotic			Frigid	Xeric	
Royal	23.5	Aridisols	Cambids	Haplocambids	Xeric Haplocambids	Coarse-loamy	Mixed	Superactive		Mesic	Aridic	
Royst	259.4	Mollisols	Xerolls	Argixerolls	Pachic Argixerolls	Clayey-skeletal	Smectitic			Frigid	Xeric	
Ruch	86.0	Alfisols	Xeralfs	Palexeralfs	Mollic Palexeralfs	Fine-loamy	Mixed	Superactive		Mesic	Xeric	
Ruckles	629.7	Mollisols	Xerolls	Argixerolls	Aridic Lithic Argixerolls	Clayey-skeletal	Smectitic			Mesic	Aridic	
Ruclick	411.1	Mollisols	Xerolls	Argixerolls	Aridic Argixerolls	Clayey-skeletal	Smectitic			Mesic	Aridic	
Ruddley	28.2	Mollisols	Xerolls	Argixerolls	Ultic Argixerolls	Loamy	Mixed	Superactive		Frigid	Xeric	
Rustlerpeak	25.5	Inceptisols	Cryepts	Humicryepts	Haploxerandic Humicryepts	Loamy-skeletal	Isotic			Cryic	Xeric	
Rustybutte	7.7	Mollisols	Udolls	Hapludolls	Typic Hapludolls	Loamy-skeletal	Magnesic			Isomesic	Udic	
Rutab	6.8	Aridisols	Cambids	Haplocambids	Xeric Haplocambids	Loamy-skeletal	Mixed	Superactive		Frigid	Aridic	
Sach	4.7	Andisols	Udands	Hapludands	Alic Hapludands	Medial over loamy-skeletal	Ferrihydritic over isotic			Frigid	Udic	
Saddlepeak	24.0	Inceptisols	Udepts	Dystrudepts	Typic Dystrudepts	Loamy-skeletal	Mixed	Superactive		Frigid	Udic	
Sag	28.9	Mollisols	Xerolls	Argixerolls	Pachic Argixerolls	Fine-loamy	Mixed	Superactive		Frigid	Xeric	
Sagehen	78.7	Aridisols	Cambids	Haplocambids	Lithic Xeric Haplocambids	Loamy-skeletal	Mixed	Superactive		Frigid	Aridic	
Sagehill	255.1	Aridisols	Calcids	Haplocalcids	Xeric Haplocalcids	Coarse-loamy	Mixed	Superactive		Mesic	Aridic	
Sagemoor	13.9	Aridisols	Cambids	Haplocambids	Xeric Haplocambids	Coarse-silty	Mixed	Superactive		Mesic	Aridic	
Sagley	4.1	Mollisols	Xerolls	Argixerolls	Pachic Argixerolls	Loamy-skeletal	Mixed	Superactive		Mesic	Xeric	
Sahaptin	6.2	Mollisols	Xerolls	Haploxerolls	Lithic Ultic Haploxerolls	Clayey-skeletal	Mixed	Superactive		Mesic	Xeric	

(continued)

Series name	Area (km^2)	Order	Suborder	Great group	Subgroup	Particle-size class	Mineralogy class	CEC activity class	Reaction class	Soil temp. regime	Soil moisture regime	Other family
Salander	153.2	Andisols	Udands	Fulvudands	Typic Fulvudands	Medial	Ferrihydritic			Isomesic	Udic	
Salem	108.1	Mollisols	Xerolls	Argixerolls	Pachic Ultic Argixerolls	Fine-loamy over sandy or sandy-skeletal	Mixed	Superactive		Mesic	Xeric	
Salhouse	66.6	Entisols	Psamments	Torripsamments	Vitrandic Torripsamments	Ashy	Glassy		Calcareous	Frigid	Aridic	
Salisbury	47.3	Mollisols	Xerolls	Durixerolls	Palexerollic Durixerolls	Fine	Smectitic			Mesic	Xeric	
Salkum	120.5	Ultisols	Humults	Palehumults	Xeric Palehumults	Fine	Kaolinitic			Mesic	Udic	
Sandgap	24.3	Entisols	Psamments	Torripsamments	Haploduridic Torripsamments	Sandy	Mixed			Frigid	Aridic	
Sandrock	7.1	Aridisols	Argids	Haplargids	Lithic Xeric Haplargids	Ashy	Glassy			Frigid	Aridic	
Sankey	4.5	Inceptisols	Udepts	Humudepts	Lithic Humudepts	Loamy-skeletal	Isotic			Mesic	Udic	
Santiam	67.7	Alfisols	Xeralfs	Haploxeralfs	Aquultic Haploxeralfs	Fine	Mixed	Superactive		Mesic	Xeric	
Saturn	33.7	Inceptisols	Udepts	Humudepts	Fluventic Humudepts	Fine-loamy over sandy or sandy-skeletal	Mixed	Superactive		Mesic	Udic	
Saum	158.5	Alfisols	Xeralfs	Palexeralfs	Ultic Palexeralfs	Fine	Mixed	Active		Mesic	Xeric	
Sauvie	100.3	Mollisols	Aquolls	Endoaquolls	Fluvaquentic Endoaquolls	Fine-silty	Mixed	Superactive		Mesic	Aquic	
Sawtell	12.3	Mollisols	Xerolls	Argixerolls	Oxyaquic Argixerolls	Loamy-skeletal	Mixed	Superactive		Mesic	Xeric	
Scalerock	8.2	Inceptisols	Udepts	Dystrudepts	Lithic Dystrudepts	Loamy-skeletal	Mixed	Superactive		Frigid	Udic	
Scaponia	173.6	Inceptisols	Udepts	Dystrudepts	Humic Dystrudepts	Fine-loamy	Isotic			Mesic	Udic	
Scaredman	27.0	Inceptisols	Udepts	Humudepts	Typic Humudepts	Loamy-skeletal	Isotic			Frigid	Udic	
Scherrard	11.1	Mollisols	Aquolls	Duraquolls	Natric Duraquolls	Fine	Smectitic			Mesic	Aquic	
Schnipper	0.7	Mollisols	Xerolls	Durixerolls	Argiduridic Durixerolls	Fine-loamy	Mixed	Superactive		Mesic	Aridic	
Schrier	19.0	Mollisols	Xerolls	Haploxerolls	Calcic Pachic Haploxerolls	Fine-loamy	Mixed	Superactive		Mesic	Xeric	
Schuelke	7.4	Mollisols	Xerolls	Argixerolls	Calcic Argixerolls	Loamy-skeletal	Mixed	Superactive		Mesic	Xeric	
Searles	117.9	Mollisols	Xerolls	Argixerolls	Aridic Argixerolls	Loamy-skeletal	Mixed	Superactive		Mesic	Aridic	
Sebastian	6.5	Mollisols	Udolls	Hapludolls	Lithic Hapludolls	Loamy-skeletal	Magnesic			Isomesic	Udic	
Segundo	58.8	Inceptisols	Xerepts	Haploxerepts	Typic Haploxerepts	Loamy-skeletal	Mixed	Superactive		Frigid	Xeric	
Seharney	122.8	Aridisols	Durids	Haplodurids	Xereptic Haplodurids	Loamy-skeletal	Mixed	Superactive		Frigid	Aridic	Shallow
Selmac	23.7	Alfisols	Xeralfs	Haploxeralfs	Ultic Haploxeralfs	Fine-loamy over clayey	Mixed over smectitic	Superactive		Mesic	Xeric	
Semiahmoo	5.0	Histosols	Saprists	Haplosaprists	Typic Haplosaprists				Euic	Mesic	Aquic	
Senra	171.9	Mollisols	Xerolls	Durixerolls	Vitritorrandic Durixerolls	Ashy	Glassy			Frigid	Aridic	Shallow
Serpentano	42.3	Inceptisols	Udepts	Eutrudepts	Dystric Eutrudepts	Loamy-skeletal	Magnesic			Mesic	Udic	
Sevencedars	20.8	Andisols	Cryands	Fulvicryands	Typic Fulvicryands	Medial-skeletal	Ferrihydritic			Cryic	Udic	
Sevenoaks	2.2	Mollisols	Xerolls	Haploxerolls	Psammentic Haploxerolls		Mixed			Mesic	Xeric	
Shanahan	332.6	Andisols	Cryands	Vitricryands	Xeric Vitricryands	Ashy over loamy	Glassy over isotic			Cryic	Xeric	
Shangland	11.5	Mollisols	Xerolls	Haploxerolls	Ultic Haploxerolls	Coarse-loamy	Mixed	Superactive		Mesic	Xeric	

(continued)

Series name	Area (km^2)	Order	Suborder	Great group	Subgroup	Particle-size class	Mineralogy class	CEC activity class	Reaction class	Soil temp. regime	Soil moisture regime	Other family
Shano	207.2	Aridisols	Cambids	Haplocambids	Xeric Haplocambids	Coarse-silty	Mixed	Superactive		Mesic	Aridic	
Sharesnout	242.7	Mollisols	Xerolls	Argixerolls	Typic Argixerolls	Clayey-skeletal	Smectitic			Frigid	Xeric	
Sharpshooter	61.1	Mollisols	Xerolls	Haploxerolls	Ultic Haploxerolls	Fine-loamy	Mixed	Superactive		Mesic	Xeric	
Shastacosta	6.7	Ultisols	Xerults	Palexerults	Typic Palexerults	Loamy-skeletal	Mixed	Superactive		Mesic	Xeric	
Shawave	5.3	Aridisols	Argids	Haplargids	Xeric Haplargids	Fine-loamy	Mixed	Superactive		Mesic	Aridic	
Sheepcreek	2.2	Mollisols	Xerolls	Haploxerolls	Andic Haploxerolls	Fine-loamy	Isotic			Frigid	Xeric	
Shefflein	56.8	Alfisols	Xeralfs	Haploxeralfs	Mollic Haploxeralfs	Fine-loamy	Mixed	Superactive		Mesic	Xeric	
Sherar	21.0	Mollisols	Xerolls	Argixerolls	Aridic Argixerolls	Fine	Smectitic			Mesic	Aridic	
Sherod	14.0	Inceptisols	Xerepts	Haploxerepts	Lithic Haploxerepts	Loamy-skeletal	Mixed	Superactive		Frigid	Xeric	
Sherval	19.0	Mollisols	Xerolls	Argixerolls	Pachic Ultic Argixerolls	Loamy-skeletal	Mixed	Superactive		Frigid	Xeric	
Shippa	27.2	Inceptisols	Xerepts	Dystroxerepts	Lithic Dystroxerepts	Loamy-skeletal	Mixed	Superactive		Mesic	Xeric	
Shiva	14.5	Mollisols	Xerolls	Haploxerolls	Vitrandic Haploxerolls	Ashy	Glassy			Mesic	Xeric	
Shivigny	32.6	Ultisols	Humults	Palehumults	Typic Palehumults	Clayey-skeletal	Mixed	Active		Mesic	Udic	
Shoat	4.0	Inceptisols	Xerepts	Haploxerepts	Typic Haploxerepts	Fine-loamy	Mixed	Active		Mesic	Xeric	
Shoepeg	0.3	Mollisols	Xerolls	Haploxerolls	Cumulic Haploxerolls	Fine-loamy	Mixed	Superactive		Mesic	Xeric	
Shroyton	9.0	Andisols	Cryands	Vitricryands	Humic Xeric Vitricryands	Ashy	Glassy			Cryic	Xeric	
Shukash	603.0	Andisols	Cryands	Vitricryands	Xeric Vitricryands	Ashy over loamy-skeletal	Glassy over isotic			Cryic	Xeric	
Sibannac	7.2	Mollisols	Aquolls	Endoaquolls	Cumulic Endoaquolls	Fine-loamy	Mixed	Superactive		Frigid	Aquic	
Sibold	14.4	Mollisols	Xerolls	Argixerolls	Aquultic Argixerolls	Fine-loamy	Mixed	Superactive		Mesic	Xeric	
Sidlake	120.9	Aridisols	Argids	Haplargids	Xeric Haplargids	Fine-loamy	Mixed	Superactive		Mesic	Aridic	
Sifton	32.0	Andisols	Xerands	Melanoxerands	Typic Melanoxerands	Medial over sandy or sandy-skeletal	Mixed			Mesic	Xeric	
Siletz	10.2	Andisols	Udands	Fulvudands	Typic Fulvudands	Medial over loamy	Ferrihydritic over isotic			Isomesic	Udic	
Silverash	63.9	Alfisols	Xeralfs	Palexeralfs	Aquandic Palexeralfs	Fine	Smectitic			Frigid	Xeric	
Silverlake	9.5	Mollisols	Xerolls	Argixerolls	Calcic Argixerolls	Fine	Smectitic			Frigid	Xeric	
Silverton	12.1	Mollisols	Xerolls	Argixerolls	Pachic Ultic Argixerolls	Fine	Mixed	Superactive		Mesic	Xeric	
Silvies	87.4	Mollisols	Aquolls	Cryaquolls	Vertic Cryaquolls	Fine	Smectitic			Cryic	Aquic	
Simas	772.5	Mollisols	Xerolls	Palexerolls	Vertic Palexerolls	Fine	Smectitic			Mesic	Xeric	
Simnasho	110.8	Andisols	Xerands	Vitrixerands	Alfic Vitrixerands	Ashy-skeletal over loamy-skeletal	Glassy over isotic			Frigid	Xeric	
Simon	5.7	Mollisols	Xerolls	Argixerolls	Aridic Argixerolls	Fine-loamy	Mixed	Superactive		Frigid	Aridic	
Sinamox	14.1	Mollisols	Xerolls	Haploxerolls	Pachic Haploxerolls	Fine-loamy	Mixed	Superactive		Mesic	Xeric	
Sinker	100.1	Mollisols	Xerolls	Haploxerolls	Pachic Haploxerolls	Loamy-skeletal	Mixed	Superactive		Frigid	Xeric	
Siskiyou	94.2	Inceptisols	Xerepts	Dystroxerepts	Typic Dystroxerepts	Coarse-loamy	Mixed	Superactive		Mesic	Xeric	
Sisley	17.1	Entisols	Orthents	Xerorthents	Typic Xerorthents	Loamy-skeletal	Mixed	Superactive	Nonacid	Frigid	Xeric	
Sisters	47.5	Andisols	Xerands	Vitrixerands	Humic Vitrixerands	Ashy over loamy	Glassy over mixed	Superactive		Frigid	Xeric	

(continued)

Series name	Area (km^2)	Order	Suborder	Great group	Subgroup	Particle-size class	Mineralogy class	CEC activity class	Reaction class	Soil temp. regime	Soil moisture regime	Other family
Sitkum	15.2	Inceptisols	Xerepts	Dystroxerepts	Typic Dystroxerepts	Coarse-loamy	Mixed	Superactive		Mesic	Xeric	
Sitton	3.5	Alfisols	Xeralfs	Haploxeralfs	Ultic Haploxeralfs	Fine-loamy	Mixed	Active		Mesic	Xeric	
Sixes	1.6	Inceptisols	Udepts	Humudepts	Pachic Humudepts	Fine-loamy	Mixed	Active		Mesic	Udic	
Skedaddle	184.8	Entisols	Orthents	Torriorthents	Lithic Xeric Torriorthents	Loamy-skeletal	Mixed	Superactive	Nonacid	Mesic	Aridic	
Skidbrackle	0.5	Mollisols	Xerolls	Argixerolls	Lithic Argixerolls	Ashy-skeletal	Glassy			Frigid	Xeric	
Skidoosprings	104.3	Inceptisols	Aquepts	Halaquepts	Duric Halaquepts	Coarse-loamy	Mixed	Superactive	Calcareous	Frigid	Aquic	
Skinner	10.1	Inceptisols	Udepts	Dystrudepts	Typic Dystrudepts	Fine-loamy	Isotic			Mesic	Udic	
Skipanon	215.3	Inceptisols	Udepts	Humudepts	Andic Humudepts	Fine-loamy	Isotic			Isomesic	Udic	
Skooker	14.3	Mollisols	Xerolls	Argixerolls	Vitrandic Argixerolls	Loamy-skeletal	Mixed	Superactive		Mesic	Xeric	
Skookum	40.4	Mollisols	Xerolls	Argixerolls	Pachic Ultic Argixerolls	Clayey-skeletal	Smectitic			Mesic	Xeric	
Skookumhouse	62.1	Ultisols	Humults	Haplohumults	Typic Haplohumults	Fine	Mixed	Active		Mesic	Udic	
Skoven	6.5	Mollisols	Xerolls	Argixerolls	Aridic Argixerolls	Clayey-skeletal	Smectitic			Mesic	Aridic	Shallow
Skull Creek	38.4	Aridisols	Durids	Haplodurids	Vitrixerandic Haplodurids	Coarse-loamy	Mixed	Superactive		Mesic	Aridic	
Skullgulch	57.8	Mollisols	Xerolls	Palexerolls	Pachic Palexerolls	Fine	Smectitic			Frigid	Xeric	
Skunkfarm	71.7	Mollisols	Aquolls	Endoaquolls	Typic Endoaquolls	Fine-loamy	Mixed	Superactive		Frigid	Aquic	
Skyline	99.7	Mollisols	Xerolls	Haploxerolls	Typic Haploxerolls	Loamy	Mixed	Superactive		Mesic	Xeric	Shallow
Slayton	12.3	Aridisols	Cambids	Haplocambids	Lithic Xeric Haplocambids	Loamy	Mixed	Superactive		Mesic	Aridic	
Slicklog	1.9	Andisols	Xerands	Vitrixerands	Humic Vitrixerands	Ashy-skeletal	Mixed			Frigid	Xeric	
Slickrock	380.3	Andisols	Udands	Hapludands	Alic Hapludands	Medial over loamy	Ferrihydritic over isotic			Mesic	Udic	
Sliptrack	19.3	Mollisols	Xerolls	Durixerolls	Vitritorrandic Durixerolls	Ashy	Glassy			Frigid	Aridic	
Smiling	236.4	Andisols	Xerands	Vitrixerands	Alfic Vitrixerands	Ashy over loamy	Glassy over isotic			Frigid	Xeric	
Snakepit	9.8	Mollisols	Xerolls	Durixerolls	Cambidic Durixerolls	Sandy	Mixed			Frigid	Aridic	
Snaker	62.5	Entisols	Orthents	Torriorthents	Lithic Xeric Torriorthents	Loamy-skeletal	Mixed	Superactive	Nonacid	Mesic	Aridic	
Snell	335.9	Mollisols	Xerolls	Argixerolls	Pachic Argixerolls	Clayey-skeletal	Smectitic			Frigid	Xeric	
Snellby	50.6	Mollisols	Xerolls	Argixerolls	Aridic Argixerolls	Clayey-skeletal	Smectitic			Frigid	Aridic	
Snow	22.4	Mollisols	Xerolls	Haploxerolls	Cumulic Haploxerolls	Fine-silty	Mixed	Superactive		Mesic	Xeric	
Snowbrier	5.5	Inceptisols	Xerepts	Dystroxerepts	Humic Dystroxerepts	Loamy-skeletal	Mixed	Superactive		Frigid	Xeric	
Snowcamp	21.1	Inceptisols	Udepts	Eutrudepts	Dystric Eutrudepts	Loamy-skeletal	Magnesic			Frigid	Udic	
Snowlin	2.1	Inceptisols	Cryepts	Humicryepts	Andic Humicryepts	Fine-loamy	Isotic			Cryic	Xeric	
Snowmore	1381.0	Aridisols	Durids	Argidurids	Xeric Argidurids	Fine-loamy	Mixed	Superactive		Mesic	Aridic	
Softscrabble	3.3	Mollisols	Xerolls	Argixerolls	Pachic Argixerolls	Loamy-skeletal	Mixed	Superactive		Frigid	Xeric	
Solarview	23.2	Entisols	Psamments	Torripsamments	Xeric Torripsamments		Mixed			Mesic	Aridic	Shallow
Sonoma	11.6	Entisols	Aquents	Fluvaquents	Aeric Fluvaquents	Fine-silty	Mixed	Superactive	Calcareous	Mesic	Aquic	
Soosap	16.2	Andisols	Cryands	Haplocryands	Typic Haplocryands	Medial	Amorphic			Cryic	Udic	
Sopher	36.1	Alfisols	Xeralfs	Haploxeralfs	Vitrandic Haploxeralfs	Clayey-skeletal	Smectitic			Mesic	Xeric	

(continued)

Series name	Area (km^2)	Order	Suborder	Great group	Subgroup	Particle-size class	Mineralogy class	CEC activity class	Reaction class	Soil temp. regime	Soil moisture regime	Other family
Sorefoot	26.3	Aridisols	Argids	Haplargids	Xerertic Haplargids	Fine	Smectitic			Mesic	Aridic	
Sorf	118.0	Aridisols	Argids	Paleargids	Vertic Paleargids	Fine	Smectitic			Mesic	Aridic	
Soughe	86.6	Aridisols	Argids	Haplargids	Lithic Xeric Haplargids	Loamy-skeletal	Mixed	Superactive		Mesic	Aridic	
Southcat	41.5	Aridisols	Cambids	Haplocambids	Sodic Haplocambids	Sandy	Mixed			Mesic	Aridic	
Spangenburg	470.6	Aridisols	Argids	Paleargids	Xeric Paleargids	Fine	Smectitic			Mesic	Aridic	
Speaker	469.8	Alfisols	Xeralfs	Haploxeralfs	Ultic Haploxeralfs	Fine-loamy	Mixed	Active		Mesic	Xeric	
Spiderhole	23.5	Aridisols	Durids	Argidurids	Xeric Argidurids	Loamy-skeletal	Mixed	Superactive		Frigid	Aridic	Shallow
Spilyay	7.0	Mollisols	Xerolls	Palexerolls	Ultic Palexerolls	Fine	Smectitic			Mesic	Xeric	
Springwater	23.3	Inceptisols	Xerepts	Haploxerepts	Humic Haploxerepts	Fine-loamy	Mixed	Superactive		Mesic	Xeric	
Srednic	19.1	Aridisols	Durids	Haplodurids	Vitrixerandic Haplodurids	Ashy	Glassy			Frigid	Aridic	
Stackyards	25.9	Inceptisols	Udepts	Humudepts	Typic Humudepts	Loamy-skeletal	Isotic			Frigid	Udic	
Stampede	74.3	Mollisols	Xerolls	Durixerolls	Vertic Durixerolls	Fine	Smectitic			Frigid	Xeric	
Stanfield	37.3	Aridisols	Durids	Haplodurids	Aquic Haplodurids	Coarse-silty	Mixed	Superactive		Mesic	Aridic	
Stanflow	9.9	Inceptisols	Aquepts	Halaquepts	Typic Halaquepts	Coarse-silty	Mixed	Superactive	Calcareous	Mesic	Aquic	
Starbuck	11.9	Aridisols	Cambids	Haplocambids	Lithic Xeric Haplocambids	Loamy	Mixed	Superactive		Mesic	Aridic	
Starkey	99.0	Mollisols	Xerolls	Argixerolls	Typic Argixerolls	Clayey-skeletal	Smectitic			Mesic	Xeric	Shallow
Statz	71.4	Mollisols	Xerolls	Durixerolls	Vitritorrandic Durixerolls	Loamy	Mixed	Superactive		Mesic	Aridic	Shallow
Stauffer	1.3	Mollisols	Xerolls	Argixerolls	Vitritorrandic Argixerolls	Ashy	Glassy			Frigid	Xeric	
Stavely	1.2	Inceptisols	Xerepts	Haploxerepts	Typic Haploxerepts	Coarse-loamy	Mixed	Superactive		Frigid	Xeric	
Stayton	13.8	Andisols	Xerands	Haploxerands	Lithic Haploxerands	Medial	Mixed			Mesic	Xeric	
Stearns	5.1	Mollisols	Aquolls	Endoaquolls	Cumulic Endoaquolls	Fine-loamy	Mixed	Superactive	Calcareous	Mesic	Aquic	
Steiger	537.9	Andisols	Cryands	Vitricryands	Xeric Vitricryands	Ashy	Glassy			Cryic	Xeric	
Steinmetz	39.3	Inceptisols	Xerepts	Dystroxerepts	Typic Dystroxerepts	Coarse-loamy	Mixed	Superactive		Mesic	Xeric	
Steiwer	74.5	Mollisols	Xerolls	Haploxerolls	Ultic Haploxerolls	Fine-loamy	Mixed	Superactive		Mesic	Xeric	
Stices	15.5	Andisols	Xerands	Vitrixerands	Typic Vitrixerands	Ashy-skeletal	Glassy			Frigid	Xeric	
Stinger	3.8	Inceptisols	Udepts	Dystrudepts	Typic Dystrudepts	Coarse-loamy	Mixed	Superactive		Mesic	Udic	
Stirfry	2.7	Histosols	Saprists	Cryosaprists	Typic Cryosaprists				Euic	Cryic	Aquic	
Stockdrive	19.5	Alfisols	Xeralfs	Natrixeralfs	Typic Natrixeralfs	Fine-loamy	Mixed	Superactive		Mesic	Xeric	
Stockel	5.7	Alfisols	Xeralfs	Haploxeralfs	Aquultic Haploxeralfs	Fine-loamy	Mixed	Superactive		Mesic	Xeric	
Stookmoor	213.0	Mollisols	Xerolls	Haploxerolls	Vitritorrandic Haploxerolls	Ashy	Glassy			Frigid	Aridic	
Stovepipe	3.2	Mollisols	Aquolls	Cryaquolls	Typic Cryaquolls	Coarse-silty over sandy or sandy-skeletal	Mixed	Superactive		Cryic	Aquic	
Straight	81.3	Inceptisols	Xerepts	Dystroxerepts	Typic Dystroxerepts	Loamy-skeletal	Mixed	Superactive		Mesic	Xeric	
Stukel	260.0	Mollisols	Xerolls	Haploxerolls	Aridic Lithic Haploxerolls	Loamy	Mixed	Superactive		Mesic	Aridic	
Sturgill	3.7	Mollisols	Aquolls	Endoaquolls	Fluvaquentic Endoaquolls	Fine-silty	Mixed	Superactive		Frigid	Aquic	

(continued)

Series name	Area (km^2)	Order	Suborder	Great group	Subgroup	Particle-size class	Mineralogy class	CEC activity class	Reaction class	Soil temp. regime	Soil moisture regime	Other family
Succor	16.2	Mollisols	Xerolls	Palexerolls	Typic Palexerolls	Fine	Smectitic			Mesic	Xeric	
Suckerflat	161.0	Mollisols	Xerolls	Haploxerolls	Aridic Lithic Haploxerolls	Ashy	Glassy			Frigid	Aridic	
Suilotem	7.9	Andisols	Xerands	Vitrixerands	Aquic Vitrixerands	Ashy	Glassy			Frigid	Xeric	
Sumine	7.3	Mollisols	Xerolls	Argixerolls	Aridic Argixerolls	Loamy-skeletal	Mixed	Superactive		Frigid	Aridic	
Sumpley	4.6	Mollisols	Xerolls	Haploxerolls	Aquic Haploxerolls	Fine-loamy over sandy or sandy-skeletal	Mixed	Superactive		Frigid	Xeric	
Sunnotch	15.9	Andisols	Cryands	Vitricryands	Typic Vitricryands	Ashy-skeletal	Amorphic			Cryic	Udic	
Sunriver	37.6	Andisols	Cryands	Vitricryands	Aquic Vitricryands	Ashy over loamy	Glassy over mixed	Superactive		Cryic	Udic	
Suppah	7.6	Aridisols	Cambids	Haplocambids	Vitrixerandic Haplocambids	Ashy-pumiceous	Glassy			Mesic	Aridic	
Sutherlin	112.9	Alfisols	Xeralfs	Haploxeralfs	Ultic Haploxeralfs	Fine-loamy over clayey	Mixed	Superactive		Mesic	Xeric	
Suttle	12.5	Andisols	Xerands	Vitrixerands	Humic Vitrixerands	Ashy	Glassy			Frigid	Xeric	
Suver	26.1	Ultisols	Humults	Haplohumults	Aquic Haplohumults	Fine	Mixed	Superactive		Mesic	Xeric	
Svensen	119.1	Inceptisols	Udepts	Humudepts	Andic Humudepts	Fine-loamy	Isotic			Isomesic	Udic	
Swaler	90.7	Aridisols	Argids	Paleargids	Xeric Paleargids	Fine	Smectitic			Frigid	Aridic	
Swalesilver	204.7	Alfisols	Xeralfs	Palexeralfs	Aquic Palexeralfs	Fine	Smectitic			Frigid	Xeric	
Swartz	1.5	Alfisols	Xeralfs	Palexeralfs	Vertic Palexeralfs	Fine	Smectitic			Mesic	Xeric	
Swedeheaven	7.0	Inceptisols	Udepts	Humudepts	Typic Humudepts	Loamy-skeletal	Isotic			Mesic	Udic	
Sweetbriar	25.1	Alfisols	Xeralfs	Haploxeralfs	Ultic Haploxeralfs	Fine	Mixed	Superactive		Mesic	Xeric	
Sweitberg	13.9	Mollisols	Xerolls	Argixerolls	Pachic Argixerolls	Fine	Smectitic			Frigid	Xeric	
Sweiting	1.0	Mollisols	Xerolls	Argixerolls	Pachic Ultic Argixerolls	Fine	Smectitic			Frigid	Xeric	
Sycan	2.4	Andisols	Cryands	Vitricryands	Xeric Vitricryands	Ashy	Glassy			Cryic	Xeric	
Syrupcreek	787.9	Andisols	Vitrands	Udivitrands	Alfic Udivitrands	Ashy over loamy-skeletal	Amorphic over isotic			Frigid	Udic	
Tablerock	6.1	Mollisols	Xerolls	Argixerolls	Pachic Argixerolls	Clayey-skeletal	Smectitic			Mesic	Xeric	
Tahkenitch	2.5	Inceptisols	Udepts	Humudepts	Typic Humudepts	Coarse-loamy	Isotic			Mesic	Udic	
Takilma	25.8	Mollisols	Xerolls	Haploxerolls	Entic Ultic Haploxerolls	Loamy-skeletal	Mixed	Superactive		Mesic	Xeric	
Talapus	17.4	Andisols	Cryands	Fulvicryands	Typic Fulvicryands	Medial-skeletal	Amorphic			Cryic	Udic	
Tallowbox	107.4	Inceptisols	Xerepts	Haploxerepts	Typic Haploxerepts	Coarse-loamy	Mixed	Superactive		Mesic	Xeric	
Tamara	144.9	Andisols	Vitrands	Udivitrands	Alfic Udivitrands	Ashy over loamy	Amorphic over isotic			Frigid	Udic	
Tamarack	0.1	Inceptisols	Udepts	Eutrudepts	Vitrandic Eutrudepts	Coarse-loamy	Mixed	Superactive		Frigid	Udic	
Tamarackcanyon	65.1	Alfisols	Xeralfs	Haploxeralfs	Vitrandic Haploxeralfs	Clayey-skeletal	Smectitic			Frigid	Xeric	
Tandy	15.0	Entisols	Aquents	Fluvaquents	Aeric Fluvaquents	Sandy over loamy	Mixed	Superactive	Calcareous	Mesic	Aquic	
Tanksel	28.4	Mollisols	Xerolls	Argixerolls	Vitrandic Argixerolls	Clayey-skeletal	Smectitic			Mesic	Xeric	
Tannahill	3.6	Mollisols	Xerolls	Argixerolls	Calcic Argixerolls	Loamy-skeletal	Mixed	Superactive		Mesic	Xeric	
Taterpa	39.5	Mollisols	Xerolls	Haploxerolls	Pachic Haploxerolls	Coarse-loamy	Mixed	Superactive		Frigid	Xeric	
Tatouche	78.3	Mollisols	Xerolls	Argixerolls	Typic Argixerolls	Fine	Smectitic			Frigid	Xeric	
Taunton	60.0	Aridisols	Durids	Haplodurids	Xeric Haplodurids	Coarse-loamy	Mixed	Superactive		Mesic	Aridic	

(continued)

Series name	Area (km^2)	Order	Suborder	Great group	Subgroup	Particle-size class	Mineralogy class	CEC activity class	Reaction class	Soil temp. regime	Soil moisture regime	Other family
Teeters	21.6	Inceptisols	Aquepts	Endoaquepts	Aquandic Endoaquepts	Ashy	Glassy		Calcareous	Mesic	Aquic	
Teewee	63.4	Mollisols	Xerolls	Argixerolls	Vitrandic Argixerolls	Fine-loamy	Mixed	Superactive		Mesic	Xeric	
Teguro	240.0	Mollisols	Xerolls	Argixerolls	Lithic Argixerolls	Loamy	Mixed	Superactive		Frigid	Xeric	
Telemon	2.6	Ultisols	Humults	Palehumults	Aquic Palehumults	Fine	Mixed	Superactive		Frigid	Udic	
Templeton	666.9	Inceptisols	Udepts	Humudepts	Andic Humudepts	Fine-silty	Isotic			Isomesic	Udic	
Tenmile	131.6	Aridisols	Argids	Haplargids	Xeric Haplargids	Clayey-skeletal	Smectitic			Mesic	Aridic	
Tenpin	19.0	Aridisols	Argids	Paleargids	Xeric Paleargids	Clayey-skeletal	Smectitic			Mesic	Aridic	
Tenwalter	33.1	Mollisols	Xerolls	Durixerolls	Palexerollic Durixerolls	Clayey-skeletal	Smectitic			Mesic	Xeric	Shallow
Terrabella	6.8	Mollisols	Aquolls	Argiaquolls	Vertic Argiaquolls	Fine	Smectitic			Mesic	Aquic	
Terwilliger	0.2	Alfisols	Xeralfs	Haploxeralfs	Typic Haploxeralfs	Fine	Smectitic			Mesic	Xeric	
Tetherow	13.1	Mollisols	Xerolls	Haploxerolls	Vitritorrandic Haploxerolls	Loamy over pumiceous or cindery	Mixed	Superactive		Mesic	Aridic	
Tethrick	47.6	Inceptisols	Xerepts	Dystroxerepts	Typic Dystroxerepts	Coarse-loamy	Mixed	Superactive		Mesic	Xeric	
Thader	5.3	Spodosols	Cryods	Humicryods	Andic Humicryods	Loamy-skeletal	Isotic			Cryic	Udic	
Thatuna	13.2	Mollisols	Xerolls	Argixerolls	Oxyaquic Argixerolls	Fine-silty	Mixed	Superactive		Mesic	Xeric	
Thenarrows	63.6	Inceptisols	Aquepts	Halaquepts	Typic Halaquepts	Coarse-loamy	Mixed	Superactive	Calcareous	Frigid	Aquic	
Thiessen	23.5	Mollisols	Xerolls	Argixerolls	Pachic Argixerolls	Clayey-skeletal	Smectitic			Mesic	Xeric	
Thirstygulch	43.3	Mollisols	Xerolls	Haploxerolls	Lithic Ultic Haploxerolls	Loamy-skeletal	Isotic			Frigid	Xeric	
Thistleburn	9.6	Ultisols	Humults	Palehumults	Typic Palehumults	Fine	Mixed	Superactive		Frigid	Udic	
Thompsoncabin	13.5	Aridisols	Argids	Natrargids	Lithic Natrargids	Loamy-skeletal	Mixed	Superactive		Mesic	Aridic	
Thorn	13.7	Alfisols	Xeralfs	Haploxeralfs	Lithic Mollic Haploxeralfs	Loamy-skeletal	Isotic			Frigid	Xeric	
Thornlake	156.3	Aridisols	Cambids	Haplocambids	Sodic Xeric Haplocambids	Ashy	Glassy			Frigid	Aridic	
Threebuck	76.0	Andisols	Xerands	Vitrixerands	Alfic Vitrixerands	Ashy over clayey-skeletal	Glassy over smectitic			Frigid	Xeric	
Threecreeks	6.0	Mollisols	Xerolls	Haploxerolls	Cumulic Haploxerolls	Coarse-loamy	Mixed	Superactive		Mesic	Xeric	
Threeforks	3.6	Inceptisols	Udepts	Dystrudepts	Humic Dystrudepts	Fine-loamy	Isotic			Mesic	Udic	
Threetrees	20.8	Inceptisols	Udepts	Dystrudepts	Typic Dystrudepts	Loamy-skeletal	Mixed	Superactive		Frigid	Udic	
Thunderegg	34.6	Mollisols	Aquolls	Natraquolls	Typic Natraquolls	Fine	Smectitic			Mesic	Aquic	
Ticino	36.5	Mollisols	Xerolls	Argixerolls	Typic Argixerolls	Fine-loamy	Mixed	Superactive		Frigid	Xeric	
Tillamook	2.1	Andisols	Udands	Melanudands	Aquic Melanudands	Medial over loamy	Ferrihydritic over isotic			Isomesic	Udic	
Timbercrater	102.1	Andisols	Cryands	Vitricryands	Typic Vitricryands	Ashy-pumiceous	Amorphic			Cryic	Udic	
Tincan	6.1	Mollisols	Xerolls	Haploxerolls	Aridic Haploxerolls	Loamy	Mixed	Superactive		Mesic	Aridic	Shallow
Tincup	6.7	Inceptisols	Udepts	Dystrudepts	Humic Dystrudepts	Loamy-skeletal	Isotic			Frigid	Udic	
Tippett	21.4	Mollisols	Xerolls	Palexerolls	Vertic Palexerolls	Fine	Smectitic			Frigid	Xeric	
Tishar	5.8	Ultisols	Xerults	Haploxerults	Typic Haploxerults	Clayey-skeletal	Mixed	Active		Mesic	Xeric	
Tolany	65.0	Andisols	Udands	Hapludands	Alic Hapludands	Medial	Ferrihydritic			Frigid	Udic	
Tolfork	5.2	Inceptisols	Udepts	Humudepts	Pachic Humudepts	Loamy-skeletal	Isotic			Frigid	Udic	

(continued)

Series name	Area (km^2)	Order	Suborder	Great group	Subgroup	Particle-size class	Mineralogy class	CEC activity class	Reaction class	Soil temp. regime	Soil moisture regime	Other family
Tolius	16.1	Mollisols	Xerolls	Argixerolls	Vitrandic Argixerolls	Fine-loamy	Mixed	Superactive		Mesic	Xeric	
Tolke	118.4	Andisols	Udands	Hapludands	Alic Hapludands	Medial	Ferrihydritic			Mesic	Udic	
Toll	23.3	Entisols	Psamments	Torripsamments	Xeric Torripsamments	Sandy	Mixed			Mesic	Aridic	
Tolo	608.0	Andisols	Xerands	Vitrixerands	Alfic Vitrixerands	Ashy over loamy	Amorphic over isotic			Frigid	Xeric	
Tolovana	220.3	Andisols	Udands	Fulvudands	Typic Fulvudands	Medial over loamy	Ferrihydritic over isotic			Isomesic	Udic	
Tonor	76.7	Aridisols	Cambids	Haplocambids	Sodic Xeric Haplocambids	Ashy	Glassy			Frigid	Aridic	
Top	83.9	Mollisols	Xerolls	Argixerolls	Vertic Argixerolls	Fine	Smectitic			Frigid	Xeric	
Topper	17.3	Mollisols	Xerolls	Haploxerolls	Vitrandic Haploxerolls	Fine-silty	Mixed	Superactive		Frigid	Xeric	
Trask	17.8	Inceptisols	Udepts	Humudepts	Typic Humudepts	Loamy-skeletal	Isotic			Mesic	Udic	
Treharne	20.9	Alfisols	Udalfs	Hapludalfs	Aquultic Hapludalfs	Fine-silty	Isotic			Mesic	Udic	
Troutmeadows	290.8	Andisols	Cryands	Vitricryands	Typic Vitricryands	Ashy over loamy-skeletal	Amorphic over isotic			Cryic	Udic	
Truax	0.0	Mollisols	Xerolls	Argixerolls	Aridic Argixerolls	Fine-loamy	Mixed	Superactive		Mesic	Aridic	
Truesdale	5.1	Aridisols	Durids	Haplodurids	Xereptic Haplodurids	Coarse-loamy	Mixed	Superactive		Mesic	Aridic	
Trunk	1.7	Aridisols	Argids	Haplargids	Xeric Haplargids	Fine	Smectitic			Mesic	Aridic	
Tub	629.6	Mollisols	Xerolls	Argixerolls	Vertic Argixerolls	Fine	Smectitic			Mesic	Xeric	
Tucker	26.1	Mollisols	Xerolls	Haploxerolls	Cumulic Haploxerolls	Fine	Smectitic			Frigid	Xeric	
Tuckerdowns	4.1	Mollisols	Xerolls	Haploxerolls	Calcic Haploxerolls	Loamy-skeletal	Mixed	Superactive		Frigid	Xeric	
Tuffcabin	1.4	Mollisols	Xerolls	Haploxerolls	Vitritorrandic Haploxerolls	Ashy	Glassy			Frigid	Aridic	
Tuffo	19.2	Entisols	Orthents	Torriorthents	Vitrandic Torriorthents	Ashy	Glassy		Nonacid	Mesic	Aridic	Shallow
Tulana	107.6	Inceptisols	Aquepts	Humaquepts	Aquandic Humaquepts	Fine-silty	Mixed	Superactive	Nonacid	Mesic	Aquic	
Tumalo	26.7	Andisols	Torrands	Vitritorrands	Duric Vitritorrands	Ashy	Glassy			Mesic	Aridic	
Tumtum	312.2	Aridisols	Durids	Argidurids	Typic Argidurids	Loamy	Mixed	Superactive		Mesic	Aridic	Shallow
Turbyfill	27.7	Entisols	Orthents	Torriorthents	Xeric Torriorthents	Coarse-loamy	Mixed	Superactive	Calcareous	Mesic	Aridic	
Turpin	203.0	Aridisols	Cambids	Haplocambids	Sodic Xeric Haplocambids	Fine-loamy	Mixed	Superactive		Mesic	Aridic	
Tusel	5.2	Mollisols	Cryolls	Argicryolls	Vitrandic Argicryolls	Loamy-skeletal	Mixed	Superactive		Cryic	Xeric	
Tutni	83.4	Andisols	Aquands	Cryaquands	Typic Cryaquands	Ashy	Glassy		Nonacid	Cryic	Aquic	
Tutuilla	8.9	Mollisols	Xerolls	Palexerolls	Typic Palexerolls	Fine	Smectitic			Mesic	Xeric	
Tweener	93.6	Mollisols	Xerolls	Argixerolls	Lithic Argixerolls	Loamy-skeletal	Mixed	Superactive		Frigid	Xeric	
TWELVEMILE	54.0	Andisols	Xerands	Vitrixerands	Typic Vitrixerands	Ashy-skeletal	Glassy			Frigid	Xeric	
Twickenham	41.9	Aridisols	Argids	Paleargids	Vertic Paleargids	Fine	Smectitic			Mesic	Aridic	
Twinbridge	0.4	Inceptisols	Xerepts	Haploxerepts	Lithic Haploxerepts	Loamy-skeletal	Isotic			Frigid	Xeric	
Tygh	16.6	Mollisols	Xerolls	Haploxerolls	Fluvaquentic Haploxerolls	Coarse-loamy	Mixed	Superactive		Mesic	Xeric	
Ukiah	84.4	Mollisols	Xerolls	Argixerolls	Vertic Argixerolls	Fine	Smectitic			Mesic	Xeric	
Umak	49.5	Andisols	Cryands	Vitricryands	Typic Vitricryands	Ashy-pumiceous	Amorphic			Cryic	Udic	

(continued)

Series name	Area (km^2)	Order	Suborder	Great group	Subgroup	Particle-size class	Mineralogy class	CEC activity class	Reaction class	Soil temp. regime	Soil moisture regime	Other family
Umapine	151.5	Inceptisols	Aquepts	Halaquepts	Typic Halaquepts	Coarse-silty	Mixed	Superactive	Calcareous	Mesic	Aquic	
Umatilla	139.7	Mollisols	Xerolls	Haploxerolls	Vitrandic Haploxerolls	Loamy-skeletal	Isotic			Frigid	Xeric	
Umpcoos	646.2	Inceptisols	Udepts	Eutrudepts	Lithic Eutrudepts	Loamy-skeletal	Isotic			Mesic	Udic	
Unionpeak	145.6	Andisols	Cryands	Duricryands	Typic Duricryands	Ashy	Amorphic			Cryic	Udic	
Upcreek	3.8	Mollisols	Xerolls	Haploxerolls	Cumulic Haploxerolls	Fine-loamy over sandy or sandy-skeletal	Mixed	Superactive		Frigid	Xeric	
Uptmor	2.6	Mollisols	Xerolls	Argixerolls	Ultic Argixerolls	Fine	Smectitic			Frigid	Xeric	
Utley	35.8	Mollisols	Xerolls	Haploxerolls	Vitrandic Haploxerolls	Fine-loamy	Mixed	Superactive		Frigid	Xeric	
Valby	464.4	Mollisols	Xerolls	Haploxerolls	Calcic Haploxerolls	Fine-silty	Mixed	Superactive		Mesic	Xeric	
Valmy	34.5	Entisols	Orthents	Torriorthents	Duric Torriorthents	Coarse-loamy	Mixed	Superactive	Calcareous	Mesic	Aridic	
Valsetz	155.8	Andisols	Cryands	Haplocryands	Alic Haplocryands	Medial-skeletal	Ferrihydritic			Cryic	Udic	
Van Horn	10.0	Mollisols	Xerolls	Argixerolls	Ultic Argixerolls	Fine-loamy	Mixed	Superactive		Mesic	Xeric	
Vandamine	68.5	Inceptisols	Cryepts	Haplocryepts	Andic Haplocryepts	Loamy-skeletal	Isotic			Cryic	Udic	
Vannoy	425.5	Alfisols	Xeralfs	Haploxeralfs	Mollic Haploxeralfs	Fine-loamy	Mixed	Superactive		Mesic	Xeric	
Vanwyper	61.8	Aridisols	Argids	Haplargids	Xeric Haplargids	Clayey-skeletal	Smectitic			Mesic	Aridic	
Veazie	46.3	Mollisols	Xerolls	Haploxerolls	Cumulic Haploxerolls	Coarse-loamy over sandy or sandy-skeletal	Mixed	Superactive		Mesic	Xeric	
Vena	13.4	Inceptisols	Xerepts	Dystroxerepts	Typic Dystroxerepts	Loamy-skeletal	Mixed	Superactive		Mesic	Xeric	
Venator	135.4	Mollisols	Xerolls	Haploxerolls	Lithic Haploxerolls	Loamy-skeletal	Mixed	Superactive		Mesic	Xeric	
Veneta	59.7	Alfisols	Xeralfs	Haploxeralfs	Ultic Haploxeralfs	Fine	Mixed	Superactive		Mesic	Xeric	
Verboort	1.6	Mollisols	Albolls	Argialbolls	Xerertic Argialbolls	Fine	Mixed	Superactive		Mesic	Xeric	
Verdico	26.3	Aridisols	Argids	Paleargids	Vertic Paleargids	Fine	Smectitic			Mesic	Aridic	
Vergas	249.1	Aridisols	Argids	Haplargids	Durinodic Xeric Haplargids	Fine-loamy over sandy or sandy-skeletal	Mixed	Superactive		Frigid	Aridic	
Vermisa	542.7	Inceptisols	Xerepts	Dystroxerepts	Lithic Dystroxerepts	Loamy-skeletal	Mixed	Superactive		Mesic	Xeric	
Vernonia	90.4	Alfisols	Udalfs	Hapludalfs	Ultic Hapludalfs	Fine-silty	Isotic			Mesic	Udic	
Veta	11.7	Aridisols	Cambids	Haplocambids	Xeric Haplocambids	Loamy-skeletal	Mixed	Superactive		Mesic	Aridic	
Vil	31.6	Mollisols	Xerolls	Durixerolls	Argiduridic Durixerolls	Loamy	Mixed	Superactive		Frigid	Aridic	Shallow
Vining	49.6	Aridisols	Cambids	Haplocambids	Xeric Haplocambids	Coarse-loamy	Mixed	Superactive		Mesic	Aridic	
Virtue	225.9	Aridisols	Durids	Argidurids	Xeric Argidurids	Fine-silty	Mixed	Superactive		Mesic	Aridic	
Vitale	270.4	Mollisols	Xerolls	Argixerolls	Typic Argixerolls	Loamy-skeletal	Mixed	Superactive		Frigid	Xeric	
Voats	21.8	Mollisols	Xerolls	Haploxerolls	Fluventic Haploxerolls	Sandy-skeletal	Mixed			Mesic	Xeric	
Volstead	14.9	Mollisols	Xerolls	Argixerolls	Vitrandic Argixerolls	Fine	Smectitic			Frigid	Xeric	
Voltage	60.9	Aridisols	Calcids	Haplocalcids	Xeric Haplocalcids	Coarse-loamy	Mixed	Superactive		Frigid	Aridic	
Vondergreen	1.0	Ultisols	Udults	Hapludults	Aquic Hapludults	Fine	Isotic			Isomesic	Udic	
Voorhies	92.5	Alfisols	Xeralfs	Haploxeralfs	Mollic Haploxeralfs	Loamy-skeletal	Mixed	Superactive		Mesic	Xeric	
Wabuska	16.7	Inceptisols	Aquepts	Halaquepts	Aeric Halaquepts	Coarse-loamy	Mixed	Superactive	Calcareous	Mesic	Aquic	
Wadecreek	8.0	Ultisols	Humults	Haplohumults	Oxyaquic Haplohumults	Fine	Isotic			Isomesic	Udic	

(continued)

Series name	Area (km^2)	Order	Suborder	Great group	Subgroup	Particle-size class	Mineralogy class	CEC activity class	Reaction class	Soil temp. regime	Soil moisture regime	Other family
Wagontire	19.8	Mollisols	Xerolls	Durixerolls	Argiduridic Durixerolls	Clayey	Smectitic			Frigid	Aridic	Shallow
Waha	504.8	Mollisols	Xerolls	Argixerolls	Pachic Argixerolls	Fine-loamy	Mixed	Superactive		Mesic	Xeric	
Wahkeena	6.9	Mollisols	Udolls	Hapludolls	Pachic Hapludolls	Fragmental	Mixed			Mesic	Udic	
Wahstal	68.2	Mollisols	Xerolls	Durixerolls	Palexerollic Durixerolls	Clayey-skeletal	Smectitic			Frigid	Xeric	Shallow
Wakamo	6.0	Mollisols	Xerolls	Argixerolls	Lithic Ultic Argixerolls	Clayey-skeletal	Smectitic			Mesic	Xeric	
Waldo	160.9	Mollisols	Aquolls	Endoaquolls	Fluvaquentic Vertic Endoaquolls	Fine	Smectitic			Mesic	Aquic	
Waldport	91.9	Entisols	Psamments	Udipsamments	Typic Udipsamments	Sandy	Mixed			Isomesic	Udic	
Walla Walla	1270.0	Mollisols	Xerolls	Haploxerolls	Typic Haploxerolls	Coarse-silty	Mixed	Superactive		Mesic	Xeric	
Wallowa	83.3	Mollisols	Xerolls	Haploxerolls	Vitrandic Haploxerolls	Fine-loamy	Mixed	Superactive		Frigid	Xeric	
Walluski	27.4	Inceptisols	Udepts	Humudepts	Andic Oxyaquic Humudepts	Fine-silty	Isotic			Isomesic	Udic	
Wamic	247.4	Inceptisols	Xerepts	Haploxerepts	Vitrandic Haploxerepts	Fine-loamy	Mixed	Superactive		Mesic	Xeric	
Wanoga	332.1	Andisols	Xerands	Vitrixerands	Humic Vitrixerands	Ashy	Glassy			Frigid	Xeric	
Wanser	21.3	Entisols	Aquents	Psammaquents	Typic Psammaquents		Mixed			Mesic	Aquic	
Wapato	164.2	Mollisols	Aquolls	Endoaquolls	Fluvaquentic Endoaquolls	Fine-silty	Mixed	Superactive		Mesic	Aquic	
Wapinitia	40.6	Mollisols	Xerolls	Argixerolls	Pachic Argixerolls	Fine-loamy	Mixed	Superactive		Mesic	Xeric	
Warden	280.5	Aridisols	Cambids	Haplocambids	Xeric Haplocambids	Coarse-silty	Mixed	Superactive		Mesic	Aridic	
Warnermount	0.2	Mollisols	Xerolls	Argixerolls	Vitrandic Argixerolls	Ashy-skeletal	Glassy			Frigid	Xeric	
Warrenton	3.6	Inceptisols	Aquepts	Humaquepts	Typic Humaquepts	Sandy	Mixed			Isomesic	Aquic	
Wasson	3.0	Inceptisols	Aquepts	Humaquepts	Fluvaquentic Humaquepts	Coarse-loamy	Mixed	Superactive	Nonacid	Mesic	Aquic	
Watama	261.5	Mollisols	Xerolls	Haploxerolls	Pachic Haploxerolls	Fine-loamy	Mixed	Superactive		Mesic	Xeric	
Watches	17.5	Inceptisols	Udepts	Dystrudepts	Typic Dystrudepts	Fine-loamy	Mixed	Active		Isomesic	Udic	
Waterbury	170.2	Mollisols	Xerolls	Argixerolls	Lithic Argixerolls	Clayey-skeletal	Smectitic			Mesic	Xeric	
Wato	54.5	Mollisols	Xerolls	Haploxerolls	Typic Haploxerolls	Coarse-loamy	Mixed	Superactive		Mesic	Xeric	
Wauld	18.2	Inceptisols	Udepts	Humudepts	Typic Humudepts	Loamy-skeletal	Isotic			Mesic	Udic	
Wauna	24.7	Inceptisols	Aquepts	Endoaquepts	Fluvaquentic Endoaquepts	Fine-silty	Mixed	Superactive	Acid	Mesic	Aquic	
Weash	4.9	Aridisols	Cambids	Haplocambids	Vitrixerandic Haplocambids	Ashy	Glassy			Mesic	Aridic	Shallow
Webfoot	1.2	Mollisols	Xerolls	Haploxerolls	Pachic Haploxerolls	Loamy-skeletal	Mixed	Superactive		Frigid	Xeric	
Wedderburn	6.0	Inceptisols	Udepts	Humudepts	Pachic Humudepts	Fine-loamy	Isotic			Isomesic	Udic	
Wegert	190.5	Mollisols	Xerolls	Haploxerolls	Vitritorrandic Haploxerolls	Ashy	Glassy			Frigid	Aridic	
Weglike	81.0	Mollisols	Xerolls	Haploxerolls	Vitritorrandic Haploxerolls	Fine-loamy	Mixed	Superactive		Frigid	Aridic	
Welch	61.5	Mollisols	Aquolls	Endoaquolls	Cumulic Endoaquolls	Fine-loamy	Mixed	Superactive		Frigid	Aquic	

(continued)

Series name	Area (km^2)	Order	Suborder	Great group	Subgroup	Particle-size class	Mineralogy class	CEC activity class	Reaction class	Soil temp. regime	Soil moisture regime	Other family
Wellsdale	22.9	Alfisols	Xeralfs	Haploxeralfs	Aquultic Haploxeralfs	Fine-loamy	Mixed	Active		Mesic	Xeric	
Wenas	9.4	Mollisols	Aquolls	Endoaquolls	Cumulic Endoaquolls	Fine-loamy	Mixed	Superactive		Mesic	Aquic	
Westbutte	642.6	Mollisols	Xerolls	Haploxerolls	Pachic Haploxerolls	Loamy-skeletal	Mixed	Superactive		Frigid	Xeric	
Westside	23.6	Aridisols	Argids	Paleargids	Durinodic Xeric Paleargids	Fine	Smectitic			Frigid	Aridic	
Whaleshead	39.6	Inceptisols	Udepts	Humudepts	Andic Humudepts	Loamy-skeletal	Isotic			Isomesic	Udic	
Whetstone	53.7	Spodosols	Cryods	Haplocryods	Typic Haplocryods	Loamy-skeletal	Isotic			Cryic	Udic	Ortstein
Whisk	2.9	Mollisols	Xerolls	Haploxerolls	Lithic Ultic Haploxerolls	Loamy	Mixed	Superactive		Mesic	Xeric	
Whiteface	2.3	Mollisols	Cryolls	Duricryolls	Argic Duricryolls	Loamy	Mixed	Active		Cryic	Xeric	Shallow
Whiteson	9.6	Mollisols	Aquolls	Endoaquolls	Fluvaquentic Vertic Endoaquolls	Fine-loamy over clayey	Mixed over smectitic	Superactive		Mesic	Aquic	
Whobrey	95.9	Inceptisols	Udepts	Eutrudepts	Aquertic Eutrudepts	Fine-silty over clayey	Isotic over smectitic			Mesic	Udic	
Wickahoney	3.8	Alfisols	Xeralfs	Haploxeralfs	Lithic Mollic Haploxeralfs	Clayey-skeletal	Smectitic			Frigid	Xeric	
Wickiup	13.2	Andisols	Aquands	Cryaquands	Typic Cryaquands	Ashy-pumiceous	Glassy		Nonacid	Cryic	Aquic	
Widowspring	98.6	Mollisols	Xerolls	Haploxerolls	Cumulic Haploxerolls	Fine-silty	Mixed	Superactive		Frigid	Xeric	
Wieland	59.3	Aridisols	Argids	Haplargids	Durinodic Xeric Haplargids	Fine	Smectitic			Mesic	Aridic	
Wildcatbutte	14.6	Mollisols	Xerolls	Haploxerolls	Vitritorrandic Haploxerolls	Ashy-skeletal	Glassy			Frigid	Aridic	
Wildhill	35.4	Aridisols	Argids	Haplargids	Durinodic Xeric Haplargids	Loamy-skeletal	Mixed	Superactive		Mesic	Aridic	
Wilhoit	75.9	Inceptisols	Udepts	Humudepts	Andic Humudepts	Fine-loamy	Isotic			Frigid	Udic	
Wilkins	21.9	Mollisols	Albolls	Argialbolls	Xerertic Argialbolls	Fine	Smectitic			Frigid	Xeric	
Willakenzie	157.4	Alfisols	Xeralfs	Haploxeralfs	Ultic Haploxeralfs	Fine-loamy	Mixed	Active		Mesic	Xeric	
Willamette	155.1	Mollisols	Xerolls	Argixerolls	Pachic Ultic Argixerolls	Fine-silty	Mixed	Superactive		Mesic	Xeric	
Willanch	15.2	Inceptisols	Aquepts	Humaquepts	Fluvaquentic Humaquepts	Coarse-loamy	Mixed	Superactive	Nonacid	Isomesic	Aquic	
Willis	148.8	Mollisols	Xerolls	Durixerolls	Haploduridic Durixerolls	Coarse-silty	Mixed	Superactive		Mesic	Xeric	
Willowdale	62.8	Mollisols	Xerolls	Haploxerolls	Cumulic Haploxerolls	Fine-loamy	Mixed	Superactive		Mesic	Xeric	
Wilt	16.2	Mollisols	Xerolls	Argixerolls	Vitrandic Argixerolls	Loamy-skeletal	Mixed	Superactive		Frigid	Xeric	
Winberry	0.7	Inceptisols	Cryepts	Dystrocryepts	Lithic Dystrocryepts	Loamy-skeletal	Isotic			Cryic	Udic	
Winchester	58.1	Entisols	Psamments	Torripsamments	Xeric Torripsamments	Sandy	Mixed			Mesic	Aridic	
Winchuck	4.1	Ultisols	Humults	Haplohumults	Typic Haplohumults	Fine	Isotic			Isomesic	Udic	
Wind River	11.1	Mollisols	Xerolls	Haploxerolls	Ultic Haploxerolls	Coarse-loamy	Mixed	Superactive		Mesic	Xeric	
Windego	70.4	Andisols	Xerands	Vitrixerands	Alfic Vitrixerands	Ashy over loamy-skeletal	Glassy over isotic			Frigid	Xeric	
Windybutte	47.5	Mollisols	Xerolls	Argixerolls	Argiduridic Argixerolls	Fine-silty	Mixed	Superactive		Frigid	Xeric	

(continued)

Series name	Area (km^2)	Order	Suborder	Great group	Subgroup	Particle-size class	Mineralogy class	CEC activity class	Reaction class	Soil temp. regime	Soil moisture regime	Other family
Windygap	313.2	Ultisols	Humults	Haplohumults	Xeric Haplohumults	Fine	Mixed	Active		Mesic	Xeric	
Windypoint	20.4	Aridisols	Argids	Haplargids	Xeric Haplargids	Fine-loamy	Mixed	Superactive		Mesic	Aridic	
Winema	5.1	Andisols	Udands	Fulvudands	Typic Fulvudands	Medial over clayey	Ferrihydritic over isotic			Isomesic	Udic	
Wingdale	8.8	Mollisols	Aquolls	Endoaquolls	Cumulic Endoaquolls	Fine-silty	Mixed	Superactive	Calcareous	Mesic	Aquic	
Wingville	78.4	Mollisols	Xerolls	Haploxerolls	Pachic Haploxerolls	Fine-silty	Mixed	Superactive		Mesic	Xeric	
Winlo	31.4	Mollisols	Aquolls	Duraquolls	Typic Duraquolls	Clayey-skeletal	Smectitic			Mesic	Aquic	Shallow
Winom	7.3	Vertisols	Uderts	Hapluderts	Oxyaquic Hapluderts	Fine	Smectitic			Frigid	Udic	
Wintercanyon	16.4	Mollisols	Xerolls	Haploxerolls	Lithic Ultic Haploxerolls	Loamy-skeletal	Isotic			Frigid	Xeric	
Winterim	117.9	Mollisols	Xerolls	Argixerolls	Pachic Argixerolls	Clayey-skeletal	Smectitic			Frigid	Xeric	
Wintley	23.6	Ultisols	Humults	Haplohumults	Typic Haplohumults	Fine	Isotic			Mesic	Udic	
Wiskan	1.3	Aridisols	Argids	Haplargids	Xeric Haplargids	Loamy-skeletal	Mixed	Superactive		Frigid	Aridic	
Witham	44.1	Mollisols	Xerolls	Haploxerolls	Vertic Haploxerolls	Fine	Smectitic			Mesic	Xeric	
Witzel	113.3	Mollisols	Xerolls	Haploxerolls	Lithic Ultic Haploxerolls	Loamy-skeletal	Mixed	Superactive		Mesic	Xeric	
Wizard	13.4	Andisols	Xerands	Vitrixerands	Aquic Vitrixerands	Ashy	Glassy			Frigid	Xeric	
Wolfer	2.1	Andisols	Udands	Fulvudands	Typic Fulvudands	Medial over sandy or sandy-skeletal	Ferrihydritic over mixed			Isomesic	Udic	
Wolfpeak	56.3	Alfisols	Xeralfs	Palexeralfs	Ultic Palexeralfs	Fine-loamy	Mixed	Superactive		Mesic	Xeric	
Wollent	11.8	Inceptisols	Aquepts	Humaquepts	Typic Humaquepts	Fine-silty	Mixed	Superactive	Nonacid	Mesic	Aquic	
Wolot	14.6	Andisols	Xerands	Vitrixerands	Alfic Vitrixerands	Ashy over loamy	Glassy over isotic			Mesic	Xeric	
Wolverine	11.7	Entisols	Psamments	Torripsamments	Xeric Torripsamments	Sandy	Mixed			Frigid	Aridic	
Woodburn	911.3	Mollisols	Xerolls	Argixerolls	Aquultic Argixerolls	Fine-silty	Mixed	Superactive		Mesic	Xeric	
Woodchopper	77.0	Mollisols	Xerolls	Argixerolls	Pachic Ultic Argixerolls	Fine	Isotic			Frigid	Xeric	
Woodcock	576.8	Andisols	Xerands	Haploxerands	Alfic Humic Haploxerands	Medial-skeletal	Amorphic			Frigid	Xeric	
Woodseye	77.5	Inceptisols	Xerepts	Dystroxerepts	Humic Lithic Dystroxerepts	Loamy-skeletal	Mixed	Superactive		Frigid	Xeric	
Woodspoint	5.8	Andisols	Cryands	Fulvicryands	Typic Fulvicryands	Medial	Ferrihydritic			Cryic	Udic	
Wrentham	333.9	Mollisols	Xerolls	Haploxerolls	Pachic Haploxerolls	Loamy-skeletal	Mixed	Superactive		Mesic	Xeric	
Wrightman	17.5	Mollisols	Xerolls	Haploxerolls	Vitrandic Haploxerolls	Fine-loamy	Mixed	Superactive		Frigid	Xeric	
Wuksi	2.2	Andisols	Cryands	Vitricryands	Xeric Vitricryands	Ashy-skeletal	Glassy			Cryic	Xeric	
Wyeast	6.0	Inceptisols	Aquepts	Fragiaquepts	Aeric Fragiaquepts	Coarse-silty	Mixed	Superactive		Mesic	Aquic	
Wyeth	26.6	Inceptisols	Xerepts	Humixerepts	Pachic Humixerepts	Loamy-skeletal	Mixed	Superactive		Mesic	Xeric	
Xanadu	72.8	Ultisols	Humults	Palehumults	Typic Palehumults	Fine	Kaolinitic			Mesic	Udic	
Yachats	9.3	Inceptisols	Udepts	Humudepts	Fluventic Humudepts	Coarse-loamy	Mixed	Superactive		Isomesic	Udic	
Yainax	36.1	Alfisols	Xeralfs	Haploxeralfs	Mollic Haploxeralfs	Fine-loamy	Mixed	Superactive		Frigid	Xeric	
Yakima	14.7	Mollisols	Xerolls	Haploxerolls	Cumulic Haploxerolls	Coarse-loamy over sandy or sandy-skeletal	Mixed	Superactive		Mesic	Xeric	

(continued)

Series name	Area (km^2)	Order	Suborder	Great group	Subgroup	Particle-size class	Mineralogy class	CEC activity class	Reaction class	Soil temp. regime	Soil moisture regime	Other family
Yakus	0.5	Mollisols	Xerolls	Haploxerolls	Lithic Haploxerolls	Loamy	Mixed	Superactive		Mesic	Xeric	
Yallani	46.8	Andisols	Xerands	Vitrixerands	Typic Vitrixerands	Ashy-skeletal	Glassy			Frigid	Xeric	
Yancy	52.4	Mollisols	Xerolls	Durixerolls	Palexerollic Durixerolls	Clayey	Smectitic			Frigid	Xeric	Shallow
Yankeewell	169.7	Aridisols	Durids	Natridurids	Xeric Natridurids	Loamy	Mixed	Superactive		Frigid	Aridic	Shallow
Yapoah	27.9	Andisols	Xerands	Vitrixerands	Humic Vitrixerands	Ashy-skeletal	Glassy			Frigid	Xeric	
Yaquina	14.3	Spodosols	Aquods	Endoaquods	Typic Endoaquods	Sandy	Isotic			Isomesic	Aquic	
Yawhee	144.3	Andisols	Vitrands	Udivitrands	Alfic Udivitrands	Ashy-skeletal over loamy-skeletal	Glassy over isotic			Frigid	Udic	
Yawkey	84.3	Mollisols	Xerolls	Palexerolls	Vertic Palexerolls	Clayey-skeletal	Smectitic			Frigid	Xeric	
Yawkola	8.8	Mollisols	Xerolls	Palexerolls	Pachic Palexerolls	Clayey-skeletal	Smectitic			Mesic	Xeric	
Yellowstone	81.2	Andisols	Cryands	Haplocryands	Lithic Haplocryands	Medial-skeletal	Ferrihydritic			Cryic	Udic	
Yoncalla	19.5	Mollisols	Xerolls	Palexerolls	Aquic Palexerolls	Fine	Smectitic			Mesic	Xeric	
Yonna	27.6	Inceptisols	Cryepts	Haplocryepts	Aquandic Haplocryepts	Loamy over ashy or ashy-pumiceous	Mixed over glassy	Superactive		Cryic	Xeric	
Yorel	18.1	Inceptisols	Udepts	Dystrudepts	Typic Dystrudepts	Fine-loamy	Isotic			Frigid	Udic	
Youtlkue	6.6	Aridisols	Cambids	Aquicambids	Vitrixerandic Aquicambids	Ashy	Glassy			Frigid	Aridic	
Yuko	4.1	Aridisols	Argids	Haplargids	Xeric Haplargids	Loamy	Mixed	Superactive		Mesic	Aridic	
Zalea	11.8	Ultisols	Humults	Haplohumults	Typic Haplohumults	Fine-loamy	Isotic			Frigid	Udic	
Zango	12.6	Inceptisols	Udepts	Dystrudepts	Lithic Dystrudepts	Loamy-skeletal	Isotic			Mesic	Udic	
Zevadez	200.2	Aridisols	Argids	Haplargids	Durinodic Xeric Haplargids	Fine-loamy	Mixed	Superactive		Mesic	Aridic	
Zing	64.1	Alfisols	Xeralfs	Haploxeralfs	Aqultic Haploxeralfs	Fine	Mixed	Superactive		Mesic	Xeric	
Zola	26.0	Mollisols	Xerolls	Haploxerolls	Cumulic Haploxerolls	Fine-loamy	Mixed	Superactive		Frigid	Xeric	
Zorravista	16.2	Entisols	Psamments	Torripsamments	Xeric Torripsamments	Sandy	Mixed			Mesic	Aridic	
Zuman	17.9	Inceptisols	Aquepts	Halaquepts	Typic Halaquepts	Fine-loamy over sandy or sandy-skeletal	Mixed	Superactive	Calcareous	Mesic	Aquic	
Zumwalt	34.7	Mollisols	Xerolls	Palexerolls	Vertic Palexerolls	Fine	Smectitic			Frigid	Xeric	
Zwagg	4.8	Inceptisols	Udepts	Humudepts	Pachic Humudepts	Coarse-loamy	Isotic			Isomesic	Udic	
Zygore	221.9	Inceptisols	Udepts	Humudepts	Andic Humudepts	Loamy-skeletal	Isotic			Frigid	Udic	
Zyzzug	7.1	Inceptisols	Aquepts	Humaquepts	Typic Humaquepts	Fine-silty	Isotic		Acid	Mesic	Aquic	

Appendix D
Benchmark, Endemic, Rare, and Endangered Soil Series in Oregon

Series name	Area (km^2)	Benchmark[1]	Endemic[2]	Rare[3]	Endangered[4]
Abegg	78.4	No	No	Yes	No
Abert	144.9	No	No	No	No
Abin	6.7	No	No	Yes	No
Abiqua	57.3	No	Yes	Yes	Yes
Absaquil	54.0	No	No	Yes	No
Acanod	6.6	No	No	Yes	No
Acker	217.6	No	Yes	No	No
Actem	749.7	No	No	No	No
Ada	84.9	No	No	Yes	No
Adieux	0.0	No	No	Yes	No
Adkins	107.6	No	No	No	No
Agate	50.3	No	Yes	Yes	Yes
Agency	135.9	No	No	No	No
Agness	2.0	No	No	Yes	No
Ahtanum	2.8	No	No	Yes	No
Akerite	18.4	No	Yes	Yes	Yes
Albee	123.7	No	No	No	No
Alcot	0.4	No	Yes	Yes	Yes
Alding	63.1	No	No	Yes	No
Algoma	29.4	No	No	Yes	No
Alicel	39.6	No	No	Yes	No
Alley	2.1	Yes	No	Yes	No
Allingham	18.8	No	Yes	Yes	Yes
Almota	1.2	No	No	Yes	No
Aloha	236.5	No	Yes	No	No
Als	8.1	No	No	Yes	No
Alsea	6.0	No	No	Yes	No
Alspaugh	130.6	No	Yes	No	No
Alstony	82.6	No	No	Yes	No
Althouse	59.9	No	No	Yes	No
Alvodest	263.4	No	No	No	No
Alyan	27.8	No	No	Yes	No
Amity	434.7	Yes	Yes	No	No
Analulu	265.0	No	No	No	No
Anatone	1589.5	No	No	No	No
Anawalt	1280.1	No	No	No	No
Anderly	185.2	No	No	No	No
Anders	4.9	No	No	Yes	No
Angelbasin	78.8	No	No	Yes	No

(continued)

T. Thorson et al., *The Soils of Oregon*, World Soils Book Series,
https://doi.org/10.1007/978-3-030-90091-5

Series name	Area (km^2)	Benchmark[1]	Endemic[2]	Rare[3]	Endangered[4]
Angelpeak	127.1	No	No	No	No
Anniecreek	3.4	No	No	Yes	No
Antelopepeak	7.9	No	Yes	Yes	Yes
Antoken	33.3	No	No	Yes	No
Anunde	32.4	No	No	Yes	No
Applegate	13.1	No	No	Yes	No
Apt	148.9	No	No	No	No
Arbidge	8.0	No	No	Yes	No
Arcia	92.6	No	No	Yes	No
Arness	8.6	No	No	Yes	No
Ascar	120.9	No	No	No	No
Aschoff	311.0	Yes	No	No	No
Aspenlake	3.5	No	Yes	Yes	Yes
Astoria	145.9	No	No	No	No
Ateron	1288.7	No	No	No	No
Athena	210.4	Yes	No	No	No
Atlow	357.3	No	No	No	No
Atring	480.2	No	No	No	No
Ausmus	235.3	No	Yes	No	No
Aval	9.5	No	No	Yes	No
Averlande	40.3	No	No	Yes	No
Awbrig	45.7	No	No	Yes	No
Axford	4.2	No	No	Yes	No
Ayres	86.6	No	No	Yes	No
Ayresbutte	40.0	No	Yes	Yes	Yes
Babbington	0.8	No	No	Yes	No
Bacona	212.6	No	No	No	No
Baconcamp	373.2	No	No	No	No
Bagness	0.8	No	Yes	Yes	Yes
Bakeoven	1225.0	Yes	No	No	No
Baker	77.9	Yes	No	Yes	No
Bald	42.7	No	No	Yes	No
Balder	18.7	No	Yes	Yes	Yes
Baldock	46.4	Yes	No	Yes	No
Baldridge	17.9	No	No	Yes	No
Balloontree	3.6	No	Yes	Yes	Yes
Balm	12.8	No	No	Yes	No
Bandarrow	17.8	No	No	Yes	No
Bandon	46.2	No	Yes	Yes	Yes
Banning	23.7	No	No	Yes	No
Barbermill	16.5	No	No	Yes	No
Barhiskey	7.6	No	No	Yes	No
Barkley	4.1	No	Yes	Yes	Yes
Barkshanty	69.7	No	Yes	Yes	Yes
Barnard	43.1	No	No	Yes	No
Barneycreek	4.9	No	No	Yes	No
Barron	24.0	No	No	Yes	No
Bashaw	171.9	Yes	No	No	No
Bata	5.7	No	No	Yes	No
Bateman	204.6	No	No	No	No
Bayside	0.2	No	Yes	Yes	Yes
Beal	36.7	No	No	Yes	No
Bearcamp	19.1	No	No	Yes	No
Bearpawmeadow	58.3	No	No	Yes	No
Bearspring	1.7	No	No	Yes	No
Beden	106.5	No	No	No	No

(continued)

Series name	Area (km^2)	Benchmark[1]	Endemic[2]	Rare[3]	Endangered[4]
Bedner	7.0	No	No	Yes	No
Beekman	647.3	Yes	No	No	No
Beeman	3.0	No	No	Yes	No
Beetville	0.3	No	No	Yes	No
Bellpine	769.6	No	No	No	No
Belrick	35.9	No	No	Yes	No
Benderly	11.9	No	No	Yes	No
Bennettcreek	107.4	No	No	No	No
Bensley	24.1	No	No	Yes	No
Bentilla	5.4	No	Yes	Yes	Yes
Beoska	55.9	Yes	No	Yes	No
Berdugo	114.7	No	No	No	No
Bergsvik	4.5	No	No	Yes	No
Bickford	1.0	No	No	Yes	No
Bicondoa	5.8	No	Yes	Yes	Yes
Bigbouldercreek	0.0	No	No	Yes	No
Bigcow	47.2	No	No	Yes	No
Bigdutch	3.1	No	No	Yes	No
Bigelk	35.4	No	No	Yes	No
Bigelow	12.4	No	No	Yes	No
Bigfrog	11.9	No	No	Yes	No
Bigriver	0.7	No	Yes	Yes	Yes
Bindle	34.8	No	Yes	Yes	Yes
Bingville	68.8	No	Yes	Yes	Yes
Bins	47.6	Yes	Yes	Yes	Yes
Bittercreek	1.8	No	Yes	Yes	Yes
Blachly	510.1	No	No	No	No
Blackgulch	17.3	No	Yes	Yes	Yes
Blackhills	2.5	No	Yes	Yes	Yes
Blacklock	21.5	Yes	No	Yes	No
Blalock	14.0	No	No	Yes	No
Blayden	44.8	No	No	Yes	No
Bler	12.6	No	No	Yes	No
Blizzard	17.6	No	No	Yes	No
Blodgett	3.3	No	Yes	Yes	Yes
Bluecanyon	67.0	No	No	Yes	No
Bluesters	9.4	No	No	Yes	No
Bly	60.8	No	No	Yes	No
Boardflower	8.3	No	No	Yes	No
Boardtree	112.2	No	Yes	No	No
Bobbitt	16.5	No	No	Yes	No
Bobsgarden	25.6	No	No	Yes	No
Bocker	1473.9	No	No	No	No
Bodale	2.2	No	No	Yes	No
Bodell	83.9	No	No	Yes	No
Bogus	6.5	No	No	Yes	No
Bohannon	2045.6	Yes	No	No	No
Boiler	5.8	No	No	Yes	No
Boilout	52.8	No	Yes	Yes	Yes
Bolobin	66.3	No	No	Yes	No
Bolony	16.7	No	No	Yes	No
Bombadil	62.3	No	No	Yes	No
Bonnick	96.7	No	No	Yes	No
Booten	39.1	No	Yes	Yes	Yes
Booth	593.7	No	No	No	No
Boravall	36.5	No	No	Yes	No

(continued)

Series name	Area (km^2)	Benchmark[1]	Endemic[2]	Rare[3]	Endangered[4]
Bordengulch	71.1	No	No	Yes	No
Borges	5.6	No	No	Yes	No
Bornstedt	67.7	No	Yes	Yes	Yes
Borobey	215.2	No	No	No	No
Bosland	44.1	No	Yes	Yes	Yes
Bott	2.0	No	Yes	Yes	Yes
Boulder lake	81.4	No	No	Yes	No
Bouldrock	56.0	No	No	Yes	No
Bowlus	24.5	No	Yes	Yes	Yes
Boyce	31.0	No	No	Yes	No
Brabble	42.6	No	No	Yes	No
Brace	1571.4	No	Yes	No	No
Brader	50.5	No	No	Yes	No
Bragton	8.9	No	Yes	Yes	Yes
Brallier	12.6	No	Yes	Yes	Yes
Brand	10.4	No	Yes	Yes	Yes
Brandypeak	28.6	No	No	Yes	No
Brannan	8.2	No	No	Yes	No
Braun	167.7	No	No	No	No
Bravo	188.9	No	No	No	No
Breadloaf	0.6	No	No	Yes	No
Bregar	40.4	No	No	Yes	No
Brenner	35.7	Yes	No	Yes	No
Brezniak	24.7	No	No	Yes	No
Bridgecreek	75.6	No	No	Yes	No
Bridgewater	1.7	No	No	Yes	No
Bridgewell	27.7	No	No	Yes	No
Briedwell	44.0	No	No	Yes	No
Brightwood	14.3	No	No	Yes	No
Brisbois	66.6	No	No	Yes	No
Broadycreek	4.3	No	No	Yes	No
Brock	10.6	No	No	Yes	No
Brockman	28.8	No	Yes	Yes	Yes
Brownlee	18.8	No	No	Yes	No
Brownscombe	27.3	No	No	Yes	No
Broyles	97.4	Yes	Yes	Yes	Yes
Bruncan	40.5	No	No	Yes	No
Brunzell	3.5	No	No	Yes	No
Btree	116.6	No	Yes	No	No
Bubus	13.7	Yes	No	Yes	No
Buckbert	7.2	No	No	Yes	No
Buckcreek	79.6	No	No	Yes	No
Bucketlake	120.9	No	No	No	No
Buckeye	13.9	No	No	Yes	No
Bucklake	26.0	No	No	Yes	No
Buckshot	13.7	No	No	Yes	No
Buckwilder	22.6	No	Yes	Yes	Yes
Budlewis	2.4	No	No	Yes	No
Buffaran	21.3	No	No	Yes	No
Buford	8.5	No	No	Yes	No
Bulgar	26.1	No	No	Yes	No
Bull run	78.1	No	Yes	Yes	Yes
Bullards	104.2	No	Yes	No	No
Bullgulch	21.3	No	No	Yes	No
Bullroar	22.4	No	No	Yes	No
Bullump	103.9	No	No	No	No

(continued)

Series name	Area (km^2)	Benchmark[1]	Endemic[2]	Rare[3]	Endangered[4]
Bullvaro	35.0	No	No	Yes	No
Bully	14.5	No	No	Yes	No
Bunchpoint	14.3	No	No	Yes	No
Bunyard	2.4	No	Yes	Yes	Yes
Burbank	46.5	No	No	Yes	No
Burgerbutte	61.2	No	No	Yes	No
Burke	62.6	Yes	No	Yes	No
Burkemont	21.4	No	Yes	Yes	Yes
Burlington	8.0	No	No	Yes	No
Burningman	0.1	No	Yes	Yes	Yes
Burnthill	12.4	No	Yes	Yes	Yes
Burntriver	25.4	No	No	Yes	No
Burntwoods	6.4	No	Yes	Yes	Yes
Burrita	2.1	No	No	Yes	No
Bybee	57.5	No	Yes	Yes	Yes
Bycracky	3.2	No	Yes	Yes	Yes
Cabell	4.6	No	No	Yes	No
Cabincreek	3.5	No	No	Yes	No
Cabinspring	9.9	No	No	Yes	No
Calder	12.5	No	No	Yes	No
Calderwood	155.8	No	No	No	No
Calfranch	27.6	No	No	Yes	No
Calimus	157.0	No	No	No	No
Camas	111.3	No	No	No	No
Camaspatch	0.1	No	No	Yes	No
Campcreek	56.1	No	No	Yes	No
Campfour	14.6	No	No	Yes	No
Camptank	1.0	No	Yes	Yes	Yes
Canderly	10.4	No	No	Yes	No
Canest	353.5	No	No	No	No
Cant	11.9	No	No	Yes	No
Cantala	165.4	No	No	No	No
Capeblanco	18.0	No	No	Yes	No
Caphealy	28.1	No	No	Yes	No
Capona	45.4	No	No	Yes	No
Caris	208.1	No	No	No	No
Carlton	39.8	No	Yes	Yes	Yes
Carney	235.3	Yes	No	No	No
Carpenterville	2.6	No	Yes	Yes	Yes
Carryback	1189.8	No	No	No	No
Carvix	151.1	No	Yes	No	No
Cascade	178.8	Yes	No	No	No
Cashner	2.7	No	Yes	Yes	Yes
Cassiday	183.3	No	No	No	No
Castlecrest	265.8	No	No	No	No
Catchell	36.7	No	No	Yes	No
Caterl	253.1	No	No	No	No
Catherine	89.2	No	No	Yes	No
Catlow	171.7	No	No	No	No
Catnapp	3.2	No	No	Yes	No
Cazadero	82.0	No	No	Yes	No
Cedarcamp	22.5	No	No	Yes	No
Cedargrove	14.3	No	No	Yes	No
Cencove	9.3	No	No	Yes	No
Central point	33.8	No	No	Yes	No
Chamate	15.7	No	No	Yes	No

(continued)

Series name	Area (km^2)	Benchmark[1]	Endemic[2]	Rare[3]	Endangered[4]
Chambeam	38.4	No	No	Yes	No
Chancelakes	7.0	No	No	Yes	No
Chapman	58.1	No	Yes	Yes	Yes
Chard	0.4	No	No	Yes	No
Chehalem	49.5	No	No	Yes	No
Chehalis	238.7	Yes	Yes	No	No
Chehulpum	42.8	No	Yes	Yes	Yes
Chen	111.9	Yes	No	No	No
Chenoweth	14.8	No	No	Yes	No
Cherry spring	14.5	No	No	Yes	No
Cherrycreek	71.3	No	No	Yes	No
Cherryhill	30.8	No	No	Yes	No
Chesebro	18.5	No	No	Yes	No
Chesnimnus	12.5	No	No	Yes	No
Chetco	14.1	No	Yes	Yes	Yes
Cheval	9.2	No	Yes	Yes	Yes
Chewaucan	13.8	No	Yes	Yes	Yes
Chilcott	93.3	Yes	No	Yes	No
Chiloquin	3.4	No	Yes	Yes	Yes
Chimneyrock	23.4	No	No	Yes	No
Chinarise	23.4	No	No	Yes	No
Chintimini	8.0	No	No	Yes	No
Chismore	16.9	No	No	Yes	No
Chitwood	22.2	No	No	Yes	No
Chock	23.7	No	No	Yes	No
Chocktoot	26.3	No	No	Yes	No
Choptie	64.6	No	No	Yes	No
Chug	24.9	No	Yes	Yes	Yes
Cinderfall	3.4	No	No	Yes	No
Circle	14.2	No	Yes	Yes	Yes
Clackamas	82.5	No	No	Yes	No
Clamp	134.3	No	No	No	No
Clarkscreek	34.0	No	No	Yes	No
Clatsop	29.6	No	Yes	Yes	Yes
Clawson	18.2	No	Yes	Yes	Yes
Clearline	42.7	No	No	Yes	No
Cleavage	29.2	Yes	No	Yes	No
Cleet	35.3	No	No	Yes	No
Cleetwood	28.8	No	Yes	Yes	Yes
Clevescove	8.3	No	No	Yes	No
Cleymor	0.5	No	No	Yes	No
Climax	17.0	No	Yes	Yes	Yes
Clinefalls	7.3	No	Yes	Yes	Yes
Cloquato	159.8	No	Yes	No	No
Clovercreek	54.5	No	No	Yes	No
Cloverland	22.6	No	Yes	Yes	Yes
Clovkamp	50.0	No	No	Yes	No
Clurde	22.6	No	No	Yes	No
Coburg	178.9	No	No	No	No
Coglin	100.1	No	No	No	No
Coker	40.6	No	No	Yes	No
Colbar	76.9	No	No	Yes	No
Coleman	15.0	No	No	Yes	No
Colepoint	16.9	No	No	Yes	No
Colestine	180.3	No	No	No	No
Collegecreek	7.1	No	No	Yes	No

(continued)

Series name	Area (km^2)	Benchmark[1]	Endemic[2]	Rare[3]	Endangered[4]
Collier	411.4	No	No	No	No
Concord	123.1	No	No	No	No
Condon	1171.4	No	No	No	No
Condorbridge	6.3	No	No	Yes	No
Conley	47.5	No	No	Yes	No
Connleyhills	122.1	No	Yes	No	No
Conser	93.3	No	No	Yes	No
Cookcreek	27.8	No	No	Yes	No
Cooperdraw	15.5	No	Yes	Yes	Yes
Cooperopolis	7.8	No	No	Yes	No
Copperfield	34.0	No	No	Yes	No
Copsey	5.1	No	No	Yes	No
Coquille	108.5	No	Yes	No	No
Cornelius	101.4	No	Yes	No	No
Cornutt	79.6	No	No	Yes	No
Corral	113.0	No	No	No	No
Cotant	10.1	No	No	Yes	No
Cotay	3.5	No	No	Yes	No
Cottrell	47.8	No	No	Yes	No
Cougarrock	38.6	No	No	Yes	No
Coughanour	29.3	No	No	Yes	No
Court	9.6	No	Yes	Yes	Yes
Courtney	9.9	No	Yes	Yes	Yes
Courtrock	3.4	No	No	Yes	No
Cove	83.2	No	No	Yes	No
Cowsly	269.9	No	No	No	No
Coyata	54.6	No	No	Yes	No
Coyotebluff	11.8	No	No	Yes	No
Coztur	117.5	No	No	No	No
Crabtree	19.6	No	No	Yes	No
Crackedground	60.2	No	No	Yes	No
Crackercreek	99.7	No	No	Yes	No
Crackler	18.9	No	No	Yes	No
Crannlcr	23.1	No	No	Yes	No
Crater lake	4.5	No	No	Yes	No
Crawfish	9.8	No	No	Yes	No
Cressler	13.8	No	Yes	Yes	Yes
Crims	7.8	No	Yes	Yes	Yes
Crofland	1.5	No	Yes	Yes	Yes
Crooked	7.3	No	Yes	Yes	Yes
Croquib	3.0	No	Yes	Yes	Yes
Crowcamp	108.6	No	No	No	No
Cruiser	51.5	No	No	Yes	No
Crume	7.3	No	No	Yes	No
Crump	70.6	No	No	Yes	No
Crutch	7.6	No	Yes	Yes	Yes
Crutchfield	16.0	No	No	Yes	No
Culbertson	22.7	No	No	Yes	No
Cullius	79.4	No	No	Yes	No
Cumley	136.1	No	No	No	No
Cunniff	26.5	No	No	Yes	No
Cupola	18.7	No	Yes	Yes	Yes
Curant	203.6	No	No	No	No
Curtin	55.5	No	Yes	Yes	Yes
Dabney	5.1	No	No	Yes	No
Dacker	50.1	Yes	No	Yes	No

(continued)

Series name	Area (km^2)	Benchmark[1]	Endemic[2]	Rare[3]	Endangered[4]
Dahl	9.7	No	Yes	Yes	Yes
Damewood	180.3	No	Yes	No	No
Damore	10.6	No	Yes	Yes	Yes
Darby	7.4	No	No	Yes	No
Dardry	17.8	No	Yes	Yes	Yes
Darkcanyon	26.1	No	No	Yes	No
Darow	20.8	No	No	Yes	No
Davey	77.0	No	No	Yes	No
Daxty	13.7	No	No	Yes	No
Day	53.5	No	No	Yes	No
Dayton	417.9	Yes	No	No	No
Dayville	29.2	No	No	Yes	No
Deadline	107.8	No	No	No	No
Deardorf	36.2	No	No	Yes	No
Debenger	38.1	No	No	Yes	No
Deck	25.9	No	No	Yes	No
Dee	13.5	No	Yes	Yes	Yes
Defenbaugh	58.1	No	Yes	Yes	Yes
Degarmo	22.3	No	No	Yes	No
Degner	11.4	No	No	Yes	No
Dehill	0.3	No	No	Yes	No
Dehlinger	58.9	No	No	Yes	No
Delena	22.0	No	Yes	Yes	Yes
Demasters	0.0	No	No	Yes	No
Dement	43.5	No	No	Yes	No
Dentdraw	2.7	No	No	Yes	No
Depoe	11.6	No	No	Yes	No
Deppy	308.6	No	No	No	No
Derallo	11.0	No	No	Yes	No
Derapter	8.8	No	No	Yes	No
Derringer	23.5	No	No	Yes	No
Deschutes	245.3	Yes	No	No	No
Deseed	64.6	No	No	Yes	No
Deschler	22.2	No	No	Yes	No
Deskamp	172.4	No	No	No	No
Desolation	4.1	No	No	Yes	No
Desons	5.1	No	No	Yes	No
Dester	365.1	No	Yes	No	No
Deter	39.5	No	No	Yes	No
Devada	5.5	No	No	Yes	No
Deven	0.9	No	No	Yes	No
Devnot	65.6	No	No	Yes	No
Devoy	3.8	No	No	Yes	No
Dewar	117.1	No	No	No	No
Diablopeak	47.8	No	Yes	Yes	Yes
Diaz	39.5	No	No	Yes	No
Dicecreek	7.2	No	No	Yes	No
Dickerson	33.2	No	Yes	Yes	Yes
Dickle	25.4	No	No	Yes	No
Diffin	6.0	No	Yes	Yes	Yes
Digger	786.8	Yes	No	No	No
Digit	3.4	No	No	Yes	No
Dilman	24.0	No	Yes	Yes	Yes
Divers	51.0	No	No	Yes	No
Dixiejett	23.3	No	No	Yes	No
Dixon	61.7	No	No	Yes	No

(continued)

Series name	Area (km²)	Benchmark[1]	Endemic[2]	Rare[3]	Endangered[4]
Dixonville	266.3	No	No	No	No
Diyou	0.5	No	No	Yes	No
Dobbins	11.2	No	No	Yes	No
Dodes	18.4	No	No	Yes	No
Dogmountain	24.3	No	Yes	Yes	Yes
Dogtown	37.9	No	No	Yes	No
Dollarlake	3.4	No	No	Yes	No
Dompier	2.0	No	Yes	Yes	Yes
Donegan	20.1	No	No	Yes	No
Donnybrook	23.5	No	Yes	Yes	Yes
Doublecreek	15.3	No	No	Yes	No
Doubleo	20.4	No	Yes	Yes	Yes
Douthit	29.4	No	No	Yes	No
Dowde	37.7	No	Yes	Yes	Yes
Downards	6.7	No	Yes	Yes	Yes
Downeygulch	30.5	No	No	Yes	No
Doyn	248.6	No	No	No	No
Drakesflat	49.4	No	No	Yes	No
Drakespeak	7.8	No	No	Yes	No
Drews	97.8	No	No	Yes	No
Drewsey	258.2	No	No	No	No
Drewsgap	9.4	No	No	Yes	No
Drinkwater	74.6	No	No	Yes	No
Droval	113.2	No	No	No	No
Drybed	19.6	No	No	Yes	No
Dryck	27.3	No	No	Yes	No
Dryhollow	30.1	No	No	Yes	No
Duart	45.7	No	No	Yes	No
Dubakella	187.7	Yes	Yes	No	No
Duckclub	46.6	No	Yes	Yes	Yes
Ducklake	48.1	No	No	Yes	No
Duff	72.0	No	No	Yes	No
Dufur	62.0	No	No	Yes	No
Dulandy	12.7	No	Yes	Yes	Yes
Dumont	179.4	No	No	No	No
Dunnlake	0.0	No	No	Yes	No
Dunres	111.6	No	No	No	No
Dunstan	52.4	No	No	Yes	No
Dupee	167.4	No	No	No	No
Dupratt	126.5	No	No	No	No
Durkee	223.4	No	No	No	No
Dyarock	1.4	No	Yes	Yes	Yes
Ead	26.0	No	No	Yes	No
Eaglespring	11.0	No	No	Yes	No
Eastlakesbasin	7.9	No	No	Yes	No
Eastpine	17.7	No	No	Yes	No
Ecola	42.2	No	No	Yes	No
Edemaps	132.2	No	No	No	No
Edenbower	27.3	No	Yes	Yes	Yes
Edson	22.0	No	No	Yes	No
Eggleson	5.6	No	Yes	Yes	Yes
Eglirim	51.5	No	No	Yes	No
Egyptcreek	111.1	No	No	No	No
Eightlar	62.7	No	No	Yes	No
Eilertsen	59.7	No	No	Yes	No
Ekoms	4.1	No	Yes	Yes	Yes

(continued)

Series name	Area (km^2)	Benchmark[1]	Endemic[2]	Rare[3]	Endangered[4]
Elbowcreek	31.7	No	No	Yes	No
Elijah	284.8	No	No	No	No
Elkhorncrest	47.5	No	No	Yes	No
Ellisforde	36.4	No	No	Yes	No
Ellum	5.9	No	No	Yes	No
Elsie	14.2	No	Yes	Yes	Yes
Embal	29.7	No	No	Yes	No
Emerson	1.9	No	No	Yes	No
Emily	35.4	No	No	Yes	No
Encina	148.5	No	No	No	No
Endcreek	22.5	No	No	Yes	No
Endersby	36.8	No	No	Yes	No
Enko	506.7	Yes	No	No	No
Era	106.0	No	No	No	No
Erakatak	495.1	No	No	No	No
Ermabell	1.8	No	No	Yes	No
Escondia	18.8	No	No	Yes	No
Esquatzel	13.4	Yes	No	Yes	No
Etelka	192.0	No	Yes	No	No
Ettersburg	1.2	No	No	Yes	No
Euchrand	13.0	No	No	Yes	No
Euchre	2.0	No	Yes	Yes	Yes
Evans	38.8	No	No	Yes	No
Evick	6.2	No	No	Yes	No
Exfo	201.7	No	Yes	No	No
Fairylawn	24.4	No	Yes	Yes	Yes
Faloma	8.9	No	No	Yes	No
Fantz	30.5	No	No	Yes	No
Farmell	44.4	No	No	Yes	No
Farva	218.2	No	No	No	No
Fawceter	34.6	No	No	Yes	No
Fawnspring	10.1	No	No	Yes	No
Feaginranch	0.7	No	Yes	Yes	Yes
Felcher	694.2	No	No	No	No
Feltham	14.6	Yes	No	Yes	No
Fendall	86.3	No	No	Yes	No
Ferguson	10.4	No	No	Yes	No
Fernhaven	77.9	No	No	Yes	No
Fernwood	71.1	No	No	Yes	No
Ferrelo	23.5	No	Yes	Yes	Yes
Fertaline	84.6	No	No	Yes	No
Final	3.0	No	No	Yes	No
Fireball	2.9	No	No	Yes	No
Firelake	20.7	No	No	Yes	No
Fitzwater	186.8	No	No	No	No
Fivebeaver	548.7	No	No	No	No
Fivebit	160.7	No	No	No	No
Fiverivers	5.7	No	No	Yes	No
Flagstaff	100.9	No	Yes	No	No
Flane	36.8	No	No	Yes	No
Flank	9.9	No	Yes	Yes	Yes
Flarm	1.1	No	No	Yes	No
Floke	247.1	No	No	No	No
Floras	48.8	No	Yes	Yes	Yes
Flowerpot	9.6	No	No	Yes	No
Flybow	0.9	No	No	Yes	No

(continued)

Series name	Area (km²)	Benchmark[1]	Endemic[2]	Rare[3]	Endangered[4]
Flycatcher	9.6	No	Yes	Yes	Yes
Flycreek	41.5	No	No	Yes	No
Flyvalley	3.3	No	Yes	Yes	Yes
Foehlin	41.2	No	No	Yes	No
Foleylake	11.3	No	No	Yes	No
Fopiano	121.3	No	Yes	No	No
Fordice	4.8	No	No	Yes	No
Fordney	124.3	Yes	No	No	No
Formader	154.4	No	No	No	No
Forshey	5.6	No	No	Yes	No
Fort rock	97.9	No	Yes	Yes	Yes
Fossilake	4.2	No	Yes	Yes	Yes
Fourbeaver	34.9	No	No	Yes	No
Fourthcreek	0.6	No	Yes	Yes	Yes
Fourwheel	168.7	No	No	No	No
Frailey	47.6	No	No	Yes	No
Frankport	5.7	No	No	Yes	No
Freels	13.9	No	No	Yes	No
Freewater	26.3	No	No	Yes	No
Freezener	264.7	No	No	No	No
Fremkle	86.9	No	Yes	Yes	Yes
Freznik	344.5	No	No	No	No
Fritsland	192.3	No	No	No	No
Frizzelcreek	1.3	No	Yes	Yes	Yes
Frohman	172.0	No	No	No	No
Fruitcreek	18.7	No	No	Yes	No
Fryrear	10.7	No	No	Yes	No
Fuego	6.1	No	No	Yes	No
Gacey	19.5	No	No	Yes	No
Gaib	169.4	No	No	No	No
Gamelake	6.1	No	No	Yes	No
Gamgee	24.5	No	No	Yes	No
Gance	49.7	No	Yes	Yes	Yes
Gap	38.3	No	No	Yes	No
Gapcot	4.6	No	No	Yes	No
Garbutt	24.0	No	No	Yes	No
Gardiner	7.2	No	Yes	Yes	Yes
Gardone	127.7	No	No	No	No
Gauldy	7.2	No	Yes	Yes	Yes
Gearhart	17.6	No	Yes	Yes	Yes
Geebarc	1.0	No	No	Yes	No
Geisel	13.0	No	Yes	Yes	Yes
Geisercreek	63.3	No	No	Yes	No
Gelderman	22.8	No	No	Yes	No
Gellatly	29.4	No	Yes	Yes	Yes
Gelsinger	4.0	No	No	Yes	No
Gem	6.9	Yes	No	Yes	No
Genaw	7.8	No	No	Yes	No
Geppert	97.0	No	No	Yes	No
Getaway	194.8	No	No	No	No
Gilispie	27.5	No	No	Yes	No
Ginger	3.1	No	Yes	Yes	Yes
Ginsberg	79.6	No	Yes	Yes	Yes
Ginser	63.5	No	No	Yes	No
Giranch	5.4	No	Yes	Yes	Yes
Giveout	6.6	No	No	Yes	No

(continued)

Series name	Area (km^2)	Benchmark[1]	Endemic[2]	Rare[3]	Endangered[4]
Glasgow	1.3	No	No	Yes	No
Glassbutte	5.4	No	No	Yes	No
Glaze	17.2	No	No	Yes	No
Glencabin	57.7	No	No	Yes	No
Gleneden	9.8	No	No	Yes	No
Glide	0.9	No	No	Yes	No
Glohm	31.3	No	No	Yes	No
Goble	261.5	No	No	No	No
Gochea	13.9	No	Yes	Yes	Yes
Goldbeach	1.3	No	No	Yes	No
Goldrun	87.5	Yes	No	Yes	No
Golfer	12.7	No	No	Yes	No
Goodin	51.2	No	No	Yes	No
Gooding	169.6	Yes	No	No	No
Goodlow	14.0	No	No	Yes	No
Goodrich	21.3	No	No	Yes	No
Goodtack	597.0	No	No	No	No
Goodwin	26.2	No	No	Yes	No
Goolaway	35.0	No	Yes	Yes	Yes
Goose creek	20.7	Yes	No	Yes	No
Goose lake	16.7	No	Yes	Yes	Yes
Gooserock	62.6	No	No	Yes	No
Gorhamgulch	32.2	No	No	Yes	No
Gosney	235.6	No	No	No	No
Gradon	81.1	Yes	No	Yes	No
Granitemountain	27.5	No	No	Yes	No
Grassmountain	4.9	No	No	Yes	No
Grassyknob	11.1	No	No	Yes	No
Gravden	12.4	No	No	Yes	No
Gravecreek	57.8	No	Yes	Yes	Yes
Greengulch	16.1	No	No	Yes	No
Greenleaf	24.9	Yes	No	Yes	No
Greenmountain	54.9	No	No	Yes	No
Greenscombe	45.6	No	No	Yes	No
Greggo	30.0	No	Yes	Yes	Yes
Gregory	13.5	No	No	Yes	No
Grell	12.5	No	No	Yes	No
Grenet	13.4	No	No	Yes	No
Greystoke	66.4	No	No	Yes	No
Gribble	59.9	No	Yes	Yes	Yes
Grindbrook	43.0	No	No	Yes	No
Grousehill	194.4	No	Yes	No	No
Grouslous	50.2	No	No	Yes	No
Grubcreek	19.0	No	No	Yes	No
Guano	40.0	No	Yes	Yes	Yes
Guerin	4.2	No	Yes	Yes	Yes
Gulliford	12.2	No	No	Yes	No
Gumble	173.2	No	No	No	No
Gurdane	190.0	No	No	No	No
Gurlidawg	2.4	No	No	Yes	No
Gustin	69.0	No	Yes	Yes	Yes
Gutridge	216.7	No	No	No	No
Gwin	649.5	No	No	No	No
Gwinly	990.8	No	No	No	No
Haar	279.0	No	No	No	No
Habenome	5.2	No	No	Yes	No

(continued)

Series name	Area (km²)	Benchmark[1]	Endemic[2]	Rare[3]	Endangered[4]
Hack	55.9	No	No	Yes	No
Hackwood	91.9	No	No	Yes	No
Haflinger	7.1	No	No	Yes	No
Hager	36.7	No	No	Yes	No
Haines	35.6	No	Yes	Yes	Yes
Halfway	3.1	No	No	Yes	No
Hall ranch	107.7	No	Yes	No	No
Hallihan	14.7	No	Yes	Yes	Yes
Hammersley	3.1	No	No	Yes	No
Hankins	362.0	Yes	No	No	No
Hanning	76.6	No	No	Yes	No
Hapgood	39.8	Yes	No	Yes	No
Happus	2.4	No	Yes	Yes	Yes
Harana	21.2	No	No	Yes	No
Harcany	119.5	No	No	No	No
Hardscrabble	30.9	No	No	Yes	No
Hardtrigger	57.7	No	No	Yes	No
Harl	126.3	No	No	No	No
Harlow	394.7	No	No	No	No
Harriman	61.7	No	No	Yes	No
Harrington	206.5	No	No	No	No
Harslow	190.1	No	No	No	No
Hart	119.4	No	Yes	No	No
Hart camp	39.4	No	No	Yes	No
Hasshollow	19.5	No	No	Yes	No
Hayespring	73.6	No	Yes	Yes	Yes
Haynap	13.9	No	No	Yes	No
Haystack	36.9	No	No	Yes	No
Hazelair	175.4	No	No	No	No
Hazelcamp	52.2	No	No	Yes	No
Hebo	25.2	No	No	Yes	No
Heceta	45.1	No	Yes	Yes	Yes
Hehe	115.6	No	No	No	No
Helmick	30.8	No	Yes	Yes	Yes
Helphenstein	85.8	No	Yes	Yes	Yes
Helter	10.1	Yes	No	Yes	No
Helvetia	76.8	No	No	Yes	No
Hembre	205.5	Yes	No	No	No
Hemcross	403.3	No	No	No	No
Henkle	159.0	No	No	No	No
Henley	63.1	Yes	Yes	Yes	Yes
Henline	86.1	No	No	Yes	No
Heppsie	38.6	No	No	Yes	No
Hermiston	67.1	No	No	Yes	No
Hershal	19.0	No	Yes	Yes	Yes
Hesslan	93.8	No	No	Yes	No
Hezel	12.3	No	Yes	Yes	Yes
Hibbard	38.3	No	Yes	Yes	Yes
Highcamp	86.5	No	No	Yes	No
Highhorn	10.9	No	No	Yes	No
Hillsboro	15.2	No	No	Yes	No
Hilltish	24.4	No	No	Yes	No
Hinton	3.6	No	Yes	Yes	Yes
Hobit	10.4	No	No	Yes	No
Hoffer	18.8	No	No	Yes	No
Hogranch	23.8	No	No	Yes	No

(continued)

Series name	Area (km²)	Benchmark[1]	Endemic[2]	Rare[3]	Endangered[4]
Holcomb	83.7	No	No	Yes	No
Holderman	14.1	No	No	Yes	No
Holland	66.6	Yes	No	Yes	No
Holmzie	43.4	No	No	Yes	No
Homefield	6.5	No	No	Yes	No
Hondu	77.2	No	No	Yes	No
Honeygrove	696.8	Yes	No	No	No
Honeymooncan	19.6	No	No	Yes	No
Hood	14.7	No	No	Yes	No
Hooly	18.8	No	Yes	Yes	Yes
Hoopal	11.8	No	Yes	Yes	Yes
Hooskanaden	24.6	No	Yes	Yes	Yes
Horeb	28.4	No	No	Yes	No
Horning	17.6	No	Yes	Yes	Yes
Horseprairie	10.0	No	No	Yes	No
Hosley	16.3	No	Yes	Yes	Yes
Hot lake	80.7	No	Yes	Yes	Yes
Housefield	45.9	No	No	Yes	No
Houstake	71.2	No	No	Yes	No
Houstenader	3.0	No	No	Yes	No
Howash	163.0	No	No	No	No
Howcan	4.0	No	No	Yes	No
Howmeadows	15.4	No	Yes	Yes	Yes
Hoxie	7.4	No	No	Yes	No
Huberly	13.5	No	No	Yes	No
Hudspeth	12.6	No	No	Yes	No
Huffling	0.5	No	Yes	Yes	Yes
Hukill	30.3	No	Yes	Yes	Yes
Hullt	21.5	No	No	Yes	No
Humarel	107.1	No	No	No	No
Hummington	119.7	No	No	No	No
Hunewill	28.2	No	No	Yes	No
Hunsaker	38.4	No	No	Yes	No
Hunterscove	13.7	No	No	Yes	No
Huntley	2.2	No	No	Yes	No
Huntrock	6.4	No	No	Yes	No
Hurryback	1.8	No	No	Yes	No
Hurwal	85.5	No	No	Yes	No
Hutchinson	17.0	No	No	Yes	No
Hutchley	80.9	No	No	Yes	No
Hutson	18.9	No	No	Yes	No
Hyall	30.1	No	Yes	Yes	Yes
Icedee	9.6	No	No	Yes	No
Icene	95.4	No	No	Yes	No
Igert	505.6	No	No	No	No
Illahee	59.2	No	No	Yes	No
Imbler	49.2	No	No	Yes	No
Immiant	2.6	No	No	Yes	No
Immig	223.5	No	No	No	No
Imnaha	173.9	No	No	No	No
Inkler	66.7	No	No	Yes	No
Ipsoot	10.3	No	No	Yes	No
Iris	3.9	No	No	Yes	No
Irma	7.8	No	No	Yes	No
Ironside	78.7	No	No	Yes	No
Irrigon	32.0	No	No	Yes	No

(continued)

Series name	Area (km^2)	Benchmark[1]	Endemic[2]	Rare[3]	Endangered[4]
Itca	16.1	Yes	No	Yes	No
Izee	28.7	No	No	Yes	No
Jacksplace	82.7	No	No	Yes	No
James canyon	0.5	No	No	Yes	No
Jayar	150.2	No	No	No	No
Jenny	0.0	No	No	Yes	No
Jerome	6.2	No	No	Yes	No
Jesse camp	8.4	No	Yes	Yes	Yes
Jett	112.9	No	No	No	No
Jimbo	34.2	No	No	Yes	No
Jimgreen	13.3	No	No	Yes	No
Joeney	5.9	No	No	Yes	No
Jojo	56.2	No	Yes	Yes	Yes
Jorn	14.3	No	Yes	Yes	Yes
Jory	872.9	Yes	No	No	No
Josephine	521.4	Yes	No	No	No
Josset	3.4	No	No	Yes	No
Jumpoff	22.0	No	No	Yes	No
Kahler	113.1	No	No	No	No
Kahneeta	5.4	No	Yes	Yes	Yes
Kamela	249.9	No	No	No	No
Kanid	249.9	No	No	No	No
Kanlee	5.7	No	No	Yes	No
Kanutchan	6.1	No	Yes	Yes	Yes
Kaskela	63.4	No	Yes	Yes	Yes
Keating	60.9	No	No	Yes	No
Kecko	13.8	No	No	Yes	No
Keel	156.6	Yes	No	No	No
Kegler	43.2	No	Yes	Yes	Yes
Kenusky	8.1	No	Yes	Yes	Yes
Kerby	21.7	No	No	Yes	No
Kerrfield	75.0	No	No	Yes	No
Ketchly	51.1	No	No	Yes	No
Kettenbach	83.1	No	No	Yes	No
Kettlecreek	9.9	No	No	Yes	No
Kewake	131.5	No	Yes	No	No
Kiesel	4.6	No	No	Yes	No
Kilchis	121.8	No	No	No	No
Killam	10.6	No	No	Yes	No
Killet	4.7	No	Yes	Yes	Yes
Kilmerque	33.1	No	No	Yes	No
Kilowan	34.3	No	No	Yes	No
Kimberly	75.2	No	No	Yes	No
Kingbolt	124.6	No	No	No	No
Kingsriver	3.0	No	No	Yes	No
Kinney	787.7	Yes	No	No	No
Kinton	60.6	No	Yes	Yes	Yes
Kinzel	75.2	No	No	Yes	No
Kiona	108.6	No	No	No	No
Kirk	107.6	No	No	No	No
Kirkendall	54.1	No	No	Yes	No
Kishwalk	191.7	No	No	No	No
Kittleson	6.5	No	Yes	Yes	Yes
Klamath	96.0	No	No	Yes	No
Klicker	1200.7	Yes	No	No	No
Klickitat	1019.5	No	No	No	No

(continued)

Series name	Area (km^2)	Benchmark[1]	Endemic[2]	Rare[3]	Endangered[4]
Klickson	185.5	No	No	No	No
Klistan	383.7	No	No	No	No
Klooqueh	4.9	No	No	Yes	No
Klootchie	495.3	No	No	No	No
Knapke	17.0	No	No	Yes	No
Knappa	33.5	No	No	Yes	No
Koehler	95.4	No	No	Yes	No
Kosh	3.5	No	Yes	Yes	Yes
Krackle	59.7	No	No	Yes	No
Krebs	24.1	No	No	Yes	No
Kubli	12.5	No	Yes	Yes	Yes
Kuck	0.4	No	No	Yes	No
Kuckup	1.4	No	Yes	Yes	Yes
Kuhl	10.0	No	No	Yes	No
Kunaton	56.8	No	No	Yes	No
Kunceider	147.0	No	Yes	No	No
Kusu	18.0	No	No	Yes	No
Kutcher	68.2	No	Yes	Yes	Yes
Kweo	8.5	No	Yes	Yes	Yes
La grande	66.9	No	No	Yes	No
Labish	9.9	No	No	Yes	No
Labuck	11.6	No	Yes	Yes	Yes
Lackeyshole	110.2	No	No	No	No
Lacy	1.1	No	No	Yes	No
Ladd	23.7	No	No	Yes	No
Laderly	167.8	No	No	No	No
Ladycomb	14.5	No	No	Yes	No
Lafollette	9.1	No	No	Yes	No
Laidlaw	53.2	No	No	Yes	No
Lakefork	31.4	No	No	Yes	No
Lakeview	95.4	Yes	No	Yes	No
Laki	81.8	No	No	Yes	No
Lalos	2.7	No	No	Yes	No
Lamath	0.7	No	No	Yes	No
Lambranch	2.7	No	No	Yes	No
Lambring	349.2	No	No	No	No
Lamonta	93.7	No	Yes	Yes	Yes
Lamulita	47.1	No	No	Yes	No
Landermeyer	11.3	No	No	Yes	No
Langellain	35.1	No	No	Yes	No
Langlois	29.1	No	No	Yes	No
Langrell	37.8	No	No	Yes	No
Langslet	13.5	No	Yes	Yes	Yes
Lapham	32.0	No	Yes	Yes	Yes
Lapine	2513.1	Yes	No	No	No
Larabee	370.6	No	No	No	No
Larmine	59.0	No	No	Yes	No
Lasere	86.5	No	No	Yes	No
Lassen	0.6	No	No	Yes	No
Lastance	23.9	Yes	No	Yes	No
Lastcall	70.2	No	No	Yes	No
Lather	101.1	Yes	No	No	No
Latourell	61.6	No	No	Yes	No
Laufer	17.9	No	No	Yes	No
Laurelwood	155.5	No	No	No	No
Lavey	17.2	No	No	Yes	No

(continued)

Series name	Area (km²)	Benchmark[1]	Endemic[2]	Rare[3]	Endangered[4]
Lawen	131.1	No	Yes	No	No
Lawyer	8.2	No	No	Yes	No
Laycock	16.6	No	No	Yes	No
Leathers	39.5	No	Yes	Yes	Yes
Lebam	46.2	Yes	No	Yes	No
Leemorris	32.3	No	No	Yes	No
Leepcreek	2.0	No	No	Yes	No
Leespeak	1.1	No	No	Yes	No
Leevan	33.3	No	No	Yes	No
Legler	122.8	No	No	No	No
Lemoncreek	8.9	No	No	Yes	No
Lemonex	33.1	No	No	Yes	No
Lempira	33.8	No	No	Yes	No
Leopold	2.6	No	Yes	Yes	Yes
Lequieu	0.0	No	No	Yes	No
Lerrow	12.3	No	No	Yes	No
Lettia	140.7	No	No	No	No
Lickskillet	1936.4	Yes	No	No	No
Limberjim	995.5	No	No	No	No
Limpy	0.4	No	No	Yes	No
Linecreek	36.2	No	No	Yes	No
Linkletter	47.7	No	No	Yes	No
Linksterly	38.0	No	No	Yes	No
Linslaw	23.5	No	No	Yes	No
Lint	38.0	No	No	Yes	No
Linville	0.0	No	No	Yes	No
Lithgow	179.2	No	No	No	No
Littlefawn	13.2	No	No	Yes	No
Littlesand	20.2	No	No	Yes	No
Lizard	15.2	No	No	Yes	No
Llaorock	100.9	No	No	No	No
Lobert	76.8	No	No	Yes	No
Locane	42.6	No	No	Yes	No
Locoda	33.7	No	No	Yes	No
Locolake	39.5	No	Yes	Yes	Yes
Loeb	15.3	No	No	Yes	No
Lofftus	7.8	No	Yes	Yes	Yes
Logdell	18.6	No	No	Yes	No
Logsden	21.4	No	Yes	Yes	Yes
Logsprings	5.3	No	Yes	Yes	Yes
Lolak	45.5	No	Yes	Yes	Yes
Lonely	336.5	No	No	No	No
Loneranch	21.2	No	No	Yes	No
Loneridge	17.1	No	Yes	Yes	Yes
Longbranch	66.5	No	No	Yes	No
Longcreek	21.1	No	No	Yes	No
Longjohn	9.5	No	No	Yes	No
Lookingglass	131.0	No	No	No	No
Lookout	217.4	Yes	No	No	No
Loomis	3.0	No	No	Yes	No
Lorella	479.4	Yes	No	No	No
Lostbasin	132.6	No	No	No	No
Lostforest	9.3	No	Yes	Yes	Yes
Lostine	10.0	No	No	Yes	No
Loupence	58.3	No	No	Yes	No
Loveboldt	40.4	No	No	Yes	No

(continued)

Series name	Area (km²)	Benchmark[1]	Endemic[2]	Rare[3]	Endangered[4]
Lovline	49.3	No	No	Yes	No
Lowerbluff	34.5	No	Yes	Yes	Yes
Luckiamute	25.1	No	No	Yes	No
Ludi	16.9	No	Yes	Yes	Yes
Lundgren	40.8	No	No	Yes	No
Lurnick	12.8	No	Yes	Yes	Yes
Lyeflat	27.5	No	No	Yes	No
Macdunn	29.7	No	No	Yes	No
Mackatie	102.6	No	No	No	No
Mackey	14.1	No	No	Yes	No
Macklyn	10.8	No	No	Yes	No
Macyflet	4.4	No	Yes	Yes	Yes
Madeline	69.9	No	No	Yes	No
Madras	301.6	No	No	No	No
Mahogee	55.0	No	No	Yes	No
Mahoon	152.3	No	No	No	No
Maklak	150.6	No	No	No	No
Malabon	167.3	No	No	No	No
Malheur	2.8	No	No	Yes	No
Malin	58.2	No	No	Yes	No
Mallory	58.3	No	No	Yes	No
Manita	123.0	Yes	No	No	No
Manlywham	4.5	No	Yes	Yes	Yes
Marack	129.9	No	No	No	No
Marblepoint	69.7	No	No	Yes	No
Marcola	8.4	No	No	Yes	No
Mariel	7.8	No	No	Yes	No
Marty	7.4	No	No	Yes	No
Mary	0.0	No	No	Yes	No
Maryspeak	0.6	No	No	Yes	No
Mascamp	0.7	No	No	Yes	No
Maset	230.2	No	No	No	No
Matheny	0.1	No	No	Yes	No
Matterhorn	2.8	No	Yes	Yes	Yes
Maupin	97.1	No	No	Yes	No
Mayger	72.0	No	No	Yes	No
Mcalpin	138.5	No	Yes	No	No
Mcbain	20.8	No	Yes	Yes	Yes
Mcbee	124.6	No	No	No	No
Mccalpinemeadow	2.1	No	No	Yes	No
Mccartycreek	80.5	No	No	Yes	No
Mccoin	19.6	No	No	Yes	No
Mccomas	3.8	No	Yes	Yes	Yes
Mcconnel	265.6	Yes	No	No	No
Mccully	292.7	No	No	No	No
Mccurdy	6.2	No	No	Yes	No
Mcduff	167.0	No	No	No	No
Mcewen	13.9	No	No	Yes	No
Mcgarr	36.7	No	No	Yes	No
Mcginnis	3.1	No	No	Yes	No
Mcivey	21.4	Yes	No	Yes	No
Mckay	66.0	No	No	Yes	No
Mcloughlin	41.3	No	No	Yes	No
Mcmeen	38.6	No	No	Yes	No
Mcmille	8.9	No	No	Yes	No
Mcmullin	286.0	No	Yes	No	No

(continued)

Series name	Area (km²)	Benchmark[1]	Endemic[2]	Rare[3]	Endangered[4]
Mcmurdie	38.6	No	No	Yes	No
Mcnab	13.1	No	No	Yes	No
Mcnamee	47.9	No	No	Yes	No
Mcnull	323.4	No	No	No	No
Mcnulty	16.2	No	Yes	Yes	Yes
Mcnye	22.2	No	No	Yes	No
Mcwillar	89.8	No	No	Yes	No
Meadowridge	123.5	No	No	No	No
Meda	76.3	No	No	Yes	No
Medco	298.3	No	Yes	No	No
Medford	64.9	No	No	Yes	No
Melbourne	129.4	Yes	No	No	No
Melby	137.2	No	No	No	No
Meld	23.1	No	No	Yes	No
Melhorn	88.9	No	No	Yes	No
Melloe	19.4	No	No	Yes	No
Mellowmoon	22.1	No	Yes	Yes	Yes
Memaloose	3.3	No	Yes	Yes	Yes
Menbo	59.9	No	No	Yes	No
Merlin	936.7	No	No	No	No
Mershon	21.2	No	Yes	Yes	Yes
Mesman	158.8	No	No	No	No
Metolius	19.0	No	Yes	Yes	Yes
Middlebox	96.2	No	Yes	Yes	Yes
Mikkalo	387.0	No	No	No	No
Milbury	265.3	No	No	No	No
Milcan	45.7	No	No	Yes	No
Milldam	24.4	No	No	Yes	No
Millenium	6.6	No	No	Yes	No
Millerflat	5.1	No	No	Yes	No
Millicoma	190.7	No	Yes	No	No
Minam	39.6	No	No	Yes	No
Minkwell	8.4	No	Yes	Yes	Yes
Minniece	8.2	No	No	Yes	No
Minveno	194.0	No	No	No	No
Mippon	3.8	No	No	Yes	No
Mislatnah	31.6	No	No	Yes	No
Moag	7.6	No	No	Yes	No
Modoc	46.2	Yes	No	Yes	No
Moe	63.2	No	Yes	Yes	Yes
Molalla	29.2	No	No	Yes	No
Mondovi	15.7	No	No	Yes	No
Monroe	4.3	No	No	Yes	No
Monumentrock	29.2	No	No	Yes	No
Moodybasin	13.1	No	No	Yes	No
Moonbeam	475.4	No	No	No	No
Moonstone	2.4	No	No	Yes	No
Morehouse	231.5	No	No	No	No
Morfitt	51.2	No	No	Yes	No
Morganhills	82.3	No	Yes	Yes	Yes
Morningstar	14.2	No	No	Yes	No
Morrow	707.9	Yes	No	No	No
Mosscreek	29.1	No	No	Yes	No
Mound	49.6	No	No	Yes	No
Mountemily	227.1	No	No	No	No
Mountireland	54.2	No	No	Yes	No

(continued)

Series name	Area (km²)	Benchmark[1]	Endemic[2]	Rare[3]	Endangered[4]
Mowako	7.5	No	No	Yes	No
Muddycreek	16.7	No	No	Yes	No
Mudlakebasin	70.4	No	No	Yes	No
Mudpot	62.0	No	Yes	Yes	Yes
Mues	8.9	No	Yes	Yes	Yes
Mugwump	7.6	No	Yes	Yes	Yes
Mulkey	6.0	No	Yes	Yes	Yes
Multnomah	61.9	No	No	Yes	No
Multorpor	4.2	No	No	Yes	No
Muni	435.4	No	No	No	No
Munsoncreek	42.3	No	No	Yes	No
Murlose	14.4	No	Yes	Yes	Yes
Murnen	22.9	No	No	Yes	No
Murtip	201.7	No	No	No	No
Musty	11.0	No	No	Yes	No
Mutt	0.7	No	No	Yes	No
Mutton	51.1	No	Yes	Yes	Yes
Nagle	21.1	No	No	Yes	No
Nailkeg	62.0	No	No	Yes	No
Nansene	110.0	No	No	No	No
Natal	6.6	No	Yes	Yes	Yes
Natroy	61.6	No	No	Yes	No
Necanicum	356.6	No	No	No	No
Needhill	9.9	No	No	Yes	No
Needle peak	0.2	No	No	Yes	No
Nehalem	57.6	Yes	Yes	Yes	Yes
Nekia	603.4	No	No	No	No
Nekoma	54.9	No	No	Yes	No
Nelscott	40.2	No	Yes	Yes	Yes
Neotsu	38.5	No	No	Yes	No
Neskowin	43.4	No	No	Yes	No
Nestucca	50.1	No	No	Yes	No
Netarts	22.3	No	Yes	Yes	Yes
Nevador	1112.8	No	No	No	No
Newanna	19.2	No	No	Yes	No
Newberg	270.7	No	No	No	No
Ninemile	1808.4	Yes	No	No	No
Ninetysix	80.5	No	No	Yes	No
Noidee	41.1	No	Yes	Yes	Yes
Noname	24.2	No	Yes	Yes	Yes
Nonpareil	87.4	No	No	Yes	No
Norad	144.2	No	No	No	No
Norcross	42.2	No	No	Yes	No
Norling	126.4	No	No	No	No
North powder	35.0	No	No	Yes	No
Northrup	3.6	No	Yes	Yes	Yes
Noti	15.2	No	No	Yes	No
Notus	4.7	No	No	Yes	No
Nuss	214.5	No	No	No	No
Nyssa	117.8	Yes	No	No	No
Oak grove	42.8	No	No	Yes	No
Oakland	147.4	No	No	No	No
Oatman	134.0	No	No	No	No
Oatmanflat	14.8	No	No	Yes	No
Observation	429.3	No	No	No	No
Ochoco	28.2	No	Yes	Yes	Yes

(continued)

Series name	Area (km^2)	Benchmark[1]	Endemic[2]	Rare[3]	Endangered[4]
Offenbacher	84.9	No	No	Yes	No
Olac	42.1	No	No	Yes	No
Olallie	4.5	No	Yes	Yes	Yes
Olaton	3.8	No	No	Yes	No
Old camp	51.6	Yes	No	Yes	No
Oldblue	7.4	No	No	Yes	No
Oldsferry	10.8	No	No	Yes	No
Olex	96.8	No	No	Yes	No
Oliphant	41.1	No	Yes	Yes	Yes
Olot	342.3	No	No	No	No
Olyic	227.5	No	No	No	No
Omahaling	3.9	No	No	Yes	No
Oneonta	26.6	No	No	Yes	No
Ontko	48.1	No	No	Yes	No
Onyx	24.7	No	No	Yes	No
Opie	62.7	No	No	Yes	No
Oreanna	8.8	No	Yes	Yes	Yes
Oreneva	122.3	No	No	No	No
Orford	304.5	No	No	No	No
Orhood	0.1	No	No	Yes	No
Ornea	2.6	No	No	Yes	No
Orovada	143.7	Yes	No	No	No
Osoll	12.0	No	No	Yes	No
Otoole	3.9	No	No	Yes	No
Otwin	9.9	No	No	Yes	No
Outerkirk	184.1	No	No	No	No
Overallflat	26.5	No	Yes	Yes	Yes
Owsel	65.7	No	Yes	Yes	Yes
Owyhee	76.9	Yes	No	Yes	No
Oxbow	9.6	No	No	Yes	No
Oxley	8.9	No	Yes	Yes	Yes
Oxman	20.5	No	No	Yes	No
Oxwall	116.1	No	No	No	No
Ozamis	181.2	No	No	No	No
Packard	8.5	No	No	Yes	No
Padigan	11.1	No	No	Yes	No
Pait	42.2	No	No	Yes	No
Palouse	122.5	Yes	No	No	No
Panther	59.6	No	Yes	Yes	Yes
Paragon	9.7	No	No	Yes	No
Parkdale	34.3	No	No	Yes	No
Parrego	47.1	No	No	Yes	No
Parsnip	97.2	No	No	Yes	No
Patit creek	3.2	No	No	Yes	No
Patron	0.3	No	No	Yes	No
Paulina	36.0	No	No	Yes	No
Paynepeak	0.9	No	No	Yes	No
Peahke	5.2	No	No	Yes	No
Pearlwise	60.2	No	No	Yes	No
Pearsoll	232.5	No	Yes	No	No
Peasley	0.6	No	No	Yes	No
Peavine	851.7	No	No	No	No
Pedigo	23.0	No	No	Yes	No
Peel	3.3	No	Yes	Yes	Yes
Pelton	2.8	No	No	Yes	No
Pengra	106.3	No	Yes	No	No

(continued)

Series name	Area (km^2)	Benchmark[1]	Endemic[2]	Rare[3]	Endangered[4]
Perdin	88.0	No	Yes	Yes	Yes
Perla	38.0	No	No	Yes	No
Pernog	7.9	No	No	Yes	No
Pernty	347.1	No	No	No	No
Pervina	33.7	No	No	Yes	No
Philomath	233.0	No	Yes	No	No
Phoenix	3.8	No	Yes	Yes	Yes
Phys	31.3	No	No	Yes	No
Picturerock	5.4	No	No	Yes	No
Piersonte	55.7	No	No	Yes	No
Pilchuck	19.8	No	No	Yes	No
Piline	73.7	No	No	Yes	No
Pilot rock	142.2	No	No	No	No
Pinehurst	148.1	No	No	No	No
Pineval	15.2	No	No	Yes	No
Pinhead	141.5	No	No	No	No
Pipp	91.7	No	No	Yes	No
Pistolriver	0.6	No	Yes	Yes	Yes
Pit	31.7	No	No	Yes	No
Pitcheranch	7.1	No	Yes	Yes	Yes
Piumpsha	21.7	No	Yes	Yes	Yes
Plainview	20.7	No	Yes	Yes	Yes
Plush	17.4	No	No	Yes	No
Poall	322.6	No	No	No	No
Pocan	1.4	No	No	Yes	No
Poden	3.7	No	No	Yes	No
Podus	1.6	No	No	Yes	No
Poe	24.3	No	Yes	Yes	Yes
Pokegema	317.8	No	No	No	No
Polander	40.6	No	Yes	Yes	Yes
Pollard	172.9	No	Yes	No	No
Polly	21.7	No	No	Yes	No
Pomerening	20.5	No	Yes	Yes	Yes
Ponina	46.3	No	Yes	Yes	Yes
Poorjug	24.7	No	No	Yes	No
Porterfield	39.5	No	No	Yes	No
Potamus	3.5	No	No	Yes	No
Poujade	290.8	No	Yes	No	No
Powder	132.1	Yes	No	No	No
Powell	49.9	No	No	Yes	No
Power	8.6	No	No	Yes	No
Powval	43.6	No	No	Yes	No
Powwatka	49.3	No	No	Yes	No
Prag	162.5	No	No	No	No
Prairie	35.7	No	No	Yes	No
Preacher	1470.2	No	No	No	No
Price	38.2	No	Yes	Yes	Yes
Prill	241.2	No	No	No	No
Prineville	23.4	No	No	Yes	No
Pritchard	1.9	No	No	Yes	No
Prosser	34.0	No	No	Yes	No
Prouty	67.8	No	No	Yes	No
Provig	10.7	No	No	Yes	No
Puderbaugh	1.2	No	No	Yes	No
Puderbaughridge	2.2	No	No	Yes	No
Puls	17.8	No	No	Yes	No

(continued)

Series name	Area (km²)	Benchmark[1]	Endemic[2]	Rare[3]	Endangered[4]
Purple	2.2	No	Yes	Yes	Yes
Puzzlebark	2.3	No	Yes	Yes	Yes
Puzzlecreek	36.8	No	No	Yes	No
Pyburn	10.3	No	No	Yes	No
Pyrady	9.0	No	Yes	Yes	Yes
Pyropatti	0.4	No	No	Yes	No
Quafeno	15.4	No	No	Yes	No
Quailprairie	5.8	No	No	Yes	No
Quartzville	33.0	No	No	Yes	No
Quatama	85.3	No	No	Yes	No
Quillamook	14.5	No	No	Yes	No
Quincy	545.6	Yes	No	No	No
Quinton	14.3	No	No	Yes	No
Quirk	110.3	No	No	No	No
Quosatana	12.6	No	Yes	Yes	Yes
Rabbitcreek	2.8	No	No	Yes	No
Rabbithills	198.5	No	No	No	No
Racing	13.4	No	Yes	Yes	Yes
Rafton	49.6	No	No	Yes	No
Rail	7.8	No	No	Yes	No
Rainey	0.5	No	No	Yes	No
Ramo	48.6	No	No	Yes	No
Randcore	8.0	No	No	Yes	No
Rastus	35.6	No	No	Yes	No
Ratsnest	0.1	No	No	Yes	No
Ratto	186.6	No	No	No	No
Raz	2279.5	No	No	No	No
Raztack	1.4	No	Yes	Yes	Yes
Reallis	240.6	No	No	No	No
Reavis	14.8	No	No	Yes	No
Redbell	22.1	No	Yes	Yes	Yes
Redcanyon	6.2	No	No	Yes	No
Redcliff	143.3	No	No	No	No
Redcone	5.4	No	Yes	Yes	Yes
Redflat	14.2	No	Yes	Yes	Yes
Redmond	99.6	No	No	Yes	No
Redmount	21.6	No	No	Yes	No
Redslide	11.4	No	Yes	Yes	Yes
Reedsport	254.3	No	No	No	No
Reese	165.7	No	No	No	No
Reinecke	7.3	No	No	Yes	No
Reinhart	6.4	No	Yes	Yes	Yes
Reluctan	343.0	Yes	No	No	No
Remote	233.8	No	No	No	No
Reston	12.5	No	No	Yes	No
Reuter	22.5	No	Yes	Yes	Yes
Reywat	16.9	No	No	Yes	No
Rhea	122.8	No	No	No	No
Ricco	6.3	No	No	Yes	No
Riceton	1.1	No	No	Yes	No
Rickreall	20.3	No	No	Yes	No
Riddleranch	114.7	No	No	No	No
Ridenbaugh	24.3	No	No	Yes	No
Ridley	43.2	No	Yes	Yes	Yes
Rilea	45.6	No	No	Yes	No
Rinconflat	119.6	No	No	No	No

(continued)

Series name	Area (km²)	Benchmark[1]	Endemic[2]	Rare[3]	Endangered[4]
Rinearson	402.4	No	No	No	No
Rio king	54.5	No	No	Yes	No
Risle	199.2	No	No	No	No
Ritner	291.8	No	No	No	No
Ritzville	1133.4	Yes	No	No	No
Roanhide	4.2	No	No	Yes	No
Robinette	29.6	No	No	Yes	No
Robson	666.1	No	No	No	No
Roca	103.4	Yes	Yes	No	Yes
Rocconda	7.5	No	No	Yes	No
Rockford	9.5	No	No	Yes	No
Rockly	774.9	No	No	No	No
Rogerson	37.8	No	No	Yes	No
Rogger	68.1	No	No	Yes	No
Rogue	50.6	No	No	Yes	No
Roloff	52.1	No	No	Yes	No
Romanose	7.1	No	No	Yes	No
Rondowa	19.9	No	No	Yes	No
Roostercomb	178.5	No	No	No	No
Roschene	44.1	No	No	Yes	No
Roseburg	22.8	No	No	Yes	No
Rosehaven	107.5	No	No	No	No
Rouen	13.4	No	No	Yes	No
Royal	23.5	No	No	Yes	No
Royst	259.4	No	No	No	No
Ruch	86.0	No	Yes	Yes	Yes
Ruckles	629.7	No	No	No	No
Ruclick	411.1	No	No	No	No
Ruddley	28.2	No	Yes	Yes	Yes
Rustlerpeak	25.5	No	No	Yes	No
Rustybutte	7.7	No	Yes	Yes	Yes
Rutab	6.8	No	No	Yes	No
Sach	4.7	No	Yes	Yes	Yes
Saddlepeak	24.0	No	No	Yes	No
Sag	28.9	No	No	Yes	No
Sagehen	78.7	No	Yes	Yes	Yes
Sagehill	255.1	No	No	No	No
Sagemoor	13.9	No	No	Yes	No
Sagley	4.1	No	No	Yes	No
Sahaptin	6.2	No	Yes	Yes	Yes
Salander	153.2	No	No	No	No
Salem	108.1	No	Yes	No	No
Salhouse	66.6	No	Yes	Yes	Yes
Salisbury	47.3	No	No	Yes	No
Salkum	120.5	Yes	No	No	No
Sandgap	24.3	No	No	Yes	No
Sandrock	7.1	No	Yes	Yes	Yes
Sankey	4.5	No	No	Yes	No
Santiam	67.7	No	No	Yes	No
Saturn	33.7	No	Yes	Yes	Yes
Saum	158.5	No	No	No	No
Sauvie	100.3	No	No	No	No
Sawtell	12.3	No	Yes	Yes	Yes
Scalerock	8.2	No	No	Yes	No
Scaponia	173.6	No	No	No	No
Scaredman	27.0	No	No	Yes	No

(continued)

Series name	Area (km^2)	Benchmark[1]	Endemic[2]	Rare[3]	Endangered[4]
Scherrard	11.1	No	No	Yes	No
Schnipper	0.7	No	No	Yes	No
Schrier	19.0	No	No	Yes	No
Schuelke	7.4	No	No	Yes	No
Searles	117.9	No	No	No	No
Sebastian	6.5	No	Yes	Yes	Yes
Segundo	58.8	No	No	Yes	No
Seharney	122.8	No	Yes	No	No
Selmac	23.7	No	No	Yes	No
Semiahmoo	5.0	Yes	No	Yes	No
Senra	171.9	No	No	No	No
Serpentano	42.3	No	No	Yes	No
Sevencedars	20.8	No	No	Yes	No
Sevenoaks	2.2	No	No	Yes	No
Shanahan	332.6	No	No	No	No
Shangland	11.5	No	No	Yes	No
Shano	207.2	Yes	No	No	No
Sharesnout	242.7	No	No	No	No
Sharpshooter	61.1	No	No	Yes	No
Shastacosta	6.7	No	Yes	Yes	Yes
Shawave	5.3	No	No	Yes	No
Sheepcreek	2.2	No	Yes	Yes	Yes
Shefflein	56.8	No	No	Yes	No
Sherar	21.0	No	No	Yes	No
Sherod	14.0	No	No	Yes	No
Sherval	19.0	No	No	Yes	No
Shippa	27.2	No	No	Yes	No
Shiva	14.5	No	No	Yes	No
Shivigny	32.6	No	Yes	Yes	Yes
Shoat	4.0	No	No	Yes	No
Shoepeg	0.3	No	No	Yes	No
Shroyton	9.0	No	No	Yes	No
Shukash	603.0	No	No	No	No
Sibannac	7.2	No	No	Yes	No
Sibold	14.4	No	No	Yes	No
Sidlake	120.9	No	No	No	No
Sifton	32.0	No	No	Yes	No
Siletz	10.2	No	No	Yes	No
Silverash	63.9	No	Yes	Yes	Yes
Silverlake	9.5	No	No	Yes	No
Silverton	12.1	No	No	Yes	No
Silvies	87.4	No	No	Yes	No
Simas	772.5	Yes	No	No	No
Simnasho	110.8	No	Yes	No	No
Simon	5.7	No	No	Yes	No
Sinamox	14.1	No	No	Yes	No
Sinker	100.1	No	No	No	No
Siskiyou	94.2	No	No	Yes	No
Sisley	17.1	No	Yes	Yes	Yes
Sisters	47.5	No	No	Yes	No
Sitkum	15.2	No	No	Yes	No
Sitton	3.5	No	No	Yes	No
Sixes	1.6	No	No	Yes	No
Skedaddle	184.8	No	No	No	No
Skidbrackle	0.5	No	No	Yes	No
Skidoosprings	104.3	No	No	No	No

(continued)

Series name	Area (km^2)	Benchmark[1]	Endemic[2]	Rare[3]	Endangered[4]
Skinner	10.1	No	No	Yes	No
Skipanon	215.3	No	No	No	No
Skooker	14.3	No	No	Yes	No
Skookum	40.4	No	Yes	Yes	Yes
Skookumhouse	62.1	No	No	Yes	No
Skoven	6.5	No	Yes	Yes	Yes
Skull creek	38.4	No	Yes	Yes	Yes
Skullgulch	57.8	No	No	Yes	No
Skunkfarm	71.7	No	No	Yes	No
Skyline	99.7	No	No	Yes	No
Slayton	12.3	No	No	Yes	No
Slicklog	1.9	No	No	Yes	No
Slickrock	380.3	No	No	No	No
Sliptrack	19.3	No	No	Yes	No
Smiling	236.4	No	No	No	No
Snakepit	9.8	No	Yes	Yes	Yes
Snaker	62.5	No	No	Yes	No
Snell	335.9	No	No	No	No
Snellby	50.6	No	No	Yes	No
Snow	22.4	No	No	Yes	No
Snowbrier	5.5	No	No	Yes	No
Snowcamp	21.1	No	No	Yes	No
Snowlin	2.1	No	Yes	Yes	Yes
Snowmore	1381.0	No	No	No	No
Softscrabble	3.3	Yes	No	Yes	No
Solarview	23.2	No	No	Yes	No
Sonoma	11.6	Yes	No	Yes	No
Soosap	16.2	No	No	Yes	No
Sopher	36.1	No	Yes	Yes	Yes
Sorefoot	26.3	No	No	Yes	No
Sorf	118.0	No	No	No	No
Soughe	86.6	No	No	Yes	No
Southcat	41.5	No	No	Yes	No
Spangenburg	470.6	No	No	No	No
Speaker	469.8	No	No	No	No
Spiderhole	23.5	No	No	Yes	No
Spilyay	7.0	No	No	Yes	No
Springwater	23.3	No	No	Yes	No
Srednic	19.1	No	Yes	Yes	Yes
Stackyards	25.9	No	No	Yes	No
Stampede	74.3	Yes	Yes	Yes	Yes
Stanfield	37.3	No	No	Yes	No
Stanflow	9.9	No	No	Yes	No
Starbuck	11.9	No	No	Yes	No
Starkey	99.0	No	Yes	Yes	Yes
Statz	71.4	No	No	Yes	No
Stauffer	1.3	No	No	Yes	No
Stavely	1.2	No	No	Yes	No
Stayton	13.8	No	Yes	Yes	Yes
Stearns	5.1	No	No	Yes	No
Steiger	537.9	No	No	No	No
Steinmetz	39.3	No	No	Yes	No
Steiwer	74.5	No	No	Yes	No
Stices	15.5	No	No	Yes	No
Stinger	3.8	No	No	Yes	No
Stirfry	2.7	No	No	Yes	No

(continued)

Series name	Area (km²)	Benchmark[1]	Endemic[2]	Rare[3]	Endangered[4]
Stockdrive	19.5	No	Yes	Yes	Yes
Stockel	5.7	No	No	Yes	No
Stookmoor	213.0	No	No	No	No
Stovepipe	3.2	No	Yes	Yes	Yes
Straight	81.3	No	No	Yes	No
Stukel	260.0	No	No	No	No
Sturgill	3.7	No	No	Yes	No
Succor	16.2	No	No	Yes	No
Suckerflat	161.0	No	No	No	No
Suilotem	7.9	No	No	Yes	No
Sumine	7.3	Yes	No	Yes	No
Sumpley	4.6	No	No	Yes	No
Sunnotch	15.9	No	No	Yes	No
Sunriver	37.6	No	Yes	Yes	Yes
Suppah	7.6	No	Yes	Yes	Yes
Sutherlin	112.9	No	No	No	No
Suttle	12.5	No	No	Yes	No
Suver	26.1	No	No	Yes	No
Svensen	119.1	No	No	No	No
Swaler	90.7	No	No	Yes	No
Swalesilver	204.7	No	Yes	No	No
Swartz	1.5	No	Yes	Yes	Yes
Swedeheaven	7.0	No	No	Yes	No
Sweetbriar	25.1	No	No	Yes	No
Sweitberg	13.9	No	No	Yes	No
Sweiting	1.0	No	No	Yes	No
Sycan	2.4	No	No	Yes	No
Syrupcreek	787.9	No	No	No	No
Tablerock	6.1	No	No	Yes	No
Tahkenitch	2.5	No	No	Yes	No
Takilma	25.8	No	No	Yes	No
Talapus	17.4	No	No	Yes	No
Tallowbox	107.4	No	No	No	No
Tamara	144.9	No	No	No	No
Tamarack	0.1	No	No	Yes	No
Tamarackcanyon	65.1	No	Yes	Yes	Yes
Tandy	15.0	No	Yes	Yes	Yes
Tanksel	28.4	No	No	Yes	No
Tannahill	3.6	No	No	Yes	No
Taterpa	39.5	No	No	Yes	No
Tatouche	78.3	No	No	Yes	No
Taunton	60.0	No	No	Yes	No
Teeters	21.6	No	No	Yes	No
Teewee	63.4	No	No	Yes	No
Teguro	240.0	No	No	No	No
Telemon	2.6	No	Yes	Yes	Yes
Templeton	666.9	Yes	No	No	No
Tenmile	131.6	No	No	No	No
Tenpin	19.0	No	No	Yes	No
Tenwalter	33.1	No	Yes	Yes	Yes
Terrabella	6.8	No	Yes	Yes	Yes
Terwilliger	0.2	No	No	Yes	No
Tetherow	13.1	No	Yes	Yes	Yes
Tethrick	47.6	No	No	Yes	No
Thader	5.3	No	Yes	Yes	Yes
Thatuna	13.2	No	No	Yes	No

(continued)

Series name	Area (km²)	Benchmark[1]	Endemic[2]	Rare[3]	Endangered[4]
Thenarrows	63.6	No	Yes	Yes	Yes
Thiessen	23.5	No	No	Yes	No
Thirstygulch	43.3	No	No	Yes	No
Thistleburn	9.6	No	No	Yes	No
Thompsoncabin	13.5	No	No	Yes	No
Thorn	13.7	No	No	Yes	No
Thornlake	156.3	No	No	No	No
Threebuck	76.0	No	Yes	Yes	Yes
Threecreeks	6.0	No	No	Yes	No
Threeforks	3.6	No	No	Yes	No
Threetrees	20.8	No	No	Yes	No
Thunderegg	34.6	No	Yes	Yes	Yes
Ticino	36.5	No	No	Yes	No
Tillamook	2.1	No	Yes	Yes	Yes
Timbercrater	102.1	No	No	No	No
Tincan	6.1	No	No	Yes	No
Tincup	6.7	No	No	Yes	No
Tippett	21.4	No	No	Yes	No
Tishar	5.8	No	No	Yes	No
Tolany	65.0	No	No	Yes	No
Tolfork	5.2	No	Yes	Yes	Yes
Tolius	16.1	No	No	Yes	No
Tolke	118.4	No	No	No	No
Toll	23.3	No	No	Yes	No
Tolo	608.0	Yes	No	No	No
Tolovana	220.3	No	No	No	No
Tonor	76.7	No	No	Yes	No
Top	83.9	No	No	Yes	No
Topper	17.3	No	No	Yes	No
Trask	17.8	No	No	Yes	No
Treharne	20.9	No	Yes	Yes	Yes
Troutmeadows	290.8	No	No	No	No
Truax	0.0	No	No	Yes	No
Truesdale	5.1	No	No	Yes	No
Turnk	1.7	No	No	Yes	No
Tub	629.6	Yes	No	No	No
Tucker	26.1	No	Yes	Yes	Yes
Tuckerdowns	4.1	No	No	Yes	No
Tuffcabin	1.4	No	No	Yes	No
Tuffo	19.2	Yes	No	Yes	No
Tulana	107.6	No	Yes	No	No
Tumalo	26.7	No	Yes	Yes	Yes
Tumtum	312.2	No	No	No	No
Turbyfill	27.7	Yes	No	Yes	No
Turpin	203.0	No	Yes	No	No
Tusel	5.2	Yes	No	Yes	No
Tutni	83.4	No	No	Yes	No
Tutuilla	8.9	No	No	Yes	No
Tweener	93.6	No	No	Yes	No
Twelvemile	54.0	No	No	Yes	No
Twickenham	41.9	No	No	Yes	No
Twinbridge	0.4	No	No	Yes	No
Tygh	16.6	No	No	Yes	No
Ukiah	84.4	No	No	Yes	No
Umak	49.5	No	No	Yes	No
Umapine	151.5	Yes	No	No	No

(continued)

Series name	Area (km²)	Benchmark[1]	Endemic[2]	Rare[3]	Endangered[4]
Umatilla	139.7	No	No	No	No
Umpcoos	646.2	No	Yes	No	No
Unionpeak	145.6	No	No	No	No
Upcreek	3.8	No	Yes	Yes	Yes
Uptmor	2.6	No	No	Yes	No
Utley	35.8	No	No	Yes	No
Valby	464.4	No	No	No	No
Valmy	34.5	No	No	Yes	No
Valsetz	155.8	No	Yes	No	No
Van horn	10.0	No	No	Yes	No
Vandamine	68.5	No	No	Yes	No
Vannoy	425.5	No	No	No	No
Vanwyper	61.8	No	No	Yes	No
Veazie	46.3	No	No	Yes	No
Vena	13.4	No	No	Yes	No
Venator	135.4	No	No	No	No
Veneta	59.7	No	No	Yes	No
Verboort	1.6	No	Yes	Yes	Yes
Verdico	26.3	No	No	Yes	No
Vergas	249.1	No	Yes	No	No
Vermisa	542.7	No	No	No	No
Vernonia	90.4	No	No	Yes	No
Veta	11.7	No	No	Yes	No
Vil	31.6	No	No	Yes	No
Vining	49.6	No	No	Yes	No
Virtue	225.9	No	No	No	No
Vitale	270.4	No	No	No	No
Voats	21.8	No	No	Yes	No
Volstead	14.9	No	Yes	Yes	Yes
Voltage	60.9	No	No	Yes	No
Vondergreen	1.0	No	No	Yes	No
Voorhies	92.5	No	No	Yes	No
Wabuska	16.7	No	No	Yes	No
Wadccrcck	8.0	No	Yes	Yes	Yes
Wagontire	19.8	No	No	Yes	No
Waha	504.8	No	No	No	No
Wahkeena	6.9	No	No	Yes	No
Wahstal	68.2	No	Yes	Yes	Yes
Wakamo	6.0	No	Yes	Yes	Yes
Waldo	160.9	No	No	No	No
Waldport	91.9	No	No	Yes	No
Walla walla	1270.0	Yes	No	No	No
Wallowa	83.3	No	No	Yes	No
Walluski	27.4	No	Yes	Yes	Yes
Wamic	247.4	No	Yes	No	No
Wanoga	332.1	No	No	No	No
Wanser	21.3	No	No	Yes	No
Wapato	164.2	No	No	No	No
Wapinitia	40.6	No	No	Yes	No
Warden	280.5	Yes	No	No	No
Warnermount	0.2	No	No	Yes	No
Warrenton	3.6	No	Yes	Yes	Yes
Wasson	3.0	No	Yes	Yes	Yes
Watama	261.5	No	No	No	No
Watches	17.5	No	No	Yes	No
Waterbury	170.2	No	No	No	No

(continued)

Series name	Area (km²)	Benchmark[1]	Endemic[2]	Rare[3]	Endangered[4]
Wato	54.5	No	No	Yes	No
Wauld	18.2	No	No	Yes	No
Wauna	24.7	No	No	Yes	No
Weash	4.9	No	Yes	Yes	Yes
Webfoot	1.2	No	No	Yes	No
Wedderburn	6.0	No	Yes	Yes	Yes
Wegert	190.5	No	No	No	No
Weglike	81.0	No	Yes	Yes	Yes
Welch	61.5	Yes	No	Yes	No
Wellsdale	22.9	No	Yes	Yes	Yes
Wenas	9.4	No	No	Yes	No
Westbutte	642.6	No	No	No	No
Westside	23.6	No	Yes	Yes	Yes
Whaleshead	39.6	No	No	Yes	No
Whetstone	53.7	No	No	Yes	No
Whisk	2.9	No	No	Yes	No
Whiteface	2.3	No	Yes	Yes	Yes
Whiteson	9.6	No	No	Yes	No
Whobrey	95.9	No	Yes	Yes	Yes
Wickahoney	3.8	No	Yes	Yes	Yes
Wickiup	13.2	Yes	No	Yes	No
Widowspring	98.6	No	Yes	Yes	Yes
Wieland	59.3	Yes	No	Yes	No
Wildcatbutte	14.6	No	No	Yes	No
Wildhill	35.4	No	No	Yes	No
Wilhoit	75.9	No	Yes	Yes	Yes
Wilkins	21.9	No	No	Yes	No
Willakenzie	157.4	No	No	No	No
Willamette	155.1	No	No	No	No
Willanch	15.2	No	No	Yes	No
Willis	148.8	No	No	No	No
Willowdale	62.8	No	No	Yes	No
Wilt	16.2	No	No	Yes	No
Winberry	0.7	No	No	Yes	No
Winchester	58.1	No	No	Yes	No
Winchuck	4.1	No	No	Yes	No
Wind river	11.1	No	No	Yes	No
Windego	70.4	No	No	Yes	No
Windybutte	47.5	No	Yes	Yes	Yes
Windygap	313.2	No	No	No	No
Windypoint	20.4	No	No	Yes	No
Winema	5.1	No	No	Yes	No
Wingdale	8.8	No	No	Yes	No
Wingville	78.4	No	No	Yes	No
Winlo	31.4	No	Yes	Yes	Yes
Winom	7.3	No	No	Yes	No
Wintercanyon	16.4	No	No	Yes	No
Winterim	117.9	No	No	No	No
Wintley	23.6	No	No	Yes	No
Wiskan	1.3	No	No	Yes	No
Witham	44.1	No	No	Yes	No
Witzel	113.3	No	No	No	No
Wizard	13.4	No	No	Yes	No
Wolfer	2.1	No	No	Yes	No
Wolfpeak	56.3	No	No	Yes	No
Wollent	11.8	No	No	Yes	No

(continued)

Series name	Area (km^2)	Benchmark[1]	Endemic[2]	Rare[3]	Endangered[4]
Wolot	14.6	No	Yes	Yes	Yes
Wolverine	11.7	No	No	Yes	No
Woodburn	911.3	Yes	Yes	No	No
Woodchopper	77.0	No	Yes	Yes	Yes
Woodcock	576.8	No	Yes	No	No
Woodseye	77.5	No	Yes	Yes	Yes
Woodspoint	5.8	No	Yes	Yes	Yes
Wrentham	333.9	No	No	No	No
Wrightman	17.5	No	No	Yes	No
Wuksi	2.2	No	No	Yes	No
Wyeast	6.0	No	No	Yes	No
Wyeth	26.6	No	No	Yes	No
Xanadu	72.8	No	Yes	Yes	Yes
Yachats	9.3	No	Yes	Yes	Yes
Yainax	36.1	No	No	Yes	No
Yakima	14.7	No	No	Yes	No
Yakus	0.5	No	No	Yes	No
Yallani	46.8	No	No	Yes	No
Yancy	52.4	No	No	Yes	No
Yankeewell	169.7	No	Yes	No	No
Yapoah	27.9	No	No	Yes	No
Yaquina	14.3	No	Yes	Yes	Yes
Yawhee	144.3	No	Yes	No	No
Yawkey	84.3	No	No	Yes	No
Yawkola	8.8	No	No	Yes	No
Yellowstone	81.2	No	No	Yes	No
Yoncalla	19.5	No	Yes	Yes	Yes
Yonna	27.6	No	Yes	Yes	Yes
Yorel	18.1	No	Yes	Yes	Yes
Youtlkue	6.6	No	Yes	Yes	Yes
Yuko	4.1	No	No	Yes	No
Zalea	11.8	No	Yes	Yes	Yes
Zango	12.6	No	No	Yes	No
Zevadez	200.2	No	No	No	No
Zing	64.1	No	No	Yes	No
Zola	26.0	No	No	Yes	No
Zorravista	16.2	No	No	Yes	No
Zuman	17.9	No	Yes	Yes	Yes
Zumwalt	34.7	No	No	Yes	No
Zwagg	4.8	No	No	Yes	No
Zygore	221.9	No	No	No	No
Zyzzug	7.1	No	No	Yes	No
Total	157,754.1				

[1]Benchmark soils
[2]Endemic soils are soil series that are the only one in a family
[3]Rare soils have an area <100 km^2
[4]Endangered soils are endemic and rare

Appendix E
Land Use, Yield, and Key Soil Characteristics Influencing Yield in Oregon

T. Thorson et al., *The Soils of Oregon*, World Soils Book Series,
https://doi.org/10.1007/978-3-030-90091-5

Series name	Area (km^2)	Subgroup	Land use[1]	Range site quality[2]	Forest indicator species	Forest site quality[3]	Depth class[4]	Drainage class[5]	Slope range (%)	Salinity class[6]	Particle-size class	Soil temp. regime	Soil moisture regime	Land capability class	
														No mgmt	With mgmt
Abegg	78.4	Ultic Haploxeralfs	F		Douglas-fir	L, M	VD	WD	2–30	L	Loamy-skeletal	Mesic	Xeric	4s	3s, 4s
Abert	144.9	Sodic Xeric Haplocambids					VD	WD	0–2	H	Ashy	Frigid	Aridic	6s	
Abin	6.7	Cumulic Haploxerolls	C, R		Or ash		D	MWD	0–3	L	Fine	Mesic	Xeric	3w	2w
Abiqua	57.3	Cumulic Ultic Haploxerolls	C, F		Douglas-fir	M, H	VD	WD	0–5	L	Fine	Mesic	Xeric	1, 2e, 2w	1e-2e
Absaquil	54.0	Typic Haplohumults	F		Douglas-fir	H	D	WD	3–60	L	Fine	Mesic	Udic	6e	
Acanod	6.6	Oxyaquic Humudepts	F		Douglas-fir	H	VD	MWD	2–25	L	Fine	Mesic	Udic	6e	
Acker	217.6	Typic Palexerults	F		Douglas-fir	M, H	VD	WD	0–65	L	Fine-loamy	Mesic	Xeric	6e	
Actem	749.7	Xeric Argidurids					S	WD	2–20	L	Clayey	Frigid	Aridic	6e, 7 s	
Ada	84.9	Typic Argixerolls	R				VD	WD	4–65		Clayey-skeletal	Mesic	Xeric	4e	
Adieux	0.0	Pachic Argixerolls					MD	WD	0–5		Fine-loamy	Mesic	Xeric	4e	
Adkins	107.6	Xeric Haplocalcids	C, R	H			D	WD	0–45	L	Coarse-loamy	Mesic	Aridic	1, 2c, 2w, 4e	2e, 2w
Agate	50.3	Typic Durixerepts	F		Douglas-fir	M	MD	WD	0–15	L	Fine-loamy	Mesic	Xeric	4s, 6s	
Agency	135.9	Aridic Haploxerolls	C, R	H			MD	WD	0–70	L	Fine-loamy	Mesic	Xeric	4e	
Agness	2.0	Pachic Humudepts					VD	WD	0–60		Fine-loamy	Mesic	Udic	6e	
Ahtanum	2.8	Typic Duraquolls	C				MD	SPD	0–5		Coarse-silty	Mesic	Aquic	4w, 7s	
Akerite	18.4	Aquic Vitrixerands	F		Ponderosa pine	M	VD	MWD	0–15		Ashy over loamy	Frigid	Xeric	4e	
Albee	123.7	Vitrandic Haploxerolls	R	H			MD	WD	2–60	L	Fine-loamy	Frigid	Xeric	4e	
Alcot	0.4	Typic Vitrixerands	F		Douglas-fir	M	VD	SED	1–35	L	Ashy-pumiceous	Mesic	Xeric	6e	
Alding	63.1	Lithic Argixerolls					S	WD	2–70		Clayey	Frigid	Xeric	7s	
Algoma	29.4	Aquandic Endoaquolls	C, R	H			D	PD	0–1	L	Ashy over sandy or sandy-skeletal	Mesic	Aquic	3w	
Alicel	39.6	Pachic Haploxerolls	C, R	H			D	WD	1–15	L	Fine-loamy	Mesic	Xeric	2e	2e
Alley	2.1	Durinodic Xeric Haplargids					VD	WD	2–75		Fine-loamy	Mesic	Aridic	7s	
Allingham	18.8	Alfic Vitrixerands	F		Ponderosa pine	M	VD	WD	0–30		Ashy over medial-skeletal	Frigid	Xeric	6e	
Almota	1.2	Calcic Haploxerolls					MD	WD	3–90		Fine-loamy	Mesic	Xeric	6e	
Aloha	236.5	Aquic Haploxerepts	C, F		Douglas-fir		VD	SPD	0–8	L	Fine-silty	Mesic	Xeric	2w	

(continued)

Series name	Area (km²)	Subgroup	Land use[1]	Range site quality[2]	Forest indicator species	Forest site quality[3]	Depth class[4]	Drainage class[5]	Slope range (%)	Salinity class[6]	Particle-size class	Soil temp. regime	Soil moisture regime	Land capability class	
														No mgmt	With mgmt
Als	8.1	Typic Torripsamments					VD	ED	1–15	L	Sandy	Mesic	Aridic	6s	
Alsea	6.0	Cumulic Ultic Haploxerolls	C, F		Douglas-fir		VD	MWD	0–5	L	Fine-loamy	Mesic	Xeric	2w, 3w	
Alspaugh	130.6	Typic Paleudults	F		Douglas-fir	H	D	WD	2–50	L	Fine	Mesic	Udic	3e, 4e, 6e	
Alstony	82.6	Alic Hapludands	F		Douglas-fir	M, H	D	WD	0–90		Medial-skeletal	Mesic	Udic	6e, 7e	
Althouse	59.9	Typic Dystroxerepts	F		Douglas-fir	L–H	D	WD	30–90	L	Loamy-skeletal	Frigid	Xeric	6e, 6s	
Alvodest	263.4	Sodic Aquicambids					VD	MWD-SPD	0–3	H	Fine	Mesic	Aridic	6s	
Alyan	27.8	Aridic Argixerolls					MD	WD	2–50	L	Fine	Frigid	Xeric	6e	
Amity	434.7	Argiaquic Xeric Argialbolls	C, F		Douglas-fir	H	VD	SPD	0–3	L	Fine-silty	Mesic	Xeric	2w	
Analulu	265.0	Vitrandic Haploxerepts	F		Ponderosa pine	M	MD	WD	0–90	L	Loamy-skeletal	Frigid	Xeric	6e	
Anatone	1589.5	Lithic Haploxerolls					S	WD	0–90	L	Loamy-skeletal	Frigid	Xeric	7e, 7s	
Anawalt	1280.1	Lithic Xeric Haplargids					S	WD	0–50	M	Clayey	Frigid	Aridic	6e	
Anderly	185.2	Typic Haploxerolls	R	H			MD	WD	1–35	L	Coarse-silty	Mesic	Xeric	3e, 3s	
Anders	4.9	Typic Haploxerolls	R	H			MD	WD	0–35	L	Coarse-loamy	Mesic	Xeric	3e, 4e	
Angelbasin	78.8	Andic Dystrocryepts	F		Douglas-fir	L	MD	WD	0–90		Loamy-skeletal	Cryic	Udic	7e	
Angelpeak	127.1	Typic Vitricryands	F		Douglas-fir	L	D-VD	WD	30–80		Ashy over loamy-skeletal	Cryic	Udic	7e	
Anniecreek	3.4	Vitrandic Haplocryolls	F		Engelmann spruce		VD	SPD	0–2		Ashy	Cryic	Udic	6w	
Antelopepeak	7.9	Typic Vitrixerands	F		Douglas-fir	L	VD	WD	0–30		Ashy over loamy	Frigid	Xeric	6c	
Antoken	33.3	Aridic Palexerolls					VD	WD	2–55		Clayey-skeletal	Mesic	Xeric	7s	
Anunde	32.4	Alic Hapludands	F		Douglas-fir	M, H	VD	WD	0–30		Medial	Mesic	Udic	6e	
Applegate	13.1	Ultic Argixerolls	C, F		Ponderosa pine	L, M	D	WD	2–15	L	Fine	Mesic	Xeric	2e, 3e	
Apt	148.9	Typic Haplohumults	F		Douglas-fir	M, H	VD	WD	2–50	L	Fine	Mesic	Udic	4e, 6e	
Arbidge	8.0	Xeric Argidurids	C, R				MD	WD	1–15		Fine-loamy	Mesic	Aridic	6c	4e
Arcia	92.6	Vitrandic Argixerolls	F		Ponderosa pine		MD	WD	2–50		Fine	Frigid	Xeric	6e, 7s	

(continued)

Series name	Area (km^2)	Subgroup	Land use[1]	Range site quality[2]	Forest indicator species	Forest site quality[3]	Depth class[4]	Drainage class[5]	Slope range (%)	Salinity class[6]	Particle-size class	Soil temp. regime	Soil moisture regime	Land capability class	
														No mgmt	With mgmt
Arness	8.6	Argiduridic Durixerolls					S	WD	1–20		Loamy	Frigid	Xeric	6e	
Ascar	120.9	Typic Fulvudands	F		Douglas-fir	M, H	MD	WD	20–100	L	Medial-skeletal	Isomesic	Udic	7s	
Aschoff	311.0	Andic Humudepts	F		Douglas-fir	H	VD	WD	5–90	L	Loamy-skeletal	Mesic	Udic	6s, 7s	
Aspenlake	3.5	Typic Duricryolls	F		Ponderosa pine	L, M	MD	WD	1–12		Coarse-loamy	Cryic	Xeric	4s	4s
Astoria	145.9	Andic Humudepts	F		Douglas-fir	M, H	VD-D	WD	0–90	L	Fine	Mesic	Udic	6e	
Ateron	1288.7	Lithic Argixerolls					S	WD	2–90	L	Clayey-skeletal	Frigid	Xeric	7s	
Athena	210.4	Pachic Haploxerolls	C, R	H			VD-D	WD	0–70	L	Fine-silty	Mesic	Xeric	2e, 3e	
Atlow	357.3	Lithic Xeric Haplargids					S	WD	2–75	L	Loamy-skeletal	Mesic	Aridic	7e, 7s	
Atring	480.2	Typic Dystroxerepts	F		Douglas-fir	L, M	MD	WD	12–90	L	Loamy-skeletal	Mesic	Xeric	6e, 7s	
Ausmus	235.3	Aquic Natrargids					VD	SPD	0–2	M	Fine-silty	Frigid	Aridic	6s, 7s	
Aval	9.5	Lithic Xeric Haplocambids					S	WD	2–20		Ashy	Frigid	Aridic	6e	
Averlande	40.3	Lithic Hapludults	F		Douglas-fir	M	S	WD	0–30		Loamy-skeletal	Mesic	Udic	7e	
Awbrig	45.7	Vertic Albaqualfs	R	H			VD	PD	0–2	L	Fine	Mesic	Aquic	4w	
Axford	4.2	Calciargidic Argixerolls	R	H			D	WD	2–55		Fine-loamy	Mesic	Aridic	4e	
Ayres	86.6	Argiduridic Durixerolls	C, R	H			S	WD	3–20	L	Loamy-skeletal	Mesic	Xeric	2e	
Ayresbutte	40.0	Vitritorrandic Durixerolls					MD	WD	0–8		Loamy-skeletal	Mesic	Xeric	6s	4s
Babbington	0.8	Calciargidic Argixerolls	C, R				VD	MWD	0–3		Fine-loamy	Mesic	Aridic	6c	2e
Bacona	212.6	Typic Palehumults	F		Douglas-fir	H	D	WD	3–30	L	Fine-silty	Mesic	Udic	6e	
Baconcamp	373.2	Pachic Haplocryolls					MD	WD	3–80	L	Loamy-skeletal	Cryic	Xeric	6e	
Bagness	0.8	Cumulic Humudepts	C, F		Douglas-fir		VD	WD	0–3		Fine-loamy	Isomesic	Udic	2w	
Bakeoven	1225.0	Aridic Lithic Haploxerolls					VS	WD	2–20	L	Loamy-skeletal	Mesic	Xeric	6e, 7s	
Baker	77.9	Haploduridic Durixerolls	R	L			MD	WD	0–7	L	Coarse-loamy	Mesic	Xeric	4s	3s
Bald	42.7	Ultic Haploxerolls	F		Ponderosa pine	L	MD	WD	5–75		Loamy-skeletal	Mesic	Xeric	6s, 7s	
Balder	18.7	Vitrandic Haploxerolls					S	WD	2–70		Loamy	Mesic	Xeric	7s	
Baldock	46.4	Typic Calciaquolls	R	H			VD	PD	0–4	L	Fine-loamy	Mesic	Aquic	3w	
Baldridge	17.9	Pachic Haploxerolls					D	WD	12–70		Loamy-skeletal	Frigid	Xeric	6e, 7e	
Balloontree	3.6	Aquic Vitricryands					VD	SPD	0–15		Ashy over loamy	Cryic	Udic	6c	

(continued)

Series name	Area (km^2)	Subgroup	Land use[1]	Range site quality[2]	Forest indicator species	Forest site quality[3]	Depth class[4]	Drainage class[5]	Slope range (%)	Salinity class[6]	Particle-size class	Soil temp. regime	Soil moisture regime	Land capability class	
														No mgmt	With mgmt
Balm	12.8	Fluvaquentic Haploxerolls	R	H			VD	SPD	0–3	L	Coarse-loamy over sandy or sandy-skeletal	Mesic	Xeric	3w	
Bandarrow	17.8	Typic Cryaquolls					VD	PD	0–30		Coarse-loamy	Cryic	Aquic	6c	
Bandon	46.2	Typic Haplorthods	F		Douglas-fir	L, M	MD	WD	0–50		Coarse-loamy	Isomesic	Udic	3e, 4e	
Banning	23.7	Pachic Argixerolls	C, R				D	SPD	0–20	L	Fine-loamy	Mesic	Xeric	2w, 3c	2w
Barbermill	16.5	Aridic Argixerolls					S	WD	2–90		Clayey	Mesic	Aridic	7e	
Barhiskey	7.6	Vitrandic Dystroxerepts	F		Douglas-fir	L	D	ED	0–3	L	Sandy	Mesic	Xeric	4s	
Barkley	4.1	Duric Argixerolls	C, R	H			VD	WD	0–30		Fine-loamy	Frigid	Xeric	4c, 4e	
Barkshanty	69.7	Typic Palehumults	F		Douglas-fir	M, H	VD	WD	0–40		Loamy-skeletal	Mesic	Udic	6e	
Barnard	43.1	Argiduridic Durixerolls	R	L			MD	WD	0–45	L	Fine	Mesic	Xeric	4s	3s, 3e
Barneycreek	4.9	Vitrandic Argixerolls	F		Ponderosa pine	L, M	VD	WD	15–60		Loamy-skeletal	Frigid	Xeric	6c	
Barron	24.0	Typic Haploxerepts	F		Ponderosa pine	M	D	SED	0–12	L	Coarse-loamy	Mesic	Xeric	4e	2e, 3e
Bashaw	171.9	Xeric Endoaquerts	C				VD	PD	0–12	L	Very-fine	Mesic	Aquic	4w	
Bata	5.7	Andic Glossocryalfs	F		Douglas-fir		D	WD	2–60		Loamy-skeletal	Cryic	Udic	6c	
Bateman	204.6	Ultic Palexeralfs	C, F		Douglas-fir	M, H	VD	WD	3–60	L	Fine	Mesic	Xeric	2e-4e,-6e	
Bayside	0.2	Aeric Fluvaquents	C				VD	PD	0–3		Fine	Isomesic	Aquic	3w	
Beal	36.7	Ultic Haploxeralfs	F, R		Douglas-fir	M	VD	WD	3–60		Fine-loamy	Mesic	Xeric	4e	
Bearcamp	19.1	Typic Humixerepts	F		Douglas-fir	M	D	WD	0–90		Loamy-skeletal	Frigid	Xeric	6e	
Bearpawmeadow	58.3	Andic Haplocryepts	F		Grand fir	M	MD	WD	0–90		Loamy-skeletal	Cryic	Udic	6c	
Bearspring	1.7	Vitrandic Haploxerolls	F		Ponderosa pine		VD	WD	20–65		Loamy-skeletal	Frigid	Xeric	7e	
Beden	106.5	Aridic Lithic Argixerolls					S	WD	0–50	L	Loamy	Frigid	Xeric	6e	
Bedner	7.0	Haplic Durixerolls	C, R	M			MD	MWD	0–1		Fine	Mesic	Xeric	3w	
Beekman	647.3	Typic Dystroxerepts	F		Douglas-fir	L, M	MD	WD	30–100	L	Loamy-skeletal	Mesic	Xeric	6e, 7e, 7s	

(continued)

Series name	Area (km^2)	Subgroup	Land use[1]	Range site quality[2]	Forest indicator species	Forest site quality[3]	Depth class[4]	Drainage class[5]	Slope range (%)	Salinity class[6]	Particle-size class	Soil temp. regime	Soil moisture regime	Land capability class	
														No mgmt	With mgmt
Beeman	3.0	Xerertic Haplocambids	C, R				VD	WD	0–2		Fine	Mesic	Aridic	6s	
Beetville	0.3	Torrifluventic Haploxerolls	C				VD	MWD	0–2		Coarse-loamy	Mesic	Aridic	6c	2e
Bellpine	769.6	Xeric Haplohumults	C, F		Douglas-fir	M, H	MD	WD	2–75	L	Fine	Mesic	Udic	2e-4e, 6e, 7e	
Belrick	35.9	Humic Vitricryands	F		Ponderosa pine	L, M	VD	WD	0–50		Ashy	Cryic	Udic	6e	
Benderly	11.9	Entic Haploxerolls	C, R	L			D	SED	0–7		Sandy-skeletal	Mesic	Xeric	4s	4s
Bennettcreek	107.4	Vitrandic Haploxeralfs	F		Ponderosa pine	L	MD	WD	0–60		Loamy-skeletal	Frigid	Xeric	7s	
Bensley	24.1	Typic Dystrocryepts	F		Douglas-fir	M	VD	WD	2–75		Loamy-skeletal	Cryic	Udic	6s, 7s	
Bentilla	5.4	Typic Palehumults	F		Douglas-fir	H	D	MWD	3–12	L	Fine	Isomesic	Udic	3e	
Beoska	55.9	Durinodic Natrargids	R				VD	WD	0–15		Fine-loamy	Mesic	Aridic	7s	3e
Berdugo	114.7	Xeric Paleargids					VD	WD	0–5	L	Fine	Mesic	Aridic	6e, 6s	
Bergsvik	4.5	Terric Haplohemists	C, F		Western hemlock		VD	VPD	0–1		Sandy or sandy-skeletal	Isomesic	Aquic	5w	
Bickford	1.0	Typic Epiaquolls	F		Douglas-fir	H	VD	SPD	3–12	L	Fine-silty over clayey	Mesic	Aquic	6w	
Bicondoa	5.8	Fluvaquentic Vertic Endoaquolls	C, R	H			VD	PD	0–2		Fine	Frigid	Aquic	4w	
Bigbouldercreek	0.0	Typic Udivitrands	F		Grand fir		VD	WD	0–15		Ashy	Frigid	Udic	6c	
Bigcow	47.2	Andic Haploxerepts	F		Grand fir		VD	WD	0–60		Loamy-skeletal	Frigid	Xeric	6c	
Bigdutch	3.1	Humic Dystrudepts	F		Douglas-fir	M, H	MD	WD	3–60		Loamy-skeletal	Frigid	Udic	6e	
Bigelk	35.4	Vitrandic Haploxerolls	F		Ponderosa pine		VD	WD	0–90		Loamy-skeletal	Frigid	Xeric	6s	
Bigelow	12.4	Typic Humicryepts	F		White fir	L	D	WD	5–65		Loamy-skeletal	Cryic	Udic	6s	
Bigfrog	11.9	Xeric Argidurids					VS-S	WD	8–40	L	Loamy	Mesic	Aridic	6s	
Bigriver	0.7	Typic Udifluvents	F		Coastal redwood		VD	WD	0–5		Coarse-loamy	Isomesic	Udic	3w	
Bindle	34.8	Vitrandic Humixerepts	F		Douglas-fir	M	MD	WD	1–70		Loamy-skeletal	Frigid	Xeric	6s	

(continued)

Series name	Area (km^2)	Subgroup	Land use[1]	Range site quality[2]	Forest indicator species	Forest site quality[3]	Depth class[4]	Drainage class[5]	Slope range (%)	Salinity class[6]	Particle-size class	Soil temp. regime	Soil moisture regime	Land capability class	
														No mgmt	With mgmt
Bingville	68.8	Pachic Palexerolls					MD	WD	2–30		Clayey-skeletal	Frigid	Xeric	6c	
Bins	47.6	Vitrandic Humixerepts	F		Douglas-fir	M	VD	WD	0–70		Fine-loamy	Frigid	Xeric	6e	
Bittercreek	1.8	Aquandic Endoaquolls	R	H			VD	PD	0–3		Coarse-loamy over sandy or sandy-skeletal	Frigid	Aquic	3w	
Blachly	510.1	Humic Dystrudepts	F		Douglas-fir	M, H	VD	WD	0–75	L	Fine	Mesic	Udic	6e, 7e	
Blackgulch	17.3	Lithic Ultic Haploxerolls	F		Ponderosa pine		S	WD	0–90		Loamy-skeletal	Frigid	Xeric	7s	
Blackhills	2.5	Aridic Lithic Haploxerolls					S	SED	15–55		Ashy-skeletal	Mesic	Xeric	7s	
Blacklock	21.5	Typic Duraquods					S	PD	0–7		Sandy	Isomesic	Aquic	6w	
Blalock	14.0	Haploduridic Durixerolls					S	WD	0–15		Loamy	Mesic	Xeric	6e	
Blayden	44.8	Argiduridic Durixerolls	R				S	WD	0–12		Loamy	Frigid	Xeric	6 s	
Bler	12.6	Vitrandic Palexerolls	F		Ponderosa pine		MD	WD	0–60		Clayey-skeletal	Frigid	Xeric	6c	
Blizzard	17.6	Lithic Argicryolls					S	WD	0–15	L	Clayey	Cryic	Xeric	6e	
Blodgett	3.3	Typic Dystrudepts	F		Douglas-fir	L, M	S	WD	30–90		Loamy-skeletal	Frigid	Udic	7e	
Bluecanyon	67.0	Lithic Haploxerolls	F		Ponderosa pine		S	WD	0–90		Loamy-skeletal	Frigid	Xeric	7s	
Bluesters	9.4	Humic Vitrixerands					S-MD	ED	15–50		Ashy over pumiceous or cindery	Frigid	Xeric	6e	
Bly	60.8	Vitrandic Argixerolls	F		Ponderosa pine	L, M	MD-D	WD	0–35		Fine-loamy	Frigid	Xeric	4c, 4e	
Boardflower	8.3	Vitrandic Haploxeralfs	F		Douglas-fir	L	VD	WD	2–20		Fine	Mesic	Xeric	3e	
Boardtree	112.2	Alfic Vitrixerands	F		Douglas-fir	L, M	VD	WD	2–70	L	Ashy over clayey	Frigid	Xeric	6e, 7e	
Bobbitt	16.5	Vitrandic Argixerolls	F		Douglas-fir		MD	WD	0–65		Loamy-skeletal	Mesic	Xeric	6e	
Bobsgarden	25.6	Humic Dystrudepts	F		Douglas-fir	M	VD	WD	0–90		Loamy-skeletal	Frigid	Udic	6e	
Bocker	1473.9	Lithic Haploxerolls	F		Ponderosa pine	M	VS	WD	0–90	L	Loamy-skeletal	Frigid	Xeric	7s	
Bodale	2.2	Cumulic Haplocryolls	F		Engelmann spruce		VD	MWD	0–30		Coarse-loamy	Cryic	Udic	6c	

(continued)

Series name	Area (km²)	Subgroup	Land use[1]	Range site quality[2]	Forest indicator species	Forest site quality[3]	Depth class[4]	Drainage class[5]	Slope range (%)	Salinity class[6]	Particle-size class	Soil temp. regime	Soil moisture regime	Land capability class	
														No mgmt	With mgmt
Bodell	83.9	Lithic Haploxerolls					S	WD	0–75		Loamy-skeletal	Mesic	Xeric	6s, 7s	
Bogus	6.5	Pachic Ultic Argixerolls	F		Douglas-fir	L	D	WD	15–50		Fine	Mesic	Xeric	4e, 6e	
Bohannon	2045.6	Andic Humudepts	F		Douglas-fir	M, H	MD	WD	2–90	L	Fine-loamy	Mesic	Udic	6e, 7e	
Boiler	5.8	Ultic Palexerolls	F		Ponderosa pine	L	D	WD	35–60		Clayey-skeletal	Frigid	Xeric	6e	
Boilout	52.8	Vitrixerandic Argidurids					S	WD	2–10		Ashy	Mesic	Aridic	6e	
Bolobin	66.3	Vitrandic Argixerolls	F		Ponderosa pine		MD	WD	0–60		Fine-loamy	Frigid	Xeric	6e	
Bolony	16.7	Vitrandic Argixerolls	F		Ponderosa pine		MD	WD	15–60		Fine-loamy	Frigid	Xeric	6c	
Bombadil	62.3	Lithic Xeric Haplargids					VS-S	WD	2–50		Loamy	Mesic	Aridic	7s	
Bonnick	96.7	Vitritorrandic Haploxerolls					VD	SED	0–5	L	Ashy	Frigid	Xeric	6e	
Booten	39.1	Vitrandic Argixerolls	F		Ponderosa pine	L	VD	WD	0–65		Ashy	Mesic	Xeric	3e, 4e, 6e	
Booth	593.7	Vertic Palexerolls	F		Ponderosa pine	L, M	MD	WD	0–65	L	Fine	Frigid	Xeric	6e, 7s	
Boravall	36.5	Aeric Halaquepts					VD	PD	0–3	H	Fine	Mesic	Aquic	6s	
Bordengulch	71.1	Andic Haplocryepts	F		Engelmann spruce		MD	WD	0–90		Loamy-skeletal	Cryic	Udic	6e	
Borges	5.6	Typic Humaquepts	C				D	PD	0–8		Fine	Mesic	Aquic	4w	
Bornstedt	67.7	Typic Palexerults	C, F		Douglas-fir	H	D	MWD	0–30	L	Fine-silty	Mesic	Xeric	2e, 3e	
Borobey	215.2	Vitritorrandic Haploxerolls					VD	SED	0–15	L	Ashy	Frigid	Xeric	6e, 6s	
Bosland	44.1	Typic Humudepts	F		Douglas-fir	H	MD	WD	0–90	L	Fine-loamy	Isomesic	Udic	6e, 7e	
Bott	2.0	Alfic Vitricryands	F		Douglas-fir	M	VD	WD	0–50		Ashy over loamy-skeletal	Cryic	Udic	6e	
Boulder lake	81.4	Xeric Epiaquerts					VD	SPD	0–2	L	Fine	Frigid	Aquic	6w	4w
Bouldrock	56.0	Humic Haploxerepts	F		Douglas-fir		MD	WD	12–80		Coarse-loamy	Frigid	Xeric	6e	
Bowlus	24.5	Pachic Ultic Haploxerolls					D	WD	40–70		Fine-silty	Frigid	Xeric	7e	
Boyce	31.0	Cumulic Endoaquolls	C, R	H			D	PD	0–2	L	Fine-silty over sandy or sandy-skeletal	Mesic	Aquic	3w, 4w	3w
Brabble	42.6	Xeric Haplodurids					MD	WD	5–25		Fine-loamy	Mesic	Aridic	6e	
Brace	1571.4	Xeric Argidurids					MD	WD	1–20	L	Fine-loamy	Frigid	Aridic	6e	
Brader	50.5	Typic Haploxerepts					S	WD	1–60	L	Loamy	Mesic	Xeric	6e	
Bragton	8.9	Terric Haplohemists					VD	VPD	0–2		Loamy	Isomesic	Aquic	5w	
Brallier	12.6	Typic Haplohemists	C, R	M			VD	VPD	0–1			Isomesic	Aquic	5w	4w

(continued)

Series name	Area (km^2)	Subgroup	Land use[1]	Range site quality[2]	Forest indicator species	Forest site quality[3]	Depth class[4]	Drainage class[5]	Slope range (%)	Salinity class[6]	Particle-size class	Soil temp. regime	Soil moisture regime	Land capability class	
														No mgmt	With mgmt
Brand	10.4	Typic Endoaquepts	R				VD	PD	0–3		Fine	Mesic	Aquic	3w	
Brandypeak	28.6	Typic Humixerepts	F		Douglas-fir	M	MD	WD	0–90		Loamy-skeletal	Frigid	Xeric	6e	
Brannan	8.2	Typic Vitrixerands	F		Douglas-fir	L	D	WD	2–70		Ashy over loamy-skeletal	Frigid	Xeric	6e, 7e	
Braun	167.7	Dystric Eutrudepts	F		Douglas-fir	M, H	MD	WD	0–90	L	Fine-loamy	Mesic	Udic	7e	
Bravo	188.9	Humic Dystrudepts	F		Douglas-fir	M, H	MD	WD	30–90	L	Fine-loamy	Mesic	Udic	7e	
Breadloaf	0.6	Leptic Haploxererts					VD	SPD	0–3		Fine	Mesic	Xeric	7e	
Bregar	40.4	Lithic Xeric Haplargids					VS-S	WD	2–75		Loamy-skeletal	Frigid	Aridic	7s	
Brenner	35.7	Fluvaquentic Humaquepts	R				VD	PD	0–3		Fine-silty	Isomesic	Aquic	3w, 4w	
Brezniak	24.7	Aridic Lithic Argixerolls					VS-S	WD	2–65	L	Clayey	Mesic	Xeric	6e, 7s	
Bridgecreek	75.6	Typic Palexerolls	R	H, M			MD	WD	1–35		Fine	Frigid	Xeric	4e, 6e	
Bridgewater	1.7	Cumulic Haploxerolls					VD	WD	0–15	L	Loamy-skeletal	Mesic	Xeric	7s	
Bridgewell	27.7	Aquandic Endoaquolls					VD	PD	0–2		Ashy	Frigid	Aquic	6w	
Briedwell	44.0	Ultic Haploxerolls	C, F		Ponderosa pine	M	VD	WD	0–20	L	Loamy-skeletal	Mesic	Xeric	2s, 3e, 4e	
Brightwood	14.3	Typic Humudepts	F		Douglas-fir	M, H	MD	WD	69–90		Loamy-skeletal	Mesic	Udic	7e	
Brisbois	66.6	Xeric Haplargids					VS	WD	2–90		Clayey	Mesic	Aridic	6e	
Broadycreek	4.3	Aquic Cumulic Haplocryolls	F		Engelmann spruce		VD	SPD	0–30	L	Coarse-loamy	Cryic	Udic	6c	
Brock	10.6	Xeric Argidurids					S	WD	4–50	L	Loamy-skeletal	Mesic	Aridic	6s	
Brockman	28.8	Vertic Haploxerepts	R				D	MWD	2–20		Fine	Mesic	Xeric	4e, 4s	3e, 3s
Brownlee	18.8	Ultic Argixerolls	C, R	H			D-VD	WD	0–60		Fine-loamy	Mesic	Xeric	3e	
Brownscombe	27.3	Aridic Argixerolls	R	L, M			MD	WD	1–60	L	Fine	Mesic	Xeric	4s, 4e, 6e	
Broyles	97.4	Durinodic Haplocambids	C, R				VD	WD	0–8		Ashy over loamy	Mesic	Aridic	7c	2s
Bruncan	40.5	Xeric Argidurids					S	WD	0–15	L	Loamy	Mesic	Aridic	6e	
Brunzell	3.5	Typic Haploxerolls	R	M			VD	WD	0–30		Loamy-skeletal	Frigid	Xeric	4s	
Btree	116.6	Alfic Udivitrands	F		Grand fir	L, M	VD-D	WD	0–90		Ashy over clayey-skeletal	Frigid	Udic	6e, 7e	
Bubus	13.7	Duric Torriorthents					VD	WD	0–8		Coarse-loamy	Mesic	Aridic	7s	
Buckbert	7.2	Vitritorrandic Haploxerolls	C, R	H			VD	WD	0–3	L	Fine-loamy	Mesic	Xeric	4c	

(continued)

Series name	Area (km^2)	Subgroup	Land use[1]	Range site quality[2]	Forest indicator species	Forest site quality[3]	Depth class[4]	Drainage class[5]	Slope range (%)	Salinity class[6]	Particle-size class	Soil temp. regime	Soil moisture regime	Land capability class	
														No mgmt	With mgmt
Buckcreek	79.6	Pachic Ultic Haploxerolls					MD	WD	45–70		Loamy-skeletal	Frigid	Xeric	7e	
Bucketlake	120.9	Typic Vitricryands	F		Engelmann spruce		VD	WD	0–90		Ashy over loamy-skeletal	Cryic	Udic	6e	
Buckeye	13.9	Pachic Ultic Argixerolls	C, F		Douglas-fir	M	MD	WD	2–60		Fine-loamy	Mesic	Xeric	3e, 4e, 6e	
Bucklake	26.0	Aridic Argixerolls					MD	WD	2–50	L	Fine	Mesic	Xeric	6e	
Buckshot	13.7	Typic Paleudults	F		Douglas-fir	H	VD	WD	3–90	L	Fine-loamy	Mesic	Udic	6e, 7e	
Buckwilder	22.6	Vertic Argicryolls					MD	WD	3–35	L	Very-fine	Cryic	Xeric	6e	
Budlewis	2.4	Typic Durixerolls					MD	WD	1–10		Fine	Frigid	Xeric	4s	4e
Buffaran	21.3	Xeric Argidurids					S	WD	0–30	L	Clayey	Mesic	Aridic	6s	
Buford	8.5	Vitrandic Haploxerolls	F		Ponderosa pine		D	WD	2–15		Fine-loamy	Frigid	Xeric	3e	
Bulgar	26.1	Typic Udivitrands	F		Grand fir		VD	WD	0–90		Ashy over loamy-skeletal	Frigid	Udic	6c	
Bull run	78.1	Eutric Fulvudands	F		Douglas-fir	H	VD	WD	3–80	L	Medial	Mesic	Udic	3e, 6e, 7e	
Bullards	104.2	Typic Haplorthods	F		Douglas-fir	M, H	VD	WD	0–60	L	Coarse-loamy	Isomesic	Udic	3e, 4e	
Bullgulch	21.3	Typic Haplohumults	F		Douglas-fir	H	VD	WD	0–60	L	Fine	Isomesic	Udic	6e	
Bullroar	22.4	Typic Udivitrands	F		Grand fir		VD	WD	0–15		Ashy over loamy-skeletal	Frigid	Udic	6c	
Bullump	103.9	Pachic Argixerolls					D-VD	WD	5–75		Loamy-skeletal	Frigid	Xeric	6e	
Bullvaro	35.0	Pachic Argixerolls					VD	WD	30–75		Loamy-skeletal	Frigid	Xeric	7e	
Bully	14.5	Xeric Torrifluvents	C, R	H			D	WD	0–2	L	Coarse-silty	Mesic	Aridic	2e	
Bunchpoint	14.3	Vitrandic Haploxerolls					MD	WD	0–15		Coarse-loamy	Frigid	Xeric	4e, 6e	
Bunyard	2.4	Durinodic Natrargids					VD	WD	0–1		Ashy	Frigid	Aridic	6s	
Burbank	46.5	Xeric Torriorthents	C, R				VD	ED	0–60		Sandy-skeletal	Mesic	Aridic	7e	4e
Burgerbutte	61.2	Lithic Humicryepts	F		Subalpine fir		S	WD	0–90		Loamy-skeletal	Cryic	Xeric	6s	
Burke	62.6	Xeric Haplodurids	R	H			MD	WD	0–65	L	Coarse-silty	Mesic	Aridic	4e	3e
Burkemont	21.4	Typic Halaquepts	R	H			D	PD	0–2	M	Fine	Mesic	Aquic	3w	
Burlington	8.0	Entic Ultic Haploxerolls	C, F		Douglas-fir		D	SED	0–15	L	Sandy	Mesic	Xeric	2e, 3e	
Burningman	0.1	Lithic Argixerolls					S	WD	4–15		Ashy over clayey	Frigid	Xeric	6s	

(continued)

Series name	Area (km^2)	Subgroup	Land use[1]	Range site quality[2]	Forest indicator species	Forest site quality[3]	Depth class[4]	Drainage class[5]	Slope range (%)	Salinity class[6]	Particle-size class	Soil temp. regime	Soil moisture regime	Land capability class	
														No mgmt	With mgmt
Burnthill	12.4	Typic Palehumults	F		Douglas-fir	H	VD	WD	0–30	L	Fine-loamy	Isomesic	Udic	3e	
Burntriver	25.4	Xeric Haplocambids	R	H			VD	WD	0–12	L	Fine-loamy	Frigid	Aridic	4c, 4e	3c
Burntwoods	6.4	Typic Fulvudands	F		Douglas-fir	M, H	VD	WD	5–60		Medial-skeletal over loamy-skeletal	Frigid	Udic	6e	
Burrita	2.1	Lithic Xeric Haplargids					S	WD	4–75		Clayey-skeletal	Mesic	Aridic	7s	
Bybee	57.5	Typic Haploxerolls	F		Douglas-fir	L	D	SPD	1–35		Fine	Frigid	Xeric	6e	
Bycracky	3.2	Terric Cryofibrists					VD	VPD	0–5		Loamy	Cryic	Aquic	6c	
Cabell	4.6	Andic Haplocryalfs	F		Engelmann spruce		VD	WD	0–50		Fine-loamy	Cryic	Udic	6c	
Cabincreek	3.5	Vitrandic Haploxerolls					D	WD	60–90		Coarse-loamy	Mesic	Xeric	7e	
Cabinspring	9.9	Vitritorrandic Argixerolls					MD	WD	20–50		Ashy-skeletal	Frigid	Xeric	6e	
Calder	12.5	Abruptic Haplic Durixeralfs	C, R	H			S	MWD	0–1		Clayey	Mesic	Xeric	4w	
Calderwood	155.8	Lithic Xeric Haplocambids					S	WD	0–25	L	Loamy-skeletal	Mesic	Aridic	7s	
Calfranch	27.6	Typic Dystrudepts	F		Douglas-fir	H	VD	WD	0–90	L	Loamy-skeletal	Isomesic	Udic	6e, 7e	
Calimus	157.0	Pachic Haploxerolls	C, R	H			VD	WD	0–35	L	Fine-loamy	Mesic	Xeric	2c, 2e, 3e	
Camas	111.3	Fluventic Haploxerolls	C, F		Or ash		VD	ED	0–5	L	Sandy-skeletal	Mesic	Xeric	4w	4w
Camaspatch	0.1	Lithic Argixerolls					S	WD	0–70		Clayey-skeletal	Mesic	Xeric	7s	
Campcreek	56.1	Vertic Palexerolls	F		Ponderosa pine		VD	WD	12–60		Fine	Frigid	Xeric	4e, 6e	
Campfour	14.6	Pachic Ultic Argixerolls	F		Ponderosa pine	M	D	WD	1–35		Fine-loamy	Mesic	Xeric	4e	
Camptank	1.0	Xeric Paleargids					D	WD	15–45		Clayey over loamy-skeletal	Frigid	Aridic	6e	
Canderly	10.4	Ultic Haploxerolls	C, F		Ponderosa pine	H	D	SED	0–8	L	Coarse-loamy	Mesic	Xeric	2e, 2s	
Canest	353.5	Aridic Lithic Argixerolls					VS	WD	1–20	L	Clayey-skeletal	Frigid	Aridic	7s	
Cant	11.9	Xeric Paleargids					MD	WD	15–40		Clayey-skeletal	Mesic	Aridic	7s	
Cantala	165.4	Typic Haploxerolls	C, R	H			D	WD	1–35	L	Fine-silty	Mesic	Xeric	2e-4e	
Capeblanco	18.0	Typic Dystrudepts	F		Douglas-fir	H	MD	WD	30–90	L	Loamy-skeletal	Isomesic	Udic	6e	

(continued)

Series name	Area (km^2)	Subgroup	Land use[1]	Range site quality[2]	Forest indicator species	Forest site quality[3]	Depth class[4]	Drainage class[5]	Slope range (%)	Salinity class[6]	Particle-size class	Soil temp. regime	Soil moisture regime	Land capability class	
														No mgmt	With mgmt
Caphealy	28.1	Vitritorrandic Haploxerolls	C, R				MD	WD	0–30		Coarse-loamy	Mesic	Xeric	6s	3s
Capona	45.4	Aridic Haploxerolls	R	H			MD	WD	0–35	L	Fine-loamy	Mesic	Xeric	3s, 3e, 4e	
Caris	208.1	Typic Haploxerepts	F		Douglas-fir	L, M	MD	WD	50–90	L	Loamy-skeletal	Mesic	Xeric	7s	
Carlton	39.8	Aquultic Haploxerolls	C, F		OR white oak		VD	MWD	0–20	L	Fine-silty	Mesic	Xeric	2w, 3e, 4e	
Carney	235.3	Udic Haploxererts	F		Ponderosa pine	L	MD	MWD	1–35	L	Fine	Mesic	Xeric	4e	3s
Carpenterville	2.6	Aquic Argiudolls	R				MD	SPD	0–60		Clayey-skeletal	Mesic	Udic	6e	
Carryback	1189.8	Vertic Palexerolls	F		Ponderosa pine		MD	WD	0–50	L	Fine	Frigid	Xeric	6e	
Carvix	151.1	Aridic Haploxerolls					VD	WD	0–8	L	Fine-loamy	Frigid	Xeric	6e, 6c	4c
Cascade	178.8	Humic Fragixerepts	F		Douglas-fir	M, H	MD	SPD	3–60	L	Fine-silty	Mesic	Xeric	3w, 3e-4e	
Cashner	2.7	Typic Duraquods					MD	PD	0–7		Coarse-loamy	Isomesic	Aquic	4w	
Cassiday	183.3	Humic Dystrudepts	F		Douglas-fir	M, H	MD	WD	30–90	L	Loamy-skeletal	Mesic	Udic	7e	
Castlecrest	265.8	Typic Vitricryands	F		Douglas-fir	M	VD	SED	0–80	L	Ashy	Cryic	Udic	7s	
Catchell	36.7	Abruptic Xeric Argidurids	R				MD	WD	0–30		Fine	Mesic	Aridic	6c	4s
Caterl	253.1	Alic Hapludands	F		Douglas-fir	M, H	VD-D	WD	3–90	L	Medial-skeletal	Frigid	Udic	6e, 7e	
Catherine	89.2	Cumulic Endoaquolls	C, R	H			VD	SPD	0–3	L	Fine-silty	Mesic	Aquic	2w	2w
Catlow	171.7	Durinodic Xeric Haplocambids	R	L, M			VD	WD	0–20	M	Loamy-skeletal	Mesic	Aridic	6e	3e
Catnapp	3.2	Xeric Natrargids					MD	WD	2–15		Fine	Frigid	Aridic	7s	
Cazadero	82.0	Rhodic Paleudults	C, F		Douglas-fir	H	D	WD	0–60	L	Fine	Mesic	Udic	2e-4e, 6e	
Cedarcamp	22.5	Dystric Eutrudepts	F		Douglas-fir	H	VD	WD	0–90	L	Loamy-skeletal	Frigid	Udic	7e	
Cedargrove	14.3	Ultic Haploxeralfs	C, F		Douglas-fir	M	VD	WD	3–60		Fine	Mesic	Xeric	3e, 4e, 6e	
Cencove	9.3	Xeric Torriorthents	C, R	H			D	WD	0–12	L	Coarse-loamy over sandy or sandy-skeletal	Mesic	Aridic	3e, 3s	
Central point	33.8	Pachic Haploxerolls	C				VD	WD	0–3	L	Coarse-loamy	Mesic	Xeric	4c	2s

(continued)

Series name	Area (km^2)	Subgroup	Land use[1]	Range site quality[2]	Forest indicator species	Forest site quality[3]	Depth class[4]	Drainage class[5]	Slope range (%)	Salinity class[6]	Particle-size class	Soil temp. regime	Soil moisture regime	Land capability class	
														No mgmt	With mgmt
Chamate	15.7	Typic Dystrudepts	F		Douglas-fir	M, H	VD	WD	3–90		Loamy-skeletal	Mesic	Udic	6e	
Chambeam	38.4	Pachic Haploxerolls	R	H, M			D	WD	12–50		Loamy-skeletal	Frigid	Xeric	4e, 6e	
Chancelakes	7.0	Xeric Epiaquerts					VD	PD	0–2		Fine	Frigid	Aquic	6w	
Chapman	58.1	Cumulic Ultic Haploxerolls	C				VD	WD	0–3	L	Fine-loamy	Mesic	Xeric	1	
Chard	0.4	Calcic Haploxerolls	C, R	H			VD	WD	0–65		Coarse-loamy	Mesic	Xeric	4e	
Chehalem	49.5	Cumulic Vertic Endoaquolls	C				VD	SPD	0–12		Fine	Mesic	Aquic	3e	
Chehalis	238.7	Cumulic Ultic Haploxerolls	F		Douglas-fir	M, H	VD	WD	0–10	L	Fine-silty	Mesic	Xeric	2w, 3e	
Chehulpum	42.8	Ultic Haploxerolls	R				S	WD	3–60		Loamy	Mesic	Xeric	6e, 6s, 7s	
Chen	111.9	Aridic Lithic Argixerolls					S	WD	2–75	L	Clayey-skeletal	Frigid	Xeric	7s	
Chenoweth	14.8	Typic Haploxerolls	C, R	H			D	WD	0–55	L	Coarse-loamy	Mesic	Xeric	2e, 3e	
Cherry spring	14.5	Haploxeralfic Argidurids					MD	WD	2–8		Fine-loamy	Mesic	Aridic	7s	
Cherrycreek	71.3	Vitrandic Haploxerolls	F		Ponderosa pine		D	WD	2–90		Loamy-skeletal	Frigid	Xeric	4e, 6e	
Cherryhill	30.8	Ultic Haploxeralfs	C, F		Douglas-fir		D	WD	0–50	L	Fine-loamy	Mesic	Xeric	2e-4e, 6e	
Chesebro	18.5	Vitrandic Argixerolls					VD	WD	15–65		Ashy-skeletal	Frigid	Xeric	6e	
Chesnimnus	12.5	Calcic Argixerolls	R	H			VD	WD	0–3	L	Fine-loamy	Frigid	Xeric	3c	
Chetco	14.1	Fluvaquentic Humaquepts	C				VD	VPD	0–3		Fine	Isomesic	Aquic	4w	
Cheval	9.2	Aquic Cumulic Haploxerolls	F		Douglas-fir		VD	SPD	0–2	L	Coarse-loamy over sandy or sandy-skeletal	Frigid	Xeric	3w	
Chewaucan	13.8	Argiduridic Argixerolls	R	H			D	WD	2–15	L	Fine	Mesic	Xeric	4e	
Chilcott	93.3	Abruptic Xeric Argidurids	R	H			MD	WD	0–30	L	Fine	Mesic	Aridic	4e	
Chiloquin	3.4	Oxyaquic Vitrandic Haploxerepts	C, R	H			VD	MWD	0–1		Fine-loamy	Frigid	Xeric	4w	
Chimneyrock	23.4	Ultic Haploxeralfs	F		Douglas-fir	M	D	WD	30–90		Loamy-skeletal	Mesic	Xeric	6e, 7e	

(continued)

Series name	Area (km^2)	Subgroup	Land use[1]	Range site quality[2]	Forest indicator species	Forest site quality[3]	Depth class[4]	Drainage class[5]	Slope range (%)	Salinity class[6]	Particle-size class	Soil temp. regime	Soil moisture regime	Land capability class	
														No mgmt	With mgmt
Chinarise	23.4	Vitrandic Haploxerolls	R				VD	SPD	0–4		Ashy	Frigid	Xeric	6e	
Chintimini	8.0	Andic Dystrudepts	F		Douglas-fir	L, M	D	WD	20–90		Loamy-skeletal	Frigid	Udic	6e, 7e	
Chismore	16.9	Aquic Palehumults	F		Douglas-fir	H	VD	MWD	0–12	L	Fine	Mesic	Udic	3w, 3e, 4e	
Chitwood	22.2	Aquandic Humudepts	C, F		Western hemlock	M	VD	SPD	0–15		Fine	Isomesic	Udic	2w, 3e, 4e	
Chock	23.7	Typic Cryaquands	C, R	M			VD	PD	0–1		Ashy	Cryic	Aquic	5w	
Chocktoot	26.3	Vitrandic Argicryolls	F		White fir		VD	WD	15–60	L	Loamy-skeletal	Cryic	Xeric	6e	
Choptie	64.6	Lithic Haploxerolls	F		Ponderosa pine		S	WD	1–60		Loamy	Frigid	Xeric	4e	
Chug	24.9	Vitrandic Haploxerolls	R				VD	WD	0–15		Fine-loamy	Frigid	Xeric	6c	6e
Cinderfall	3.4	Vitritorrandic Haploxerolls					VD	SED	1–8		Ashy-skeletal	Frigid	Xeric	6e	
Circle	14.2	Alfic Vitrixerands	F		Douglas-fir	M	VD	WD	0–30		Ashy	Frigid	Xeric	6e	
Clackamas	82.5	Typic Argiaquolls	F		Douglas-fir		VD	SPD	0–3	L	Fine-loamy	Mesic	Aquic	3w	
Clamp	134.3	Lithic Haplocryolls					VS-S	WD	5–70	L	Loamy-skeletal	Cryic	Xeric	7e, 7s	
Clarkscreek	34.0	Haploxerandic Humicryepts					MD	WD	0–90		Loamy-skeletal	Cryic	Xeric	6e	
Clatsop	29.6	Histic Humaquepts	R	H, L			D	VPD	0–3		Fine-silty	Isomesic	Aquic	4w, 6w	
Clawson	18.2	Typic Endoaquepts	R	H			VD	PD	0–7	L	Coarse-loamy	Mesic	Aquic	3w	3w
Clearline	42.7	Vitrandic Haploxerolls	F		Ponderosa pine		D	WD	30–90		Ashy-skeletal	Frigid	Xeric	7e	
Cleavage	29.2	Aridic Lithic Argixerolls					S	WD	2–75	L	Loamy-skeletal	Frigid	Xeric	7s	
Cleet	35.3	Xeric Argidurids					S	WD	2–15	L	Loamy-skeletal	Mesic	Aridic	7s	
Cleetwood	28.8	Vitrandic Cryopsamments					VD	ED	0–30		Ashy	Cryic	Udic	6e	
Clevescove	8.3	Andic Humudepts	F		Douglas-fir	H	D	WD	30–70	L	Coarse-loamy	Isomesic	Udic	6a	
Cleymor	0.5	Vertic Argixerolls	F		Douglas-fir		VD	WD	4–35		Fine	Frigid	Xeric	4e	
Climax	17.0	Leptic Haploxererts	C, R	H, M			MD	MWD	12–60		Very-fine	Mesic	Xeric	4e, 6e	
Clinefalls	7.3	Typic Vitritorrands					VD	WD	0–3	L	Ashy over sandy or sandy-skeletal	Mesic	Aridic	6s	
Cloquato	159.8	Cumulic Ultic Haploxerolls	C, F		Douglas-fir	M, H	VD	WD	0–5	L	Coarse-silty	Mesic	Xeric	2w	

(continued)

Series name	Area (km^2)	Subgroup	Land use[1]	Range site quality[2]	Forest indicator species	Forest site quality[3]	Depth class[4]	Drainage class[5]	Slope range (%)	Salinity class[6]	Particle-size class	Soil temp. regime	Soil moisture regime	Land capability class	
														No mgmt	With mgmt
Clovercreek	54.5	Lithic Argixerolls	R	H			S	WD	2–35		Loamy-skeletal	Mesic	Xeric	3s	
Cloverland	22.6	Xeric Argialbolls	F		Ponderosa pine	L	VD	MWD	2–45	L	Fine-silty	Frigid	Xeric	3e	
Clovkamp	50.0	Vitritorrandic Haploxerolls	C, R				VD, D	SED	0–25	L	Ashy over sandy or sandy-skeletal	Mesic	Xeric	6s	3s
Clurde	22.6	Durinodic Xeric Haplocambids	C, R				VD	WD	0–15		Fine-loamy	Mesic	Aridic	6e	
Coburg	178.9	Oxyaquic Argixerolls	C, F		Douglas-fir		VD	MWD	0–5	L	Fine	Mesic	Xeric	2w	
Coglin	100.1	Xeric Paleargids					VD	WD	2–15	L	Fine	Frigid	Aridic	7s	
Coker	40.6	Xeric Endoaquerts	R				VD	SPD	0–12		Very-fine	Mesic	Aquic	4e, 4w	
Coleman	76.9	Typic Palexerolls	C, R				D	MWD	0–7	L	Fine	Mesic	Xeric	4e	2e
Colepoint	15.0	Typic Humudepts	F		Douglas-fir	H	D	WD	0–60	L	Fine-loamy	Mesic	Udic	6e	
Colestine	16.9	Typic Dystroxerepts	F		Douglas-fir	L, M	MD	WD	20–90	L	Fine-loamy	Mesic	Xeric	6e, 7e	
Collegecreek	180.3	Typic Vitrixerands	F		Douglas-fir		VD	WD	0–30		Ashy over loamy	Mesic	Xeric	3e	
Collier	7.1	Xeric Vitricryands	F		Ponderosa pine	M	VD	SED	0–80		Ashy	Cryic	Xeric	6c	
Combsflat	411.4	Alfic Humic Vitrixerands	F		Ponderosa pine		VD	WD	5–40		Ashy over clayey	Frigid	Xeric	4e	
Concord	123.1	Typic Endoaqualfs	R	H			VD	PD	0–2	L	Fine	Mesic	Aquic	3w	
Condon	1171.4	Typic Haploxerolls	R	H, M			MD	WD	0–40	L	Fine-silty	Mesic	Xeric	3e, 4e, 6e	
Condorbridge	6.3	Andic Humudepts	C, F		Douglas-fir	H	VD	WD	3–15	L	Fine-loamy	Isomesic	Udic	2e, 3e	
Conley	47.5	Xerertic Argialbolls	R	H			VD	SPD	0–15	L	Fine	Mesic	Xeric	3w	
Connleyhills	122.1	Vitritorrandic Argixerolls					MD	WD	1–20		Clayey-skeletal	Frigid	Xeric	6e	
Conser	93.3	Vertic Argiaquolls	C				VD	PD	0–3	L	Fine	Mesic	Aquic	3w	
Cookcreek	27.8	Vitrandic Haploxerolls					MD	WD	30–90		Loamy-skeletal	Mesic	Xeric	7e	
Cooperdraw	15.5	Xeric Argidurids					MD	WD	1–5	L	Loamy-skeletal	Frigid	Aridic	6e	
Cooperopolis	7.8	Aridic Lithic Argixerolls					S	WD	2–50		Loamy	Mesic	Aridic	6e	
Copperfield	34.0	Pachic Argixerolls					VD	WD	30–90		Loamy-skeletal	Mesic	Xeric	6e, 7e	
Copsey	5.1	Vertic Endoaquolls	C, R	H			D	PD	0–9		Fine	Mesic	Aquic	4w	3w

(continued)

Series name	Area (km^2)	Subgroup	Land use[1]	Range site quality[2]	Forest indicator species	Forest site quality[3]	Depth class[4]	Drainage class[5]	Slope range (%)	Salinity class[6]	Particle-size class	Soil temp. regime	Soil moisture regime	Land capability class	
														No mgmt	With mgmt
Coquille	108.5	Fluvaquentic Endoaquepts	R	H, L			VD	VPD	0–1	L	Fine-silty	Isomesic	Aquic	3w, 4w, 6w	
Cornelius	101.4	Mollic Fragixeralfs	C, F		Douglas-fir	H	MD	WD	2–60	L	Fine-silty	Mesic	Xeric	2e, 4e	
Cornutt	79.6	Ultic Haploxeralfs	F		Ponderosa pine	M	D	WD	7–55		Fine	Mesic	Xeric	4e, 6e	
Corral	113.0	Xeric Haplargids					S	WD	0–50	L	Loamy	Mesic	Aridic	6e, 7s	
Cotant	10.1	Aridic Argixerolls					S	WD	2–50	L	Clayey	Frigid	Xeric	6e	
Cotay	3.5	Vitrandic Haploxeralfs	F		Grand fir		MD	WD	0–90		Clayey-skeletal	Frigid	Xeric	6c	
Cottrell	47.8	Aquic Haplohumults	C, F		Douglas-fir	H	D	MWD	2–30	L	Fine	Mesic	Udic	3e, 3w, 4e	
Cougarrock	38.6	Vitrandic Haploxeralfs	F		Grand fir		MD	WD	0–60		Clayey-skeletal	Frigid	Xeric	6e	
Coughanour	29.3	Argiduridic Durixerolls	R	H			MD	WD	0–12	L	Fine-silty	Mesic	Xeric	4c, 4e	3e, 3s
Court	9.6	Vitritorrandic Haploxerolls	R	H			MD	WD	0–8	L	Coarse-loamy over sandy or sandy-skeletal	Mesic	Xeric	4e	
Courtney	9.9	Abruptic Argiaquolls	C, R	H			VD	PD	0–3	L	Fine	Mesic	Aquic	4w	
Courtrock	3.4	Calcidic Haploxerolls	C				VD	WD	2–20		Coarse-loamy	Mesic	Aridic	3c, 3e	
Cove	83.2	Vertic Endoaquolls	R	H			VD	PD-VPD	0–3	L	Fine	Mesic	Aquic	3e, 4w	
Cowsly	269.9	Xerertic Argialbolls	F		Ponderosa pine	L, M	VD-D	WD	2–30	L	Fine	Frigid	Xeric	3e, 4e	3c
Coyata	54.6	Humic Dystroxerepts	F		Douglas-fir	L–H	MD	WD	1–80	L	Loamy-skeletal	Mesic	Xeric	3 s, 4e	
Coyotebluff	11.8	Humic Vitrixerands	F		Grand fir		D	WD	30–90		Ashy-skeletal over loamy-skeletal	Frigid	Xeric	6e	
Coztur	117.5	Lithic Xeric Haplargids					S	WD	2–30	L	Loamy	Frigid	Aridic	6e, 7s	
Crabtree	19.6	Aquic Dystrocryepts	F		Douglas-fir	L, M	D	MWD	2–75		Loamy-skeletal	Cryic	Udic	6s, 7s	
Crackedground	60.2	Vitritorrandic Haploxerolls	F		Douglas-fir	M	D	WD	1–6		Ashy-skeletal	Frigid	Xeric	6e	
Crackler	99.7	Typic Vitrixerands	F		Douglas-fir	L	D	WD	2–50		Ashy over loamy-skeletal	Frigid	Xeric	4e	
Crannler	18.9	Typic Humicryepts	F		White fir	L	MD	SED	50–90		Loamy-skeletal	Cryic	Udic	7s	

(continued)

Series name	Area (km^2)	Subgroup	Land use[1]	Range site quality[2]	Forest indicator species	Forest site quality[3]	Depth class[4]	Drainage class[5]	Slope range (%)	Salinity class[6]	Particle-size class	Soil temp. regime	Soil moisture regime	Land capability class	
														No mgmt	With mgmt
Crater lake	23.1	Typic Vitrixerands	F		Douglas-fir	M, H	VD	WD	0–70		Ashy	Mesic	Xeric	3e, 4e	
Crawfish	4.5	Lithic Humicryepts					VS	WD	0–60		Loamy-skeletal	Cryic	Udic	7s	
Cressler	9.8	Fluvaquentic Endoaquolls	R	M			VD	SPD	0–2		Fine	Frigid	Aquic	5w	
Crims	13.8	Terric Haplohemists	C, R	H			D	VPD	0–3		Loamy	Mesic	Aquic	3w	
Crofland	7.8	Aquic Haplohumults	F		Douglas-fir	H	VD	SPD	0–3	L	Fine	Isomesic	Udic	3w	3w
Crooked	1.5	Sodic Aquicambids	R	H			VD	SPD	0–2	H	Coarse-loamy	Mesic	Aridic	3w, 4w	
Croquib	7.3	Alic Epiaquands	C, R	H			D	PD	0–3		Medial over loamy-skeletal	Isomesic	Aquic	4w	
Crosswhite	3.0	Alfic Vitrixerands	F		Grand fir	M	MD	WD	5–50		Ashy over loamy-skeletal	Frigid	Xeric	6e	
Crowcamp	108.6	Vertic Palexerolls	C, R				VD	SPD	0–2	L	Fine	Frigid	Xeric	6w	4w
Cruiser	51.5	Typic Haplocryands	F		Douglas-fir	M, H	D-VD	WD	2–70		Medial	Cryic	Udic	6e, 7e	
Crume	7.3	Typic Haploxerolls	C, R	H			D	WD	0–8		Fine-loamy	Frigid	Xeric	4c, 4e	
Crump	70.6	Histic Humaquepts	R	H, M			VD	VPD	0–1	L	Fine-silty	Mesic	Aquic	3w, 5w	
Crutch	7.6	Aquentic Haplorthods	F		Douglas-fir	L	MD	MWD	0–10		Sandy-skeletal	Mesic	Udic	6s, 6w	
Crutchfield	16.0	Typic Humudepts	F		Douglas-fir	H	MD	WD	0–60	L	Fine-loamy	Mesic	Udic	6e	
Culbertson	22.7	Typic Humixerepts	C, F		Douglas-fir	M	VD, D	WD	0–50	L	Fine-loamy	Mesic	Xeric	2e, 3e, 6e	
Cullius	79.4	Aridic Lithic Argixerolls	F		Subalpine fir		S	WD	0–15	L	Clayey	Mesic	Xeric	6e	
Cumley	136.1	Oxyaquic Palehumults	F		Douglas-fir	M, H	D	MWD	2–20	L	Fine	Mesic	Udic	6e	
Cunniff	26.5	Typic Palehumults	F		Douglas-fir	H	VD	WD	0–30	L	Fine	Isomesic	Udic	3e, 4e	
Cupola	18.7	Typic Udivitrands	F		Douglas-fir	M	VD	WD	3–30		Ashy-skeletal	Mesic	Udic	6s	
Curant	203.6	Calcic Pachic Haploxerolls					VD	WD	8–70	L	Fine-silty	Mesic	Xeric	6e, 7e	
Curtin	55.5	Aquic Haploxererts	R				VD	SPD	3–20		Very-fine	Mesic	Xeric	3w	
Dabney	5.1	Typic Udipsamments	F		Douglas-fir	L	D	SED	0–3		Sandy	Mesic	Udic	6s	

(continued)

Series name	Area (km²)	Subgroup	Land use[1]	Range site quality[2]	Forest indicator species	Forest site quality[3]	Depth class[4]	Drainage class[5]	Slope range (%)	Salinity class[6]	Particle-size class	Soil temp. regime	Soil moisture regime	Land capability class	
														No mgmt	With mgmt
Dacker	50.1	Xeric Argidurids	R				MD	WD	0–15		Fine-loamy	Mesic	Aridic	6s	3e
Dahl	9.7	Entic Haploxererts	R	H			VD	MWD	0–3		Fine	Mesic	Xeric	3w	
Damewood	180.3	Andic Humudepts	F		Douglas-fir	H	MD	WD	30–90	L	Loamy-skeletal	Mesic	Udic	7e	
Damore	10.6	Fluvaquentic Haploxerolls	C, R	H			D	SPD	0–3		Fine	Frigid	Xeric	3e	
Darby	7.4	Pachic Ultic Argixerolls	F		Douglas-fir	H	D	WD	12–60	L	Fine	Mesic	Xeric	4e, 6e	
Dardry	17.8	Cumulic Ultic Haploxerolls	F		Ponderosa pine		VD	WD	0–10		Loamy-skeletal	Frigid	Xeric	6c	
Darkcanyon	26.1	Xeric Haplocambids					MD	WD	30–80		Loamy-skeletal	Mesic	Aridic	7e	
Darow	20.8	Vertic Argixerolls	R				MD	MWD	1–35	L	Fine	Mesic	Xeric	4e	3s
Davey	77.0	Xeric Haplocambids					VD	SED	0–45		Sandy	Mesic	Aridic	6e	
Daxty	13.7	Vitrandic Haploxerepts	F		Ponderosa pine		MD	WD	2–50		Loamy-skeletal	Frigid	Xeric	6e	
Day	53.5	Chromic Haploxererts					D-VD	WD	0–70		Very-fine	Mesic	Xeric	6e, 7e	
Dayton	417.9	Vertic Albaqualfs	R	H			VD	PD	0–2	L	Fine	Mesic	Aquic	4w	
Dayville	29.2	Cumulic Endoaquolls	R				VD	SPD	0–2		Fine-silty over sandy or sandy-skeletal	Mesic	Aquic	3w, 4w	
Deadline	107.8	Humic Dystrudepts	F		Douglas-fir	M, H	D	WD	0–90	L	Loamy-skeletal	Mesic	Udic	6e	
Deardorf	36.2	Typic Udivitrands	F		Grand fir		MD	WD	0–60		Ashy over loamy-skeletal	Frigid	Udic	6c	
Debenger	38.1	Typic Haploxerepts	R				MD	WD	1–60		Fine-loamy	Mesic	Xeric	6e	4e
Deck	25.9	Vitrandic Haploxerolls	F		Ponderosa pine		MD	WD	0–90		Loamy-skeletal	Frigid	Xeric	6e	
Dee	13.5	Aquic Vitrixerands	C, F		Douglas-fir		VD	SPD	0–12	L	Ashy	Mesic	Xeric	2w, 3e	
Defenbaugh	58.1	Typic Haplocambids					VD	WD	0–4	L	Fine-loamy	Mesic	Aridic	6 s	
Degarmo	22.3	Cumulic Endoaquolls	R	H			VD	SPD-PD	0–2		Fine-loamy over sandy or sandy-skeletal	Frigid	Aquic	4w	4w
Degner	11.4	Calcic Argixerolls					MD	WD	12–40		Clayey-skeletal	Mesic	Xeric	6e	
Dehill	0.3	Pachic Haploxerolls	C				VD	WD	0–15		Coarse-loamy	Mesic	Xeric	4e	

(continued)

Series name	Area (km^2)	Subgroup	Land use[1]	Range site quality[2]	Forest indicator species	Forest site quality[3]	Depth class[4]	Drainage class[5]	Slope range (%)	Salinity class[6]	Particle-size class	Soil temp. regime	Soil moisture regime	Land capability class	
														No mgmt	With mgmt
Dehlinger	58.9	Pachic Haploxerolls					D	WD	15–65		Loamy-skeletal	Mesic	Xeric	7s	
Delena	22.0	Humic Fragiaquepts	F		Douglas-fir		MD	PD	3–15		Fine-silty	Mesic	Aquic	4w, 6e	
Demasters	0.0	Pachic Ultic Argixerolls	F		Douglas-fir		D-VD	WD	3–90		Fine-loamy	Frigid	Xeric	7e	
Dement	43.5	Humic Dystrudepts	F		Douglas-fir	M, H	D	WD	2–70		Fine	Mesic	Udic	3e-4e, 6e-7e	
Dentdraw	2.7	Fluvaquentic Endoaquolls					VD	PD	0–2		Fine-loamy	Frigid	Aquic	6s	
Denvnot	65.6	Lithic Argixerolls					S	WD	2–60		Clayey	Mesic	Xeric	6s	
Depoe	11.6	Typic Duraquods	F		Western hemlock	L	S	PD	0–8		Loamy	Isomesic	Aquic	6w	
Deppy	308.6	Argidic Argidurids					VS-S	WD	2–50	L	Loamy	Mesic	Aridic	6e, 7s	
Derallo	11.0	Vitritorrandic Argixerolls					D	WD	15–60		Ashy-skeletal	Frigid	Xeric	6e	
Derapter	8.8	Calciargidic Argixerolls					D	WD	30–70		Loamy-skeletal	Frigid	Aridic	7s	
Derringer	23.5	Typic Argixerolls	R	H			MD	WD	12–60		Clayey-skeletal	Frigid	Xeric	4e	
Deschutes	245.3	Vitritorrandic Haploxerolls	C, R	H, M, L			MD	WD	0–30	L	Coarse-loamy	Mesic	Xeric	2e, 6s, 6e	3 s, 3e
Deseed	64.6	Xeric Haplargids					MD	WD	2–50		Fine	Frigid	Aridic	6e	
Deshler	22.2	Pachic Argixerolls	R				MD	WD	0–60		Fine	Mesic	Xeric	6e	
Deskamp	172.4	Vitritorrandic Haploxerolls	C, R				MD	SED	0–15	L	Ashy	Mesic	Xeric	6s	3s
Desolation	4.1	Humic Dystrudepts	F		Douglas-fir	M	D	WD	10–35		Fine	Mesic	Udic	6e	
Desons	5.1	Typic Palehumults	F		Douglas-fir	H	VD	WD	0–30	L	Fine	Isomesic	Udic	6e	
Dester	365.1	Vitritorrandic Argixerolls					MD	WD	0–8	L	Fine-loamy	Frigid	Xeric	6e	
Deter	39.5	Pachic Argixerolls	C, R	H			VD	WD	0–30	L	Fine	Mesic	Xeric	2e, 2w	
Devada	5.5	Aridic Lithic Argixerolls					S	WD	0–50	L	Clayey	Mesic	Xeric	7s	
Deven	0.9	Lithic Argixerolls					S	WD	0–50		Clayey	Mesic	Xeric	6e	
Devoy	3.8	Xeric Argicryolls	F		Pinyon-juniper		MD	WD	2–50	L	Clayey-skeletal	Cryic	Xeric	6e	
Dewar	117.1	Xeric Argidurids					S	WD	0–50		Loamy	Mesic	Aridic	7s	
Diablopeak	47.8	Lithic Natrargids					S	WD	2–20		Clayey	Frigid	Aridic	7s	
Diaz	39.5	Xeric Haplargids					MD	WD	2–30		Fine	Mesic	Aridic	6e	
Dicecreek	7.2	Lithic Dystroxerepts	F		Douglas-fir	M	S	WD	30–60		Loamy	Mesic	Xeric	6e	

(continued)

Series name	Area (km^2)	Subgroup	Land use[1]	Range site quality[2]	Forest indicator species	Forest site quality[3]	Depth class[4]	Drainage class[5]	Slope range (%)	Salinity class[6]	Particle-size class	Soil temp. regime	Soil moisture regime	Land capability class	
														No mgmt	With mgmt
Dickerson	33.2	Lithic Xerorthents					VS	WD	3–90		Loamy	Mesic	Xeric	7s	
Dickle	25.4	Lithic Haplocryolls					S	WD	3–35		Loamy	Cryic	Xeric	6e	
Diffin	6.0	Vertic Palexerolls					D-VD	WD	30–60		Clayey-skeletal	Frigid	Xeric	6e	
Digger	786.8	Dystric Eutrudepts	F		Douglas-fir	M, H	MD	WD	3–90	L	Loamy-skeletal	Mesic	Udic	6e	
Digit	3.4	Typic Vitricryands	F		Engelmann spruce		VD	WD	0–30		Ashy over loamy	Cryic	Udic	6c	
Dilman	24.0	Aquandic Cryaquolls	C, R	H			VD	PD	0–1		Loamy over ashy or ashy-pumiceous	Cryic	Aquic	4w	
Divers	51.0	Typic Haplocryands	F		Douglas-fir	L, M	VD	WD	3–90	L	Medial-skeletal	Cryic	Udic	6s	
Dixiejett	23.3	Typic Argixerolls					D	WD	30–90		Loamy-skeletal	Mesic	Xeric	6e	
Dixon	61.7	Xeric Haplocambids					VD	WD	0–15	L	Fine-loamy over sandy or sandy-skeletal	Mesic	Aridic	6e, 6s, 7s	
Dixonville	266.3	Pachic Ultic Argixerolls	F		Douglas-fir	L–H	MD	WD	3–60	L	Fine	Mesic	Xeric	3e-4e, 6e	
Diyou	0.5	Fluvaquentic Haploxerolls	C				VD	SPD	0–2		Fine-loamy	Mesic	Xeric	3w	2w
Dobbins	11.2	Humic Dystrudepts	F		Douglas-fir	M, H	MD	WD	69–90		Clayey-skeletal	Mesic	Udic	7e	
Dodes	18.4	Aridic Argixerolls	C, R	H			MD	WD	0–15	L	Fine-loamy	Mesic	Xeric	3s, 4e	
Dogmountain	24.3	Vitrixerandic Haplodurids					MD	WD	4–20		Ashy-skeletal	Frigid	Aridic	6e	
Dogtown	37.9	Vitrandic Haploxerolls	F		Ponderosa pine		D-VD	WD	12–80		Loamy-skeletal	Frigid	Xeric	6e	
Dollarlake	3.4	Vitrixerandic Humicryepts					VD	WD	0–60		Loamy-skeletal	Cryic	Xeric	6s	
Dompier	2.0	Aquic Fragixeralfs	F		Douglas-fir	M	MD	SPD	30–60		Fine-silty	Mesic	Xeric	6e	
Donegan	20.1	Humic Dystroxerepts	F		Douglas-fir	L–H	MD	WD	3–65		Loamy-skeletal	Frigid	Xeric	6e	
Donnybrook	23.5	Calcidic Haploxerolls					S	WD	10–60		Loamy	Mesic	Xeric	6e	
Doublecreek	15.3	Vitrandic Haploxerolls	C, F		Ponderosa pine		VD	WD	2–60		Coarse-loamy	Mesic	Xeric	3e-4e, 6e	
Doubleo	20.4	Fluvaquentic Vertic Endoaquolls	R	M			VD	PD	0–2		Clayey over loamy	Frigid	Aquic	5w	
Douthit	29.4	Typic Vitricryands	F		Douglas-fir	M	VD	WD	0–65		Ashy-skeletal	Cryic	Udic	6s	
Dowde	37.7	Humic Dystrudepts	F		Douglas-fir	M, H	D	WD	30–90		Fine-silty	Mesic	Udic	6e	

(continued)

Series name	Area (km²)	Subgroup	Land use[1]	Range site quality[2]	Forest indicator species	Forest site quality[3]	Depth class[4]	Drainage class[5]	Slope range (%)	Salinity class[6]	Particle-size class	Soil temp. regime	Soil moisture regime	Land capability class	
														No mgmt	With mgmt
Downards	6.7	Andic Hapludolls	F		Grand fir		VD	WD	15–90		Loamy-skeletal	Frigid	Udic	6e, 7e	
Downeygulch	30.5	Vitrandic Haploxerepts	F		Ponderosa pine		MD	WD	0–30		Coarse-loamy	Frigid	Xeric	4e	
Doyn	248.6	Aridic Lithic Haploxerolls					VS	WD	2–30	L	Loamy	Frigid	Xeric	7s	
Drakesflat	49.4	Calciargidic Argixerolls					MD	WD	2–30		Fine	Frigid	Aridic	6e	
Drakespeak	7.8	Xeric Humicryepts					VD	WD	20–50		Loamy-skeletal	Cryic	Xeric	6e	
Drews	97.8	Pachic Argixerolls	R	H			VD	WD	0–35	L	Fine-loamy	Mesic	Xeric	3e, 4e	3e, 4e
Drewsey	258.2	Xeric Haplocambids					VD	WD	1–20	L	Coarse-loamy	Mesic	Aridic	6e	
Drewsgap	9.4	Typic Durixerolls	C, R				MD	WD	0–15	L	Fine-loamy	Mesic	Xeric	7s	3s
Drinkwater	74.6	Xeric Haplocambids					VD	WD	2–90		Loamy-skeletal	Mesic	Aridic	7s	
Droval	113.2	Sodic Aquicambids					VD	SPD	0–3	H	Fine	Mesic	Aridic	6s	
Drybed	19.6	Calciargidic Argixerolls	R	H			VD	WD	0–8	L	Fine-loamy	Mesic	Aridic	4e	3e
Dryck	27.3	Cumulic Haploxerolls	R	H			VD	WD	0–2	L	Coarse-loamy over sandy or sandy-skeletal	Mesic	Xeric	3c	3c
Dryhollow	30.1	Vitritorrandic Haploxerolls	R	H, M			VD	WD	0–55		Ashy	Mesic	Xeric	4e, 6e	
Duart	45.7	Typic Haploxerolls	R	H, M			MD	WD	1–55	L	Coarse-loamy	Mesic	Xeric	3e, 6e	
Dubakella	187.7	Mollic Haploxeralfs	F		Douglas-fir	M	MD	WD	5–75	L	Clayey-skeletal	Mesic	Xeric	7s	
Duckclub	46.6	Sodic Aquicambids					VD	SPD	0–2	H	Coarse-loamy	Frigid	Aridic	6s	
Ducklake	48.1	Typic Vitricryands	F		Engelmann spruce		D	WD	5–60		Ashy over loamy-skeletal	Cryic	Udic	6c	
Duff	72.0	Pachic Haplocryolls					D	WD	15–75		Fine-loamy	Cryic	Xeric	6e	
Dufur	62.0	Calcic Haploxerolls	C, R	H, M			D	WD	0–40	L	Coarse-loamy	Mesic	Xeric	2e, 3e, 6e	
Dulandy	42.7	Humic Dystrudepts	F		Douglas-fir	M, H	MD	WD	0–90		Loamy-skeletal	Isomesic	Udic	7e	
Dumont	179.4	Typic Palexerults	C, F		Douglas-fir	L–H	VD	WD	0–60	L	Fine	Mesic	Xeric	2e-4e	
Dunnlake	0.0	Aridic Lithic Argixerolls	R				S	WD	0–50		Clayey	Mesic	Aridic	6e	
Dunres	0.0	Vitrandic Durixerolls					S	WD	1–20	L	Clayey	Frigid	Xeric	6e	
Dunstan	111.6	Vitrandic Haploxeralfs	F		Douglas-fir		D	WD	0–90		Clayey-skeletal	Frigid	Xeric	6e	
Dupee	52.4	Aquultic Haploxeralfs	C, R				VD-D	SPD-MWD	2–30	L	Fine	Mesic	Xeric	3e, 4e	
Dupratt	167.4	Vitrandic Argixerolls	F		Ponderosa pine		MD	WD	20–50		Clayey-skeletal	Frigid	Xeric	6e	

(continued)

Series name	Area (km^2)	Subgroup	Land use[1]	Range site quality[2]	Forest indicator species	Forest site quality[3]	Depth class[4]	Drainage class[5]	Slope range (%)	Salinity class[6]	Particle-size class	Soil temp. regime	Soil moisture regime	Land capability class	
														No mgmt	With mgmt
Durkee	126.5	Calcic Argixerolls	F		Douglas-fir	M	MD	WD	1–60	L	Fine	Frigid	Xeric	4e, 6e	
Dyarock	1.4	Oxyaquic Vitricryands					VD	MWD	2–20		Ashy	Cryic	Udic	6e	
Ead	26.0	Pachic Humudepts	F		Douglas-fir		MD	WD	5–60		Fine	Mesic	Udic	6e	
Eaglespring	11.0	Vitrandic Haploxerolls					D	WD	12–80		Loamy-skeletal	Mesic	Xeric	4e, 6e, 7e	
Eastlakesbasin	7.9	Lithic Humicryepts	F		Grand fir		S	WD	0–90		Loamy-skeletal	Cryic	Udic	7s	
Eastpine	17.7	Vitrandic Haploxerolls	F		Ponderosa pine		MD	WD	15–90		Loamy-skeletal	Frigid	Xeric	6s, 7e	
Ecola	42.2	Andic Humudepts	F		Douglas-fir	H	MD	WD	0–90	L	Fine-silty	Isomesic	Udic	6e, 7e	
Edemaps	132.2	Argiduridic Durixerolls					MD	WD	2–20	L	Fine	Frigid	Xeric	6e	
Edenbower	27.3	Lithic Ultic Haploxerolls					VS	WD	3–60		Clayey	Mesic	Xeric	7s	
Edson	22.0	Typic Palehumults	F		Douglas-fir	M, H	VD	WD	0–30		Fine	Mesic	Udic	6e	
Eggleson	5.6	Oxyaquic Haploxerolls	C, F		Ponderosa pine		VD	MWD	0–2		Sandy-skeletal	Frigid	Xeric	4s	
Eglirim	51.5	Aridic Argixerolls					VD	WD	20–50	L	Clayey-skeletal	Mesic	Xeric	6e	
Egyptcreek	111.1	Vitrandic Haploxerolls	F		Ponderosa pine		MD	WD	20–60		Loamy-skeletal	Frigid	Xeric	6e	
Eightlar	62.7	Typic Dystroxerepts	F		Douglas-fir	L, M	VD	WD	3–90	L	Clayey-skeletal	Mesic	Xeric	3e, 7s	
Eilertsen	59.7	Ultic Hapludalfs	C, F		Douglas-fir	H	VD	WD	0–7	L	Fine-silty	Mesic	Udic	2c	
Ekoms	4.1	Typic Haplohumults	F		Douglas-fir	H	VD	WD	0–12	L	Fine-loamy	Isomesic	Udic	3e	3e
Elbowcreek	31.7	Alfic Udivitrands	F		Grand fir		VD	WD	0–15		Ashy over loamy	Frigid	Udic	6c	
Elijah	284.8	Xeric Argidurids	C				MD	WD	0–20		Fine-silty	Mesic	Aridic	6c	4e
Elkhorncrest	47.5	Lithic Dystrocryepts	F		Subalpine fir		S	WD	0–90		Loamy-skeletal	Cryic	Udic	7e	
Ellisforde	36.4	Calcidic Haploxerolls	C, R	H			VD	WD	0–65	L	Coarse-silty	Mesic	Xeric	3e	2e
Ellum	5.9	Xereptic Haplodurids					MD	WD	2–12		Loamy-skeletal	Mesic	Aridic	6e	
Elsie	14.2	Typic Haplohumults	C, F		Douglas-fir	H	VD	WD	0–15	L	Fine-silty	Mesic	Udic	2c, 3e, 4e	
Embal	29.7	Vitritorrandic Haploxerolls	R				VD	WD	0–5		Ashy	Frigid	Xeric	6c	
Emerson	1.9	Xeric Torriorthents	C				D	WD	0–12		Coarse-loamy over sandy or sandy-skeletal	Mesic	Aridic	6c	3e

(continued)

Series name	Area (km^2)	Subgroup	Land use[1]	Range site quality[2]	Forest indicator species	Forest site quality[3]	Depth class[4]	Drainage class[5]	Slope range (%)	Salinity class[6]	Particle-size class	Soil temp. regime	Soil moisture regime	Land capability class	
														No mgmt	With mgmt
Emily	35.4	Vitrandic Haploxerolls	F, R		Ponderosa pine		D-VD	WD	2–60		Loamy-skeletal	Mesic	Xeric	3s, 4e, 6e	4e
Encina	148.5	Calciargidic Argixerolls	R	H, M			D	WD	1–65	L	Fine	Mesic	Aridic	3e, 4e, 6e	4e
Endcreek	22.5	Typic Vitrixerands	F		Douglas-fir		MD	WD	0–60		Ashy over loamy-skeletal	Frigid	Xeric	6c	
Endersby	36.8	Cumulic Haploxerolls	C, R	H			D	SED	0–3	L	Coarse-loamy	Mesic	Xeric	2c, 2e, 3s	1, 3s
Enko	506.7	Durinodic Xeric Haplocambids					VD	WD	0–30	M	Coarse-loamy	Mesic	Aridic	6e	3e
Era	106.0	Vitritorrandic Haploxerolls	R	H, L			VD-D	WD	0–40	L	Coarse-loamy	Mesic	Xeric	4e, 6s	3e
Erakatak	495.1	Vitrandic Argixerolls					MD	WD	2–80	L	Clayey-skeletal	Frigid	Xeric	6e, 7e	
Ermabell	1.8	Humic Vitrixerands	F		Ponderosa pine		VD	WD	0–3		Ashy	Frigid	Xeric	6c	
Escondia	18.8	Alfic Vitrixerands	F		Grand fir		VD	WD	30–60		Ashy-skeletal over loamy-skeletal	Frigid	Xeric	6e	
Esquatzel	13.4	Torrifluventic Haploxerolls	C, R	H			VD, D	WD	0–5	L	Coarse-silty	Mesic	Xeric	3c	1
Etelka	192.0	Oxyaquic Dystrudepts	F		Douglas-fir	M, H	VD	MWD	7–70	L	Fine	Mesic	Udic	7e	
Ettersburg	1.2	Humic Dystrudepts	C, F		Douglas-fir	M, H	VD	WD	0–3		Fine-loamy	Isomesic	Udic	2e	2e
Euchrand	13.0	Lithic Dystrudepts	F		Douglas-fir	L	S	WD	30–90		Loamy-skeletal	Frigid	Udic	7e	
Euchre	2.0	Alic Endoaquands	C				VD	SPD	0–7		Medial over loamy	Isomesic	Aquic	3e, 3w	
Evans	38.8	Cumulic Haploxerolls	F		Ponderosa pine		D	WD	0–3	L	Coarse-loamy	Mesic	Xeric	3w, 4w	2w
Evick	6.2	Lithic Xeropsamments	F		Ponderosa pine	L	S	SED	30–65		Sandy	Mesic	Xeric	6s	
Exfo	201.6	Lithic Torriorthents	R				VS	SED	2–90		Sandy-skeletal	Mesic	Aridic	7s	
Fairylawn	24.4	Abruptic Durixeralfs	R				MD	WD	1–25		Fine	Mesic	Xeric	4e	
Faloma	8.9	Fluvaquentic Endoaquolls	R	H, L			VD	PD-VPD	0–3	L	Coarse-silty over sandy or sandy-skeletal	Mesic	Aquic	4w, 6w	2w
Fantz	30.5	Pachic Ultic Haploxerolls	F		Douglas-fir	L	MD	WD	30–90		Loamy-skeletal	Mesic	Xeric	6 s	
Farva	44.4	Typic Haploxerepts	F		Douglas-fir	L, M	MD	WD	3–70	L	Loamy-skeletal	Frigid	Xeric	6e, 7e	

(continued)

Series name	Area (km^2)	Subgroup	Land use[1]	Range site quality[2]	Forest indicator species	Forest site quality[3]	Depth class[4]	Drainage class[5]	Slope range (%)	Salinity class[6]	Particle-size class	Soil temp. regime	Soil moisture regime	Land capability class	
														No mgmt	With mgmt
Fawceter	218.2	Pachic Fulvudands	F		Douglas-fir	M	D	WD	5–90		Medial-skeletal	Isofrigid	Udic	7e	
Fawnspring	34.6	Vitrandic Palexerolls	F		Ponderosa pine	L	D	WD	2–30		Fine	Mesic	Xeric	4e	
Feaginranch	10.1	Aquandic Cryaquolls	R	H			VD	VPD	0–2		Fine	Cryic	Aquic	4w	
Felcher	0.7	Xeric Haplocambids					MD	WD	5–70	L	Loamy-skeletal	Mesic	Aridic	6e, 6s, 7s	
Feltham	694.2	Xeric Torriorthents	C, R	H			D	WD-SED	0–25		Sandy	Mesic	Aridic	3e, 3s	
Fendall	14.6	Andic Humudepts	F		Douglas-fir	M, H	MD	WD	3–85		Fine	Isomesic	Udic	4e, 6e	
Ferguson	86.3	Typic Udivitrands	F		Douglas-fir	M	VD	WD	2–60		Ashy over loamy-skeletal	Frigid	Udic	6e	
Fernhaven	10.4	Typic Paleudults	F		Douglas-fir	H	VD	WD	3–75	L	Fine-loamy	Mesic	Udic	6e	
Fernwood	77.9	Andic Humudepts	F		Douglas-fir	M	MD	WD	5–90		Loamy-skeletal	Frigid	Udic	6s	
Ferrelo	71.1	Typic Humudepts	F		Douglas-fir	H	VD	WD	0–60	L	Coarse-loamy	Isomesic	Udic	3e, 4e, 6e	
Fertaline	23.5	Abruptic Xeric Argidurids	R				MD	WD	0–15	L	Fine	Frigid	Aridic	6e	
Fiddler	84.6	Typic Argixerolls					MD	WD	2–50		Clayey-skeletal	Mesic	Xeric	6e	
Final	3.0	Xerertic Natrargids					VD	SPD	0–2		Fine	Mesic	Aridic	6s	
Finsel	2.9	Vitrandic Argixerolls					MD	WD	12–70		Clayey-skeletal	Frigid	Xeric	6e	
Firelake	20.7	Lithic Xeric Torriorthents					VS	SED	2–20		Loamy	Mesic	Aridic	7s	
Fitzwater	186.8	Aridic Haploxerolls					VD-D	WD	0–70	L	Loamy-skeletal	Frigid	Xeric	6e, 6s	
Fivebeaver	548.7	Lithic Ultic Haploxerolls	F		Ponderosa pine	M	S	WD	0–90	L	Loamy-skeletal	Frigid	Xeric	6s, 7s	
Fivebit	160.7	Lithic Ultic Haploxerolls	F		Ponderosa pine	L	S	WD	0–90		Loamy-skeletal	Frigid	Xeric	7s	
Fiverivers	5.7	Andic Dystrudepts	F		Douglas-fir	M	MD	WD	5–60		Fine-loamy	Frigid	Udic	6e	
Flagstaff	100.9	Typic Aquisalids					VD	MWD	0–1	H	Ashy	Frigid	Aridic	6e, 6s	
Flane	36.8	Humic Dystrudepts	F		Douglas-fir	M	D	WD	3–75		Clayey-skeletal	Frigid	Udic	6e	
Flank	9.9	Lithic Xeric Torriorthents					VS-S	WD	1–20		Ashy-skeletal	Frigid	Aridic	7s	
Flarm	1.1	Ultic Palexeralfs	F		Douglas-fir		VD	SPD	0–15		Fine	Frigid	Xeric	6w	
Floke	247.1	Abruptic Xeric Argidurids					S	WD	2–15	L	Clayey	Frigid	Aridic	7s	
Floras	48.8	Humic Dystrudepts	F		Douglas-fir	M, H	D	WD	30–90		Fine	Isomesic	Udic	7e	

(continued)

Series name	Area (km^2)	Subgroup	Land use[1]	Range site quality[2]	Forest indicator species	Forest site quality[3]	Depth class[4]	Drainage class[5]	Slope range (%)	Salinity class[6]	Particle-size class	Soil temp. regime	Soil moisture regime	Land capability class	
														No mgmt	With mgmt
Flowerpot	9.6	Aquandic Humudepts	F		Western hemlock		VD	SPD	5–30		Fine	Isomesic	Udic	6e	
Flybow	0.9	Lithic Xerorthents					VS	WD	4–100		Loamy-skeletal	Mesic	Xeric	7s	
Flycatcher	9.6	Lithic Eutrudepts	F		Douglas-fir		S	WD	0–90		Loamy-skeletal	Frigid	Udic	7e	
Flycreek	41.5	Alfic Udivitrands	F		Grand fir	L, M	MD	WD	0–90		Ashy over clayey	Frigid	Udic	6e, 7e	
Flyvalley	3.3	Lithic Udivitrands	F		Grand fir		S	WD	2–60		Ashy	Frigid	Udic	6e	
Foehlin	41.2	Typic Argixerolls	C				VD	WD	0–12	L	Fine-loamy	Mesic	Xeric	4c, 4e	2 s, 3e
Foleylake	11.3	Abruptic Xeric Argidurids					MD	WD	2–15	L	Fine	Frigid	Aridic	6e	
Fopiano	121.3	Vitrandic Argixerolls					S	WD	2–50	L	Clayey	Frigid	Xeric	6e	
Fordice	4.8	Ultic Argixerolls	C				VD	WD	0–12	L	Loamy-skeletal	Mesic	Xeric	6s	6s
Fordney	124.3	Torripsammentic Haploxerolls	R	H			VD	ED	0–20	L		Mesic	Xeric	3e, 3w, 4e	
Formader	154.4	Alic Hapludands	F		Douglas-fir	M, H	MD	WD	3–80	L	Medial over loamy	Mesic	Udic	6e	
Forshey	5.6	Vitrandic Argixerolls	F		Ponderosa pine		VD	WD	0–60		Fine-loamy	Frigid	Xeric	6c	
Fort rock	97.9	Vitritorrandic Haploxerolls					VD	SED	0–8	L	Ashy over sandy or sandy-skeletal	Frigid	Xeric	6e, 6s	
Fossilake	4.2	Aquandic Halaquepts					VD	SPD	0–1	H	Ashy	Frigid	Aquic	6s	
Fourbeaver	34.9	Lithic Ultic Argixerolls					S	WD	0–60		Clayey-skeletal	Frigid	Xeric	7s	
Fourthcreek	0.6	Typic Vitrixerands	F		Douglas-fir		MD	WD	15–30		Ashy over loamy	Frigid	Xeric	6c	
Fourwheel	168.7	Xeric Paleargids					MD	WD	3–40	L	Fine	Frigid	Aridic	6e, 7s	
Frailey	47.6	Typic Haploxerepts	F		Douglas-fir	L, M	D	WD	3–70		Coarse-loamy	Mesic	Xeric	6e, 7e	
Frankport	5.7	Typic Udipsamments					VD	ED	0–30		Sandy	Isomesic	Udic	7e	
Freels	13.9	Cumulic Haploxerolls	R	H			VD	WD	0–3	L	Coarse-loamy	Frigid	Xeric	3c	
Freewater	26.3	Fluventic Haploxerolls	C, R	L			D	SED	0–3		Sandy-skeletal	Mesic	Xeric	4s, 6s	
Freezener	264.7	Ultic Haploxeralfs	F		Douglas-fir	M	D	WD	0–65	L	Fine	Mesic	Xeric	4e	
Fremkle	86.9	Lithic Vitrixerands	R				S	WD	0–30		Ashy	Frigid	Xeric	6e	4e
Freznik	344.5	Xeric Paleargids					MD	MWD	2–30	L	Fine	Frigid	Aridic	6e	
Fritsland	192.3	Humic Dystrudepts	F		Douglas-fir	M, H	D	WD	30–90	L	Fine-loamy	Mesic	Udic	6e	
Frizzelcreek	1.3	Andic Humicryepts	F		Grand fir		MD	WD	15–30		Coarse-loamy	Cryic	Xeric	6e	

(continued)

Series name	Area (km^2)	Subgroup	Land use[1]	Range site quality[2]	Forest indicator species	Forest site quality[3]	Depth class[4]	Drainage class[5]	Slope range (%)	Salinity class[6]	Particle-size class	Soil temp. regime	Soil moisture regime	Land capability class	
														No mgmt	With mgmt
Frohman	172.0	Xeric Haplodurids	R	H			S	WD	0–20	L	Loamy	Mesic	Aridic	4e	
Fruitcreek	18.7	Vitrixerandic Humicryepts					MD	WD	0–90		Loamy-skeletal	Cryic	Xeric	7e	
Fryrear	10.7	Humic Vitrixerands					MD	WD	0–50		Ashy-skeletal	Frigid	Xeric	6e	
Fuego	6.1	Typic Haploxerolls	F		Ponderosa pine	L	MD	SED	3–40		Loamy-skeletal	Frigid	Xeric	7s	
Gacey	19.5	Argiduridic Durixerolls	R				S	WD	1–12		Clayey-skeletal	Mesic	Aridic	7s	4s
Gaib	169.4	Lithic Ultic Argixerolls					S	WD	0–80	L	Loamy-skeletal	Frigid	Xeric	7 s	
Gamelake	6.1	Typic Humudepts	F		Douglas-fir	M, H	VD	WD	0–90		Loamy-skeletal	Frigid	Udic	6e, 7e	
Gamgee	24.5	Haplic Natrargids					MD	WD	1–25		Fine-loamy	Mesic	Aridic	7s	
Gance	49.7	Durinodic Xeric Haplargids					VD	WD	2–50		Clayey-skeletal	Mesic	Aridic	7s	
Gap	38.3	Xeric Vitricryands	F		Ponderosa pine	M	D	WD	0–30		Ashy over loamy	Cryic	Xeric	6e	
Gapcot	4.6	Typic Haploxerepts	C, F		Douglas-fir	M	S	WD	3–60		Loamy	Mesic	Xeric	6s, 6e	
Garbutt	24.0	Typic Torriorthents	C, R	H			VD	WD	0–12	L	Coarse-silty	Mesic	Aridic	1, 2e	
Gardiner	7.2	Typic Udipsamments	C, R	H			VD	WD	0–3		Sandy	Mesic	Udic	4w	
Gardone	127.7	Vitritorrandic Haploxerolls					VD	ED	0–20	L	Ashy	Frigid	Xeric	6e	
Gauldy	7.2	Fluventic Humudepts	C, F		Douglas-fir		VD	SED	0–7		Coarse-loamy over sandy or sandy-skeletal	Isomesic	Udic	2s-4 s, 4w	
Gearhart	17.6	Psammentic Humudepts					VD	SED	0–30			Isomesic	Udic	4e, 6e	
Geebarc	1.0	Oxyaquic Haplocryolls	F		Engelmann spruce		VD	SPD	0–5		Sandy-skeletal	Cryic	Udic	6s	
Geisel	13.0	Typic Humudepts	F		Douglas-fir	H	D	WD	2–50	L	Fine	Isomesic	Udic	3e, 4e, 6e	
Geisercreek	63.3	Alfic Udivitrands	F		Grand fir	M	VD	WD	0–30		Ashy over clayey	Frigid	Udic	6e	
Gelderman	22.8	Xeric Haplohumults	C, F		Douglas-fir	M	MD	WD	2–30		Fine	Mesic	Udic	3e, 4e	
Gellatly	29.4	Pachic Argixerolls	C, F, R		Douglas-fir	M	VD	WD	3–60		Fine	Mesic	Xeric	2e, 4e, 6e	
Gelsinger	4.0	Calcic Pachic Argixerolls	C, R	H			VD	WD	2–25	L	Fine	Mesic	Xeric	2e, 3e	
Gem	6.9	Calcic Argixerolls	C, R				MD	WD	0–60		Fine	Mesic	Xeric	7e	
Genaw	7.8	Xeric Haplargids					S	WD	2–30		Loamy	Mesic	Aridic	7s	
Geppert	97.0	Typic Dystroxerepts	F		Douglas-fir	L, M	MD	WD	1–70	L	Loamy-skeletal	Mesic	Xeric	7e	

(continued)

Series name	Area (km^2)	Subgroup	Land use[1]	Range site quality[2]	Forest indicator species	Forest site quality[3]	Depth class[4]	Drainage class[5]	Slope range (%)	Salinity class[6]	Particle-size class	Soil temp. regime	Soil moisture regime	Land capability class	
														No mgmt	With mgmt
Getaway	194.8	Vitrandic Argixerolls	F		Ponderosa pine	L, M	D	WD	15–90		Loamy-skeletal	Frigid	Xeric	4e, 6e, 7e	
Gilispie	27.5	Lithic Argicryolls					S	WD	0–60	L	Loamy	Cryic	Ustic	7s	
Ginger	3.1	Typic Melanaquands	C, R	H			VD	SPD	0–15		Medial over clayey	Isomesic	Aquic	3e	
Ginsberg	79.6	Alic Hapludands	F		Douglas-fir	M, H	VD	WD	5–60	L	Medial over clayey	Mesic	Udic	6e	
Ginser	63.5	Pachic Haploxerolls					MD	WD	12–70		Loamy-skeletal	Frigid	Xeric	6e, 7e	
Giranch	5.4	Vitritorrandic Durixerolls					MD	WD	2–20		Ashy-skeletal	Frigid	Xeric	6e	
Giveout	6.6	Alic Hapludands	F		Douglas-fir	M	MD	WD	5–60		Medial	Frigid	Udic	6e	
Glasgow	1.3	Xeric Haplargids					MD	WD	2–60	L	Fine	Mesic	Aridic	6e	
Glassbutte	5.4	Vitritorrandic Argixerolls					VD	WD	20–65		Ashy-skeletal	Frigid	Xeric	6e	
Glaze	17.2	Xeric Vitricryands	F		Douglas-fir	M	D	WD	15–50		Ashy over loamy-skeletal	Cryic	Xeric	6e	
Glencabin	57.7	Vitrandic Haploxerolls	F		Ponderosa pine		MD	WD	15–65		Ashy-skeletal	Frigid	Xeric	6e	
Gleneden	9.8	Aquic Humudepts	F		Western hemlock	M	VD	SPD	0–20		Fine	Isomesic	Udic	3e	
Glide	0.9	Humic Vitrixerands	C, F		Ponderosa pine		VD	WD	0–3	L	Ashy	Mesic	Xeric	1	
Glohm	31.3	Typic Fragiudepts	F		Douglas-fir	H	MD	WD	3–30	L	Fine-silty	Mesic	Udic	6e	
Goble	261.5	Andic Fragiudepts	F		Douglas-fir	H	MD	MWD	2–80	L	Fine-silty	Mesic	Udic	6e	
Gochea	13.9	Argiduridic Argixerolls					VD	WD	2–30	L	Fine-loamy	Frigid	Xeric	6e	
Goldbeach	1.3	Lithic Humudepts					S	WD	0–60		Loamy-skeletal	Mesic	Udic	7s	
Goldrun	87.5	Xeric Torripsamments					VD	SED	0–30		Sandy	Mesic	Aridic	6e	
Golfer	12.7	Vitrandic Haploxerolls	F		Ponderosa pine		MD	WD	0–90		Loamy-skeletal	Frigid	Xeric	7s	
Goodin	51.2	Ultic Haploxeralfs	C, F, R		Douglas-fir		MD	WD	2–60		Fine	Mesic	Xeric	3e	
Gooding	169.6	Vertic Paleargids	R				D	WD	0–20		Fine	Mesic	Aridic	4e	
Goodlow	14.0	Andic Humicryepts	F		Douglas-fir	H	VD	WD	5–60	L	Loamy-skeletal	Cryic	Udic	7s	
Goodrich	21.3	Entic Haploxerolls	C, R	H			D	WD	0–12	L	Coarse-loamy	Mesic	Xeric	4e	
Goodtack	597.0	Vitritorrandic Durixerolls					S	WD	0–20	L	Ashy	Frigid	Xeric	6e	
Goodwin	26.2	Humic Dystroxerepts	F		Douglas-fir	H	D	WD	5–80	L	Loamy-skeletal	Frigid	Xeric	6s	

(continued)

Series name	Area (km^2)	Subgroup	Land use[1]	Range site quality[2]	Forest indicator species	Forest site quality[3]	Depth class[4]	Drainage class[5]	Slope range (%)	Salinity class[6]	Particle-size class	Soil temp. regime	Soil moisture regime	Land capability class	
														No mgmt	With mgmt
Goolaway	35.0	Typic Dystroxerepts	F		Douglas-fir	L, M	MD	WD	12–50		Fine-silty	Mesic	Xeric	4e, 6e	
Goose creek	20.7	Cumulic Haploxerolls	C				VD	MWD-SPD	0–6		Fine-loamy	Mesic	Xeric	6c	2e
Goose lake	16.7	Aquandic Argialbolls	R	H, M			VD	PD	0–1	L	Fine	Mesic	Udic	3w, 5w	3w
Gooserock	62.6	Vitritorrandic Haploxerolls					VD	WD	2–90		Loamy-skeletal	Mesic	Aridic	7s	
Gorhamgulch	32.2	Typic Udivitrands	F		Grand fir		VD	WD	0–90		Ashy over loamy	Frigid	Udic	6e	
Gosney	235.6	Lithic Torripsamments	C, R				S	SED	0–15	L	Ashy	Mesic	Aridic	7e	4e
Gradon	81.1	Argiduridic Durixerolls					MD	WD	0–8		Fine-loamy	Frigid	Xeric	6e	
Granitemountain	27.5	Vitrixerandic Humicryepts					VD	WD	15–90		Loamy-skeletal	Cryic	Xeric	6s	
Grassmountain	4.9	Andic Dystrudepts	F		Douglas-fir	M	VD	WD	5–60		Fine-loamy	Frigid	Udic	6e	
Grassyknob	11.1	Andic Humudepts	F		Douglas-fir	H	MD	WD	0–60	L	Fine-loamy	Isomesic	Udic	6e	
Gravden	12.4	Xeric Haplodurids					S	WD	4–40		Loamy-skeletal	Mesic	Aridic	6e	
Gravecreek	57.8	Typic Dystroxerepts	F		Douglas-fir	L, M	MD	WD	3–90		Loamy-skeletal	Mesic	Xeric	6s, 7e	
Greengulch	16.1	Ultic Haploxeralfs	F		Douglas-fir		MD	WD	0–60		Fine	Mesic	Xeric	4e	
Greenleaf	24.9	Xeric Calciargids	C, R	H			VD	WD	0–30	L	Fine-silty	Mesic	Xeric	1, 2e	
Greenmountain	54.9	Vitritorrandic Durixerolls					MD	WD	1–15		Ashy	Frigid	Xeric	6e	
Greenscombe	45.6	Typic Haploxerolls	R	H			MD	WD	2–35		Fine-loamy	Mesic	Xeric	4e	
Greggo	30.0	Lithic Eutrudepts	F		Douglas-fir		S	WD	0–90		Loamy-skeletal	Mesic	Udic	7e	
Gregory	13.5	Typic Argiaquolls	C, R	H			D	PD	0–3	L	Fine	Mesic	Aquic	4w	2w
Grell	12.5	Lithic Haploxerolls					S	WD	7–50		Loamy-skeletal	Mesic	Xeric	7s	
Grenet	13.4	Typic Vitrixerands	F		Douglas-fir	L	MD	SED	0–65		Ashy-skeletal	Frigid	Xeric	6s	
Greystoke	66.4	Pachic Ultic Argixerolls	F		Douglas-fir	L	D	WD	1–75		Loamy-skeletal	Frigid	Xeric	6e, 7e	
Gribble	59.9	Haplic Durixerolls					MD	WD	5–20		Clayey-skeletal	Mesic	Xeric	6e	
Grindbrook	43.0	Oxyaquic Humudepts	C, F		Douglas-fir	H	VD	MWD	0–30	L	Fine-silty	Isomesic	Udic	2e, 3e, 4e	2e
Grousehill	194.4	Oxyaquic Duricryands	F		White fir		MD	MWD	0–35		Medial-skeletal	Cryic	Udic	7s	
Grouslous	50.2	Lithic Dystrudepts	F		Douglas-fir	M	S	WD	60–90		Loamy-skeletal	Mesic	Udic	7s	

(continued)

Series name	Area (km^2)	Subgroup	Land use[1]	Range site quality[2]	Forest indicator species	Forest site quality[3]	Depth class[4]	Drainage class[5]	Slope range (%)	Salinity class[6]	Particle-size class	Soil temp. regime	Soil moisture regime	Land capability class	
														No mgmt	With mgmt
Grubcreek	19.0	Vitrandic Haploxerolls	F		Ponderosa pine		MD	WD	0–60		Loamy-skeletal	Frigid	Xeric	6s	
Guano	40.0	Xeric Haplargids					S	WD	2–15		Loamy	Frigid	Aridic	6e	
Guerin	4.2	Lithic Humudepts	F		Douglas-fir	M	S	WD	0–90		Loamy-skeletal	Isomesic	Udic	7e	
Gulliford	12.2	Oxyaquic Udorthents					VD	PD	0–5		Sandy-skeletal	Frigid	Udic	7 s	
Gumble	173.2	Xeric Haplargids					S	WD	2–50	L	Clayey	Mesic	Aridic	6e	
Gurdane	190.0	Pachic Argixerolls	R	H, M			MD	WD	0–45	L	Clayey-skeletal	Mesic	Xeric	3e, 4e, 6e	3e
Gurlidawg	2.4	Xeric Vitricryands	F		Lodgepole pine	M	MD	WD	4–50		Ashy-skeletal	Cryic	Xeric	6e	
Gustin	69.0	Aquic Palehumults	F		Douglas-fir	H	VS	WD	2–90		Fine	Mesic	Udic	5e, 6e	
Gutridge	216.7	Typic Udivitrands	F		Grand fir	L	D	WD	0–90		Ashy over loamy-skeletal	Frigid	Udic	6e	
Gwin	649.5	Lithic Argixerolls					S	WD	0–90	L	Loamy-skeletal	Mesic	Xeric	7e, 7s	
Gwinly	990.8	Lithic Argixerolls					S	WD	2–120	L	Clayey-skeletal	Mesic	Xeric	7e, 7s	
Haar	279.0	Xeric Torriorthents					VS	WD	8–50		Loamy	Mesic	Aridic	7s	
Habenome	5.2	Typic Cryaquolls					VD	PD	0–30		Coarse-loamy	Cryic	Aquic	6c	
Hack	55.9	Calcic Argixerolls	C, R				VD	WD	2–20	L	Fine-loamy	Mesic	Xeric	7s	2e
Hackwood	91.9	Pachic Haplocryolls	F		Trembling aspen		VD	WD	2–80		Fine-loamy	Cryic	Xeric	6e	
Haflinger	7.1	Entic Humudepts	F		Douglas-fir	H	VD	ED	0–5		Sandy-skeletal	Mesic	Udic	6s	
Hager	36.7	Xeric Argidurids					MD	WD	1–15	L	Fine-loamy	Mesic	Aridic	6e, 7s	
Haines	35.6	Typic Endoaquepts	C, R	H			D	PD	0–2		Coarse-silty	Mesic	Aquic	4w	
Halfway	3.1	Typic Haploxererts	C, R	H			VD	MWD	0–3		Very-fine	Mesic	Xeric	3s	
Hall ranch	107.6	Vitrandic Haploxerolls	F		Ponderosa pine		MD	WD	2–65	L	Fine-loamy	Frigid	Xeric	6c, 6e, 7e	
Hallihan	14.7	Xeric Vitricryands	F		White fir	M	VD	SED	0–60		Ashy-skeletal	Cryic	Xeric	6s	
Hammersley	3.1	Pachic Argicryolls					D	WD	15–70	L	Fine	Cryic	Xeric	6e	
Hankins	361.7	Vertic Palexerolls	F		Ponderosa pine	L, M	VD-D	WD	1–70	L	Fine	Frigid	Xeric	6e, 7e	
Hanning	76.6	Pachic Argixerolls	C				VD, D	WD	0–60		Fine-silty	Mesic	Xeric	6e	

(continued)

Series name	Area (km²)	Subgroup	Land use[1]	Range site quality[2]	Forest indicator species	Forest site quality[3]	Depth class[4]	Drainage class[5]	Slope range (%)	Salinity class[6]	Particle-size class	Soil temp. regime	Soil moisture regime	Land capability class	
														No mgmt	With mgmt
Hapgood	39.8	Pachic Haplocryolls					D	WD	2–75		Loamy-skeletal	Cryic	Xeric	6e	
Happus	2.4	Vitrandic Haploxerepts	F, R		Ponderosa pine	L	MD	SED	12–30		Ashy-pumiceous	Mesic	Xeric	6s	
Harana	21.2	Cumulic Haploxerolls	C, R	H			D	MWD	0–3	L	Fine-silty	Mesic	Xeric	2c, 3s	
Harcany	119.4	Pachic Haplocryolls					VD	WD	2–75	L	Loamy-skeletal	Cryic	Xeric	6e	
Hardscrabble	30.9	Aquic Palexeralfs	F		Douglas-fir		D	SPD	2–20		Fine	Mesic	Xeric	3w, 4e	
Hardtrigger	57.6														
Harl	126.3	Typic Udivitrands	F		Grand fir		VD	WD	30–90		Ashy-skeletal over loamy-skeletal	Frigid	Udic	6e, 7e	
Harlow	394.7	Lithic Argixerolls					S	WD	2–90	L	Clayey-skeletal	Frigid	Xeric	7s	
Harriman	61.7	Pachic Argixerolls	C, R	H			VD, D	WD	0–35	L	Fine-loamy	Mesic	Xeric	2c, 3e	
Harrington	206.5	Typic Humudepts	F		Douglas-fir	M, H	MD	WD	3–90	L	Loamy-skeletal	Mesic	Udic	6e, 7e	
Harslow	190.1	Alic Hapludands	F		Douglas-fir	M, H	MD	WD	5–90	L	Medial-skeletal	Mesic	Udic	6s, 7s, 7e	
Hart	119.4	Duric Palexerolls					D	WD	2–50	L	Fine	Frigid	Xeric	6e, 7s	
Hart camp	39.4	Aridic Argixerolls					S	WD	4–50	L	Loamy	Frigid	Xeric	6e	
Hasshollow	19.5	Alfic Vitricryands	F		Grand fir	M	VD	WD	0–90		Ashy-skeletal over loamy-skeletal	Cryic	Udic	6c	
Hayespring	73.6	Vitritorrandic Durixerolls					MD	WD	0–20		Fine-loamy	Frigid	Xeric	6e	
Haynap	13.9	Humic Vitricryands	F		Ponderosa pine		VD	SED	0–70		Ashy-skeletal	Cryic	Udic	6s	
Haystack	36.9	Calcidic Haploxerolls					VD	WD	0–15		Loamy-skeletal	Mesic	Xeric	6s	
Hazelair	175.4	Vertic Haploxerolls	R				MD	SPD	2–35	L	Very-fine	Mesic	Xeric	3e, 4e	
Hazelcamp	52.2	Typic Haplohumults	F		Douglas-fir	H	MD	WD	0–30		Fine	Mesic	Udic	6e	
Hebo	25.2	Typic Humaquepts	C, F		Douglas-fir	H	VD	PD	0–7		Fine	Isomesic	Aquic	4w	
Heceta	45.1	Typic Psammaquents	R	H			VD	PD	0–3			Isomesic	Aquic	4w	
Hehe	115.6	Vitrandic Argixerolls	F		Ponderosa pine	L, M	MD	WD	0–65	L	Loamy-skeletal	Mesic	Xeric	6e	
Helmick	30.8	Vertic Haploxerepts	C, R	H, M			VD	SPD	3–50		Very-fine	Mesic	Xeric	3e, 4e, 6e	
Helphenstein	85.8	Sodic Aquicambids					VD	SPD	0–5	H	Fine-loamy	Mesic	Aridic	6w, 6s	
Helter	10.1	Xeric Vitricryands	F		Grand fir	L, M	VD-D	WD	2–65	L	Ashy over loamy	Cryic	Xeric	6e, 7e	

(continued)

Series name	Area (km^2)	Subgroup	Land use[1]	Range site quality[2]	Forest indicator species	Forest site quality[3]	Depth class[4]	Drainage class[5]	Slope range (%)	Salinity class[6]	Particle-size class	Soil temp. regime	Soil moisture regime	Land capability class	
														No mgmt	With mgmt
Helvetia	76.8	Ultic Argixerolls	C, F		Douglas-fir	H	VD	MWD	0–30	L	Fine	Mesic	Xeric	2e, 3e, 4e	
Hembre	205.5	Andic Humudepts	F		Douglas-fir	M,H	D	WD	3–90	L	Fine-loamy	Mesic	Udic	6e, 7e	
Hemcross	403.3	Alic Hapludands	F		Douglas-fir	M, H	VD	WD	3–90	L	Medial	Mesic	Udic	6e, 7e	
Henkle	159.0	Lithic Vitrixerands	F		Ponderosa pine	L	S	SED	0–65	L	Ashy-skeletal	Frigid	Xeric	6e, 7e	
Henley	63.1	Aquic Haplodurids	R	H			S	SED	0–15		Coarse-loamy	Mesic	Aridic	3w, 4w	
Henline	86.1	Typic Humicryepts	F		Douglas-fir	M	MD	WD	6–90		Loamy-skeletal	Cryic	Udic	6s	
Heppsie	38.6	Vertic Haploxerolls					MD	WD	35–70		Fine	Mesic	Xeric	6e	
Hermiston	67.1	Cumulic Haploxerolls	C, R	H			D	WD	0–3	L	Coarse-silty	Mesic	Xeric	2c, 2e	1
Hershal	19.0	Cumulic Endoaquolls	R	H			VD	PD	0–2	L	Coarse-silty over sandy or sandy-skeletal	Mesic	Aquic	3w	3w
Hesslan	93.8	Typic Haploxerolls					MD	WD	5–70	L	Coarse-loamy	Mesic	Xeric	7s	
Hezel	12.3	Xeric Torriorthents	C				VD	SED	0–60		Sandy over loamy	Mesic	Aridic	7e	4e
Hibbard	38.3	Palexerollic Durixerolls	R	H			MD	WD	2–12	L	Fine	Mesic	Xeric	4e	3e
Highcamp	86.5	Typic Haplocryands	F		Douglas-fir	M	D	WD	5–90		Medial-skeletal	Cryic	Udic	7s	
Highhorn	10.9	Vitrandic Haploxerepts	F		Ponderosa pine	L	D	WD	12–75		Loamy-skeletal	Frigid	Xeric	6e, 7e	
Hillsboro	15.2	Ultic Argixerolls	C, F		Douglas-fir		D	WD	0–20	L	Fine-silty	Mesic	Xeric	1, 2e, 3e	
Hilltish	24.4	Typic Dystroxerepts	F		Douglas-fir	M	MD	WD	30–90		Loamy-skeletal	Mesic	Xeric	6e, 7e	
Hinton	3.6	Duric Torriorthents					VD	WD	0–5		Coarse-loamy over sandy or sandy-skeletal	Mesic	Aridic	4s	
Hobit	10.4	Typic Fulvicryands	F		Douglas-fir	L, M	MD	WD	12–60		Medial	Cryic	Udic	7e	
Hoffer	18.8	Typic Vitricryands	F		Engelmann spruce		D-VD	WD	15–60		Ashy over loamy	Cryic	Udic	4e	
Holcomb	23.8	Typic Argialbolls	C				VD	SPD	0–3	L	Fine	Mesic	Aquic	3w	3w
Holderman	83.7	Typic Haplocryands	F		Douglas-fir	M	MD	WD	0–90		Medial-skeletal	Cryic	Udic	6s	
Holland	14.1	Ultic Haploxeralfs	F		Douglas-fir	M	VD	WD	2–75	L	Fine-loamy	Mesic	Xeric	4e, 6e	2e

(continued)

Series name	Area (km^2)	Subgroup	Land use[1]	Range site quality[2]	Forest indicator species	Forest site quality[3]	Depth class[4]	Drainage class[5]	Slope range (%)	Salinity class[6]	Particle-size class	Soil temp. regime	Soil moisture regime	Land capability class	
														No mgmt	With mgmt
Holmzie	66.6	Aridic Argixerolls					MD	WD	0–30	L	Fine	Mesic	Xeric	6e	
Homefield	43.4	Cumulic Endoaquolls	R	M			VD	VPD	0–1		Fine-silty	Frigid	Aquic	5w	
Homehollow	6.5	Vitritorrandic Durixerolls					MD	WD	0–20	L	Fine-loamy	Mesic	Xeric	6s	
Hondu	77.2	Andic Haploxerepts	F		Ponderosa pine		VD	WD	0–90		Loamy-skeletal	Frigid	Xeric	6s	
Honeygrove	696.8	Typic Palehumults	F		Douglas-fir	M, H	VD	WD	0–75	L	Fine	Mesic	Udic	4e, 6e, 7e	
Honeymooncan	19.6	Alfic Udivitrands	F		Douglas-fir	M	D	WD	0–60		Ashy over loamy-skeletal	Frigid	Udic	6c	
Hood	14.7	Ultic Haploxeralfs	C, F		Douglas-fir		VD	WD	0–65	L	Fine-loamy	Mesic	Xeric	1, 2e, 3e, 4e	
Hooly	18.8	Typic Endoaquands	C, R	H			VD	SPD	0–2		Medial over loamy	Mesic	Aquic	4w	3w
Hoopal	11.8	Typic Duraquolls	C, R	L			MD	SPD	0–2		Coarse-loamy	Mesic	Aquic	4s	3s
Hooskanaden	24.6	Andic Hapludalfs					VD	SPD	0–60		Fine	Isomesic	Udic	6e	
Horeb	28.4	Oxyaquic Humudepts	F		Douglas-fir		VD	MWD	0–35		Fine-loamy	Mesic	Udic	3e	
Horning	17.6	Vitrandic Torriorthents					VD	SED	0–20		Ashy	Frigid	Aridic	6e, 6s	
Horseprairie	10.0	Andic Humudepts	F		Douglas-fir	H	VD	WD	0–30	L	Fine-loamy	Isomesic	Udic	3e	
Hosley	16.3	Natric Duraquolls	C, R	H			MD	SPD	0–1		Fine-loamy	Mesic	Aquic	4w	
Hot lake	80.7	Aquic Haploxerands	C, R	H			VD	SPD	0–2	L	Medial over loamy	Mesic	Xeric	2w	2w
Housefield	45.9	Cumulic Endoaquolls	R	M			VD	VPD	0–1		Fine-silty	Frigid	Aquic	5w	
Houstake	71.2	Vitritorrandic Haploxerolls	R	L			VD	WD	0–8	L	Coarse-loamy	Mesic	Xeric	6e	3e
Houstenader	3.0	Aquic Argiudolls	R				VD	SPD	0–60		Fine-loamy	Mesic	Udic	6e	
Howash	163.0	Humic Udivitrands	F		Ponderosa pine	M	VD	SED	12–65	L	Ashy-skeletal	Frigid	Udic	6e	
Howcan	4.0	Typic Argixerolls					D-VD	WD	12–60		Loamy-skeletal	Frigid	Xeric	7e	
Howmeadows	15.4	Vertic Epiaqualfs	R	H			MD	PD	0–3		Fine	Frigid	Aquic	4w	
Hoxie	7.4	Humic Cryaquepts	C, R	M			VD	PD	0–1		Fine-silty	Cryic	Aquic	5w	5w
Huberly	13.5	Typic Fragiaquepts	R	H			VD	PD	0–3	L	Fine-silty	Mesic	Aquic	3w	
Hudspeth	12.6	Ultic Palexerolls					MD	WD	12–60		Clayey-skeletal	Frigid	Xeric	6s	
Huffling	0.5	Typic Umbraquults	C				D	PD	0–3		Fine	Isomesic	Aquic	4w	

(continued)

Series name	Area (km^2)	Subgroup	Land use[1]	Range site quality[2]	Forest indicator species	Forest site quality[3]	Depth class[4]	Drainage class[5]	Slope range (%)	Salinity class[6]	Particle-size class	Soil temp. regime	Soil moisture regime	Land capability class	
														No mgmt	With mgmt
Hukill	30.3	Typic Rhodoxeralfs	F		Ponderosa pine	M	D	WD	1–12	L	Fine	Mesic	Xeric	4e	3e
Hullt	21.5	Typic Humixerepts	F		Douglas-fir	H	D	WD	2–60	L	Fine-loamy	Mesic	Xeric	3e, 4e, 6e	
Humarel	107.1	Vitrandic Argixerolls	F		Ponderosa pine		MD	WD	0–90		Clayey-skeletal	Frigid	Xeric	6e, 6s	
Hummington	119.7	Typic Haplocryands	F		Douglas-fir	M, H	MD	WD	5–90		Medial-skeletal	Cryic	Udic	6s, 7e	
Hunewill	28.2	Xeric Haplargids	C, R				VD	WD	0–30		Loamy-skeletal	Mesic	Aridic	6c	3e
Hunsaker	38.4	Andic Hapludalfs	F		Grand fir		VD	WD	0–90		Fine-loamy	Frigid	Udic	6e	
Hunterscove	13.7	Typic Haplohumults	F		Douglas-fir	H	MD	WD	0–60		Fine	Isomesic	Udic	6e	
Huntley	2.2	Lithic Hapludolls	F		Douglas-fir	H	S	WD	0–60		Loamy	Mesic	Udic	7e	
Huntrock	6.4	Vitrandic Haploxerepts	F		Ponderosa pine	L	MD	WD	12–75		Loamy-skeletal	Frigid	Xeric	6e	
Hurryback	1.8	Pachic Argixerolls					VD	WD	8–40		Fine-loamy	Frigid	Xeric	4e	
Hurwal	85.5	Vitrandic Argixerolls	R	H			VD, D	WD	2–60	L	Fine-silty	Frigid	Xeric	3e, 4e, 6e	
Hutchinson	17.0	Typic Durixerolls	F		Ponderosa pine	M	MD	WD	0–20	L	Fine	Mesic	Xeric	4e	3e
Hutchley	80.9	Lithic Argixerolls					S	WD	2–75		Loamy-skeletal	Frigid	Xeric	7s	
Hutson	18.9	Typic Vitricryands	F		Douglas-fir	M	VD	WD	0–65		Ashy	Cryic	Udic	6e, 7e	
Hyall	30.1	Torrertic Argixerolls					MD	WD	12–60		Clayey-skeletal	Mesic	Xeric	6e	
Icedee	9.6	Alfic Vitricryands	F		Grand fir	M	D-VD	WD	0–30		Ashy over loamy	Cryic	Udic	6c	
Icene	95.4	Typic Aquisalids					VD	MWD	0–1	H	Fine-loamy	Mesic	Aridic	6s	
Igert	505.6	Fine-loamy					MD	WD	2–20		Mixed	Aridic	Aridic	6c	
Illahee	59.2	Typic Humudepts	F		Douglas-fir	M, H	VD	WD	3–90		Loamy-skeletal	Frigid	Udic	6e	
Imbler	49.2	Pachic Haploxerolls	C, R	H			D	WD	1–5	L	Coarse-loamy	Mesic	Xeric	3e	2e
Immiant	2.6	Typic Argixerolls					MD	WD	0–70		Fine-loamy	Mesic	Xeric	4s	
Immig	223.5	Typic Argixerolls	R	L, M			MD	WD	2–80		Clayey-skeletal	Mesic	Xeric	4s, 6e, 6s	4e
Imnaha	173.9	Vitrandic Argixerolls	F		Ponderosa pine		MD	WD	2–90	L	Loamy-skeletal	Frigid	Xeric	4e, 6e, 7e	
Inkler	66.7	Andic Haploxerepts	F		Ponderosa pine	L	D-VD	WD	0–75		Loamy-skeletal	Frigid	Xeric	4e, 6e, 7e	
Ipsoot	10.3	Xeric Vitricryands	F		Ponderosa pine	M	MD	ED	15–65		Ashy over pumiceous or cindery	Cryic	Xeric	6e	

(continued)

Series name	Area (km²)	Subgroup	Land use[1]	Range site quality[2]	Forest indicator species	Forest site quality[3]	Depth class[4]	Drainage class[5]	Slope range (%)	Salinity class[6]	Particle-size class	Soil temp. regime	Soil moisture regime	Land capability class	
														No mgmt	With mgmt
Iris	3.9	Duridic Haploxerolls	R	H			VD	WD	0–1	L	Fine-silty	Mesic	Xeric	3c	3c
Irma	7.8	Humic Dystrudepts	F		Douglas-fir	M, H	VD	WD	0–30		Fine-loamy	Mesic	Udic	6e	
Ironside	78.7	Vitrandic Haploxerolls	F		Ponderosa pine		MD	WD	0–90		Loamy-skeletal	Frigid	Xeric	6s	
Irrigon	32.0	Xeric Haplocambids	R	L			MD	WD	2–12	L	Coarse-loamy	Mesic	Aridic	6e	3e
Itca	16.1	Lithic Argixerolls					S	WD	2–75		Clayey-skeletal	Frigid	Xeric	7s	
Izee	28.7	Vitrandic Haploxerolls					MD	WD	0–65		Fine-loamy	Frigid	Xeric	4e, 6e	
Jacksplace	82.7	Vitritorrandic Argixerolls	R				MD	WD	1–20		Ashy-skeletal	Frigid	Xeric	6e	
James canyon	0.5	Cumulic Endoaquolls	C, R				VD	PD	0–8		Fine-loamy	Mesic	Aquic	5w	3w
Jayar	150.2	Typic Dystroxerepts	F		Douglas-fir	L, M	MD	WD	12–90	L	Loamy-skeletal	Frigid	Xeric	6e, 6s, 7s	
Jenny	0.0	Entic Calcixererts	C				VD	WD	0–15		Fine	Mesic	Xeric	3e	3e
Jerome	6.2	Typic Epiaquepts	C, R	H			D	PD	0–3	L	Coarse-loamy over clayey	Mesic	Aquic	3w	3w
Jesse camp	8.4	Xeric Haplocambids					VD	WD	0–8		Fine-silty	Frigid	Aridic	6e	
Jett	112.9	Cumulic Haploxerolls	C, R	H			D	WD	0–3	L	Fine-silty	Mesic	Xeric	2c, 4c	2c
Jimbo	34.2	Andic Humudepts	C, F		Douglas-fir	H	VD	WD	0–15	L	Coarse-loamy	Mesic	Udic	1	
Jimgreen	13.3	Hemic Haplosaprists	C, R	M			VD	VPD	0–1			Frigid	Aquic	5w	
Joeney	5.9	Typic Duraquods	C, F		Douglas-fir	M	S	PD	0–7		Loamy	Isomesic	Aquic	4w	
Jojo	56.2	Typic Vitricryands	F		Western hemlock	L	MD	SED	0–70		Medial-skeletal	Cryic	Udic	6 s	
Jorn	14.3	Vitrandic Argixerolls	F		Ponderosa pine	L	MD	WD	2–55		Fine	Mesic	Xeric	6e	
Jory	872.9	Xeric Palehumults	C, F		Douglas-fir	M, H	VD	WD	2–90	L	Fine	Mesic	Udic	2e-4e, 6e	
Josephine	521.4	Typic Haploxerults	C, F		Douglas-fir	L–H	D	WD	2–75	L	Fine-loamy	Mesic	Xeric	2e-4e, 6e	
Josset	3.4	Cumulic Haploxerolls	R	H			VD	MWD	0–2	L	Coarse-loamy over sandy or sandy-skeletal	Frigid	Xeric	3w	
Jumpoff	22.0	Ultic Haploxeralfs	F		Douglas-fir	L, M	D	MWD	7–50		Fine	Mesic	Xeric	4e, 6e	3e

(continued)

Series name	Area (km²)	Subgroup	Land use[1]	Range site quality[2]	Forest indicator species	Forest site quality[3]	Depth class[4]	Drainage class[5]	Slope range (%)	Salinity class[6]	Particle-size class	Soil temp. regime	Soil moisture regime	Land capability class	
														No mgmt	With mgmt
Kahler	113.1	Vitrandic Haploxerolls	F		Ponderosa pine		VD-D	WD	2–90	L	Fine-loamy	Frigid	Xeric	3e, 4e, 6e, 7e	
Kahneeta	5.4	Argic Duraquolls	R	L			MD	SPD	0–3		Clayey-skeletal	Mesic	Aquic	4s	
Kamela	249.9	Vitrandic Haploxerepts	F		Ponderosa pine	L, M	MD	WD	0–90	L	Loamy-skeletal	Frigid	Xeric	6e, 6s	
Kanid	249.9	Typic Dystroxerepts	F		Douglas-fir	L, M	D	WD	12–90	L	Loamy-skeletal	Mesic	Xeric	4e, 6e, 7e	
Kanlee	5.7	Typic Argixerolls					MD	SED	1–55		Fine-loamy	Frigid	Xeric	4e	
Kanutchan	6.1	Xeric Endoaquerts	C, R				D	SPD	1–8		Fine	Frigid	Aquic	6w, 7s	
Kaskela	63.4	Typic Haploxererts	R	H			VD	WD	0–30		Very-fine	Mesic	Xeric	4e	
Keating	60.9	Typic Argixerolls	R	H, L			MD	WD	2–35	L	Fine	Mesic	Xeric	4e, 4s	
Kecko	13.8	Xeric Haplocalcids	C, R				VD, D	WD	0–20		Coarse-loamy	Mesic	Aridic	6s	4e
Keel	156.6	Typic Haplocryands	F		Douglas-fir	M	MD	WD	2–75	L	Medial	Cryic	Udic	6e, 7e	
Kegler	43.2	Haploduridic Durixerolls					MD	WD	2–5		Fine-loamy	Frigid	Xeric	6e	
Kenusky	8.1	Umbric Paleaquults	F		Douglas-fir	M	VD	PD	0–15		Fine	Mesic	Aquic	6w	
Kerby	21.7	Typic Haploxerepts	F		Douglas-fir	M	D	WD, MWD	0–3	L	Fine-loamy	Mesic	Xeric	4c	1
Kerrfield	75.0	Durinodic Xeric Haplocambids					MD	WD	3–20		Coarse-loamy	Mesic	Aridic	6e	
Ketchly	51.1	Vitrandic Haploxeralfs	F		Douglas-fir	L, M	VD-D	WD	3–65		Fine-loamy	Frigid	Xeric	6e, 7e	
Kettenbach	83.1	Pachic Argixerolls					MD	WD	3–90		Loamy-skeletal	Mesic	Xeric	6e, 7e	
Kettlecreek	9.9	Vitrandic Haploxerolls	F		Ponderosa pine		D	WD	0–90		Loamy-skeletal	Frigid	Xeric	6e	
Kewake	131.5	Vitrandic Torripsamments					VD	ED	1–65	M	Sandy	Mesic	Aridic	6e	
Kiesel	4.6	Xeric Natrargids	C, R	L			D	WD	0–2		Fine	Mesic	Aridic	4s	
Kilchis	121.8	Lithic Humudepts	F		Douglas-fir	M	S	WD	3–100	L	Loamy-skeletal	Mesic	Udic	6s, 7s	
Killam	10.6	Pachic Fulvudands	F		Western hemlock		MD	WD	60–90		Medial-skeletal	Isofrigid	Udic	7e	
Killet	4.7	Humic Dystroxerepts	F		Douglas-fir	M, H	D	WD	3–35		Fine-loamy	Frigid	Xeric	4e	
Kilmerque	33.1	Vitrandic Haploxerolls	F		Ponderosa pine		MD	WD	1–80		Coarse-loamy	Frigid	Xeric	4e, 6e, 7e	
Kilowan	34.3	Humic Dystrudepts	F		Douglas-fir	M, H	MD	WD	3–75		Fine	Mesic	Udic	6e, 7e	

(continued)

Series name	Area (km^2)	Subgroup	Land use[1]	Range site quality[2]	Forest indicator species	Forest site quality[3]	Depth class[4]	Drainage class[5]	Slope range (%)	Salinity class[6]	Particle-size class	Soil temp. regime	Soil moisture regime	Land capability class	
														No mgmt	With mgmt
Kimberly	75.2	Torrifluventic Haploxerolls	C, R	H			VD	WD	0–3	L	Coarse-loamy	Mesic	Xeric	1, 3e	1
Kingbolt	124.6	Typic Vitrixerands	F		Douglas-fir		MD	WD	0–60		Ashy over loamy-skeletal	Frigid	Xeric	6c	
Kingsriver	3.0	Cumulic Endoaquolls	C, R				VD	VPD	0–2		Coarse-loamy	Mesic	Aquic	6w	3w
Kinney	787.7	Andic Humudepts	F		Douglas-fir	H	D-VD	WD	0–90	L	Fine-loamy	Mesic	Udic	6e, 7e	
Kinton	60.6	Typic Fragixerepts	C, F		Douglas-fir	H	D	MWD	2–60	L	Fine-silty	Mesic	Xeric	2e-4e	
Kinzel	75.2	Typic Fulvicryands	F		Douglas-fir	M	VD	WD	5–90		Medial-skeletal	Cryic	Udic	6e	
Kiona	108.6	Xeric Haplocambids					VD	WD	0–120		Loamy-skeletal	Mesic	Aridic	7e	
Kirk	107.6	Typic Cryaquands	R	M			VD	PD	0–1		Ashy-pumiceous	Cryic	Aquic	5w	
Kirkendall	54.1	Oxyaquic Humudepts	C, F		Douglas-fir	H	VD	WD-MWD	0–3		Fine-silty	Mesic	Udic	2w, 3w	
Kishwalk	191.7	Pachic Argixerolls					MD	WD	2–80		Clayey-skeletal	Mesic	Xeric	6e, 7e	
Kittleson	6.5	Xeric Vitricryands	F		White fir	L	D	WD	15–75		Ashy	Cryic	Xeric	6e	
Klamath	96.0	Cumulic Cryaquolls	R	H			VD	VPD	0–1	L	Fine-silty	Cryic	Aquic	4w	
Klicker	1200.7	Vitrandic Argixerolls	F		Ponderosa pine	L, M	MD	WD	0–90	L	Loamy-skeletal	Frigid	Xeric	6s, 7s, 7e	
Klickitat	1019.5	Typic Humudepts	F		Douglas-fir	M, H	D	WD	3–90	L	Loamy-skeletal	Mesic	Udic	6e, 6s, 7s	
Klickson	185.5	Vitrandic Argixerolls	F		Ponderosa pine	L	VD-D	WD	0–90		Loamy-skeletal	Frigid	Xeric	6e	
Klistan	383.7	Alic Hapludands	F		Douglas-fir	M, H	VD-D	WD	3–90	L	Medial-skeletal	Mesic	Udic	6e	
Klooqueh	4.9	Typic Palehumults	C, F		Douglas-fir	H	VD	WD	0–8		Fine	Isomesic	Udic	2c, 3e	2e, 3e
Klootchie	495.3	Typic Fulvudands	F		Douglas-fir	M, H	VD-D	WD	3–90	L	Medial	Isomesic	Udic	6s	
Knapke	17.0	Entic Ultic Haploxerolls	F		Douglas-fir	L	VD	WD	30–90		Loamy-skeletal	Mesic	Xeric	6e	
Knappa	33.5	Andic Humudepts	F, R		Douglas-fir	M, H	VD	WD	0–30		Fine-silty	Isomesic	Udic	2c, 2e, 3e	
Koehler	95.4	Xeric Haplodurids	C, R				MD	SED	0–15	L	Sandy	Mesic	Aridic	7e	4e
Kosh	3.5	Lithic Ultic Haploxerolls					S	ED	8–90		Sandy-skeletal	Frigid	Xeric	7e	

(continued)

Series name	Area (km^2)	Subgroup	Land use[1]	Range site quality[2]	Forest indicator species	Forest site quality[3]	Depth class[4]	Drainage class[5]	Slope range (%)	Salinity class[6]	Particle-size class	Soil temp. regime	Soil moisture regime	Land capability class	
														No mgmt	With mgmt
Krackle	59.7	Xeric Haplocryolls					MD	WD	3–65		Loamy-skeletal	Cryic	Xeric	6e	
Krebs	24.1	Calciargidic Argixerolls	C, R	M			D	WD	2–40	L	Fine	Mesic	Aridic	6e	1
Kubli	12.5	Aquic Haploxerolls	C, R	H			D	SPD	0–7	L	Fine-loamy over clayey	Mesic	Xeric	4e, 4w	3w
Kuck	0.4	Vertic Argixerolls	C, R				MD	WD	2–50		Fine	Mesic	Xeric	6e	
Kuckup	1.4	Vitric Haplocryands	F		Pacific silver fir	M	VD	ED	15–40		Medial over pumiceous or cindery	Cryic	Udic	6e	
Kuhl	10.0	Lithic Haploxerolls					S	WD	0–65		Loamy	Mesic	Xeric	7s	
Kunaton	56.8	Abruptic Xeric Argidurids	C, R				S	WD	0–12		Clayey	Mesic	Aridic	6s	4s
Kunceider	147.0	Aridic Lithic Haploxerolls					S	SED	0–30	L	Ashy-skeletal	Frigid	Xeric	6e	
Kusu	18.0	Typic Vitrixerands	F		Douglas-fir	L	D	WD	0–65		Ashy-skeletal	Frigid	Xeric	6s	
Kutcher	68.2	Alfic Udivitrands	F		Douglas-fir	L, M	MD	WD	0–30		Ashy-skeletal over loamy-skeletal	Frigid	Udic	6s	
Kweo	8.5	Xeric Vitricryands	F		Douglas-fir	M	MD	ED	8–50		Ashy over pumiceous or cindery	Cryic	Xeric	6e	
La grande	66.9	Pachic Haploxerolls	C, R	H			D	MWD	0–3	L	Fine-silty	Mesic	Xeric	2c	2c
Labish	9.9	Cumulic Humaquepts	C, R	H			D	PD	0–1		Fine	Mesic	Aquic	3w	
Labuck	11.6	Vitrandic Haploxerepts	F		Ponderosa pine	L	MD	WD	5–35		Coarse-loamy	Frigid	Xeric	6e	
Lackeyshole	110.2	Typic Vitricryands	F		Engelmann spruce		D	WD	0–90		Ashy over loamy-skeletal	Cryic	Udic	6c	
Lacy	1.1	Lithic Ultic Argixerolls	F		Ponderosa pine		S	WD	0–90		Loamy-skeletal	Mesic	Xeric	7e	
Ladd	23.7	Typic Argixerolls	C, R	H			D	WD	0–65	L	Fine-loamy	Mesic	Xeric	2e, 3e	2e, 3e
Laderly	167.8	Alic Hapludands	F		Douglas-fir	M, H	MD	WD	3–90	L	Medial-skeletal	Frigid	Udic	6e, 7e	
Ladycomb	14.5	Aridic Lithic Haploxerolls					VS	WD	8–25		Loamy	Mesic	Xeric	7 s	
Lafollette	9.1	Vitritorrandic Haploxerolls	C, R	L, M			MD	WD	0–8	L	Coarse-loamy	Mesic	Xeric	6s, 6e	3s, 3e
Laidlaw	53.2	Humic Vitrixerands	F		Ponderosa pine		VD	WD	0–40		Ashy	Frigid	Xeric	6e	4e
Lakefork	31.4	Typic Udivitrands	F		Grand fir		VD	WD	5–90		Ashy-skeletal over loamy-skeletal	Frigid	Udic	6c	

(continued)

Series name	Area (km^2)	Subgroup	Land use[1]	Range site quality[2]	Forest indicator species	Forest site quality[3]	Depth class[4]	Drainage class[5]	Slope range (%)	Salinity class[6]	Particle-size class	Soil temp. regime	Soil moisture regime	Land capability class	
														No mgmt	With mgmt
Lakeview	95.4	Cumulic Haploxerolls	C, R	H			VD	MWD-SPD	0–2	L	Fine-loamy	Mesic	Xeric	2w	
Laki	81.8	Typic Haploxerolls	C, R	H			D	MWD	0–2		Fine-loamy	Mesic	Xeric	3s	
Lalos	2.7	Sodic Xeric Haplocambids	C, R				D	WD	2–15		Coarse-silty	Mesic	Aridic	4e	4e
Lamath	0.7	Aquandic Endoaquepts	C, R				D	PD	0–1		Ashy over sandy or sandy-skeletal	Mesic	Aquic	4w	4w
Lambranch	2.7	Xeric Haplargids	C, R				VD	WD	2–8		Clayey-skeletal	Mesic	Aridic	6e	
Lambring	349.2	Pachic Haploxerolls					VD	WD	5–70	L	Loamy-skeletal	Frigid	Xeric	6e	
Lamonta	93.7	Abruptic Argiduridic Durixerolls	C, R	H			MD	WD	0–20		Fine	Mesic	Xeric	2e, 4e	
Lamulita	47.1	Vitrandic Argixerolls	F		Ponderosa pine		D	WD	0–60		Clayey-skeletal	Frigid	Xeric	7s	
Landermeyer	11.3	Vitritorrandic Haploxerolls	C, R				VD	WD	0–15		Fine-loamy	Mesic	Aridic	6c	3c
Langellain	35.1	Ultic Haploxeralfs	F		Ponderosa pine	L, M	MD	MWD	1–40		Fine-loamy over clayey	Mesic	Xeric	4e	3e
Langlois	29.1	Typic Fluvaquents	C, R				VD	VPD	0–3		Fine	Isomesic	Aquic	4w	
Langrell	37.8	Pachic Haploxerolls	R	H, L			VD	WD	0–3	L	Loamy-skeletal	Mesic	Xeric	3s, 4s	3s, 4s
Langslet	13.5	Xeric Aquicambids					VD	SPD	0–2		Fine	Frigid	Aridic	6s	
Lapham	32.0	Vitritorrandic Haploxerolls	R				VD	SED	0–30		Ashy over loamy-skeletal	Frigid	Xeric	6e, 6s	
Lapine	2513.1	Xeric Vitricryands	F		Ponderosa pine	L, M	VD	ED	0–70	L	Ashy-pumiceous	Cryic	Xeric	6e, 6s	
Larabee	370.6	Vitrandic Argixerolls	F		Ponderosa pine		MD	WD	0–90		Loamy-skeletal	Frigid	Xeric	6e, 7e	
Larmine	59.0	Lithic Haploxerepts	F		Douglas-fir	M	S	WD	12–90		Loamy-skeletal	Mesic	Xeric	6e, 7e	
Lasere	86.5	Typic Palexerolls	R	H, L			MD	WD	2–50		Fine	Mesic	Xeric	4e, 7s	4e
Lassen	0.6	Leptic Haploxererts	C, R				MD	WD	2–50		Fine	Mesic	Xeric	6e	
Lastance	23.9	Typic Haplocryods	F		Noble fir	L	D	WD	5–60		Loamy-skeletal	Cryic	Udic	7s	
Lastcall	70.2	Vitritorrandic Argixerolls					MD	WD	0–15		Ashy	Frigid	Xeric	6e	
Lather	101.1	Limnic Haplohemists	C, R	H			D	VPD	0–1	L		Frigid	Aquic	4w	

(continued)

Series name	Area (km^2)	Subgroup	Land use[1]	Range site quality[2]	Forest indicator species	Forest site quality[3]	Depth class[4]	Drainage class[5]	Slope range (%)	Salinity class[6]	Particle-size class	Soil temp. regime	Soil moisture regime	Land capability class	
														No mgmt	With mgmt
Latourell	61.6	Ultic Haploxeralfs	C, F		Douglas-fir		D	WD	0–30	L	Fine-loamy	Mesic	Xeric	1, 2e, 3e, 4e	
Laufer	17.9	Lithic Argixerolls					S	WD	2–120		Clayey-skeletal	Mesic	Xeric	7e, 7s	
Laurelwood	155.5	Ultic Haploxeralfs	C, F		Douglas-fir	H	VD	WD	3–60	L	Fine-silty	Mesic	Xeric	2e-4e, 6e	
Lavey	17.2	Calciargidic Argixerolls	R	H			MD	WD	2–15		Fine	Mesic	Aridic	4e	
Lawen	131.1	Calciargidic Argixerolls					VD	WD	2–5	L	Coarse-loamy	Frigid	Aridic	6e	4e
Lawyer	8.2	Pachic Ultic Argixerolls					D-VD	WD	15–90		Loamy-skeletal	Mesic	Xeric	7e	
Laycock	16.6	Lithic Ultic Haploxerolls	F		Ponderosa pine		S	WD	15–75		Loamy-skeletal	Frigid	Xeric	7s	
Leathers	39.5	Sodic Xeric Haplocambids					VD	WD	0–3	H	Coarse-loamy	Mesic	Aridic	6s	
Lebam	46.2	Typic Fulvudands	F		Douglas-fir	M, H	VD	WD	1–90		Medial over clayey	Isomesic	Udic	4e, 6e, 7e	
Leemorris	32.3	Pachic Argicryolls					MD	WD	3–35	L	Fine	Cryic	Xeric	6e	
Leepcreek	2.0	Vertic Argixerolls					D	WD	15–60		Fine	Mesic	Xeric	6e	
Leespeak	1.1	Vitrixerandic Humicryepts					MD	WD	60–90		Loamy-skeletal	Cryic	Xeric	7e	
Leevan	33.3	Typic Argixerolls					MD	WD	4–60		Clayey-skeletal	Frigid	Xeric	6e	
Legler	122.8	Xeric Haplocambids					VD	WD	0–20	L	Fine-loamy	Mesic	Aridic	6e	
Lemoncreek	8.9	Vitrandic Haploxerepts	F		Ponderosa pine		MD	WD	0–60		Loamy-skeletal	Frigid	Xeric	6c	
Lemonex	33.1	Vitrandic Argixerolls	F		Ponderosa pine		MD	WD	3–50		Fine	Frigid	Xeric	7s	
Lempira	33.8	Typic Hapludands	F		Douglas-fir	H	VD	WD	3–60	L	Medial	Frigid	Udic	6e	
Leopold	2.6	Andic Dystrudepts	F		Douglas-fir	M	MD	WD	3–60		Fine-loamy	Frigid	Udic	6e	
Lequieu	0.0	Lithic Xeric Torriorthents					VS	WD	0–9		Loamy-skeletal	Mesic	Aridic	7e	
Lerrow	12.3	Vitritorrandic Argixerolls					MD	WD	4–50		Fine	Frigid	Xeric	7e	
Lettia	140.7	Ultic Haploxeralfs	F		Douglas-fir	M, H	D	WD	2–70	L	Fine-loamy	Mesic	Xeric	3e, 4e, 6e	
Lickskillet	1936.4	Aridic Lithic Haploxerolls					S	WD	1–120	L	Loamy-skeletal	Mesic	Xeric	7s	
Limberjim	995.5	Alfic Udivitrands	F		Grand fir	L, M	D	WD	0–90	L	Ashy over loamy-skeletal	Frigid	Udic	6e	
Limpy	0.4	Lithic Humudepts	F		Douglas-fir	M	S	SED	60–100		Loamy-skeletal	Frigid	Udic	7e	
Linecreek	36.2	Vitrandic Haploxerolls	F		Ponderosa pine		VD	WD	15–90		Ashy-skeletal	Frigid	Xeric	7e	

(continued)

Series name	Area (km²)	Subgroup	Land use[1]	Range site quality[2]	Forest indicator species	Forest site quality[3]	Depth class[4]	Drainage class[5]	Slope range (%)	Salinity class[6]	Particle-size class	Soil temp. regime	Soil moisture regime	Land capability class	
														No mgmt	With mgmt
Linkletter	47.7	Duric Xeric Petroargids					D	WD	8–25		Fine	Mesic	Aridic	6c	6e
Linksterly	38.0	Humic Vitricryands	F		White fir		VD	WD	0–50		Ashy	Cryic	Udic	6e	
Linslaw	23.5	Aquultic Haploxeralfs	F		Douglas-fir		VD	SPD	0–8	L	Fine	Mesic	Xeric	3w	
Lint	38.0	Typic Fulvudands	C, F		Douglas-fir	M, H	VD	WD	0–40		Medial	Isomesic	Udic	2e-4e, 6e	
Linville	0.0	Pachic Haploxerolls					D-VD	WD	0–90		Fine-loamy	Mesic	Xeric	7e	
Lithgow	179.2	Xeric Haplargids					MD	WD	20–70		Loamy-skeletal	Mesic	Aridic	6e	
Littlefawn	13.2	Vitrandic Haploxeralfs	F		Douglas-fir	L	MD	WD	2–55		Fine	Mesic	Xeric	4e	
Littlesand	20.2	Typic Dystroxerepts	F		Douglas-fir	M, H	MD	WD	60–90		Fine-loamy	Mesic	Xeric	7e	
Lizard	15.2	Vitrandic Palexerolls					VD	WD	0–60		Clayey-skeletal	Frigid	Xeric	6c	
Llaorock	100.9	Vitric Haplocryands	F		Ponderosa pine	L	VD	SED	0–80		Medial-skeletal	Cryic	Udic	7s	
Lobert	76.8	Pachic Haploxerolls	F		Ponderosa pine	L, M	VD	WD	0–25		Coarse-loamy	Frigid	Xeric	4c, 4e	
Locane	42.6	Lithic Xeric Haplargids					S	WD	2–75		Clayey-skeletal	Frigid	Aridic	6e, 7s	
Locoda	33.7	Typic Fluvaquents	C, R	H, L			D	VPD	0–3		Fine-silty	Mesic	Aquic	3w, 4w, 6w	
Locolake	39.5	Typic Natridurids					S	WD	2–15		Loamy	Frigid	Aridic	6e, 7s	
Loeb	15.3	Typic Haplohumults	F		Douglas-fir	M, H	D	WD	0–30		Fine	Isomesic	Udic	6e	
Lofftus	7.8	Aquicambidic Haplodurids					MD	SPD	0–1	L	Ashy	Mesic	Aridic	7s	
Logdell	18.6	Lithic Ultic Haploxerolls					VS	WD	15–75		Loamy-skeletal	Frigid	Xeric	7s	
Logsden	21.4	Typic Humudepts	C, F		Douglas-fir	H	VD	WD	0–3	L	Fine-silty	Isomesic	Udic	2c, 2e	
Logsprings	5.3	Vitrandic Haploxeralfs	F		Douglas-fir	L	VD	MWD	0–8		Fine-loamy over clayey	Mesic	Xeric	3e	
Lolak	45.5	Vertic Halaquepts	R				VD	PD	0–2	H	Fine	Frigid	Aquic	6s	
Lonely	336.5	Xeric Haplocambids					MD	WD	2–30	L	Fine-loamy	Frigid	Aridic	6e	
Loneranch	21.2	Aquandic Humudepts					MD	MWD	0–60		Fine-loamy	Isomesic	Udic	6e	
Loneridge	17.1	Vertic Palexeralfs	F		Douglas-fir	L	VD	WD	0–65		Clayey-skeletal	Frigid	Xeric	6e	
Longbranch	66.5	Pachic Argixerolls	R	H			D	WD	12–50		Clayey-skeletal	Frigid	Xeric	4e	
Longcreek	21.1	Aridic Lithic Argixerolls					S	WD	2–75	L	Clayey-skeletal	Mesic	Xeric	7s	
Longjohn	9.5	Humic Vitricryands					VD	SED	15–50		Ashy-skeletal	Cryic	Udic	6s	
Lookingglass	131.0	Xerertic Argialbolls	F		Ponderosa pine	L, M	VD	MWD	2–30	L	Fine	Mesic	Xeric	3e, 4e	

(continued)

Series name	Area (km^2)	Subgroup	Land use[1]	Range site quality[2]	Forest indicator species	Forest site quality[3]	Depth class[4]	Drainage class[5]	Slope range (%)	Salinity class[6]	Particle-size class	Soil temp. regime	Soil moisture regime	Land capability class	
														No mgmt	With mgmt
Lookout	217.4	Abruptic Xeric Argidurids	C, R	H, M, L			MD	WD	0–45	L	Fine	Mesic	Aridic	2e, 6e, 6s	
Loomis	3.0	Lithic Xeric Haplargids					VS-S	WD	2–35		Clayey-skeletal	Mesic	Aridic	7s	
Lorella	479.4	Lithic Argixerolls	R				S	WD	0–75	L	Clayey-skeletal	Mesic	Xeric	6e, 7s	4e
Lostbasin	132.6	Typic Haploxerepts					MD	WD	12–80	L	Loamy-skeletal	Frigid	Xeric	6e	
Lostforest	9.3	Vitrixerandic Haplocambids					MD	WD	0–5		Ashy	Frigid	Aridic	6e	
Lostine	10.0	Pachic Haploxerolls	R	H			VD	WD	0–30	L	Coarse-silty	Frigid	Xeric	3c	
Loupence	58.3	Cumulic Haploxerolls	R	H			VD	WD, MWD	0–3	L	Fine-silty	Mesic	Xeric	3c	3c
Loveboldt	40.4	Typic Torrifluvents					VD	WD	0–2		Sandy	Mesic	Aridic	7w	
Lovline	49.3	Aridic Haploxerolls					MD	WD	2–70		Fine-loamy	Mesic	Xeric	6e, 7e	
Lowerbluff	34.5	Lithic Vitrixerands	F		Douglas-fir		S	WD	0–15		Ashy	Frigid	Xeric	6e	
Luckiamute	25.1	Lithic Dystrocryepts	F		Douglas-fir	L, M	S	WD	3–90		Loamy-skeletal	Cryic	Udic	6s, 7s	
Ludi	16.9	Vitrandic Haploxerolls					VD	SED	15–65		Ashy-skeletal over fragmental or cindery	Frigid	Xeric	6e	
Lundgren	40.8	Humic Vitrixerands	F, R		Ponderosa pine		VD	WD	0–3		Ashy	Frigid	Xeric	6s	
Lurnick	12.8	Andic Dystrocryepts	F		Douglas-fir	M	MD	WD	3–90		Clayey-skeletal	Cryic	Udic	6s, 7s	
Lyeflat	27.5	Lithic Haplocambids					S	WD	2–50		Loamy	Mesic	Aridic	7e	
Macdunn	29.7	Humic Haploxerepts	F		Douglas-fir	M	D	WD	30–90		Clayey-skeletal	Mesic	Xeric	7s	
Mackatie	102.6	Alfic Udivitrands	F		Douglas-fir	M	D	WD	0–30		Ashy over loamy	Frigid	Udic	6e	
Mackey	14.1	Xeric Haplocalcids					MD	WD	8–60		Loamy-skeletal	Mesic	Aridic	7e	
Macklyn	10.8	Typic Haplohumults	F		Douglas-fir	H	MD	WD	0–30		Fine	Isomesic	Udic	6e	
Macyflet	4.4	Vertic Paleargids					VD	MWD	0–2		Very-fine	Frigid	Aridic	7s	
Madeline	69.9	Aridic Lithic Argixerolls					S	WD	0–50	L	Clayey	Frigid	Xeric	6e	
Madras	301.6	Aridic Argixerolls	C,R	H			MD	WD	0–40	L	Fine-loamy	Mesic	Xeric	4e	3e
Mahogee	55.0	Lithic Argicryolls					S	WD	15–75		Loamy	Cryic	Xeric	7s	
Mahoon	152.3	Aridic Palexerolls					MD	WD	2–40	L	Fine	Mesic	Xeric	6e	
Maklak	150.6	Xeric Vitricryands	F		Ponderosa pine	M	VD	ED	0–15		Ashy-pumiceous	Cryic	Xeric	6s	
Malabon	167.3	Pachic Ultic Argixerolls	C, F		Douglas-fir		VD	WD	0–3	L	Fine	Mesic	Xeric	1, 2w, 2s	

(continued)

Series name	Area (km²)	Subgroup	Land use[1]	Range site quality[2]	Forest indicator species	Forest site quality[3]	Depth class[4]	Drainage class[5]	Slope range (%)	Salinity class[6]	Particle-size class	Soil temp. regime	Soil moisture regime	Land capability class	
														No mgmt	With mgmt
Malheur	2.8	Xeric Natridurids	C, R	H			MD	WD	0–8		Fine-silty	Mesic	Aridic	3e, 3s	
Malin	58.2	Fluvaquentic Endoaquolls	R	H			VD	PD	0–1	M	Fine	Mesic	Aquic	4w	
Mallory	58.3	Pachic Argixerolls					MD	WD	2–90		Clayey-skeletal	Mesic	Xeric	6e, 6s	
Manita	123.0	Mollic Haploxeralfs	F		Douglas-fir	L, M	D	WD	2–50	L	Fine	Mesic	Xeric	4e, 6e	2e
Marack	4.5	Calciargidic Argixerolls	R	M			D	WD	2–35	L	Fine	Frigid	Aridic	6e	3e
Marblepoint	129.9	Andic Haplocryepts	F		Engelmann spruce		VD	WD	0–90		Loamy-skeletal	Cryic	Udic	6e	
Marcola	69.7	Pachic Ultic Argixerolls	F		Douglas-fir		D	MWD	2–7		Clayey-skeletal	Mesic	Xeric	4s	
Mariel	8.4	Typic Cryohemists					VD	VPD	0–10			Cryic	Aquic	5w	
Markscreek	7.8	Cumulic Haploxerolls	R	L			VD	WD	0–3	L	Coarse-loamy over sandy or sandy-skeletal	Frigid	Xeric	4s	3s
Marty	7.4	Alic Hapludands	F		Douglas-fir	H	VD	WD	0–60		Medial	Frigid	Udic	6e	
Maryspeak	0.0	Andic Dystrocryepts	F		Douglas-fir		VD	WD	5–60		Sandy-skeletal	Cryic	Udic	6e	
Mascamp	0.6	Aridic Lithic Argixerolls					S	WD	0–50	L	Loamy-skeletal	Frigid	Xeric	7e	
Maset	0.7	Alfic Vitrixerands	F		Ponderosa pine	L, M	MD	WD	1–45	L	Ashy over loamy-skeletal	Frigid	Xeric	6e	
Matheny	230.2	Calcic Pachic Argixerolls					D	WD	10–90	L	Loamy-skeletal	Mesic	Xeric	7e	
Matterhorn	0.1	Typic Calcixerolls	F		Ponderosa pine		VD	SED	0–3	L	Sandy-skeletal	Frigid	Xeric	4s	
Maule	2.8	Vitrandic Argixerolls	F		Ponderosa pine		MD	WD	5–70		Clayey-skeletal	Frigid	Xeric	6e	
Maupin	97.1	Haploduridic Durixerolls	C, R	H			MD	WD	0–35	L	Fine-loamy	Mesic	Xeric	2e, 3e	2s
Mayger	72.0	Aquic Palehumults	F		Douglas-fir	H	D	SPD	0–30		Fine	Mesic	Udic	6e	
Mcalpin	138.5	Aquic Cumulic Haploxerolls	C, F		Douglas-fir	H	VD	MWD	0–6	L	Fine	Mesic	Xeric	2e, 2w	2w
Mcbain	20.8	Sodic Xeric Haplocalcids	R				VD	MWD	0–2	M	Fine-loamy	Frigid	Aridic	6s	
Mcbee	124.6	Aquic Cumulic Haploxerolls	C, F		Douglas-fir	H	VD	MWD	0–3	L	Fine-silty	Mesic	Xeric	2w, 3w	
Mccalpinemeadow	2.1	Typic Vitricryands	F		Grand fir		MD	WD	0–60		Ashy over loamy-skeletal	Cryic	Udic	6c	
Mccartycreek	80.5	Vitrandic Haploxerolls					MD	WD	0–90		Loamy-skeletal	Frigid	Xeric	7s	
Mccoin	19.6	Aridic Haploxerolls	C, R	H			S	WD	2–60		Loamy	Mesic	Xeric	4e	

(continued)

Series name	Area (km^2)	Subgroup	Land use[1]	Range site quality[2]	Forest indicator species	Forest site quality[3]	Depth class[4]	Drainage class[5]	Slope range (%)	Salinity class[6]	Particle-size class	Soil temp. regime	Soil moisture regime	Land capability class	
														No mgmt	With mgmt
Mccomas	3.8	Aquic Palexeralfs	F		Douglas-fir		VD	SPD	3–30		Clayey-skeletal	Mesic	Xeric	4e	
Mcconnel	265.6	Xeric Haplocambids	C, R				VD	SED	0–50	L	Sandy-skeletal	Mesic	Aridic	6e	4e
Mccully	292.7	Typic Humudepts	F		Douglas-fir	H	D	WD	2–70	L	Fine	Mesic	Udic	3e, 4e, 6e, 7e	
Mccurdy	6.2	Oxyaquic Palehumults	F		Douglas-fir	H	VD	MWD	0–30		Fine	Mesic	Udic	3e, 4e	
Mcduff	167.0	Typic Haplohumults	F		Douglas-fir	M, H	MD	WD	0–75	L	Fine	Mesic	Udic	6e, 7e	
Mcewen	13.9	Vitrandic Haploxeralfs	C, F		Ponderosa pine	L	VD	WD	2–20		Fine-loamy	Frigid	Xeric	3c, 3e	
Mcgarr	36.7	Vitrandic Haploxerolls	F		Ponderosa pine		MD	WD	5–75		Fine-loamy	Frigid	Xeric	6s, 7s	
Mcginnis	3.1	Typic Haploxerults	F		Douglas-fir	M	MD	WD	3–80		Clayey-skeletal	Mesic	Xeric	4e, 6e, 7e	
Mcivey	21.4	Typic Argixerolls	R				VD	WD	2–75		Clayey-skeletal	Frigid	Xeric	7e	
Mckay	66.0	Calcic Argixerolls	C, R	H			D	WD	0–25		Fine-silty	Mesic	Xeric	3e	
Mcloughlin	41.3	Xeric Haplocambids	R	H			D	WD	0–8		Fine-silty	Mesic	Aridic	3s	
Mcmeen	38.6	Haplic Haploxerollic Durixerolls	R	H			MD	WD	0–12		Fine-loamy	Mesic	Xeric	3e	
Mcmille	8.9	Andic Humudepts	F		Douglas-fir	H	D-VD	WD	0–60		Fine-silty	Frigid	Udic	6e	
Mcmullin	286.0	Lithic Ultic Haploxerolls					S	WD-SED	1–75	L	Loamy	Mesic	Xeric	6e	
Mcmurdie	38.6	Calcic Pachic Argixerolls	R	H			VD	WD	1–10		Fine	Mesic	Xeric	3e, 4e	
Mcnab	13.1	Aquic Palexeralfs	F		Douglas-fir	M	VD	SPD	0–20	L	Fine	Mesic	Xeric	3e	
Mcnamee	47.9	Andic Haplocryalfs	F		Engelmann spruce		VD	WD	0–60		Loamy-skeletal	Cryic	Udic	6c, 6e	
Mcnull	323.4	Ultic Argixerolls	F		Douglas-fir	L, M	MD	WD	12–60	L	Fine	Mesic	Xeric	6e	
Mcnulty	16.2	Fluventic Dystrudepts	C, R	H			VD	WD	0–3		Coarse-loamy	Mesic	Udic	2w	
Mcnye	22.2	Xeric Haplocambids					D	SED	2–50		Sandy-skeletal	Mesic	Aridic	6e	
Mcwillar	89.8	Alfic Vitrixerands	F		Grand fir	M	D	WD	0–90		Ashy over loamy-skeletal	Frigid	Xeric	6e	
Meadowridge	123.5	Vitritorrandic Argixerolls	R	H			VD	WD	2–60		Fine-loamy	Mesic	Xeric	4e	
Meda	76.3	Typic Humudepts	F		Douglas-fir	H	VD	WD	2–20		Fine-loamy	Mesic	Udic	3e	

(continued)

Series name	Area (km^2)	Subgroup	Land use[1]	Range site quality[2]	Forest indicator species	Forest site quality[3]	Depth class[4]	Drainage class[5]	Slope range (%)	Salinity class[6]	Particle-size class	Soil temp. regime	Soil moisture regime	Land capability class	
														No mgmt	With mgmt
Medco	298.3	Ultic Haploxerolls	C, F, R		Ponderosa pine	L	MD	MWD	1–50	L	Fine	Mesic	Xeric	6e	4e
Medford	64.9	Pachic Argixerolls	R	H			VD	MWD	0–15	L	Fine	Mesic	Xeric	4c, 4e	
Melbourne	129.4	Ultic Palexeralfs	C, F		Douglas-fir	H	VD	WD	0–65	L	Fine	Mesic	Xeric	2e, 3e, 6e	
Melby	137.2	Humic Dystrudepts	F		Douglas-fir	M, H	D	WD	3–90	L	Fine	Mesic	Udic	6e	
Meld	23.1	Vitritorrandic Durixerolls					MD	WD	2–20		Ashy	Frigid	Xeric	6e	
Melhorn	88.9	Vitrandic Argixerolls	F		Grand fir		VD	WD	0–60		Fine-loamy	Frigid	Xeric	4e, 6e	
Melloe	19.4	Typic Cryaquolls					VD	PD	0–5		Loamy-skeletal	Cryic	Aquic	6c	
Mellowmoon	22.1	Typic Humudepts	F		Douglas-fir	H	VD	WD	3–60		Fine-loamy	Frigid	Udic	6e	
Memaloose	3.3	Andic Dystrudepts	F		Douglas-fir	M	MD	WD	5–30		Fine-loamy	Frigid	Udic	6e	
Menbo	59.9	Vitrandic Argixerolls					MD	WD	4–75		Clayey-skeletal	Frigid	Xeric	6s	
Merlin	936.7	Lithic Argixerolls					S	WD	0–40	L	Clayey	Frigid	Xeric	6e, 7s	
Mershon	21.2	Aquic Humudepts	F		Douglas-fir	M	D	MWD	0–30	L	Fine-silty	Mesic	Udic	3e, 4e	
Mesman	158.8	Xeric Natrargids					VD	WD	0–15	H	Fine-loamy	Mesic	Aridic	6s	
Metolius	19.0	Vitrixerandic Haplocambids	C, R	H			VD	MWD	0–2	L	Coarse-loamy	Mesic	Aridic	2s, 4e	
Middlebox	96.2	Vitrandic Torriorthents					MD	WD	5–40	L	Ashy-skeletal	Frigid	Aridic	6e	
Mikkalo	387.0	Calcidic Haploxerolls	C, R	H			MD	WD	0–60	L	Coarse-silty	Mesic	Xeric	3e, 4e	1
Milbury	265.3	Typic Humudepts	F		Douglas-fir	H	MD	WD	30–80	L	Loamy-skeletal	Mesic	Udic	6e	
Milcan	45.7	Vitritorrandic Durixerolls					MD	SED	0–5		Ashy	Frigid	Xeric	6e	
Milldam	24.4	Typic Durixerolls	C, R	L			MD	WD	0–3		Fine-loamy	Mesic	Xeric	4s	
Millenium	6.6	Vitritorrandic Argixerolls					VD	WD	0–2		Ashy	Frigid	Xeric	6c	
Millerflat	5.1	Vitrandic Argixerolls	F		Ponderosa pine	M	VD	MWD	0–3		Loamy-skeletal	Frigid	Xeric	4w	
Millicoma	190.7	Andic Humudepts	F		Douglas-fir	H	MD	WD	10–90	L	Loamy-skeletal	Isomesic	Udic	6e, 7e	
Minam	39.6	Vitrandic Haploxerolls	F		Ponderosa pine		VD	WD	0–15	L	Fine-loamy	Frigid	Xeric	3e, 4e	
Minkwell	8.4	Alfic Vitricryands	F		Douglas-fir	M	VD	WD	0–50		Ashy over medial	Cryic	Udic	6e	
Minniece	8.2	Vertic Endoaqualfs	F		Douglas-fir	M, H	V-VD	SPD-PD	0–8		Fine	Mesic	Aquic	6w	

(continued)

Series name	Area (km^2)	Subgroup	Land use[1]	Range site quality[2]	Forest indicator species	Forest site quality[3]	Depth class[4]	Drainage class[5]	Slope range (%)	Salinity class[6]	Particle-size class	Soil temp. regime	Soil moisture regime	Land capability class	
														No mgmt	With mgmt
Minveno	194.0	Xeric Haplodurids					S	WD	0–20		Loamy	Mesic	Aridic	6s	
Mippon	3.8	Fluventic Haploxerolls	C, F		Grand fir	L, M	VD	MWD	0–5		Sandy-skeletal	Frigid	Xeric	4s	
Mislatnah	31.6	Dystric Eutrudepts	F		Douglas-fir	H	MD	WD	0–90		Loamy-skeletal	Mesic	Udic	7e	
Moag	7.6	Vertic Endoaquepts					VD	VPD	0–2	L	Fine	Mesic	Aquic	6w	
Modoc	46.2	Vitritorrandic Durixerolls	R	H, M			MD	WD	0–15	L	Fine-loamy	Mesic	Xeric	3s, 3e, 6e	
Moe	63.2	Andic Humudepts	F		Douglas-fir	H	VD	WD	2–75		Fine	Frigid	Udic	6e, 7e	
Molalla	29.2	Typic Humudepts	F, R		Douglas-fir	H	D	WD	2–30		Fine-loamy	Mesic	Udic	3e, 4e	
Mondovi	15.7	Cumulic Haploxerolls	C, R	H			VD	WD	0–8	L	Coarse-silty	Mesic	Xeric	2c	1
Monroe	4.3	Cumulic Haploxerolls	C, R				VD	WD	0–4		Fine-loamy	Mesic	Xeric	6c	3e
Monumentrock	29.2	Andic Haplocryepts	F		Engelmann spruce		D	WD	15–90		Loamy-skeletal	Cryic	Udic	6c	
Moodybasin	13.1	Oxyaquic Humicryepts					MD	MWD	0–90		Loamy-skeletal	Cryic	Udic	7s	
Moonbeam	475.4	Vitritorrandic Durixerolls					S	WD	0–20	L	Clayey	Frigid	Xeric	6e, 7s	
Moonstone	2.4	Pachic Ultic Haploxerolls					MD	WD	8–60		Coarse-loamy	Frigid	Xeric	7e	
Morehouse	231.5	Vitrandic Torripsamments					VD	SED	0–35	L	Ashy	Frigid	Aridic	6e	
Morfitt	51.2	Xeric Haplargids	C				VD	WD	0–25	L	Fine-loamy	Mesic	Aridic	6c	3c
Morganhills	82.3	Vitrandic Torriorthents					S	WD	2–35		Ashy	Frigid	Aridic	7s	
Morningstar	14.2	Ultic Argixerolls	F		Ponderosa pine	L	VD	WD	12–60		Loamy-skeletal	Frigid	Xeric	6s	
Morrow	707.9	Calcic Argixerolls	R	H, M			MD	WD	0–40	L	Fine-silty	Mesic	Xeric	3e, 4e, 6e	
Mosscreek	29.1	Pachic Fulvudands	F		Douglas-fir	M	VD	WD	5–90		Medial	Isofrigid	Udic	6e	
Mound	49.6	Pachic Ultic Argixerolls	F		Douglas-fir	M	D	WD	0–70		Clayey-skeletal	Frigid	Xeric	4e	
Mountemily	227.1	Typic Vitricryands	F		Ponderosa pine	L	VD	WD	0–90	L	Ashy over loamy-skeletal	Cryic	Udic	6e, 7e	
Mountireland	54.2	Alfic Vitricryands	F		Grand fir	M	D	WD	0–30		Ashy over loamy	Cryic	Udic	6c	
Mowako	7.5	Ultic Haploxerolls	F		Ponderosa pine	L	MD	WD	30–80		Loamy-skeletal	Mesic	Xeric	6s, 7e	
Muddycreek	16.7	Andic Humicryepts	F		Engelmann spruce		VD	SED	15–90	L	Loamy-skeletal	Cryic	Udic	7e	
Mudlakebasin	70.4	Typic Vitricryands	F		Engelmann spruce		MD	WD	0–90		Ashy over loamy-skeletal	Cryic	Udic	6c	

(continued)

Series name	Area (km[2])	Subgroup	Land use[1]	Range site quality[2]	Forest indicator species	Forest site quality[3]	Depth class[4]	Drainage class[5]	Slope range (%)	Salinity class[6]	Particle-size class	Soil temp. regime	Soil moisture regime	Land capability class	
														No mgmt	With mgmt
Mudpot	62.0	Chromic Endoaquerts	R	M			VD	PD	0–2		Fine	Frigid	Aquic	5w	
Mues	8.9	Aquic Fulvudands	C, F		Western hemlock		VD	MWD	0–3		Medial over loamy-skeletal	Isomesic	Udic	2w	
Mugwump	7.6	Cumulic Hapludolls	F		Grand fir		VD	WD	0–30		Loamy-skeletal	Frigid	Udic	6c	
Mulkey	6.0	Pachic Fulvicryands					MD	WD	3–60		Medial	Cryic	Udic	6e	
Multnomah	61.9	Humic Dystroxerepts	F		Douglas-fir	H	D	WD	0–60	L	Coarse-loamy over sandy or sandy-skeletal	Mesic	Xeric	3s, 3e, 4e	
Multorpor	4.2	Typic Udorthents					D	ED	0–10		Sandy-skeletal	Mesic	Udic	7w	
Muni	435.4	Haploxeralfic Argidurids	R				S	WD	2–8		Loamy	Mesic	Aridic	7s	4e
Munsoncreek	42.3	Andic Humudepts	F		Douglas-fir	H	D	WD	5–60		Fine	Isomesic	Udic	6e	
Murlose	14.4	Vitrandic Durixerolls					S	WD	1–20		Ashy	Frigid	Xeric	6e	
Murnen	22.9	Typic Fulvudands	F		Douglas-fir	H	VD	WD	3–65		Medial	Frigid	Udic	6e	
Murtip	201.7	Alic Hapludands	F		Douglas-fir	M, H	D	WD	3–90	L	Medial	Frigid	Udic	6e, 7e	
Musty	11.0	Humic Dystroxerepts	F		Douglas-fir	L–H	MD	WD	12–50		Loamy-skeletal	Mesic	Xeric	4e, 6e	
Mutt	0.7	Andic Humudepts	F		Douglas-fir	H	MD	WD	5–30		Fine-silty	Frigid	Udic	6e	
Mutton	51.1	Vitrandic Haploxeralfs	F		Douglas-fir	L	VD	WD	12–80		Ashy-skeletal	Mesic	Xeric	6e, 7e	
Nagle	21.1	Pachic Argixerolls					D	WD	12–50		Fine-loamy	Frigid	Xeric	6e	
Nailkeg	62.0	Typic Dystrudepts	F		Douglas-fir	M	MD	WD	0–90		Loamy-skeletal	Mesic	Udic	6e	
Nansene	110.0	Pachic Haploxerolls					VD-D	WD	2–70	L	Coarse-silty	Mesic	Xeric	7e	
Natal	6.6	Umbric Endoaqualfs	C, R	H			VD	PD	0–4		Fine	Mesic	Aquic	3w, 4w	
Natroy	61.6	Xeric Endoaquerts	R	H			VD	PD	0–2		Very-fine	Mesic	Aquic	4w	
Nebopeak	356.6	Andic Haploxeralfs	F		Grand fir	M	VD	WD	30–60		Loamy-skeletal	Frigid	Xeric	6c	
Necanicum	9.9	Typic Fulvudands	F		Douglas-fir	M, H	VD-D	WD	0–90	L	Medial-skeletal	Isomesic	Udic	6e, 7e	
Needhill	0.2	Vitrandic Argixerolls	R	H			D	WD	0–30		Loamy-skeletal	Frigid	Xeric	4e	
Nehalem	57.6	Fluventic Humudepts	C, F		Douglas-fir	H	VD	WD	0–3	L	Fine-silty	Isomesic	Udic	2c, 2w, 3w	

(continued)

Series name	Area (km²)	Subgroup	Land use[1]	Range site quality[2]	Forest indicator species	Forest site quality[3]	Depth class[4]	Drainage class[5]	Slope range (%)	Salinity class[6]	Particle-size class	Soil temp. regime	Soil moisture regime	Land capability class	
														No mgmt	With mgmt
Nekia	603.4	Xeric Haplohumults	C, F		Douglas-fir	M, H	MD	WD	2–60	L	Fine	Mesic	Udic	2e-4e, 6e	
Nekoma	54.9	Fluventic Humudepts	F		Douglas-fir	H	VD	MWD-WD	0–3	L	Coarse-loamy	Mesic	Udic	3w	
Nelscott	40.2	Typic Durorthods	F		Douglas-fir	H	MD	MWD	0–50	L	Fine-loamy over sandy or sandy-skeletal	Isomesic	Udic	3e	
Neotsu	38.5	Typic Fulvudands	F		Douglas-fir	H	MD	WD	0–90		Medial	Isomesic	Udic	6e	
Neskowin	43.4	Typic Fulvudands	F		Douglas-fir	M, H	MD	WD	5–99		Medial	Isomesic	Udic	6e	
Nestucca	50.1	Fluvaquentic Humaquepts	C, F		Red alder		VD	SPD	0–3		Fine-silty	Isomesic	Aquic	2w, 3w	
Netarts	22.3	Entic Haplorthods					VD	WD	0–60		Sandy	Isomesic	Udic	6e	4e
Nevador	1112.8	Durinodic Xeric Haplargids					VD	WD	0–15		Fine-loamy	Mesic	Aridic	6e	
Newanna	19.2	Typic Fulvicryands	F		Douglas-fir	L	MD	WD	0–100		Medial-skeletal	Cryic	Udic	6e	
Newberg	270.7	Fluventic Haploxerolls	C, F		Douglas-fir	H	VD	SED	0–4	L	Coarse-loamy	Mesic	Xeric	2w, 3w	2w
Ninemile	1808.4	Aridic Lithic Argixerolls					S	WD	0–70	L	Clayey	Frigid	Xeric	6e, 6s	
Ninetysix	80.5	Calcic Haploxerolls					VD	WD	2–90		Loamy-skeletal	Mesic	Xeric	6e	
Noidee	41.1	Lithic Xeric Natrargids					S	WD	2–15		Clayey	Frigid	Aridic	7s	
Noname	24.2	Lithic Haplocryepts					VS-S	WD	3–80		Loamy	Cryic	Xeric	7s	
Nonpareil	87.4	Typic Dystroxerepts					S	WD	3–90		Loamy	Mesic	Xeric	7e	
Norad	144.2	Xeric Haplargids					VD	MWD-WD	0–2	M	Fine-silty	Mesic	Aridic	6c	
Norcross	42.2	Vitrandic Durixerolls					S	WD	0–10		Clayey	Frigid	Xeric	6e, 7s	
Norling	126.4	Ultic Haploxeralfs	F		Douglas-fir	L, M	MD	WD	30–60	L	Fine-loamy	Mesic	Xeric	6e	
North powder	35.0	Xeric Haplocambids	C, R				MD	WD	2–35	L	Fine-loamy	Mesic	Aridic	6e	4e
Northrup	3.6	Oxyaquic Haplohumults	C, F		Douglas-fir	H	D	SPD	0–20		Fine-silty	Mesic	Udic	3w, 4e	
Noti	15.2	Typic Humaquepts	C, R	H			D	PD	0–3		Coarse-loamy over sandy or sandy-skeletal	Mesic	Aquic	4w	
Notus	4.7	Aquic Xerofluvents	C, R	H			VD	MWD-SPD	0–4		Sandy-skeletal	Mesic	Xeric	4w	
Nuss	214.5	Lithic Haploxerolls	F		Ponderosa pine	L	S	WD	0–70	L	Loamy	Frigid	Xeric	6e, 7s	
Nyssa	117.8	Xeric Haplodurids	R	H			MD	WD	0–20	L	Coarse-silty	Mesic	Aridic	3e, 3s	

(continued)

Series name	Area (km²)	Subgroup	Land use[1]	Range site quality[2]	Forest indicator species	Forest site quality[3]	Depth class[4]	Drainage class[5]	Slope range (%)	Salinity class[6]	Particle-size class	Soil temp. regime	Soil moisture regime	Land capability class	
														No mgmt	With mgmt
Oak grove	42.8	Ultic Palexeralfs	C, F		Douglas-fir	M	VD	WD	0–60	L	Fine	Mesic	Xeric	2e, 3e, 4e, 6e	
Oakland	147.4	Ultic Haploxeralfs	F		Douglas-fir	M	MD	WD	3–60	L	Fine	Mesic	Xeric	3e, 4e, 6e	
Oatman	134.0	Typic Haplocryands	F		Ponderosa pine	L	VD-D	WD	0–65	L	Medial-skeletal	Cryic	Udic	6e	
Oatmanflat	14.8	Vitritorrandic Haploxerolls					D	WD	0–5		Ashy	Frigid	Xeric	6e	
Observation	429.3	Typic Argixerolls					MD	WD	2–50	L	Fine	Frigid	Xeric	4e, 6s	
Ochoco	28.2	Vitrixerandic Argidurids	C, R	H			MD	WD	0–8		Fine-loamy	Mesic	Aridic	2e	
Offenbacher	84.9	Typic Haploxerepts	F		Douglas-fir	L, M	MD	WD	50–80		Fine-loamy	Mesic	Xeric	7e	
Olac	42.1	Lithic Xeric Haplargids					VS-S	WD	2–75		Loamy-skeletal	Mesic	Aridic	7s	
Olallie	4.5	Cumulic Endoaquolls	R	H			VD	PD	0–3	L	Loamy-skeletal	Mesic	Aquic	3w	
Olaton	3.8	Pachic Ultic Haploxerolls					VD	SED	4–90		Coarse-loamy	Mesic	Xeric	7e	
Old camp	51.6	Lithic Xeric Haplargids					S	WD	2–75		Loamy-skeletal	Mesic	Aridic	7s	
Oldblue	7.4	Andic Humudepts	F		Douglas-fir	M, H	VD	WD	5–60		Fine-loamy	Frigid	Udic	6e	
Oldsferry	10.8	Typic Haploxerolls					MD	WD	25–85	L	Loamy-skeletal	Mesic	Xeric	6s	
Olex	96.8	Calcidic Haploxerolls					VD	WD	0–65	L	Loamy-skeletal	Mesic	Xeric	6e, 7e	
Oliphant	41.1	Calcic Pachic Haploxerolls	C, R	H			VD, D	WD	0–55	L	Coarse-silty	Mesic	Xeric	2c	1
Olot	342.3	Typic Vitrixerands	F		Douglas-fir	L, M	MD	WD	2–90	L	Ashy over loamy-skeletal	Frigid	Xeric	4e, 6e, 7e	
Olyic	227.5	Typic Haplohumults	F		Douglas-fir	H	D	WD	0–90	L	Fine-loamy	Mesic	Udic	6e, 7e	
Omahaling	3.9	Fluvaquentic Haploxerolls	C				VD	SPD	0–5		Coarse-loamy	Frigid	Xeric	6e	
Oneonta	26.6	Typic Haplocryands	F		Douglas-fir	M	VD	WD	0–60		Medial	Cryic	Udic	6e	
Ontko	48.1	Aquandic Cryaquolls	R	H			VD	VPD	0–1		Fine-loamy	Cryic	Aquic	4w	
Onyx	24.7	Cumulic Haploxerolls	C, R	H			VD	WD	0–5	L	Coarse-silty	Mesic	Xeric	2c	1
Opie	62.7	Cumulic Endoaquolls					VD	PD	0–2		Fine-silty	Frigid	Aquic	6s	6s
Oreanna	8.8	Typic Haplocambids					VD	WD	0–3		Fine-loamy over sandy or sandy-skeletal	Mesic	Aridic	6e, 6s	
Oreneva	122.3	Xeric Haplocambids	C				MD	WD	0–60	L	Loamy-skeletal	Frigid	Aridic	6c	2e

(continued)

Series name	Area (km^2)	Subgroup	Land use[1]	Range site quality[2]	Forest indicator species	Forest site quality[3]	Depth class[4]	Drainage class[5]	Slope range (%)	Salinity class[6]	Particle-size class	Soil temp. regime	Soil moisture regime	Land capability class	
														No mgmt	With mgmt
Orford	304.5	Typic Palehumults	F		Douglas-fir	H	VD	WD	0–75	L	Fine	Mesic	Udic	6e	
Orhood	0.1	Aridic Lithic Argixerolls	F		Westerm juniper		S	WD	2–50		Loamy-skeletal	Mesic	Aridic	7e	
Ornea	2.6	Typic Haplargids					VD	WD	1–40		Fine-loamy over sandy or sandy-skeletal	Mesic	Aridic	6c	3e
Orovada	143.7	Durinodic Xeric Haplocambids	C				VD	WD	0–15		Coarse-loamy	Mesic	Aridic	6c	2e
Osoll	12.0	Typic Haplodurids					S	WD	2–50		Loamy-skeletal	Mesic	Aridic	7s	
Otoole	3.9	Xeric Haplodurids	C, R	H			S	SPD	0–2		Loamy	Mesic	Aridic	4w	
Otwin	9.9	Typic Haplocryands	F		White fir	L, M	MD	WD	0–12		Medial-skeletal	Cryic	Udic	6e	
Outerkirk	184.1	Durinodic Haplocalcids					VD	WD	1–6	L	Coarse-loamy	Mesic	Aridic	6e	
Overallflat	26.5	Aquic Paleargids	C				VD	MWD	0–2		Ashy	Frigid	Aridic	6s	
Owsel	65.7	Durinodic Xeric Haplargids	R				VD	WD	0–20		Fine-silty	Mesic	Aridic	6c	4e
Owyhee	76.9	Xeric Haplocambids	C, R	H			VD	WD	0–30	L	Coarse-silty	Mesic	Xeric	1, 2e, 3e	
Oxbow	9.6	Palexerollic Durixerolls	C				MD	WD	0–10		Fine	Mesic	Xeric	7s	4s
Oxley	8.9	Typic Argiaquolls	C, R	H			D	SPD	0–3	L	Loamy-skeletal	Mesic	Aquic	3w	3w
Oxman	20.5	Xeric Haplocambids					MD	WD	2–35		Fine-loamy	Mesic	Aridic	6e	
Oxwall	116.1	Palexerollic Durixerolls					S	WD	0–15		Clayey	Mesic	Xeric	7s	6s
Ozamis	184.2	Fluvaquentic Endoaquolls	C				VD	PD	0–2	M	Fine-loamy	Mesic	Aquic	5w, 6s	6s
Packard	8.5	Pachic Haploxerolls	F		Ponderosa pine		VD	WD	0–5	L	Loamy-skeletal	Mesic	Xeric	4 s	
Padigan	11.1	Xeric Endoaquerts	C, R	H			VD	PD	0–3		Very-fine	Mesic	Aquic	4w	4w
Pait	42.2	Aridic Haploxerolls	C, R				VD	WD	1–30		Loamy-skeletal	Mesic	Xeric	6e, 6s, 7s	4e
Palouse	122.5	Pachic Ultic Haploxerolls	C, R	H, M			D	WD	0–60	L	Fine-silty	Mesic	Xeric	2e-4e, 6e	
Panther	59.6	Vertic Epiaquolls	C, R				VD-D	PD	2–20	L	Very-fine	Mesic	Aquic	6w	6w
Paragon	9.7	Pachic Ultic Argixerolls	F		Douglas-fir	L	MD	WD	1–35		Fine-loamy	Mesic	Xeric	4e	
Parkdale	34.3	Humic Vitrixerands	C, F		Ponderosa pine	M	VD	WD	0–40	L	Ashy	Mesic	Xeric	2e, 3e, 4e	

(continued)

Series name	Area (km^2)	Subgroup	Land use[1]	Range site quality[2]	Forest indicator species	Forest site quality[3]	Depth class[4]	Drainage class[5]	Slope range (%)	Salinity class[6]	Particle-size class	Soil temp. regime	Soil moisture regime	Land capability class	
														No mgmt	With mgmt
Parrego	47.1	Ultic Haploxeralfs	F		Douglas-fir		MD	WD	0–50		Fine-loamy	Frigid	Xeric	6e	
Parsnip	97.2	Lithic Argixerolls					s	WD	0–30		Loamy	Frigid	Xeric	6e	
Patit creek	3.2	Cumulic Haploxerolls	C, R	H			D	WD	0–5		Coarse-loamy	Mesic	Xeric	2e	
Patron	0.3	Vitrandic Palexerolls	F		Ponderosa pine		VD	WD	5–70		Fine	Mesic	Xeric	6e	
Paulina	36.0	Aquandic Endoaquolls	R				VD	VPD	0–1		Ashy	Frigid	Aquic	6w	
Paynepeak	0.9	Vitrandic Argicryolls					D	WD	4–50	L	Ashy-skeletal	Cryic	Xeric	6e	
Peahke	5.2	Vitrandic Haploxerolls	F		Ponderosa pine		MD	WD	2–55		Loamy-skeletal	Frigid	Xeric	6s	
Pearlwise	60.2	Pachic Haploxerolls					MD	WD	2–65		Fine-loamy	Frigid	Xeric	6e	
Pearsoll	232.5	Lithic Dystroxerepts	F		Jeffrey pine		S	WD	3–90	L	Clayey-skeletal	Mesic	Xeric	7s	
Peasley	0.6	Haplic Durixererts					MD	WD	2–6		Fine	Mesic	Xeric	4e	3e
Peavine	851.7	Typic Haplohumults	F		Douglas-fir	M, H	MD	WD	2–75	L	Fine	Mesic	Udic	6e, 7e	
Pedigo	23.0	Cumulic Haploxerolls	C, R	H			D	SPD	0–3	L	Coarse-silty	Mesic	Xeric	2w, 3w	2w
Peel	3.3	Vertic Argixerolls	F		Douglas-fir	M	MD	MWD	3–30		Fine	Mesic	Xeric	4e	
Pelton	2.8	Torrifluventic Haploxerolls	R	L			VD	WD	0–3		Loamy-skeletal	Mesic	Xeric	4s	
Pengra	106.3	Vertic Epiaquolls	R				VD	SPD	1–30	L	Fine-silty over clayey	Mesic	Aquic	3w, 4w	
Perdin	88.0	Ultic Haploxeralfs	F		Jeffrey pine		MD	WD	5–90		Fine	Frigid	Xeric	7e	
Perla	38.0	Aridic Argixerolls	R				MD	WD	0–60		Fine	Mesic	Aridic	4s	3e
Pernog	7.9	Lithic Argixerolls					S	WD	15–50		Loamy-skeletal	Frigid	Xeric	7s	
Pernty	347.1	Aridic Lithic Argixerolls					S	WD	2–75	L	Loamy-skeletal	Frigid	Xeric	7s	
Pervina	33.7	Typic Hapludults	F		Douglas-fir		D	WD	2–60		Fine	Mesic	Udic	3e, 4e, 6e	
Philomath	233.0	Vertic Haploxerolls					S	WD	3–70	L	Clayey	Mesic	Xeric	6e, 6s	
Phoenix	3.8	Xeric Epiaquerts	C, R	H			MD	PD	0–3		Very-fine	Mesic	Aquic	4w	
Phys	31.3	Typic Argixerolls	R	H, L			VD	WD	0–30	L	Loamy-skeletal	Mesic	Xeric	3s, 4s	
Picturerock	5.4	Vitritorrandic Haploxerolls					VD	MWD	1–3		Ashy	Frigid	Xeric	6s	
Piersonte	55.7	Vitrandic Haploxerolls	F		Ponderosa pine		VD	WD	35–75		Loamy-skeletal	Frigid	Xeric	6e, 7e, 7s	
Pilchuck	19.8	Dystric Xeropsamments	F		Douglas-fir	M	VD	ED	0–8		Sandy	Mesic	Xeric	4w, 6w	

(continued)

Series name	Area (km^2)	Subgroup	Land use[1]	Range site quality[2]	Forest indicator species	Forest site quality[3]	Depth class[4]	Drainage class[5]	Slope range (%)	Salinity class[6]	Particle-size class	Soil temp. regime	Soil moisture regime	Land capability class	
														No mgmt	With mgmt
Pilot rock	73.7	Haploxerollic Durixerolls	R	H			MD	WD	1–40	L	Coarse-silty	Mesic	Xeric	3e, 4e	
Pilotbutte	142.2	Humic Vitrixerands	F		Ponderosa pine		VD	SED	12–70		Ashy	Frigid	Xeric	6e	
Pinehurst	148.1	Pachic Ultic Argixerolls	F		Douglas-fir	L, M	D	WD	1–50	L	Fine-loamy	Frigid	Xeric	6e	
Pinhead	15.2	Vitric Haplocryands	F		Douglas-fir	L	VD	SED	0–65	L	Medial-skeletal	Cryic	Udic	6s	
Pinuscreek	141.5	Andic Haploxeralfs	F		Grand fir	M	D	WD	0–60	L	Loamy-skeletal	Frigid	Xeric	6c	
Pipp	91.7	Humic Vitrixerands	F		Ponderosa pine	M	D	SED	12–65	L	Ashy-skeletal	Frigid	Xeric	6s	
Pistolriver	0.6	Typic Humaquepts	R	H			VD	SPD	0–3		Coarse-loamy over sandy or sandy-skeletal	Isomesic	Aquic	3w	
Pit	31.7	Xeric Endoaquerts	R	H, M			VD	PD	0–5		Fine	Mesic	Aquic	3w, 5w	4w
Pitcheranch	7.1	Aquandic Endoaquepts					VD	PD	0–1		Ashy	Frigid	Aquic	6w	
Piumpsha	21.7	Alfic Vitricryands					VD	WD	2–40		Medial	Cryic	Udic	6e	
Plainview	20.7	Typic Vitritorrands	C, R	L			MD	WD	0–8	L	Ashy over loamy-skeletal	Mesic	Aridic	6e, 6s	3e, 3s
Plush	17.4	Xeric Haplargids					D	WD	25–50		Loamy-skeletal	Mesic	Aridic	6e	
Poall	322.6	Xeric Paleargids					VD	WD	2–40	L	Fine	Mesic	Aridic	6e	6e
Poden	1.4	Cumulic Haploxerolls	C, R	H			D	WD	0–5	L	Fine-silty over sandy or sandy-skeletal	Mesic	Xeric	2s	
Poe	3.7	Typic Durixerepts	R	H			MD	SPD	0–2	L	Sandy	Mesic	Xeric	3w, 4w	
Pokegema	1.6	Humic Haploxerands	F		Douglas-fir	M	VD-D	WD	1–35	L	Medial	Frigid	Xeric	6e	
Polander	24.3	Typic Vitrixerands	F		White fir		D	WD	0–70		Ashy	Frigid	Xeric	6e	
Pollard	317.8	Typic Palexerults	F		Douglas-fir	L, M	VD	WD	2–60	L	Fine	Mesic	Xeric	4e, 6e	2e
Polly	40.6	Calciargidic Argixerolls	R	H			VD	WD	1–15		Fine-loamy	Mesic	Aridic	2e	
Pomerening	172.9	Vitrandic Torriorthents					VD	ED	2–20		Ashy	Frigid	Aridic	6e	
Ponina	21.7	Abruptic Durixeralfs					S	WD	1–8		Clayey	Frigid	Xeric	7s	
Poorjug	20.5	Lithic Xeric Haplocambids					S	WD	0–15		Loamy-skeletal	Mesic	Aridic	6e	
Porch	46.3	Vitrandic Haploxerolls	F		Ponderosa pine		MD	WD	15–60		Loamy-skeletal	Frigid	Xeric	6s	

(continued)

Series name	Area (km^2)	Subgroup	Land use[1]	Range site quality[2]	Forest indicator species	Forest site quality[3]	Depth class[4]	Drainage class[5]	Slope range (%)	Salinity class[6]	Particle-size class	Soil temp. regime	Soil moisture regime	Land capability class	
														No mgmt	With mgmt
Porterfield	24.7	Vitrandic Torriorthents					S	WD	2–60		Ashy	Mesic	Aridic	6e	
Potamus	39.5	Typic Haploxerolls	R	H			VD	WD	0–2		Loamy-skeletal	Frigid	Xeric	4c	
Poujade	3.5	Durinodic Xeric Natrargids					VD	MWD	0–5	H	Fine-loamy	Frigid	Aridic	6s	
Powder	290.8	Cumulic Haploxerolls	C, R	H			VD	WD	0–3	L	Coarse-silty	Mesic	Xeric	2c, 2w	1
Powderriver	132.1	Lithic Haploxerepts	F		Grand fir		S	WD	15–60		Loamy-skeletal	Frigid	Xeric	7s	
Powell	49.9	Humic Fragixerepts	C, F		Douglas-fir	M	D	SPD	0–30	L	Fine-silty	Mesic	Xeric	3w, 3e, 4e	
Powellbutte	8.6	Vitrandic Argixerolls	R	H, M			MD	WD	0–30		Fine-loamy	Frigid	Xeric	4e, 6e	
Powval	43.6	Pachic Haploxerolls	C, R	H			D	WD	0–3	L	Coarse-silty	Mesic	Xeric	4c	2c
Powwatka	49.3	Vitrandic Argixerolls	R	H			MD	WD	2–30		Fine-silty	Frigid	Xeric	4e	
Prag	162.5	Pachic Palexerolls					MD	WD	2–70	L	Fine	Frigid	Xeric	6e, 7e	
Prairie	35.7	Xeric Vitricryands	F		Ponderosa pine	M	MD	WD	0–50		Ashy over loamy	Cryic	Xeric	6e	
Preacher	1470.2	Andic Humudepts	F		Douglas-fir	M, H	VD	WD	0–90	L	Fine-loamy	Mesic	Udic	6e, 7e	
Price	38.2	Humic Haploxerepts	C, F		Douglas-fir	M	VD	WD	30–90		Fine	Mesic	Xeric	2e, 3e	
Prill	241.2	Pachic Palexerolls					MD	WD	2–55		Fine	Mesic	Xeric	6e	
Prineville	23.4	Durinodic Xeric Haplocambids	C, R	H			D	WD	0–8		Coarse-loamy	Mesic	Aridic	2e	
Pritchard	1.9	Typic Palexerolls	R	H			D	WD	2–20	L	Fine	Mesic	Xeric	3c	3e
Prosser	34.0	Xeric Haplocambids	C, R	H, M			MD	WD	0–60	L	Coarse-loamy	Mesic	Aridic	4e, 6e	2 s
Prouty	67.8	Andic Dystrocryepts	F		Douglas-fir		MD	WD	30–65		Loamy-skeletal	Cryic	Udic	4e	
Provig	10.7	Typic Argixerolls	R	H, M			D	WD	3–35		Clayey-skeletal	Mesic	Xeric	4e, 6e	
Puderbaugh	1.2	Pachic Argixerolls					MD	WD	15–90		Loamy-skeletal	Frigid	Xeric	7e	
Puderbaughridge	2.2	Aridic Lithic Haploxerolls					S	WD	30–90		Loamy-skeletal	Mesic	Aridic	7s	
Puls	17.8	Abruptic Xeric Argidurids					S	WD	0–30		Clayey	Mesic	Aridic	6s	
Purple	2.2	Histic Cryaquepts					VD	PD	0–15		Coarse-loamy over sandy or sandy-skeletal	Cryic	Aquic	4c	
Puzzlebark	2.3	Vitrixerandic Haplodurids					S	WD	0–5		Ashy	Frigid	Aridic	6e	
Puzzlecreek	36.8	Haploxerandic Humicryepts					MD	WD	15–60		Loamy-skeletal	Cryic	Xeric	6e	

(continued)

Series name	Area (km²)	Subgroup	Land use[1]	Range site quality[2]	Forest indicator species	Forest site quality[3]	Depth class[4]	Drainage class[5]	Slope range (%)	Salinity class[6]	Particle-size class	Soil temp. regime	Soil moisture regime	Land capability class	
														No mgmt	With mgmt
Pyburn	10.3	Typic Umbraquults					VD	PD	0–8		Fine	Mesic	Aquic	4w	
Pyrady	9.0	Oxyaquic Palehumults	F		Douglas-fir	L	VD	MWD	0–30		Fine	Frigid	Udic	6e	
Pyropatti	0.4	Vitrandic Argicryolls					D	WD	2–30	L	Ashy-skeletal	Cryic	Xeric	6e	
Quafeno	15.4	Aquultic Haploxerolls	C, F		Douglas-fir		D	MWD	0–15	L	Coarse-loamy	Mesic	Xeric	2e, 2w	
Quailprairie	5.8	Pachic Humudepts					VD	WD	0–60		Fine-loamy	Mesic	Udic	6e	
Quartzville	33.0	Andic Humudepts	F		Douglas-fir	H	VD	WD	2–75		Fine	Mesic	Udic	6e	
Quatama	85.3	Aquultic Haploxeralfs	C, F		Douglas-fir		VD	MWD	0–30	L	Fine-loamy	Mesic	Xeric	2w, 2e, 3e	
Quillamook	14.5	Pachic Melanudands	C, F		Western hemlock		VD	WD	0–15	L	Medial	Isomesic	Udic	2c, 2e	3e
Quincy	545.6	Xeric Torripsamments	R	H, M			VD	ED	0–65	L	Sandy	Mesic	Aridic	3s, 3w, 7e	4e, 4s
Quinton	14.3	Xeric Torripsamments	C				MD	ED	0–30		Sandy	Mesic	Aridic	7e	4e
Quirk	110.3	Vitrandic Palexerolls	F		Ponderosa pine		MD	WD	0–30		Fine	Frigid	Xeric	4e	
Quosatana	12.6	Fluvaquentic Humaquepts	C				VD	PD	0–3		Fine-silty	Mesic	Aquic	3w	
Rabbitcreek	2.8	Typic Haplocambids					VD	WD	0–3		Fine-loamy	Mesic	Aridic	6s	
Rabbithills	198.5	Xereptic Haplodurids					S	WD	0–20	L	Loamy	Mesic	Aridic	6e	
Racing	13.4	Aquandic Cryaquepts					VD	PD	0–3		Fine-loamy	Cryic	Aquic	6w	
Rafton	49.6	Fluvaquentic Endoaquepts	R	H, L			VD	VPD	0–2	L	Fine-silty	Mesic	Aquic	3w, 4w, 6w	
Rail	7.8	Xeric Endoaquerts	C, R	H			VD	SPD	0–2		Fine	Mesic	Aquic	4w	
Rainey	0.5	Entic Haploxerolls	C, R				MD	WD	15–40		Coarse-loamy	Mesic	Xeric	6e	
Ramo	48.6	Typic Argixerolls	C, R	H, L			VD	WD	2–35	L	Fine	Mesic	Xeric	2e, 3e, 4e, 6s	
Randcore	8.0	Lithic Xerorthents					VS-S WD	MWD	0–5		Loamy-skeletal	Mesic	Xeric	7s	
Rastus	35.6	Palexerollic Durixerolls	R	H			MD	WD	1–10		Fine	Frigid	Xeric	4e	
Ratsnest	0.1	Typic Haplargids					MD	WD	3–12		Fine	Mesic	Aridic	6c	6e
Ratto	186.6	Xeric Argidurids					S	WD	2–15	L	Clayey	Frigid	Aridic	6e	
Raz	2279.5	Xeric Haplodurids					S	WD	1–20	L	Loamy	Frigid	Aridic	6e	
Raztack	1.4	Vitrandic Palexeralfs					D	WD	0–1		Fine	Frigid	Xeric	6s	

(continued)

Series name	Area (km²)	Subgroup	Land use[1]	Range site quality[2]	Forest indicator species	Forest site quality[3]	Depth class[4]	Drainage class[5]	Slope range (%)	Salinity class[6]	Particle-size class	Soil temp. regime	Soil moisture regime	Land capability class	
														No mgmt	With mgmt
Reallis	240.6	Durinodic Xeric Haplocambids					VD	WD	0–8	M	Coarse-loamy	Frigid	Aridic	6e, 6s	4e
Reavis	14.8	Calcic Haploxerolls	R	H			VD	WD	0–3	L	Fine-loamy	Frigid	Xeric	3c	
Redbell	22.1	Aquultic Argixerolls	C				D	SPD	0–5	L	Fine	Mesic	Xeric	2w	
Redcanyon	6.2	Calcidic Haploxerolls					MD	WD	30–50		Loamy-skeletal	Mesic	Xeric	7s	
Redcliff	143.3	Aridic Haploxerolls					MD	WD	0–75	L	Loamy-skeletal	Mesic	Xeric	6e, 7e	
Redcone	5.4	Typic Duricryands	F		Lodgepole pine		MD	SED	30–60		Ashy-skeletal	Cryic	Udic	7e	
Redflat	14.2	Dystric Eutrudepts	F		Douglas-fir	H	VD	WD	0–60		Fine-loamy	Mesic	Udic	6e	
Redmond	99.6	Vitritorrandic Haploxerolls	C, R	H, L			MD	WD	0–15	L	Fine-loamy	Mesic	Xeric	2e, 6s	3s
Redmount	21.6	Pachic Haploxerolls	R	H			VD	WD	0–8	L	Coarse-loamy	Frigid	Xeric	3c, 3e	
Redslide	11.4	Vitritorrandic Haploxerolls					MD	WD	15–50		Loamy-skeletal	Mesic	Xeric	6e	
Reedsport	254.3	Andic Humudepts	F		Douglas-fir	H	MD	WD	0–90	L	Fine-loamy	Isomesic	Udic	7e	
Reese	165.7	Duric Halaquepts					VD	PD	0–2	H	Fine-loamy	Mesic	Aquic	6s	
Reinecke	7.3	Typic Vitrixerands	F		Douglas-fir	M	VD	WD	0–5		Ashy over loamy	Mesic	Xeric	4e	
Reinhart	6.4	Lithic Humudepts					S	WD	0–60		Loamy-skeletal	Isomesic	Udic	7e	
Reluctan	343.0	Aridic Argixerolls					MD	WD	0–50	L	Fine-loamy	Frigid	Xeric	6e	
Remote	233.8	Typic Dystrudepts	F		Douglas-fir	L–H	VD	WD	3–90	L	Loamy-skeletal	Mesic	Udic	6e, 7e	
Reston	12.5	Lithic Ultic Haploxerolls					VS	WD	3–75		Loamy	Mesic	Xeric	7e	
Reuter	22.5	Vitritorrandic Haploxerolls	C				S	WD	0–30		Loamy	Mesic	Xeric	6s	3s
Reywat	16.9	Aridic Lithic Argixerolls					S	WD	0–90		Loamy-skeletal	Mesic	Aridic	6s	
Rhea	122.8	Calcic Haploxerolls	C, R	H			D	WD	1–50	L	Fine-silty	Mesic	Xeric	2e, 3e, 4e	
Ricco	6.3	Fluvaquentic Vertic Endoaquolls	C				VD	PD	0–3		Fine	Mesic	Aquic	4w	
Riceton	1.1	Pachic Ultic Haploxerolls	C, R				VD	WD	0–12		Coarse-loamy	Frigid	Xeric	3e	3e
Rickreall	20.3	Xeric Haplohumults					S	WD	3–75	L	Clayey	Mesic	Udic	6e, 7e	
Riddleranch	114.7	Aridic Haploxerolls					MD	WD	15–70	L	Loamy-skeletal	Frigid	Xeric	6e	

(continued)

Series name	Area (km²)	Subgroup	Land use[1]	Range site quality[2]	Forest indicator species	Forest site quality[3]	Depth class[4]	Drainage class[5]	Slope range (%)	Salinity class[6]	Particle-size class	Soil temp. regime	Soil moisture regime	Land capability class	
														No mgmt	With mgmt
Ridenbaugh	24.3	Abruptic Xeric Argidurids	C, R				S	WD	0–4		Clayey	Mesic	Aridic	6s	4s
Ridley	43.2	Pachic Palexerolls	R	H			D	WD	2–12		Fine	Mesic	Xeric	3e	3e
Rilea	45.6	Typic Dystrudepts	F		Douglas-fir	L	MD	WD	0–90		Loamy-skeletal	Frigid	Udic	6e	
Rinconflat	119.6	Xeric Haplocambids					VD	WD	3–10	L	Loamy-skeletal	Frigid	Aridic	6e	
Rinearson	402.4	Typic Humudepts	F		Douglas-fir	H	D	WD	0–90	L	Fine-silty	Mesic	Udic	6e, 7e	
Rio king	54.5	Aridic Haploxerolls	C, R	L, M			VD	MWD	0–6	L	Coarse-loamy	Mesic	Xeric	6c, 6e	3c, 3e
Risley	199.2	Xeric Haplargids					MD	WD	2–50		Fine	Mesic	Aridic	6e	
Ritner	291.8	Humic Haploxerepts	F		Douglas-fir	L–H	MD	WD	2–90	L	Clayey-skeletal	Mesic	Xeric	4s, 6s, 7s	
Ritzville	1133.4	Calcidic Haploxerolls	R	H, M			D-VD	WD	0–70	L	Coarse-silty	Mesic	Xeric	3e, 4e, 6e	2e, 3e
Roanhide	4.2	Ultic Haploxerolls					MD	WD	4–60		Coarse-loamy	Frigid	Xeric	7e	
Robinette	29.6	Pachic Argixerolls	R	H			D	WD	2–12		Fine-loamy	Mesic	Xeric	3e	3e
Robson	666.1	Lithic Xeric Haplargids					S	WD	2–75	L	Clayey-skeletal	Frigid	Aridic	7s	
Roca	103.4	Xeric Haplargids					MD	WD	4–75		Clayey-skeletal	Frigid	Aridic	6e	
Rocconda	7.5	Lithic Xeric Haplargids					VS-S	WD	2–50		Clayey-skeletal	Mesic	Aridic	7s	
Rockford	9.5	Ultic Haploxerolls	F		Ponderosa pine	M	VD	WD	0–30		Loamy-skeletal	Mesic	Xeric	6s, 7s	4s
Rockly	774.9	Lithic Haploxerolls					VS-S	WD	0–120	L	Loamy-skeletal	Mesic	Xeric	7s	
Rogerson	37.8	Abruptic Xeric Argidurids	R				S	WD	2–6		Clayey	Mesic	Aridic	6s	4e
Rogger	68.1	Ultic Haploxerolls	F		Ponderosa pine		MD	WD	0–60		Loamy-skeletal	Frigid	Xeric	6e	
Rogue	50.6	Typic Dystroxerepts	F		Douglas-fir	L, M	D	SED	12–80	L	Coarse-loamy	Frigid	Xeric	6e	
Roloff	52.1	Aridic Haploxerolls	R	H			MD	WD	0–70	L	Coarse-loamy	Mesic	Xeric	3c, 3e	
Romanose	7.1	Lithic Hapludands	F		Douglas-fir	L, M	S	WD	30–90		Medial-skeletal	Frigid	Udic	7e	
Rondowa	19.9	Pachic Haploxerolls	F		Douglas-fir	L	VD	WD	2–60		Loamy-skeletal	Frigid	Xeric	3s, 4s, 6s	
Roostercomb	178.5	Typic Argixerolls	R	H			MD	WD	2–50	L	Clayey-skeletal	Frigid	Xeric	4e	
Roschene	44.1	Cumulic Haploxerolls	R	H			VD	MWD	0–3		Fine-loamy	Frigid	Xeric	4c	4c
Roseburg	22.8	Pachic Ultic Argixerolls	C				VD	WD	0–3	L	Fine-loamy	Mesic	Xeric	1	

(continued)

Series name	Area (km^2)	Subgroup	Land use[1]	Range site quality[2]	Forest indicator species	Forest site quality[3]	Depth class[4]	Drainage class[5]	Slope range (%)	Salinity class[6]	Particle-size class	Soil temp. regime	Soil moisture regime	Land capability class	
														No mgmt	With mgmt
Rosehaven	107.5	Ultic Haploxeralfs	C, F		Douglas-fir	M, H	VD	WD	3–90	L	Fine-loamy	Mesic	Xeric	2e-4e, 6e, 7e	
Rouen	13.4	Typic Vitrixerands	F		Douglas-fir	L, M	MD	WD	2–50		Ashy over loamy-skeletal	Frigid	Xeric	6e	
Royal	23.5	Xeric Haplocambids	C, R	L			VD	WD	0–40	L	Coarse-loamy	Mesic	Aridic	6e	3e
Royst	259.4	Pachic Argixerolls	F		Douglas-fir	M	MD	WD	0–70	L	Clayey-skeletal	Frigid	Xeric	4e, 6e, 7s	
Ruch	86.0	Mollic Palexeralfs	F		Douglas-fir	M, H	VD	WD	2–60	L	Fine-loamy	Mesic	Xeric	4e	2e
Ruckles	629.7	Aridic Lithic Argixerolls					S	WD	0–80	L	Clayey-skeletal	Mesic	Xeric	7e, s	
Ruclick	411.1	Aridic Argixerolls					MD	WD	0–70	L	Clayey-skeletal	Mesic	Xeric	6s	
Ruddley	28.2	Ultic Argixerolls	F		Ponderosa pine		S	WD	2–30		Loamy	Frigid	Xeric	6e	
Rustlerpeak	25.5	Haploxerandic Humicryepts	F		White fir	M	MD	WD	3–70		Loamy-skeletal	Cryic	Xeric	6e, 7e	
Rustybutte	7.7	Typic Hapludolls	F		Douglas-fir		MD	WD	0–60		Loamy-skeletal	Isomesic	Udic	6e	
Rutab	6.8	Xeric Haplocambids					VD	WD	0–15		Loamy-skeletal	Frigid	Aridic	6e	
Sach	4.7	Alic Hapludands	F		Douglas-fir	M	MD	WD	0–90		Medial over loamy-skeletal	Frigid	Udic	6e	
Saddlepeak	24.0	Typic Dystrudepts	F		Douglas-fir	H	VD	WD	0–90		Loamy-skeletal	Frigid	Udic	6e	
Sag	28.9	Pachic Argixerolls	R	H, M			D-VD	WD	12–90		Fine-loamy	Frigid	Xeric	4e, 6e, 7e	
Sagehen	78.7	Lithic Xeric Haplocambids					S	WD	0–70		Loamy-skeletal	Frigid	Aridic	7s	
Sagehill	255.1	Xeric Haplocalcids	C, R	H, M			VD-D	WD	0–60	L	Coarse-loamy	Mesic	Aridic	2e, 2s, 3s, 4e, 6e	2e, 3e
Sagemoor	13.9	Xeric Haplocambids	C, R				VD	WD	0–60		Coarse-silty	Mesic	Aridic	6e	
Sagley	4.1	Pachic Argixerolls	R	H			D	WD	12–55		Loamy-skeletal	Mesic	Xeric	4e	
Sahaptin	6.2	Lithic Ultic Haploxerolls	F		Douglas-fir	M	S	WD	30–90		Clayey-skeletal	Mesic	Xeric	6s, 7e	
Salander	153.2	Typic Fulvudands	F		Douglas-fir	M, H	VD	WD	0–90	L	Medial	Isomesic	Udic	6e, 7e	
Salem	108.1	Pachic Ultic Argixerolls	C, F		Douglas-fir	H	VD	WD	0–12	L	Fine-loamy over sandy or sandy-skeletal	Mesic	Xeric	2s	2s

(continued)

Series name	Area (km^2)	Subgroup	Land use[1]	Range site quality[2]	Forest indicator species	Forest site quality[3]	Depth class[4]	Drainage class[5]	Slope range (%)	Salinity class[6]	Particle-size class	Soil temp. regime	Soil moisture regime	Land capability class	
														No mgmt	With mgmt
Salhouse	66.6	Vitrandic Torripsamments					VD	SED	0–20	M	Ashy	Frigid	Aridic	6e	
Salisbury	47.3	Palexerollic Durixerolls	C, R	H			MD	WD	0–45	L	Fine	Mesic	Xeric	2e	
Salkum	120.5	Xeric Palehumults	C, F		Douglas-fir	H	VD	WD	0–65	L	Fine	Mesic	Udic	2e, 2w, 3e	
Sandgap	24.3	Haploduridic Torripsamments					VD	SED	1–8		Sandy	Frigid	Aridic	6e	
Sandrock	7.1	Lithic Xeric Haplargids	R				S	WD	0–5		Ashy	Frigid	Aridic	6e, 6s	
Sankey	4.5	Lithic Humudepts					S	WD	0–60		Loamy-skeletal	Mesic	Udic	7e	
Santiam	67.7	Aquultic Haploxeralfs	C, F		Douglas-fir	M	VD	MWD	2–20	L	Fine	Mesic	Xeric	2e, 2w, 3e	
Saturn	33.7	Fluventic Humudepts	F		Douglas-fir	H	VD	WD	0–5	L	Fine-loamy over sandy or sandy-skeletal	Mesic	Udic	3s	
Saum	158.5	Ultic Palexeralfs	C, F		Douglas-fir	M, H	VD	WD	2–90	L	Fine	Mesic	Xeric	2e-4e, 6e	
Sauvie	100.3	Fluvaquentic Endoaquolls	C				D	PD	0–3	L	Fine-silty	Mesic	Aquic	6e	2w
Sawtell	12.3	Oxyaquic Argixerolls	C, F		Douglas-fir	H	VD	MWD	0–15	L	Loamy-skeletal	Mesic	Xeric	2w	Hw
Scalerock	8.2	Lithic Dystrudepts	F		Douglas-fir	L, M	S	WD	39–90		Loamy-skeletal	Frigid	Udic	7e	
Scaponia	173.6	Humic Dystrudepts	F		Douglas-fir	M, H	D	WD	0–90	L	Fine-loamy	Mesic	Udic	6e	
Scaredman	27.0	Typic Humudepts	F		Douglas-fir	M, H	MD	WD	30–90		Loamy-skeletal	Frigid	Udic	6e	
Scarpal	11.1	Humic Vitrixerands	F		Grand fir		VD	WD	20–70		Ashy-skeletal over loamy-skeletal	Frigid	Xeric	7s	
Scherrard	0.7	Natric Duraquolls	C, R	H			MD	SPD	0–1		Fine	Mesic	Aquic	4w	
Schrier	19.0	Calcic Pachic Haploxerolls	C, R	H, M			VD, D	WD	2–90	L	Fine-loamy	Mesic	Xeric	2e-4e, 6e	
Schuelke	7.4	Calcic Argixerolls					MD	WD	2–90		Loamy-skeletal	Mesic	Xeric	6e	
Searles	117.9	Aridic Argixerolls					MD	WD	0–80	L	Loamy-skeletal	Mesic	Xeric	6e, 7s	
Sebastian	6.5	Lithic Hapludolls					S	WD	0–60		Loamy-skeletal	Isomesic	Udic	7e	
Segundo	58.8	Typic Haploxerepts	F		Ponderosa pine	L, M	D	WD	2–75		Loamy-skeletal	Frigid	Xeric	4e, 6e, 7e	
Seharney	122.8	Xereptic Haplodurids					S	WD	3–20	L	Loamy-skeletal	Frigid	Aridic	7s	
Selmac	23.7	Ultic Haploxeralfs	F		Douglas-fir	M	VD	MWD	2–20	L	Fine-loamy over clayey	Mesic	Xeric	4e	3e

(continued)

Series name	Area (km^2)	Subgroup	Land use[1]	Range site quality[2]	Forest indicator species	Forest site quality[3]	Depth class[4]	Drainage class[5]	Slope range (%)	Salinity class[6]	Particle-size class	Soil temp. regime	Soil moisture regime	Land capability class	
														No mgmt	With mgmt
Semiahmoo	5.0	Typic Haplosaprists	C, R	H			VD	VPD	0–3			Mesic	Aquic	3w	
Senra	171.9	Vitritorrandic Durixerolls					S	WD	0–20	L	Ashy	Frigid	Xeric	6e	
Serpentano	42.3	Dystric Eutrudepts	F		Douglas-fir	M, H	D	WD	3–70		Loamy-skeletal	Mesic	Udic	6e, 7e	
Sevencedars	20.8	Typic Fulvicryands	F		Douglas-fir	H	VD	WD	5–100		Medial-skeletal	Cryic	Udic	6e, 7e	
Sevenoaks	2.2	Psammentic Haploxerolls	C, R	L			D	SED	0–3	L		Mesic	Xeric	4s	3s
Shanahan	332.6	Xeric Vitricryands	F		Ponderosa pine	L, M	VD	SED	0–45	L	Ashy over loamy	Cryic	Xeric	6s, 6e	
Shangland	11.5	Ultic Haploxerolls	C, R	H			MD	WD	2–35		Coarse-loamy	Mesic	Xeric	4e	
Shano	207.2	Xeric Haplocambids	C, R	H			VD	WD	0–65	L	Coarse-silty	Mesic	Aridic	4e	2e, 3e
Sharesnout	242.7	Typic Argixerolls					MD	WD	3–35		Clayey-skeletal	Frigid	Xeric	7s	
Sharpshooter	61.1	Ultic Haploxerolls	F		Ponderosa pine	H	D	WD	3–90		Fine-loamy	Mesic	Xeric	3e-4e, 6e, 7e	
Shastacosta	6.7	Typic Palexerults	F		Douglas-fir	M	VD	WD	2–60		Loamy-skeletal	Mesic	Xeric	6e	
Shawave	5.3	Xeric Haplargids					VD	WD	2–30		Fine-loamy	Mesic	Aridic	6c	3e
Sheepcreek	2.2	Andic Haploxerolls					MD	MWD	0–15		Fine-loamy	Frigid	Xeric	6c	
Shefflein	56.8	Mollic Haploxeralfs	F		Ponderosa pine	M	D	WD	2–35	L	Fine-loamy	Mesic	Xeric	4e, 6e	2e
Sherar	21.0	Aridic Argixerolls					MD	WD	0–70	L	Fine	Mesic	Xeric	6e, 7e	
Sherod	14.0	Lithic Haploxerepts					S	SPD	0–3		Loamy-skeletal	Frigid	Xeric	6w	
Sherval	19.0	Pachic Ultic Argixerolls	F		Trembling aspen		VD	WD	5–20		Loamy-skeletal	Frigid	Xeric	6e	
Shippa	27.2	Lithic Dystroxerepts	F		Douglas-fir	L	S	WD	30–90		Loamy-skeletal	Mesic	Xeric	6e	
Shiva	14.5	Vitrandic Haploxerolls	R	H, M			VD	WD	0–55		Ashy	Mesic	Xeric	3e, 4e, 6e	
Shivigny	32.6	Typic Palehumults	F		Douglas-fir	M, H	VD	WD	3–90		Clayey-skeletal	Mesic	Udic	6e	
Shoat	4.0	Typic Haploxerepts	R	H			MD	WD	0–5		Fine-loamy	Mesic	Xeric	4e	
Shoepeg	0.3	Cumulic Haploxerolls	C				D-VD	PD	0–3		Fine-loamy	Mesic	Xeric	3w	3w
Shroyton	9.0	Humic Xeric Vitricryands	F		Ponderosa pine		D	WD	30–50		Ashy	Cryic	Xeric	6e	

(continued)

Series name	Area (km^2)	Subgroup	Land use[1]	Range site quality[2]	Forest indicator species	Forest site quality[3]	Depth class[4]	Drainage class[5]	Slope range (%)	Salinity class[6]	Particle-size class	Soil temp. regime	Soil moisture regime	Land capability class	
														No mgmt	With mgmt
Shukash	603.0	Xeric Vitricryands	F		Grand fir	M	VD	SED	0–65	L	Ashy over loamy-skeletal	Cryic	Xeric	6e	
Sibannac	7.2	Cumulic Endoaquolls	R	H			D	PD	0–7		Fine-loamy	Frigid	Aquic	4w	
Sibold	14.4	Aquultic Argixerolls	C				VD	SPD	0–5	L	Fine-loamy	Mesic	Xeric	3w	3w
Sidlake	120.9	Xeric Haplargids					MD	WD	1–30		Fine-loamy	Mesic	Aridic	6c	
Sifton	32.0	Typic Melanoxerands	C, R	H, L			D	WD	0–8	L	Medial over sandy or sandy-skeletal	Mesic	Xeric	3s, 4s, 6e	
Siletz	10.2	Typic Fulvudands	C, F		Douglas-fir	M, H	VD	WD	0–7	L	Medial over loamy	Isomesic	Udic	2c, 2e, 3s	
Silverash	63.9	Aquandic Palexeralfs					VD	PD	0–2		Fine	Frigid	Xeric	6w	
Silverlake	9.5	Calcic Argixerolls	R	H			VD	WD	0–3	L	Fine	Frigid	Xeric	3c	
Silverton	12.1	Pachic Ultic Argixerolls	F		Douglas-fir		MD	WD	2–20		Fine	Mesic	Xeric	3e	
Silvies	87.4	Vertic Cryaquolls	R	M			VD	PD	0–7		Fine	Cryic	Aquic	5w	
Simas	772.5	Vertic Palexerolls					VD	WD	0–80	L	Fine	Mesic	Xeric	6e, 7s	
Simnasho	110.8	Alfic Vitrixerands	F		Douglas-fir	L, M	MD	WD	0–40	L	Ashy-skeletal over loamy-skeletal	Frigid	Xeric	6s	
Simon	5.7	Aridic Argixerolls					VD	WD	2–50	L	Fine-loamy	Frigid	Xeric	6e	
Sinamox	14.1	Pachic Haploxerolls	C, R	H			D	WD	0–70	L	Fine-loamy	Mesic	Xeric	3e, 4e	
Sinker	100.1	Pachic Haploxerolls					MD	WD	12–80		Loamy-skeletal	Frigid	Xeric	7e	
Siskiyou	94.2	Typic Dystroxerepts	F		Douglas-fir	L, M	MD	SED	20–90	L	Coarse-loamy	Mesic	Xeric	6e, 7e	
Sisley	17.1	Typic Xerorthents	F		Ponderosa pine	L	MD	WD	2–70		Loamy-skeletal	Frigid	Xeric	4e, 6e, 7e	
Sisters	47.5	Humic Vitrixerands	F		Ponderosa pine		D-VD	WD	0–50		Ashy over loamy	Frigid	Xeric	6e	
Sitkum	15.2	Typic Dystroxerepts	F		Douglas-fir	M	MD	SED	30–90		Coarse-loamy	Mesic	Xeric	7e	
Sitton	3.5	Ultic Haploxeralfs	F		Douglas-fir		VD	WD	30–60		Fine-loamy	Mesic	Xeric	7e	
Sixes	1.6	Pachic Humudepts					MD	WD	0–60		Fine-loamy	Mesic	Udic	6e	
Skedaddle	184.8	Lithic Xeric Torriorthents					VS-S	WD	4–75	L	Loamy-skeletal	Mesic	Aridic	7s	
Skidbrackle	0.5	Lithic Argixerolls					S	WD	2–30		Ashy-skeletal	Frigid	Xeric	6s	
Skidoosprings	104.3	Duric Halaquepts					D	MWD	0–3	H	Coarse-loamy	Frigid	Aquic	6s, 7e	
Skinner	10.1	Typic Dystrudepts	F		Douglas-fir		MD	WD	25–75		Fine-loamy	Mesic	Udic	6e	

(continued)

Series name	Area (km^2)	Subgroup	Land use[1]	Range site quality[2]	Forest indicator species	Forest site quality[3]	Depth class[4]	Drainage class[5]	Slope range (%)	Salinity class[6]	Particle-size class	Soil temp. regime	Soil moisture regime	Land capability class	
														No mgmt	With mgmt
Skipanon	215.3	Andic Humudepts	F		Douglas-fir	M, H	VD-D	WD	3–60	L	Fine-loamy	Isomesic	Udic	6e	
Skooker	14.3	Vitrandic Argixerolls	F		Ponderosa pine	L	D	WD	0–20		Loamy-skeletal	Mesic	Xeric	4e	
Skookum	40.4	Pachic Ultic Argixerolls					MD	WD	1–70		Clayey-skeletal	Mesic	Xeric	7e	
Skookumhouse	62.1	Typic Haplohumults	F		Douglas-fir	H	D	WD	0–30		Fine	Mesic	Udic	6e	
Skoven	6.5	Aridic Argixerolls					S	WD	0–15	L	Clayey-skeletal	Mesic	Xeric	6s	
Skull creek	38.4	Vitrixerandic Haplodurids	R				MD	WD	0–8		Coarse-loamy	Mesic	Aridic	7s	3e
Skullgulch	57.8	Pachic Palexerolls	R	H, M			VD	WD	7–60		Fine	Frigid	Xeric	4e, 6e	
Skunkfarm	71.7	Typic Endoaquolls	R	M			VD	SPD	0–2		Fine-loamy	Frigid	Aquic	5w	5w
Skyline	99.7	Typic Haploxerolls					S	WD	5–70	L	Loamy	Mesic	Xeric	7s	
Slayton	12.3	Lithic Xeric Haplocambids	R	H			S	WD	1–40		Loamy	Mesic	Aridic	4e	
Slicklog	1.9	Humic Vitrixerands	F		Ponderosa pine		VD	WD	15–90		Ashy-skeletal	Frigid	Xeric	6e, 7e	
Slickrock	380.3	Alic Hapludands	F		Douglas-fir	M, H	VD-D	WD	0–75	L	Medial over loamy	Mesic	Udic	6e	
Sliptrack	19.3	Vitritorrandic Durixerolls					MD	WD	0–8		Ashy	Frigid	Xeric	6e	
Smiling	236.4	Alfic Vitrixerands	F		Ponderosa pine	M	VD-D	WD	0–70	L	Ashy over loamy	Frigid	Xeric	6e	
Snakepit	9.8	Cambidic Durixerolls					MD	SED	0–3		Sandy	Frigid	Xeric	6s, 7e	
Snaker	62.5	Lithic Xeric Torriorthents					S	WD	30–80		Loamy-skeletal	Mesic	Aridic	7e	
Snell	335.9	Pachic Argixerolls					MD	WD	0–90	L	Clayey-skeletal	Frigid	Xeric	7e, 7s	
Snellby	50.6	Aridic Argixerolls					MD	WD	12–80	L	Clayey-skeletal	Frigid	Xeric	7e	
Snow	22.4	Cumulic Haploxerolls	C, R	H			VD	WD	0–30	L	Fine-silty	Mesic	Xeric	2c	1
Snowbrier	5.5	Humic Dystroxerepts	F		Douglas-fir	L–H	MD	WD	25–50		Loamy-skeletal	Frigid	Xeric	6e	
Snowcamp	21.1	Dystric Eutrudepts	F		Douglas-fir	H	MD	WD	0–90		Loamy-skeletal	Frigid	Udic	6e	
Snowlin	2.1	Andic Humicryepts	F		White fir	M	D	WD	0–60		Fine-loamy	Cryic	Xeric	6e	
Snowmore	1381.0	Xeric Argidurids	R				MD	WD	0–30		Fine-loamy	Mesic	Aridic	6C	3s
Softscrabble	3.3	Pachic Argixerolls					VD	WD	4–75		Loamy-skeletal	Frigid	Xeric	6e	
Solarview	23.2	Xeric Torripsamments					S	ED	35–65			Mesic	Aridic	7e	

(continued)

Series name	Area (km^2)	Subgroup	Land use[1]	Range site quality[2]	Forest indicator species	Forest site quality[3]	Depth class[4]	Drainage class[5]	Slope range (%)	Salinity class[6]	Particle-size class	Soil temp. regime	Soil moisture regime	Land capability class	
														No mgmt	With mgmt
Sonoma	11.6	Aeric Fluvaquents	C				VD	PD	0–2		Fine-silty	Mesic	Aquic	7w	6w
Soosap	16.2	Typic Haplocryands	F		Douglas-fir	M	MD	WD	5–30		Medial	Cryic	Udic	6e	
Sopher	36.1	Vitrandic Haploxeralfs	F		Douglas-fir	L, M	D	WD	15–90		Clayey-skeletal	Mesic	Xeric	6e	
Sorefoot	26.3	Xerertic Haplargids	C				VD	WD	1–8		Fine	Mesic	Aridic	6e, 7e	
Sorf	118.0	Vertic Paleargids					MD	WD	2–60		Fine	Mesic	Aridic	6e, 7e	
Southcat	86.6	Sodic Haplocambids					VD	SED	0–10	H	Sandy	Mesic	Aridic	6e	
Spangenburg	41.5	Xeric Paleargids	C, R				VD	MWD-WD	0–2	L	Fine	Mesic	Aridic	6s	3s
Spartabutte	470.6	Vitrandic Argixerolls	F		Ponderosa pine		MD	WD	0–60		Fine-loamy	Frigid	Xeric	6c	
Speaker	469.8	Ultic Haploxeralfs	F		Douglas-fir	L–H	MD	WD	2–75	L	Fine-loamy	Mesic	Xeric	6e	
Spiderhole	23.5	Xeric Argidurids	R				S	WD	2–15	L	Loamy-skeletal	Frigid	Aridic	7s	
Spilyay	7.0	Ultic Palexerolls	F		Ponderosa pine	L	VD	WD	0–30		Fine	Mesic	Xeric	4e	
Springwater	23.3	Humic Haploxerepts	F		Douglas-fir	M, H	MD	WD	2–60	L	Fine-loamy	Mesic	Xeric	3e, 4e, 6e	
Srednic	19.1	Vitrixerandic Haplodurids					MD	WD	2–20		Ashy	Frigid	Aridic	6e	
Stackyards	25.9	Typic Humudepts	F		Douglas-fir	M, H	D	WD	30–90		Loamy-skeletal	Frigid	Udic	7e	
Stampede	74.3	Vertic Durixerolls					MD	WD	1–20		Fine	Frigid	Xeric	6e	
Stanfield	37.3	Aquic Haplodurids	R	L			MD	MWD	0–3		Coarse-silty	Mesic	Aridic	4s, 6s	3s, 4s
Stanflow	9.9	Typic Halaquepts	C, R	L			MD	WD	0–2	H	Coarse-silty	Mesic	Aquic	4s	4s
Starbuck	11.9	Lithic Xeric Haplocambids					S	WD	0–65		Loamy	Mesic	Aridic	6e, 7s	
Starkey	99.0	Typic Argixerolls					S	WD	2–50		Clayey-skeletal	Mesic	Xeric	7s	
Statz	71.4	Vitritorrandic Durixerolls					S	WD	0–30		Loamy	Mesic	Xeric	6e	
Stauffer	1.3	Vitritorrandic Argixerolls					VD	WD	0–2		Ashy	Frigid	Xeric	6c	
Stavely	1.2	Typic Haploxerepts	F		Ponderosa pine		VD	WD	10–60		Coarse-loamy	Frigid	Xeric	6e	
Stayton	13.8	Lithic Haploxerands					S	WD	1–40		Medial	Mesic	Xeric	6e	
Stearns	5.1	Cumulic Endoaquolls	C, R	H			VD	PD	0–2		Fine-loamy	Mesic	Aquic	4w	
Steiger	537.9	Xeric Vitricryands	F		Ponderosa pine	L, M	VD	SED	0–65	L	Ashy	Cryic	Xeric	3s, 6e, 6s	
Steinmetz	39.3	Typic Dystroxerepts	F		Douglas-fir	M, H	VD	SED	30–90		Coarse-loamy	Mesic	Xeric	6e	

(continued)

Series name	Area (km^2)	Subgroup	Land use[1]	Range site quality[2]	Forest indicator species	Forest site quality[3]	Depth class[4]	Drainage class[5]	Slope range (%)	Salinity class[6]	Particle-size class	Soil temp. regime	Soil moisture regime	Land capability class	
														No mgmt	With mgmt
Steiwer	74.5	Ultic Haploxerolls	F		Ponderosa pine		MD	WD	3–60	L	Fine-loamy	Mesic	Xeric	3e, 4e, 6e	
Stices	15.5	Typic Vitrixerands	F		Douglas-fir	L	VD	WD	35–80		Ashy-skeletal	Frigid	Xeric	6e, 7e	
Stinger	3.8	Typic Dystrudepts	F		Douglas-fir	H	MD	SED	60–90		Coarse-loamy	Mesic	Udic	7e	
Stirfry	2.7	Typic Cryosaprists					VD	VPD	0–15			Cryic	Aquic	6w	
Stockdrive	19.5	Typic Natrixeralfs	C, R				VD	SPD	0–1	L	Fine-loamy	Mesic	Xeric	6s	
Stockel	5.7	Aquultic Haploxeralfs	C, F		Douglas-fir		VD	SPD	3–12		Fine-loamy	Mesic	Xeric	3w	3w
Stookmoor	213.0	Vitritorrandic Haploxerolls					MD	SED	1–50	L	Ashy	Frigid	Xeric	6e	
Stovepipe	3.2	Typic Cryaquolls	R	M			D	PD	0–3		Coarse-silty over sandy or sandy-skeletal	Cryic	Aquic	5w	
Straight	81.3	Typic Dystroxerepts	F		Douglas-fir	L, M	MD	WD	12–80		Loamy-skeletal	Mesic	Xeric	6e, 6s	
Stukel	260.0	Aridic Lithic Haploxerolls	C, R				S	WD	0–40	L	Loamy	Mesic	Xeric	6e	4e
Sturgill	3.7	Fluvaquentic Endoaquolls	C, R	H			VD	PD	0–2		Fine-silty	Frigid	Aquic	4w	
Succor	16.2	Typic Palexerolls					VD	WD	3–35		Fine	Mesic	Xeric	4e	
Suckerflat	161.0	Aridic Lithic Haploxerolls					S	WD	0–40	L	Ashy	Frigid	Xeric	6e	
Suilotem	7.9	Aquic Vitrixerands	F		Ponderosa pine		VD	SPD	0–8		Ashy	Frigid	Xeric	6e	
Sumine	7.3	Aridic Argixerolls					MD	WD	8–75		Loamy-skeletal	Frigid	Aridic	7s	
Sumpley	4.6	Aquic Haploxerolls	C, R	H			D	SPD	0–3		Fine-loamy over sandy or sandy-skeletal	Frigid	Xeric	3c	3c
Sunnotch	15.9	Typic Vitricryands					VD	SED	0–35		Ashy-skeletal	Cryic	Udic	6e	
Sunriver	37.6	Aquic Vitricryands	F		Engelmann spruce		VD	SPD	0–3		Ashy over loamy	Cryic	Udic	6c	
Suppah	7.6	Vitrixerandic Haplocambids					MD	SED	12–30		Ashy-pumiceous	Mesic	Aridic	7s	
Sutherlin	112.9	Ultic Haploxeralfs	F		Douglas-fir	M	VD	MWD	3–60	L	Fine-loamy over clayey	Mesic	Xeric	4e, 6e	
Suttle	12.5	Humic Vitrixerands	F		Ponderosa pine		VD	SPD	0–15		Ashy	Frigid	Xeric	6e	
Suver	26.1	Aquic Haplohumults	C, R		Or white oak	H	D	SPD	3–50		Fine	Mesic	Xeric	3e, 4e	
Svensen	119.1	Andic Humudepts	F		Douglas-fir	H	VD-D	WD	0–90	L	Fine-loamy	Isomesic	Udic	6e, 7e	

(continued)

Series name	Area (km^2)	Subgroup	Land use[1]	Range site quality[2]	Forest indicator species	Forest site quality[3]	Depth class[4]	Drainage class[5]	Slope range (%)	Salinity class[6]	Particle-size class	Soil temp. regime	Soil moisture regime	Land capability class	
														No mgmt	With mgmt
Swaler	90.7	Xeric Paleargids					VD	MWD	0–2		Fine	Frigid	Aridic	6s, 6w	
Swalesilver	204.7	Aquic Palexeralfs					VD	SPD	0–2	L	Fine	Frigid	Xeric	6w	
Swartz	1.5	Vertic Palexeralfs	C, R	H			VD	SPD	0–3		Fine	Mesic	Xeric	4w	
Swedeheaven	7.0	Typic Humudepts					MD	WD	0–60		Loamy-skeletal	Mesic	Udic	6e	
Sweetbriar	25.1	Ultic Haploxeralfs	F		Douglas-fir	H	VD	WD	3–60		Fine	Mesic	Xeric	3e, 4e, 6e	
Sweitberg	13.9	Pachic Argixerolls	C, R	H			MD	WD	2–30		Fine	Frigid	Xeric	4e	
Sweiting	1.0	Pachic Ultic Argixerolls	C, F		Ponderosa pine		MD	WD	2–30		Fine	Frigid	Xeric	4e	
Sycan	2.4	Xeric Vitricryands	C, R	L, H			VD	SED	0–2		Ashy	Cryic	Xeric	4s, 4w	
Syrupcreek	787.9	Alfic Udivitrands	F		Grand fir	L, M	MD	WD	0–60	L	Ashy over loamy-skeletal	Frigid	Udic	6e	
Tablerock	6.1	Pachic Argixerolls	R	H			D	MWD	20–50		Clayey-skeletal	Mesic	Xeric	4e	
Tahkenitch	2.5	Typic Humudepts	F		Douglas-fir	H	D	WD	3–75		Coarse-loamy	Mesic	Udic	6e, 7e	
Takilma	25.8	Entic Ultic Haploxerolls	R				VD	WD	0–3		Loamy-skeletal	Mesic	Xeric	4s	4s
Talapus	17.4	Typic Fulvicryands	F		Douglas-fir	L	VD	WD	2–60		Medial-skeletal	Cryic	Udic	7s	
Tallowbox	107.4	Typic Haploxerepts	F		Ponderosa pine	M	MD	SED	20–70	L	Coarse-loamy	Mesic	Xeric	6e, 7e	
Tamara	144.9	Alfic Udivitrands	F		Grand fir	M	VD	WD	0–60		Ashy over loamy	Frigid	Udic	6e	
Tamarack	0.1	Vitrandic Eutrudepts	F		Ponderosa pine		VD	WD	0–60		Coarse-loamy	Frigid	Udic	6s	
Tamarackcanyon	65.1	Vitrandic Haploxeralfs	F		Douglas-fir	L	MD	WD	2–90		Clayey-skeletal	Frigid	Xeric	6e, 7e	
Tandy	15.0	Aeric Fluvaquents					VD	SPD	0–2		Sandy over loamy	Mesic	Aquic	6w	
Tanksel	28.4	Vitrandic Argixerolls					MD	WD	15–65		Clayey-skeletal	Mesic	Xeric	7e	
Tannahill	3.6	Calcic Argixerolls					D	WD	7–90		Loamy-skeletal	Mesic	Xeric	6e, 7e	
Taterpa	39.5	Pachic Haploxerolls	R	H, M			D	WD	12–60		Coarse-loamy	Frigid	Xeric	4e, 6e	
Tatouche	78.3	Typic Argixerolls	F		Douglas-fir	L	D	WD	12–65		Fine	Frigid	Xeric	6e	
Taunton	60.0	Xeric Haplodurids	C, R	L			MD	WD	0–45	L	Coarse-loamy	Mesic	Aridic	6e	3e, 4e
Teeters	21.6	Aquandic Endoaquepts	C, R	H			D	PD	0–1		Ashy	Mesic	Aquic	3w	
Teewee	63.4	Vitrandic Argixerolls	F		Ponderosa pine	L	D	WD	0–75		Fine-loamy	Mesic	Xeric	4e	

(continued)

Series name	Area (km²)	Subgroup	Land use[1]	Range site quality[2]	Forest indicator species	Forest site quality[3]	Depth class[4]	Drainage class[5]	Slope range (%)	Salinity class[6]	Particle-size class	Soil temp. regime	Soil moisture regime	Land capability class	
														No mgmt	With mgmt
Teguro	240.0	Lithic Argixerolls					S	WD	2–50	L	Loamy	Frigid	Xeric	6e	
Telemon	2.6	Aquic Palehumults	F		Douglas-fir	M	VD	SPD-MWD	3–60		Fine	Frigid	Udic	6e	
Templeton	666.9	Andic Humudepts	F		Douglas-fir	M, H	D	WD	0–90	L	Fine-silty	Isomesic	Udic	3e, 4e, 6e, 7e	
Tenwalter	131.6	Palexerollic Durixerolls					S	WD	0–3		Clayey-skeletal	Mesic	Xeric	6e	
Terlough	19.0	Oxyaquic Hapludolls	F		Grand fir		VD	SPD	0–5		Loamy-skeletal	Frigid	Udic	6c	
Terrabella	33.1	Vertic Argiaquolls	C, R	H			D	PD	0–3	L	Fine	Mesic	Aquic	3w	3w
Terrodd	6.8	Aquic Cumulic Hapludolls	F		Grand fir		VD	SPD	0–5		Fine-loamy	Frigid	Udic	6c	
Tertoo	0.2	Typic Vitrixerands	F		Douglas-fir		VD	WD	0–10		Ashy over loamy-skeletal	Frigid	Xeric	6c	
Tetherow	13.1	Vitritorrandic Haploxerolls	C				MD-S	ED	0–50	L	Loamy over pumiceous or cindery	Mesic	Xeric	6e, 6s	4e, 4s
Tethrick	47.6	Typic Dystroxerepts	F		Douglas-fir	L, M	D	WD	35–90		Coarse-loamy	Mesic	Xeric	6e, 7e	
Thader	5.3	Andic Humicryods	F		Douglas-fir	L	MD	WD	5–90		Loamy-skeletal	Cryic	Udic	7s	
Thatuna	13.2	Oxyaquic Argixerolls	C, R	H			VD	MWD	0–55	L	Fine-silty	Mesic	Xeric	2e, 3e	
Thenarrows	63.6	Typic Halaquepts					VD	PD	0–2	H	Coarse-loamy	Frigid	Aquic	6s	
Thiessen	23.5	Pachic Argixerolls					MD	WD	2–90		Clayey-skeletal	Mesic	Xeric	6s, 7e	
Thirstygulch	43.3	Lithic Ultic Haploxerolls	F		Ponderosa pine		S	WD	2–60		Loamy-skeletal	Frigid	Xeric	7s	
Thistleburn	9.6	Typic Palehumults	F		Douglas-fir	M, H	VD	WD	3–60		Fine	Frigid	Udic	6e	
Thompsoncabin	13.5	Lithic Natrargids					S	WD	15–70		Loamy-skeletal	Mesic	Aridic	7s	
Thorn	13.7	Lithic Mollic Haploxeralfs	F		Ponderosa pine		S	WD	15–50		Loamy-skeletal	Frigid	Xeric	7e	
Thornlake	156.3	Sodic Xeric Haplocambids					VD	WD	0–5	H	Ashy	Frigid	Aridic	6s	
Threebuck	76.0	Alfic Vitrixerands	F		Douglas-fir	L, M	D	WD	2–90		Ashy over clayey-skeletal	Frigid	Xeric	4e, 6e, 7e	
Threecreeks	6.0	Cumulic Haploxerolls	C				VD	MWD	0–3		Coarse-loamy	Mesic	Xeric	3w	3w
Threeforks	3.6	Humic Dystrudepts	F		Douglas-fir	M, H	VD	WD	30–90		Fine-loamy	Mesic	Udic	6e, 7e	
Threetrees	20.8	Typic Dystrudepts	F		Douglas-fir	M	MD	WD	0–90		Loamy-skeletal	Frigid	Udic	6e	

(continued)

Series name	Area (km²)	Subgroup	Land use[1]	Range site quality[2]	Forest indicator species	Forest site quality[3]	Depth class[4]	Drainage class[5]	Slope range (%)	Salinity class[6]	Particle-size class	Soil temp. regime	Soil moisture regime	Land capability class	
														No mgmt	With mgmt
Thunderegg	34.6	Typic Natraquolls	R				VD	PD	0–1		Fine	Mesic	Aquic	6w	
Ticino	36.5	Typic Argixerolls	R	H			MD	WD	4–30		Fine-loamy	Frigid	Xeric	4e	
Tillamook	2.1	Aquic Melanudands	C, F		Sitka spruce		VD	MWD	0–15	L	Medial over loamy	Isomesic	Udic	2e	
Timbercrater	102.1	Typic Vitricryands	F		Lodgepole pine		VD	ED	0–80		Ashy-pumiceous	Cryic	Udic	6e	
Tincan	6.1	Aridic Haploxerolls					S	WD	20–60		Loamy	Mesic	Xeric	6e	
Tincup	6.7	Humic Dystrudepts	F		Douglas-fir	L, M	MD	WD	0–90		Loamy-skeletal	Frigid	Udic	6s	
Tippett	21.4	Vertic Palexerolls	R	H			D-VD	WD	0–3		Fine	Frigid	Xeric	3c	
Tishar	5.8	Typic Haploxerults	F		Douglas-fir	M	D	WD	3–80		Clayey-skeletal	Mesic	Xeric	6e	
Tolany	65.0	Alic Hapludands	F		Douglas-fir	M	VD	WD	0–60		Medial	Frigid	Udic	6e	
Tolfork	5.2	Pachic Humudepts	F		Douglas-fir	M	D	WD	30–90		Loamy-skeletal	Frigid	Udic	6s, 7e	
Tolius	16.1	Vitrandic Argixerolls	F		Ponderosa pine	L	VD	WD	0–8		Fine-loamy	Mesic	Xeric	4e	
Tolke	118.4	Alic Hapludands	F		Douglas-fir	M, H	VD	WD	0–60	L	Medial	Mesic	Udic	6e	
Toll	23.3	Xeric Torripsamments					D	SED	0–15		Sandy	Mesic	Aridic	6e	
Tolo	608.0	Alfic Vitrixerands	F		Douglas-fir	L, M	D-VD	WD	2–65	L	Ashy over loamy	Frigid	Xeric	3e, 4e, 6e, 7e	
Tolovana	220.3	Typic Fulvudands	F		Douglas-fir	M, H	VD	WD	3–85	L	Medial over loamy	Isomesic	Udic	6e	
Tonor	76.7	Sodic Xeric Haplocambids					VD	WD	0–15	M	Ashy	Frigid	Aridic	6s	
Top	83.9	Vertic Argixerolls	F		Douglas-fir	M	D-VD	WD	2–75		Fine	Frigid	Xeric	4e, 6e, 7e	
Topper	17.3	Vitrandic Haploxerolls	R	H			VD	WD	2–30	L	Fine-silty	Frigid	Xeric	3e, 4e	
Trask	17.8	Typic Humudepts	F		Douglas-fir	M, H	MD	WD	3–90		Loamy-skeletal	Mesic	Udic	6s, 7s	
Treharne	20.9	Aquultic Hapludalfs	C, F		Douglas-fir		VD	MWD	0–3	L	Fine-silty	Mesic	Udic	2c	
Troutmeadows	290.8	Typic Vitricryands	F		Grand fir	L, M	MD	WD	0–90	L	Ashy over loamy-skeletal	Cryic	Udic	6e, 7e	
Truax	0.0	Aridic Argixerolls	C				D	WD	0–15		Fine-loamy	Mesic	Aridic	4e	
Truesdale	5.1	Xereptic Haplodurids	R	H			MD	WD	0–12	L	Coarse-loamy	Mesic	Aridic	3 s	
Trunk	1.7	Xeric Haplargids					MD	WD	4–50		Fine	Mesic	Aridic	7s	
Tub	629.6	Vertic Argixerolls	R	H			VD-D	WD	1–70	L	Fine	Mesic	Xeric	3e	7e

(continued)

Series name	Area (km^2)	Subgroup	Land use[1]	Range site quality[2]	Forest indicator species	Forest site quality[3]	Depth class[4]	Drainage class[5]	Slope range (%)	Salinity class[6]	Particle-size class	Soil temp. regime	Soil moisture regime	Land capability class	
														No mgmt	With mgmt
Tuckerdowns	26.1	Calcic Haploxerolls	C, R	H			VD	WD	2–30		Loamy-skeletal	Frigid	Xeric	3e, 4e	
Tuffcabin	4.1	Vitritorrandic Haploxerolls					D	WD	0–10		Ashy	Frigid	Xeric	6e	
Tuffo	1.4	Vitrandic Torriorthents					VS-S	SED	4–50		Ashy	Mesic	Aridic	7s	
Tulana	19.2	Aquandic Humaquepts	R	H			VD	PD-VPD	0–1	L	Fine-silty	Mesic	Aquic	3w	
Tumalo	107.6	Duric Vitritorrands	C, R	L			MD	WD	0–8	L	Ashy	Mesic	Aridic	6e	3e
Tumtum	26.7	Typic Argidurids					S	WD	2–15	L	Loamy	Mesic	Aridic	6e, 7s	
Turbyfill	312.2	Xeric Torriorthents	C, R	H			D-VD	WD	0–35	L	Coarse-loamy	Mesic	Xeric	1, 2e, 3e	
Turpentine	27.7	Alfic Udivitrands	F		Grand fir	M	D-VD	WD	0–60		Ashy over loamy	Frigid	Udic	6c	
Turpin	203.0	Sodic Xeric Haplocambids					VD	WD-MWD	0–15	M	Fine-loamy	Mesic	Aridic	6e, 6s	
Tuscor	5.2	Vertic Argixerolls					D	WD	12–65		Clayey-skeletal	Frigid	Xeric	6e	
Tutni	83.4	Typic Cryaquands	F		Ponderosa pine	M	VD	SPD	0–3		Ashy	Cryic	Aquic	6s	
Tutuilla	8.9	Typic Palexerolls	C, R	H			D	WD	1–35		Fine	Mesic	Xeric	3e, 4e	
Tweener	93.6	Lithic Argixerolls					VS-S	WD	4–75		Loamy-skeletal	Frigid	Xeric	7s	
Twelvemile	54.0	Typic Vitrixerands	F		White fir		VD	SED	0–60		Ashy-skeletal	Frigid	Xeric	6s	
Twickenham	41.9	Vertic Paleargids					VD	WD	2–60		Fine	Mesic	Aridic	4e, 6e	
Twinbridge	0.4	Lithic Haploxerepts	F		Ponderosa pine		S	WD	15–60		Loamy-skeletal	Frigid	Xeric	s	
Tygh	16.6	Fluvaquentic Haploxerolls	R	H			VD	SPD	0–2	L	Coarse-loamy	Mesic	Xeric	3w	
Ukiah	84.4	Vertic Argixerolls	R	H, L			MD	WD	2–40		Fine	Mesic	Xeric	4e, 4s, 6e, 7s	
Umak	49.5	Typic Vitricryands	F		Lodgepole pine		VD	ED	0–10		Ashy-pumiceous	Cryic	Udic	6e	
Umapine	151.5	Typic Halaquepts	C, R	H, L			VD	SPD	0–5	H	Coarse-silty	Mesic	Aquic	3c, 3w, 6w	2 s
Umatilla	139.7	Vitrandic Haploxerolls	F		Ponderosa pine		VD	WD	15–70	L	Loamy-skeletal	Frigid	Xeric	6e, 7e	
Umpcoos	646.2	Lithic Eutrudepts	F		Douglas-fir	L, M	S	WD	30–99	L	Loamy-skeletal	Mesic	Udic	7e	
Unionpeak	145.6	Typic Duricryands	F		Lodgepole pine		MD	SED	0–35		Ashy	Cryic	Udic	6e	
Upcreek	3.8	Cumulic Haploxerolls	C				VD	SPD	1–3		Fine-loamy over sandy or sandy-skeletal	Frigid	Xeric	4w	4w

(continued)

Series name	Area (km^2)	Subgroup	Land use[1]	Range site quality[2]	Forest indicator species	Forest site quality[3]	Depth class[4]	Drainage class[5]	Slope range (%)	Salinity class[6]	Particle-size class	Soil temp. regime	Soil moisture regime	Land capability class	
														No mgmt	With mgmt
Uptmor	2.6	Ultic Argixerolls	F		Douglas-fir		D-VD	WD	3–40		Fine	Frigid	Xeric	6e	
Utley	35.8	Vitrandic Haploxerolls					D-VD	WD	0–40		Fine-loamy	Frigid	Xeric	6e	
Valby	464.4	Calcic Haploxerolls	R	H			MD	WD	1–30	L	Fine-silty	Mesic	Xeric	3s, 3e, 4e	
Valmy	34.5	Duric Torriorthents	C, R				VD	WD	0–8		Coarse-loamy	Mesic	Aridic	6w	2w
Valsetz	155.8	Alic Haplocryands	F		Douglas-fir	M	MD	WD	3–90	L	Medial-skeletal	Cryic	Udic	6s, 7s	
Van horn	10.0	Ultic Argixerolls	C, F		Ponderosa pine		D	WD	0–35	L	Fine-loamy	Mesic	Xeric	2e-4e	
Vandamine	68.5	Andic Haplocryepts	F		Engelmann spruce		VD	WD	15–90		Loamy-skeletal	Cryic	Udic	6e	
Vannoy	425.5	Mollic Haploxeralfs	F		Douglas-fir	L, M	MD	WD	2–60	L	Fine-loamy	Mesic	Xeric	6e	
Vanwyper	61.8	Xeric Haplargids					MD	WD	8–70		Clayey-skeletal	Mesic	Aridic	7e	
Veazie	46.3	Cumulic Haploxerolls	C, R	H, L			VD	WD	0–5	L	Coarse-loamy over sandy or sandy-skeletal	Mesic	Xeric	2w, 3w, 4s	2s, 3s, 3w
Vena	13.4	Typic Dystroxerepts	F		Douglas-fir	M	MD	WD	20–100		Loamy-skeletal	Mesic	Xeric	7e	
Venator	135.4	Lithic Haploxerolls	F		Ponderosa pine		S	WD	2–80		Loamy-skeletal	Mesic	Xeric	6e, 6s, 7s	
Veneta	59.7	Ultic Haploxeralfs	C, F		Douglas-fir	M, H	VD	MWD	0–20	L	Fine	Mesic	Xeric	2e, 3e	
Verboort	1.6	Xerertic Argialbolls	C, R				VD	PD	0–3	L	Fine	Mesic	Xeric	3w	3w
Verdico	26.3	Vertic Paleargids					MD	WD	4–30		Fine	Mesic	Aridic	7s	
Vergas	249.1	Durinodic Xeric Haplargids					VD	WD	0–8	L	Fine-loamy over sandy or sandy-skeletal	Frigid	Aridic	6e, 6s	
Vermisa	542.7	Lithic Dystroxerepts	F		Douglas-fir	L, M	S	SED	12–100	L	Loamy-skeletal	Mesic	Xeric	6e, 7e, 7s	
Vernonia	90.4	Ultic Hapludalfs	F		Douglas-fir	H	D	WD	3–30		Fine-silty	Mesic	Udic	6e	
Veta	11.7	Xeric Haplocambids	C				VD	WD	0–15		Loamy-skeletal	Mesic	Aridic	6c	
Vil	31.6	Argiduridic Durixerolls	R				S	WD	2–20		Loamy	Frigid	Xeric	6e	
Vining	49.6	Xeric Haplocambids					MD	WD	0–30		Coarse-loamy	Mesic	Aridic	6e	
Virtue	225.9	Xeric Argidurids	R	H, L			MD	WD	0–20	L	Fine-silty	Mesic	Aridic	3s, 4s	
Vitale	270.4	Typic Argixerolls					MD	WD	2–75	L	Loamy-skeletal	Frigid	Xeric	6e, 6s	
Voats	21.8	Fluventic Haploxerolls	R	L			VD	WD	0–5		Sandy-skeletal	Mesic	Xeric	4s	

(continued)

Series name	Area (km^2)	Subgroup	Land use[1]	Range site quality[2]	Forest indicator species	Forest site quality[3]	Depth class[4]	Drainage class[5]	Slope range (%)	Salinity class[6]	Particle-size class	Soil temp. regime	Soil moisture regime	Land capability class	
														No mgmt	With mgmt
Volstead	14.9	Vitrandic Argixerolls	F		Ponderosa pine		D	WD	0–60		Fine	Frigid	Xeric	4e, 6e	
Voltage	60.9	Xeric Haplocalcids					VD	WD	0–2		Coarse-loamy	Frigid	Aridic	6s	
Vondergreen	1.0	Aquic Hapludults	F		Douglas-fir	H	D	SPD	0–30		Fine	Isomesic	Udic	6e	
Voorhies	92.5	Mollic Haploxeralfs	F		Douglas-fir	L, M	MD	WD	35–55	L	Loamy-skeletal	Mesic	Xeric	7s	
Wabuska	16.7	Aeric Halaquepts	C, R				VD	SPD	0–2		Coarse-loamy	Mesic	Aquic	6w	3w
Wadecreek	8.0	Oxyaquic Haplohumults	F		Douglas-fir	H	VD	MWD	0–20		Fine	Isomesic	Udic	3e	
Wagontire	19.8	Argiduridic Durixerolls	R				S	WD	2–20		Clayey	Frigid	Xeric	6e	
Waha	504.8	Pachic Argixerolls	F		Ponderosa pine		MD	WD	0–65	L	Fine-loamy	Mesic	Xeric	3e, 4e, 6e	
Wahkeena	6.9	Pachic Hapludolls	F		Douglas-fir		D	WD	69–90		Fragmental	Mesic	Udic	7s	
Wahstal	68.2	Palexerollic Durixerolls					S	WD	2–12		Clayey-skeletal	Frigid	Xeric	7s	
Wakamo	6.0	Lithic Ultic Argixerolls	F		Ponderosa pine	L	S	WD	2–30		Clayey-skeletal	Mesic	Xeric	6s	
Waldo	160.9	Fluvaquentic Vertic Endoaquolls	C				VD	PD	0–3	L	Fine	Mesic	Aquic	3w	
Waldport	91.9	Typic Udipsamments	F		Douglas-fir	L	VD	SED	0–70		Sandy	Isomesic	Udic	6e, 7e	
Walla walla	1270.0	Typic Haploxerolls	C, R	H, M			D-VD	WD	0–65	L	Coarse-silty	Mesic	Xeric	2e, 3e, 3w, 4e, 6e	2e, 3e
Wallowa	83.3	Vitrandic Haploxerolls	F		Ponderosa pine		MD	WD	1–15		Fine-loamy	Frigid	Xeric	4e	
Walluski	27.4	Andic Oxyaquic Humudepts	C, F		Western hemlock		VD	MWD	0–20		Fine-silty	Isomesic	Udic	2e, 3e	3e
Wamic	247.4	Vitrandic Haploxerepts	F		Ponderosa pine	L	D	WD	0–70	L	Fine-loamy	Mesic	Xeric	3e, 4e, 6e, 7e	
Wanoga	332.1	Humic Vitrixerands	F		Ponderosa pine	M	MD	WD	0–65	L	Ashy	Frigid	Xeric	6e	4e
Wanser	21.3	Typic Psammaquents	R				D	PD	0–12			Mesic	Aquic	6w	4w
Wapato	164.2	Fluvaquentic Endoaquolls	R	H			VD	PD	0–3	L	Fine-silty	Mesic	Aquic	3w	
Wapinitia	40.6	Pachic Argixerolls	C, R	H			D	WD	0–35	L	Fine-loamy	Mesic	Xeric	3e	
Warden	280.5	Xeric Haplocambids	C, R	H, M			VD-D	WD	0–65	L	Coarse-silty	Mesic	Aridic	4c, 4e, 6e	2e

(continued)

Series name	Area (km^2)	Subgroup	Land use[1]	Range site quality[2]	Forest indicator species	Forest site quality[3]	Depth class[4]	Drainage class[5]	Slope range (%)	Salinity class[6]	Particle-size class	Soil temp. regime	Soil moisture regime	Land capability class	
														No mgmt	With mgmt
Warnermount	0.2	Vitrandic Argixerolls					MD	WD	4–50		Ashy-skeletal	Frigid	Xeric	6e	
Warrenton	3.6	Typic Humaquepts	C				D	VPD	0–3		Sandy	Isomesic	Aquic	4w	
Wasson	3.0	Fluvaquentic Humaquepts					VD	PD	0–3		Coarse-loamy	Mesic	Aquic	3w	
Watama	261.5	Pachic Haploxerolls	R	H			MD	WD	0–35	L	Fine-loamy	Mesic	Xeric	4e	
Watches	17.5	Typic Dystrudepts	F		Douglas-fir	H	VD	WD	0–90		Fine-loamy	Isomesic	Udic	6e, 7e	
Waterbury	170.2	Lithic Argixerolls					S	WD	2–80	L	Clayey-skeletal	Mesic	Xeric	7s	
Wato	54.5	Typic Haploxerolls	R	H			VD	WD	0–35	L	Coarse-loamy	Mesic	Xeric	3e, 4e	
Wauld	18.2	Typic Humudepts	F		Douglas-fir	M, H	MD	WD	30–70		Loamy-skeletal	Mesic	Udic	7s	
Wauna	24.7	Fluvaquentic Endoaquepts	C, R	H, L			D	PD	0–3		Fine-silty	Mesic	Aquic	2w, 3w, 4s	
Webbgulch	4.9	Vitrandic Haploxerolls	F		Ponderosa pine		MD	WD	0–90		Loamy-skeletal	Frigid	Xeric	6c	
Webfoot	1.2	Pachic Haploxerolls	C, F		Ponderosa pine	M	VD	SPD	0–7		Loamy-skeletal	Frigid	Xeric	3e	
Wedderburn	6.0	Pachic Humudepts	F		Douglas-fir		D	WD	0–90		Fine-loamy	Isomesic	Udic	6s	
Wegert	190.5	Vitritorrandic Haploxerolls					MD	SED	0–15	L	Ashy	Frigid	Xeric	6e, 6s	
Weglike	81.0	Vitritorrandic Haploxerolls					MD	WD	0–6		Fine-loamy	Frigid	Xeric	6e	
Welch	61.5	Cumulic Endoaquolls	R	M			VD	PD-VPD	0–15		Fine-loamy	Frigid	Aquic	5w	
Wellsdale	22.9	Aquultic Haploxeralfs	C, F		Douglas-fir	M	VD	MWD	2–30	L	Fine-loamy	Mesic	Xeric	2e, 3e 4e	
Wenas	9.4	Cumulic Endoaquolls	R	M			VD	SPD, PD	0–3	L	Fine-loamy	Mesic	Aquic	5w	5w
Westbutte	642.6	Pachic Haploxerolls					MD	WD	2–75	L	Loamy-skeletal	Frigid	Xeric	6e, 7s	
Westside	23.6	Durinodic Xeric Paleargids					VD	WD	2–15		Fine	Frigid	Aridic	6s	
Whaleshead	39.6	Andic Humudepts	F		Douglas-fir	M, H	VD	WD	30–90		Loamy-skeletal	Isomesic	Udic	6e	
Whetstone	53.7	Typic Haplocryods	F		Western hemlock		MD	WD	3–75		Loamy-skeletal	Cryic	Udic	7e	
Whisk	2.9	Lithic Ultic Haploxerolls					S	SED	8–90		Loamy	Mesic	Xeric	7s	
Whiteface	2.3	Argic Duricryolls	F		Ponderosa pine	L	S	WD	1–12		Loamy	Cryic	Xeric	6s	
Whiteson	9.6	Fluvaquentic Vertic Endoaquolls	R	H			D	SPD	0–3		Fine-loamy over clayey	Mesic	Aquic	4w	
Whobrey	95.9	Aquertic Eutrudepts	F		Douglas-fir	L, M	VD	SPD	7–60	L	Fine-silty over clayey	Mesic	Udic	6e	
Wickahoney	3.8	Lithic Mollic Haploxeralfs					S	WD	1–45		Clayey-skeletal	Frigid	Xeric	7s	7s

(continued)

Series name	Area (km²)	Subgroup	Land use[1]	Range site quality[2]	Forest indicator species	Forest site quality[3]	Depth class[4]	Drainage class[5]	Slope range (%)	Salinity class[6]	Particle-size class	Soil temp. regime	Soil moisture regime	Land capability class	
														No mgmt	With mgmt
Wickiup	13.2	Typic Cryaquands	F		Lodgepole pine		VD	PD	0–3		Ashy-pumiceous	Cryic	Aquic	6w	
Widowspring	98.6	Cumulic Haploxerolls	R	H, L			VD	MWD	0–2	L	Fine-silty	Frigid	Xeric	4c, 6c	
Wieland	59.3	Durinodic Xeric Haplargids					VD	WD	0–30		Fine	Mesic	Aridic	6s	
Wildcatbutte	14.6	Vitritorrandic Haploxerolls					VD	WD	15–65		Ashy-skeletal	Frigid	Xeric	6e, 7s	
Wildhill	35.4	Durinodic Xeric Haplargids					MD	WD	2–60		Loamy-skeletal	Mesic	Aridic	7s	
Wilhoit	75.9	Andic Humudepts	F		Douglas-fir	H	D	WD	5–60		Fine-loamy	Frigid	Udic	6e	
Wilkins	21.9	Xerertic Argialbolls	C, R	H			VD	SPD	0–5	L	Fine	Frigid	Xeric	3w, 4w	
Willakenzie	157.4	Ultic Haploxeralfs	F		Douglas-fir	M, H	MD	WD	2–60	L	Fine-loamy	Mesic	Xeric	3e, 4e, 6e	
Willamette	155.1	Pachic Ultic Argixerolls	C, F		Douglas-fir	M	VD	WD	0–20	L	Fine-silty	Mesic	Xeric	1, 2e, 3e	
Willanch	15.2	Fluvaquentic Humaquepts	C				VD	PD	0–3		Coarse-loamy	Isomesic	Aquic	3w	
Willis	148.8	Haploduridic Durixerolls	C, R	H			MD	WD	0–65	L	Coarse-silty	Mesic	Xeric	3e, 4e	2e
Willowdale	62.8	Cumulic Haploxerolls	C, R	H			VD	WD	0–3	L	Fine-loamy	Mesic	Xeric	2w, 3w, 4s, 4c	3c
Wilt	16.2	Vitrandic Argixerolls	F		Ponderosa pine		MD	WD	0–15		Loamy-skeletal	Frigid	Xeric	6e	
Winberry	0.7	Lithic Dystrocryepts					S	SED	10–70		Loamy-skeletal	Cryic	Udic	7s	
Winchester	58.1	Xeric Torripsamments	C, R				VD	ED	0–65		Sandy	Mesic	Aridic	7e	4e, 4s
Winchuck	4.1	Typic Haplohumults	C, F		Douglas-fir	H	VD	WD	0–30		Fine	Isomesic	Udic	2e-4e	2e-3e
Wind river	11.1	Ultic Haploxerolls	C, F		Douglas-fir	M	VD	WD	0–30	L	Coarse-loamy	Mesic	Xeric	2e, 3e, 3s	
Windego	70.4	Alfic Vitrixerands	F		Douglas-fir	M	VD	WD	0–70		Ashy over loamy-skeletal	Frigid	Xeric	6e	
Windybutte	47.5	Argiduridic Argixerolls	C, R				VD	WD	2–5	L	Fine-silty	Frigid	Xeric	6e	4e
Windygap	313.2	Xeric Haplohumults	C, F		Douglas-fir	M, H	D	WD	2–60	L	Fine	Mesic	Xeric	2e-4e, 6e	
Windypoint	20.4	Xeric Haplargids	C, R				VD	WD	1–4		Fine-loamy	Mesic	Aridic	6c	3e
Winema	5.1	Typic Fulvudands	C, F		Douglas-fir	M, H	VD	WD	3–70		Medial over clayey	Isomesic	Udic	4e	
Wingdale	8.8	Cumulic Endoaquolls	R	H			VD	PD	0–2	L	Fine-silty	Mesic	Aquic	3w	3w

(continued)

Series name	Area (km²)	Subgroup	Land use[1]	Range site quality[2]	Forest indicator species	Forest site quality[3]	Depth class[4]	Drainage class[5]	Slope range (%)	Salinity class[6]	Particle-size class	Soil temp. regime	Soil moisture regime	Land capability class	
														No mgmt	With mgmt
Wingville	78.4	Pachic Haploxerolls	C, R	H			D	SPD	0–2	L	Fine-silty	Mesic	Xeric	2w	2w
Winlo	31.4	Typic Duraquolls	R				S	SPD	0–3		Clayey-skeletal	Mesic	Aquic	6s	
Winom	7.3	Oxyaquic Hapluderts	C, R	L			D	MWD	0–3		Fine	Frigid	Udic	4s	
Wintercanyon	16.4	Lithic Ultic Haploxerolls	F		Douglas-fir		S	WD	30–90		Loamy-skeletal	Frigid	Xeric	7e	
Winterim	117.9	Pachic Argixerolls	F		Ponderosa pine	L	D	WD	0–60	L	Clayey-skeletal	Frigid	Xeric	6e	
Wintley	23.6	Typic Haplohumults	C				VD	WD	0–30		Fine	Mesic	Udic	3e, 4e	
Wiskan	1.3	Xeric Haplargids					MD	WD	15–75		Loamy-skeletal	Frigid	Aridic	6e	
Witham	44.1	Vertic Haploxerolls	F, R		Or white oak		VD	SPD	2–20		Fine	Mesic	Xeric	3e	
Witzel	113.3	Lithic Ultic Haploxerolls	F		Douglas-fir	L, M	S	WD	3–75	L	Loamy-skeletal	Mesic	Xeric	6s, 7s	
Wizard	13.4	Aquic Vitrixerands	F		Ponderosa pine		VD	SPD	0–15		Ashy	Frigid	Xeric	6e	
Wolfer	2.1	Typic Fulvudands	C				VD	SED	0–3		Medial over sandy or sandy-skeletal	Isomesic	Udic	3s	
Wolfpeak	56.3	Ultic Palexeralfs	F		Douglas-fir	L, M	VD	WD	3–60	L	Fine-loamy	Mesic	Xeric	3e, 4e	2e, 3w
Wollent	11.8	Typic Humaquepts	R	H			VD	PD	0–3	L	Fine-silty	Mesic	Aquic	3w	
Wolot	14.6	Alfic Vitrixerands	C, F		Ponderosa pine	M	VD	WD	0–30		Ashy over loamy	Mesic	Xeric	2e, 3e, 4e	
Wolverine	11.7	Xeric Torripsamments	C				D-VD	SED	0–30		Sandy	Frigid	Aridic	6e, 7s	7s
Woodburn	911.3	Aquultic Argixerolls	C, F		Douglas-fir	H	VD	MWD	0–55	L	Fine-silty	Mesic	Xeric	2w, 2e, 3e	
Woodchopper	77.0	Pachic Ultic Argixerolls	F		Ponderosa pine		VD	WD	0–40		Fine	Frigid	Xeric	6e	
Woodcock	576.8	Alfic Humic Haploxerands	F		Ponderosa pine	L, M	VD-D	WD	1–60	L	Medial-skeletal	Frigid	Xeric	6e	
Woodseye	77.5	Humic Lithic Dystroxerepts	F		Douglas-fir	L	S	WD	2–90		Loamy-skeletal	Frigid	Xeric	7s	
Woodspoint	5.8	Typic Fulvicryands	F		Douglas-fir		D	WD	5–60		Medial	Cryic	Udic	6e, 7e	
Wrentham	333.9	Pachic Haploxerolls					MD	WD	35–70	L	Loamy-skeletal	Mesic	Xeric	7s	
Wrightman	17.5	Vitrandic Haploxerolls					MD	WD	2–25		Fine-loamy	Frigid	Xeric	7s	

(continued)

Series name	Area (km²)	Subgroup	Land use[1]	Range site quality[2]	Forest indicator species	Forest site quality[3]	Depth class[4]	Drainage class[5]	Slope range (%)	Salinity class[6]	Particle-size class	Soil temp. regime	Soil moisture regime	Land capability class	
														No mgmt	With mgmt
Wuksi	2.2	Xeric Vitricryands	F		Ponderosa pine	M	VD	SED	2–70		Ashy-skeletal	Cryic	Xeric	6e	
Wyeast	6.0	Aeric Fragiaquepts	C, R	H			MD	SPD	0-	L	Coarse-silty	Mesic	Aquic	2w	
Wyeth	26.6	Pachic Humixerepts	F		Douglas-fir	M	VD	WD	5–75		Loamy-skeletal	Mesic	Xeric	6s	
Xanadu	72.8	Typic Palehumults	F		Douglas-fir	H	VD	WD	3–60		Fine	Mesic	Udic	6e, 7s	
Yachats	9.3	Fluventic Humudepts	C, F		Douglas-fir		VD	WD	0–3		Coarse-loamy	Isomesic	Udic	2w, 3w	
Yainax	36.1	Mollic Haploxeralfs	F, R		Ponderosa pine	L	MD	WD	1–15		Fine-loamy	Frigid	Xeric	4e	
Yakima	14.7	Cumulic Haploxerolls	R	L			VD	WD	0–3	L	Coarse-loamy over sandy or sandy-skeletal	Mesic	Xeric	4s	2s
Yakus	0.5	Lithic Haploxerolls					S	WD	12–90		Loamy	Mesic	Xeric	7e	
Yallani	46.8	Typic Vitrixerands	F		Douglas-fir	M	VD	WD	8–75	L	Ashy-skeletal	Frigid	Xeric	6s	
Yancy	52.4	Palexerollic Durixerolls	R	L, M			S	WD	0–8		Clayey	Frigid	Xeric	4s, 6e	
Yankeewell	169.7	Xeric Natridurids					S	WD	2–20	M	Loamy	Frigid	Aridic	7s	
Yapoah	27.9	Humic Vitrixerands	F		Ponderosa pine		VD	SED	0–75		Ashy-skeletal	Frigid	Xeric	6e	
Yaquina	14.3	Typic Endoaquods	R				VD	SPD	0–5		Sandy	Isomesic	Aquic	4w	
Yawhee	144.3	Alfic Udivitrands	F		Ponderosa pine	L, M	VD	SED	3–40	L	Ashy-skeletal over loamy-skeletal	Frigid	Udic	6e	
Yawkey	84.3	Vertic Palexerolls	F		Ponderosa pine		D-VD	WD	1–70		Clayey-skeletal	Frigid	Xeric	7e	
Yawkola	8.8	Pachic Palexerolls	F		Ponderosa pine	L	VD	WD	2–55		Clayey-skeletal	Mesic	Xeric	4e	
Yellowstone	81.2	Lithic Haplocryands	F		Douglas-fir	L	S	SED	3–90		Medial-skeletal	Cryic	Udic	6s, 7s	
Yoncalla	19.5	Aquic Palexerolls	R				VD	SPD	2–30		Fine	Mesic	Xeric	3w, 4w	
Yonna	27.6	Aquandic Haplocryepts	C, R	H			VD	SPD	0–2		Loamy over ashy or ashy-pumiceous	Cryic	Xeric	4w	
Yorel	18.1	Typic Dystrudepts	F		Douglas-fir	M	MD	WD	0–60		Fine-loamy	Frigid	Udic	6e	
Youtlkue	6.6	Vitrixerandic Aquicambids					MD	SPD	0–2		Ashy	Frigid	Aridic	6s	
Yuko	4.1	Xeric Haplargids					VS-S	WD	2–50		Loamy	Mesic	Aridic	7s	

(continued)

Series name	Area (km²)	Subgroup	Land use[1]	Range site quality[2]	Forest indicator species	Forest site quality[3]	Depth class[4]	Drainage class[5]	Slope range (%)	Salinity class[6]	Particle-size class	Soil temp. regime	Soil moisture regime	Land capability class	
														No mgmt	With mgmt
Zalea	11.8	Typic Haplohumults	F		Douglas-fir	M, H	MD	WD	0–30		Fine-loamy	Frigid	Udic	6e	
Zango	12.6	Lithic Dystrudepts	F		Douglas-fir	M	S	ED	60–90		Loamy-skeletal	Mesic	Udic	7e	
Zevadez	200.2	Durinodic Xeric Haplargids					VD	WD	0–50		Fine-loamy	Mesic	Aridic	6c	
Zing	64.1	Aquultic Haploxeralfs	F		Ponderosa pine		VD	MWD	0–45	L	Fine	Mesic	Xeric	5e, 6e	
Zola	26.0	Cumulic Haploxerolls	R				VD	MWD	0–2		Fine-loamy	Frigid	Xeric	3c	3c
Zorravista	16.2	Xeric Torripsamments					VD	ED	0–15	M	Sandy	Mesic	Aridic	7s	
Zuman	17.9	Typic Halaquepts	R	H, L			D	PD	0–1	H	Fine-loamy over sandy or sandy-skeletal	Mesic	Aquic	3w, 6w	
Zumwalt	34.7	Vertic Palexerolls	C, R	H			MD	WD	0–20		Fine	Frigid	Xeric	4e	
Zwagg	4.8	Pachic Humudepts					MD	WD	0–90		Coarse-loamy	Isomesic	Udic	6e	
Zygore	221.9	Andic Humudepts	F		Douglas-fir	H	VD	WD	5–90	L	Loamy-skeletal	Frigid	Udic	6e, 6s	
Zyzzug	7.1	Typic Humaquepts	C				VD	PD	0–3		Fine-silty	Mesic	Aquic	3w	3w

[1]Land use: C = crops and pasture (dry or irrigated); F = forests; R = rangeland; areas left blank are devoid of forest vegetation and currently are not used for crops, but may be suitable for range, watersheds, recreation and wildlife habitat.

[2]Rangeland yield: L = low, <1340 kg dry matter/ha; M = 1340–1785; H = >1785.

[3]Forest site quality: site index for Douglas-fir (df); L = <30; M = 30–40; H = >130 (base age = 100 year); site index for ponderosa pine (pp); L = <27; M = 27–37; H = >37 (base age = 100 year); site index for grand fir (gf): L = <21; M = 21–27; H = >27 (base age 50 year).

[4]Depth class: VD = very deep (>150 cm); D = deep (100–150 cm); MD = moderately deep (50–100 cm); S = shallow (50–25 cm); VS = very shallow (<25 cm).

[5]Drainage class: ED = excessively drained; SED = somewhat excessively drained; WD = well drained; MWD = moderately well drained; SPD = somewhat poorly drained; VPD = very poorly drained.

[6]Salinity class: H = Salids suborder, Halaquepts great group, sodic subgroup (EC > 16 dS/m); M = Natr- great groups (EC = 4 to <16); L = EC < 4.

Yellow highlighting = estimated by author.

Orange highlighting = columns to be removed.

Index

A

Accreted terrane, 37–39, 323
Adiabatic lapse rate, 43, 46, 50
Agricultural Water Quality Act of 1993, 270
Albaqualfs, 56, 70–72, 92, 201, 203, 204, 281, 334, 368, 381, 392, 468, 482
Albic horizon, 13, 91, 161, 162, 173, 201, 203, 207, 209, 211
Alder, red, 44, 50, 54, 60, 68, 81, 88, 105, 141, 143, 149, 151, 175, 178, 188, 192, 197, 199, 201, 202, 236, 302–304, 326–333, 335–338, 340–343, 346–353, 355, 356, 358–363, 511
Alfisols, 14, 15, 64, 67, 70, 91, 92, 96, 109, 201–204, 215, 216, 219, 285, 324, 380–384, 387–390, 392, 394–405, 407–414, 416–418, 420–428, 430–432
Alkali sacaton, 53, 55, 350, 351, 353, 359
Alluvial fan, 54, 88, 91, 104, 106, 117, 147, 148, 155, 157, 158, 163, 164, 181
Alluvium, 40, 53, 54, 55, 57, 59, 65, 67, 70, 73–78, 82, 84–86, 88, 91, 104–106, 111, 114, 115, 117, 118, 128, 129, 132, 133, 141, 147, 148, 155–158, 161–168, 171, 175, 177, 179, 181, 183, 185, 187, 192, 205, 206, 210, 251, 323, 326, 328–340, 342–363
Alpine zone, 48, 50
Alvord Desert, 21, 40, 86, 220
Andisolization, 190, 215, 216, 218, 324
Andisols, 14, 15, 60, 62, 67, 68, 81, 89, 91, 92, 96, 108, 109, 187–190, 192, 194, 215, 216, 219, 324, 380–432
Animal Unit Months (AUM), 225, 268, 299, 300
Antelope bitterbrush, 29, 54, 61, 67, 77, 136, 305, 326, 329–338, 340–347, 350, 352–359, 361, 362
Aquicambids, 56, 86, 92, 216, 327, 335, 340, 366, 369, 371, 380, 391, 394, 402, 407, 408, 432, 467, 481, 485, 494, 502, 504, 536
Aquic soil moisture regime, 155, 161, 209
Aquisalids, 56, 86, 92, 185, 216, 217, 337, 341, 369, 371, 397, 403, 488, 497
Aquolls, 94, 165, 168, 380, 382, 383, 385, 388–394, 396, 400–405, 410, 411, 414–416, 419, 421–424, 426, 429–431
Argialbolls, 56, 63, 70–72, 78, 91, 94, 106–108, 161, 162, 164, 165, 172, 327, 333, 341, 345, 366, 368, 371, 372, 381, 389–391, 400, 402, 409, 428, 430, 467, 479, 480, 492, 495, 504, 531, 534
Argicryolls, 57, 94, 346, 373, 384, 386, 389, 399, 401, 407, 409, 416, 418, 427, 471, 474, 478, 483, 491, 493, 503, 505, 514, 517
Argids, 91, 92, 96, 181, 185, 215, 216, 324, 380, 381, 383–388, 390, 391, 393, 396–401, 403, 405, 407–409, 411, 412, 414–422, 424–428, 430–432
Argidurids, 53, 56, 57, 74, 85, 88, 91, 92, 96, 104, 107, 108, 117, 118, 164, 181, 184, 215, 324, 326, 329, 330, 332–334, 336, 337, 344, 345, 348, 352, 357, 359, 361, 366–369, 372, 374–377, 380, 381, 383–386, 388–390, 392, 393, 395, 397, 401, 406, 409, 413, 414, 418–420, 423, 424, 427, 428, 466, 467, 470, 472–474, 476–479, 482, 483, 486, 488, 489, 493, 501, 505, 510, 512, 516, 517, 519, 524, 525, 530, 531
Argillic horizon, 13, 14, 89, 91, 96, 111–113, 116–119, 124–126, 132–134, 136, 138, 139, 141, 142, 146–148, 161–165, 168–171, 173, 182, 190, 197, 199, 200, 201, 203, 204, 217
Argilluviation, 168, 185, 198, 201, 215, 216, 218, 324
Argixerolls, 12, 13, 45, 47, 49, 50, 53, 56, 57, 61–64, 70–72, 74, 75, 78–80, 91, 94, 96, 103, 104, 107, 108, 111–113, 164, 165, 170, 172, 215, 233, 236, 270, 277, 281, 282, 289, 292, 296, 299, 300, 308, 324, 326–351, 353–355, 357–363, 366–378, 380–431, 466–469, 471–475, 477–535
Aridic soil moisture regime, 14, 61, 82, 85, 86, 110, 112, 116, 117, 122, 124, 128, 132, 147, 165, 208
Aridisols, 14–16, 53, 61, 73, 82, 85, 86, 91, 92, 96, 181–185, 215, 219, 324, 380, 381, 383–432
Ash, Oregon, 54, 70, 105, 326–328, 331, 332, 334, 349, 351, 352, 361
Aspen, quaking, 49, 305
Association, soil, 13, 61

B

Baker City, 5, 6, 9, 60, 323
Baker County, 9, 61, 230, 258, 299–301
Basin and Range, 1, 23, 26, 32, 34, 35, 40, 50, 52, 53, 77, 86, 323
Benchmark soil, 219, 324
Benton County, 9, 10, 12, 29, 31, 38, 39, 68–72, 82, 84, 113, 135, 142, 146, 150, 156, 298
Black greasewood, 29, 53, 55, 86, 175, 178, 181, 327, 328, 334, 335, 337, 340–342, 344, 347, 352, 353, 355, 356, 359, 360
Bluebunch wheatgrass, 29, 32, 49, 50, 53, 55, 74, 77, 79, 82, 85, 89, 110, 112, 116, 117, 124, 128, 132, 146, 153, 166, 181, 215, 268, 269, 300, 326–363
Bluegrass, alkali, 53, 55, 351
Bluegrass, Sandberg, 29, 49, 50, 53, 55, 73, 74, 79, 82, 85, 112, 132, 153, 166, 181, 215, 300, 326–342, 344–363
Blue Mountains, 1, 21, 22, 32, 34, 35, 39–41, 43, 49, 50, 52, 53–56, 60–62, 110, 112, 119, 124, 126, 132, 136, 138, 140, 147, 153, 157, 165, 168, 175, 187, 192, 201, 205, 216, 258, 268, 299–301, 305, 310, 323
Bottlebrush squirreltail, 328–330, 334
Bureau of Land Management (BLM), 5, 55, 64, 74, 85, 223–226, 232, 258, 300
Bureau of Soils, 5

C

Calcic horizon, 13, 91, 111, 113, 128, 147, 148, 181–183, 217
Calcification, 215–218

T. Thorson et al., *The Soils of Oregon*, World Soils Book Series,
https://doi.org/10.1007/978-3-030-90091-5

Cambic horizon, 14, 89, 110, 111, 114, 116, 119, 120, 122, 123, 128–130, 136, 138–141, 143, 145, 149, 151, 152, 155, 157, 159, 160, 165–169, 175–180, 182, 187–193, 201, 207, 215, 323
Cambids, 91, 92, 181, 185, 216, 217, 380, 383, 385, 387, 388, 390–394, 396, 402, 404, 405, 407–411, 414, 415, 417–420, 422–429, 432
Cambisolization, 168, 177, 180, 185, 190, 208, 215, 216, 218, 324
Cascade Mountains, 1, 22, 32, 40, 43, 52–54, 56, 59, 66, 67, 114, 119, 126, 134, 136, 138, 140, 141, 143, 159, 175, 178, 187, 192, 197, 199, 201, 205, 206, 213, 216, 217, 233, 269, 283–286, 302, 304, 305, 308, 309, 311, 323
Cascade Range, 25, 35, 43, 47, 49, 201, 248
Cascadia Subduction Zone, 248, 249
Cation-Exchange Capacity (CEC), 99, 190, 200, 204
Catlow Valley, 57, 58
CEC activity class, 96, 110, 112, 116, 117, 122, 126, 128, 130, 132, 134, 141–143, 147, 155, 157, 161, 163, 168, 176, 197, 208, 380–432
Cedar, Incense, 27, 30, 64, 68
Cedar, Port Orford, 27, 30, 54, 88, 206
Cedar, western red, 54, 255, 338, 347
Cheatgrass, 29, 269, 300
Climate change, 223, 241–243, 252, 255, 257–259, 296, 312, 324
Coastal erosion, 223, 241, 249, 254, 259, 324
Coast Range, 1, 21–23, 25, 32, 35, 37, 40, 41, 43, 44, 50, 52–54, 56, 67–71, 81, 114, 134, 141, 143, 145, 149, 175, 178, 187, 189, 191, 192, 216, 220, 235, 247, 249, 268, 302, 303, 323
Colluvium, 17, 39–41, 53–55, 57, 59–61, 64–71, 73–79, 82, 84, 91, 104–106, 110, 111, 113, 114, 116–118, 122, 124–126, 128, 130, 131, 133–146, 149–152, 157, 159, 163, 165, 166, 169, 170, 175, 178, 180, 181, 187, 188, 191, 192, 197, 199, 201, 203, 205–207, 210, 269, 294, 323, 326–363
Columbia Basin, 52–54, 56, 82, 84, 122, 123, 153, 175, 178, 181, 185, 205, 213, 215, 217, 229, 279, 283–288, 299, 311, 323
Columbia County, 9, 12, 299
Columbia Plateau, 34, 35, 47, 52–54, 56, 74–76, 110, 122, 123, 128, 129, 165, 168, 175, 178, 217, 278–280, 297
Columbia River, 1, 2, 21, 23, 26, 32, 35, 40, 43, 46, 47, 49, 70, 82–85, 154, 244, 245, 247, 279, 299, 309, 311–313
Community Wildfire Protection Plan (CWPP), 258
Complex, soil, 13, 66
Confined Animal Feeding Operation (CAFO), 230, 241, 270, 273, 314, 324
Conservation Reserve Program, 225, 231, 258, 270, 275, 277–280, 314
Consociation, soil, 13
Coos County, 9, 68, 70, 272, 292
Coquille River, 35, 43
Crater Lake, 1, 37, 43, 65, 66, 120, 187, 190, 223, 224, 249, 253, 266, 305, 391, 439, 481
Crater Lake National Park, 9, 21, 29, 30, 37, 59, 120, 121
Cropland, 212, 226, 229–232, 241–243, 255, 258, 265, 268, 269, 272–279, 285, 289, 290, 297, 302, 314
Cryands, 50, 91, 92, 96, 187, 192, 324, 381–383, 385, 386, 388, 390, 391, 393–395, 398–410, 412–417, 419, 421–428, 431, 432
Cryaquands, 45, 47, 92, 343, 359, 372, 377, 389, 405, 427, 430, 478, 500, 530, 534
Cryaquolls, 45, 50, 94, 343, 356, 372, 376, 382, 393, 396, 401, 405, 411, 415, 422, 424, 469, 484, 488, 493, 500, 508, 512, 523, 526
Cryic soil temperature regime, 60, 62, 67, 119, 159, 208, 209
Cumulic subgroup, 89
Curry County, 9, 10, 66, 197, 199, 205, 211, 248, 250, 304, 305

D

Department of Defense (DOD), 223
Deschutes-Umatilla Plateau, 1, 32, 34, 35, 74, 79, 82
Developed land, 226, 229, 230, 246, 247, 255, 257, 258, 265, 302
Diagnostic horizons, 13, 89, 90, 103, 107, 215, 365
Douglas County, 9, 38, 127, 144, 165, 168, 257, 294
Douglas-fir, 12, 17, 27, 29, 40, 43, 46, 48–50, 60, 61, 64, 67, 68, 70, 77, 81, 88, 89, 91, 112, 114, 126, 130, 133, 136, 140, 141, 143, 145, 149, 151, 159, 161, 163, 166, 168, 175, 178, 188, 192, 197, 199, 201, 202, 233, 235, 236, 255, 258, 269, 270, 288, 289, 302–305, 315, 323, 324
Duraquods, 81, 91, 94, 205, 211, 384, 388, 392, 404, 471, 476, 483, 498
Durids, 91, 92, 96, 181, 185, 216, 217, 324, 380, 381, 383–386, 388–390, 392–395, 397, 398, 400–402, 406, 408–410, 412–415, 418–421, 423–425, 427, 428, 432
Duripan, 13, 14, 89–91, 107, 108, 117, 118, 128, 129, 132, 147, 148, 165, 168, 169, 173, 175, 179, 182, 183, 185, 187, 190, 217, 220, 323, 366–378
Durixerolls, 45, 49, 53, 56, 75, 78, 82, 84, 91, 94, 104–108, 132, 133, 164, 165, 169, 172, 217, 220, 328, 335, 338, 340, 344, 346, 348, 351, 355, 357, 361–363, 366, 369, 370, 372–376, 378, 381–384, 386, 390, 394, 395, 398–403, 405, 407, 410–416, 418, 421, 423, 424, 426, 428–430, 432, 468, 469, 471, 474, 480, 485, 486, 490–492, 494–497, 499, 502, 506–511, 513, 515, 517, 521, 522, 524, 525, 528, 531, 532, 534, 536
Dystrocryepts, 43, 44, 56, 63, 93, 175, 178, 327, 352, 366, 375, 381, 383, 391, 395, 409, 410, 418, 430, 467, 470, 480, 486, 505, 506, 516, 534
Dystroxerepts, 45, 50, 56, 64–66, 71, 91, 93, 104, 107, 108, 130, 131, 164, 175, 178, 179, 215, 296, 327, 328, 332, 333, 335, 337, 338, 342, 348, 349, 351, 354, 356, 357, 360, 363, 366, 368–371, 373–378, 380–383, 390, 391, 394, 395, 399, 400, 402, 404, 405, 408, 413, 414, 416, 420, 422–424, 426, 428, 431, 467–469, 479, 480, 483, 484, 486, 490–492, 495, 498, 499, 504, 510, 511, 514, 519, 522–526, 528, 531, 535
Dystrudepts, 43, 44, 56, 60, 68, 69, 81, 88, 93, 107, 108, 143, 144, 164, 175, 178, 180, 215, 329–331, 334, 336, 337, 339, 347, 349, 353, 355, 367–370, 373–376, 383–385, 387–389, 392, 394, 396–398, 400, 404, 405, 408, 411, 413, 419–421, 420, 423, 424, 426, 429, 432, 470, 471, 473, 475–478, 482–485, 487–489, 492, 498, 499, 503, 507, 508, 510, 518–521, 523, 526, 528, 529, 533, 536, 537

E

Earthquakes, 223, 241, 247–249, 252, 253, 255, 259, 324
Ecoregions, 29, 32, 33, 220
Endangered soil, 219, 220, 433, 463
Endemic soil, 219
Endoaquolls, 53, 56, 57, 73, 94, 105, 107, 108, 155, 156, 164, 165, 173, 331, 333, 346, 350, 351, 355, 356, 361, 362, 367, 368, 373, 374, 376, 378, 380, 383, 385, 388, 390–392, 394, 396, 402–405, 410, 414, 416, 419, 421–424, 429–431, 466, 470–473, 476, 477, 479–484, 487, 495, 496, 498, 500, 506, 512–514, 518, 521, 523–526, 532–534
Entisols, 14–16, 81, 82, 88, 93, 96, 205–209, 212, 213, 216, 218, 219, 250, 382, 383, 386, 388, 389, 392, 393, 395–403, 405, 407–409, 411–414, 416–418, 421–423, 425, 427–432
Eocene, 37–39
Eolian deposits, 82, 338
Epiaquerts, 57, 58, 94, 205, 329, 351, 367, 374, 385, 388, 416, 472, 477, 514
Epipedons, 12–14, 89, 90, 108, 176, 190, 197, 201, 221, 323
Eugene, 1, 9, 230
Eutrudepts, 56, 68, 70, 93, 105, 107, 108, 149, 150, 164, 175, 180, 330, 334, 360, 362, 367, 369, 377, 378, 385, 388, 393, 397, 400, 412, 419, 421, 423, 425, 428, 430, 473, 476, 484, 489, 492, 509, 518, 522, 524, 527, 530, 533

Exclusive farm use, 237, 240
Extragrade subgroups, 91

F

Family, soil, 91
Farm Service Agency (FSA), 225, 228, 274, 275, 280
Farmsteads, 229, 230, 265
Fescue, Idaho, 29, 32, 48–50, 53, 64, 67, 73, 74, 77, 79, 89, 110, 112, 116, 124, 132, 136, 166, 215, 268, 300, 305, 326–363
Fir, grand, 27, 45, 46, 48, 50, 88, 104, 105, 133, 138, 143, 175, 178, 188, 192, 206, 236, 269, 302, 305, 326, 328, 330, 331, 333–337, 339–356, 358–363, 469, 470, 473, 474, 480–482, 485–487, 489, 490, 492–494, 497, 501, 503, 506–510, 515, 516, 521, 523, 527–530, 537
Fir, noble, 27, 29, 43–46, 50, 54, 60, 106, 159, 206, 288, 333, 334, 341–344, 348, 352, 360, 362, 363, 502
Fir, Pacific silver, 29, 43, 44, 50, 159, 206, 340–344, 348, 352, 360, 362, 363, 501
Fir, subalpine, 27, 45, 48–50, 61, 104, 119, 175, 178, 188, 192, 201, 327, 328, 330, 342–344, 346–348, 352, 358–360, 474, 481, 486
Fir, white, 27, 45, 46, 50, 54, 64, 77, 175, 178, 188, 192, 327, 330, 333, 335–338, 342, 344, 346, 350–352, 354, 357–360, 363, 470, 478, 480, 492, 493, 500, 504, 513, 515, 520, 524, 530
Flooding, 40, 41, 223, 237, 241, 243–247, 255, 259, 265, 268, 284, 285, 292, 324
Fluvaquents, 56, 81, 93, 299, 382, 407, 408, 423, 425, 469, 502, 504, 525, 527
Foredunes, 250, 251, 254, 255
Forestland, 77, 225, 226, 228, 229, 233, 234, 236, 237, 242, 255, 258, 265, 269, 272, 297, 298, 301, 302, 305, 314, 324
Fragipan, 13, 14, 89, 90, 366–378
Fragixerepts, 56, 70, 93, 308, 331, 343, 367, 371, 387, 405, 417, 476, 500, 516
Fremont Mountains, 49, 50
Frigid soil temperature regime, 53, 61, 68, 73, 78, 79, 110, 112, 114, 116, 117, 122, 124, 126, 128, 130, 134, 138, 140, 141, 143, 147, 149, 153, 155, 161, 163, 190, 197, 204
Fulvicryands, 43, 44, 92, 187, 191, 343, 372, 402, 405, 413, 414, 421, 425, 431, 495, 500, 510, 511, 522, 527, 535
Fulvudands, 43, 44, 50, 56, 81, 92, 105, 107, 108, 151, 152, 164, 187, 189, 192, 194, 327, 330, 343, 349, 354, 359, 366, 367, 372, 374, 375, 377, 381, 386, 387, 396, 405–408, 412, 413, 421, 422, 427, 431, 468, 474, 475, 488, 499, 500, 503, 504, 509–511, 520, 523, 529, 534, 535

G

Glaciolacustrine, 54, 70, 71, 82, 84, 106, 161, 166, 201, 203, 326, 327, 330, 332, 334, 340, 341, 345, 348, 352, 355, 362, 363
Gleization, 168, 177, 180, 190, 201, 208, 209, 215, 216, 218, 324
Glossic horizon, 90, 366–378
Goose Lake Valley, 78
Great group, soil, 12, 44, 49, 52, 88, 89, 91, 96, 103, 104, 108, 109, 110, 215, 220

H

Halaquepts, 53, 56, 57, 86, 93, 217, 353, 356, 359, 360, 375–377, 384, 386, 397, 409, 419, 423, 424, 426, 428, 432, 472, 474, 489, 504, 518, 523, 525, 528, 530, 532, 537
Haplargids, 53, 56–58, 74, 91, 92, 96, 104, 107, 108, 116, 164, 181, 184, 324, 327, 329, 330, 332–334, 339, 341, 345, 348–350, 353, 354, 356–358, 360, 362, 363, 368, 370–378, 380, 381, 384, 385, 387, 390, 391, 393, 396–401, 403, 407–409, 412, 414–422, 424, 426–428, 430–432, 466–468, 472, 473, 475, 480, 483, 490, 491, 493, 497, 502, 504, 505, 509, 511–513, 515, 517, 519, 521–523, 525, 529, 531, 534–537
Haplocalcids, 56, 74, 82, 85, 88, 92, 185, 217, 283, 326, 350, 351, 354, 361, 366, 374, 375, 378, 380, 405, 409, 410, 415, 420, 428, 466, 499, 505, 506, 513, 520, 532
Haplocambids, 45, 49, 53, 56–58, 75, 76, 82, 84, 89, 91, 92, 104, 107, 108, 122, 123, 164, 181, 184, 215, 326, 330, 331, 333–336, 342, 343, 345, 347, 350, 353–355, 359, 361, 366–369, 371–378, 380, 381, 383, 385, 387, 388, 390, 392–394, 396, 404, 405, 407–411, 414, 415, 417–420, 422–429, 466, 468, 470, 473, 475, 476, 479, 482, 484, 485, 487, 488, 498–500, 502–505, 507, 508, 511–513, 515–520, 522, 524–526, 528–532
Haplocryands, 45, 47, 50, 92, 106–108, 159, 160, 164, 187, 192, 194, 215, 333, 334, 341, 342, 345, 350, 352, 360, 363, 368, 369, 371, 372, 374, 375, 377, 378, 391, 393, 402, 403, 405, 406, 408, 414–416, 423, 428, 432, 481, 484, 495, 497, 499, 501, 504, 512, 513, 515, 525, 531, 536
Haplocryepts, 45, 49, 93, 328, 329, 346, 360, 366, 373, 367, 377, 382, 384, 410, 412, 414, 428, 432, 469, 472, 506, 509, 511, 531, 536
Haplocryods, 49, 60, 91, 94, 205, 311, 362, 378, 407, 430, 502, 533
Haplocryolls, 45, 49, 53, 56, 57, 94, 328, 332, 335, 339, 343, 366, 368, 369, 370, 372, 381, 382, 384, 385, 389, 393, 394, 398, 401, 406, 467, 468, 471, 473, 478, 484, 485, 490, 493, 494, 501
Haplodurids, 45, 49, 53, 56, 82, 85, 91, 92, 104, 107, 108, 128, 129, 164, 181, 184, 296, 330, 337, 340, 343, 348, 350, 352, 353, 355, 358, 367, 370, 371–377, 385, 386, 394, 395, 398, 400, 402, 406, 408, 412, 414, 415, 418, 421, 423–425, 427, 472, 474, 484, 486, 490, 492, 495, 500, 504, 509, 511, 513, 516, 517, 521, 524, 525, 527, 529
Haplohemists, 56, 78, 81, 93, 205, 218, 345, 372, 383, 385, 391, 407, 470, 472, 481, 502
Haplohumults, 44, 56, 60, 66, 68–72, 88, 89, 91, 94, 104, 107, 108, 133–135, 164, 197–200, 216, 220, 284, 296, 326–328, 340, 347, 349–351, 356, 362, 366, 370, 373, 374, 376, 378, 380, 381, 383, 386, 390, 391, 395, 399, 401, 403, 408–410, 413–416, 419, 423, 425, 428, 430–432, 466, 467, 470, 474, 480, 481, 486, 490, 494, 497, 504, 505, 507, 511, 512, 514, 518, 524, 526, 532, 534, 535, 537
Haplorthods, 81, 91, 94, 205, 211, 212, 311, 330, 367, 382, 386, 391, 413, 469, 474, 481, 511
Haploxeralfs, 45, 50, 56, 64, 66, 91, 92, 104, 107, 108, 126, 127, 164, 201, 202, 204, 215, 284, 288, 289, 296, 304, 308, 326, 328, 333, 335, 337, 338, 341, 342, 345, 346, 349–352, 354, 355, 357, 358, 360–363, 366, 368–378, 380, 382–384, 388–390, 394, 395, 398–400, 402, 404, 405, 407–411, 413, 414, 416, 418, 420–426, 428, 430–432, 466, 469–471, 476, 477, 480, 485, 489, 491, 492, 495, 496, 498, 499, 502–504, 506, 507, 510–512, 514, 515, 517, 520–523, 525–528, 531–534, 536, 537
Haploxerands, 45, 50, 56, 78, 92, 187, 188, 193, 341, 352, 363, 371, 375, 378, 403, 417, 424, 431, 496, 515, 525, 535
Haploxerepts, 45, 50, 56, 67, 73, 93, 105, 107, 108, 140, 141, 164, 175, 178, 180, 215, 326, 327, 330, 331, 336, 341, 342, 344, 345, 350, 353, 355, 358, 361, 366, 367, 369, 371, 372, 374–378, 380–383, 385, 387, 389, 392, 394, 396, 398, 401–409, 414, 417, 419, 421, 422, 424, 425, 427, 429, 466, 467, 469, 470, 472, 473, 476, 477, 482, 485, 487, 489, 490, 494–497, 499, 501–503, 505, 512, 516, 519, 521, 522, 525, 527, 530, 532
Haploxerolls, 45, 47, 49, 50, 53, 56, 57, 61–64, 71–76, 78–80, 82, 84, 89, 91, 94, 96, 104, 107, 108, 110, 111, 164, 165, 167, 168, 173, 215, 217, 218, 220, 233, 278–280, 283, 284, 296, 299, 324, 326–357, 360–363, 366–378, 380–432, 466–524, 526–537
Hapludands, 43, 44, 50, 56, 68, 92, 107, 108, 145, 146, 164, 187, 192, 194, 215, 220, 326, 331, 337, 338, 340, 343, 344, 348, 356, 359, 366–367, 369–372, 374, 376, 377, 380, 381, 388, 397, 399, 401, 402, 405, 406, 408, 410, 413, 420, 423, 426, 427, 467, 476, 489, 491, 494, 495, 500, 501, 503, 506, 510, 519, 520, 524, 529

Harney County, 1, 9, 30, 32, 49, 57, 58, 86, 133, 181, 257, 278
Hellgate Canyon, 64
Hemlock, mountain, 30, 46, 50, 54, 60, 106, 159, 206, 305, 338, 340–343, 345, 352, 359, 360, 362
Hemlock, western, 27, 29, 30, 40, 43, 46, 50, 60, 68, 81, 89, 104, 114, 133, 141, 143, 145, 151, 175, 178, 188, 192, 197, 199, 201, 202, 233, 235, 302–305, 323, 326–331, 333–356, 358–363, 470, 478, 483, 489, 491, 498, 499, 510, 517, 532, 533
High Lava Plains, 1, 22, 32, 34, 35, 52, 53
Highly Erodible Land (HEL), 273, 274, 280, 314
Histic epipedon, 13, 208, 323
Histosols, 14, 78, 81, 93, 96, 109, 205, 207–209, 212, 213, 216, 218, 242, 383, 385, 387, 391, 404, 407, 410, 421, 424
Holocene, 40, 41, 188, 192, 206, 249
Humaquepts, 45, 50, 56, 70, 78, 81, 93, 175, 177, 218, 333, 349, 359, 368, 374, 377, 384, 385, 388, 389, 391, 401, 406, 413, 414, 417, 418, 427, 429–432, 472, 473, 477, 478, 481, 494, 501, 511, 515, 517, 530, 533–535, 537
Humboldt Area, 52, 53, 55, 56, 86, 122, 157, 215, 217
Humicryepts, 45, 49, 50, 56, 63, 89, 93, 330, 367, 371, 383, 386, 389, 391, 394, 395, 398–400, 402, 408, 412, 413, 418, 420, 423, 470, 474, 478, 480, 481, 484–486, 489–492, 495, 503, 509, 516, 520, 524
Humification, 177, 180, 198, 209, 215, 216, 218, 324
Humudepts, 43, 44, 45, 47, 50, 56, 68–70, 81, 89, 91, 93, 96, 104, 107, 108, 114, 115, 164, 175, 178, 180, 215, 220, 236, 303, 304, 324, 327, 329, 333, 336, 340–343, 347–349, 352, 353, 356, 358, 362, 363, 366–378, 380–382, 384, 385, 389–392, 395–406, 408–414, 417–419, 421, 423–427, 429–432, 466, 468, 472, 473, 478, 479, 481, 482, 486, 488–500, 503, 504, 507–512, 516–519, 521, 523–529, 532–534, 536, 537
Humults, 91, 94, 96, 197–199, 216, 380–384, 389–391, 393, 395, 399–404, 406, 408–410, 413–416, 418, 419, 421–423, 425, 426, 428, 430–432
Hydric soils, 237, 239, 243, 258, 274

I

Idaho Batholith, 40
Inceptisols, 14, 15, 60, 62, 64, 67, 68, 70, 78, 81, 86, 88, 89, 91, 93, 96, 109, 175–179, 180, 215, 216, 219, 250, 251, 324, 380–414, 416–432
Indian ricegrass, 29, 54, 82, 181, 215, 326–330, 333, 334, 336–339, 341–343, 345, 347–350, 352, 353, 355, 357, 359, 361–363
Inland saltgrass, 29, 53, 55, 86, 175, 178, 181, 215, 326–328, 337, 340, 344, 346, 350, 351, 353, 355, 356, 359, 360
Intergrade, subgroups, 91
Isofrigid soil temperature regime, 25, 81, 151
Isomesic soil temperature regime, 88, 197, 199, 209

J

Jackson County, 9, 65, 66, 237, 283–285
John Day, formation, 38, 39
John Day, fossil beds, 38, 39
Jory soil series, 17, 142, 323
Josephine County, 9, 12, 39, 269, 291, 294, 299
Juniper, western, 27, 29, 30, 48–50, 54, 55, 61, 67, 77, 89, 166, 168, 232, 305, 326, 328, 332, 335–338, 342, 344, 345, 349, 351, 352, 354, 355, 357, 361, 362

K

Klamath County, 5, 9, 37, 120, 205, 282, 283, 297
Klamath Mountains, 1, 25, 32, 35, 37, 39, 41, 43, 50, 52, 63, 88, 125, 163, 164, 323
Klamath Valleys and Basins, 40, 77, 112, 155, 165, 168, 187, 192, 216, 218, 243, 284, 305

L

Lacustrine, 40, 41, 53–55, 57, 58, 70, 77, 78, 82, 84–86, 91, 104, 105, 116, 122, 128, 146, 153, 166, 181, 183, 185, 206, 209, 300, 323, 326–329, 333, 335–337, 340, 341, 344–359, 361
Lake County, 1, 9, 53, 57, 78, 209, 257, 266, 277, 278
Lake Missoula, 25
Lake Owyhee State Park, 85
Land capability classes, 232, 236, 238, 258, 284, 285, 299, 466–537
Land Conservation and Development Commission (LCDC), 226, 238, 241
Landslides, 223, 241, 247, 248, 250, 252, 254, 259, 272, 324
Larch, western, 55, 62, 105, 305, 327–330, 333, 335, 337, 339, 341–345, 347, 348, 350, 352, 358–360
Late Wisconsin, 1
Liquefaction, 249
Lithic contact, 143, 168, 183
Lithic subgroup, 168
Loess, 41, 54, 55, 61, 63, 67, 74–76, 78–80, 82, 84, 85, 91, 104, 106, 110, 111, 118, 122, 123, 139, 147, 148, 157, 158, 161, 162, 165, 166, 169, 175, 180, 181, 217, 247, 300, 323, 326, 327, 330–333, 335, 336, 338, 339, 341–357, 360–363

M

Major Land Resource Areas (MLRAs), 45, 51–54, 56, 59–64, 66–68, 70, 71, 73, 74, 77–79, 81, 82, 85, 86, 88, 123, 165, 170, 175, 178, 181, 185, 187, 192, 197, 199, 201, 202, 205, 215, 219, 220, 268, 278, 280, 281, 283–286, 288, 289, 293, 295, 298, 299, 301, 304, 305, 308, 309, 323
Malheur County, 9, 16, 118, 148, 225, 278, 283, 287, 291
Malheur High Plateau, 49, 52, 53, 55–57, 110, 112, 116, 117, 122, 124, 128, 132, 147, 153, 155, 157, 163, 165, 168, 170, 181, 185, 205, 213, 216–218, 323
Maple, bigleaf, 50, 60, 68, 143, 149, 175, 178, 188, 192, 197, 199, 201, 202
Marion County, 9, 10, 12, 244, 257, 283, 294
Marys Peak, 35, 43
Measure 49, 240, 259
Melanic epipedon, 13, 151, 187, 192
Mesic soil temperature regime, 62, 64, 70, 74, 82, 85, 86, 96, 132, 136, 145, 157, 168, 197
Miocene, 37–39, 74, 79, 82
Mollic epipedon, 13, 14, 89, 96, 110–113, 124, 125, 132, 136, 155, 161, 162, 165–173, 188, 190, 193, 201, 323
Mollisols, 12–15, 53, 61, 62, 64, 67, 70, 73, 74, 78, 79, 89, 91, 94, 96, 108, 165, 166, 168, 169, 172, 215–219, 251, 270, 285, 324, 380–432
Moraine, 63, 207, 338
Mountain mahogany, 300
Mt Ashland, 35, 50
Mt Hood, 1, 32, 35, 43, 59, 242, 249, 253, 257
Mt Mazama, 1, 22, 32, 37, 41, 43, 62, 67, 187, 188, 192, 249, 253

N
National inventory groupings, 236, 237
National Park Service (NPS), 5, 223, 224, 257, 304
Natrargids, 53, 56, 57, 92, 181, 183, 217, 328, 347, 352, 366, 373, 375, 381, 383, 386, 388, 393, 397, 398, 405, 411, 414, 417, 426, 468, 470, 474, 476, 483, 488, 490, 499, 508, 511, 516, 528
Natric horizon, 14, 181, 183, 217, 218
Needlegrass, Thurber, 29, 53–55, 64, 73, 82, 85, 116, 117, 128, 146, 166, 181, 215
Newberry Crater, 249, 253
Northern spotted owl, 223, 233, 258, 272
Nursery crops, 258, 289, 290, 324

O
Oak, California black, 201, 202, 304, 305, 326, 331, 342, 346, 347, 350, 352, 354–358, 360, 361
Oak, Oregon white, 17, 31, 46, 50, 64, 70, 89, 91, 112, 140, 141, 161, 163, 166, 197, 199, 201, 202, 304, 309, 315
Ochric epipedon, 13, 14, 116–120, 122, 123, 126, 128–130, 136, 138–143, 145–147, 149, 153, 154, 157, 159, 160, 163, 164, 175, 177–179, 181–183, 185, 187, 190, 201, 203, 207, 209, 323
Oligocene, 38, 39
Orchards, 17, 192, 231, 238, 274, 283–287, 314
Order, soil, 14–16, 19, 20, 53, 60–62, 64, 67, 68, 70, 73, 74, 78, 79, 81, 82, 85, 86, 88, 89, 95, 103, 108–110, 165, 168, 175, 178, 181, 185, 187, 189, 192, 197, 199, 205, 284
Oregon Climate Change Research Institute (OCCRI), 241
Oregon Forest Practices Act (OFPA), 270, 272, 303, 304, 314, 324
Oregon Land Use Act of 1973, 237, 259, 324
Oregon Land Use Planning, 237
Oregon State University, 29, 226, 230, 270, 277, 281, 282, 285, 286, 289, 290, 297, 299, 314
Ortstein, 13, 14, 89, 90, 205, 206, 208, 209, 211–213, 220, 366–378, 382, 384, 388, 392, 404, 413, 430
Outwash, 59, 351, 362, 363
Owyhee High Plateau, 52, 53, 55, 56, 73, 74, 112, 116, 117, 122, 128, 165, 170, 175, 178, 181, 185, 216, 217

P
Pachic subgroup, 89, 187, 192, 216
Pacific madrone, 45, 54, 64, 68, 91, 104, 126, 130, 133, 175, 178, 197, 199, 201, 202, 304, 326, 328, 330–335, 337, 339–342, 345–350, 352, 354–358, 360–363
Paleargids, 53, 56–58, 92, 107, 108, 146–148, 164, 181, 184, 329, 332, 337, 338, 352, 357, 358, 366, 368–370, 375–377, 383, 387, 390, 398, 399, 409, 415, 417, 424–428, 430, 470, 475, 479, 489, 491, 505, 513, 515, 525, 527, 530, 531, 533
Palehumults, 17, 20, 45, 47, 56, 60, 68–73, 94, 103, 105, 107, 108, 141, 142, 164, 197, 199, 200, 215, 216, 220, 294, 296, 323, 328, 333, 339, 341, 342, 346, 350, 355, 363, 366, 368, 370, 371, 373, 374, 376, 378, 382, 383, 386, 389, 391, 393, 395, 400, 402, 404, 406, 410, 415, 418, 421, 422, 426, 431, 468–470, 475, 478, 481, 483, 486, 493, 496, 498, 500, 506, 507, 513, 517, 521, 522, 528, 536
Palexeralfs, 92, 106–108, 163, 164, 201, 202, 204, 296, 328, 347, 354–356, 358, 363, 366, 373, 375–378, 382, 397, 401, 409–411, 414, 418, 420–422, 425, 431, 469, 488, 494, 504, 507, 508, 512, 517, 520, 521, 523, 527, 535
Palexerolls, 45, 49, 53, 56, 61, 62, 78, 91, 104, 107, 108, 124, 125, 164, 165, 173, 215, 216, 220, 329–331, 333, 339, 340, 344, 346, 352, 356, 363, 366–368, 370, 372, 373, 375, 376, 378, 381, 383–385, 387, 390, 391, 393, 396, 401, 403, 407–409, 416–419, 422–427, 432, 467, 471–473, 475, 476, 479, 481, 484, 488, 493, 494, 496, 502, 504, 505, 514, 516, 517, 519, 523–526, 529, 530, 536, 537
Palexerults, 45, 50, 65, 66, 94, 326, 329, 335, 352, 366, 367, 369, 375, 380, 384, 395, 417, 422, 466, 472, 485, 515, 522
Paludization, 209, 215, 216, 218
Paralithic contact, 91, 113, 116, 126, 176, 180, 220
Pasture, 212, 226, 229, 231–233, 236, 237, 243, 245, 247, 258, 265, 266, 268, 272, 275, 276, 281, 297–302, 314, 324, 537
Pedodiversity, 220
Pine, lodgepole, 22, 27, 29, 40, 45–48, 50, 54, 60, 62, 67, 119, 138, 175, 178, 188, 192, 201, 206, 305, 315
Pine, ponderosa, 22, 27, 29, 40, 45–50, 54, 55, 61, 64, 67, 77, 89, 91, 104–106, 112, 119, 126, 130, 136, 140, 166, 168, 175, 178, 188, 192, 197, 201, 202, 236, 269, 304–306, 308, 315, 323, 335, 344, 346, 355, 357, 359
Pine, shore, 54, 81, 206, 304, 330, 361
Pine, sugar, 27, 54, 64, 67, 77, 197, 199, 326, 331, 337, 338, 341, 345, 352, 356, 363
Pine, whitebark, 45, 50, 55, 62
Playa, 37, 40, 77, 86, 217
Pleistocene, 1, 25, 58, 70, 77, 85, 181, 197, 199, 201, 203, 244
Pliocene, 40, 41
Pluvial lake, 1, 26, 40
Podzolization, 215–217
Portland, 1, 71, 226, 230, 233, 234, 241, 244, 245, 247, 249, 258, 279, 302, 311, 323
Prime farmland, 237, 258, 277, 299

Q
Quaternary, 35, 37, 53, 54, 60, 323

R
Rabbitbrush, 29, 53, 55, 73, 86, 300, 326
Rangeland, 77, 223, 225, 226, 229, 231–233, 237, 241, 255–258, 268–270, 297, 299–302, 314, 315, 324, 537
Rare soil, 219
Redwood, coastal, 40, 43, 44, 52–54, 56, 67, 87, 88, 114, 134, 141, 175, 470
Residuum, 17, 39–41, 53–55, 57, 60, 61, 64–71, 73–79, 82, 84, 88, 91, 104–106, 110, 111, 113, 114, 116, 124, 126, 128, 130–135, 139, 141–143, 145, 146, 149–152, 159, 161, 165, 166, 169, 170, 175, 178, 180, 181, 187–189, 191, 192, 197, 199, 201, 203, 206, 269, 294, 323, 326–363, 366
Rocky Mountains, 2, 60

S
Sagebrush, bud, 55, 86, 181, 328, 330, 334, 337, 341, 349, 350, 359
Sagebrush, low, 29, 5–55, 73, 77, 89, 181, 331, 336, 337, 340, 344, 347–349, 354, 355, 361, 363
Sagebrush, mountain big, 29, 32, 48–50, 53–55, 61, 67, 89, 181, 326–347, 349–351, 353, 354, 356–359, 361, 362
Sagebrush Rebellion, 225
Sagebrush, Wyoming big, 29, 48, 49, 53, 61, 73, 74, 82, 85, 86, 89, 104, 110, 116, 117, 128, 132, 146, 153, 181, 300, 326–339, 341, 342–363
Salem, 1, 5, 7, 10, 230, 323, 354, 376, 421, 456, 520
Salic horizon, 14, 89, 182, 185, 217, 323
Salinization, 185, 215–217, 218
Serpentinite, 1, 35, 37, 201, 203, 335
Seventh Approximation, 5, 10
Shadscale, 55, 86, 181, 328, 330, 333, 334, 337, 341, 359, 363
Shallow soil, 220, 272
Sherman County, 10, 75, 76
Silicification, 185, 190, 209, 215–217, 218, 324

Siskiyou Area, 64
Site index, 236, 269, 270, 272, 303–305, 315, 537
Snake River, 1, 21, 32, 52–54, 56, 73, 85, 122, 128, 157, 181, 185, 205, 284, 287, 288
Snowberry, 53, 55, 304, 326–330, 332–339, 341–345, 347, 349–352, 354–356, 358–363
Soil and Water Conservation Districts, 226, 256, 270, 275
Soil, definition, 5
Soil health, 229, 242, 246, 270, 275–277, 286, 288, 289, 293, 295, 312, 314
Soil map, 6, 7, 13, 15–18, 103, 230, 236, 253, 269, 272, 274, 280, 284, 285, 323
Soil series, 8, 10–13, 17, 20, 21, 40, 53, 56, 57, 61, 68, 69, 73, 76, 78, 81, 82, 84, 89–91, 96–101, 103, 110–124, 126–161, 163–169, 171, 175, 178, 179, 181, 183, 187, 189–192, 197–199, 201–203, 205, 207–211, 215, 217–221, 250, 269, 296, 323, 324, 463
Soil survey, 5, 8–10, 12, 17, 22, 51, 103, 109, 135, 142, 144, 146, 150, 156, 225, 229, 230, 236–239, 250, 251, 253, 265, 268–270, 272, 277, 283, 285, 294, 303–305, 314, 323
Soil survey management groups, 236, 268
Soil Taxonomy, 5, 10, 11, 13, 14, 17, 22, 89, 185, 190, 219, 221, 269, 270, 296
Solonization, 215–217, 218
Spiny hopsage, 55, 86, 181, 327, 330, 333–335, 338, 341, 342, 347, 348, 350, 355, 357, 359
Spodic horizon, 14, 91, 190, 205–207, 209, 211–213, 218
Spodosols, 14, 15, 60, 81, 91, 94, 96, 109, 190, 205–209, 212, 213, 216, 217, 382, 384, 386, 388, 391, 392, 404, 407, 413, 426, 430, 432
Spruce, Engelmann, 27, 48–50, 61, 119, 175, 178, 188, 192, 201
Spruce, Sitka, 27, 30, 40, 43, 44, 52–54, 56, 67, 81, 91, 105, 114, 141, 151, 175, 177, 178, 187, 188, 192, 205, 206, 211, 213, 216–218, 288, 292, 299, 304, 311, 314, 323, 327, 330, 336, 343, 348, 349, 353, 354, 356, 358, 359, 361, 529
SSURGO database, 314, 323
STATSGO2 database, 103
Steens Mountain, 25, 32, 35, 49, 86
Subalpine Zone, 50
Subgroup, soil, 12, 91, 97
Suborder, soil, 12, 95

T

Tanoak, 27, 68, 305, 326–328, 330–332, 334, 336, 337, 339, 340, 342, 348, 349, 360, 363
Till, 54, 55, 59, 91, 104, 118, 160, 187, 190, 207, 242, 243, 271, 274, 276, 280, 282, 283, 288, 327, 330, 333, 338, 343, 346, 348, 350, 363
Tillamook County, 5, 10, 25, 29, 30, 38, 39, 81, 115, 187, 235, 246, 251, 252, 254, 257, 277, 288, 297, 299
Timber Wars, 225
Tolerable erosion, 237, 274
Torric (aridic) soil moisture regime, 23, 25, 105, 108, 153
Torriorthents, 82, 84, 91, 93, 105, 107, 108, 157, 158, 164, 205, 336, 339, 348, 356, 369, 370, 373, 376, 386, 388, 395–398, 401–403, 408, 411–413, 417, 423, 427, 428, 473, 474, 476, 486–488, 490, 493, 495, 496, 503, 508, 509, 515, 516, 523, 524, 530, 531
Torripsamments, 56, 82, 84, 91, 93, 105, 107, 108, 153, 154, 164, 205, 209, 212, 283, 338, 342, 348, 352, 355, 362, 370, 371, 373, 375, 376, 378, 380, 399, 400, 405, 412, 418, 421, 423, 427, 430–432, 467, 491, 492, 499, 509, 517, 521, 524, 529, 534, 535, 537
Tsunamis, 223, 241, 248, 249, 253, 255, 259, 324

U

Udands, 89, 92, 187, 192, 380, 381, 386–388, 396, 397, 399, 401, 402, 405, 406, 408, 410, 412, 413, 418, 420–423, 426, 427, 431
Udepts, 91, 93, 96, 175, 178, 324, 380–385, 387–406, 408–414, 417–421, 423–432
Udic soil moisture regime, 60, 62, 68, 81, 88, 114, 119, 134, 138, 141, 143, 145, 149, 151, 176, 197, 199, 204
Udivitrands, 45, 47, 49, 92, 105, 107, 108, 138, 139, 164, 187, 192, 194, 220, 330, 337, 339, 341, 344, 345, 358, 363, 367, 370–372, 377, 378, 383, 386, 391, 392, 395–398, 400–403, 406, 408, 409, 425, 432, 470, 473, 474, 481, 482, 486, 488–490, 492–494, 496, 501, 503, 505, 527, 530, 536
Ultisols, 14–16, 60, 64, 68, 70, 88, 91, 94, 96, 103, 197–200, 215, 216, 251, 285, 324, 380–384, 386, 388–391, 393, 395, 397, 399–406, 408–411, 413–419, 421–423, 425, 426, 428, 430–432
Umatilla County, 10, 74, 75, 79, 80, 169, 282–284, 286
Umbric epipedon, 13, 89, 114, 133, 134, 136, 145, 151, 152, 159, 175, 177, 180, 187, 189, 191, 197, 199, 201
Umpqua River, 64, 165, 167
Unique farmland, 237, 238, 258
Universal Soil Loss Equation (USLE), 273
University of Oregon, 294
US Fish and Wildlife Service (USFWS), 5, 223, 230, 233, 243
US Forest Service (USFS), 5, 30, 32, 223, 225, 228, 233, 257

V

Vertisols, 14–16, 94, 96, 205–210, 212, 213, 216, 217, 285, 382, 385, 387–390, 392, 401, 404, 405, 407, 413, 415–418, 431
Vertization, 208, 209, 215–218
Vitrands, 92, 187, 192, 383, 386, 389, 392, 395–398, 400–403, 406, 408, 409, 425, 432
Vitricryands, 45, 47, 49, 50, 56, 60, 63, 67, 91, 92, 96, 104, 107, 108, 118–121, 164, 187, 192, 194, 215, 220, 305, 327, 330–332, 342, 344, 346, 348, 355, 357, 359, 366–368, 371–373, 376, 377, 381–383, 385, 386, 388, 390, 393–395, 398–410, 412, 413, 417, 421, 422, 424–427, 431, 467, 468, 470, 472, 474, 476, 479, 484–486, 490, 491, 493–495, 497, 498, 500–502, 504–506, 508, 509, 515, 516, 522, 523, 525–527, 529, 530, 536
Vitrixerands, 45, 47, 49, 56, 65–67, 91, 92, 105, 107, 108, 136, 137, 164, 187, 192, 195, 305, 310, 329, 333, 337, 340, 343, 344, 346, 347, 350, 352, 356, 359–362, 367, 368, 370–378, 380, 381, 384, 385, 389–392, 396, 398–400, 402, 405, 406, 409–411, 415–417, 419, 420, 422–427, 429–432, 466, 467, 471, 478–482, 487, 489–492, 495, 500, 501, 505–507, 512, 513, 515, 518, 520, 521, 523, 524, 526, 528–530, 532, 534–536
Volcanic ash, 32, 40, 41, 54, 55, 60–63, 65, 66, 74, 78, 91, 110, 111, 118, 136–139, 153, 161, 162, 166, 169, 175, 180, 181, 183, 187, 188, 190, 192, 207, 209, 217, 248, 323, 325–330, 332–334, 337–341, 343–348, 350, 352, 355–363
Volcanoes, 35, 43, 223, 241, 248, 249, 251, 253, 259, 324

W

Wallowa Batholith, 34, 47
Wallowa County, 38, 39, 47, 61, 63, 79, 111, 162
Wallowa Mountains, 21, 25, 32, 34, 40, 47–49, 61–63, 305
Web Soil Survey (WSS), 5, 8, 103, 225, 236, 253, 265, 266, 268–270, 272, 274, 277, 299, 314, 315, 323
Wetland loss, 223, 241, 243, 259, 324
Wildfires, 223, 241–243, 255–259, 296, 304–306, 324
Wildrye, basin, 29, 32, 53, 73, 86, 181, 215, 326–328, 331, 333, 334–340, 342–345, 347, 348, 350–353, 356, 359–362

Willamette Valley, 2, 3, 17, 21, 22, 26, 35, 37, 40, 52–54, 56, 59, 70–73, 110, 112, 113, 126, 134, 140, 141, 155, 161, 163, 165, 175, 201–203, 205, 210, 213, 216, 217, 228–230, 237, 243, 244, 247, 258, 269, 279, 281–289, 292–299, 302, 304, 307–309, 323
Working lands, 265, 270, 271, 273, 275–277, 297, 304, 314, 324

X

Xeralfs, 91, 92, 201, 202, 380, 382–384, 387–390, 394–405, 407–411, 413, 414, 416–418, 420–426, 428, 430–432
Xerands, 92, 187, 192, 380, 381, 384, 389–392, 396, 398–400, 402, 403, 405, 406, 409–411, 415–417, 419, 420, 422–427, 429–432
Xerepts, 91, 93, 96, 175, 178, 380–383, 385, 387, 389–396, 398–409, 413, 414, 416, 417, 419–429, 431
Xeric soil moisture regime, 53, 64, 67, 70, 73, 74, 78, 79, 96, 112, 126, 130, 136, 140, 159, 163, 166, 168, 180, 197, 201, 217
Xerolls, 13, 91, 94, 96, 165, 168, 324, 380, 382–432

Y

Yamhill County, 5, 7, 10, 68, 69, 71, 294, 323

Z

Zoning, 226, 237, 257, 258
Zumwalt Prairie, 53, 79, 125, 207